MECHANOBIOLOGY HANDBOOK

MECHANOBIOLOGY HANDBOOK

Jiro Nagatomi

CRC Press
Taylor & Francis Group
Boca Raton   London   New York

CRC Press is an imprint of the
Taylor & Francis Group, an **informa** business

Cover image: Bladder smooth muscle cell, courtesy of Dr. Rosalyn Adam, Children's Hospital

CRC Press
Taylor & Francis Group
6000 Broken Sound Parkway NW, Suite 300
Boca Raton, FL 33487-2742

First issued in paperback 2017

© 2011 by Taylor & Francis Group, LLC
CRC Press is an imprint of Taylor & Francis Group, an Informa business

No claim to original U.S. Government works

Version Date: 20110701

ISBN 13: 978-1-138-07226-8 (pbk)
ISBN 13: 978-1-4200-9121-2 (hbk)

Library of Congress Cataloging-in-Publication Data

Mechanobiology handbook / editor, Jiro Nagatomi.
    p. cm.
    Includes bibliographical references and index.
    ISBN 978-1-4200-9122-9 (hardcover : alk. paper)
    1. Biomechanics. 2. Molecular biology. I. Nagatomi, Jiro. II. Title.

QH513.M44 2011
612'.014--dc23   2011017668

Visit the Taylor & Francis Web site at
http://www.taylorandfrancis.com

and the CRC Press Web site at
http://www.crcpress.com

Contents

SECTION I   Basic Science Tools for Mechanobiology Research

SECTION II (Part 1)   Literature Review of Mechanobiology Research Findings and Theories: Cardiovascular Systems

SECTION II (Part 2)   *Literature Review of Mechanobiology Research Findings and Theories: Musculoskeletal Systems*

SECTION II (Part 3)   *Literature Review of Mechanobiology Research Findings and Theories: Other Organ Systems*

SECTION II (Part 4)　Literature Review of Mechanobiology Research Findings and Theories: Frontiers of Mechanobiology

Foreword

When we observe muscle mass increasing as a consequence of lifting weights or when we note large mass losses in musculoskeletal tissues of astronauts returning after a sojourn in space, we observe manifestations of the mechanism of mechanotransduction, which is at the heart of mechanobiology. The interdisciplinary field of mechanobiology has grown naturally and inexorably from traditional mechanics, or more precisely biomechanics, by incorporating strong elements of molecular and cell biology. Thus, while the term "biomechanics" came to represent mechanical characterization of tissues or kinematic analysis of biological systems, the modern term "mechanobiology" encompasses mechanistic cascades of biological events initiated or governed by mechanical forces. So, for example, the mechanosensing ability of pain-sensing nociceptors to detect strains that are then transmitted as pain signals to the brain is under the umbrella of this field, as are the effects of Wolff's law. The interrelationships between mechanical signals (forces or stresses, and deformations or strains) and biological processes are pivotal in understanding health and disease processes in systems as diverse as the cardiovascular and the musculoskeletal, as well as in pulmonary and sensory organs.

In my own group's research efforts, which deal with biomechanics, biomaterials, and tissue engineering, I have observed a shift in the problems that we study from the classical continuum-based treatment of stress–strain relationships to the elucidation of biological cascades, often necessitating the use of cellular or molecular pathways. For example, we have quantified strong relationships between either hydrostatic pressure or direct compression and the biomechanical properties of tissue-engineered articular cartilage. While these observations were initially made phenomenologically, it is imperative that we understand the ion pumps and channels that seem to be involved in the pathways of these mechanobiological phenomena if we are to understand the mechanisms involved. This gradual linking of basic mechanics and basic biology is a phenomenon observed by numerous investigators in biomechanics, and as exemplified by the creation of journals that are almost specific to mechanobiology, such as *Cellular and Molecular Bioengineering* and *Biomechanics and Modeling in Mechanobiology*. But even more established journals devote a significant number of their papers on mechanobiology. Examples include the *Annals of Biomedical Engineering* (the flagship journal of the Biomedical Engineering Society for which I happen to serve as editor in chief), the *Journal of Biomechanics*, and the *Journal of Biomechanical Engineering*.

The timely book by Jiro Nagatomi comes about as an endorsement of this new scientific approach. I have known Professor Nagatomi ever since he was a graduate student in biomedical engineering at Rensselaer Polytechnic Institute (RPI). I remember meeting him and asking him about his work, which I have followed through the years as I perceive him to be an up-and-coming investigator. Indeed, since that time he has become a well-recognized principal investigator and a solid faculty member working on mechanobiology. So when he asked me if I could write a short foreword for this book, I was delighted to do so. This is a handbook that encompasses both basic and applied aspects of mechanobiology. Since understanding of mechanics is central to this field, the book covers both solid and fluid mechanics. In addition, due to the fact that mechanics is coupled with biological mechanisms and corresponding assays, the book presents that aspect of analysis as well. These background sections are followed by sections on mechanobiology of the cardiovascular system, the musculoskeletal system, and other systems and organs. I feel that the book's format is appropriate and sufficiently comprehensive.

It is my expectation that this book will serve as a compendium of comprehensive reviews covering the entire spectrum of mechanobiology. Not only students, but also seasoned research investigators and scientists in industry will find this collection to be of immense importance. The book, organized and edited magnificently by Professor Jiro Nagatomi, is at the heart of mechanics, biology, and medicine, as it presents fundamental aspects of the emerging and exciting field of mechanobiology. I know that it will be read widely.

Kyriacos A. Athanasiou
University of California, Davis, California

Preface

The initial concept for *Mechanobiology Handbook* spawned from the insightful vision of a CRC Press editor, Michael Slaughter, who found potential in the growth of this field and took a chance to embark on this project with me. While the definition of mechanobiology may vary depending on whom you ask, in this book we will simply define it as "studies of the effects of mechanical environments on the biological processes of cells." Of course, "mechanical environments" can be interpreted as the forces that result from gravity, locomotion, weight-bearing, pumping of the heart, blood flow, containment of urine, and contraction of the muscles, or the mechanical properties of the surroundings of cells such as stiffness of the extracellular matrix or the viscosity of the blood and other fluids inside the body. "Biological processes" can be just as broadly interpreted to be morphological shift, cell growth or death, differential expression of genes and proteins, reorganization of the cytoskeleton, changes in cell membrane potential or intracellular ion concentrations, release of soluble compounds, or phosphorylation of kinases. Researchers have demonstrated that all of the stated mechanical factors influence all of these and a number of other cellular and molecular biological events, which, in turn, influence health, disease, and injury. Considering the complexity of both the mechanics and the biology of the human body, there is so much more to be studied than what we have already discovered. Aided by the continuous advancement of research tools in both mechanics and biology, more sophisticated experiments and analyses are now possible, and the field of mechanobiology is expected to continue growing.

A broad range of organ systems are currently studied by investigators including surgeons, physicians, basic scientists, and engineers. Due to this diversity in the field of mechanobiology, even if a discovery in one organ system or in one discipline may be applicable to other researchers, it may be overlooked and not fully appreciated. This is the reason I decided to put together a single-volume handbook that collects cutting-edge research findings from multiple laboratories in the hope that it will become a good reference to investigators in the field of mechanobiology from different technical communities. I was fortunate enough to receive many contributions from my own professional contacts, and their contacts, with whom I truly enjoyed working. Since mechanobiology is a growing field and many of the research questions are still unsettled, the viewpoints of the authors presented here may be contradictory or conflicting with each other. At the same time, there may be some overlapping information on issues that they all agree on. The goal of this book was not to force one unified theory, but to bring out many different viewpoints and approaches to stimulate further research questions. The handbook may then be updated to incorporate newer discoveries and theories in future editions.

The main goals of research in the field of mechanobiology are threefold: (1) to quantify or estimate the mechanical environment to which cells are subjected in health and disease, (2) to identify and quantify mechanosensitive responses and the molecular mechanisms of mechanically induced pathological conditions, and (3) to ultimately apply the knowledge obtained to the development of new therapies. To achieve these goals, investigators must be familiar with both the basic concepts of mechanics and the modern tools of cellular/molecular biology. However, the current literature contains numerous studies that incorrectly estimate or mimic the in vivo mechanical environments of interest, or misuse standard mechanics terminology. At the same time, those who are well versed in mechanics may be able to design elegant experimental setups, but may not be able to come up with appropriate molecular analyses. The aim of this handbook, thus, is not only to present the cutting-edge research findings in various fields, but also to provide the elementary chapters on mechanics and molecular analysis techniques that will hopefully help the reader plan their experiments better or understand the findings previously reported.

Acknowledgments

I would like to thank each and every chapter author and coauthor for their contributions. Readers will find that these are well-recognized, up-and-coming, and/or future star researchers of mechanobiology. I sincerely hope that the authors have experienced as much excitement and satisfaction in bringing together this useful information as I have had in editing this book. I would also like to thank Professor Kyriacos (Kerry) Athanasiou for his foreword and accept it as a stamp of approval by one of the greatest in biomedical engineering. Last but not least, this project could not have been completed without the help of Sravya Durugu, who assisted me in managing the chapters, and the timely support provided by Kari A. Budyk and Robert Sims, CRC Press project coordinator and project editor, respectively. I hope that *Mechanobiology Handbook* will serve as a useful reference for those who are new to the field and will stimulate new ways of thinking among the researchers of mechanobiology.

Jiro Nagatomi
Clemson, South Carolina

Editor

Jiro Nagatomi is an assistant professor of bioengineering and the director of Cell Mechanics and Mechanobiology Laboratory at Clemson University (South Carolina). He completed his BS followed by a PhD in biomedical engineering with Rena Bizios at Rensselaer Polytechnic Institute (New York). His doctoral thesis was on an in vitro investigation of the effects of hydrostatic pressure on bone cell functions. He worked as a postdoctoral research associate under Michael Sacks at the University of Pittsburgh (Pennsylvania) in the field of soft tissue biomechanics before assuming his current faculty position at Clemson University. His research group is interested in ion channels involved in cellular mechanotransduction of hydrostatic pressure and the development of micro-devices for research in the field of mechanobiology.

Contributors

Bethany Acampora
Department of Bioengineering
Clemson University
Clemson, South Carolina

Rosalyn M. Adam
Department of Urology
Children's Hospital Boston and Harvard
    Medical School
Boston, Massachusetts

Sarah C. Baxter
Department of Mechanical Engineering
University of South Carolina
Columbia, South Carolina

Rena Bizios
Department of Biomedical Engineering
University of Texas
San Antonio, Texas

Francis Boudreault
Department of Environmental Health
Harvard School of Public Health
Boston, Massachusetts

Karen J.L. Burg
Department of Bioengineering
Clemson University
Clemson, South Carolina

Tiffany Camp
Department of Mechanical Engineering
The Johns Hopkins University
Baltimore, Maryland

Natasha Case
Department of Medicine
University of North Carolina at Chapel Hill
Chapel Hill, North Carolina

Alesha B. Castillo
Rehabilitation Research and Development
Veterans Affairs Palo Alto Health Care System
Palo Alto, California

Kevin Champaigne
Department of Bioengineering
Clemson University
Clemson, South Carolina

Wen Li Kelly Chen
Institute of Biomaterials and Biomedical
    Engineering
University of Toronto
Toronto, Ontario, Canada

Naomi C. Chesler
Department of Biomedical Engineering
University of Wisconsin-Madison
Madison, Wisconsin

Adam J. Engler
Department of Bioengineering
University of California
La Jolla, California

Richard Figliola
Department of Mechanical Engineering and
    Bioengineering
Clemson University
Clemson, South Carolina

Peter A. Galie
Department of Biomedical Engineering and
    Bioengineering
University of Michigan
Ann Arbor, Michigan

Wilda Helen
Department of Bioengineering
University of California
La Jolla, California

Jong Wook Hong
Department of Mechanical Engineering
Auburn University
Auburn, Alabama

Joshua D. Hutcheson
Department of Biomedical Engineering
Vanderbilt University
Nashville, Tennessee

Christopher R. Jacobs
Department of Biomedical Engineering
Columbia University
New York, New York

Sachin Jambovane
Department of Mechanical Engineering
Auburn University
Auburn, Alabama

Ramaswamy Krishnan
Department of Environmental Health
Harvard School of Public Health
Boston, Massachusetts

Sanjay Kumar
Department of Bioengineering
UCSF/UC Berkeley Joint Graduate Group in
    Bioengineering
University of California
Berkeley, California

Martine LaBerge
Department of Bioengineering
Clemson University
Clemson, South Carolina

Jeoung Soo Lee
Department of Bioengineering
Clemson University
Clemson, South Carolina

Bin Li
Department of Orthopaedic Surgery
University of Pittsburgh
Pittsburgh, Pennsylvania

Jeen-Shang Lin
Department of Civil and Environmental
    Engineering
University of Pittsburgh
Pittsburgh, Pennsylvania

Fei Liu
Department of Environmental Health
Harvard School of Public Health
Boston, Massachusetts

Xin L. Lu
Department of Mechanical Engineering
University of Delaware
Newark, Delaware

Aya Makino
Department of Medicine
University of California
La Jolla, California

Brittany Ho McGowan
Department of Bioengineering
Clemson University
Clemson, South Carolina

W. David Merryman
Department of Biomedical Engineering
Vanderbilt University
Nashville, Tennessee

Jiro Nagatomi
Department of Bioengineering
Clemson University
Clemson, South Carolina

Sheila Nagatomi
Poly-Med, Inc.
Anderson, South Carolina

Michael P. Nilo
Department of Biomedical Engineering
University of Alabama
Birmingham, Alabama

Shawn Olsen
Department of Bioengineering
Clemson University
Clemson, South Carolina

Darryl R. Overby
Department of Bioengineering
Imperial College London
London, United Kingdom

Shawn J. Peniston
Poly-Med, Inc.
Anderson, South Carolina

Aruna Ramachandran
Department of Urology
Children's Hospital Boston and Harvard
    Medical School
Boston, Massachusetts

Janet Rubin
Division of Endocrinology/Metabolism,
    Medicine
University of North Carolina at Chapel Hill
Chapel Hill, North Carolina

Geert W. Schmid-Schönbein
Department of Bioengineering
University of California
La Jolla, California

Shalaby W. Shalaby
Poly-Med, Inc.
Anderson, South Carolina

Hainsworth Y. Shin
Center for Biomedical Engineering
University of Kentucky
Lexington, Kentucky

Vassilios I. Sikavitsas
Bioengineering Center
School of Chemical, Biological, and Materials
    Engineering
University of Oklahoma
Norman, Oklahoma

Craig A. Simmons
Institute of Biomaterials and Biomedical
    Engineering
University of Toronto
Toronto, Ontario, Canada

Harold A. Singer
The Center for Cardiovascular Sciences
Albany Medical College
Albany, New York

Jan P. Stegemann
Department of Biomedical Engineering
University of Michigan
Ann Arbor, Michigan

Diana M. Tabima
Department of Biomedical Engineering
University of Wisconsin-Madison
Madison, Wisconsin

Daniel J. Tschumperlin
Department of Environmental Health
Harvard School of Public Health
Boston, Massachusetts

Theresa A. Ulrich
Department of Bioengineering
UCSF/UC Berkeley Joint Graduate Group in
    Bioengineering
University of California
Berkeley, California

Samuel B. VanGordon
Bioengineering Center
School of Chemical, Biological, and Materials
    Engineering
University of Oklahoma
Norman, Oklahoma

James H.-C. Wang
Departments of Orthopaedic Surgery,
    Bioengineering, and Mechanical
    Engineering and Materials Science
University of Pittsburgh
Pittsburgh, Pennsylvania

Ken Webb
Department of Bioengineering
Clemson University
Clemson, South Carolina

Brad Winn
Department of Bioengineering
Clemson University
Clemson, South Carolina

Yongren Wu
Department of Bioengineering
Clemson University
Clemson-MUSC Joint Bioengineering Program
Charleston, South Carolina

Hai Yao
Department of Bioengineering
Clemson University
Clemson-MUSC Joint Bioengineering Program
Charleston, South Carolina

Warren Yates
Bioengineering Center
School of Chemical, Biological, and Materials
    Engineering
University of Oklahoma
Norman, Oklahoma

Xiaoyan Zhang
Center for Biomedical Engineering
University of Kentucky
Lexington, Kentucky

Section I

Basic Science Tools for Mechanobiology Research

1 An Introductory Guide to Solid Mechanics

Sarah C. Baxter

CONTENTS

1.1 INTRODUCTION

The goal of this chapter is to present an overview of solid mechanics to an audience that is expert in another field, primarily biology. In the formation of interdisciplinary research teams, it is important that everyone learn some of the "other" field—but the tendency is to try to prescribe a complete education. For mechanics, this would be to suggest that other team members complete the statics, solids, dynamics, continuum sequence, as well as the prerequisite math courses for each. This is not a realistic solution, and takes no advantage of the fact that the "students" are experts in their own field; they are independently good at seeing parallels, asking good questions, and analyzing new and unexpected results. With respect to mechanics, what a nonmechanician needs to know is a basic vocabulary, what kinds of questions those trained in mechanics ask and answer, what kinds of problems they know how to set up, and what tools are available to them to solve these problems.

With these goals in mind, this chapter provides, first, a broad definition of mechanics; second, a list of the fundamental concepts and terms; third, some of the fundamental equations associated with mechanics and the resulting constitutive models; and finally, a few of the classic model problems from mechanics that are well established in the literature, offered as starting points for more advanced research. The differential equations are, without exception, those whose general solutions are either known or easily obtained. Very little is formally derived; the emphasis is more on the mechanisms than the math. The notation is mostly drawn from the engineering side of mechanics; again the goal is not to turn biologists into mechanicians, but to provide them with some sense of the available tools, and information as to what kinds of questions mechanics could be used to answer.

1.2 MECHANICS: A BROAD DEFINITION

Continuum mechanics is arguably the most mature field of engineering science. Its parallel course work in physics is more often called Newtonian mechanics. Both assume that all events occur at speeds considerably less than the speed of light, i.e., no relativity. It has also been for many years the core subject in applied mathematics, sometimes under the title of rational mechanics. Continuum mechanics is the mathematical theory associated with the mechanical behavior of both solids and fluids, as they are subjected to forces or displacements considered on a macroscopic scale. At the finest scale, matter is discontinuous, composed of molecules, atoms, and smaller particles with significant space between them. Most engineering applications, however, deal with matter on a much larger, most often observable scale. At these scales, the concern is with the average response of the collection of the bits and pieces of matter rather than a detailed description of individual particles. While the mathematics of continuum mechanics often considers limits that approach a value at a single point in space, the point is still assumed to have the properties of the bulk material. Continuum mechanics can be further divided into kinematics, which is the study of motion, displacement, velocity, and acceleration, without a specific consideration of the forces required to affect the motion, and mechanics of materials, which is the study of the forces and variables that relate to them, forces per unit area, or per unit volume that can be linked to the energy of deformation as well as displacements and the gradients of displacement, which describe deformation. The most significant difference between solid and fluid mechanics is the idea of a reference state. In solids, even those that can exhibit large deformation, descriptions are with respect to a defined reference state, i.e., position, place, or shape of the material. In contrast, it is extremely difficult to define an original material-based shape for a fluid; fluids assume the shape of their external containers. Thus, the emphasis in fluid mechanics is on tracking what is observed with respect to a global, rather than material, position, i.e., what is the speed of the fluid past a specific point.

Quantum mechanics also attempts to describe the physical behavior of matter, but at smaller, subatomic or atomic scales. Based on statistical probabilities, it includes the nondeterministic behavior of matter and energy; often a probabilistic analysis. While "nanoscale" has as its popular definition any object that has at least one of its dimensions ~100 nm, a more useful definition may be to consider that nano is the dividing scale line between quantum and continuum; below the nanoscale, quantum effects must be considered, above it, there is validity to a continuum approach.

1.2.1 Mechanics of Materials: Fundamental Terms

1.2.1.1 Concepts and Descriptors

1.2.1.1.1 Load and Displacement

Most mechanical testing is done in terms of load and displacement. Either a load or force is applied and the corresponding displacements are observed, or a displacement is effected that results in forces in the structural element. Two visual examples are as follows: hang a weight at the end of a bar and see how much it stretches downward, or stretch a beam so that its ends are attached to two walls and notice if it is in tension or compression, as a result. The advantage of both load and displacement is that they can each be measured. The disadvantage is that the relationship depend on the size and shape of the sample you are testing.

1.2.1.1.2 Stress and Strain

In order to generalize load and displacement, the mechanical behavior of materials is more often expressed in terms of stress and strain. Stress is the force per unit area, in units of pressure. Mathematically, it is a limit, the force at a single point, which means that it cannot really be measured. For traditional engineering materials that hopefully do not deform a great deal when in use, stress is presented in terms of force divided by the undeformed, or reference, cross-sectional area. In highly deformable materials, stress can be compared to the current cross-sectional area; this is more difficult to do at the same time as the mechanical testing. Strains are gradients or the rate of change of displacements; strains are dimensionless. The traditional notations for stress and strain are σ and ε for normal stresses and strains, and τ and γ for shear stresses and strains, respectively. Formally, both stress and strain are tensor quantities. This means that they are invariant under coordinate transformation, or that the physical effect is not changed if you choose a different coordinate system. Position vectors are tensors in this sense; if you pick a point in space, the point itself does not move if you rotate your x–y axis $90°$, but the numbers that describe the position do.

The stress and strain tensors look like symmetric matrices. Each element of each tensor is assigned two subscripts, denoting the row and column in the matrix, and symmetric means that $\sigma_{ij} = \sigma_{ji}$. So,

$$\sigma = \begin{bmatrix} \sigma_{xx} & \sigma_{xy} & \sigma_{xz} \\ \sigma_{xy} & \sigma_{yy} & \sigma_{yz} \\ \sigma_{xz} & \sigma_{yz} & \sigma_{zz} \end{bmatrix}, \quad \varepsilon = \begin{bmatrix} \varepsilon_{xx} & \varepsilon_{xy} & \varepsilon_{xz} \\ \varepsilon_{xy} & \varepsilon_{yy} & \varepsilon_{yz} \\ \varepsilon_{xz} & \varepsilon_{yz} & \varepsilon_{zz} \end{bmatrix}.$$

Imagine stresses applied to a cube positioned with one corner at the origin of an (x, y, z) coordinate system; one subscript tells you what face of the cube you are on by indicating the perpendicular or normal direction to that face of the cube; the second subscript tells you the direction of the stress. So σ_{xx} is the stress on the side of the cube that faces the positive x-axis, and the stress is in that same direction. The shear stress, σ_{xy}, is the stress on the same face of the cube, but along or in the plane of the face rather than normal to it; or equivalently, σ_{yx} is the stress on the side of the cube facing the y-axis, with the stress acting in the x direction. Strains match stresses by similar notation. Engineers most often use a contracted notation that presents stress and strain as vector quantities. The convention is that each double subscript, using $(1, 2, 3)$ for (x, y, z), is contracted to a single subscript, $(11 \rightarrow 1, 22 \rightarrow 2, 33 \rightarrow 3, 23 \rightarrow 4, 13 \rightarrow 5, 12 \rightarrow 6)$. The normal stresses and strains are 1, 2, and 3, the shears are 4, 5, and 6. By labeling the Cartesian axes as $(1, 2, 3)$, referring to (x, y, z), the pattern for the shears is to remove 1, $\sigma_{23} = \sigma_4$, then 2, $\sigma_{13} = \sigma_5$, then 3, $\sigma_{12} = \sigma_5$. The stress vector is then $\sigma = [\sigma_{xx}, \sigma_{yy}, \sigma_{zz}, \sigma_{yz}, \sigma_{xz}, \sigma_{xy}]^T$, where the superscript T indicates the transpose, i.e., that the column vector has been written as a row vector.

1.2.1.1.3 Boundary Conditions

In solid mechanics *boundary value* problems are defined on physical bodies consisting of an interior surrounded by a boundary. Loads and displacements are prescribed on the boundaries, and a differential equation, usually an equilibrium, is solved for the whole physical body, subject to the boundary conditions. The order of the differential equation, the highest derivative present, determines the number of boundary conditions that are needed, e.g., a second-order differential equation requires two boundary conditions.

1.2.1.1.4 Constitutive Models

Constitutive models describe relationships between stress and strain due to material properties.

1.2.1.1.5 Linear Elastic Materials

The simplest, most well-developed and validated theories and models are those for linear elastic materials. Elastic means that however a material is loaded, when the load is removed the material returns to its original shape. Linear refers to the relationship between stress and strain; stress is proportional to strain. At small strains, most materials will have linear stress–strain curves, so the small strain theories can be used in many cases.

1.2.1.1.6 Stiffness

Stiffness describes how strongly a material resists being deformed. For linear elastic materials, and most materials at very small strains, stiffness is the slope of the stress–strain curve. Isotropic linear elastic materials exhibit the same mechanical response regardless of the direction of the loading. If an element is placed in uniaxial tension, this slope is called the elastic modulus (E), the axial modulus, or Young's modulus. If a material is deformed in shear, its shear stiffness is called a shear modulus, modulus of rigidity, or G. Isotropic materials can be completely characterized by three material properties, E, G, and Poisson's ratio, ν. Poisson's ratio, again for an isotropic material, describes how much the material narrows in one direction when it is stretched in the perpendicular direction, as

$$\nu = -\frac{\varepsilon_{lateral}}{\varepsilon_{longitudinal}}. \tag{1.1}$$

Some foams with novel microstructures have been shown to widen when stretched, but most materials decrease in width. Only two of these three properties are independent. They are related as

$$G = \frac{E}{2(1+\nu)}. \tag{1.2}$$

A fourth material parameter, K, the bulk modulus, also often appears in the literature. It describes a material's resistance to uniform compression. It can be written in terms of the others as

$$K = \frac{EG}{3(3G-E)} \tag{1.3}$$

The inverse of stiffness is *compliance*. For anisotropic materials, e.g., wood, whose properties vary with direction, observable by the fact that it can be more easily cut with the grain, it is important to describe the connections between directions. The axial tension along the x-axis may produce different compressions in the y than in the z directions. For materials with more anisotropy, a full tensor description of the stiffnesses/compliance is required. In totally nonintuitive notation, mechanics uses **C** for the stiffness tensor and **S** for the compliance tensor. Stiffness is a material property and does not depend on geometry.

1.2.1.1.7 Strength

Strength is the stress at which a material breaks; it is characterized by an ultimate stress. What is probably most important to remember is that strength and stiffness are not the same, and do not necessarily correlate. A rubber band is not stiff, hang a light weight on it and it stretches a lot. A cotton string, hung with the same weight is stiffer; it does not stretch as much as the rubber band under the same load. However, it is possible that both the rubber band and the cotton string have roughly the same strength, i.e., they might break under the same load.

1.2.1.1.8   Incompressible

Compressibility is a measure of the change in volume in response to pressure. An incompressible material does not change volume when loaded. A Poisson's ratio of $\nu = 0.5$ is known as the incompressible limit; at this value $G = E/3$ and the bulk modulus becomes infinite. Most rubber-like materials have Poisson ratios of ~0.5.

1.2.1.1.9   Viscoelastic Materials

Viscoelastic materials exhibit strain rate effects in response to an applied strain. Often, they have a region of loading that is relatively unaffected by strain rate and behaves as an elastic material, and a second region that is dominated by strain rate and behaves more as a viscous fluid.

1.2.1.1.10   Creep

When a constant load (stress) is applied and held constant, but the displacement (strain) continues to increase with time, the material is said to exhibit creep; think of Silly Putty©.

1.2.1.1.11   Relaxation

When a constant stretch (strain) is applied and held, but over time the internal stress lessens, then the material is said to relax. Stretch a rubber band around a deck of cards for several weeks, the tightness of the rubber band will lessen over time.

1.2.1.1.12   Hysteresis

A material exhibits hysteresis when there is a time lag in the mechanical response to a change in load. It is a manifestation of what is described as internal friction in the material and is characteristic of both viscoelastic and nonlinearly elastic materials. Usually, this is presented in the form of a stress–strain curve that shows both loading and unloading; the unloading curve follows a different path on the graph to that of the loading.

1.2.1.1.13   Hardness

Hardness is defined as a material's (usually a metal) resistance to plastic deformation, but is not considered a fundamental material property. It is measured using indentation tests and generally provides an empirical and comparative measure.[1] Correlations between hardness and yield stress do not generalize well, although they may be useful for a particular experiment.

1.2.1.1.14   Yield

Beyond the yield point, generally a yield stress, a material is no longer deforming elastically. If the load is removed, the material will not return to its original shape. Most often, the term is applied to metals. For crystalline solids, it means that the regular rows of atoms in the lattice have gotten out of line, been dislocated to such an extent that they cannot assume their original position on unloading. These dislocations cause plastic deformation. Yield stress has no absolute definition. For traditional engineering materials, yield stress is defined as 0.02% offset. At a strain of 0.02%, a line is drawn with slope $= E$. The yield stress is then defined as the point where this line intersects the stress–strain curve.

1.2.1.1.15   Ductile

A ductile material is one that yields, deforms, or shows a change in material properties—a material that shows damage before it breaks.

1.2.1.1.16   Brittle

A brittle material is one that does not yield before it breaks. There are no warnings of failure.

1.2.1.1.17   Fatigue, Fatigue Strength

This refers to the ability of a material to resist cyclic loading. It is usually presented as a graph of stress (*y*-axis) vs. cycles to failure (*x*-axis). Fatigue occurs as the material accumulates damage.

1.2.1.1.18   Failure

This generally refers to the point at which a material can no longer carry a prescribed load; it can be specifically associated with a length scales.

1.2.1.1.19   Fracture

When a physical body (material) separates into two or more parts as a result of loading, it is said to fracture. It is also the result of damage accumulating to a critical level.

1.2.1.1.20   Residual Stress

This effect can also be characterized through a residual strain. When the material that you are testing is already under some loading, then any measurements and observations should reflect the existence of these underlying states if they are to accurately model the properties of the material itself. Structures, including those in biological systems, are often mechanically improved by designing in residual stress–strain, e.g., the spokes in a bicycle wheel are loaded in radial tension as they are put into place to counteract/match some of the compression load that occurs in use.

1.2.1.1.21   Homogeneous

If a material's mechanical response is independent of position within the structure, then it is said to be homogeneous. This designation can be a function of the scale that is being considered; biological tissue may consist of several types of polymer chains, fiber, and/or a cross-linked network, but it might still be most useful to treat it mathematically as a homogeneous material.

1.2.1.1.22   Micromechanics

Micromechanics is a relatively mature subfield of mechanics that was developed to predict the properties of composite materials. Homogenization and averaging techniques are used to extend models of homogeneous phases to generate the effective properties of the composite, as if it were an equivalent homogeneous material.

1.2.1.2   Deformations

1.2.1.2.1   Tension

Stretching and elongation, by convention positive stresses and strains, are tensile (Figure 1.1a).

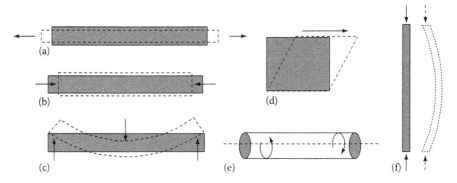

FIGURE 1.1   Modes of deformation: (a) tension, (b) compression, (c) bending, (d) simple shear, (e) torsion, and (f) buckling.

1.2.1.2.2 Compression

The opposite direction from tension, by convention negative stresses and strains, indicate compression (Figure 1.1b).

1.2.1.2.3 Bending

This can also be referred to as flexure. This generally applies to slender beams or thin plates and is the result of a load applied perpendicularly to the long axis or the plane of a sheet (Figure 1.1c). Flexural stiffness is EI, the product of the elastic modulus and the cross-sectional moment of inertia, and depends on geometry. A meter stick is "stiffer" when it is bent with the wider side facing out than if the wider side is facing down.

1.2.1.2.4 Shear

There are several types of shear, but fundamentally a shearing effect is caused by two planes in a material shifting with respect to one another (Figure 1.1d).

1.2.1.2.5 Torsion

This is the result of a twisting motion, or an applied torque. Torsional stiffness depends not only on the elastic modulus but also on the polar moment of inertia. It is strongly affected by geometry. A thick-walled, large radius tube will have large torsional stiffness (Figure 1.1e).

1.2.1.2.6 Buckling

True buckling is an instability, a mathematically and energetically unstable configuration. It is common, however, to hear the term buckling used to describe rippling or bunching that results from compression (Figure 1.1f).

1.2.1.3 Mechanisms

There are many powerful models describing material behavior in the literature, including those developed from first principles, empirical models that capture experimental observation, and those that model damage and failure mechanisms. The two regimes that probably have the most relevance in biology, however, are elasticity and viscoelasticity. The essence of each can be illustrated by two very simple and intuitive mechanical mechanisms, springs and dashpots.

Elastic deformation, in its most basic form, can be described by the simple model of a mass on a spring. A force can be applied to stretch the spring, and when the force is quasi-statically released, the spring returns to its original shape and position. *Viscous* materials behave as dashpots or damping elements. An example of damped motion is the device on screen doors that slows their closing, or shock absorbers on your car. The force in viscous materials is proportional to the rate at which they are being stretched, rather than how far. This is the simple model for a fluid. *Viscoelastic* materials exhibit properties of both solids and fluids, which include observed effects, such as rate dependent elasticity, stiffness depending on the speed at which the load is applied, and hysteresis due to creep and relaxation. Combining springs and dashpots in varying arrangements can create models that capture these effects. The most elaborate of them can capture very complex deformations, and although it is often difficult to determine the appropriate material parameters associated with a specific system, the usefulness of these models is that they illustrate, in an intuitive way, the combined mechanisms; spring/dashpot, elastic/viscous. In what follows are formal presentations of three basic models—the spring mass model for linear elasticity, and the Maxwell and Kelvin models for viscoelasticity.

1.2.1.3.1 Linear Elasticity

1.2.1.3.1.1 The Spring Mass Model The simplest mechanistic model for an elastic material is to imagine that they behave as springs; they resist being stretched with a force that is proportional to their constitutive makeup. The force on a spring is equal to the distance it is stretched, x, times

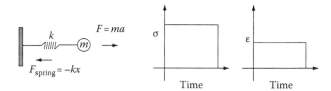

FIGURE 1.2 A simple model of elastic behavior, spring-mass. An applied force produces a resistance that is proportional to the distance it is stretched; the constant of proportionality is the spring constant, a material property. The spring model responds and recovers instantaneously when the load is applied or removed.

the spring constant, k, the material property of the spring (see schematic in Figure 1.2). It shows instantaneous elastic response and an instantaneous recovery (see graphs in Figure 1.2).

When the stretching load is removed, the spring returns to its original length. If the material property of the spring is a constant over all stretches, then the model is for a linear elastic material. If the spring constant varies with the stretch, it is a nonlinear elastic model. The spring mass model is a more of a kinematic than a constitutive model as it is formulated in terms of displacements, but it illustrates our intuitive concept of the response of an elastic material. If no damping is included in the model, then from Newton's second law, $F = ma$. Force is mass times acceleration:

$$F = -kx$$

$$ma = -kx \qquad (1.4)$$

$$m\frac{d^2x}{dt^2} + kx = 0.$$

The final differential equation describes undamped oscillation. It can be solved for the displacement, x, as a function of time:

$$x(t) = A\cos\left(\sqrt{k}t\right) + B\sin\left(\sqrt{k}t\right). \qquad (1.5)$$

The constants A and B are determined by initial conditions, such as the initial position, $x(t_0)$, and initial velocity, $\dot{x}Y(t_0)$, where the dotted variable denotes the first derivative with respect to time. This model can also include an external, time-dependent forcing function. Because this model is phrased in terms of time, it is not a boundary value problem, but an initial value problem. Again, its solution provides more of a kinematic model rather than a constitutive model, but as one of the most well-studied and well-understood models, it can be a useful tool.

For completeness, the model can be extended to include damping, or drag forces, by assuming that damping is proportional to velocity, here by the damping constant, β. In this case, the forces and differential equations becomes

$$F = -kx - \gamma\dot{x}$$

$$ma = -kx - \beta\dot{x} \qquad (1.6)$$

$$m\ddot{x} + \beta\dot{x} + kx = 0.$$

where the double dot indicates a second derivative with respect to time.

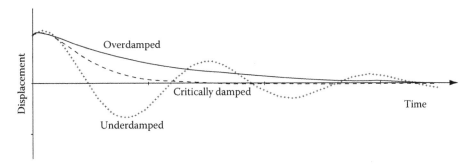

FIGURE 1.3   Oscillation of mass with damping; overdamped, underdamped, and critically damped.

The solutions to this equation are also well known (see Figure 1.3).

$$\text{if } \beta > 2\sqrt{km} \text{ then } x(t) = Ae^{r_1 t} + Be^{r_2 t}, r_1, r_2 = \frac{-\beta \pm \sqrt{\gamma^2 - 4mk}}{2m}, \text{ overdamped,} \qquad (1.7)$$

$$\text{if } \beta < 2\sqrt{km} \text{ then } x(t) = e^{-\lambda t}\left[A\cos(\mu t) + B\sin(\mu t)\right], \lambda = \frac{\beta}{2m}, \mu = \frac{\sqrt{|\beta^2 - 4mk|}}{2m}, \text{ underdamped,}$$

$$(1.8)$$

and

$$\text{if } \beta = 2\sqrt{km} \text{ then } x(t) = e^{-\lambda t}\left[A + Bt\right], \text{ critically damped.} \qquad (1.9)$$

1.2.1.3.2   Viscoelasticity

The Maxwell model consists of a linear spring and a linear dashpot in series (see schematic in Figure 1.4). The Maxwell model can capture time-dependent effects, like creep. For example, since the spring and dashpot are connected in series in the Maxwell model, the total strain is the sum of the strain in the spring and the strain in the dashpot (in direct analogy to circuits in series). This results in the differential equation

$$\dot{\varepsilon} = \frac{\dot{\sigma}}{k} + \frac{\sigma}{\eta} \qquad (1.10)$$

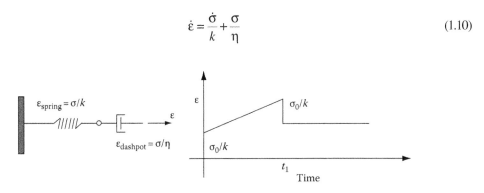

FIGURE 1.4   Maxwell model of viscoelastic response of a material under constant stress load; spring and dashpot in series. The material displays instantaneous elasticity due to the spring response, followed by the creep–continued strain without additional load. Upon unloading, the elastic strain is recovered immediately, but the time-delayed response results in permanent deformation.

where the dot again implies a derivative with respect to time, k is the spring (elastic) constant, and η is the viscosity. The first order differential equation can be solved as

$$\varepsilon(t) = \frac{\sigma(t) - \sigma(0)}{k} + \int_0^t \frac{\sigma(0)}{\eta} dt. \qquad (1.11)$$

If a constant stress, σ_0, is applied at time zero, the strain $\varepsilon(t)$ is

$$\varepsilon(t) = \frac{\sigma_0}{k} + \frac{\sigma_0}{\eta} t, \qquad (1.12)$$

consisting of an elastic part and a viscous part. As time goes on, even though no additional load is applied, the strain continues to increase; the material creeps (see graph in Figure 1.4).

If the stress is removed at time t_1, the solution past this point is

$$\varepsilon(t) = \frac{\sigma(t) - \sigma(t_1)}{k} + \int_0^{t_1} \frac{\sigma_0}{\eta} dt + \int_{t_1}^t \frac{0}{\eta} dt$$

$$\varepsilon(t) = \int_0^{t_1} \frac{\sigma_0}{\eta} dt = \frac{\sigma_0}{\eta} t_1 \qquad (1.13)$$

where in the last equation, the elastic strain has been recovered, its contribution is now zero, but there is a permanent strain that remains, $(\sigma_0/\eta)t_1$.

The Kelvin model consists of a linear spring and a linear dashpot in parallel (see schematic in Figure 1.5). Since the spring and dashpot are in parallel, the total stress in the model is equal to sum of the stresses in each component, resulting in

$$\dot{\varepsilon} + \frac{k}{\eta} \varepsilon = \frac{\sigma}{\eta}. \qquad (1.14)$$

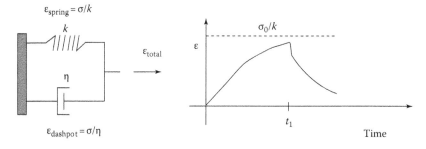

FIGURE 1.5 Kelvin model of viscoelastic response of a material under constant stress load; spring and dashpot are connected in parallel. The material exhibits a gradual creep response up to an asymptotic level. For unloading, the strain can be calculated by imagining an equal but opposite stress applied. The material recovers slowly, but eventually returns to an original state.

If a constant stress is applied to this model, the solution is

$$\varepsilon = \frac{\sigma_0}{k}(1 - e^{-kt/\eta}) \qquad (1.15)$$

In this case, the strain increases to an asymptotic value of σ_0/k. If the stress is removed at time t_1, we can determine the subsequent strain by imagining that an equal but opposite stress is applied (see graph in Figure 1.5). The sum of the positive and negative constant stresses then gives a net zero stress. This is the principle of superposition. If the material is elastic, then multiple loadings can be considered separately, and the various solutions added together to capture the net effect. So, if the strain due to the constant stress σ_0 is

$$\varepsilon^+ = \frac{\sigma_0}{k}(1 - e^{-kt/\eta}), \qquad (1.16)$$

and the strain due to the addition at time t_1 of $-\sigma_0$, is

$$\varepsilon^- = -\frac{\sigma_0}{k}(1 - e^{-k(t-t_1)/\eta}), \qquad (1.17)$$

then the total strain is $\varepsilon^+ + \varepsilon^-$, or

$$\varepsilon = \frac{\sigma_0}{k}e^{-kt/\eta}(e^{-kt_1/\eta} - 1). \qquad (1.18)$$

Neither of these models is a great predictive model for most viscoelastic materials, but the basic elements do provide an intuitive sense of the mechanisms.

1.3 MODELING

1.3.1 FUNDAMENTAL EQUATIONS

Mechanics is largely based on Newton's second law, also known as the conservation of momentum. This law says that the net result of all the forces acting on a body is balanced by a change in momentum, or

$$F = \frac{d(mv)}{dt}, \qquad (1.19)$$

where
 F is force
 m is mass
 v is velocity
 mass times velocity, mv is momentum

At speeds much slower than the speed of light, mass does not change with time, so the equation becomes the more familiar $F = ma$. Undergraduates begin their study of mechanics in a class called statics. This class deals with the mechanics of things that are not accelerating—they could be in motion, but only at a constant velocity, ($a = 0$). In this case Newton's second law says that all the forces, usually drawn as vectors (arrows) with both magnitude (length of arrow) and direction, (angle of arrow) must add up to zero. Vectors can be added visually, if they are drawn to scale, by putting the tail of the second to the head of first (see Figure 1.6a). The class extends the concept to forces that produce moments, e.g., forces applied in opposite directions at the end of a bar will try to spin it around a center point and create what is called a moment (see Figure 1.6b). The complete

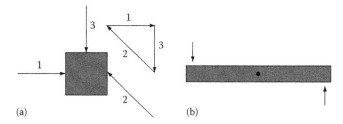

FIGURE 1.6 The first class in mechanics, statics, deals with forces and moments that balance each other. In (a), forces, drawn as arrows, applied to the body add to zero, since the summed arrows (arrow 1 plus arrow 2 plus arrow 3; tail to head) return to the starting position of arrow 1. In (b), forces applied in opposite directions will try to spin the object about the center point, which, if they are resisted, produce a moment.

course in statics can be concisely stated as, for an object with zero acceleration, the sum of all the forces and the sum of all the moments must be zero.

1.3.1.1 Strain

Engineering normal strains are defined as the change in length of a line segment, $\overline{AB}$, divided by its original length:

$$\varepsilon = \frac{\Delta L}{L}. \tag{1.20}$$

Engineering shear strains are defined as the angular change between two line segments that are originally 90° apart. The angle after being deformed is θ, and the shear strain is

$$\gamma = \frac{\pi}{2} - \theta. \tag{1.21}$$

The generalized form for linearized (engineering) strains, valid in small strain regimes or for linear elastic materials is, for the in-plane normal strains, ε, and shear strains, γ.

$$\varepsilon_x = \frac{du}{dx}, \quad \varepsilon_y = \frac{dv}{dx} \quad \gamma_{xy} = \left(\frac{dv}{dx} + \frac{du}{dy} \right) \tag{1.22}$$

where the displacements in the (x, y, z) directions are denoted as (u, v, w). (This is a very engineering notation, and often other symbols will be used for displacement, primarily as vectors.)

$$u = \left(u_x, u_y, u_z \right) \quad \text{or} \quad u = \left(u_1, u_2, u_3 \right).$$

1.3.1.2 Stretch

Stretch is the ratio of the deformed length of a line segment to its original length

$$\lambda = \frac{L + \Delta L}{L} = 1 + \varepsilon. \tag{1.23}$$

1.3.2 Constitutive Laws

Constitutive laws describe the relationship between stress and strain. The fundamental parameters are material properties; however, stress and strain can be functions of time and/or a spatial position (within the material, as for example in a composite material). The constitutive laws that follow are samples of the most well-developed theories. Each has validity under various, usually fairly significant, assumptions. If the assumptions are not valid, then the models are not good. These stress–strain relationships are presented as valuable for back of the envelope calculations, potentially immediately useful in application, and certainly as a starting place for more complex models.

1.3.2.1 Linear Elasticity

The constitutive law for a linear elastic material is known as Hooke's law or generalized Hooke's law, $\sigma = C\varepsilon$. For isotropic materials, this can be written in terms of the elastic constants as

$$\sigma_x = \frac{vE}{(1+v)(1-2v)}\left(\varepsilon_x + \varepsilon_y + \varepsilon_z\right) + \frac{E}{1+v}\varepsilon_x$$

$$\sigma_y = \frac{vE}{(1+v)(1-2v)}\left(\varepsilon_x + \varepsilon_y + \varepsilon_z\right) + \frac{E}{1+v}\varepsilon_y. \tag{1.24}$$

$$\sigma_z = \frac{vE}{(1+v)(1-2v)}\left(\varepsilon_x + \varepsilon_y + \varepsilon_z\right) + \frac{E}{1+v}\varepsilon_z$$

The shear stresses can be written as

$$\tau_{yz} = G\gamma_{yz}$$

$$\tau_{xz} = G\gamma_{xz}. \tag{1.25}$$

$$\tau_{xy} = G\gamma_{xy}$$

Normal and shear stresses in a linear elastic material are independent of each other (normals from shears).

1.3.2.2 Viscoelasticity

For biological materials, constitutive laws that capture nonlinear elastic and/or time-dependent effects are of interest. The classic descriptive model of a viscoelastic material, and the infinite extension of multiple springs and dashpots is in the form of an integral equation that describes the stress relaxation under an arbitrarily prescribed strain, and strain rate as,

$$\sigma(t) = \int_0^t E(t-\tau)\frac{\partial\varepsilon(\tau)}{\partial(\tau)}d\tau \tag{1.26}$$

In this equation, $E(t)$ is the relaxation modulus, a function of the history of the loading up to the time t, and the partial derivative in the integrand is the strain rate. The inverse relationship for strain as a function of an arbitrarily prescribed stress is

$$\varepsilon(t) = \int_0^t J(t-\tau)\frac{\partial\sigma(\tau)}{\partial(\tau)}d\tau, \tag{1.27}$$

where $J(t)$ is the creep compliance.

Linear viscoelastic materials are often characterized under oscillating loads. If the input is an oscillating stress at frequency ω, then the strain response will be an oscillation at the same frequency but lagging behind by a phase angle δ, or

$$\varepsilon = \varepsilon_0 \cos(\omega t - \delta) \tag{1.28}$$

where ε_0 is the strain magnitude. Similarly, if the input is an oscillating strain, the stress response, with a magnitude of σ_0, will lead the strain by a phase angle δ, or

$$\sigma = \sigma_0 \cos(\omega t + \delta). \tag{1.29}$$

The mathematics of this loading requires that, in this case, the dynamic creep compliance and relaxation modulus functions are expressed as complex numbers. The complex relaxation modulus, for example, takes the form of

$$E^* = \frac{\sigma_0}{\varepsilon_0} e^{i\delta} = E_1 + iE_2 \tag{1.30}$$

or

$$G^* = \frac{\tau_0}{\gamma_0} e^{i\delta} G_1 + iG_2, \tag{1.31}$$

if the loading is an applied shear.

The real part of these moduli is called the storage modulus, e.g.,

$$E_1 = \frac{\sigma_0}{\varepsilon_0} \cos(\delta), \tag{1.32}$$

and the complex part is called the loss modulus, e.g.,

$$E_2 = \frac{\sigma_0}{\varepsilon_0} \sin(\delta). \tag{1.33}$$

Instead of stress–strain plots for these material, their mechanical behavior is characterized by the mechanical loss, which is defined as $\tan(\delta)$, and by the magnitude of the complex modulus $E = \sigma_0/\varepsilon_0$ (see Figure 1.7).

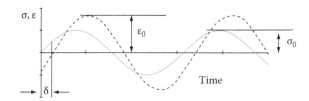

FIGURE 1.7   Oscillating stress and strain with phase lag, δ, due to viscoelasticity.

1.3.2.3 Nonlinear Elasticity

Linear elasticity is formulated on the assumption of small strains. The theory of large deformations is called finite elasticity. One of the classic models in the literature is the Mooney–Rivlin model,[2,3] which is used to model the behavior of plastic or rubber-like materials. Nonlinear elasticity theories are most often formulated in terms of a deformation tensor, which is beyond the scope of this chapter. However, for the case of uniaxial tension, the model predicts the following axial stresses as a function of the axial stretch λ_1. First, the stress with respect to the deformed cross-sectional area (true stress)

$$\sigma_{11} = \left(2C_1 + \frac{2C_2}{\lambda_1}\right)\left(\lambda_1^2 - \frac{1}{\lambda_1}\right) \tag{1.34}$$

and with respect to the original and undeformed, reference cross-sectional area

$$\sigma_{11} = \left(2C_1 + \frac{2C_2}{\lambda_1}\right)\left(\lambda_1 - \frac{1}{\lambda_1^2}\right) \tag{1.35}$$

with the constants C_1, C_2 fit experimentally.

1.3.2.4 Pseudoelasticity

Fung (1993) introduced the idea of pseudoelasticity for biological materials. He pointed out that biological materials are not perfectly elastic and so cannot really be considered elastic materials in a thermodynamic sense, since all the energy is not recovered on unloading. However, researchers had noticed that under the initial cycles cyclic loading, called preconditioning of the material, the stress–strain curve stopped varying with the strain rate. If strain rate effects can be ignored, then Fung (1993) suggested considering the loading and unloading curves separately, each modeled as an elastic material with a separate constitutive law.

For nonlinear constitutive laws, stresses and/or strains are coupled. Therefore, uniaxial testing is not sufficient to determine material properties, and biaxial testing is needed. There are two general approaches to constitutive models for biaxial testing; polynomial and exponential.

Fung et al.[4] proposed the following exponential model for a pseudo-elastic stress–strain relationship [article] for blood vessels. He formulated the constitutive law in terms of a pseudo-strain energy function, given in radial coordinates (r, θ, z) as

$$\rho_o W^{(2)} = \frac{C}{2}\exp\left[a_1\left(\varepsilon_\theta^2 - \varepsilon_\theta^{*2}\right) + a_2\left(\varepsilon_z^2 - \varepsilon_z^{*2}\right) + 2a_4\left(\varepsilon_\theta\varepsilon_z - \varepsilon_\theta^*\varepsilon_z^*\right)\right] \tag{1.36}$$

where C, a_1, a_2, and a_4 are material constants, and the starred quantities are arbitrary stress–strain pairs, constants measured in the physiological range. For elastic materials, the potential energy that is stored in a body as a result of mechanical work is called strain energy; it depends only on deformation. The assumption is that the radial stresses are significantly smaller than either the circumferential or axial. The stress–strain relationships are found by taking the derivative of the strain energy function with reference to the circumferential and longitudinal strains; this results in two coupled equations for the stresses in terms of the strains:

$$\sigma_{\theta\theta} = \frac{C}{2}\exp\left[a_1\left(\varepsilon_\theta^2 - \varepsilon_\theta^{*2}\right) + a_2\left(\varepsilon_z^2 - \varepsilon_z^{*2}\right) + 2a_4\left(\varepsilon_\theta\varepsilon_z - \varepsilon_\theta^*\varepsilon_z^*\right)\right]\left[2a_1\left(\varepsilon_\theta\right) + 2a_4\left(\varepsilon_z\right)\right]$$

$$\sigma_{zz} = \frac{C}{2}\exp\left[a_1\left(\varepsilon_\theta^2 - \varepsilon_\theta^{*2}\right) + a_2\left(\varepsilon_z^2 - \varepsilon_z^{*2}\right) + 2a_4\left(\varepsilon_\theta\varepsilon_z - \varepsilon_\theta^*\varepsilon_z^*\right)\right]\left[2a_2\left(\varepsilon_z\right) + 2a_4\left(\varepsilon_\theta\right)\right]$$

$$\tag{1.37}$$

Measured data from, for example, axial loading in combination with internal pressure is fit to these functional forms, in a least squares sense, to determine the material constants.

1.3.2.5 Inelastic Behavior

There are many constitutive models for inelastic effects, mostly for traditional engineering materials, but no consensus on a "best" model. Many of these models can, however, be formulated with respect to the linear elastic model. If the total strain is assumed to be the sum of all the strains, i.e., elastic strain, plastic strain, thermal strain, etc., $\varepsilon^{\text{total}} = (\varepsilon^{\text{elastic}} + \varepsilon^{\text{plastic}} + \varepsilon^{\text{thermal}} + \cdots)$ then by solving the elastic strain, the linear elastic constitutive law can be used as a framework.

$$\varepsilon^{\text{elastic}} = \varepsilon^{\text{total}} - (\varepsilon^{\text{plastic}} + \varepsilon^{\text{thermal}} + \cdots)$$

$$\sigma = \mathbf{C} \left[\varepsilon^{\text{total}} - (\varepsilon^{\text{plastic}} + \varepsilon^{\text{thermal}} + \cdots) \right].$$

(1.38)

Thermal strains are defined as $\varepsilon = \alpha \Delta T$, where α is the coefficient of the thermal expansion and ΔT is the change in temperature. Materials can also respond anisotropically to changes in temperature, thus there can be subscripted α_x, α_y, α_z. There is less consensus on how to model inelastic strains, although well-established models do exist for plasticity as it is associated with metals.

1.4 ENGINEERING MECHANICS: CLASSIC PROBLEMS

The following problems are classic problems in mechanics. These problems and solutions are only valid in small strain regimes for elastic homogeneous materials. However, they represent elementary and fundamental deformations that are valid for study of either traditional engineering or biological materials.

1.4.1 AXIAL EXTENSION

The deformation of an elastic beam or rod, of length L, with a constant cross-sectional area, A, under an axial load, P, which can be described by normal stress and strain are

$$\sigma = \frac{P}{A}, \quad \varepsilon = \frac{\Delta L}{L}.$$

(1.39)

If the cross-sectional area varies along the length, then the force due to an external load will also vary along the length, and the stress and strain at any point are

$$\sigma(x) = \frac{P(x)}{A(x)}, \quad \varepsilon = \frac{\partial u}{\partial x}$$

(1.40)

where the displacement, u, at any point along the beam can be calculated by

$$u = \int_0^L \frac{P(x)}{A(x)} \frac{dx}{E}, \quad \text{where for an elastic body, } \sigma = E\varepsilon.$$

(1.41)

1.4.2 BENDING

These formulae (Euler–Bernoulli theory for thin elastic beams) can be applied to beams, with various boundary conditions, where small strains are expected. The deflection of a horizontal beam, with a constant cross-sectional area, and homogeneous properties, in the y direction is given by

$$EI\frac{d^4y}{dx^4} = -w(x) \tag{1.42}$$

where
  w is the distributed transverse (vertical) load as a function of x
  E is the elastic modulus
  I is the cross-sectional moment of inertia

This fourth order differential equation requires four boundary conditions that are determined by the boundary conditions associated with the position of the beam. Often, instead of an axial load, a moment or shear load is known, and an established relationship between axial loads, shears forces, V, and moments, M, can be used to rewrite the equation as either

$$EI\frac{d^3y}{dx^3} = V(x)$$

$$EI\frac{d^2y}{dx^2} = M(x). \tag{1.43}$$

The tensile stress in the beam is given by

$$\sigma = \frac{Mc}{I} = Ec\frac{d^2y}{dx^2}, \tag{1.44}$$

where c is the distance in the y direction from the center axis of the beam.

1.4.3 BENDING OF A CANTILEVER BEAM

The solution for any of the differential equations above for a cantilevered beam, of length L, with a force F, applied at the free end, is given by

$$y(x) = \frac{F}{EI}\left[\frac{Lx^2}{2} - \frac{x^3}{6}\right], \tag{1.45}$$

where x is measured from the fixed end. The boundary conditions for this problem are first, that it is completely fixed at $x=0$; completely fixed means that the displacement (deflection) and the slope are both zero, or

$$y(0) = 0,$$

$$\left.\frac{dy}{dx}\right|_{x=0} = 0, \tag{1.46}$$

and second, that it is completely free at the other end, $x = L$, which is specified by requiring that both the shear and moment to be zero or

$$\frac{d^3 y}{dx}\bigg|_{x=L} = 0,$$

$$\frac{d^2 y}{dx^2}\bigg|_{x=L} = 0. \tag{1.47}$$

1.4.4 TORSION OF CIRCULAR SHAFTS

The elementary theory of torsion has been developed for circular elastic beams and assumes that there is no warping of the transverse planes in the cylinder. It assumes that the shear stress (there are no normal stresses) is proportional to the angle of twist, θ, the distance from the center axis of the shaft, r, and the shear modulus of the material, G, or

$$\tau_{yz} = G\theta x$$

$$\tau_{xz} = G\theta y. \tag{1.48}$$

1.4.5 THIN-WALLED PRESSURE VESSELS

The calculation of stress in a thin-walled pressure vessel, either spherical or cylindrical (like soft drink cans), is also a well-known problem. These stresses are presented in a radial geometry, so that σ_1 is the normal stress in the circumferential direction around the curvature of the vessel, and σ_2 is the stress in the longitudinal or axial direction, perpendicular to the thickness of the wall in a cylinder, equal to σ_1 in a sphere. The stresses in a cylindrical vessel and a spherical vessel, with wall thickness, t, and internal radius, r, due to an internal pressure, p, are

$$\sigma_1 = \frac{pr}{t}$$

$$\sigma_2 = \frac{pr}{2t} \tag{1.49}$$

and

$$\sigma_1 = \sigma_2 = \frac{pr}{2t} \tag{1.50}$$

respectively. Each stress is assumed to be constant through the thickness of the wall and each produces a tensile load. The thin wall condition requires that $r/t \geq 10$.

1.4.6 BUCKLING

The curvature (second derivative) of an axial-loaded beam can be described by

$$\frac{d^2 y}{dx^2} = \frac{M}{EI} \tag{1.51}$$

where

   M is the bending moment
   E is the elastic modulus
   I is the cross-sectional moment of inertia

If the beam is loaded by an axial load P, then the bending moment is $M = -Py$, and the equation becomes

$$\frac{d^2y}{dx^2} + p^2y = 0, \quad p = \frac{P}{EI}. \tag{1.52}$$

The general solution for this problem, with boundary conditions, is

$$y(x) = A\cos(px) + B\sin(px), \quad y(0) = 0, \quad y(L) = 0. \tag{1.53}$$

Applying the boundary conditions results in the condition that

$$\sin(pL) = 0, \text{ or } p = \frac{n\pi}{L}, \tag{1.54}$$

which is satisfied for all integer values of n. Each value of n defines a load and the associated shape of the beam under that load; each is referred to as a buckling load—loads at which the column moves to a different conformation. The first of these, $n = 1$, has practical significance because failure will occur at the first buckling load, thus the critical load is

$$p = \frac{\pi^2 EI}{L^2}, \tag{1.55}$$

also known as the Euler buckling load.

1.5 CONCLUSION

This chapter was not designed to provide a formal education in mechanics. The goal was to provide those who work with mechanicians a primer that would allow them to contribute constructively to identifying questions and problems in biology whose solutions might be approached using a mechanics of material perspective. The perspective is that of using mechanics as a methodology and as a tool, not on developing new mechanics models, which while also valid and necessary in biological and biomechanics research, is not the scope of this work. While this chapter, for the most part, lacks derivations, it does mention, in context, most of the things that mechanicians think are important to their field. Once an area has been identified, more detail and/or solutions to the simple presentations in this work will be easy to find, as well as collaborators, mathematicians, or mechanicians who can provide information on solving and extending. All the information presented here is widely available in textbooks and articles, but I have added a list of references, each representing the broad areas sketched out here and all offering additional details and scope. The choice of references is based on the texts with which I am most familiar, but are certainly not the only accurate sources of information.

REFERENCES

1. Chawla, K. and Meyers, M., 2009. *Mechanical Behavior of Materials*. Cambridge University Press.
2. Mooney, M., 1940. A theory of large elastic deformation. *J. Appl. Phys.* 11, 582–592.

3. Rivlin, R. S., 1948. Large elastic deformations of isotropic materials, I. Fundamental concepts. *Philos. Trans. R. Soc. Lond. A* 240, 459–490.
4. Fung, Y. C., Fronek, K., and Patitucci, P., 1979. Pseudoelasticity of arteries and the choice of its mathematical expression. *Am. J. Physiol.* 237(5), H620–H631.

FURTHER READING

Findlay, W. N., Lai, J. S., and Onaran, K., 1976. *Creep and Relaxation of Nonlinear Viscoelastic Materials.* Dover Publications, New York.

Fung, Y.-C., 1993. *Biomechanics: Mechanical Properties of Living Tissue*, 2nd edn. Springer-Verlag, New York.

Howard, J., 2001. *Mechanics of Motor Proteins and the Cytoskeleton.* Sinauer Associates, Inc, Sunderland, MA.

Humphery, J. D. and Delange, S. L., 2004. *An Introduction to Biomechanics: Solids and Fluids, Analysis and Design.* Springer, New York.

Meyers, M. A. and Chawla, K. K., 1999. *Mechanical Behavior of Materials.* Prentice Hall, Upper Saddle River, NJ.

Spencer, A. J. M., 2004. *Continuum Mechanics.* Dover Publications, New York.

Timoshenko, S. P. and Goodier, J. N., 1970. *Theory of Elasticity.* McGraw-Hill, Auckland, New Zealand.

2 Fluid Mechanics

Tiffany Camp and Richard Figliola

CONTENTS

2.1 INTRODUCTION

Fluid mechanics is the study of fluids at rest and in motion. This concept plays a critical role in the area of biomedical engineering. To better understand this, let us first discuss "what is a fluid?" A fluid can be described as something that does not hold its form or shape. It is something that is easily deformed (as compared to a solid). Liquids, such as water and oils, and gases, such as air and oxygen, all fit into this category. The classical definition of a fluid is a substance that deforms continuously under constant shearing stress.

Now knowing the definition of a fluid, one can identify many within the human body. They include blood and blood plasma in veins and arteries, airflow within the lungs, and even the flow of waste through the kidneys and urinary system. As engineers, we need to understand the physiological behavior of the fluid systems in the body so that we can further develop solutions to medical problems. In order to develop medical devices like artificial hearts, prosthetic heart valves, and ventricular-assist devices, knowledge of biofluid mechanics is needed. The same is true for the machines such as heart/lung bypass machines and dialysis machines. Therefore, this chapter aims to introduce the fundamentals of fluid mechanics and how it relates to the human body.

The governing equations used in the analyses of fluid mechanics are based on the laws of physics including Newton's second law, the conservation of mass, and the first law of thermodynamics. This discussion will focus on the fundamentals of fluid mechanics and how they apply to biofluids. We will review the following: fluid dynamics, fluid kinematics, finite control volume analysis, differential control volume analysis, and blood characteristics.

2.2 FLUID PROPERTIES AND VISCOSITY

First, it is necessary to go over some properties of fluids that help describe how fluids behave.

There are a few measures that are important to understand in fluid studies. The *density* of a fluid is defined as the mass per unit of volume (kg/m³). The density of water is 999 kg/m³. The density of human blood is 1060 kg/m³. A fluid is considered incompressible if its density change is negligible for large changes in pressure. Mathematically, density is treated as a constant in these cases.

Specific weight is another property that is used in fluid mechanics. It is the weight per unit volume (N/m³). The specific weight can be related to the density of a fluid with the equation

$$\gamma = \rho g \tag{2.1}$$

where g is the local acceleration due to gravity.

The *specific gravity* of a fluid is the dimensionless ratio of the density of a given fluid to the density of a reference fluid. Water is typically used for other liquids and air is used for gases. The specific gravity of a liquid is

$$SG = \frac{\rho}{\rho_{H_2O @ 4°C}} \tag{2.2}$$

where the density of water is taken at a specific temperature, usually 4°C.

One can decipher how "heavy" a fluid is by its density and specific weight. But still there is information missing about how the fluid flows. A fluid with the same density can still flow quite differently. Consider blood and water as an example. Their densities are close; yet, if you have ever gotten a small cut and watched the blood drip, the difference in how the two substances flow is visible. This is not attributed to the density. Instead it is a function of another property used in fluid mechanics known as viscosity.

Viscosity is the measure of the resistance of a fluid to shearing stress. It can best be illustrated by an example. The standard example is to consider a material between two infinitely long and wide plates, as shown in Figure 2.1. The coordinate system is arranged as shown.

In this example, the bottom plate is fixed and the top plate is pulled with a force of magnitude F at a velocity of U m/s. The velocity of the fluid in between the plates is dependent on the vertical position y, and is thus written as $u(y)$. The fluid in contact with the plates is known to "stick" to the surface. This is an experimental observation called the "no-slip" condition that all liquids and gases fulfill. Therefore, the fluid velocities at the "boundaries" are established by conditions of the plate in which they touch. By applying these "no-slip" conditions, one finds that the fluid positioned at the bottom plate is at rest, $u(y=0)=0$ and the fluid touching the top plate moves with a speed of U or $u(y=b)=U$. The velocity profile in this particular case would be linear, given as $u(y)=Uy/b$.

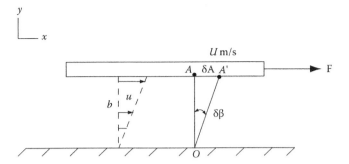

FIGURE 2.1 Fluid placed between two parallel plates.

Figure 2.1 shows the line *OA* that represents the plate's position at $t=0$. At $t=t+\delta t$, the top plate has displaced a distance δa. The new position is denoted by dotted line *OA'*. The distance $\delta a = U\delta t$. Thus, the subsequent angle of rotation, $\delta\beta$, is as follows:

$$\tan\delta\beta = \frac{\delta a}{b} = \frac{U\delta t}{b} \approx \delta\beta$$

In this situation if the top plate is pulled with a higher force, the velocity of the plate will increase. As a result, the angle is a function of both force and time. So the shear stress is related to the rate of shearing strain which is represented by the Greek letter $\dot{\gamma}$. By definition the shearing rate is $\dot{\gamma} = \lim_{\delta t \to 0} \delta\beta/\delta t$. In this case, it is equal to

$$\dot{\gamma} = \frac{U}{b} = \frac{du}{dy}.$$

Thus, the shear stress is directly proportional to the rate of shearing strain:

$$\tau \propto \frac{du}{dy}$$

The constant of proportionality is μ, the *absolute viscosity*. It is also referred to as the dynamic viscosity or just viscosity. Fluids that have this linear relationship between shear stress and the rate of shearing strain are called Newtonian fluids. Water, oils, and air are all Newtonian fluids. For these fluids, the shear stress can be defined as the following equation:

$$\tau = \mu\frac{du}{dy} \tag{2.3}$$

Fluids for which the shearing strain rate is not linearly related to the shear stress are referred to as non-Newtonian fluids. Shear thinning fluids are non-Newtonian fluids whose apparent viscosity decreases as the shear rate increases. A common shear thinning fluid is latex paint. Blood also behaves as a shear thinning fluid at low shear rates. In contrast, shear thickening fluids have apparent viscosities that increase as the shearing strain rate increases. Quicksand is the most commonly presented example of this. Figure 2.2 shows how the rate of shearing strain relates to shear stress for different types of fluids.

Viscosity also appears in another form that includes density. This is the *kinematic viscosity*:

$$n = \frac{\mu}{\rho} \tag{2.4}$$

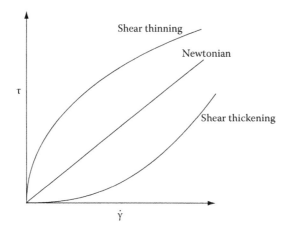

FIGURE 2.2 Shearing stress as a function of rate of shearing strain for different types of fluids.

2.3 CHARACTERISTICS OF BLOOD

Blood is a suspension made up of blood plasma and cells. Blood plasma exhibits Newtonian behavior and has a viscosity of 1.2 cP at 37°C. The types of cells in blood are erythrocytes, leukocytes, and platelets. Erythrocytes, or red blood cells (RBCs), make up ~95% of the cellular content. The RBCs have higher density due to the oxygen-carrying component hemoglobin.

The fraction of RBCs in the blood is given by the hematocrit. It typically ranges from 40% to 50%.

The presence of RBCs causes whole blood to behave as a non-Newtonian fluid at lower shear rates. For shear rates less than $100\,s^{-1}$, blood is shear thinning. However, at shear rates higher than $100\,s^{-1}$, blood behaves as a Newtonian fluid. Figure 2.3 shows the shear stress as a function of rate of shearing strain for blood. The viscosity of blood ranges from 3 to 6 cP or 0.003 to 0.006 N/m² as compared to the viscosity of water which is 0.7 cP at 37°C. The viscosity of blood is also a function of the hematocrit as shown in Figure 2.4. Expectedly as the hemotocrit increases, the viscosity also increases. The viscosity of a fluid is also temperature dependent, but blood is maintained at 37°C.

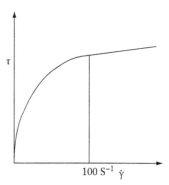

FIGURE 2.3 Shearing stress as a function of rate of shearing strain for blood.

2.4 FLUID KINEMATICS

2.4.1 VELOCITY AND ACCELERATION FIELDS

The position for a particle at time t can be written in vector form in Cartesian coordinates as

$$\mathbf{x} = x(t)\,\hat{\mathbf{i}} + y(t)\,\hat{\mathbf{j}} + z(t)\,\hat{\mathbf{k}} \qquad (2.5)$$

Vector notation is marked by boldface characters.

The velocity vector is the time rate of change of position. This is written as

$$\mathbf{V}(x,\,y,\,z,\,t) = \frac{d\mathbf{x}}{dt} = u\,\hat{\mathbf{i}} + v\,\hat{\mathbf{j}} + w\,\hat{\mathbf{k}} \qquad (2.6)$$

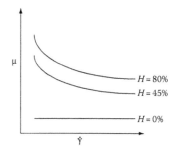

FIGURE 2.4 Viscosity of blood as a function of rate of shearing strain for varying hematocrit levels.

where $\hat{\mathbf{i}}$, $\hat{\mathbf{j}}$, and $\hat{\mathbf{k}}$ are the unit vectors in the x, y, and z directions, respectively. u, v, and w are

$$u = \frac{dx}{dt}, \quad v = \frac{dy}{dt}, \quad w = \frac{dz}{dt}$$

Cylindrical polar coordinates are commonly used in fluid mechanics. They are especially useful in biomedical applications when observed flow takes place through arteries, veins, and other vessels with circular cross sections. Thus, the velocity vector can be written in the following form:

$$\mathbf{V}(r, \theta, z, t) = v_r \hat{\mathbf{e}}_r + v_\theta \hat{\mathbf{e}}_\theta + v_z \hat{\mathbf{e}}_z \tag{2.7}$$

Acceleration is the time rate of change of velocity:

$$a = \frac{\partial \mathbf{V}}{\partial t} + \frac{\partial \mathbf{V}}{\partial x} + \frac{\partial \mathbf{V}}{\partial y} + \frac{\partial \mathbf{V}}{\partial z} \tag{2.8}$$

This can be broken up into components and shown as follows:

$$a_x = \frac{\partial u}{\partial t} + u \frac{\partial u}{\partial x} + v \frac{\partial u}{\partial y} + w \frac{\partial u}{\partial z} \tag{2.8a}$$

$$a_y = \frac{\partial v}{\partial t} + u \frac{\partial v}{\partial x} + v \frac{\partial v}{\partial y} + w \frac{\partial v}{\partial z} \tag{2.8b}$$

$$a_z = \frac{\partial w}{\partial t} + u \frac{\partial w}{\partial x} + v \frac{\partial w}{\partial y} + w \frac{\partial w}{\partial z} \tag{2.8c}$$

The time-dependent term $\partial u/\partial t$ is the local acceleration term. It is the acceleration that a particle experiences relative to time. For example, a particle in pulsatile flow accelerates and decelerates over time. The latter terms (i.e., $u(\partial u/\partial t)$) are the convective acceleration. This is the change in velocity with respect to position. An example of this is flow through a narrowed vessel. The particle's velocity will increase in the region with a smaller cross section.

Acceleration can also be expressed in cylindrical coordinates as shown in Equations 2.9:

$$a_r = \frac{\partial v_r}{\partial t} + v_r \frac{\partial v_r}{\partial r} + \frac{v_\theta}{r} \frac{\partial v_r}{\partial \theta} - \frac{v_\theta^2}{r} + v_z \frac{\partial v_r}{\partial z} \tag{2.9a}$$

$$a_\theta = \frac{\partial v_\theta}{\partial t} + v_r \frac{\partial v_\theta}{\partial r} + \frac{v_\theta}{r} \frac{\partial v_\theta}{\partial \theta} + \frac{v_r v_\theta}{r} + v_z \frac{\partial v_\theta}{\partial z} \tag{2.9b}$$

$$a_z = \frac{\partial v_z}{\partial t} + v_r \frac{\partial v_z}{\partial r} + \frac{v_\theta}{r} \frac{\partial v_z}{\partial \theta} + v_z \frac{\partial v_z}{\partial z} \tag{2.9c}$$

2.4.1.1 Steady and Unsteady Flows

The terms steady flow and unsteady flow relay how the flow and its properties change with respect to time. Real flows are almost always unsteady. However, an assumption of steady flow can be made in many instances and simplifies problems. Steady flows are those whose properties (velocity, pressure, temperature, etc.) at a point in space do not change over time. Unsteady flows may be periodic, non-periodic, or random. The randomness is associated with *turbulent* flow in which mixing occurs. *Laminar* flow is void of this randomness.

2.4.1.2   Eulerian versus Lagrangian

Two methods are used in analyzing fluid mechanics problems. The first is the Eulerian approach. This is examining flow and its properties at a fixed point. It is the equivalent of watching snow from a window and counting the number of flakes that are observed over time. The second is the Lagrangian approach. This the manner of tracking a specific particle as it moves. In contrast to the prior method, in this case the path of a particular snow flake would be followed. The Eulerian approach is easier, though there are some situations in which the Lagrangian method is more beneficial.

2.4.1.3   Control Volume and System

Fluid mechanics problems are solved by applying the fundamental laws of physics. The ways in which this is done varies. This section focuses on two such approaches. One way is to treat the fluid as a system, or a collection of matter of fixed identity. This matter moves, flows, and interacts with its surroundings. The system can be large or small. It can undergo changes in volume, pressure, and temperature. However, the mass of the system remains the same. For example, in studying a problem involving air escaping a tank, the system could be the original mass of air contained.

Another approach to solving fluid mechanics problems is to use a control volume. A control volume is a volume of a certain size through which flow can enter and exit. It can move or deform. The matter within the control volume can change with time as can the amount of mass. The surface of a control volume is referred to as a control surface.

2.4.1.4   Bernoulli Equation

A very familiar equation in fluid mechanics is the Bernoulli equation (Daniel Bernoulli, 1700–1782) shown in Equation 2.10:

$$p_1 + \frac{1}{2}\rho V^2 + \gamma z = \text{constant along a streamline} \tag{2.10}$$

It is obtained by summing the forces on fluid particles along a streamline. This is a very useful tool; however, there were several assumptions that were made during the derivation of this equation that must be taken into account before using this equation. The assumptions are as follows: (1) the flow is incompressible, (2) the flow is steady, (3) the viscous effects can be neglected, and (4) the equation is applied along a streamline.

2.5   FINITE CONTROL VOLUME ANALYSIS

There are three balance relationships that are important in fluid mechanics: the conservation of mass, the conservation of momentum, and the conservation of energy (the first law of thermodynamics).

2.5.1   Reynolds Transport Theorem

The governing equations of motion for a fluid are intended to be used in a system approach. Therefore, a mathematical expression known as the Reynolds transport theorem is used to move from system analysis to control volume analysis. The general form of the Reynolds transport theorem for a fixed, non-deforming control volume is

$$\frac{dB_{sys}}{dt} = \frac{\partial}{\partial t} \int_{CV} \rho b d\mathcal{V} + \int_{CS} \rho b \mathbf{V} \cdot \hat{\mathbf{n}} dA \qquad (2.11)$$

B represents any flow parameter, including velocity, acceleration, temperature, mass and momentum. The lowercase letter b represents the parameter per unit mass such that $B = mb$. The volume is defined with a $\mathcal{V}$ so that it is not confused with the symbol for velocity.

B is the extensive property and b is the intensive property. The left side of the equation is the time rate of change of the parameter B. The first term on the right side is the time rate of change of B in the control volume. The second term on the right is the net effect of the flow into and out of the control volume through each of the control surfaces.

2.5.2   Conservation of Mass

We consider a system to be a collection of unchanging contents. The conservation of mass states that the time rate of change of mass in a system is 0. This is written mathematically as

$$\frac{dM_{sys}}{dt} = 0 \qquad (2.12)$$

where M_{sys} is the mass of the system. It can be written as

$$M_{sys} = \int_{sys} \rho d\mathcal{V} \qquad (2.13)$$

The integration takes places over the volume of the system. Using the Reynolds transport theorem with B representing the mass of the system, the control volume expression for the conservation of mass for a fixed, non-deforming control volume is written as

$$\frac{\partial}{\partial t} \int_{cv} \rho d\mathcal{V} + \int_{cs} \rho \mathbf{V} \cdot \hat{\mathbf{n}} dA = 0 \qquad (2.14)$$

where the first term is the time rate of change of the mass in the control volume and the second term is net rate flow of mass through the control surface. Equation 2.14 is also called the *continuity equation*.

Let us look at the two terms more closely. The first integral is to be carried out over the entire control volume. Moreover, if the flow in question was steady, the first term would be equal to zero as follows:

$$\int_{cs} \rho \mathbf{V} \cdot \hat{\mathbf{n}} dA = 0 \qquad (2.15)$$

The integrand $\rho \mathbf{V} \cdot \hat{\mathbf{n}} dA$ in the second term is the mass flow rate through the surface area dA.

It also known that mass flow rate, $\dot{m}$ (kg/s), through a surface area is the product of the density and the volumetric flow rate, Q (m³/s). The average velocity, $\bar{V}$, is used to calculate Q.

$$\dot{m} = \rho Q = \rho \bar{V} A \tag{2.16}$$

Also,

$$\dot{m} = \int_{Area} \rho \mathbf{V} \cdot \hat{\mathbf{n}} \, dA \tag{2.17}$$

Thus, the average velocity, $\bar{V}$, can also be defined as

$$\bar{V} = \frac{\displaystyle\int_{Area} \rho \mathbf{V} \cdot \hat{\mathbf{n}} \, dA}{\rho A} \tag{2.18}$$

The conservation of mass for a finite control volume can be used to find global information about the flow velocity and flow rates.

2.5.3 CONSERVATION OF MOMENTUM

Everyone is familiar with Newton's second law of motion in the form of

$$F = ma \tag{2.19}$$

Newton's second law for a system, in words, states that the time rate of change of the linear momentum of the system = sum of external forces acting on the system:

$$\frac{d\mathbf{P}_{sys}}{dt} = \sum \mathbf{F}_{sys} \tag{2.20}$$

where $\mathbf{P} = m\mathbf{V}$, the linear momentum. The system forces, F_{sys}, are equal to the forces acting on a control volume when said control volume is coincident with the system. So, in the same manner as the conservation of mass for a finite control volume was expressed, the conservation of linear momentum for a fixed, non-deforming control volume is

$$\frac{\partial}{\partial t} \int_{cv} \mathbf{V} \rho \, d\mathcal{V} + \int_{cs} \mathbf{V} \rho \mathbf{V} \cdot \hat{\mathbf{n}} \, dA = \sum \mathbf{F} \tag{2.21}$$

The sign notation of the forces should be consistent with the coordinate system chosen for the problem. By using the conservation of momentum along with the conservation of mass, information about flow-induced forces can be determined.

2.5.4 CONSERVATION OF ENERGY

Finally, the conservation of energy is stated in Equation 2.22.

$$\frac{dE_{sys}}{dt} = \dot{Q} + \dot{W} \tag{2.22}$$

E_{sys} is the total energy and can also be written as the product of the density and the volume

$$E_{sys} = e\rho d\forall \tag{2.23}$$

The total energy per unit mass is represented by the symbol e, and is equal to the sum of the internal energy, $\breve{u}$, potential energy, gz, and kinetic energy, $V^2/2$.

$$e = \breve{u} + gz + \frac{V^2}{2} \tag{2.24}$$

The conservation of energy for a fixed, non-deforming system

$$\frac{\partial}{\partial t} \int_{cv} e\rho d\forall + \int_{cs} e\rho \mathbf{V} \cdot \hat{\mathbf{n}} \, dA = \left(\dot{Q} + \dot{W} \right)_{cv} \tag{2.25}$$

Combining Equations 2.24 and 2.25 yields

$$\frac{\partial}{\partial t} \int_{cv} \rho \left(\breve{u} + gz + \frac{V^2}{2} \right) d\forall + \int_{cs} \rho \left(\breve{u} + gz + \frac{V^2}{2} \right) \mathbf{V} \cdot \hat{\mathbf{n}} \, dA = \left(\dot{Q} + \dot{W} \right)_{cv} \tag{2.26}$$

The conservation of energy has many applications. One specific application is how it is used in flow through pipes or other conduits. This will be discussed later.

2.5.5 APPLICATIONS IN BIOMEDICAL ENGINEERING

It is noteworthy that the above relations provide global information about a system such as the average velocity exiting through a volume or the net force acting on a body. This information can be gathered without considering if a fluid is Newtonian or non-Newtonian. The conservation of mass can be used to evaluate flows through vessels with different cross-sectional area. Consider an in vitro experiment that models flow through a pulmonary artery branch where one branch has a smaller cross-sectional area than the other. The conservation of mass can be used to determined flow velocities in each branch for a given flow rate. Similar applications include blood flow through stenosed vessels or airflow through restricted bronchial tubes. The conservation of momentum, when coupled with the continuity equation, can be used to find the forces created by the flow. In the lab environment, this can be beneficial when planning a setup involving piping that needs to be supported. The forces that result from the flow through any bends in the system would need to be sufficiently compensated.

While control volume analysis can be applied to many fluids mechanics problems, it is many times necessary to know more detailed information about the flow. For instance, in cardiovascular flows, the levels of shear stresses that the artery walls are exposed to are of particular interest. Therefore, another type of analysis is needed.

2.6 DIFFERENTIAL FLUID ELEMENT ANALYSIS

Analysis of a differential fluid element yields more detailed information about the flow and how it varies with time and position as opposed to the averaged information that the finite control volume analysis produces.

2.6.1 Continuity Equation

The differential form of the continuity equation can be found by applying the conservation of mass to a differential element.

Figure 2.5a shows the differential element, a small cube of size δx, δy, and δz and density ρ. The first term of the continuity equation (Equation 2.14) can be written as

$$\frac{\partial}{\partial t} \int_{cv} \rho \, d\mathcal{V} \approx \frac{\partial \rho}{\partial t} \delta x \delta y \delta z \qquad (2.27)$$

In Figure 2.5b, the mass flow rate through the cube is graphically explained. The x-direction flow rate at the center of the element is ρu. Thus, the flow rate exiting the element is

$$\rho u \Big|_{x+\left(\frac{\delta x}{2}\right)} = \rho u + \frac{\partial(\rho u)}{\partial x} \frac{\delta x}{2}$$

and the flow entering the element is

$$\rho u \Big|_{x-\left(\frac{\delta x}{2}\right)} = \rho u - \frac{\partial(\rho u)}{\partial x} \frac{\delta x}{2}$$

These previous two equations are in actuality a Taylor series expansion of ρu with the higher order terms neglected. We can multiply the above relations by the surface area $\delta y \delta z$ to get net rate of mass flow in the x-direction through the surfaces.

$$\left(\rho u + \frac{\partial(\rho u)}{\partial x} \frac{\delta x}{2} \right) \delta y \delta z - \left(\rho u - \frac{\partial(\rho u)}{\partial x} \frac{\delta x}{2} \right) \delta y \delta z = \frac{\partial(\rho u)}{\partial x} \delta x \delta y \delta z$$

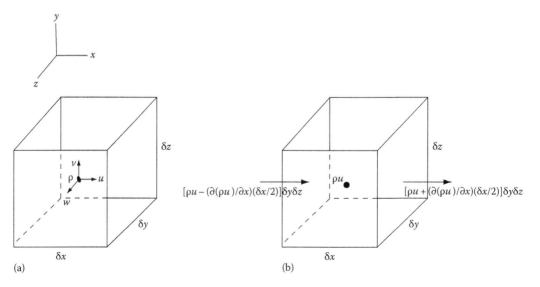

FIGURE 2.5 A differential element used in the derivation of the conservation of mass (a) with dimension size δx, δy, δz and density ρ (b) with the mass flow rate terms shown.

Following the same methods for the y and z directions shows that net rate of mass flow in those directions are $(\delta(\rho v)/dy)\delta x\delta y\delta z$ and $(\delta(\rho w)/\partial z)\delta x\delta y\delta z$.

Thus,

$$\text{Net rate of mass flow in all directions} = \left(\frac{\partial(\rho u)}{\partial x} + \frac{\partial(\rho v)}{\partial y} + \frac{\partial(\rho w)}{\partial z}\right)\delta x\delta y\delta z \tag{2.28}$$

Using Equations 2.14, 2.27, and 2.28, the conservation of mass for a differential element is

$$\frac{\partial \rho}{\partial t} + \frac{\partial(\rho u)}{\partial x} + \frac{\partial(\rho v)}{\partial y} + \frac{\partial(\rho w)}{\partial z} = 0 \tag{2.29}$$

As stated before, cylindrical coordinates can be used in cases with flow through a circular conduit. Thus, it is important to know the differential forms of the balances of equations in this alternate coordinate system. So, the continuity equation is

$$\frac{\partial \rho}{\partial t} + \frac{1}{r}\frac{\partial(r\rho v_r)}{\partial r} + \frac{1}{r}\frac{\partial(\rho v_\theta)}{\partial \theta} + \frac{\partial(\rho v_z)}{\partial z} = 0 \tag{2.30}$$

2.6.2 Conservation of Linear Momentum

We can develop the differential linear momentum equation by the system approach and applying Equation 2.20 to a differential mass $\delta m = \rho\delta x\delta y\delta z$. The equation can be expressed as

$$\delta\mathbf{F} = \frac{d(\mathbf{V}\delta m)}{dt}$$

The mass, δm, can be treated as a constant and the definition of acceleration can be applied so the above equation becomes

$$\delta\mathbf{F} = \delta m\mathbf{a} \tag{2.31}$$

Figure 2.6 shows an element is of volume $\delta x\delta y\delta z$. For simplicity, only the x-direction forces are shown in Figure 2.6. Body forces and surface forces act on the element. The body forces considered to be gravitational forces are given as

$$\delta\mathbf{F}_b = \delta m\mathbf{g} \tag{2.32}$$

where $\mathbf{g} = g_x\hat{\mathbf{i}} + g_y\hat{\mathbf{j}} + g_z\hat{\mathbf{k}}$.

The surface forces are shear and normal stresses (σ) and hydrostatic forces. The stresses σ_{xx}, σ_{yy}, and σ_{zz} are at the center of the element. The Taylor series expansion excluding the higher order terms is used to get the surface forces. The sum of the surface forces in the x-direction are

$$\delta F_{sx} = \left(\frac{\partial\sigma_{xx}}{\partial x} + \frac{\partial\tau_{yx}}{\partial y} + \frac{\partial\tau_{zx}}{\partial z}\right)\delta x\delta y\delta z \tag{2.33a}$$

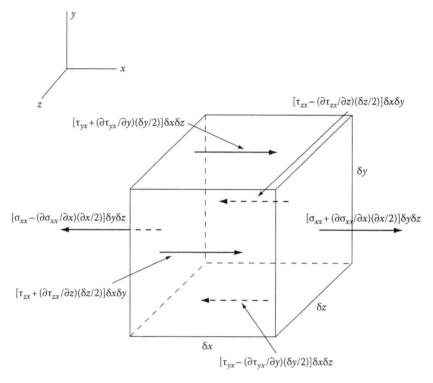

FIGURE 2.6 A differential element showing the acting surface forces in the *x*-direction.

Likewise, the *y*- and *z*-direction summed forces are

$$\delta F_{sy} = \left(\frac{\partial \tau_{xy}}{\partial x} + \frac{\partial \sigma_{yy}}{\partial y} + \frac{\partial \tau_{zy}}{\partial z} \right) \delta x \delta y \delta z \tag{2.33b}$$

$$\delta F_{sz} = \left(\frac{\partial \tau_{xz}}{\partial x} + \frac{\partial \tau_{yz}}{\partial y} + \frac{\partial \sigma_{zz}}{\partial z} \right) \delta x \delta y \delta z \tag{2.33c}$$

Thus, the total surface force is

$$\delta \mathbf{F}_s = \delta F_{sx} \hat{\mathbf{i}} + \delta F_{sy} \hat{\mathbf{j}} + \delta F_{sz} \hat{\mathbf{k}} \tag{2.34}$$

Equation 2.31 expands to

$$\rho g_x + \frac{\partial \sigma_{xx}}{\partial x} + \frac{\partial \tau_{yx}}{\partial y} + \frac{\partial \tau_{zx}}{\partial z} = \rho \left(\frac{\partial u}{\partial t} + u \frac{\partial u}{\partial x} + v \frac{\partial u}{\partial y} + w \frac{\partial u}{\partial z} \right) \tag{2.35a}$$

$$\rho g_y + \frac{\partial \tau_{xy}}{\partial x} + \frac{\partial \sigma_{yy}}{\partial y} + \frac{\partial \tau_{zy}}{\partial z} = \rho \left(\frac{\partial v}{\partial t} + u \frac{\partial v}{\partial x} + v \frac{\partial v}{\partial y} + w \frac{\partial v}{\partial z} \right) \tag{2.35b}$$

$$\rho g_z + \frac{\partial \tau_{xz}}{\partial x} + \frac{\partial \tau_{yz}}{\partial y} + \frac{\partial \sigma_{zz}}{\partial z} = \rho \left(\frac{\partial w}{\partial t} + u \frac{\partial w}{\partial x} + v \frac{\partial w}{\partial y} + w \frac{\partial w}{\partial z} \right) \tag{2.35c}$$

2.6.3 EULER'S EQUATIONS OF MOTION

Euler's equations of motion (Leonhard Euler, 1707–1783) are for flow in which the viscous effects are neglected. This is referred to as inviscid flow. Therefore, the shearing stresses are all equal to zero and

$$-p = \sigma_{xx} = \sigma_{yy} = \sigma_{zz}$$

As a result, Equations 2.35 become
x-direction

$$\rho\left(\frac{\partial u}{\partial t} + u\frac{\partial u}{\partial x} + v\frac{\partial u}{\partial y} + w\frac{\partial u}{\partial z}\right) = -\frac{\partial p}{\partial x} + \rho g_x \qquad (2.36a)$$

y-direction

$$\rho\left(\frac{\partial v}{\partial t} + u\frac{\partial v}{\partial x} + v\frac{\partial v}{\partial y} + w\frac{\partial v}{\partial z}\right) = -\frac{\partial p}{\partial y} + \rho g_y \qquad (2.36b)$$

z-direction

$$\rho\left(\frac{\partial w}{\partial t} + u\frac{\partial w}{\partial x} + v\frac{\partial w}{\partial y} + w\frac{\partial w}{\partial z}\right) = -\frac{\partial p}{\partial z} + \rho g_z \qquad (2.36c)$$

These equations contain nonlinear velocity terms that make it difficult to solve analytically. However, they can be simplified with assumptions.

2.6.3.1 Navier–Stokes Equations

The *Navier–Stokes* equations are the governing equations (Equation 2.35) in terms of velocity. This can be achieved for Newtonian incompressible fluids by using the *Navier–Poisson* (L.M.H. Navier, 1785–1836 and S.D. Poisson, 1781–1840) equations. These equations define the relationship between the stresses and rate of deformation for Newtonian incompressible fluids. The normal stresses are expressed in Cartesian coordinates as

$$\sigma_{xx} = -p + 2\mu\frac{\partial u}{\partial x} \qquad (2.37a)$$

$$\sigma_{yy} = -p + 2\mu\frac{\partial v}{\partial y} \qquad (2.37b)$$

$$\sigma_{zz} = -p + 2\mu\frac{\partial w}{\partial z} \qquad (2.37c)$$

Further, the shear stresses in Cartesian coordinates are expressed as

$$\tau_{zx} = \tau_{xz} = \mu\left(\frac{\partial w}{\partial x} + \frac{\partial u}{\partial z}\right) \qquad (2.38a)$$

$$\tau_{yz} = \tau_{zy} = \mu\left(\frac{\partial v}{\partial z} + \frac{\partial w}{\partial y}\right) \tag{2.38b}$$

$$\tau_{xy} = \tau_{yx} = \mu\left(\frac{\partial u}{\partial y} + \frac{\partial v}{\partial x}\right) \tag{2.38c}$$

The *Navier–Stokes* (L. M. H. Navier, 1785–1836 and Sir G. G. Stokes, 1819–1903) equations can be obtained by substituting the above equations with the differential form of the conservation of linear momentum (Equations 2.39) and simplifying using the continuity equation (Equation 2.32). First, the expressions for the *Navier–Stokes* in Cartesian coordinates are
x-direction

$$\rho\left(\frac{\partial u}{\partial t} + u\frac{\partial u}{\partial x} + v\frac{\partial u}{\partial y} + w\frac{\partial u}{\partial z}\right) = -\frac{\partial p}{\partial x} + \rho g_x + \mu\left(\frac{\partial^2 u}{\partial x^2} + \frac{\partial^2 u}{\partial y^2} + \frac{\partial^2 u}{\partial z^2}\right) \tag{2.39a}$$

y-direction

$$\rho\left(\frac{\partial v}{\partial t} + u\frac{\partial v}{\partial x} + v\frac{\partial v}{\partial y} + w\frac{\partial v}{\partial z}\right) = -\frac{\partial p}{\partial y} + \rho g_y + \mu\left(\frac{\partial^2 v}{\partial x^2} + \frac{\partial^2 v}{\partial y^2} + \frac{\partial^2 v}{\partial z^2}\right) \tag{2.39b}$$

z-direction

$$\rho\left(\frac{\partial w}{\partial t} + u\frac{\partial w}{\partial x} + v\frac{\partial w}{\partial y} + w\frac{\partial w}{\partial z}\right) = -\frac{\partial p}{\partial z} + \rho g_z + \mu\left(\frac{\partial^2 w}{\partial x^2} + \frac{\partial^2 w}{\partial y^2} + \frac{\partial^2 w}{\partial z^2}\right) \tag{2.39c}$$

The following are the *Navier–Stokes* equations in cylindrical polar coordinates form:
r-direction

$$\rho\left(\frac{\partial v_r}{\partial t} + v_r\frac{\partial v_r}{\partial r} + \frac{v_\theta}{r}\frac{\partial v_r}{\partial \theta} - \frac{v_\theta^2}{r} + v_z\frac{\partial v_r}{\partial z}\right)$$

$$= -\frac{\partial p}{\partial r} + \rho g_r + \mu\left[\frac{1}{r}\frac{\partial}{\partial r}\left(r\frac{\partial v_r}{\partial r}\right) - \frac{v_r}{r} + \frac{1}{r^2}\frac{\partial^2 v_r}{\partial \theta^2} - \frac{2}{r^2}\frac{\partial v_\theta}{\partial \theta} + \frac{\partial^2 v_r}{\partial z^2}\right] \tag{2.40a}$$

θ-direction

$$\rho\left(\frac{\partial v_\theta}{\partial t} + v_r\frac{\partial v_\theta}{\partial r} + \frac{v_\theta}{r}\frac{\partial v_\theta}{\partial \theta} + \frac{v_r v_\theta}{r} + v_z\frac{\partial v_\theta}{\partial z}\right)$$

$$= -\frac{1}{r}\frac{\partial p}{\partial \theta} + \rho g_\theta + \mu\left[\frac{1}{r}\frac{\partial}{\partial r}\left(r\frac{\partial v_\theta}{\partial r}\right) - \frac{v_\theta}{r^2} + \frac{1}{r^2}\frac{\partial^2 v_\theta}{\partial \theta^2} + \frac{2}{r^2}\frac{\partial v_r}{\partial \theta} + \frac{\partial^2 v_\theta}{\partial z^2}\right] \tag{2.40b}$$

z-direction

$$\rho\left(\frac{\partial v_z}{\partial t}+v_r\frac{\partial v_z}{\partial r}+\frac{v_\theta}{r}\frac{\partial v_z}{\partial\theta}+v_z\frac{\partial v_z}{\partial z}\right)$$

$$=-\frac{\partial p}{\partial z}+\rho g_z+\mu\left[\frac{1}{r}\frac{\partial}{\partial r}\left(r\frac{\partial v_z}{\partial r}\right)+\frac{1}{r^2}\frac{\partial^2 v_z}{\partial\theta^2}+\frac{\partial^2 v_z}{\partial z^2}\right]\qquad(2.40c)$$

The *Navier–Stokes* equations are complex due to their nonlinearity. They can be approximated numerically; however, no analytical solution exists. There are some special cases in which the nonlinear terms go away and these will be discussed.

2.6.3.1.1   Steady, Laminar Flow between Two Fixed Parallel Plates

For the first example, we consider flow between two infinitely long flat plates in the horizontal plane as shown in Figure 2.7. The flow is assumed to be laminar, Newtonian, and steady. It is also only in the x-direction. First, the continuity equation (Equation 2.29) is evaluated. Since the flow is steady, $\partial u/\partial t=0$. As stated, $v=w=0$; therefore, their derivatives are 0. That leaves, $\partial u/\partial x=0$ meaning that the velocity in the x-direction does not vary with position (fully developed); thus, u is only a function of y. We set $g_x=g_z=0$ and $g_y=-g$.

The Navier–Stokes equations reduce to

$$0=-\frac{\partial p}{\partial x}+\mu\left(\frac{\partial^2 u}{\partial y^2}\right)\qquad(2.41)$$

$$0=-\frac{\partial p}{\partial y}-\rho g\qquad(2.42)$$

$$0=-\frac{\partial p}{\partial z}\qquad(2.43)$$

Integrating Equation 2.42 yields a relationship between pressure and gravity.

$$p=-\rho gy+c_1\qquad(2.44)$$

This is similar to the hydrostatic pressure variation in a fluid. The constant, c_1, will be dependent on the x position since a pressure gradient in that direction, $\partial p/\partial x$ exists. Therefore, $c_1=f(x)$. Equation 2.43 simply shows that the pressure does not vary in the z-direction.

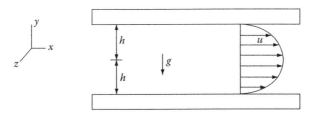

FIGURE 2.7   Viscous flow between two infinitely long, parallel plates separated by distance $2h$.

The pressure gradient $\partial p/\partial x$ is treated as a constant so that Equation 2.41 can be integrated twice as follows:

$$\frac{d^2u}{dy^2} = -\frac{1}{\mu}\frac{\partial p}{\partial x}$$

$$\frac{du}{dy} = -\frac{1}{\mu}\left(\frac{\partial p}{\partial x}\right)y + c_1$$

$$u = -\frac{1}{2\mu}\left(\frac{\partial p}{\partial x}\right)y^2 + c_1 y + c_2 \tag{2.45}$$

It is known that $u=0$ at the plates where $y=\pm h$. Using these boundary conditions, c_1 is found to be 0 and c_2 is

$$c_2 = -\frac{1}{2\mu}\left(\frac{\partial p}{\partial x}\right)h^2$$

Finally, the velocity profile between the two plates can be written as

$$u = -\frac{1}{2\mu}\left(\frac{\partial p}{\partial x}\right)\left(y^2 - h^2\right) \tag{2.46}$$

A few other relations can be derived from this problem. The volumetric flow rate on a per unit width basis can be found by integrating the velocity over the cross-sectional area between the plates:

$$q = \int_{-h}^{h} u\, dy = \int_{-h}^{h} -\frac{1}{2\mu}\left(\frac{\partial p}{\partial x}\right)\left(y^2 - h^2\right)dy$$

This is equal to

$$q = -\frac{2h^3}{3\mu}\left(\frac{dp}{dx}\right) \tag{2.47}$$

The velocity can also be expressed in terms of the pressure gradient as well:

$$V = \frac{Q}{A} = \frac{q}{2h} = -\frac{h^2}{2\mu}\left(\frac{dp}{dx}\right) \tag{2.48}$$

2.6.3.1.2 *Steady, Laminar Flow in Circular Tubes*

The next example has many applications in biofluid mechanics. It involves laminar flow through circular tubes (blood vessels, airway passages, etc.). In fact it is based on the work of a physician (J.L. Poiseuille, 1799–1869) who was interested in blood flow through capillaries. This type of

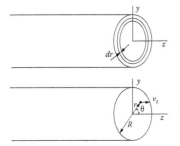

FIGURE 2.8 Viscous flow in a circular tube of radius R.

flow is often referred to as Poiseuille flow or Hagen–Poiseuille flow based on the contributions by Poiseuille and later by G.H.L. Hagen (1797–1884).

Figure 2.8 shows the schematic of the circular tube. Cylindrical coordinates are used in this problem based on the geometry involved. The flow moves only in the axial (z) direction. Just as in the previous example, the flow is also assumed to be steady, laminar, Newtonian and incompressible. Since the flow is steady and unidirectional, the continuity equation reduces to $\partial v_z/\partial z = 0$. Along the same lines, the *Navier–Stokes* equations become

$$0 = -\rho g \sin\theta - \frac{\partial p}{\partial r} \tag{2.49}$$

$$0 = -\rho g \cos\theta - \frac{1}{r}\frac{\partial p}{\partial \theta} \tag{2.50}$$

$$0 = -\frac{\partial p}{\partial z} + \mu\left[\frac{1}{r}\frac{\partial}{\partial r}\left(r\frac{\partial v_z}{\partial r}\right)\right] \tag{2.51}$$

The gravitational forces act as $g_r = -g\sin\theta$ and $g_\theta = -g\cos\theta$

Both Equations 2.49 and 2.50 can be integrated one time to yield

$$p = -\rho g r \sin\theta + c_1 \tag{2.52}$$

where $c_1 = f(z)$. It can be noted that the quantity $r\sin\theta = y$. Therefore, the pressure in this problem also only varies hydrostatically at any point in the cross section.

The final equation can be rearranged as

$$\frac{1}{r}\frac{\partial}{\partial r}\left(r\frac{\partial v_z}{\partial r}\right) = \frac{1}{\mu}\frac{\partial p}{\partial z}$$

Integrating twice yields

$$r\frac{\partial v_z}{\partial r} = \frac{1}{2\mu}\left(\frac{\partial p}{\partial z}\right)r^2 + c_1$$

so that

$$v_z = \frac{1}{4\mu}\left(\frac{\partial p}{\partial z}\right)r^2 + c_1\ln r + c_2 \tag{2.53}$$

The boundary conditions for this problem are $v_z = 0$ when $r = R$ and v_z must be finite at $r = 0$. The only solution for c_1 that fulfills this requirement is for $c_1 = 0$. That leaves

$$c_2 = -\frac{1}{4\mu}\left(\frac{\partial p}{\partial z}\right)R^2$$

So, the velocity profile in the circular tube is given as

$$v_z = \frac{1}{4\mu}\left(\frac{\partial p}{\partial z}\right)\left(r^2 - R^2\right) \tag{2.54}$$

The volumetric flow rate can be found by integrating over the cross-sectional area of the tube (Equation 2.55). We know that $dA = 2\pi dr$ where dA is the differential area. As a result

$$Q = 2\pi \int_0^R v_z r \, dr \tag{2.55}$$

Therefore,

$$Q = -\frac{\pi R^4}{8\mu}\left(\frac{\partial p}{\partial z}\right) \tag{2.56}$$

The corresponding average velocity can be found by dividing the volumetric flow rate by the area such that

$$V = \frac{Q}{A} = -\frac{R^2}{8\mu}\left(\frac{\partial p}{\partial z}\right) \tag{2.57}$$

The velocity can also be put in terms of maximum velocity, V_{max}, which occurs at $r = 0$.

$$V = V_{max}\left(1 - \frac{r^2}{R^2}\right) \tag{2.58}$$

2.6.3.1.3 Pulsatile Flow in Circular Tube

The previous solution for flow in a circular tube can be reworked to include an unsteady contribution based on Womersley's approach.

The pressure gradient is described by multiple harmonics such that

$$\frac{\partial p}{\partial z} = \phi_0 + \sum^N \left(\phi_n \cos n\omega t + \psi_n \sin n\omega t\right) \tag{2.59}$$

where

ϕ_0 is the steady portion of the pressure gradient
ϕ_n and ψ_n are Fourier coefficients for the nth harmonic
$i = \sqrt{-1}$

For the sake of space, Equation 2.59 can also be expressed as

$$\frac{\partial p}{\partial z} = \phi_0 + \sum_{n=1}^N \text{Real}\left(\Psi_n e^{in\omega t}\right) \tag{2.60}$$

where we take only the real part of

$$\Psi e^{i\omega t} = \phi \cos \omega t + \psi \sin \omega t + i\left(\phi \sin \omega t - \psi \cos \omega t\right)$$

The *Navier–Stokes* equations are similar to those used in the steady flow example, except the time term remains.

$$0 = -\rho g \sin\theta - \frac{\partial p}{\partial r} \tag{2.61}$$

$$0 = -\rho g \cos\theta - \frac{1}{r}\frac{\partial p}{\partial \theta} \tag{2.62}$$

$$\rho \frac{\partial v_z}{\partial t} = -\frac{\partial p}{\partial z} + \mu \left[\frac{1}{r}\frac{\partial}{\partial r}\left(r\frac{\partial v_z}{\partial r} \right) \right] \tag{2.63}$$

The third of these equations is of interest here. It can be rewritten as

$$\rho \frac{\partial v_z(r,t)}{\partial t} + \frac{\partial p(t)}{\partial z} = \mu \left[\frac{1}{r}\frac{\partial}{\partial r}\left(r\frac{\partial v_z(r,t)}{\partial r} \right) \right] \tag{2.64}$$

Thus, emphasizing that the velocity and pressure are both functions of time. The complete solution to this problem is the sum of the steady flow solution and the unsteady flow solution. Since, the steady flow solution has already been solved; here we will only look at the unsteady flow solution.

We will again treat $\partial p/\partial z$ as a constant in the axial direction. After combining Equations 2.60 and 2.64 we get

$$\mu \left[\frac{1}{r}\frac{\partial}{\partial r}\left(r\frac{\partial v_z(r,t)}{\partial r} \right) \right] - \rho \frac{\partial v_z(r,t)}{\partial t} = \Psi_n e^{in\omega t} \tag{2.65}$$

Separation of variables is used to get one part of the equation in terms of r and the other in terms of t.

$$\frac{d^2 V_n}{dr^2} + \frac{1}{r}\frac{dV_n}{dr} + \lambda^2 V_n = \frac{\Psi}{\mu} \tag{2.66}$$

where $\lambda^2 = i^3 n\omega\rho/\mu$. This is now in the form a standard Bessel equation with a solution in the form

$$y(x) = c_1 J_0(x) + c_2 Y_0(x)$$

$J_0(x)$ is a zeroth order Bessel function of the first kind and $Y_0(x)$ is a Bessel function of the second kind.

The final solution (Equation 2.67) is obtained by solving the homogenous and particular solutions using the following boundary conditions: $V_n(r)$ must be finite when $r=0$ and $V_n(r)=0$ when $r=R$. For brevity, the details of this derivation will not be discussed here.

$$v_z(r,t) = v_{steady}(r) + \text{Real}\left\{ \sum_{n=1}^{N} \left[\frac{\Psi}{\mu\lambda_n^2}\left(1 - \frac{J_0(\lambda_n r)}{J_0(\lambda_n R)} \right) e^{in\omega t} \right] \right\} \tag{2.67}$$

where v_{steady} is the steady flow solution given in Equation 2.54 and the number of harmonics, $n = 1, 2, \ldots, N$.

2.6.4 APPLICATIONS IN BIOMEDICAL ENGINEERING

The exact solutions presented can be analogous to many meaningful biomedical applications. As stated in the introduction to this section, differential element analysis of a problem provides detailed information including velocity profile, pressure distribution, and shear stresses in a given flow field. These examples are all based on flows with limitations (laminar, incompressible, steady, Newtonian); however, they can be useful for engineers in biomedical engineering fields. For example, parallel-plate flow chambers are used to study the effects of shear stresses on cells such as in the experiments done by Van Kooten et al. (1992). The cells can be positioned between two plates and exposed to a known flow. Measurements are made accordingly so that the shear stresses levels can be found. Poiseuille's flow can be utilized in a similar manner to find the shear stresses that act on the walls of blood vessels.

2.7 PIPE FLOW

Pipe flow is the general term used to describe flow in closed conduit. The implication is not that all pipes have round cross sections. This topic involves flow through piping, whether it be water in pipes, air through duct work, or natural gas in pipe work supplying a home. The understanding of pumping systems is commonly included in the study of pipe flow.

2.7.1 REYNOLDS NUMBER

The Reynolds number is a well-known parameter in fluid mechanics that must be included in the discussion of pipe flow. It is a dimensionless parameter named in honor of Osborne Reynolds (1842–1912):

$$Re = \frac{\rho V d}{\mu} \tag{2.68}$$

It is the ratio of inertia forces to viscous forces and is used to differentiate between laminar flow and turbulent flow. For small values of Re, the inertial effects are negligible relative to the viscous effects. On the contrary, for large Re, the inertial forces are dominant. Flow can be characterized as laminar, transitional, or turbulent. For pipe flow, flow is considered laminar if $Re < 2100$. Flow is turbulent if $Re > 4000$. If the Reynolds number is in the middle of this range, the flow is transitional.

2.7.2 ENTRANCE LENGTH/FULLY DEVELOPED FLOW

When flow enters a pipe its flow profile is disrupted. The fluid "sticks" to the pipe walls, and the velocity profile changes. A boundary layer is created. The flow continues to develop for a distance known as the entrance length. During this length the viscous effects are very important and the velocity profile continues to change. Once the velocity profile no longer changes, the flow is considered *fully developed*.

2.7.3 LAMINAR FLOW IN PIPES

A relationship between flow rate and pressure gradient in a circular tube was obtained through the analysis using the *Navier–Stokes* equations. Recall Equation 2.56.

$$Q = -\frac{\pi R^4}{8\mu}\left(\frac{\partial p}{\partial z}\right)$$

There is negative sign present because the flow moves from a region of higher pressure to lower pressure. Thus, the change in pressure with respect to distance is negative. In pipe flow, the pressure gradient term is replaced with a pressure *drop* as shown

$$-\frac{\partial p}{\partial x} = \frac{\Delta p}{\Delta x} = \frac{\Delta p}{l} \tag{2.69}$$

where l is the distance between two points at which the pressures were measured. Equation 2.56 becomes

$$Q = \frac{\pi R^4}{8\mu}\left(\frac{\Delta p}{l}\right) \tag{2.70}$$

From this relationship we can surmise a few things. First, it is clear that the flow rate is proportional to the pressure drop. If there is no pressure difference (and no gravity), then there will be no flow. Flow is also inversely proportional to the viscosity and pipe length. Most notable is the fact that the flow rate is very much dependent on the pipe size. It is directly proportional to the pipe radius (or diameter) to the fourth power. This equation is valid for steady, laminar, fully developed flow in a horizontal pipe. If the pipe is not horizontal, gravitational effects are introduced and must be taken into consideration.

2.7.4   Energy Equation

The conservation of energy was earlier derived for a control volume analysis (Equation 2.26). This relationship is useful and necessary when discussing flow in pipes. When the conservation of energy is applied to steady, incompressible flow, it is rewritten as

$$\frac{p_1}{\gamma} + \frac{V_1^2}{2g} + z_1 = \frac{p_2}{\gamma} + \frac{V_2^2}{2g} + z_2 + h_L + h_s \tag{2.71}$$

The flow properties are evaluated at two control surfaces, 1 and 2. h_s is the work done on the system by a pump or produced by the system with a turbine. Head is the amount of energy per unit weight. h_L is the head loss due to friction and other elements of resistance in the piping system.

2.7.5   Head Losses

Head loss is broken into two categories: minor and major. Major head losses are due to friction and minor losses are created by piping system components such as valves, elbows, contractions, etc. Major head loss is defined as

$$h_L = f\frac{l}{D}\frac{V^2}{2g} \tag{2.72}$$

where
   $D = 2R$
   f is a friction factor.

The friction factor is a function of the Reynolds number and the surface roughness (ε/D) of the pipe. Since f is dependent on Re, it is different for laminar and turbulent flows. For laminar flows, the friction factor is only a function of the Reynolds number. It can be calculated as

$$f = \frac{64}{Re} \tag{2.73}$$

Determining the friction factor for turbulent flows is a little more involved. The Moody chart (Moody 1944) plots the results from experiments and gives the friction factor as a function of Re for many lines of constant ε/D. Additionally, there is an equation that approximates the nonlaminar range of the Moody chart:

$$\frac{1}{\sqrt{f}} = -2.0 \log \left(\frac{\varepsilon/D}{3.7} + \frac{2.51}{Re\sqrt{f}} \right) \tag{2.74}$$

Despite the implicitness of this equation, it can be solved either through an iterative scheme or with the use of most scientific calculators.

Minor head loss is defined as

$$h_{Lminor} = K_L \frac{V^2}{2g} \tag{2.75}$$

where K_L is the loss coefficient for a piping component (elbows, tees, valves, entrances, exits, etc.). These tabulated values can be found in many sources.

2.7.6 Applications in Biomedical Engineering

Pipe flow applications are plentiful in the world of researchers, especially experimentalists. Therefore, the knowledge of selecting the proper pump and piping system is beneficial. Take for instance a lab that conducts in vitro studies on heart valves. Mock circulatory systems such as the one described by Camp et al. (2007) are commonly used to mimic the flow through heart valves. For these applications, the pumps have to be properly sized so that the flow could overcome the resistance in the piping system resulting from tubing and fittings.

REFERENCES

Camp, T. A., K. C. Stewart, R. S. Figliola, and T. McQuinn. 2007. In vitro study of flow regulation for pulmonary insufficiency. *Journal of Biomechanical Engineering* 129(2):284–288.
Moody, L. F. 1944. Friction factors for pipe flow. *Trans of the ASME* 66:671–784.
Van Kooten, T. G., J. M. Schakenraad, H. C. Van der Mei, and H. J. Busscher. 1992. Development and use of a parallel-plate flow chamber for studying cellular adhesion to solid surfaces. *Journal of Biomedical Materials Research* 26(6):725–738.

3 Molecular Analysis in Mechanobiology

Ken Webb and Jeoung Soo Lee

CONTENTS

3.1 INTRODUCTION

All the cells of a given organism possess an identical genotype, the literal sequence content of their entire genome. The unique phenotypes (observable characteristics) displayed by neurons, osteoblasts, hepatocytes, etc., within an organism derive from differential gene expression, protein synthesis, and activation. Over the last several decades, numerous studies have demonstrated the critical and cooperative role of extracellular stimuli in the development and maintenance of cellular phenotypic characteristics and functions [1,2]. Growth factors, hormones, and extracellular matrix adhesion molecules have been shown to influence cell survival, division, migration, and differentiation [3,4]. In general, these effects are mediated through transmembrane receptors that couple extracellular ligand binding to the activation of intracellular enzymes. Subsequent signal transduction cascades of cytoplasmic enzymatic reactions ultimately culminate in the activation of transcription factors, proteins that bind to regulatory sequences within DNA, activating or inhibiting transcription of specific genes.

The application or removal of externally applied mechanical loads has long been recognized to have profound effects on the macroscopic structure and biomechanics of bone, cartilage, muscle, and tendon/ligament [5–8]. Mechanobiology clearly functions at the cellular level with numerous in vitro studies using isolated cells and 3D culture models to demonstrate loading-induced changes in cell proliferation, gene expression, and biomechanical properties [9–12]. However, compared to growth factor or integrin-mediated signaling, relatively little is known about the molecular mechanisms by which cells detect external mechanical stimuli, transduce these inputs into changes in intracellular biochemistry, and ultimately alter levels of gene expression and protein synthesis/activation to achieve functional changes in cellular/tissue properties [13–17]. The effects of mechanobiology have been most widely observed in connective tissue and muscle, specifically with respect to the regulation of extracellular matrix metabolism and tissue biomechanics. An improved understanding of mechanotransduction will provide opportunities for the development of novel strategies to manipulate ECM production that may be particularly important in the treatment of various fibroproliferative disorders and chronic wounds, as well as tissue engineering of musculoskeletal and cardiovascular tissues [18–20].

The objective of this chapter is to provide researchers from traditional engineering disciplines entering the field of mechanobiology/mechanotransduction with an introduction to the theoretical principles and practical techniques of experimental molecular biology. The first section examines the analysis of mRNA expression levels with a focus on real-time quantitative reverse transcription-polymerase chain reaction (qRT-PCR), including RNA isolation, qualification/quantification, controls, normalization, and quantitative data analysis. The second section introduces antibodies and their application in analysis of protein expression using immunohistochemistry (IHC), Western blotting, and enzyme-linked immunosorbent assay (ELISA). The final section presents a brief introduction to the use of various types of inhibitors for "loss of function" assays that are useful in identifying causal roles of specific signaling mediators in mechanotransduction, including recently developed techniques of RNA interference.

3.2 ANALYSIS OF mRNA EXPRESSION

3.2.1 INTRODUCTION

The first critical factor in selecting a method of mRNA expression analysis is assessing whether a rational hypothesis can be formulated regarding specific genes (targets) that are expected to be responsive to the experimental treatment. When specific targets are unknown, the best approach is to begin with a high-throughput screening technique such as microarray analysis that can simultaneously assay all expressed transcripts from the entire genome. Alternatively, targets that are known or can be predicted with high probability may be immediately analyzed using Northern blotting,

RNase protection assay, in situ hybridization, or RT-PCR. In a review of studies that investigated "mRNA quantification," Thellin et al. reported use of a variety of techniques in 1999; however, by 2008, 88% of studies used real-time quantitative RT-PCR (qRT-PCR) [21]. Therefore, this section will present the basic principles underlying each technique, with a more detailed emphasis on qRT-PCR methods and experimental design.

3.2.2 MICROARRAY ANALYSIS

The most commonly used commercial platform for high-throughput analysis of gene expression is the Affymetrix Gene Chip® technology [22]. Gene chips are prepared by direct synthesis of short oligonucleotide sequences (probes) on the surface of small silicon wafers, providing multiple probes for each expressed sequence from the entire genome of the animal model from which the cells/tissues are derived. In the most common approach, total RNA from a control group and one or more experimental groups is reversed transcribed using oligo-dT primers incorporating a promoter sequence. In vitro transcription is then performed on this complimentary DNA (cDNA) using one labeled deoxyribonucleotide triphosphate (dNTP). The resulting complimentary RNA (cRNA) is fragmented and incubated with the gene chip to allow hybridization of the target cRNA with complimentary probes on the chip surface. The chips are then rinsed and digitally scanned to detect fluorescence derived from the labeled dNTP incorporated within the cRNA. Data are processed for background subtraction; normalization among replicates and control/experimental groups; and statistical analysis. Due to the sample sizes involved, data analysis is complex and significant differences in gene expression levels are routinely confirmed in replicate experiments using one of the conventional mRNA analysis techniques described below.

3.2.3 NORTHERN BLOTTING/RNASE PROTECTION ASSAY

Both Northern blotting and RNase protection assays use hybridization of labeled nucleic acid probes to detect and measure the amount of target mRNAs with complimentary sequence. In Northern blotting, isolated RNA samples and a "ladder" composed of oligonucleotides of known length are separated on the basis of size by agarose gel electrophoresis and transferred to a membrane support [23]. The membrane is incubated with the probe to allow hybridization, washed, processed as appropriate for the labeling method chosen, and analyzed by digital imaging and densitometry. Because of the electrophoresis step, Northern blotting is able to provide information on transcript size and alternative splicing. In RNase protection assays, samples of RNA are hybridized with one or more labeled probes in solution, treated with RNase, and then electrophoresed, processed, and analyzed by digital imaging and densitometry [24]. The resistance of double-stranded RNA to RNase degradation allows selective protection of probe/target hybrids and elimination of other transcripts. By designing probes of varying length, RNase protection assays can be performed in a "multiplex" format in which multiple targets are analyzed simultaneously and distinguished on the basis of size during the analysis steps. Normalization among replicates and experimental groups in both techniques is achieved by loading equal quantities of RNA sample and confirmed by analysis of a putatively constitutively expressed "housekeeping" gene (see Section 3.2.5.10 for further discussion of normalization in mRNA expression assays).

One benefit of Northern blotting and RNase protection assays is that both directly assay the target mRNA without additional processing steps (such as amplification in RT-PCR). The corresponding drawback of this is that both techniques require relatively large amounts of starting material and have limited sensitivity, making their application to transcripts with low expression levels difficult. In addition, densitometry analysis has a limited dynamic range, the ratio of the highest to lowest reliably detectable signal. Relative to qRT-PCR, both techniques can be performed with common laboratory equipment without the need for specialized instrumentation.

3.2.4 IN SITU HYBRIDIZATION

In situ hybridization is performed by direct probing of cultured cells or tissue sections with labeled nucleic acid probes [25]. This avoids the need for bulk sample homogenization and has been widely used to identify spatially localized patterns of gene expression in developing and pathological tissues. Most commonly, radioisotope labels are employed, detected by exposure of the sample to photographic film, and quantified by densitometry. Although not yet widely applied in mechanobiology research, in situ hybridization offers important opportunities for the spatial analysis of gene expression in cells cultured on/within anisotropic materials or subjected to variable external loading regimes.

3.2.5 REVERSE TRANSCRIPTION-POLYMERASE CHAIN REACTION

3.2.5.1 Fundamental Principles

PCR was invented by Dr. Kary Mullis and colleagues at Cetus Corp in the mid-1980s [26,27]. The reaction consists of a template DNA molecule, two short DNA oligonucleotides complimentary to the template (primers), DNA polymerase, buffer with appropriate Mg^{2+} concentration, and dNTPs. Through a process of thermal cycling, a specific region of the DNA template is selectively amplified. The use of a thermostable DNA polymerase (Taq polymerase) derived from the bacterium *Thermus aquaticus* substantially improved the efficiency and utility of PCR [28]. Reverse transcriptase, an RNA-dependent DNA polymerase that synthesizes cDNA from an RNA template allowed the application of PCR to mRNA analysis (RT-PCR).

Each PCR cycle consists of three steps: (1) denaturation—the reaction temperature is raised (typically to 95°C) such that double-stranded DNA molecules are denatured into single strands, (2) annealing—the reaction temperature is lowered to the annealing temperature (varies depending on the primer melting temperatures) to allow the primers to hybridize to the DNA template through sequence-specific base pairing, and (3) extension—the temperature is raised to the temperature for optimal activity of the DNA polymerase that synthesizes two new strands of DNA complimentary to the double-stranded template (Figure 3.1). Thus in theory, PCR provides an exponential amplification in which the number of DNA copies doubles with each cycle and the copy number measured at later stages in the reaction is mathematically related to the initial copy number (Equation 3.1)

$$N = N_0 2^n \tag{3.1}$$

where
  N is the copy number at a given cycle
  N_0 is the initial copy number
  n is the number of cycles

Denaturation   Annealing   Extension

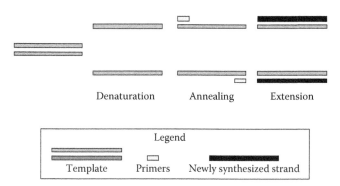

Legend

Template   Primers   Newly synthesized strand

FIGURE 3.1 Schematic representation of the three steps (denaturation, annealing, and extension) performed during each cycle of a PCR reaction.

In reality, PCR does not always achieve an exact doubling with each cycle and is more realistically represented as shown in Equation 3.2

$$N = N_0(1 + E)^n \tag{3.2}$$

where E represents the reaction "efficiency," with $E = 1$ corresponding to a perfect doubling of product per cycle. In addition, efficiency varies during the course of the reaction, with E being very close to 1 in the early stages under optimal conditions and dramatically declining at later cycles as the reaction enters a linear phase and eventually reaches a plateau. This results from depletion of reagents, decreased polymerase activity, and amplicons competing with primers for annealing.

In its early stages of development, RT-PCR was run for a fixed number of cycles and the products electrophoresed, stained, digitally imaged, and analyzed by densitometry (endpoint analysis). The limitation of this approach for quantitative analysis is that the sensitivity of the gel staining/imaging makes product detection difficult until the later cycles (linear/plateau region), at which point Equation 3.2 is no longer valid. Endpoint PCR analysis is still widely reported to semiquantitatively demonstrate relatively binary relationships (activation/deactivation). To improve the quantitative capability of RT-PCR, a number of competitive assay methods were developed (see Bustin [29]); however, their labor-intensive nature hindered widespread adoption. In 1992, Higuichi first described real-time PCR, using ethidium bromide to monitor the amount of product produced during the course of the reaction [30,31]. Figure 3.2A shows representative data from real-time amplification of a serial dilution of cDNA template. During the early cycles (typically 10–15, depending on the amount of template), the fluorescent signal is not readily discernable above background noise, establishing the baseline. The fluorescence then exponentially increases and eventually plateaus. A central parameter in qRT-PCR is the threshold cycle (C_T), a factional cycle number at which the sample fluorescent signal exceeds a specified level (typically 10 times the standard deviation of the baseline). C_T values are inversely proportional to the amount of starting template and provide the basis for absolute or relative quantitative analysis as described below (Section 3.2.5.11). In contrast to endpoint analysis, C_T values are determined relatively early in the amplification while the reaction is still in the exponential phase and efficiency is stable and approximately equal to 1 under appropriate conditions. Amplification efficiency (E) can be readily calculated from the slope of a linear regression of C_T values plotted versus the logarithm of the starting template amount (Figure 3.2B, Equation 3.3). A slope of −3.322 corresponds to an amplification efficiency of 1.

$$E = 10^{-1/\text{slope}} - 1 \tag{3.3}$$

In the last 10 years, qRT-PCR has become the dominant method of mRNA expression analysis, supported by rapid advances in methodology and instrument sophistication. Relative to alternative techniques, qRT-PCR offers significant advantages in sensitivity, dynamic range, and sample throughput. The remainder of this section will discuss critical decisions, techniques, and strategies for qRT-PCR, as well as its limitations. A recommended workflow for designing and implementing qRT-PCR experiments for the first time or under new experimental conditions is shown in Figure 3.3. In a recent review, Bustin and colleagues have provided more detailed step-by-step protocols as a recommended basis for increasing standardization among researchers and facilitating the comparison of data between labs [32].

3.2.5.2 Real-Time Detection Methods

The fundamental basis of qRT-PCR is the inclusion of a dye/probe in the reaction mixture that can be monitored during the amplification and generates a fluorescent signal proportional to the amount of product. This can be achieved using either a nonspecific DNA intercalating dye (predominantly SYBR green) or sequence-specific nucleic acid probes conjugated with fluorescent dyes. The

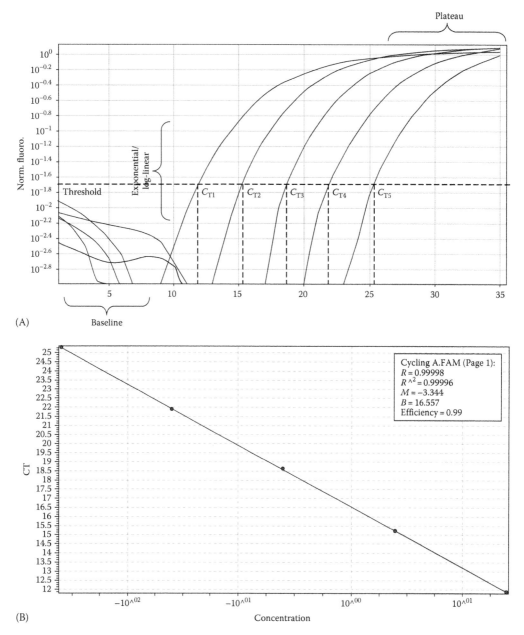

FIGURE 3.2 (A) Plot of cycle number versus normalized fluorescence (logarithmic scale) over a 35 cycle PCR reaction performed on a serial dilution of template covering five orders of magnitude concentration. Perpendicular lines drawn from the threshold level to the x-axis define threshold cycle (C_T) values for each sample. (B) Plot of template concentration (logarithmic scale) versus threshold cycle number showing linear regression line, correlation coefficient, and reaction efficiency calculated from the slope. (Data collected on Corbett Research RotorGene 3000, software version 6.1.)

weakness of SYBR green is that it binds to any dsDNA sequence, including the intended amplicon, primer-dimers, and nonspecific amplification products. The benefit of this nonspecificity is that it can be used in any PCR reaction, whereas a unique probe is required for each target. For the analysis of numerous targets, SYBR green is considerably more cost effective. SYBR green does require additional validation of amplification specificity as discussed below (Section 3.2.5.9).

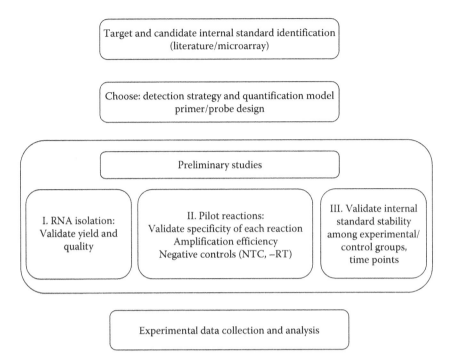

FIGURE 3.3   Recommended workflow for developing qRT-PCR assays, emphasizing appropriate planning and preliminary validation studies.

The most widely used probe chemistries are hydrolysis [33] (TaqMan) and hairpin [34] (molecular beacons) probes. Hydrolysis probes are labeled with 5′ reporter and 3′ quencher dyes. When intact, fluorescence resonance energy transfer (FRET) substantially reduces the reporter fluorescence. Hydrolysis probes hybridize to amplicons during a combined annealing/extension phase of each cycle. When the probes are reached by the Taq polymerase during DNA synthesis, they are hydrolyzed by the enzyme's 5′-exonuclease activity, releasing free reporter dye that generates a signal proportional to the number of amplicons. Molecular beacons consist of a central domain complimentary to the target and flanking domains on each side complimentary to one another with a fluorescent dye and quencher conjugated to the ends. In solution, molecular beacons adopt a hairpin structure that places the dye and quencher in close proximity resulting in fluorescence quenching by FRET. During the annealing step, molecular beacons hybridize to the amplicon, separating the dye and quencher, and generating a fluorescent signal proportional to the number of amplicons. A number of other sequence-specific detection chemistries are now available and have recently been reviewed by Wong and Medrano [35].

3.2.5.3   Primer/Probe Design

Appropriate primer/probe design is a critical aspect of effective RT-PCR. The commercial availability of "pre-designed, validated" primer/probe sets as well as public and commercial databases and software design programs (recently reviewed by VanGuilder et al. [36]) have made primer/probe design relatively automated and routine. However, knowledge of the basic principles involved is critical for troubleshooting and custom experimental design. The key features of primer design are as follows:

- *Primer length*: Primers are typically 18–25 bp in length. The probability of any given sequence is equal to 1 in 4^n, where n = the number of bases in the sequence. A length of 16 bases yields a probability of 1 in 4 billion, approximately the size of human genome.

- *Base composition*: Random, approximately 50% GC content, and limited repeated sequences of the same nucleotide.
- *Melting temperature* (T_m): Primer T_m is a function of length/composition and is a critical parameter since it affects the annealing temperature for hybridization. The guidelines above generally produce primers with T_ms ranging between 55°C and 65°C, which work well in qRT-PCR. Primer pairs should have very similar T_ms such that both hybridize effectively at the annealing temperature. Probe T_ms vary depending upon the stage of the reaction at which they hybridize to the target.
- *Amplicon size*: Amplicons for qRT-PCR are designed to be between 75 and 150 bp. Such short sequences are more easily amplified with high efficiency.
- *Complimentarity*: The presence of complimentary sequences between primers or within individual primers must be avoided to prevent the formation of secondary structure and primer-dimers that can inhibit amplification.
- *Exon/intron location*: Ideally, the two primers should bind to different exons, although for some transcripts this is impossible or impractical without compromising other design principles. This ensures that any product amplified from contaminating genomic DNA will differ in size (inclusion of the intron) from the amplicon and can be readily detected. Large introns may effectively inhibit amplification from genomic DNA due to insufficient time for replication of the larger product during the extension step.
- *Detection of splice variants*: Appropriate design of primers/probes to alternatively spliced exons can allow the selective amplification of specific splice variants [37].
- *3′ bias*: The use of oligo-dT primers and the poor processivity of reverse transcriptase can lead to incomplete cDNA synthesis from larger transcripts (>1500 bp). Therefore, as a general rule, it is preferable to design primers/amplicons located closer to the 3′ end of a large transcript when planning to use oligo-dT primers.

Primers/probes are usually received lyophilized and should be reconstituted in nuclease-free water or Tris EDTA (TE) buffer (0.1 mM Tris HCl, 1 mM EDTA, pH 8.0), aliquoted, and stored at −20°C.

3.2.5.4 Basic Principles of Nucleic Acid Handling

Proper handling of nucleic acids is essential due to the presence of degradative enzymes (DNases/RNases) in cells, biological fluids, and the environment. This is particularly important for RNA because RNases are resistant to inactivation by routine methods such as autoclaving. Today, numerous commercial vendors offer plasticware, chemicals, and liquids certified DNase-, RNase-, and pyrogen-free for RNA work. For materials that must be decontaminated "in-house," several established methods have been developed:

- Nuclease-free solutions—RNases in aqueous solutions can be inactivated by diethylpyrocarbonate (DEPC). Typically, 0.1% DEPC is added to the solution, mixed overnight, and then autoclaved, which degrades DEPC into carbon dioxide and ethanol. DEPC-treated water is commonly used for rinsing apparatus such as gel rigs, but generally avoided for enzymatic reactions that might be adversely affected by residual DEPC or ethanol.
- Glass or metal—equipment may be decontaminated by exposure to high heat (>180°C) for 4 h.
- Plasticware—tubes and other supplies can be immersed for 2 h at 37°C in 0.1 M NaOH/1 mM EDTA or 1% SDS in absolute ethanol, rinsed with nuclease-free water, and heated to 100°C for 15 min.

Several general lab practices are also important for RNase control. RNA work should ideally be performed in a dedicated, relatively isolated space. Clean cabinets designed for PCR work are commercially available, but not required. Lab benches and equipment should be cleaned with an RNase inactivating chemical such as RNase Away (Molecular Bioproducts) or RNase Zap (Ambion), and personnel should wear gloves at all times. The lab should be equipped with a dedicated set of pipettors and all work should be performed using tips containing an aerosol-resistant barrier filter to prevent pipettor contamination.

3.2.5.5 RNA Isolation, Purification, and Storage

If tissues/cells are not processed immediately at the experimental endpoint, they must be stored appropriately in order to prevent RNA degradation. Tissue samples may be flash-frozen using liquid nitrogen and stored at −80°C or stored in sulfate-containing buffers such as RNA*later*® (Ambion). Cellular samples can be stably stored for several months at −80°C in most commercially available lysis reagents.

RNA isolation and purification procedures involve cell lysis, RNase inactivation, and selective separation of RNA from other cellular macromolecules. The two most common methods are organic extraction and silica-based membranes. In the organic extraction method, cells or tissues are lysed/homogenized in an acid guanidinium thiocyanate-phenol solution then mixed with chloroform [38]. Phenol denatures proteins while guanidium thiocyanate inhibits endogenous RNases. The solution is mixed and centrifuged at high speed to separate aqueous and organic phases. At acidic pH, DNA is protonated and remains in the organic phase, while deprotonated RNA partitions to the aqueous phase, and proteins/lipids accumulate at the interface. RNA is recovered from the aqueous phase by precipitation. For silica-based methods, samples are lysed/homogenized in an aqueous buffer containing guanidinium thiocyanate. The lysate is then applied to the silica membrane, typically contained within a small column that can be used in a microcentrifuge. A combination of high salt and/or alcohol concentration (depending upon the manufacturer's formulation) facilitates selective binding of RNA to the silica membrane. The columns are rinsed to remove other macromolecules and the RNA eluted. Both methods yield purified total RNA (mRNA, tRNA, and rRNA), although some commercially available silica-membrane systems do not bind RNA sequences less than 200 nucleotides, reducing the content of small RNAs. Combined protocols are also used in which the aqueous phase obtained after phenol/chloroform extraction is applied to a silica membrane. This is beneficial in reducing the potential for sample contamination with phenol/ethanol. Total RNA purified by organic extraction and/or silica-based membranes can be reconstituted/eluted in nuclease-free water and stably stored at −80°C for several months. mRNA can be selectively isolated, typically using oligo-dT modified particles, but this is usually not required for most analytical methods.

3.2.5.6 DNase Treatment

Although current methods of RNA purification are very effective, the co-amplification of even trace amounts of contaminating genomic DNA can lead to substantial experimental errors, particularly when analyzing transcripts with low expression levels. Therefore, it is common practice to include a DNase treatment step either during or after RNA purification. One important consideration is that the DNase enzyme must be removed or inactivated prior to reverse transcription and this should be accomplished without heating the samples, which may result in RNA degradation in the presence of divalent cations contained in the DNase buffer. A variety of RNase-free DNase enzymes and kits are commercially available for this purpose.

3.2.5.7 RNA Qualification/Quantification

The final step before the actual performance of RT-PCR is to verify the RNA quality and measure its concentration. RNA degradation can be a source of large experimental error in RT-PCR. One method of RNA qualification is the assessment of ribosomal RNA integrity by denaturing agarose gel electrophoresis. rRNA accounts for approximately 75%–80% of total RNA and the 28S and 18S rRNAs are easily detected by ethidium bromide staining. Intact preparations are characterized by sharp 28S and 18S rRNA bands and an approximately 2:1 ratio in staining intensity. RNA degradation is recognized by a decrease in this ratio and "smearing," staining of a broad range of lower molecular weight fragments. More recently, electrophoresis has been miniaturized and accelerated using lab-on-chip technologies such as the Agilent Bioanalyzer. The Bioanalyzer measures 28S:18S ratios, as well as calculating an RNA integrity number (RIN) ranging from 1 to 10 that is an objective measurement derived from the entire electrophoresis profile [39,40]. While lab-on-chip technology is quicker, the costs associated with the instrument and consumables are considerably greater.

A particular benefit of the Bioanalyzer technology is that the RNA ladder can also provide a standard curve for simultaneously calculating sample RNA concentrations. However, despite the sophistication of the instrument, it still relies on accurate pipetting of 1 µL aliquots of ladder and samples, which can be a source of variability between replicates tested on separate chips. Alternatively, RNA concentrations may be measured by UV absorbance at 260 nm or RNA-specific fluorescent dyes such as RiboGreen® (Invitrogen) in conventional microplate assays. UV absorbance is simple and cost effective, but relatively limited in sensitivity and does not discriminate between DNA, RNA, and free nucleotides.

3.2.5.8 Reverse Transcription and Amplification

The RT and PCR amplification reactions can be performed in either one-step or two-step formats. In one-step RT-PCR, all necessary reagents are included within a single tube and both reactions are performed in a single, continuous process. In the two-step approach, the RT and PCR reactions are performed separately. Both methods are effective and widely used. The one-step procedure is generally more convenient and efficient when assaying large numbers of experimental samples. On the other hand, if multiple targets are to be assayed from each sample, the two-step method allows multiple PCR amplifications to be performed from a single cDNA solution generated in the previous RT step. The analysis of multiple targets from a single cDNA stock may reduce variability arising from the RT reaction and is generally more cost effective since the RT enzyme is one of the more expensive components of RT-PCR.

In two-step RT-PCR, the basic components of the RT reaction are the RNA sample, the RT enzyme, primers, RT buffer, dNTPs, and nuclease-free water as required to achieve the final recommended reaction volume. The first step of normalization for differences in the amount of biological material among experimental samples and groups is performed in the setup of the RT reaction by adding an equal amount of total RNA to each reaction (typically 0.5–1.0 µg) calculated from the previously determined sample concentrations. Most RT kits use reverse transcriptase enzymes derived from the Moloney murine leukemia virus (MMLV) of avian myeloblastosis virus (AMV). Oligo-dT or random hexamers/nonamers may be used for priming the RT reaction. Oligo-dT primers are more specific, providing a selective reverse transcription of mRNA. However, mRNA secondary structure or degradation may interfere with the synthesis of full length transcripts, introducing a 3′ bias in the cDNA that should be considered during primer design (see Section 3.2.5.3). Random primers are less sensitive to RNA quality because they bind at multiple sites; however, this generates more than one copy of cDNA per RNA template and can result in an overestimation of target copy numbers [41]. The first step in RT reactions is an optional heat denaturation, in which the RNA template, primers, and water are heated at 70°C–85°C for several minutes to remove secondary structure that can inhibit efficient cDNA synthesis. After the samples are cooled to room temperature, the remaining components are added and the reverse transcription performed at the recommended temperature for optimal enzyme activity, followed by a brief heat inactivation of the RT enzyme. Some recently developed recombinant RT enzymes with increased thermal stability also allow the RT reaction to be performed at higher temperatures, minimizing interference from secondary structure.

There are two major designs of real-time thermal cyclers, microplate-based and centrifugal systems. Both are widely and effectively used. Many microplate-based systems are capable of providing variable temperatures among the plate wells. This can be very valuable if reaction optimization, particularly of primer annealing temperatures, is required. The centrifugal design provides the highest temperature uniformity among samples. Most current instrument designs are capable of monitoring fluorescence at multiple wavelengths. Using probe-based detection methods, it is feasible to perform "multiplex" reactions in which two or more targets are amplified in the same tube and monitored using different reporter dyes. This is not widely practiced as the inclusion of multiple primer/probe sets can increase the potential for nonspecific amplification and requires more stringent design and additional optimization. More detailed descriptions of specific instruments are available in several recent reviews [42,43].

The components of real-time PCR amplification reactions are the cDNA template (a defined volume of the RT reaction), Taq polymerase, target-specific primers, detection dye/probe, buffer, and dNTPs. Today, commercially available master mixes containing the polymerase, buffer, dNTPs, and SYBR green (if desired) are widely available. Most commercial reagents now use a "hot-start" polymerase that includes antibodies or other inhibitors that block enzymatic activity at room temperature and are inactivated when heated [44]. This alleviates the need for keeping samples on ice during reaction setup and improves specificity. After calculating the required volumes based on the number of samples to be assayed for each target; the primers, probes (if required), and water are added to the commercial master mix and then aliquoted into the required number of tubes/wells. The use of such master mixes minimizes the number of pipetting steps, reducing chances of sample/reagent contamination and experimental error. Finally, the required volume (usually 1–2 µL) of each cDNA sample is added to each tube/well and the samples are loaded into the thermal cycler. The template aliquot should not exceed 10% of the final reaction volume. The salt composition of the RT buffer is different from the PCR buffer and can inhibit PCR reactions at higher concentrations. PCR amplifications are usually carried out for 30–40 cycles depending on the target transcript expression level. When using SYBR green, a melting curve should be included at the end of each procedure (Section 3.2.5.9).

In its early stages of development, efficient and sensitive qRT-PCR usually required substantial optimization of reagent concentrations (Mg^{++}, primers, dNTPs, SYBR green if used). Most of the currently available commercial master mixes are advertised as "pre-optimized" and usually work effectively with minimal optimization. Primers are usually included at concentrations ranging between 50–500 nM and optimization within this range can significantly increase reaction sensitivity [32]. The other variable that may require optimization is the annealing temperature. As a general guideline, the annealing temperature is set 4°C–5°C below the average primer Tm. If the annealing temperature is too low, primer hybridization becomes less stringent and tolerant of imperfect matches, which can lead to amplification of additional products. As noted above, this is primarily a problem when using nonspecific detection, but it can be readily observed in melting curve analysis and corrected by increasing the annealing temperature. On the other hand, if the annealing temperature is too high, primer hybridization may be inhibited, which can be recognized by poor amplification efficiency.

The final step in PCR amplification is data processing to obtain C_T values. Most current instruments include default settings and algorithms for data processing, but attentive monitoring is required to ensure these perform properly. The first step in data processing is to define the range of cycles that comprise the baseline. The baseline region defines an average background fluorescence during the early cycles (usually by default cycles 3–15) when amplification cannot be effectively detected. The baseline region should correspond to a relatively flat region of the amplification profile and may require manual adjustment for highly expressed targets that begin to exhibit amplification within the default baseline region. The threshold fluorescence level may be automatically or arbitrarily assigned. The key point is that the threshold level must be within the log-linear (exponential) phase of amplification. Although its exact placement will alter the absolute C_T values, small changes in the threshold setting should not significantly affect relative differences in C_T values among samples and groups. In some cases, it is necessary and appropriate to adjust baseline and threshold settings differently for each target. Provided controls and validation are appropriate, raw C_T values are finally analyzed to obtain quantitative information about absolute or relative expression levels.

It is important to note that the quantification approach (Section 3.2.5.11) must be determined before experimental sample analysis. For absolute quantification, a standard curve of known template concentrations must be included for each target analyzed. For relative quantification, it is critical that each experimental run include the internal standard (Section 3.2.5.10). One final point involves the use of experimental replicates. Given the sensitivity and potential variability of PCR, it is widely recommended that all samples/standards/controls be performed in duplicate. However, this does not account for numerous potential sources of error/variability such as inherent biological variability, small changes in bioreactor/sample configuration and applied forces in mechanobiology studies, sample processing, and total RNA quantification/quality. It is strongly recommended that all studies for publication be independently replicated and analyzed.

3.2.5.9 Controls and Validation

The most routine controls for qRT-PCR are no-template controls (NTC) and minus-RT controls (−RT). NTCs should be performed for every amplification target and prepared from the same master mix as the experimental samples with the final addition of nuclease-free water instead of template. In the absence of template, no product should be produced and no C_T value generated. A positive signal in an NTC is indicative of DNA contamination of reagents/buffers. Minus RT controls may be prepared by omitting the RT enzyme (most common for one step) or attempting to amplify an equivalently diluted sample of the original RNA (most cost effective for two step). For targets with primers that bind to different exons, −RT controls are unnecessary, since amplification of genomic DNA will produce a product larger than the intended amplicon that can be detected by melt curve analysis. In circumstances where both primers bind to the same exon, −RT controls are essential.

Because SYBR green detection is nonspecific, validation procedures are required to ensure the amplification of a single product with desired sequence. The simplest method of validation is melting curve analysis, which should be included at the end of every reaction [45]. To generate a melting curve after the last cycle of amplification, the instrument incrementally increases the temperature from the detection to denaturation temperature while monitoring sample fluorescence. Separation of the dsDNA product at its melting temperature leads to an abrupt decline in fluorescence. The product melting temperature is readily identified as a peak in a plot of the negative rate of change of sample fluorescence with respect to temperature. For a valid reaction, there should be one peak corresponding to one product. Melting curve analysis can also reveal the presence of primer-dimers. Due to their shorter length, primer-dimers appear as small peaks at relatively lower temperatures than amplicons. Primer-dimers may be observed in NTCs, but should never occur in template-loaded samples if appropriate guidelines for minimizing primer complimentarity have been followed. In order to confirm that the product is in fact the intended specific amplicon, products should be analyzed by Southern blotting or DNA sequencing.

3.2.5.10 Normalization (Internal Standards)

Normalization in qRT-PCR is intended to account for potential variations in the amount of biological material (quantification inaccuracies) and its quality. As presented in a recent review by Huggett et al., normalization among experimental samples and groups can be performed at three levels [46]. First, the amount of starting material (cell numbers or tissue mass) should be reasonably consistent. Second, samples are generally normalized to total RNA by adding equal amounts to each RT reaction. As sole method of normalization, total RNA has several limitations, including the assumption that the experimental conditions do not alter total RNA levels, the need for highly accurate quantification and consistent RNA quality, and the inability to account for potential variability in the efficiency of the RT step [47–49]. Therefore, it is a common but problematic procedure to finally normalize target gene expression levels to the expression levels of another endogenously expressed gene (an internal standard). Internal standards in early RT-PCR work were chosen from putative housekeeping genes whose products are involved in essential structural or metabolic functions and expression levels are not expected to be affected by the experimental conditions. Common examples include beta-actin, glyceraldehyde-3-phosphate dehydrogenase (GAPDH), hypoxanthine phosphoribosyltransferase (HPRT), and beta-2-microglobulin.

In recent years, many studies have demonstrated that mRNA expression levels of conventional housekeeping genes vary significantly among different experimental conditions, tissue origins, or pathologies [50–54]. Use of an inappropriate internal standard can produce significant errors in qRT-PCR data analysis [55]. Therefore, the selection of valid internal standards has become a subject of considerable discussion and investigation [46,56,57]. It is now widely recognized that the selection of an internal standard should be validated for the specific experimental conditions in which it is being applied. Validation involves screening of candidate internal standards and performing statistical analysis to identify one or more genes exhibiting the lowest levels of variability among the experimental conditions or time points. One method that is particularly suitable to time course studies is to statistically compare fold changes in gene expression calculated as $2^{-\Delta C_T}$, where $\Delta C_T = C_{T,\text{time } x} - C_{T,\text{time } 0}$ [58]. Another method is to statistically compare mRNA copy numbers determined by absolute quantification [53] (described in the next section). More recently, several new statistical models and software-based tools have become available for comparing the expression stability of multiple candidate internal standards and geometric averaging of two or more identified suitable genes [59–61]. While rigorous, these advanced methods can become cumbersome and costly. We recommend that during the early stages of setting up qRT-PCR protocols, investigators screen candidates from two or more independent experiments using the $2^{-\Delta C_T}$ method or absolute quantification to identify an internal standard that does not significantly vary among their specific experimental conditions. In addition, investigators should consistently be attentive to effects of internal standard normalization and aware that the smaller the changes in gene expression levels attempting to be detected, the greater the potential for internal standard normalization to significantly alter the results and conclusions. In our experience, beta-2-microglobulin has exhibited very stable expression between static and vibratory cultures across multiple time points and the effects of internal standard normalization have been minimal [62].

3.2.5.11 Quantitative Data Analysis Methods

qRT-PCR can be performed using either absolute or relative quantification methods. In absolute quantification, standards containing known amounts of the target template (typically expressed as mass or copy numbers) are amplified with experimental samples. A standard curve is generated defining a linear relationship between C_T values and the initial amount of template, from which the absolute quantity of the target in unknown samples can be calculated [29]. RNA or DNA standards may be used in absolute quantification. cRNA standards can be generated by in vitro transcription [63]. DNA standards may be chemically synthesized, cloned into plasmid vectors, or purified from large volume PCR reactions. cRNA standards are the most rigorous, as DNA standards cannot account for potential variability in the RT reaction. Accurate quantification of standards is essential and should be performed using fluorescent nucleic acid binding dyes. Standard curves for qRT-PCR should exhibit linearity over at least four orders of magnitude variation in initial template concentration. It is also important to note that all experimental sample values must fall within the range encompassed by the standard curve. Normalization relative to an internal standard (reference) is performed by calculating a normalized expression level (Equation 3.4):

$$\text{Normalized expression level} = \frac{\text{pg/copy \# (target)}}{\text{pg/copy \# (reference)}} \quad (3.4)$$

Normalized expression levels (pg/copy numbers) may be directly compared among experimental and control groups or compared using a relative expression ratio (Equation 3.5):

$$\text{Relative expression ratio} = \frac{\text{normalized expression level (experimental)}}{\text{normalized expression level (control)}} \quad (3.5)$$

Relative quantification methods use C_T values to make relative comparisons between experimental and control groups without calculating exact quantities. The $2^{-\Delta\Delta C_T}$ method is the most widely used approach to relative quantification [58]. It is important to note that this method involves two assumptions that must be experimentally confirmed for its valid application. First, it is assumed that the amplification efficiencies of the target(s) and the internal standard are approximately equal. Second, it incorporates an internal standard (reference) that is assumed to maintain a stable expression level among the various experimental and control groups. Comparable amplification efficiency can be demonstrated by performing pilot experiments with a serial dilution of template and plotting log template dilution versus ΔC_T ($C_{T(\text{target})} - C_{T(\text{reference})}$) and verifying that the absolute value of the slope is approximately equal to zero. An internal standard (reference) must be experimentally validated as described above (Section 3.2.5.10). If these assumptions can be validated, relative expression levels (corresponding to fold changes) can be calculated as $2^{-\Delta\Delta C_T}$ as shown in Equation 3.6:

$$2^{-\Delta\Delta C_T} = \left(C_{T(\text{target})} - C_{T(\text{reference})} \right)_{\text{experimental}} - \left(C_{T(\text{target})} - C_{T(\text{reference})} \right)_{\text{control}} \quad (3.6)$$

The designation of the "control" group is at the researcher's discretion, in mechanobiology it is usually defined as a static group or the initial time point. Also, statistical analysis can only be applied to $2^{-\Delta\Delta C_T}$ relative expression levels and not raw C_T values. In circumstances where the equal amplification efficiency assumption cannot be validated, Pfaffl has described an alternative model that factors in the difference in amplification efficiencies [64]. Although not yet as widely adopted, several newer mathematical models for relative quantification, which include improvements such as calculating the efficiency of each sample individually from raw fluorescence data, have been developed and recently reviewed by Wong and Medrano [35].

3.2.5.12 Limitations

Despite its capability and widespread use, several limitations common to mRNA analysis and specific to real-time qRT-PCR should be considered. As with all the transcriptional analysis techniques described above, the data obtained from qRT-PCR measure steady-state mRNA levels and due to variability in mRNA stability, may not necessarily directly correlate with transcriptional activity. Recombinant constructs consisting of the target gene promoter fused to a reporter gene that encodes a readily assayed molecule such as chloramphenicol acetyltransferase (CAT) can offer direct measurements of transcriptional activity, but such experiments require additional time and expertise. Another common limitation of all mRNA analysis techniques is that changes in mRNA expression levels do not always correlate with changes in protein synthesis/activity. Therefore, it is common practice to confirm changes in gene expression at the protein level using methods such as those described in the next section. Specific to qRT-PCR is the limitation that it does not directly measure sample mRNA and two enzymatic reactions are required to produce the amplified product that is ultimately measured. While this process provides the remarkable sensitivity of qRT-PCR, it also creates considerable opportunity for experimental error and the need for attention and experience in experimental design and quality control. Several studies have reported significant variations among identical assays with respect to lab site/operator and reagent lots [42,65]. Maintaining consistency among these external variables is important for reproducibility. Finally, as discussed in detail above, the quality of data obtained from qRT-PCR is highly dependent upon appropriate selection and validation of internal standards/references used to normalize data.

3.3 ANALYSIS OF PROTEIN EXPRESSION

3.3.1 INTRODUCTION

Antibodies are proteins expressed by B lymphocytes that contribute to specific immune responses by binding to foreign (nonself) molecules termed antigens. The basic antibody subunit is a Y-shaped structure that is composed of two identical copies of heavy and light chains connected by disulfide bridges (Figure 3.4). Five different classes of antibodies are expressed (IgM, IgG, IgD, IgA, and IgE) that differ in the number of Y subunits and the type of heavy chain. Each antibody subunit has two antigen-binding domains in the Fab regions that are identical among all the antibodies produced by an individual B cell, but highly variable among the overall population of B cells. This diversity results from recombination of the immunoglobulin genes during B cell maturation and provides each individual with an antibody repertoire capable of recognizing an estimated >10^9 different antigens [66]. Each antibody subunit also has an Fc region that is constant among all the antibodies of a particular class and provides binding sites for complement proteins and macrophages. The antigen-binding domains of each antibody recognize a relatively small region of the antigen (typically 6–10 amino acids) termed an epitope. Therefore, a typical antigen such as a protein will contain multiple epitopes, each of which may be recognized by different specific antibodies. Antibody–antigen binding is mediated by multiple, cooperative non-covalent interactions, providing very high specificity that has made antibodies the central tool for studying protein expression.

Several proteins of widespread interest such as collagen and elastin can be selectively analyzed by unique methods based upon their chemical composition. Since antibodies are generally consistent in all properties except their antigen specificity, they may be used interchangeably in common assay formats that can be applied to a broad variety of proteins. The following sections provide additional background and general guidelines for working with antibodies and describe three analytical methods: IHC, Western blotting, and enzyme-linked immunosorbent assay (ELISA).

3.3.2 ANTIBODY PRODUCTION, FORMAT, PURIFICATION, AND HANDLING

Antibodies used in research are prepared by repeated immunization of a host animal with the desired antigen. Resting B lymphocytes express cell-surface forms of IgM and IgD that function

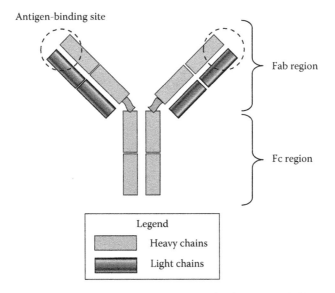

FIGURE 3.4 Schematic representation of the basic structure of an immunoglobulin subunit.

as receptors. Antigen-binding results in B cell activation and heavy chain class switching allowing the production of other antibody classes and their secretion. IgG is the predominant antibody class released into the bloodstream. Antibodies may be obtained by collecting serum during the peak of the immune response. Because most antigens are macromolecular proteins, the serum will contain a polyclonal mixture of antibodies with specificities for multiple epitopes. Alternatively, spleen cells may be harvested from the immunized animal, mixed with a myeloma cell line under conditions that induce cell fusion, and cultured in selective conditions that only allow the survival of fused B lymphocyte/myeloma cells (hybridomas). These hybridoma cells are then plated at high dilutions that allow the isolation of single cells in individual wells or with cloning rings. Multiple clones are expanded and tested to identify a clonal population with high antibody production and antigen-binding affinity. Such cells originating from a single clone are genetically identical and produce monoclonal antibodies that recognize only one specific epitope. Hybridomas can be cultured in vitro and antibody harvested from the medium or injected into the peritoneum of another animal and the antibody-enriched ascites fluid subsequently collected.

The crude antisera, hybridoma supernatants, or ascites fluid produced by these various methods can be used effectively in antibody-based assays. However, the vast majority of commercially available antibodies are purified by affinity chromatography. In this procedure, the solution containing the antibodies of interest is passed over a column packed with low protein-binding hydrogel microparticles to which the antigen has been covalently conjugated. After washing out other molecules, the bound antibodies can be eluted at high purity. Both monoclonal and polyclonal antibody formats are widely used. Monoclonal antibodies of appropriate specificity can be used to discriminate between similar proteins arising from alternative mRNA splicing or among families of proteins with high homology, as well as detect conformational changes and activation. Assays using polyclonal antibodies are generally less affected by suboptimal assay conditions in which proteins may be partially denatured.

Antibodies should be stored at 4°C or −20°C, depending on the supplier's specifications. Because antibodies are unstable in dilute solution and may lose activity after repeated freeze/thaw cycles; small, undiluted aliquots should be prepared prior to freezing. Frozen antibodies generally retain activity for at least 1 year. Working solutions of antibodies should be diluted in blocking buffers (see below) containing high overall protein concentrations and can usually be stored at 4°C for at least 1 month without significant loss of activity. Sodium azide is commonly added as a preservative (final concentration 0.05%) to working solutions that will be stored. Sodium azide inhibits the activity of horse radish peroxidase and should not be added to solutions of secondary antibodies conjugated with this enzyme.

3.3.3 Detection, Blocking, Titer, and Cross-Reactivity

Several common principles are important for all antibody-based applications. Antibody detection is achieved through the covalent conjugation of labels (fluorescent dyes, biotin) or enzymes. It should be noted that conjugated antibodies generally exhibit decreased stability and can be prone to aggregation after prolonged storage. This problem can be detected in IHC or Western blots by intense, discrete background staining. Such stocks can usually be "cleaned up" by either high-speed centrifugation (2 min, ~10,000g) of the stock solution or 0.2 µm filtration of the working solution. Labels can be directly conjugated to the "primary" antibody that recognizes the antigen of interest (Figure 3.5). More commonly, an "indirect" detection method is used in which the sample is first incubated with a primary antibody that recognizes the antigen of interest, followed by incubation with a conjugated "secondary" antibody directed against the primary antibody. Secondary antibodies are always polyclonal and the capacity for binding of multiple secondary antibodies to each primary provides signal amplification. Additional signal amplification may

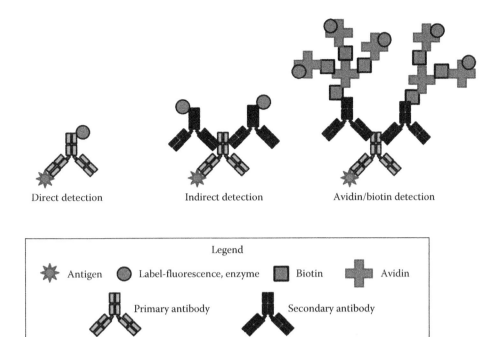

Direct detection   Indirect detection   Avidin/biotin detection

Legend

⬢ Antigen   ● Label-fluorescence, enzyme   ■ Biotin   ✚ Avidin

Primary antibody   Secondary antibody

FIGURE 3.5 Signal amplification and detection strategies for antibody-based assays. Note that most labeling procedures conjugate more than one label per antibody molecule. Single labels have been shown here for clarity.

be obtained using a biotinylated secondary and fluorescent- or enzyme-conjugated avidin or streptavidin.

Antigens are typically present in the context of a physical support (membrane, plate/slide) and therefore the measurements obtained from an assay reflect both specific signal derived from antibody–antigen binding and noise originating from nonspecific antibody adsorption. Buffers containing a high concentration of bulk protein intended to saturate any potential sites for adsorption are commonly used for "blocking" steps before antibody application and also for antibody dilution.

Almost all commercial antibodies are provided with a specified concentration. However, the effective working dilution is a function of both the antibody concentration and its binding affinity for a particular antigen that may vary among batches, suppliers, and assay conditions. Therefore, each investigator should empirically test antibodies when they are first used in the lab. The optimal titer can be determined by performing assays with positive control samples using antibody working solutions of varying dilution. The objective is to identify the lowest dilution that offers a plateau-level signal with minimum background. Although most vendors offer guidelines for working dilutions, many antibodies can be used effectively at much greater dilution.

Finally, it is important to be aware of cross-reactivity, the potential for an antibody to bind to an identical or closely related epitope in a molecule other than the antigen. Interspecies cross-reactivity (the ability of an antibody against human type I collagen to recognize mouse type I collagen) is generally beneficial and can allow a single antibody to be used with cells and tissue from multiple species. Alternatively, cross-reactivity with other closely related antigens within the same sample compromises specificity and is detrimental (such as an antibody against beta-III-tubulin used to selectively identify neurons binding to other tubulins expressed in all cells). Such problems can be avoided by careful attention to antibody specifications and detected using negative controls that do not contain the antigen of interest.

3.3.4 Immunohistochemistry

IHC is analogous to in situ hybridization and provides information about the spatial localization of protein expression and qualitative evaluation of expression levels. Staining for cell-type-specific antigens is also commonly performed to characterize stem cell differentiation or recruitment of specific cells to sites of injury/implantation. The general steps in the procedure are fixation, embedding, and sectioning (for tissue samples), detergent permeabilization of cell membranes, antigen retrieval (commonly performed on paraffin-embedded sections), blocking, and incubation with appropriate primary and secondary antibodies. Thorough washing with blocking buffer is performed between each step. In general, fluorescent labels offer better spatial resolution than enzymes that cleave soluble substrates to produce insoluble products that precipitate on/within the sample. Simultaneous staining of up to three antigens in a single sample is possible as long as each primary antibody is either a different isotype or raised in a different species so that each can be recognized by distinct secondary antibodies with different labels.

Negative controls are required in IHC to confirm specificity of staining. One option is application of the same reagents to sections of tissue known not to contain the antigen of interest. Another option is "no primary controls" in which case the primary antibody is omitted. However, this only provides insight into the level of nonspecific binding of the secondary antibody. A more rigorous method is to substitute another primary antibody produced in the same species directed against an antigen known to not be expressed in the tissue of interest. Although IHC is generally a qualitative assay, improved microscopy and imaging techniques have allowed some quantitative analysis of signal intensity [67,68]. Care must be taken to maintain constant exposure times among all samples and account for uneven sample illumination.

3.3.5 SDS-PAGE, Western Blotting, and Zymography

Sodium dodecyl sulfate-polyacrylamide gel electrophoresis (SDS-PAGE)/Western blotting is analogous to Northern blotting for mRNA analysis, providing a separation on the basis of molecular weight and identification/semi-quantitative measurements of specific proteins using antibodies. Samples may be prepared from biological fluids, cell culture supernatants, or cell lysates. The technique is particularly useful for studying intracellular and transmembrane proteins with limited solubility in aqueous solutions. Samples are mixed with a concentrated loading buffer containing SDS, heated (usually 10 min at 60°C), and loaded into wells formed in covalently cross-linked polyacrylamide gels [69]. Heating in the presence of SDS denatures the proteins, reducing secondary structure so that their migration in the gel will be primarily a function of molecular weight, while SDS binding provides a net negative charge to all proteins. It is also common to include beta-mercaptoethanol in the loading buffer to reduce disulfide bonds (reducing conditions). A "ladder" composed of proteins of known molecular weight is loaded in one well as a basis for molecular weight determination. Polyacrylamide gels are usually prepared with a top "stacking" region containing the sample wells and a bottom "resolving" region. Proteins migrate rapidly through the stacking region and are concentrated at the interface before entering the resolving region. The acrylamide concentration in the resolving gel can be varied to improve resolution within certain molecular weight ranges (8% for the 100–200 kDa range; 12% gels for the 10–40 kDa range).

After electrophoresis, proteins are transferred to a hydrophobic membrane (PVDF or nitrocellulose) and the membrane is incubated with a blocking buffer. In the Western blotting step, antibodies are used to probe the membrane; most commonly using an indirect method with an enzyme-conjugated secondary antibody. Antibody incubation steps are usually performed for 1 h; however, incubation with the primary antibody overnight at 4°C at a relatively lower dilution can sometimes increase sensitivity and reduce background. After antibody binding and washing, a developing solution is added containing either a colorimetric or chemiluminescent substrate. Enzymatic substrate cleavage is a kinetic process and care must be taken to stop the development process before plateau signal levels are reached. Development of colorimetric assays can be visually monitored. Chemiluminescence is detected using photographic film in a closed cartridge and may require more optimization of development time. However, chemiluminescence-based assays provide much greater sensitivity. Western blots can be "stripped" and re-probed several times to detect assay multiple proteins from a single blot.

Normalization among samples and experimental groups is based on the loading of each sample lane with an equal amount of total protein (normalization to total sample DNA may also be used). Sample protein concentrations can be determined using the Lowry, Bradford, or bicinchoninic acid (BCA) assays [70]. This approach involves the assumption that total cellular protein levels are unaffected by the experimental conditions. For further validation, it is common practice to re-probe the membrane for a putatively constitutively expressed protein such as beta-actin.

Zymography is a variation of SDS-PAGE that analyzes the activity of matrix degradative enzymes [71,72]. The method is most widely applied using gelatin as a substrate for MMP-2 and MMP-9, but variations have been developed for other MMPs such as collagenases, as well as hyaluronidases. A substrate for the enzyme of interest is incorporated within the resolving gel and SDS-PAGE performed under nonreducing conditions. The gel is washed in a Triton X-100 buffer to remove SDS and then incubated under appropriate conditions for enzymatic activity. The gel is then stained with Coomassie blue (for protease assays) to reveal clear bands at molecular weights corresponding to specific enzymes. Substrate degradation and corresponding enzymatic activity can be measured semi-quantitatively by densitometry. A further variation, reverse zymography, may also be performed to analyze the activity of tissue inhibitors of matrix metalloproteases (TIMPs) in which both a substrate and a protease are included within the gel and TIMP activity detected as stained bands where degradation has been inhibited.

3.3.6   ELISA

ELISA is the most sensitive and quantitative technique for protein analysis, offering high specificity even with samples such as blood containing high concentrations of other proteins. Although several different formats of ELISA have been developed, the sandwich ELISA for analyzing soluble protein concentrations is by far the most useful and widely practiced and will be the focus of this section. The most common application is the measurement of growth factor/cytokine levels in culture supernatants or biological fluids. A sandwich ELISA requires two antibodies that recognize different epitopes (either a polyclonal antibody solution or two different monoclonal antibodies) and a highly purified standard of known concentration of the antigen of interest. A variety of ELISA kits can be purchased commercially or assays may be developed in-house. Commercial kits are convenient and efficient, but are costly and have limited throughput capability (a typical kit contains one 96-well plate). If a particular antigen needs to be assayed repeatedly, it is much more cost effective to develop a custom ELISA and many suppliers offer "paired" antibodies for this purpose; however, optimal working dilutions and other assay conditions will need to be empirically optimized.

ELISA is performed in a microplate format using unmodified polystyrene or proprietary hydrophobic, high protein-binding plates. The first step in a sandwich ELISA is coating the wells with a solution of the capture antibody that adsorbs to the surface (Figure 3.6). After blocking, wells are incubated with serially diluted standards of known concentration and experimental samples. A group without sample/standards is included as a "blank" to determine the background signal of the assay. An enzyme-conjugated detection antibody is then added, followed by a substrate that is cleaved to form a soluble colorimetric, fluorescent, or chemiluminescent product. Extensive washing is performed between each step and a "stop" buffer usually added at the end to terminate the enzyme reaction after a specific time period. Raw data is collected on an appropriate plate reader, the average background signal subtracted, and a standard curve generated for calculation of unknown sample concentrations. Under appropriate conditions, ELISA can accurately measure sample concentrations in the low (<10) picogram/milliliter range. ELISA data is usually normalized

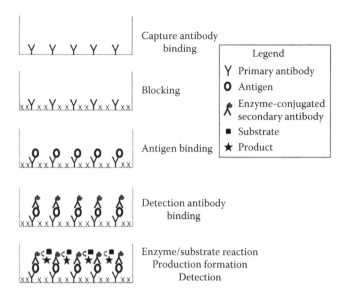

FIGURE 3.6 Procedure for sandwich ELISA assay.

to total protein or DNA (e.g., picogram antigen/nanogram DNA) in order to account for potential variability in cell number among samples and experimental groups.

3.4 MECHANISTIC TOOLS

3.4.1 INTRODUCTION

Early-stage investigations in mechanobiology usually involve the discovery and characterization of stimulus–response relationships. For example, the seminal observation that cyclic strain increases collagen expression was first reported in 1976 [73]. The next stage of investigation is elucidation of the underlying mechanisms of mechanochemical conversion and signal transduction. A central strategy is the use of experimental tools that allow selective activation or inhibition of specific molecules hypothesized to participate in signal generation/transduction. Returning to the example above, in the mid-1980s, transforming growth factor beta-1 (TGFβ-1) was found to stimulate collagen expression in static cultures [74,75] and several studies showed that strain-mediated collagen induction was accompanied by significant increases in TGFβ-1 expression [76,77]. Shortly thereafter, inhibition of TGFβ-1 activity with function blocking antibodies was shown to eliminate collagen induction by cyclic strain, providing convincing evidence that increased TGFβ-1 expression is a fundamental mechanism contributing to increased collagen expression in cells subjected to cyclic strain [78]. More recently, the transcription factor involved and its binding site in the procollagen alpha 1 (I) promoter have been identified [79]. The following sections describe various classes of inhibitory molecules and review examples of their application in the literature to gain new insight into the mechanisms responsible for mechanotransduction.

3.4.2 MOLECULAR INHIBITORS

One of the earliest identified mechanisms of mechanochemical conversion was stretch-activated (SA) ion channels that exhibit increased ion transport in response to mechanical stimulation. A variety of chemical and biological antagonists have been developed/identified that function as channel blockers. These compounds usually block transport of a specific ion, but may not discriminate between different channels capable of transporting the same ion. Chemical inhibitors are also available for many cellular enzymes; ranging from broad spectrum tyrosine kinase inhibitors to highly specific compounds targeting a specific growth factor receptor. In mechanobiology, these inhibitors have been most widely applied to modulate cytoskeletal structure/tension, receptor tyrosine kinases, and matrix metalloproteases (MMPs). The primary limitation of these compounds is that few are completely specific/selective, and there is always the possibility of "off-target" effects on other cellular enzymes that are difficult to recognize and may confound the interpretation of results.

The most specific methods of molecular inhibition utilize biologically derived recognition principles, including antibody–antigen binding and sequence-specific nucleic acid hybridization. The first approach is the use of "function-blocking" antibodies that bind to either (1) soluble extracellular signaling molecules, thereby sequestering them or (2) the extracellular ligand binding domain of transmembrane receptors, interfering with ligand binding and subsequent intracellular signal activation. Common targets in mechanobiology include growth factor and integrin receptors. The second approach is the exogenous activation of RNA interference, an evolutionarily conserved cellular mechanism for the destruction of double-stranded RNA (dsRNA) [80,81]. In the native physiological pathway, dsRNA is cleaved by the enzyme Dicer into 21–23 nucleotide small interfering RNAs (siRNAs) [82] that combine with an RNA-induced silencing complex (RISC) [83], which mediates binding and endonuclease cleavage of complementary mRNAs [84], resulting in sequence-specific posttranscriptional gene silencing. As a potential experimental tool, the RNAi pathway can be directly activated by intracellular delivery of chemically synthesized siRNA oligonucleotides [85,86] or transfection with plasmid or viral vectors encoding short-hairpin RNA (shRNA) [87] that is recognized and processed by the Dicer pathway to generate siRNA. In principle, the activation of RNAi offers a highly specific and effective method to block the expression of a selected gene product and investigate its contribution to an observed biological response. RNAi has been reported to discriminate between mutated and wild-type genes that differ by a single point mutation [88]. Vector designs for shRNA delivery are now available with inducible promoters and selectable markers that can be used to purify highly enriched populations of transfected cells [89]. Despite its promise, several caveats must be considered in the use of RNAi technology. Although RNAi can achieve complete inhibition of gene expression (silencing) in plants and some lower organisms, the process in mammalian cells and organisms only achieves a "knock-down" of expression, typically in the range of 80%–90% reduction under optimized conditions [90]. The specificity of RNAi remains an intensely studied and debated issue. Using microarray technology, conflicting studies have reported minimal and significant "off-target" effects on the expression of other genes with varying degrees of sequence similarity [91]. Finally, effective use of RNAi requires validation of inhibition at both mRNA and protein levels and corresponding negative controls using the same vector with a sequence not corresponding to any endogenously expressed gene.

3.4.3 Applications in the Literature

3.4.3.1 Mechanisms of Strain-Mediated Cell Reorientation

Cellular reorientation and alignment perpendicular to the direction of applied strain was one of the earliest observed responses to cyclic strain [92]. Only recently, has insight been gained into the fundamental mechanisms responsible for this behavior. Using the Flexcell Strain Unit to apply variable strain to fibroblasts, Wen et al. observed significant increases in cell orientation at strains between 15%–24% [93]. This response was significantly reduced when cells were pretreated with function-blocking antibodies against β1 integrins, but not α1, α2, or α3. Western blotting was performed to determine if strain-mediated reorientation was accompanied by changes in the activation of several intracellular signaling molecules associated with integrin binding. Exposure to cyclic strain stimulated rapid increases in activated focal adhesion kinase (FAK), Rho, and the p38 mitogen-activated protein kinase (MAPK) and this response was significantly reduced by β1 integrin antibodies. Fibroblasts were then pretreated with the Rho inhibitor C3 transferase or a chemical inhibitor of p38 activation and shown to exhibit reduced cell orientation in response to cyclic strain, providing further support for a mechanism involving β1 integrin-mediated signaling through Rho and p38.

Thodeti et al. also recently reported on the mechanisms of capillary endothelial (CE) cell reorientation in response to cyclic strain [94]. Similar to the study above, they showed that cyclic strain-induced CE reorientation is accompanied by increased β1 integrin, FAK, and extracellular signal-regulated kinase [ERK]1/2 activation and that the cell orientation response could be blocked using function-blocking antibodies against β1 integrin. In order to investigate potential mechanisms responsible for β1 integrin activation, they evaluated the activation of phosphatidylinositol-3-kinase (PI3K), an intracellular enzyme involved in integrin and subsequent Rho activation. Western blotting demonstrated increased activation of the PI3K substrate AKT in strained samples. The application of a chemical inhibitor of PI3K blocked both AKT and β1 integrin activation. The authors hypothesized that SA ion channels might be involved in the orientation response and demonstrated that the general SA channel blocker, gadolinium chloride, inhibited β1 integrin activation and cell reorientation in response to cyclic strain. To identify the specific SA ion channel responsible for the reorientation effect, siRNA was used to knockdown the expression of several members of the transient receptor potential vanilloid (TRPV) family of ion channels. RT-PCR analysis demonstrated that siRNAs targeting TRPV2 or TRPV4 achieved 70%–90% knockdown of mRNA expression levels, while a nontargeting control sequence did not affect their expression. siRNA-mediated knockdown of TRPV4-inhibited strain-induced calcium influx measured by a fluorescent reporter dye; activation of β1 integrin, AKT, and ERK1/2; and CE cell reorientation; suggesting a pathway of strain-mediated calcium influx through TRPV4, PI3K activation, and β1 integrin activation in initiating the endothelial reorientation response to cyclic strain.

Another fundamental response to mechanical loading observed in many different cell types is the induction of tissue-specific matrix expression and increased mechanical properties. This concept has been widely applied as a strategy for preconditioning tissue engineering constructs using in vitro bioreactors to improve their mechanical properties prior to implantation. Seliktar et al. showed that cyclic distention of aortic smooth muscle cells (SMCs) in collagen gels around silicone tubes resulted in increased circumferential cell orientation and lamellar collagen organization, accompanied by significant increases in tensile properties [95]. In a subsequent study using the same experimental model, Seliktar et al. used Western blotting and zymography to show that cyclic strain increased total MMP-2 expression and activity relative to static controls [96]. High ribose medium (HRM) and N-acetylcysteine (NAC) were used as nonspecific inhibitors of MMP activity. Both conditions were shown to antagonize strain-induced MMP-2 expression and substantially reduced (HRM) or essentially eliminated (NAC) strain-induced increases in construct mechanical properties; confirming a critical role for MMP-mediated matrix remodeling in the mechanical conditioning of collagen/SMC constructs.

3.4.3.2 The Tensegrity Model of Cellular Mechanotransduction

One of the major paradigms of mechanotransduction developed by Donald Ingber is the concept of the cytoskeleton as a tensegrity structure, possessing an internal prestress originating from actin–myosin contraction resisted by microtubules [14–16,97]. Application of external loads causes structural rearrangement, achieving mechanotransduction by solid-state biochemistry originating from the physical immobilization of numerous enzymes, substrates, and signaling molecules to the cytoskeleton [98]. One early line of evidence in support of the tensegrity model was the demonstration that changes in cell shape could switch cells between proliferation, differentiation, and apoptosis [99,100].

In a recent study, McBeath et al. investigated the effects of cell shape and cytoskeletal tension on mesenchymal stem cell (MSC) fate specification [101]. MSC cell shape was controlled by growing at high- or low-density culture or on chemically defined adhesive islands of varying size. When cultured in a mixed media formulation containing both osteogenic and adipogenic induction factors, MSCs that attained extensive spreading (low-density culture, relative large islands) exhibited increased osteogenic differentiation, while limited cell spreading favored adipocytic differentiation. In order to determine if this effect was attributable to changes in differences in cell shape per se or associated cytoskeletal tension, MSCs were cultured in mixed media with a chemical inhibitor of the Rho effector enzyme Rho kinase (ROCK). Inhibition of ROCK and cytoskeletal tension did not alter cell shape, but reduced osteogenic and increased adipogenic differentiation. To further confirm the role of cytoskeletal tension in fate selection, MSCs were transfected with adenoviral constructs encoding constitutively active RhoA or dominant-negative RhoA and cultured in conventional growth media without differentiation inducing factors. Overexpression of RhoA promoted osteogenic differentiation, while its inactivation by a dominant-negative form promoted adipogenic differentiation, suggesting that RhoA can regulate MSC differentiation independent of soluble differentiation factors.

In another study relating cytoskeletal tension to cellular differentiation, Engler et al. showed that variation of substrate elastic modulus could direct MSC differentiation [102]. MSCs cultured on polyacrylamide gels modeling the elastic modulus of brain (0.1–1 kPa), striated muscle (8–17 kPa), and osteoid collagen (25–40 kPa) in the absence of any soluble differentiation factors exhibited morphological characteristics consistent with neuronal, muscular, and osteoblastic phenotypes, respectively. Transcriptional analysis using microarrays with confirmation by Western blotting demonstrated substrate elasticity-dependent induction of lineage-specific markers at mRNA and protein levels. The authors hypothesized that nonmuscle myosin (NMM) II isoforms and resulting cytoskeletal tension mediated these effects. The addition of blebbistatin, a chemical inhibitor of all NMM II isoforms, at the time of plating inhibited cell spreading on all substrates. When blebbistatin was added 24 h after plating to MSCs cultured on gels modeling striated muscle elasticity, the adopted spindle-shaped morphology was preserved, but expression of the muscle-specific transcription factor MyoD was reduced. Subsequent removal of the drug resulted in restoration of MyoD expression within 72 h; suggesting that actin–myosin contractility is necessary for cells to detect and respond to the underlying substrate elasticity.

To investigate the importance of tensegrity principles at the tissue level, Moore et al. investigated the role of cytoskeletal tension in epithelial branching morphogenesis using an embryonic mouse lung organotypic model [103]. Three chemical inhibitors targeting different enzymes involved in the generation of cytoskeletal tension (ROCK, myosin light chain kinase, and myosin ATPase) were shown to inhibit epithelial bud formation in a dose-dependent manner. Conversely, addition of a Rho activator, cytotoxic necrotizing factor-1, significantly increased bud formation at low doses; indicating that Rho-dependent changes in cytoskeletal tension play a critical role in the generation of complex epithelial tissue structures.

3.4.3.3 A Novel Mechanism of Mechanotransduction

Tschumperlin et al. investigated the effects of compressive stress on bronchial epithelial cells, modeling conditions associated with airway constriction [104]. Epithelial cells were cultured at an air–liquid interface to produce a pseudo-stratified monolayer where the cells were connected by tight junctions at the apical surface and separated by a lateral intercellular space (LIS) at the basolateral surface. Western blotting showed that compressive stress strongly activated the epidermal growth factor receptor (EGFR) and a downstream member of the MAPK pathway, extracellular regulated kinase (ERK). To confirm the role of the EGFR in signal generation, two chemical inhibitors were applied basolaterally to confirm that inhibition of EGFR lowered stress-induced ERK activation, while inhibition of the platelet-derived growth factor receptor did not affect the response. Similarly, basolateral administration of a function-blocking antibody against EGFR blocked stress-mediated ERK activation. In order to determine which EGFR ligand was responsible, neutralizing antibodies against transforming growth factor α, EGF and heparin-binding EGF (HB-EGF) were applied during compressive loading. Anti-HB-EGF reduced ERK activation in a dose-dependent manner, while the other two antibodies had no effect even at the highest concentration. The application of a broad spectrum MMP inhibitor was shown to reduce both baseline and stress-induced ERK activation, demonstrating that the signaling activity required proteolytic cleavage and ectodomain shedding of HB-EGF. Using fluorescently labeled dextran, the author's showed that compressive stress caused a substantial reduction in the volume of the LIS. Finally, a mathematical model was developed that predicted the measured reduction in LIS volume would produce an approximately eightfold increase in the concentration of soluble HB-EGF. A subsequent Western blot confirmed that a 10-fold increase in exogenous HB-EGF from the expected baseline level yielded a similar increase in ERK activation to that produced by compressive stress; demonstrating a novel mechanism of mechanotransduction whereby mechanically induced changes in LIS volume result in an effective increase in the concentration and activity of a soluble signaling molecule.

3.5 CONCLUSION

Research in mechanotransduction is rapidly accelerating in response to growing recognition of the critical impact of mechanical factors in tissue development, homeostatic maintenance, and pathology, as well as potential applications in tissue engineering and regenerative medicine [20,105,106]. The last decade has seen substantial advances in experimental platforms for precisely manipulating the mechanical microenvironment and measuring changes in cellular mechanics [107–110]. Continued integration of engineering technology with cell and molecular analytical techniques promises rapid advances in the discovery of fundamental mechanisms by which cells detect and respond to mechanical forces and the translation of these principles into clinical therapies.

REFERENCES

1. Adams, J. C. and F. M. Watt. 1993. Regulation of development and differentiation by the extracellular matrix. *Development* 117(4):1183–1198.
2. Sastry, S. K. and A. F. Horwitz. 1996. Adhesion-growth factor interactions during differentiation: An integrated biological response. *Dev Biol* 180(2):455–467.
3. Chen, R. R. and D. J. Mooney. 2003. Polymeric growth factor delivery strategies for tissue engineering. *Pharm Res* 20(8):1103–1112.
4. Danen, E. H. and A. Sonnenberg. 2003. Integrins in regulation of tissue development and function. *J Pathol* 201(4):632–641.
5. Lanyon, L. E. 1987. Functional strain in bone tissue as an objective, and controlling stimulus for adaptive bone remodelling. *J Biomech* 20(11–12):1083–1093.
6. Vandenburgh, H. H. 1987. Motion into mass: How does tension stimulate muscle growth? *Med Sci Sports Exerc* 19(5 Suppl):S142–S149.
7. Wong, M. and D. R. Carter. 2003. Articular cartilage functional histomorphology and mechanobiology: A research perspective. *Bone* 33(1):1–13.
8. Woo, S. L.-Y., S. D. Abramowitch, J. C. Loh, V. Musahl, and J. H.-C. Wang. 2003. Ligament healing: Present status and the future of functional tissue engineering. In: *Functional Tissue Engineering*, F. Guilak, D. L. Butler, S. A. Goldstein, and D. J. Mooney (eds.). New York: Springer-Verlag.
9. Kim, B. S., J. Nikolovski, J. Bonadio, and D. J. Mooney. 1999. Cyclic mechanical strain regulates the development of engineered smooth muscle tissue. *Nat Biotechnol* 17(10):979–983.
10. Yang, G., R. C. Crawford, and J. H. Wang. 2004. Proliferation and collagen production of human patellar tendon fibroblasts in response to cyclic uniaxial stretching in serum-free conditions. *J Biomech* 37(10):1543–1550.
11. Wang, J. H. and B. P. Thampatty. 2006. An introductory review of cell mechanobiology. *Biomech Model Mechanobiol* 5(1):1–16.
12. Joshi, S. D. and K. Webb. 2008. Variation of cyclic strain parameters regulates development of elastic modulus in fibroblast/substrate constructs. *J Orthop Res* 26(8):1105–1113.
13. Banes, A. J., G. Lee, R. Graff, C. Otey, J. Archambault, M. Tsuzaki, M. Elfervig, and J. Qiu. 2001. Mechanical forces and signaling in connective tissue cells-cellular mechanisms of detection, transduction, and responses to mechanical deformation. *Curr Opin Orthop* 12:389–396.
14. Ingber, D. E. 2003. Tensegrity I. Cell structure and hierarchical systems biology. *J Cell Sci* 116 (Pt 7):1157–1173.

15. Ingber, D. E. 2003. Tensegrity II. How structural networks influence cellular information processing networks. *J Cell Sci* 116(Pt 8):1397–1408.
16. Ingber, D. E. 2006. Cellular mechanotransduction: Putting all the pieces together again. *FASEB J* 20(7):811–827.
17. Wang, N., J. D. Tytell, and D. E. Ingber. 2009. Mechanotransduction at a distance: Mechanically coupling the extracellular matrix with the nucleus. *Nat Rev Mol Cell Biol* 10(1):75–82.
18. Parker, K. K. and D. E. Ingber. 2007. Extracellular matrix, mechanotransduction and structural hierarchies in heart tissue engineering. *Philos Trans R Soc Lond B Biol Sci* 362(1484):1267–1279.
19. Kisseleva, T. and D. A. Brenner. 2008. Mechanisms of fibrogenesis. *Exp Biol Med (Maywood)* 233(2):109–122.
20. Butler, D. L., S. A. Goldstein, R. E. Guldberg, X. E. Guo, R. Kamm, C. T. Laurencin, L. V. McIntire et al. 2009. The impact of biomechanics in tissue engineering and regenerative medicine. *Tissue Eng Part B Rev* 15(4):477–484.
21. Thellin, O., B. ElMoualij, E. Heinen, and W. Zorzi. 2009. A decade of improvements in quantification of gene expression and internal standard selection. *Biotechnol Adv* 27(4):323–333.
22. Dalma-Weiszhausz, D. D., J. Warrington, E. Y. Tanimoto, and C. G. Miyada. 2006. The affymetrix GeneChip platform: An overview. *Methods Enzymol* 410:3–28.
23. Brown, T., K. Mackey, and T. Du. 2004. Analysis of RNA by northern and slot blot hybridization. *Curr Protoc Mol Biol* Chapter 4:Unit 4.9.
24. Qu, Y. and M. Boutjdir. 2007. RNase protection assay for quantifying gene expression levels. *Methods Mol Biol* 366:145–158.
25. Hicks, D. G., G. Longoria, J. Pettay, T. Grogan, S. Tarr, and R. Tubbs. 2004. In situ hybridization in the pathology laboratory: General principles, automation, and emerging research applications for tissue-based studies of gene expression. *J Mol Histol* 35(6):595–601.
26. Mullis, K., F. Faloona, S. Scharf, R. Saiki, G. Horn, and H. Erlich. 1986. Specific enzymatic amplification of DNA in vitro: The polymerase chain reaction. *Cold Spring Harb Symp Quant Biol* 51(Pt 1):263–273.
27. Saiki, R. K., S. Scharf, F. Faloona, K. B. Mullis, G. T. Horn, H. A. Erlich, and N. Arnheim. 1985. Enzymatic amplification of beta-globin genomic sequences and restriction site analysis for diagnosis of sickle cell anemia. *Science* 230(4732):1350–1354.
28. Saiki, R. K., D. H. Gelfand, S. Stoffel, S. J. Scharf, R. Higuchi, G. T. Horn, K. B. Mullis, and H. A. Erlich. 1988. Primer-directed enzymatic amplification of DNA with a thermostable DNA polymerase. *Science* 239(4839):487–491.
29. Bustin, S. A. 2000. Absolute quantification of mRNA using real-time reverse transcription polymerase chain reaction assays. *J Mol Endocrinol* 25(2):169–193.
30. Higuchi, R., G. Dollinger, P. S. Walsh, and R. Griffith. 1992. Simultaneous amplification and detection of specific DNA sequences. *Biotechnology (NY)* 10(4):413–417.
31. Higuchi, R., C. Fockler, G. Dollinger, and R. Watson. 1993. Kinetic PCR analysis: Real-time monitoring of DNA amplification reactions. *Biotechnology (NY)* 11(9):1026–1030.
32. Nolan, T., R. E. Hands, and S. A. Bustin. 2006. Quantification of mRNA using real-time RT-PCR. *Nat Protoc* 1(3):1559–1582.
33. Gibson, U. E., C. A. Heid, and P. M. Williams. 1996. A novel method for real time quantitative RT-PCR. *Genome Res* 6(10):995–1001.
34. Tyagi, S. and F. R. Kramer. 1996. Molecular beacons: Probes that fluoresce upon hybridization. *Nat Biotechnol* 14(3):303–308.
35. Wong, M. L. and J. F. Medrano. 2005. Real-time PCR for mRNA quantitation. *Biotechniques* 39(1):75–85.
36. VanGuilder, H. D., K. E. Vrana, and W. M. Freeman. 2008. Twenty-five years of quantitative PCR for gene expression analysis. *Biotechniques* 44(5):619–626.
37. Kafert, S., J. Krauter, A. Ganser, and M. Eder. 1999. Differential quantitation of alternatively spliced messenger RNAs using isoform-specific real-time RT-PCR. *Anal Biochem* 269(1):210–213.
38. Chomczynski, P. and N. Sacchi. 1987. Single-step method of RNA isolation by acid guanidinium thiocyanate-phenol-chloroform extraction. *Anal Biochem* 162(1):156–159.
39. Imbeaud, S., E. Graudens, V. Boulanger, X. Barlet, P. Zaborski, E. Eveno, O. Mueller, A. Schroeder, and C. Auffray. 2005. Towards standardization of RNA quality assessment using user-independent classifiers of microcapillary electrophoresis traces. *Nucleic Acids Res* 33(6):e56.
40. Schroeder, A., O. Mueller, S. Stocker, R. Salowsky, M. Leiber, M. Gassmann, S. Lightfoot, W. Menzel, M. Granzow, and T. Ragg. 2006. The RIN: An RNA integrity number for assigning integrity values to RNA measurements. *BMC Mol Biol* 7:3.

41. Bustin, S. A. and T. Nolan. 2004. Pitfalls of quantitative real-time reverse-transcription polymerase chain reaction. *J Biomol Tech* 15(3):155–166.

42. Bustin, S. A. 2002. Quantification of mRNA using real-time reverse transcription PCR (RT-PCR): Trends and problems. *J Mol Endocrinol* 29(1):23–39.

43. Kubista, M., J. M. Andrade, M. Bengtsson, A. Forootan, J. Jonak, K. Lind, R. Sindelka et al. 2006. The real-time polymerase chain reaction. *Mol Aspects Med* 27(2–3):95–125.

44. Sharkey, D. J., E. R. Scalice, K. G. Christy Jr., S. M. Atwood, and J. L. Daiss. 1994. Antibodies as thermolabile switches: High temperature triggering for the polymerase chain reaction. *Biotechnology (NY)* 12(5):506–509.

45. Ririe, K. M., R. P. Rasmussen, and C. T. Wittwer. 1997. Product differentiation by analysis of DNA melting curves during the polymerase chain reaction. *Anal Biochem* 245(2):154–160.

46. Huggett, J., K. Dheda, S. Bustin, and A. Zumla. 2005. Real-time RT-PCR normalisation; strategies and considerations. *Genes Immun* 6(4):279–284.

47. Stahlberg, A., M. Kubista, and M. Pfaffl. 2004. Comparison of reverse transcriptases in gene expression analysis. *Clin Chem* 50(9):1678–1680.

48. Fleige, S. and M. W. Pfaffl. 2006. RNA integrity and the effect on the real-time qRT-PCR performance. *Mol Aspects Med* 27(2–3):126–139.

49. Fleige, S., V. Walf, S. Huch, C. Prgomet, J. Sehm, and M. W. Pfaffl. 2006. Comparison of relative mRNA quantification models and the impact of RNA integrity in quantitative real-time RT-PCR. *Biotechnol Lett* 28(19):1601–1613.

50. Hamalainen, H. K., J. C. Tubman, S. Vikman, T. Kyrola, E. Ylikoski, J. A. Warrington, and R. Lahesmaa. 2001. Identification and validation of endogenous reference genes for expression profiling of T helper cell differentiation by quantitative real-time RT-PCR. *Anal Biochem* 299(1):63–70.

51. Schmittgen, T. D. and B. A. Zakrajsek. 2000. Effect of experimental treatment on housekeeping gene expression: Validation by real-time, quantitative RT-PCR. *J Biochem Biophys Methods* 46(1–2):69–81.

52. Dheda, K., J. F. Huggett, S. A. Bustin, M. A. Johnson, G. Rook, and A. Zumla. 2004. Validation of housekeeping genes for normalizing RNA expression in real-time PCR. *Biotechniques* 37(1):112–114, 116, 118–119.

53. de Kok, J. B., R. W. Roelofs, B. A. Giesendorf, J. L. Pennings, E. T. Waas, T. Feuth, D. W. Swinkels, and P. N. Span. 2005. Normalization of gene expression measurements in tumor tissues: Comparison of 13 endogenous control genes. *Lab Invest* 85(1):154–159.

54. Roge, R., J. Thorsen, C. Torring, A. Ozbay, B. K. Moller, and J. Carstens. 2007. Commonly used reference genes are actively regulated in in vitro stimulated lymphocytes. *Scand J Immunol* 65(2):202–209.

55. Dheda, K., J. F. Huggett, J. S. Chang, L. U. Kim, S. A. Bustin, M. A. Johnson, G. A. Rook, and A. Zumla. 2005. The implications of using an inappropriate reference gene for real-time reverse transcription PCR data normalization. *Anal Biochem* 344(1):141–143.

56. Hendriks-Balk, M. C., M. C. Michel, and A. E. Alewijnse. 2007. Pitfalls in the normalization of real-time polymerase chain reaction data. *Basic Res Cardiol* 102(3):195–197.

57. Guenin, S., M. Mauriat, J. Pelloux, O. Van Wuytswinkel, C. Bellini, and L. Gutierrez. 2009. Normalization of qRT-PCR data: The necessity of adopting a systematic, experimental conditions-specific, validation of references. *J Exp Bot* 60(2):487–493.

58. Livak, K. J. and T. D. Schmittgen. 2001. Analysis of relative gene expression data using real-time quantitative PCR and the 2(-Delta Delta C(T)) method. *Methods* 25(4):402–408.

59. Vandesompele, J., K. De Preter, F. Pattyn, B. Poppe, N. Van Roy, A. De Paepe, and F. Speleman. 2002. Accurate normalization of real-time quantitative RT-PCR data by geometric averaging of multiple internal control genes. *Genome Biol* 3(7):RESEARCH0034.

60. Pfaffl, M. W., A. Tichopad, C. Prgomet, and T. P. Neuvians. 2004. Determination of stable housekeeping genes, differentially regulated target genes and sample integrity: BestKeeper—Excel-based tool using pair-wise correlations. *Biotechnol Lett* 26(6):509–515.

61. Andersen, C. L., J. L. Jensen, and T. F. Orntoft. 2004. Normalization of real-time quantitative reverse transcription-PCR data: A model-based variance estimation approach to identify genes suited for normalization, applied to bladder and colon cancer data sets. *Cancer Res* 64(15):5245–5250.

62. Kutty, J. K. and K. Webb. 2010. Vibration stimulates vocal mucosa-like matrix expression by hydrogel-encapsulated fibroblasts. *J Tissue Eng Regen Med* 4(1):62–72.

63. Fronhoffs, S., G. Totzke, S. Stier, N. Wernert, M. Rothe, T. Bruning, B. Koch, A. Sachinidis, H. Vetter, and Y. Ko. 2002. A method for the rapid construction of cRNA standard curves in quantitative real-time reverse transcription polymerase chain reaction. *Mol Cell Probes* 16(2):99–110.

64. Pfaffl, M. W. 2001. A new mathematical model for relative quantification in real-time RT-PCR. *Nucleic Acids Res* 29(9):e45.

65. Bolufer, P., F. Lo Coco, D. Grimwade, E. Barragan, D. Diverio, B. Cassinat, C. Chomienne et al. 2001. Variability in the levels of PML-RAR alpha fusion transcripts detected by the laboratories participating in an external quality control program using several reverse transcription polymerase chain reaction protocols. *Haematologica* 86(6):570–576.

66. Abbas, A. K., A. H. Lichtman, and J. S. Pober. 1994. Maturation of B lymphocytes and expression of immunoglobulin genes. In: *Cellular and Molecular Immunology*. Philadelphia, PA: W. B. Saunders Company.

67. Kim, Y. T., R. W. Hitchcock, M. J. Bridge, and P. A. Tresco. 2004. Chronic response of adult rat brain tissue to implants anchored to the skull. *Biomaterials* 25(12):2229–2237.

68. Biran, R., D. C. Martin, and P. A. Tresco. 2005. Neuronal cell loss accompanies the brain tissue response to chronically implanted silicon microelectrode arrays. *Exp Neurol* 195(1):115–126.

69. Laemmli, U. K. 1970. Cleavage of structural proteins during the assembly of the head of bacteriophage T4. *Nature* 227(5259):680–685.

70. Sapan, C. V., R. L. Lundblad, and N. C. Price. 1999. Colorimetric protein assay techniques. *Biotechnol Appl Biochem* 29(Pt 2):99–108.

71. Miura, R. O., S. Yamagata, Y. Miura, T. Harada, and T. Yamagata. 1995. Analysis of glycosaminoglycan-degrading enzymes by substrate gel electrophoresis (zymography). *Anal Biochem* 225(2):333–340.

72. Snoek-van Beurden, P. A. and J. W. Von den Hoff. 2005. Zymographic techniques for the analysis of matrix metalloproteinases and their inhibitors. *Biotechniques* 38(1):73–83.

73. Leung, D. Y., S. Glagov, and M. B. Mathews. 1976. Cyclic stretching stimulates synthesis of matrix components by arterial smooth muscle cells in vitro. *Science* 191(4226):475–477.

74. Ignotz, R. A. and J. Massague. 1986. Transforming growth factor-beta stimulates the expression of fibronectin and collagen and their incorporation into the extracellular matrix. *J Biol Chem* 261(9):4337–4345.

75. Roberts, A. B., M. B. Sporn, R. K. Assoian, J. M. Smith, N. S. Roche, L. M. Wakefield, U. I. Heine et al. 1986. Transforming growth factor type beta: Rapid induction of fibrosis and angiogenesis in vivo and stimulation of collagen formation in vitro. *Proc Natl Acad Sci USA* 83(12):4167–4171.

76. Neidlinger-Wilke, C., I. Stalla, L. Claes, R. Brand, I. Hoellen, S. Rubenacker, M. Arand, and L. Kinzl. 1995. Human osteoblasts from younger normal and osteoporotic donors show differences in proliferation and TGF beta-release in response to cyclic strain. *J Biomech* 28(12):1411–1418.

77. Riser, B. L., P. Cortes, C. Heilig, J. Grondin, S. Ladson-Wofford, D. Patterson, and R. G. Narins. 1996. Cyclic stretching force selectively up-regulates transforming growth factor-beta isoforms in cultured rat mesangial cells. *Am J Pathol* 148(6):1915–1923.

78. Riser, B. L., P. Cortes, J. Yee, A. K. Sharba, K. Asano, A. Rodriguez-Barbero, and R. G. Narins. 1998. Mechanical strain- and high glucose-induced alterations in mesangial cell collagen metabolism: Role of TGF-beta. *J Am Soc Nephrol* 9(5):827–836.

79. Lindahl, G. E., R. C. Chambers, J. Papakrivopoulou, S. J. Dawson, M. C. Jacobsen, J. E. Bishop, and G. J. Laurent. 2002. Activation of fibroblast procollagen alpha 1(I) transcription by mechanical strain is transforming growth factor-beta-dependent and involves increased binding of CCAAT-binding factor (CBF/NF-Y) at the proximal promoter. *J Biol Chem* 277(8):6153–6161.

80. Fire, A., S. Xu, M. K. Montgomery, S. A. Kostas, S. E. Driver, and C. C. Mello. 1998. Potent and specific genetic interference by double-stranded RNA in *Caenorhabditis elegans*. *Nature* 391(6669):806–811.

81. Hannon, G. J. 2002. RNA interference. *Nature* 418(6894):244–251.

82. Bernstein, E., A. A. Caudy, S. M. Hammond, and G. J. Hannon. 2001. Role for a bidentate ribonuclease in the initiation step of RNA interference. *Nature* 409(6818):363–366.

83. Hammond, S. M., E. Bernstein, D. Beach, and G. J. Hannon. 2000. An RNA-directed nuclease mediates post-transcriptional gene silencing in Drosophila cells. *Nature* 404(6775):293–296.

84. Liu, J., M. A. Carmell, F. V. Rivas, C. G. Marsden, J. M. Thomson, J. J. Song, S. M. Hammond, L. Joshua-Tor, and G. J. Hannon. 2004. Argonaute2 is the catalytic engine of mammalian RNAi. *Science* 305(5689):1437–1441.

85. Caplen, N. J., S. Parrish, F. Imani, A. Fire, and R. A. Morgan. 2001. Specific inhibition of gene expression by small double-stranded RNAs in invertebrate and vertebrate systems. *Proc Natl Acad Sci USA* 98(17):9742–9747.

86. Elbashir, S. M., J. Harborth, W. Lendeckel, A. Yalcin, K. Weber, and T. Tuschl. 2001. Duplexes of 21-nucleotide RNAs mediate RNA interference in cultured mammalian cells. *Nature* 411(6836):494–498.

87. Brummelkamp, T. R., R. Bernards, and R. Agami. 2002. A system for stable expression of short interfering RNAs in mammalian cells. *Science* 296(5567):550–553.

88. Ding, H., D. S. Schwarz, A. Keene, B. Affar el, L. Fenton, X. Xia, Y. Shi, P. D. Zamore, and Z. Xu. 2003. Selective silencing by RNAi of a dominant allele that causes amyotrophic lateral sclerosis. *Aging Cell* 2(4):209–217.

89. Fewell, G. D. and K. Schmitt. 2006. Vector-based RNAi approaches for stable, inducible and genome-wide screens. *Drug Discov Today* 11(21–22):975–982.

90. Juliano, R. L., V. R. Dixit, H. Kang, T. Y. Kim, Y. Miyamoto, and D. Xu. 2005. Epigenetic manipulation of gene expression: A toolkit for cell biologists. *J Cell Biol* 169(6):847–857.

91. Dykxhoorn, D. M. and J. Lieberman. 2005. The silent revolution: RNA interference as basic biology, research tool, and therapeutic. *Annu Rev Med* 56:401–423.

92. Buckley, M. J., A. J. Banes, L. G. Levin, B. E. Sumpio, M. Sato, R. Jordan, J. Gilbert, G. W. Link, and R. Tran Son Tay. 1988. Osteoblasts increase their rate of division and align in response to cyclic, mechanical tension in vitro. *Bone Miner* 4(3):225–236.

93. Wen, H., P. A. Blume, and B. E. Sumpio. 2009. Role of integrins and focal adhesion kinase in the orientation of dermal fibroblasts exposed to cyclic strain. *Int Wound J* 6(2):149–158.

94. Thodeti, C. K., B. Matthews, A. Ravi, A. Mammoto, K. Ghosh, A. L. Bracha, and D. E. Ingber. 2009. TRPV4 channels mediate cyclic strain-induced endothelial cell reorientation through integrin-to-integrin signaling. *Circ Res* 104(9):1123–1130.

95. Seliktar, D., R. A. Black, R. P. Vito, and R. M. Nerem. 2000. Dynamic mechanical conditioning of collagen-gel blood vessel constructs induces remodeling in vitro. *Ann Biomed Eng* 28(4):351–362.

96. Seliktar, D., R. M. Nerem, and Z. S. Galis. 2001. The role of matrix metalloproteinase-2 in the remodeling of cell-seeded vascular constructs subjected to cyclic strain. *Ann Biomed Eng* 29(11):923–934.

97. Ingber, D. E. 2008. Tensegrity-based mechanosensing from macro to micro. *Prog Biophys Mol Biol* 97(2–3):163–179.

98. Janmey, P. A. 1998. The cytoskeleton and cell signaling: Component localization and mechanical coupling. *Physiol Rev* 78(3):763–781.

99. Ingber, D. E. and J. Folkman. 1989. Mechanochemical switching between growth and differentiation during fibroblast growth factor-stimulated angiogenesis in vitro: Role of extracellular matrix. *J Cell Biol* 109(1):317–330.

100. Chen, C. S., M. Mrksich, S. Huang, G. M. Whitesides, and D. E. Ingber. 1997. Geometric control of cell life and death. *Science* 276(5317):1425–1428.

101. McBeath, R., D. M. Pirone, C. M. Nelson, K. Bhadriraju, and C. S. Chen. 2004. Cell shape, cytoskeletal tension, and RhoA regulate stem cell lineage commitment. *Dev Cell* 6(4):483–495.

102. Engler, A. J., S. Sen, H. L. Sweeney, and D. E. Discher. 2006. Matrix elasticity directs stem cell lineage specification. *Cell* 126(4):677–689.

103. Moore, K. A., T. Polte, S. Huang, B. Shi, E. Alsberg, M. E. Sunday, and D. E. Ingber. 2005. Control of basement membrane remodeling and epithelial branching morphogenesis in embryonic lung by Rho and cytoskeletal tension. *Dev Dyn* 232(2):268–281.

104. Tschumperlin, D. J., G. Dai, I. V. Maly, T. Kikuchi, L. H. Laiho, A. K. McVittie, K. J. Haley et al. 2004. Mechanotransduction through growth-factor shedding into the extracellular space. *Nature* 429(6987):83–86.

105. Ingber, D. E. 2003. Mechanobiology and diseases of mechanotransduction. *Ann Med* 35(8):564–577.

106. Ghosh, K. and D. E. Ingber. 2007. Micromechanical control of cell and tissue development: Implications for tissue engineering. *Adv Drug Deliv Rev* 59(13):1306–1318.

107. Lele, T. P., J. E. Sero, B. D. Matthews, S. Kumar, S. Xia, M. Montoya-Zavala, T. Polte, D. Overby, N. Wang, and D. E. Ingber. 2007. Tools to study cell mechanics and mechanotransduction. *Methods Cell Biol* 83:443–472.

108. Addae-Mensah, K. A. and J. P. Wikswo. 2008. Measurement techniques for cellular biomechanics in vitro. *Exp Biol Med (Maywood)* 233(7):792–809.

109. Wang, Y., J. Y. Shyy, and S. Chien. 2008. Fluorescence proteins, live-cell imaging, and mechanobiology: Seeing is believing. *Annu Rev Biomed Eng* 10:1–38.

110. Kim, D. H., P. K. Wong, J. Park, A. Levchenko, and Y. Sun. 2009. Microengineered platforms for cell mechanobiology. *Annu Rev Biomed Eng* 11:203–233.

Section II (Part 1)

Literature Review of Mechanobiology Research Findings and Theories

Cardiovascular Systems

4 Effects of Endovascular Intervention on Vascular Smooth Muscle Cell Function

Brad Winn, Bethany Acampora,
Jiro Nagatomi, and Martine LaBerge

CONTENTS

4.1 INTRODUCTION

Cardiovascular disease is the leading cause of mortality in the United States and Europe, accounting for approximately half of all deaths [1,2]. The most common form of cardiovascular disease is atherosclerosis, which is commonly treated using balloon angioplasty usually in conjunction with the deployment of a stent. Stent deployment helps to hold the vessel open following the local injury caused by balloon inflation and prevents elastic recoil. Stenting has been shown to significantly reduce restenosis rates from approximately 20% to 50% without a stent to about 10%–30% with stent deployment [3]. However, this still leaves significant room for improvement.

This review will take a systematic approach to gaining an understanding of the different mechanisms of atherosclerosis as well as the potential drawbacks and limitations of vascular stenting that lead to restenosis. Several in vitro test setups reviewed may be useful for the modeling of a stented blood vessel. The use of such an in vitro simulator allows for a controlled environment that reduces variability in order to maximize potential improvement of clinical success while minimizing research expense. First, however, we will look to gaining an understanding of the basic mechanisms of atherosclerosis and intravascular stenting as a primary modality of treatment.

4.2 ATHEROSCLEROSIS

4.2.1 INTRODUCTION

Atherosclerosis is a disease that systemically affects the large and medium-sized arteries of the vasculature [1,4]. However, it is more common in the carotid, coronary, cerebral, and renal arteries, as well as the aorta. Risk factors for the development of atherosclerosis include high blood pressure, high cholesterol, diabetes, obesity, smoking, family history, stress, lack of exercise, excessive alcohol consumption, race, and gender [1,2,4–6].

4.2.2 PATHOLOGY OF ATHEROSCLEROSIS

The basic pathology of atherosclerosis is characterized by the development of an atheromatous plaque, which consists of proliferating cells and connective tissue with a soft lipid-rich core [6]. This lesion can eventually grow to occlude the lumen, causing complications such as obstruction to blood flow, thrombosis risk, loss of elasticity, and impaired contractility. Atherosclerosis is also characterized by a chronic inflammatory response [6,7]. At this point, the initial cause of atherosclerotic development can be a topic of debate with two commonly accepted theories referred to as the "fatty streak theory" and the "response to injury theory." However, these theories are not mutually exclusive. In both cases, high blood triglyceride levels, particularly low-density lipoproteins (LDLs), are the necessary catalyst for lesion development and progression [1,5–7]. The primary difference between the two theories is in the modality of monocyte recruitment to the lesion site.

Following the fatty streak theory, fatty streaks develop early in life as a result of lipid accumulation in the vessel wall caused by a high fat diet as well as genetic predisposition [6,8,9]. Early childhood lesions commonly regress but may continue to develop rapidly during the age span of 15–34 years [5,6]. In this process, LDLs are transported by transcytosis across the endothelium and deposited in the intima where a small amount may be taken up by local smooth muscle cells (SMCs) [5]. However, the LDLs may become oxidized during the process of transcytosis, thereby forming the toxic oxLDL. It is this injurious oxLDL that is thought to damage neighboring cells, thereby triggering the inflammatory response [9]. This results in local macrophage infiltration from nearby tissues as well as monocyte recruitment from the blood. The resulting macrophage population then phagocytizes the LDL. The resulting lipid-filled macrophages and SMCs form what are known as foam cells that are characteristic of atherosclerotic lesions.

The response to injury theory varies in that the inflammatory response is thought to be sparked by an initial injury to the endothelium. This initial injury is theorized to be caused by factors such as smoking, high blood pressure, infection, etc. [8]. The damaged endothelium promotes platelet adhesion and subsequent release of platelet derived growth factor (PDGF) as well as monocyte recruitment. PDGF is known to be both chemotactic and mitogenic for SMCs [10–12]. The damaged endothelium also exhibits altered permeability, leading to lipid accumulation in the vessel wall in the presence of high lipid content in the blood. The local macrophages and SMCs will then uptake this lipid, thereby forming foam cells.

Whatever the initial cause, atherosclerosis is a progressive disease marked by a chronic inflammatory response. Lesion progression takes place over the course of years through a series of well-documented stages [13]. This is illustrated in Table 4.1 reproduced from a review by Stary [13]. It is important to note that the progression of these stages with respect to time is not necessarily linear.

In addition to simply growing in size to physically occlude the lumen, the ongoing progression of a plaque often leads to fissuring and rupture, which in turn causes thrombosis [14]. These thrombi give rise to the serious risk of emboli formation and subsequent heart attack or stroke. Figure 4.1

TABLE 4.1
Classification of Atherosclerotic Lesions

	Recommended Terms	Description	Conventional Terms Based on Appearance with the Unaided Eye	Comments
Type I	(initial lesion)	Intimal lipoprotein accumulation; lipid in macrophages; changes discernible only microscopically or chemically; no tissue damage	None	Types I and II are sometimes combined as "early lesions"
Type II	(fatty streak)	Lipoprotein accumulation in intima; lipid in macrophages and smooth muscle cells; quantities large enough to be visible to the unaided eye but still no tissue damage	Fatty streak	
IIa	(progression-prone:colocalized with specific adaptive thickening			
IIb	(progression resistant)			
Type III	(preatheroma)	All type IIa changes plus multiple deposits of pooled extracellular lipid; microscopic evidence of tissue damage and disorder	None	An "intermediate" or "transitional" lesion had be suspected
Type IV	(atheroma)	All type IIa changes plus confluent mass of extracellular lipid (lipid core) with massive structural damage to intima	Fibrous plaque; fibrolipid plaque; plaque	Types IV to VII are sometimes combined as "advanced lesions"
Type V	(fibroatheroma)	All type IV changes plus development of marked collagen and smooth muscle cell increase (cap) above lipid core		
Type VI	(complicated fibroatheroma)	All type V changes plus a thrombotic deposit, and/or erosion or fissure	Complicated plaque	
VIa	(thrombo-hemmorrhagic)			
VIa	(thrombotic)			
VIb				
VIc	(hemmorrhagic)			
Type VII	(calcific lesion)	Any advanced lesion type composed predominantly of calcium; substantial structural deformity	Calcified plaque	
Type VIII	(fibrotic lesion)	Any advanced lesion type composed predominantly of collagen; lipid may be absent	Fibrous plaque	

Source: Stary, H., *Virchows Archiv A Pathological Anatomy and Histopathology,* 421, 277, 1992.

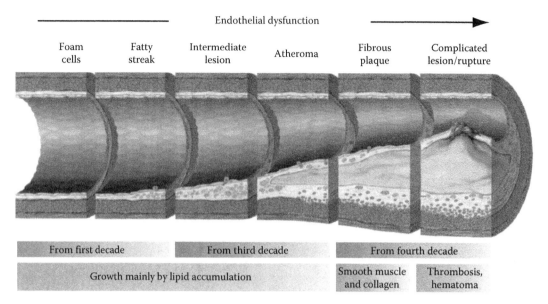

FIGURE 4.1 (See color insert.)   Progression of an atherosclerotic lesion. (From Pepine, C.J., *The American Journal of Cardiology*, 82(10), S24, 1998.)

provides a general timeline for the formation and progression of atherosclerotic plaques [15]. It is important to note that the progression of atherosclerosis can be clinically silent until the very late stages of plaque formation. This is the reason for the somewhat common incidences of sudden death from sudden coronary blockage or stroke caused by emboli arising from atherosclerotic lesions.

4.2.3 TREATMENT OF ATHEROSCLEROSIS: INDICATIONS FOR VASCULAR STENTING

Traditionally, bypass surgery was the gold standard for treatment of cardiovascular and peripheral artery occlusive disease as a result of atherosclerosis [16]. However, in recent years, balloon angioplasty procedures have become a viable and increasingly popular minimally invasive alternative. Today, either bypass surgery or percutaneous transluminal angioplasty (PCTA or PTA) may be used to treat patients based on a number of factors known to dictate the potential procedural success [16,17]. The final decision on the modality of interventional therapy by the physician is made on an individual patient basis considering all of the possible risk factors of each treatment as well as patient consultation. In general, endovascular surgical therapy is indicated for treatment of vessels in which the occlusion is not immediately life threatening, and vessel diameter is greater than 2.5 mm. It is contradicted by chronic total occlusion, small-caliber arteries, diffusely diseased arteries having long or multiple closely oriented lesions, and vessels with no collateral blood supply to distal vessels. Table 4.2 reproduced from Rutherford's *Vascular Surgery* gives basic guidelines for the use to PTA or surgery for lesions in the lower extremities.

4.3 INTRALUMINAL VASCULAR STENTS

4.3.1 INTRODUCTION

The deployment of an intravascular stent in conjunction with the angioplasty procedure reduces the risk of restenosis to within the range of 10%–30% versus the 20%–50% restenosis risk of angioplasty alone [3]. However, restenosis still remains the main cause of long-term stent failure. In order to reduce this risk, it is important to understand the basic mechanisms of in-stent restenosis. However, for the purposes of enhancing the reader's understanding of the following sections, we

TABLE 4.2
Indications for Modality of Vascular Intervention

	General Description of Disease	Clinical Usefulness Based upon Risks/Benefits		Authors' Recommendation	American Heart Association Categories of Disease
		PTA	Surgery		
Category 1	Short, focal, stenotic disease at the site of intervention. Mild or no disease in the proximal or distal arterial segments	++	0	Surgery not indicated as initial treatment	Lesions for which PTA alone is the procedure of choice. Treatment of these lesions will result in a high technical success rate and will generally result in complete relief of symptoms or normalization of pressure gradients
Category 2	Moderate-length, focal, stenotic disease at the site of intervention. Mild disease in the proximal or distal arterial segments	++	+	PTA is appropriate initial therapy. Surgery is initial therapy in selected cases, such as young, good risk patients	Lesions that are well suited for PTA. Treatment of these lesions will result in complete relief or significant improvement in symptoms, pulses, or pressure gradients. This category includes lesions that will be treated by procedures to be followed by surgical bypass to treat multilevel vascular disease
Category 3	Long stenotic disease or short occlusion at site of intervention. Moderate disease in the proximal or distal arterial segments	+	++	Surgery is initial therapy except in selected cases, such as prohibitive surgical risk	Lesions that may be treated with PTA, but because of disease extent location. Or severity have a significantly lower chance of initial technical success or long term benefit than if treated with surgery. However, PTA may be performed, generally because of patient risk factors or because of lack of suitable bypass material
Category 4	Diffuse or extensive stenotic disease or long occlusion at site of intervention. Severe disease in the proximal or distal arterial segments	0	+++	Surgery preferable; PTA not indicated	Extensive vascular disease, for which PTA has a very limited role because of low technical success rate or poor long-term benefit. In very high risk patients or in those for whom no surgical procedure is applicable, PTA may have some place

Source: Rutherford, R., *Vascular Surgery*, 5th edn. Vol. 1. W.B. Saunders, Philadelphia, PA, 2000.

will first give a brief overview of smooth muscle cells as it relates to the discussion of endovascular intervention.

4.3.2 Smooth Muscle Cells

The most abundant cell type in an artery is the smooth muscle cell. Interestingly, smooth muscle cells are the only cells of the myogenic lineage to exhibit phenotypic plasticity in response to stimuli as first confirmed in vitro by Chamley-Cambell et al. in 1979 [18–21]. The two ends of the spectrum for SMC phenotype are contractile and synthetic [6,18]. It is important to note that the differentiation state of SMCs is in fact a spectrum where cells may exist in any number of differentiation states between the two ends. It is not a simple black and white issue with cells being easily classified into one phenotype or the other. Healthy adult SMCs are primarily of the differentiated contractile phenotype [5,21]. Their contractile ability is important for the maintenance of blood vessel tone as well as regulation of blood pressure and flow [1,8,21]. They exhibit properties such as low synthetic activity, low proliferation rate, and expression of unique contractile proteins such as smooth muscle α-actin, myosin heavy chain, h-caldesmon, and calponin [5,21,22]. The most abundant of these proteins is α-actin, making up approximately 40% of the total cell protein [22].

Highly synthetic SMCs can be referred to as "immature" and exhibit characteristics such as increased proliferation, increased organelles for production of lipids and proteins, and increased excretion of extracellular matrix as well as increased migration [18,21,22]. They are also characterized by increased expression of the proteins vimentin and β-actin [23–25]. In addition, the cells lose their contractile ability, which is confirmed by cell staining to show reduced expression of contractile markers such as smooth muscle α-actin, myosin heavy chain, h-caldesmon, and calponin [21,22,24]. Synthetic SMCs also exhibit altered morphology when examined under a microscope. Contractile SMCs are smaller, elongated, and spindle-shaped when compared to a synthetic SMC's hypertrophic appearance surrounded by copious extracellular matrix production (Figure 4.2). It is important to note that SMCs cultured in vitro will eventually dedifferentiate toward a more synthetic phenotype after several passages. For this reason, experimentations must be carried out under low passage numbers.

As mentioned previously, SMCs of a contractile phenotype are generally referred to as a "healthy" phenotype because they are present in healthy adult arteries. On the other hand, SMCs of a highly synthetic phenotype present in vivo are generally indicative of diseased conditions. In addition to

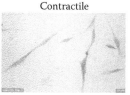

Contractile

Arterial injury, growth factors

Arteriogenesis

Synthetic

Marker of differentiation:

Decreased cell size
  • Spindle elongated morphology
Decreased ECM production
  • Predominance collagen IV, laminin
Increased contractile protein expression
  • SM-MHC2, calponin, actin
Decreased migration
  • Decreased MMPs, increased TIMPs

Marker of de-differentiation:

Increased cell size
  • Hypertrophic appearance
  • "Hill and valley" growth
Increased ECM production
  • Predominance collagen III, fibronectin
Decreased contractile protein expression
  • Increase osteopontin
Increase migration
  • Increased MMP-1 and MMP-3

FIGURE 4.2 Markers of smooth muscle cell phenotype. (From Rzucidlo, E., Martin, K., and Powell, R., *Journal of Vascular Surgery*, 45(A), A25, 2007.)

excessive extracellular matrix production, synthetic SMCs can also release cytokines, which are chemotactic for other SMCs, induce cell proliferation, and can stimulate phenotypic change from contractile to synthetic [8].

One must keep in mind when comparing and contrasting experimental results of in vitro studies presented in the literature that smooth muscle cells may react differently based on various factors involving the experimental materials and methods. For example, growth substrates are commonly coated with proteins such as collagen, fibronectin, and laminin in order to promote cell adhesion. However, it has been shown that surfaces coated with fibronectin elicit a more rapid synthetic phenotypic shift in SMC compared to laminin and collagen [25,26]. Also, it is possible that SMCs of different animal origins may react differently to similar stimuli [27].

4.3.3 Restenosis of Vascular Stents

In basic terms, a balloon angioplasty procedure is a forceful displacement of an atherosclerotic lesion serving to widen the vessel lumen to increase blood flow. In so doing, this causes stretching of the vessel wall, tears in the atherosclerotic plaques, and general damage to the vessel [28]. This in turn signals a complex cascade of elastic recoil, thrombosis, inflammation, intimal thickening, and vascular remodeling [28]. Elastic recoil and negative remodeling are the primary causative factors of restenosis when vessel expansion is done by balloon only. Indeed, the post-procedural lumen diameter is the most telling predictor of the expected long-term lumen diameter [28]. The deployment of a vascular stent greatly reduces the incidence of elastic recoil by serving to mechanically hold the vessel open. However, in the treatment of calcified and highly fibrous lesions, acute stent recoil of up to 15%–30% is commonly observed [28]. Stent deployment also further complicates the immunological response by triggering a foreign body response from the implantation of a biomaterial into the body. When performing an angioplasty procedure, particularly in conjunction with stent deployment, a certain degree of vascular injury is inevitable. However, the initial injury can be further complicated by the inflammatory response, the severity of which can ultimately dictate the degree of restenosis and subsequently affect procedural success.

Following balloon angioplasty, it is common to experience significant local damage to the endothelium or often a near complete denudation of the endothelial layer [28,29]. The acute reaction is an increased thrombogenic response due to exposure of the underlying collagen and sub-endothelial matrix [28,30]. It also triggers an inflammatory response and the extrinsic coagulation pathway due to injury to the vessel wall. To further amplify this thrombogenic response, the implanted biomaterial itself also triggers the intrinsic coagulation pathway and amplified inflammatory response [30]. With the use of current anticoagulant and antiplatelet therapy, thrombus formation to such an extent as to occlude the vessel lumen is relatively rare, and is not the primary determinant of long-term procedural success [28]. Alternately, it is the thrombus formation and recruitment of inflammatory cells, leading to high local concentration of growth factors such as cytokines and PDGF as well as fibrin that triggers a mitogenic and chemotactic response from smooth muscle cells [10,28]. The proliferative response of SMCs in turn results in the formation of often copious amounts of neo-intimal tissue generally known as intimal hyperplasia. The formation of this new tissue, primarily consisting of SMCs of the synthetic phenotype and their subsequent extracellular matrix, is the sole cause of in-stent restenosis since the stent normally serves to prevent elastic recoil and negative remodeling [28]. When the degree of restenosis progresses the extent such that follow-up interventional therapy is needed, it is given the term clinical restenosis.

An overview of the sequence of events following the placement of a bare metallic stent was established by Palmaz [31]. Due to the electropositive nature of the metallic surface in electrolyte solution, electronegative plasma proteins, primarily fibrinogen, bind to the surface of the stent immediately following implantation (Figure 4.3) [31]. Platelets and red blood cells then accumulate at the site of stent deployment within minutes and thrombus formation occurs shortly thereafter. Over the course of a few days to weeks, SMCs then respond to chemotactic factors from injury and clot formation,

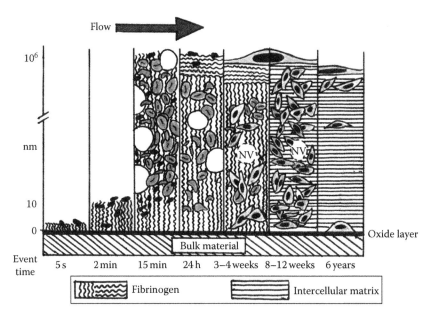

FIGURE 4.3 Sequence of events following placement of a metallic stent showing thickness of tissue deposited on surface. Vertical wavy lines represent randomly oriented fibrinogen. Horizontal wavy lines represent fibrinogen strands oriented in the direction of flow. Parallel straight lines represent extracellular matrix consisting primarily of collagen and ground substance; NV, neovessel; nm, nanometers. (From Palmaz, J., *American Journal of Roentgenology*, 160, 613, 1993.)

migrate into the area, and soon begin to proliferate replacing the thrombus with fibromuscular tissue [28,31,32]. These synthetic SMCs proliferate over a period of a few months when the majority of the restenosis takes place [28,32,33]. It is generally accepted that restenosis will present itself within the first 6–9 months if it is going to occur with restenosis initiating after 12 months being rare [28,33]. This is followed by a somewhat quiescent period several years after placement in which the neointimal material is composed primarily of extracellular matrix with a few scattered SMCs.

The degree of in-stent restenosis varies from case to case under similar circumstances although the mechanisms are not yet fully understood. However, to date, some factors such as host response to the implanted biomaterial and biomaterial thrombogenicity as well as degree of vascular injury upon stent deployment have been shown to have a significant influence [34–36]. In addition to these factors, there are others that are thought to influence restenosis, such as contact stresses and rheological factors. Low wall shear stress and wall shear stress gradients are thought to be directly linked to SMC hyperplasia and subsequent restenosis. However, these mechanical factors are influenced by variations in stent design. Therefore, careful consideration in stent design paying special attention to reduction of vessel injury, biocompatibility of materials, and hemodynamics of blood flow could lead to reduced restenosis and better long-term patency.

In addition to contributing to degree of injury to the vessel wall, stent geometry can also influence the hemodynamics of blood flow through the stented artery. It has been shown in various studies that areas having low wall shear stress (WSS) or wall shear stress gradients are conducive to the development of intimal hyperplasia [37–43]. Kohler et al. demonstrated this by implanting polytetrafluoroethylene vascular grafts in baboons [37]. In this experiment, arteriovenous fistulas were created to increase blood flow from approximately 230 to 785 mL/min and subsequently shear from approximately 26 to 78 dyn/cm^2. This increase in shear stress resulted in a significant reduction in the cross-sectional area of the neointima at 3 months (approximately 2.6–0.42 mm^2) marked by an equivalent percentage decrease in SMCs and matrix content. The ligation of the fistula 2 months post graft implantation resulted in a rapid increase in neointimal thickness of

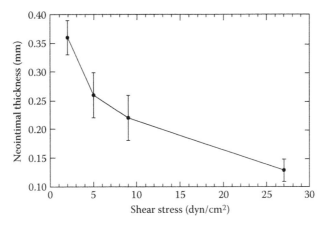

FIGURE 4.4   Neointimal hyperplasia thickening versus wall shear stress in a dog arterial graft. (From Salam, T. et al., *Journal of Vascular Investigation*, 2, 12, 1996.)

approximately $0.6–3.88\,mm^2$ 1 month later. The effect of WSS on intimal hyperplasia is clearly illustrated in Figure 4.4 from the work of Salam et al., which investigated the effect of varying levels of shear stress in PTFE grafts implanted into the femoral and carotid arteries of mongrel dogs [42,44]. The degree of intimal thickening was found to be almost closely correlated with the inverse of the mean shear with a correlation coefficient of 0.96.

The effect of shear stress on intimal thickening is thought to be due at least in part to the increased expression of vascular cell adhesion molecule-1 (VCAM-1) by endothelial cells stimulated by low wall shear stress (WSS), leading to the recruitment of monocytes [43,45]. Gonzales et al. confirmed this by observing human monocyte adhesion to endothelial cells grown in culture under varying shear levels using a parallel plate flow model [45]. The results showed that shearing ECs with stress levels between 2 and $10\,dyn/cm^2$ prior to the addition of monocytes lead to an increased VCAM-1 expression by the ECs and subsequent significant increase in monocyte adherence at $1\,dyn/cm^2$ compared to an unsheared control. However, when the endothelial cells were sheared at $30\,dyn/cm^2$ prior to the addition of the monocytes for attachment at $1\,dyn/cm^2$, monocytic attachment was unchanged compared to an unsheared control. This observation indicates that VCAM-1 expression is stimulated in ECs by low levels of shear, but not at higher shear levels.

LaDisa et al. used computational fluid dynamics to characterize the flow of blood in the stented portion of a rabbit iliac artery [46]. These data were then used to determine areas within the stented region that were exposed to the lowest WSS or had acute WSS gradients. These areas of altered WSS were further shown to directly correlate with locations experiencing the greatest levels of neointimal hyperplasia in the rabbit model [46]. Computational fluid dynamics modeling has also yielded useful results related to effect of strut thickness on flow profile. It was predicted that a thicker strut would cause a larger distal region of low WSS as compared to a thinner strut of similar design [47]. These areas of low WSS distal to the stent struts should thus cause higher incidences in neointimal hyperplasia in thicker strutted stent. The results of the various computational studies suggest that areas of higher WSS lacking WSS gradients hence areas with less disruption of native blood flow should be less prone to hyperplasia and subsequent restenosis [37]. These ideas were also evident in clinical experience.

In considering the overall effect of strut thickness on incidence of restenosis, perhaps the most useful data can be obtained from clinical trials. Comparing the use of a thin strut stent ($50\,\mu m$) to a thick strut stent ($140\,\mu m$) of similar design showed that patients who received the thinner strut device demonstrated a marked decrease in both angiographic and clinical restenosis of the coronary arteries [48]. The effect of strut thickness on restenosis frequency was further demonstrated by rates of angiographic restenosis at 15.0% for the thin versus 25.8% for the thick and clinical restenosis rates of 8.6% for the thin and 13.8% for the thick using stents of similar design (Figure 4.5) [48].

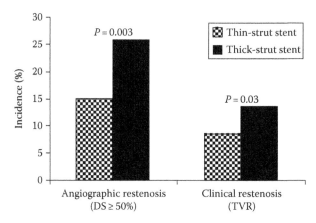

FIGURE 4.5 Incidence of restenosis when comparing thick and thin-strut stents of similar design. (From Kastrati, A. et al., *Circulation*, 103, 2816, 2001.)

This amounts to a reduction or angiographic restenosis by 42% and of clinical restenosis by nearly 38%. In a following study, the same group also found that restenosis is similarly influenced by strut thickness when comparing different strut thicknesses in stents of dissimilar design [49]. This reduced risk of restenosis in thinner strut stents in human patients is further confirmed by another study by Briguori et al. that evaluated thin strut stents of less than 0.10 mm thick ($n = 200$ lesions) versus thick strut stents greater than/equal to 0.10 mm ($n = 150$ lesions) in small coronary arteries between 2.76 and 2.99 mm in diameter [50]. This study found restenosis rates of 23.5% for the thin strut group versus 37% for the thick strut group at an average of 8 ± 2 months. These findings of these two clinical studies demonstrate that a thinner stent strut results in consistently lower rates of restenosis. This is potentially due to the reduced injury to the vessel wall caused by the thinner struts as well as a shorter extension of areas subjected to low WSS.

Another aspect of stent design that can affect the hemodynamics of blood flow is inter-strut spacing and orientation. It has been computationally shown that larger spacing between struts is beneficial from a fluid dynamics point of view [41]. The larger spacing allows for the restoration of normal flow and subsequent normal levels of WSS between struts. However, this could give rise to greater stresses (contact force/contact area) to the vessel wall from having fewer struts to impart the radial strength necessary as well as allowing tissue prolapse between struts [41,51]. This may in turn cause greater injury to the vessel wall, causing increased restenosis rates. Also, studies have shown that a minimum number of struts should be oriented perpendicular to the direction of flow with the use of longitudinal connectors only as necessary [41].

4.4 IN VITRO EXPERIMENTAL MODELS OF A STENTED BLOOD VESSEL

4.4.1 INTRODUCTION

A number of stent design attributes, such as material selection and treatment, stent geometric design, etc., are known to have a direct impact on the potential for restenosis. In order to test the effect of variations in these design criteria, some researchers have turned to the use of in vivo animal models. However, animal models carry the inherent drawbacks of uncontrollable variables causing poor reproducibility in efforts to understand what forms of stimuli will have an effect on SMC growth as it relates to potential restenosis. In vitro simulators can be invaluable in this endeavor in that they allow for testing under more repeatable controlled conditions. This allows for the testing of a single variable at a time, making it easier to evaluate its individual effect on the SMCs.

Static in vitro experimentation using SMCs is limited in its ability to closely mimic the in vivo conditions because it lacks the influence of concurrent mechanical stimuli constantly present in the body, and thus may not comprehensively represent what the cellular response will be in vivo. For this reason, many researchers are developing dynamic testing systems to simulate the native environment of the vascular cells and investigate the effects of changing variables. The most common modes of stimulating cells are flow shear, mechanical stretching of cell culture substrates, and/or compressive loading on cells. The stimulus may be applied as a one-time event to the cells to model a condition such as stretch induced injury in an angioplasty or may be applied cyclically to mimic the pulsatile flow, pressure, and mechanical stretch experienced by SMCs in vivo. The following sections review the current literature on the studies that investigated the effects of hemodynamic parameters on SMCs.

4.4.2 EFFECT OF STRAIN ON SMOOTH MUSCLE CELLS

Although a number of approaches for the exposure of cells to strains exist, the most popular experimental setups use vacuum suction to the bottom of a flexible membrane on which cells are cultured [52–55]. Some of these systems are able to produce approximately equibiaxial strain to which cells are subjected during experiments while other systems generate nonuniform distributions of applied strain [53,56,57].

The exposure of SMCs to cyclic strain has been shown to increase cell proliferation and induce a shift toward a synthetic phenotype in some experiments [52,53,55,58,59] while preventing cell proliferation and promoting a contractile phenotype expression in others [60–63]. However, efforts to explain the phenomenon governing these observations regarding SMC response has been the cause of some debate. On initial observation, the underlying reasons for these discrepancies in results may be unclear. However, upon closer evaluation, these results can be roughly separated into two groups: those stretched at high and low strain rates with the separation point being approximately 10%. For example, Hipper and Isenberg cyclically stretched SMCs of the A10 cell line derived from the aortas of embryonic rats grown on type I collagen-coated membranes at 0.5 Hz using 5% elongation [62]. This resulted in decreased DNA synthesis of 30% at 24 h and 50% at 48 h compared to a static control [62]. This observation remained unaffected when the experiment was repeated on laminin- and pronectin-coated membranes. Chapman et al. used flow cytometry to show that cyclic stretch of type I collage-coated membranes prevented rat aortic SMC proliferation via a G1/S phase transition inhibition using strain levels of 10% at 1 Hz compared to a static control [60]. In contrast, Mills et al. found that cyclic stretching of the type I collagen-coated culture substrate (24% at the edges and 7% in the center) caused increased bovine aortic SMC proliferation and cyclic adenosine monophosphate (cAMP) accumulation compared to static controls [55]. Hasaneen et al. subjected human airway SMCs grown on type I collagen-coated membranes to cyclic strain between 17% and 18.5% and observed increased SMC proliferation at 3 and 5 days compared to a static control [53]. Additionally, the conditioned media from the strain group contained elevated levels of matrix metalloproteinase-1 (MMP-1), MMP-3, and both pro and activated MMP-2, likely indicating an increase in SMC migration. This increase in cyclically strained SMC migration compared to statically cultured SMCs was then confirmed using a modified Boyden chamber assay that operates based on counting the number of cells able to migrate through a gelatin-coated membrane. Addition of the broad spectrum MMP inhibitor Prinomastat prevented this increase in proliferation and migration while also inhibiting MMP-2 activation in a dose-dependent manner.

Physiological levels of strain in vivo are generally less than 10% and may be as low as 0%–4% depending on the location in the vasculature [5,57,62,64]. The previously mentioned studies by Mills, Hu, and Sotoudeh show that SMC proliferation and dedifferentiation into a synthetic phenotype is enhanced by high strain. This may explain the fact that high blood pressure, and thus increased force on the vessel wall and increased strain by as much as 15%, causes thickening of the blood vessel wall due in part to SMC proliferation in vivo [52,65–67]. In addition, it has been observed that high levels of strain experienced during balloon angioplasty also leads to varying degrees of restenosis caused by SMC proliferation. Together, it can be concluded that the maintenance of healthy contractile SMCs with low proliferation rates is achieved by closely mimicking the physiological strain level experienced in vivo while deviations from this normal mechanical environment lead to a shift toward a synthetic phenotype.

4.4.3 Effects of Fluid Shear Stress on Endothelial Cells and Smooth Muscle Cells

A number of different custom-designed systems have been devised in order to stimulate cells with the use of fluid shear. These systems generally break down into two basic design configurations: the cone and plate system and the parallel plate flow chamber (Figure 4.6) [56]. In the cone model, variations in angular velocity of the cone provide kinematic control of fluid shear stress on the cells under investigation. The parallel plate setup uses a simple pressure differential between the inlet and outlet to drive cell culture media across the culture area. This can be achieved via either gravity heads or active pumps. The parallel plate setup is advantageous in that the entrance length can be designed to allow the development of a laminar flow proximal to the culture area, thereby providing a highly homogeneous stress stimulus.

In vivo, the artery is lined by a monolayer of highly hemophilic endothelial cells in direct contact with blood. Subsequently, SMCs are not directly exposed to fluid shear from blood flow. However, following the deployment of a vascular sent, it is common to experience a near complete denudation of the endothelial layer, thereby exposing the underlying SMCs and thrombogenic collagen and sub-endothelial matrix [28,29]. The restoration of this endothelial layer is important for the recovery of the artery to its native state preventing late thrombosis and reducing restenosis [28]. For this reason, much of the work available in the literature focuses on the exposure of endothelial cells (ECs) to fluid shear in vitro. In the stented artery model by Punchard et al., a vascular stent was deployed inside a fibronectin-coated silicone tube seeded with human umbilical cord ECs [68]. It was hypothesized that the turbulent areas of flow caused by the stent struts, along with cell injury from the stent strut, was a major contributing factor to the observed increased expression of genes by ECs for E-selectin, intercellular adhesion molecule-1 (ICAM-1), and VCAM-1 [68]. Lin et al. utilized bovine ECs exposed to laminar shear stress for 24h using a parallel plate flow chamber, and observed an increase in the expression of the tumor suppressor gene p53 [69]. The transcription factor p53 can be activated in a number of different cell types in response to various stresses such as UV and ionizing radiation, DNA-damaging agents, and hypoxia [69,70]. It is known to lead to the induction of various growth arrest genes such as p21cip1 as well as the DNA damage–inducible gene 45 (GADD45), thereby leading to growth inhibition. This increase in p53 expression was observed at shear stress levels of $3\,dyn/cm^2$ and higher, but not at $1.5\,dyn/cm^2$ [69]. These results indicate that the threshold of growth inhibition for ECs caused by shear stress may lie between 1.5 and $3\,dyn/cm^2$. This is further supported by the work of Akimoto et al. that found an inhibition of DNA synthesis in bovine aortic ECs subjected to shear stress levels of 30 and $5\,dyn/cm^2$ compared to a static control, but no significant inhibition was observed at $1\,dyn/cm^2$ [71].

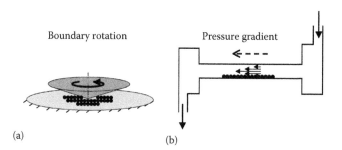

FIGURE 4.6 Schematic of (a) the cone and plate system and (b) the parallel plate system for stimulation using fluid shear. (From Brown, T., *Journal of Biomechanics*, 33, 3, 2000.)

Following an angioplasty procedure, it is common to observe a denudation of the endothelial layer, thereby directly exposing the underlying SMCs to shear stress from blood flow. The direct stimulation of SMCs with shear stress in vitro is done in order to model this situation. Asada et al. stimulated bovine SMCs using orbital shear stress levels of $11.5\,dyn/cm^2$ and observed increased proliferation as well as decreased α-actin and calponin expression along with increased expression of vimentin and β-actin compared to a static control [72]. In another study, bovine SMCs were stimulated with oscillatory shear stress of $14\,dyn/cm^2$ using an orbital shaker by Haga et al., which resulted in a significant increase in cell number and ^{3}[H]thymidine incorporation compared to a static control at 5 days [73]. Moreover, Qi et al. demonstrated a decreased expression of protein Rho-GDP dissociation inhibitor alpha (Rho-GDIα) in rat aortic SMCs exposed to low levels of shear stress ($5\,dyn/cm^2$) as compared to those exposed to higher levels of shear stress ($15\,dyn/cm^2$) [74]. Since decreased Rho-GDIα expression has been associated with enhanced vascular SMC migration and apoptosis in SMCs, this molecule may become a drug target in the development of future pharmaceutical treatments to reduce SMC migration and vascular remodeling following angioplasty in vivo.

It can be seen from evaluating these and a number of other studies that, shear stress plays an important role in the activity of both ECs and SMCs. In both cases, there is threshold level at which shear stress acts to inhibit cell proliferation. At higher levels of shear stress of approximately $15–28\,dyn/cm^2$, which is within the normal range of physiologic shear stress of $5–25\,dyn/cm^2$, SMC proliferation is reduced [75,76]. The shear level at which EC proliferation is inhibited in vitro is much lower at approximately $1.5–3\,dyn/cm^2$ [69,71].

4.4.4 EFFECT OF COMBINED STRAIN AND FLUID SHEAR ON ENDOTHELIAL AND SMOOTH MUSCLE CELLS

Exposure of SMCs to concurrent fluid shear and cyclic strain stimuli adds another level of complexity to modeling of the in vivo environment experienced by SMCs in an artery following the denudation of the endothelial layer. The development of experimental setups to apply controlled shear stress and equibiaxial strain could lead to cost-effective and time-efficient test systems with a much higher degree of reproducibility than that observed using in vivo experimentation. To date, however, few experimental models are available to evaluate the combined effect of these two stimuli applied simultaneously.

To the best of our knowledge, the earliest example of such an in vitro study was reported by Moore et al. [64]. This device made use of compliant silicone tubes coated with fibronectin and seeded with bovine pulmonary endothelial cells and allowed to reach confluence. After this process, they were subjected to pulsatile pressurized flow in order to apply both cyclic stretch and fluid shear. Although it was not reported, based on the design, it can be speculated that cell seeding inside the tube as well as visualization of adherent cells would be problematic. This general test setup, however, has been used by several groups to test the response of ECs to cyclic stretch and fluid shear [77–79]. For example, Zhao et al. observed a significantly elongated morphology as well as an orientation in the direction of flow and perpendicular to stretch in bovine aortic endothelial cells seeded in fibronectin coated Sylgard tubes subjected to shear stress greater than $2\,dyn/cm^2$ or hoop stress greater than 2% at 24 h compared to a static control [79]. The formation of stress fibers aligned with the long axis of the cell was induced by both shear stress as well as hoop stress. Stress fiber size and alignment were significantly enhanced by the concurrent application of shear and hoop stress as compared to application of either stimuli, singularly indicating a possible synergistic effect between the two stimuli on the cell morphology of ECs.

Another 3D vascular simulator that utilized a silicone compliant tube was developed by Punchard et al. [68,80]. This study differed from the previously mentioned compliant tube studies in that a Libertè bare metal stent (Boston Scientific Corporation, Natick, MA) was deployed inside the silicone tube using a balloon catheter in order to evaluate the effects of the deployed stent on human umbilical cord vein endothelial cells (HUVECs) in vitro [68]. The group observed local endothelial denudation immediately after stent deployment at 15 atm for 30 s as a result of physical damage to the cell layer [68]. Additionally, the HUVECs subjected to flow without a deployed stent oriented themselves in the direction of flow. Patterns of cell orientation were less apparent in the stented model subjected to flow and completely randomized in the stented model not subjected to flow, indicating a possible response of ECs to regions of turbulent flow caused by stent struts [68]. Furthermore, the deployed stent increased the expression of E-selectin, ICAM-1, and VCAM-1 by the ECs 24 h following stent deployment [68]. While these studies highlight the effects of fluid shear and cyclic strain on ECs, little is known about the effects of combined cyclic strain and fluid shear on SMCs.

The work of Acampora et al. conducted in our laboratory utilized a parallel plate flow chamber combined with computer-controlled vacuum pressure to strain a silicone membrane across a stationary post in order to provide independently controlled fluid shear and cyclic strain stimuli [57]. Using this setup, Acampora et al. for the first time exposed rat aortic SMC to a stain stimulus (two cycles of 12% strain) to mimic angioplasty injury followed by combined strain and fluid shear (4% and 0.5 dyn/cm², respectively) to mimic physiological hemodynamic conditions. This study demonstrated a 75% increase in SMC proliferation compared to the dynamic control not subjected to the 12% strain injury at a time point of 8 h [57]. Additionally, the authors reported a 19% decrease in the expression of the contractile marker smooth muscle α-actin in the angioplasty injury model compared to a static control [57]. In a follow-up study when cells were subjected to similar concurrent strain and shear conditions for 24 h, rat aortic SMC proliferation, expression of the synthetic marker vimentin, cellular apoptosis, and hypertrophy were all observed higher compared to the group stimulated with cyclic strain alone [81]. Interestingly, the cells exposed to concurrent cyclic strain and fluid shear exhibited similar SMC proliferation compared to the cells stimulated with cyclic tension only (Figure 4.7). This lack of effect on SMC proliferation and apoptosis is likely due to the low level of fluid shear (0.5 dyn/cm²) employed in the experiment as it was shown in previous shear only studies that shear stress levels below approximately 15 dyn/cm² do not act to inhibit SMC proliferation [72–76]. However, the increased expression of the synthetic marker vimentin in the Acampora angioplasty injury model subjected to concurrent shear and strain demonstrates that there may be a synergistic effect on synthetic marker expression from concurrent force application [81]. Previous studies in the literature have shown decreased contractile marker expression and increased vimentin expression independently in response to oscillations of shear stress and strain

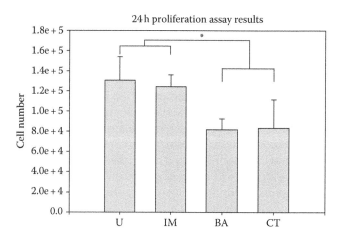

FIGURE 4.7 CyQuant proliferation assay for 24 h cell quantification. Left to right: U 13.10×10⁴ cells±2.33×10⁴ SD (n=6), IM 12.41×10⁴ cells±1.16×10⁴ SD (n=6), BA 8.15×10⁴ cells±1.10×10⁴ SD (n=5), CT 8.35×10⁴ cells±2.80×10⁴ SD (n=6). Asterisk indicated statistical significance, p=0.001. (From Acampora, K.B. et al., *Annals of Vascular Surgery*, 24(1), 116, 2010.)

compared to static controls; the Acampora study was the first to illustrate the important additive effects of shear and tensile forces.

The experimental setup developed by Acampora et al. also allowed investigation of cell response to various pharmacologic treatments encountered during vascular intervention. For example, the authors investigated the effect of combined shear and strain in the presence of heparin, an anticoagulant commonly administered during angioplasty procedures [82]. In addition to its properties as an anticoagulant, heparin has also been shown to decrease SMC migration and proliferation [82–84]. A common clinical protocol is to administer a 3–5 min 10,000 unit bolus IV injection of heparin immediately prior to balloon angioplasty [85]. When simulated in vitro, this 10,000 unit bolus injection of heparin was held in contact with the cells for 5 min followed by a 100 µg/mL heparin dosage for 24 h and resulted in decreased rat aortic SMC proliferation [82]. Thus, the aim of the study by Acampora et al. was to test the efficacy of the drug treatment in minimizing intimal hyperplasia following balloon injury. Furthermore, Fasciano et al. demonstrated a decrease in rat SMC proliferation using 100 µg/mL heparin under static conditions as compared to a negative control [86]. This was thought to be due to a block in the G_1 to S phase, caused by an inhibition of cyclin-dependent kinase 2 activity. This is in agreement with other studies reported in the literature in which SMC proliferation decreased following heparin administration [82–84,86–88]. A unique finding in the Acampora study was the differential expression of calponin, a calcium-binding protein that inhibits the ATPase activity of myosin, in response to changes in mechanical environment [82]. It was shown that the cells subjected to concurrent strain and fluid shear in the presence of heparin exhibited decreased calponin expression compared to a static control. However, heparin administration to static SMCs increased calponin expression compared to a static control [82]. The results of this study demonstrate a dependency of calponin expression by SMCs in response to heparin administration under varying culture conditions. This differential response of SMCs under static versus dynamic culture conditions clearly illustrates the importance of studying SMCs subjected to physiological levels of mechanical stimuli. A dynamic in vitro simulator is a useful tool with which to investigate SMC response to new pharmaceutical treatments when cultured under physiological mechanical conditions. The in vitro application of mechanical stimuli including cyclic strain and fluid shear is necessary in order to gather experimental data more likely to mirror the in vivo response, and thus be of significant value in the clinical realm. Future improvements to this model may include the addition of a model stent strut to represent an artery after the deployment of a stent. This would allow for rapid evaluation of various stent treatments and coatings for improved cell–biomaterial interaction as well as drug elution. Additionally, coculturing SMCs and ECs together may more closely mimic the native artery.

4.5 CONCLUSION

In summary, atherosclerosis is a progressive arterial disease characterized by the development of SMC-rich plaques that eventually grow to occlude the vessel lumen [1,4,6]. The use of balloon angioplasty in conjunction with stent deployment as an alternative treatment option to bypass surgery is a viable and increasingly popular option. However, the outcome of this procedure still suffers from varying degrees of restenosis under similar circumstances (e.g., patient, device, location, lesion classification) due to factors that are not yet fully understood. The development of an in vitro model of a stented blood vessel is an important area of current research aimed at gaining a better understanding of the phenomenon involved in restenosis. In order to yield clinically significant data, this model must account for the different facets of mechanical stimuli found in vivo such as cyclic strain and flow shear. A number of different in vitro models from the literature were reviewed in order to gain an understanding of how SMCs react to various stimuli and how these reactions are influenced by the mechanical environment.

For the development of an optimal in vitro simulator, the researcher should be able to apply cyclic strain and fluid shear either singularly or in combination. Additionally, the magnitude of these stimuli should be independently controlled by the operator in order to mimic the normal in vivo environment as well as angioplasty injury and different flow conditions based on location in the vasculature. This will allow the researcher to investigate the effect of secondary stimuli such as pharmacologic treatments, cell–biomaterial interactions, etc. under dynamic conditions. The improved understanding of the various aspects of SMC response under dynamic in vitro culture conditions will help the future development of highly effective vascular stents designs and pharmacologic treatment regimes that minimize the risk for restenosis.

REFERENCES

1. Guyton, A. and Hall, J. 2006. *Textbook of Medical Physiology*, 11th edn. Elsevier Saunders: Amsterdam, the Netherlands.
2. Copstead, L. and Banasik, J. 2005. *Pathophysiology*, 3rd edn. Elsevier Saunders: Amsterdam, the Netherlands.
3. Douglas, J. 2007. Pharmacologic approaches to restenosis prevention. *The American Journal of Cardiology* **100**(5, Supplement 1):S10–S16.
4. Topol, E. 2007. *Textbook of Cardiovascular Medicine*, 3rd edn. Lippincott Williams & Wilkins: Philadelphia, PA.
5. Fuster, V., Ross, R., and Topol, E. 1996. *Atherosclerosis and Coronary Artery Disease*, Vol. 1. Lippincott-Raven: Philadelphia, PA.
6. Dean, R. and Kelly, D. 2000. *Atherosclerosis: Gene Expression, Cell Interactions, and Oxidation*. Oxford University Press: Oxford, U.K.
7. Packard, C. and Rader, D. 2006. *Lipids and Atherosclerosis. Advances in Translational Medical Science*. Taylor & Francis: Abingdon, U.K.
8. Rutherford, R. 2000. *Vascular Surgery*, 5th edn. Vol. 1. W.B. Saunders: Philadelphia, PA.
9. Napoli, C. et al. 1997. Fatty streak formation occurs in human fetal aortas and is greatly enhanced by maternal hypercholesterolemia—Intimal accumulation of low density lipoprotein and its oxidation precede monocyte recruitment into early atherosclerotic lesions. *Journal of Clinical Investigation* **100**(11): 2680.
10. Martin, A. et al. 2008. Dual regulation of cofilin activity by LIM Kinase and slingshot-1L phosphatase controls platelet-derived growth factor-induced migration of human aortic smooth muscle cells. *Circulation Research* **102**(4): 432–438.
11. Pintucci, G. et al. 2005. PDGF-BB induces vascular smooth muscle cell expression of high molecular weight FGF-2, which accumulated in the nucleus. *Journal of Cellular Biochemistry* **95**:1292–1300.
12. Waltenberger, J. 1997. Modulation of growth factor action: Implications for the treatment of cardiovascular diseases. *Circulation* **96**:4083–4094.
13. Stary, H. 1992. Composition and classification of human atherosclerotic lesions. *Virchows Archiv A Pathological Anatomy and Histopathology* **421**:277–290.
14. Waddington, E. et al. 2003. Fatty acid oxidation products in human atherosclerotic plaque: An analysis of clinical and histopathological correlates. *Atherosclerosis* **167**:111–120.
15. Pepine, C.J. 1998. The effects of angiotensin-converting enzyme inhibition on endothelial dysfunction: Potential role in myocardial ischemia. *The American Journal of Cardiology* **82**(10):S24–S27.
16. Mardikar, H.M. and Mukherjee, D. 2007. Current endovascular treatment of peripheral arterial disease. *Progress in Cardiovascular Nursing* **22**:31–37.
17. White, C. and Gray, W. 2007. Endovascular therapies for peripheral arterial disease: An evidence-based review. *Circulation* **116**:2203–2215.
18. Halayko, A. and Solway, J. 2001. Molecular mechanisms of phenotypic plasticity in smooth muscle cells. *Journal of Applied Physiology* **90**(1):358–368.
19. Chamley-Campbell, J., Campbell, G., and Ross, R. 1979. The smooth muscle cell in culture. *Physiological Reviews* **59**(1):1–61.
20. Majno, G. and Joris, I. 2004. *Cells, Tissues, and Disease*, 2nd edn. Oxford University Press: Oxford, U.K.
21. Rzucidlo, E., Martin, K., and Powell, R. 2007. Regulation of vascular smooth muscle cell differentiation. *Journal of Vascular Surgery* **45**(A):A25–A32.
22. Owens, G. 1995. Regulation of differentiation of vascular smooth muscle cells. *Physiological Reviews* **75**(3):487–517.

23. Jones, B. et al. 1996. Phenotypic characterization of human smooth muscle cells derived from atherosclerotic tibial and peroneal arteries. *Journal of Vascular Surgery* **24**:883–891.
24. Glukhova, M. et al. 1988. Modulation of human aorta smooth muscle cell phenotype: A study of muscle-specific variants of vinculin, caldesmon, and actin expression. *Proceedings of the National Academy of Sciences of the United States of America* **85**:9542–9546.
25. Thyberg, J. et al. 1990. Regulation of differentiated properties and proliferation of arterial smooth muscle cells. *Arteriosclerosis, Thrombosis, and Vascular Biology* **10**(6):966–989.
26. Zheng, B., Duan, C., and Clemmons, D. 1998. The effect of extracellular matrix proteins on porcine smooth muscle cell insulin-like growth factor (IGF) binding protein-5 synthesis and responsiveness to IGF-I. *The Journal of Biological Chemistry* **273**(15):8994–9000.
27. Grewe, P. et al. 2000. Acute and chronic tissue response to coronary stent implantation: Pathologic findings in human specimen. *Journal of the American College of Cardiology* **35**(1):157–163.
28. Faxon, D. 2001. *Restenosis: A Guide to Therapy.* Martin Dunitz Ltd: London, U.K.
29. Welt, F. and Rogers, C. 2002. Inflammation and restenosis in the stent era. *Arteriosclerosis, Thrombosis, and Vascular Biology* **22**(11):1769.
30. Dee, K., Puleo, D., and Bizios, R. 2002. *An Introduction to Tissue–Biomaterial Interactions.* Wiley-Liss, Hoboken, NJ.
31. Palmaz, J. 1993. Intravascular stents: Tissue-stent interactions and design considerations. *American Journal of Roentgenology* **160**:613–618.
32. Rajagopal, V. and Rockson, S. 2003. Coronary restenosis: A review of mechanisms and management. *The American Journal of Medicine* **115**:547–553.
33. Serruys, P. et al. 1988. Incidence of restenosis after successful coronary angioplasty: A time-related phenomenon circulation **77**(2):361–371.
34. Morton, A., Walker, R., and Gunn, J. 2007. Current challenges in coronary stenting: From bench to bedside. *Biochemical Society Transactions* **35**(Pt 5):900–904.
35. Kornowski, R. et al. 1998. In-stent restenosis: Contributions of inflammatory responses and arterial injury to neointimal hyperplasia. *JACC* **31**(1):224–230.
36. Bertrand, O. et al. 1998. Biocompatibility aspects of new stent technology. *JACC* **32**(3):562–571.
37. Kohler, T. et al. 1991. Increased blood flow inhibits neointimal hyperplasia in endothelialized vascular grafts. *Circulation Research* **69**:1557–1565.
38. LaDisa, J. et al. 2003. Three-dimensional computational fluid dynamics modeling of alterations in coronary wall shear stress produced by stent implantation. *Annals of Biomedical Engineering* **31**:972–980.
39. Moore, J. et al. 1994. Fluid wall shear stress measurements in a model of the human abdominal aorta: Oscillatory behavior and relationship to atherosclerosis. *Atherosclerosis* **110**:225–240.
40. Sabbah, H. et al. 1986. Relation of atherosclerosis to arterial wall shear in the left anterior descending coronary artery of man. *The American Heart Journal* **112**:453–458.
41. He, Y. et al. 2005. Blood flow in stented arteries: A parametric comparison of strut design patterns in three dimensions. *Journal of Biomedical Engineering* **127**:637–647.
42. Wootton, D. and Ku, D. 1999. Fluid mechanics of vascular systems, diseases, and thrombosis. *Annual Review of Biomedical Engineering* **1**:299–329.
43. Ainsworth, S. 1999. Modeling of artery restenosis influenced by endovascular stent geometry, in Department of Bioengineering, Clemson University, Clemson, SC.
44. Salam, T. et al. 1996. Low shear stress promotes intimal hyperplasia thickening. *Journal of Vascular Investigation* **2**:12–22.
45. Gonzales, R. and Wick, T. 1996. Hemodynamic modulation of cell adherence to vascular endothelium. *Annals of Biomedical Engineering* **24**:382–393.
46. LaDisa, J.F. et al. 2005. Alterations in wall shear stress predict sites of neointimal hyperplasia after stent implantation in rabbit iliac arteries. *American Journal of Physiology. Heart and Circulatory Physiology* **288**:H2465–H2475.
47. LaDisa, J.F. et al. 2004. Stent design properties and deployment ratio influence indexes of wall shear stress: A three-dimensional computational fluid dynamics investigation within a normal artery. *Journal of Applied Physiology* **97**(1):424–430.
48. Kastrati, A. et al. 2001. Intracoronary stenting and angiographic results: Strut thickness effect on restenosis outcome (ISAR-STEREO) trial. *Circulation* **103**:2816–2821.
49. Pache, J. et al. 2003. Intracoronary stenting and angiographic results: Strut thickness effect on restenosis outcome (ISAR-STEREO-2) trial. *Journal of the American College of Cardiology* **41**:1283–1288.
50. Briguori, C. et al. 2002. In-stent restenosis in small coronary arteries: Impact of strut thickness. *Journal of the American College of Cardiology* **40**(3):403–409.

51. Rogers, C. et al. 1999. Balloon-artery interactions during stent placement: A finite element analysis approach to pressure, compliance, and stent design as contributors to vascular injury. *Circulation Research* **84**:378–383.

52. Hu, Y. et al. 1998. Activation of PDGF receptor α in vascular smooth muscle cells by mechanical stress. *FASEB Journal* **12**(12):1135–1142.

53. Hasaneen, N. et al. 2005. Cyclic mechanical strain-induced proliferation and migration of human airway smooth muscle cells: Role of EMMPRIN and MMPs. *FASEB Journal* **19**:1507–1509.

54. Grote, K. et al. Stretch-inducible expression of the angiogenic factor CCN1 in vascular smooth muscle cells in mediated by Egr-1. *The Journal of Biological Chemistry* **279**(53):55675–55681.

55. Mills, I. et al. 1997. Strain activation of bovine aortic smooth muscle cell proliferation and alignment: Study of strain dependency and the role of protein kinase A and C signaling pathways. *Journal of Cellular Physiology* **170**:228–234.

56. Brown, T. 2000. Techniques for mechanical stimulation of cells in vitro: A review. *Journal of Biomechanics* **33**:3–14.

57. Acampora, K. et al. 2007. Development of a novel vascular simulator and injury model to evaluate smooth muscle cell response following balloon angioplasty. *Annals of Vascular Surgery* **21**(6):734–741.

58. Wilson, E. et al. 1993. Mechanical strain induces growth of vascular smooth muscle cells via autocrine action of PDGF. *The Journal of Cell Biology* **123**(3):741–747.

59. Butcher, J., Barrett, B., and Nerem, R. 2006. Equibiaxial strain stimulates fibroblastic phenotype shift in smooth muscle cells in an engineered tissue model of the aortic wall. *Biomaterials* **27**:5252–5258.

60. Chapman, G. et al. 2000. Physiological cyclic stretch causes cell cycle arrest in cultured vascular smooth muscle cells. *American Journal of Physiology. Heart and Circulatory Physiology* **278**:H748–H754.

61. Grenier, G. et al. 2006. Mechanical loading modulates the differentiation state of vascular smooth muscle cells. *Tissue Engineering* **12**(11):3159–3170.

62. Hipper, A. and Isenberg, G. 2000. Cyclic mechanical strain decreases the DNA synthesis of vascular smooth muscle cells. *Pflugers Archiv European Journal of Physiology* **440**(1):19–27.

63. Schulze, P. et al. 2003. Biomechanically induced gene iex-1 inhibits vascular smooth muscle cell proliferation and neointima formation. *Circulation Research* **93**:1210–1217.

64. Moore, J. et al. 1994. A device for subjecting vascular endothelial cells to both fluid shear stress and circumferential cyclic stretch. *Annals of Biomedical Engineering* **22**:416–422.

65. Haga, J., Li, Y. and Chien, S. 2007. Molecular basis of the effects of mechanical stretch on vascular smooth muscle cells. *Cardiovascular and Interventional Radiology* **40**:947–960.

66. Sudir, K. et al. 1993. Mechanical strain and collagen potentiate mitogenic activity of angiotensin II in rat vascular smooth muscle cells. *Journal of Clinical Investigation* **92**:3003–3007.

67. Safar, M.E. et al. 1981. Pulsed Doppler: Diameter, blood flow velocity, and volumic flow of the brachial artery in sustained essential hypertension. *Circulation* **63**:393–400.

68. Punchard, M.A. et al. 2009. Evaluation of human endothelial cells post stent deployment in a cardiovascular simulator in vitro. *Annals of Biomedical Engineering* **37**(7):1322–1330.

69. Lin, K. et al. 2000. Molecular mechanism of endothelial growth arrest by laminar shear stress. *Proceedings of the National Academy of Sciences of the United States of America* **97**:9385–9389.

70. Levine, A.J. 1997. p53, the cellular gatekeeper for growth and division. *Cell* **88**(3): 323–331.

71. Akimoto, S. et al. 2000. Laminar shear stress inhibits vascular endothelial cell proliferation by inducing cyclin-dependent kinase inhibitor p21 Sdi1/Cip1/Waf1. *Circulation Research* **86**:185–190.

72. Asada, H. et al. 2005. Sustained orbital shear stress stimulates smooth muscle cell proliferation via the extracellular signal-regulated protein kinase 1/2 pathway. *Journal of Vascular Surgery* **42**:772–780.

73. Haga, M. et al. 2003. Oscillatory shear stress increases smooth muscle cell proliferation and Akt phosphorylation. *Journal of Vascular Surgery* **37**:1277–1284.

74. Qi, Y.-X. et al. 2008. Rho-GDP dissociation inhibitor alpha downregulated by low shear stress promotes vascular smooth muscle cell migration and apoptosis: A proteomic analysis. *Cardiovascular Research* **80**:114–122.

75. Ueba, H., Kawakami, M., and Yaginuma, T. 1997. Shear stress as an inhibitor of vascular smooth muscle cell proliferation. *Arteriosclerosis, Thrombosis, and Vascular Biology* **17**:1512–1516.

76. Papadaki, M., Eskin, S., and McIntire, L. 1997. Flow modulation of smooth muscle cells (SMC) proliferation and metabolism. *Cardiovascular Pathology* **5**(5):292.

77. Walsh, P. et al. 1999. Morphological response of endothelial cells to combined pulsatile shear stress plus cyclic stretch. In: *BMES/EMBS Joint Conference: Serving Humanity, Advancing Technology*, Atlanta, GA.

78. Moreno, M. et al. 2003. Combined effects of pulsatile shear stress and pressure driven cyclic strain on protein expression by the isolated endothelium. In: *Summer Bioengineering Conference*, Sonesta Beach Resort, Key Biscayne, FL.
79. Zhao, S. et al. 1995. Synergistic effects of fluid shear stress and cyclic circumferential stretch on vascular endothelial cell morphology and cytoskeleton. *Arteriosclerosis, Thrombosis, and Vascular Biology* **15**:1781–1786.
80. Punchard, M.A. et al. 2007. Endothelial cell response to biomechanical forces under simulated vascular loading conditions. *Journal of Biomechanics* **40**:3146–3154.
81. Acampora, K.B. et al. 2010. Increased synthetic phenotype behavior of smooth muscle cells in response to in vitro balloon angioplasty injury model. *Annals of Vascular Surgery* **24**(1):116–126.
82. Acampora, B. 2008. Effect of clinically relevant mechanical forces on smooth muscle cell response in model of balloon angioplasty mechanical forces on smooth muscle cell response in model of balloon angioplasty, in Department of Bioengineering, Clemson University, Clemson, SC.
83. Marcum, J., Reilly, C. and Rosenberg, R. 1986. The role of specific forms of heparin sulfate in regulating blood vessel wall function. *Progress in Hemostasis and Thrombosis* **8**:185–215.
84. Clowes, A.W. and Karnowsky, M.J. 1977. Suppression by heparin of smooth muscle cell proliferation in injured arteries. *Nature* **265**:625–626.
85. Hirsh, J. et al. 2001. Heparin and low-molecular-weight heparin: Mechanisms of action, pharmacokinetics, dosing, monitoring, efficacy, and safety. *Chest* **119**:64S–94S.
86. Fasciano, S. et al. 2005. Regulation of vascular smooth muscle proliferation by heparin. *The Journal of Biological Chemistry* **280**(16):15682–15689.
87. Reilly, C. et al. 1989. Heparin prevents vascular smooth muscle cell progression through the G1 phase of the cell cycle. *Journal of Biological Chemistry* **264**:6990–6995.
88. Castellot, J., Cochran, D., and Karnovsky, M. 1985. Effect of heparin on vascular smooth muscle cells. I. Cell metabolism. *Journal of Cellular Physiology* **124**:21–28.

5 Effects of Pressure on Vascular Smooth Muscle Cells

Sheila Nagatomi, Harold A. Singer, and Rena Bizios

CONTENTS

5.1 INTRODUCTION

Smooth muscle cells are the predominant cell constituent of the media, the middle circumferential layer of the arterial wall tissue.[1] Within this tissue, a single arterial smooth muscle cell has a volume of approximately 1200–1600 μm^3 in rats and humans.[2]

Smooth muscle cells contribute to the structure and composition of the vascular tissue since under normal physiological conditions, smooth muscle cells synthesize various chemical compounds such as elastin, collagen (types I, III, and V), fibronectin, and proteoglycans,[2,3] which are constituents of the vascular extracellular matrix. Most importantly, contraction and relaxation of smooth muscle cells induce artery vasoconstriction and vasodilation, respectively.[4] Such changes of the blood vessel wall tone induce adjustments in the internal diameter of, and thus conditions within, the arterial lumen, which affect pressure (including the transmural component) and the dynamics of blood flow in the vasculature. Smooth muscle cells are also involved in vascular pathology; specifically in atherosclerosis, smooth muscle cell migration and hyperplasia (i.e., increased number of these cells) in the intima contribute to the thickening of the arterial wall tissue, which is a clinical symptom of the disease.[5,6]

In both health and disease, vascular cells exist and function in the dynamic milieu of the vascular system. Recognition that mechanical forces affect constituent cell functions, which are critical in both the physiology and pathology of the vasculature, has motivated extensive pertinent research. Advances in cellular engineering, molecular biology, and biochemistry as well as availability of appropriate instrumentation and laboratory methodologies have facilitated and contributed to the success of such endeavors.

5.1.1 MECHANICAL FORCES IN THE VASCULATURE

In vivo, the mammalian vascular wall tissue is exposed to several mechanical forces (Figure 5.1). In the arterial network, values for the physiological fluid shear stress due to blood flow in the circulatory system are greater than 15 dyn/cm^2.[7] Pumping of the heart moves blood across a pressure difference in the direction of blood flow. Moreover, due to containment of blood in the vasculature, all layers and constituent cells in the human blood vessel wall are exposed to pressure perpendicular to the vascular tissue, which varies in a range up to 120 mm Hg in healthy individuals.[8] Finally, due to the compliance of the vascular tissue and the pulsatile nature of blood flow, the blood vessel wall tissue is exposed to mechanical strains (stretch) of up to 10%.[9]

Not all constituent cells of the vascular tissue, however, are exposed to all of the aforementioned mechanical forces. The endothelial cells lining the vascular lumen are exposed to all three of these mechanical forces. Smooth muscle cells in the medial layer of the arterial tissue do not directly experience shear forces because they are covered, and thus protected, by the normal, healthy, intact endothelial layer; for this reason, the dominant forces to which vascular smooth muscle cells are exposed are strain and pressure.

In contrast to the extensive studies of the effects of mechanical shear and tensile stresses on vascular cell function,[10–15] the effect of pressure has attracted limited, and more recent, attention from researchers. It is well accepted now that mechanical forces affect various smooth muscle cell functions such as synthesis and release of growth factors and cell proliferation both in vivo and in vitro.[4,16] Undoubtedly, further research is needed to elucidate crucial aspects of the role of mechanical forces on the function of vascular cells in health and in disease.

5.1.2 VASCULAR SMOOTH MUSCLE CELL PHENOTYPE

In the healthy adult mammalian artery, smooth muscle cells exhibit a range of characteristics that are described as either "contractile" or "synthetic." Fully differentiated, "contractile" smooth muscle cells exhibit an elongated, spindle morphology and express marker proteins such as α-smooth muscle actin, smooth muscle myosin heavy chain, smoothelin, and nestin.[17–19] "Synthetic" smooth muscle cells typically exhibit a cobblestone morphology, have either decreased or absence of contractile smooth muscle marker expression,[17] and proliferate faster than their contractile counterparts.

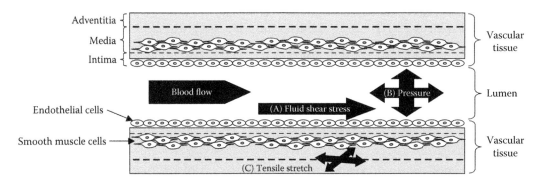

FIGURE 5.1 Schematic illustration of the mechanical forces in the vascular milieu. (A) Fluid shear due to blood flow. (B) Pressure in the direction of blood flow (due to the pumping of the heart) and perpendicular to the vascular tissue (due to containment of blood within the vasculature). (C) Tensile stretch resulting from compliance of the blood vessel wall tissue. (Modified with permission from Dr. H. Y. Shin, University of Kentucky, Lexington, KY.)

Recent reviews also discussed a "pro-inflammatory" smooth muscle cell phenotype that can trigger an inflammatory response and, thus, contribute to atherosclerosis.[20]

The phenotype variants of vascular smooth cells have been studied extensively.[17,21] There is also evidence that phenotypic shifts are sometimes reversible.[22] The most important parameters that affect the smooth muscle cell "contractile" or "synthetic" phenotype include biochemical compounds, extracellular matrix components, mechanical forces, and tissue/cell injury.[17,20,23]

5.1.3 MECHANICAL FORCES AND THE VASCULAR SMOOTH MUSCLE CELL PHENOTYPE

With each pulse of the heart and the concomitant burst/rest pattern of blood flow, strain in the circumferential direction through all layers of the arterial wall tissue is approximately 10%.[9] Such stretch promotes expression of smooth muscle cell differentiation markers and results in decreased DNA synthesis[24]; these results are characteristics of cells in the "contractile" phenotype.[25] In general, however, there is no consensus on this issue because exposure of smooth muscle cells to mechanical strain has also been associated with increased cell proliferation,[26] a characteristic of cells in the "synthetic" phenotype.

In contrast to literature reports describing the effect of strain on the smooth muscle cell phenotype plasticity,[17,27] there is no consensus regarding the effect of pressure on the phenotypic shift in vascular smooth muscle cells. There is evidence, however, that the pressure effects on vascular smooth muscle cells depend on the initial phenotype of these cells. Specifically, vascular smooth muscle cells exposed to pulsatile pressures of 64/14 mm Hg for 3 days responded differently based on the cell phenotype.[28] In contrast to results obtained with cells exhibiting a "less" differentiated phenotype (i.e., cells that expressed low mRNA levels of differentiation markers such as calponin), exposure of smooth muscle cells that highly expressed differentiation markers to pressure resulted in increased DNA synthesis and cell proliferation.[28] Furthermore, it should be noted that, while after exposure to pressure, mRNA levels for calponin were maintained in cells expressing the "more" differentiated phenotype, the low levels of calponin expression were no longer detectable in the cells of the "less" differentiated phenotype.[28]

This evidence provides an explanation for the differences in the responses of smooth muscle cells to pressure that, to date, have been reported by various studies in the literature[29,30]: smooth muscle cell populations composed of such mixed cell phenotypes may lead to erroneous conclusions regarding the effect of mechanical stimuli (such as pressure) on these cells. While the range of phenotypic characteristics in smooth muscle cells has been recognized, most experimental models used to investigate the effects of pressure on smooth muscle cell function have not taken into consideration the phenotype of smooth muscle cells prior to exposure of these cells to mechanical stimuli, including pressure. Evidence for the dependence of smooth muscle cell function on the phenotype of these cells, however, established cell phenotype as an important requirement for present and future studies of vascular smooth muscle cells. The role of smooth muscle cell phenotype (and phenotypic changes) in health and in disease must be elucidated. Further research in this field is, therefore, needed.

5.1.4 VASCULAR SMOOTH MUSCLE CELL RESPONSES TO PRESSURE IN HYPERTENSION IN VIVO

Thickening of the arterial blood vessel wall due to smooth muscle cell hyperplasia occurs in hypertension, a clinical condition of chronically elevated blood pressure[31] in the range of 140/100[32]– 170/130 mm Hg.[33] Supporting clinical evidence was provided by the observation that the intima and media layers in the femoral and carotid arteries were significantly ($p < 0.001$) thicker in hypertensive (than in non-hypertensive) human subjects.[34] Similar results were obtained in animal models; specifically, smooth muscle cell proliferation was higher in hypertensive than in normotensive rats.[35]

This evidence of the effects of pressure on vascular smooth muscle cell function pertinent to vascular physiology and pathology became the impetus for research at the cellular level. Advances in appropriate instrumentation and availability of novel mammalian cell culture methodologies provided a most useful in vitro model for such studies.

5.2 VASCULAR SMOOTH MUSCLE CELL PROLIFERATION UNDER PRESSURE

5.2.1 EFFECTS OF COMBINED PRESSURE AND STRAIN

Several studies reported in the literature used explanted vascular tissue segments and experimental setups, which exposed these tissue samples (and constituent cells including endothelial and smooth muscle cells) to both pressure and stretch,[36–38] and reported data that (due to the compliance of the vascular tissue) were the result of the combined effects of pressure and strain. One such study reported that exposure of rat aortic segments to sustained pressures of either 80 or 150 mm Hg for 3 consecutive days did not affect DNA synthesis, an index of cell proliferation.[36] Another study reported increased ERK 1/2 (extracellular signal-regulated kinase 1 and 2) activation in segments of rat aortic tissue exposed to 150 mm Hg sustained pressure for 5 min and 24 h.[37]

Undoubtedly, the combined pressure and strain models better simulate the mechanical milieu of the vasculature. It is important to note, however, that the results from these types of experiments neither identify the respective contributions of each mechanical force nor correlate the effects on vascular cells which are due to one mechanical stimulus (e.g., pressure) alone. An attempt to resolve this issue was a study in which the increased cell proliferation of rat aortic smooth muscle cells exposed simultaneously to pressure and strain (either 10 cm H_2O pressure and 4% strain or 30 cm H_2O pressure and 10% strain for 7 days) proved to be similar to that observed when these cells were exposed solely to the respective pressure levels for the same time period[39] (Figure 5.2). These results provided evidence that pressure may be the predominant mechanical stimulus that affects vascular smooth muscle cell mitogenesis. Moreover, this evidence became the rationale and impetus for further pertinent studies whose results are reviewed in the section that follows.

5.2.2 EFFECTS OF PRESSURE ALONE

Few in vitro studies have investigated the effects of pressure, without associated strain, on vascular smooth muscle cell function. One such study reported that rat aortic smooth muscle cells cultured

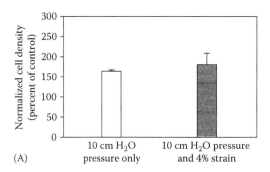
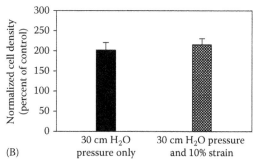

FIGURE 5.2 Comparison of smooth muscle cell proliferation under pressure alone and under simultaneous pressure with strain. Smooth muscle cell proliferation was similar under all conditions tested, that is, 10 cm H_2O sustained pressure alone (A; white bar) and under simultaneous 10 cm H_2O pressure with 4% strain (striped bar) as well as under 30 cm H_2O sustained pressure alone (B; black bar) and under simultaneous 30 cm H_2O pressure with 10% strain (checker bar). Values are means (normalized to results of cell proliferation obtained on controls [i.e., tissue culture polystyrene] under similar experimental conditions) ± SEM; n = 3. (From Dela Cruz, S., Effects of pressure on vascular smooth muscle cells function, MS thesis, Rensselaer Polytechnic Institute, Troy, NY, 2002.)

in serum-free media and exposed to either 40, 80, or 120 mm Hg sustained pressure exhibited significantly increased DNA synthesis after 24 h.[30] Exposure of rat aortic smooth muscle cells to cyclic pressure (64/14 mm Hg at 2 Hz for 24 h) applied via a transcapillary system resulted in either increased or decreased [³H]-thymidine incorporation, an index of cell proliferation, depending on the phenotype of the cells.[28] In one case, exposure of rat smooth muscle cells of a differentiated phenotype (in other words, cells that exhibited high levels of mRNA expression for typical "contractile" markers, such as calponin and myosin) to cyclic pressure (64/14 mm Hg at 2 Hz for 24 h) resulted in a significant (p<0.05) increase in cell proliferation compared to control (i.e., rat smooth muscle cells maintained at 24/18 mm Hg pressure) but otherwise similar experimental conditions.[28] In contrast, exposure of rat smooth muscle cells of a less differentiated phenotype to similar cyclic pressure conditions (specifically, 64/14 mm Hg at 2 Hz for 24 h) resulted in significant (p<0.05) decrease in cell proliferation compared to control (i.e., rat smooth muscle cells maintained at 24/18 mm Hg pressure) but otherwise similar experimental conditions.[28] Although the transcapillary system involved varying flow rates inside the capillaries, the enhanced cell proliferation was attributed to the pulse pressure because cell proliferation was not affected in controls under similar flow conditions in the absence of pressure.[28] Increased cell proliferation was also reported in rat aortic smooth muscle cells exposed to either 40 or 80 mm Hg sustained pressure for 48 h[30] and in bovine aortic smooth muscle cells exposed to either cyclic (120/90 mm Hg above atmospheric pressure; 1 Hz frequency) or sustained pressure (105 mm Hg above atmospheric pressure) for 5–9 days.[40] Exposure of rat vascular smooth muscle cells to sustained hydrostatic pressure as low as 10 cm H₂O (about 7.4 mm Hg) for 5 and 7 days resulted in significant (p<0.05) increased cell proliferation[39] (Figure 5.3). In contrast, exposure of rat aortic smooth muscle cells to high mean arterial pressure, specifically 130 mm Hg (selected based on clinically relevant hypertensive pressures) for 1 day, resulted in a significant (p<0.008), but transient, decrease in cell proliferation, which was no longer observed after 3 and 5 days of cell exposure to 130 mm Hg pressure, at which time cell proliferation was similar to that of cells exposed to either ambient or high pressure levels.[29]

Based on the results of the aforementioned studies, the effect of pressure on smooth muscle cell proliferation appears to be a function of the level, as well as of the duration, of cell exposure to pressure. The pertinent threshold values of these parameters have not yet been determined. Undoubtedly, this is an area of great potential and promise for further research.

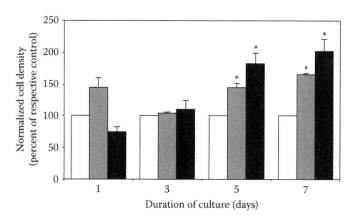

FIGURE 5.3 Time course of smooth muscle cell proliferation under sustained hydrostatic pressure. Compared to control (0.3 cm H₂O) pressure (white bars), smooth muscle cells under either 10 cm H₂O (gray bars) or 30 cm H₂O (black bars) sustained pressure for 5 and 7 days exhibited significantly increased cell density. Values are means (normalized to the respective controls) ± SEM; n = 3; *p < 0.05 (compared to the respective controls; analysis of variance; ANOVA). (From Dela Cruz, S., Effects of pressure on vascular smooth muscle cell function, MS thesis, Rensselaer Polytechnic Institute, Troy, NY, 2002.)

5.2.3 Apoptosis

Evidence that the observed effects of pressure on cell proliferation could result from apoptosis was provided by several studies. For example, exposure of vascular rat smooth muscle cells to 130–135 mm Hg pressure for 5 days exhibited decreased expression of c-myc oncoprotein (an index of cell proliferation); under these conditions a significant (15%) increase in the percentage of apoptotic cells was determined using acridine orange staining.[41] While co-culture of these rat smooth muscle cells with rat endothelial cells under ambient pressure conditions increased apoptosis (compared to results obtained with monoculture), exposure of the co-culture to pressure (130–135 mm Hg pressure for 5 days) resulted in the largest (specifically, 22%) increase of apoptosis compared to results obtained under control ambient pressure conditions.[41] Furthermore, apoptosis of smooth muscle cells exposed to ambient pressure and "conditioned media" (i.e., media from endothelial cells pre-exposed to sustained 130–135 mm Hg pressure for 5 days), was increased by 50%.[41] This result led to the hypothesis that the underlying molecular-level mechanism for the observed increased apoptosis in smooth muscle cells may involve yet unidentified paracrine chemical factors of endothelial origin.[41] This explanation is consistent with the fact that, in health and in disease, smooth muscle cell function is associated with, and complements, that of endothelial cells, the other major cell constituent of the vascular tissue.

5.3   MECHANISMS UNDERLYING VASCULAR SMOOTH MUSCLE CELL RESPONSES TO PRESSURE

Advances in cellular biology and biochemistry have provided various and novel methodologies to investigate the responses of smooth muscle cells at the molecular and genetic levels. These approaches are providing important insights of the underlying mechanisms of cell responses to biophysical and biochemical stimuli.

5.3.1 Role of Autocrine and Paracrine Chemical Compounds

The mechanism(s) underlying the enhanced cell proliferation observed when smooth muscle cells are exposed to pressure involve(s) bioactive chemicals. Attempts to identify the nature and source of such chemical compounds showed that autocrine chemicals were not involved when rat smooth muscle cells were exposed to moderately hypertensive (130–135 mm Hg) sustained pressures for 5 days[41]; instead, paracrine chemical compounds of endothelial origin were implicated in this case.[41] In another study, the mechanism underlying the significantly ($p < 0.05$) enhanced proliferative response of rat vascular smooth muscle cell proliferation after exposure to 10 cm H_2O sustained hydrostatic pressure for 5 and 7 consecutive days[39] (Figure 5.3) involved significant ($p < 0.05$) activation of ERK 1/2 (an extracellular-related kinase and a mitogen-activated protein kinase which is associated with enhanced cell proliferation); evidence of such activation was provided when these cells were exposed to 10 cm H_2O for 2 h[39] (Figure 5.4). This finding was in agreement with another study that reported pressure-induced phosphorylation of ERK in rat vascular smooth muscle cells exposed to 120 mm Hg for 2 min.[42] Further molecular-level investigation of the underlying mechanism revealed that the observed smooth muscle cell proliferation under 10 mm H_2O sustained pressure for 5 days involved soluble, transferable, bioactive mitogenic protein(s) and/or peptide(s), which had originated from the cells exposed to pressure[39] (Figure 5.5). Platelet-derived growth factor or PDGF (but not endothelial growth factor or EGF) was involved in the smooth muscle cell proliferation under 10 cm H_2O sustained pressure for 5 days; evidence supporting this conclusion was provided by significantly ($p < 0.05$) decreased rat aortic smooth muscle cell proliferation in the presence of AG 1296, a PDGF-receptor inhibitor[39] (Figure 5.6). This cell proliferation inhibition was dependent on the dose (5, 10, and 15 μM) of AG 1296, but was not completely blocked even when the highest concentration of inhibitor was used.[39]

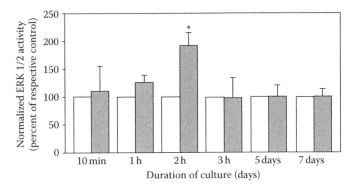

FIGURE 5.4 Effect of sustained hydrostatic pressure on smooth muscle cell ERK 1/2 activity. Compared to results obtained under standard cell culture conditions (i.e., culture media containing 10% fetal bovine serum) and control (i.e., 0.3 cm H_2O) pressure conditions (white bars), ERK 1/2 activity of smooth muscle cells increased significantly after exposure to 10 cm H_2O pressure (gray bars) for 2h. ERK 1/2 activity in smooth muscle cells exposed to 10 cm H_2O pressure, however, was similar to the respective controls at all other time points (specifically, 10 min, 1 h, 3 h, 5 days, and 7 days) tested. Values are means (normalized to the respective control) ± SEM; n = 3; *p < 0.05 (Student's t-test). (From Dela Cruz, S., Effects of pressure on vascular smooth muscle cell function, MS thesis, Rensselaer Polytechnic Institute, Troy, NY, 2002.)

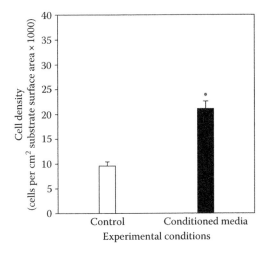

FIGURE 5.5 Evidence for the release of transferable bioactive mitogens. Smooth muscle cells cultured under "conditioned media" (i.e., supernatant media from cells exposed to 10 cm H_2O pressure for 5 days; striped bar) and 0.3 cm H_2O pressure for 3 days exhibited increased cell density (i.e., cells per cm² substrate surface area) compared to cell proliferation under control conditions (i.e., cells cultured under supernatant media from cells maintained under 0.3 cm H_2O pressure for 5 days; white bar) and 0.3 cm H_2O pressure for 3 days. Values are means ± SEM; n = 3; *p < 0.01 (Student's t-test). (From Dela Cruz, S., Effects of pressure on vascular smooth muscle cell function, MS thesis, Rensselaer Polytechnic Institute, Troy, NY, 2002.)

Further research is needed to determine what other growth factors, cytokines, and bioactive chemical compounds may be involved as well as to elucidate the underlying mechanisms of the proliferation of vascular smooth muscle cells exposed to pressure.

5.3.2 OTHER MOLECULAR-LEVEL RESPONSES

Other molecular-level responses include evidence that exposure of rat smooth muscle cells to pressure induced DNA synthesis via phospholipase C activation.[30] Evidence to support this claim was

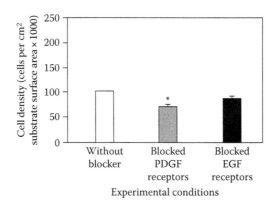

FIGURE 5.6 Identification of growth factors released from vascular smooth muscle cells exposed to pressure. Smooth muscle cell proliferation was similar in controls (i.e., smooth muscle cells exposed to 10 cm H_2O sustained pressure in the absence of any growth factor receptor kinase blocker; white bar) and in cells exposed to 10 cm H_2O sustained pressure in media containing 15 nM AG 1478, a blocker of endothelial growth factor (EGF) receptor kinase on cell membranes (black bar). In contrast, smooth muscle cells exposed to 10 cm H_2O sustained pressure in media containing 5 mM AG 1296 (gray bar), a blocker of platelet-derived growth factor (PDGF) receptor kinase on cell membranes, exhibited significantly decreased proliferation compared to that of controls (white bar). Values are means (normalized to controls) ± SEM; n = 3; *p < 0.05 (compared to results obtained using conditioned media (controls) without blocker; Student's t-test). (From Dela Cruz, S., Effects of pressure on vascular smooth muscle cell function, MS thesis, Rensselaer Polytechnic Institute, Troy, NY, 2002.)

provided by the absence of pressure-induced DNA synthesis in rat aortic smooth muscle cells pretreated with 2-nitro-4-carboxyphenyl-N,N-diphenylcarbamate (200 μM), a phospholipase C inhibitor, following exposure to sustained 80 mm Hg pressure.[30] Activation of phospholipase C triggers a signal transduction pathway leading to increased intracellular calcium, which, in turn, induces smooth muscle cell contraction.[8,30] An overstimulated contractile state of smooth muscle cells induces constriction of blood vessels that may contribute to hypertension.[43]

Exposure of rat vascular smooth muscle cells to either 30 cm H_2O hydrostatic pressure or 140/100 mm Hg cyclic pressure for 24 h also induced heat shock proteins, specifically, increased mRNA expression for Stch (stress 70 protein chaperone, AF006617), a member of the heat shock protein 70 (HSP 70) family.[39] Compared to controls (i.e., cells cultured under similar conditions but maintained at atmospheric pressure), Stch mRNA expression was upregulated in cells exposed to either 30 cm H_2O sustained (11.3-fold increase) or 140/100 cyclic (9.1-fold increase) pressure.[39] In addition, gene expression of a protein associated with microtubules (elements of the cytoskeleton) was upregulated in smooth muscle cells exposed to either 30 cm H_2O sustained or 140/100 cyclic pressure for 24 h (14.9-fold and 6.5-fold increases, respectively).[39] In contrast, gene expression for fibronectin (an extracellular matrix protein) and several growth factors was downregulated in the smooth muscle cells exposed to the aforementioned two pressure conditions tested for 24 h.[39] In the analysis of these results, a threefold difference from the respective controls was considered significant. These findings[39] provided the first genetic-level evidence that exposure to high (either 30 cm H_2O sustained or 140/100 mm Hg/mm Hg cyclic) pressure induced smooth muscle cell responses that are similar to those observed, and reported, for acute hypertension in rats in vivo.[44]

Advances in molecular biology and biochemistry as well as the availability of novel laboratory methodologies that allow investigation of events at the molecular and genetic levels have enabled research that has great, but to date mostly untapped, potential for major contributions in vascular physiology and pathology.

5.4 CLOSING REMARKS

This chapter focused on studies that exposed vascular smooth muscle cells to various types of pressure, including hypotensive,[39] normotensive,[30,42] and hypertensive sustained pressures[42] as well as normotensive and hypertensive cyclic pressures.[28,29,39–41] These results provide insights into the current understanding of smooth muscle cell responses to mechanical forces to which vascular cells are exposed in vivo and, in addition, establish that pressure (a mechanical force of the vascular milieu) affects functions of smooth muscle cells which are pertinent to the physiology and pathology of the vasculature. Specifically, the enhanced smooth muscle cell proliferation observed under either 10 or 30 cm H_2O sustained hydrostatic pressure (Figure 5.3) and the related investigation into the underlying mechanism(s) (Figures 5.4 through 5.6) provide a cellular- and molecular-level explanation for the smooth muscle cell hyperplasia observed in vivo under elevated pressure conditions.[45] In addition to elucidating aspects of the effects of pressure on smooth muscle cell function, these findings imply that pressure may be one of the stimuli that induce the compositional (and consequently structural) changes observed in vascular diseases.[34,35] Most importantly, this information provides insight into the cellular- and molecular-level events that may contribute to the etiology and progress of cardiovascular diseases such as hypertension and atherosclerosis.

Undoubtedly, further research is needed to elucidate the underlying mechanisms (especially at the molecular and genetic levels) of vascular smooth muscle cell functions in health and in disease. In pursuit of these objectives, future research must take into consideration the phenotype of the smooth muscle cell populations tested and must also establish the effects of this aspect on pressure-induced mechanotransduction. In addition, further research is necessary to determine the role of apoptosis (in conjunction with proliferation) of the various phenotypes of vascular smooth muscle cells in response to various pressure levels from the normotensive to the hypertensive range. Smooth muscle cells are a research area of great potential for major contributions in vascular physiology and pathology.

ACKNOWLEDGMENTS

The authors would like to thank Dr. Roman Ginnan, Albany Medical College, Albany, NY, for training Sheila Nagatomi to conduct the molecular-level analyses whose results are reported in this chapter. We would also like to thank Dr. Jiro Nagatomi, Clemson University, Clemson, SC, for the invitation to contribute a chapter to the present book and for his most gracious consideration and support during the long process of preparing this manuscript.

REFERENCES

1. Fung, Y.C. (1993) *Biomechanics: Mechanical Properties of Living Tissues*. New York: Springer-Verlag.
2. Schwartz, S.M. and Mecham, R.P. (1995) *The Vascular Smooth Muscle Cell: Molecular and Biological Responses to the Extracellular Matrix*. San Diego, CA: Academic Press.
3. Newby, A.C. and Zaltsman, A.B. (2000) Molecular mechanisms in intimal hyperplasia. *J Pathol* **190**:300–309.
4. Osol, G. (1995) Mechanotransduction by vascular smooth muscle. *J Vasc Res* **32**:275–292.
5. Aqel, N.M., Ball, R.Y., Waldmann, H., and Mitchinson, M.J. (1985) Identification of macrophages and smooth muscle cells in human atherosclerosis using monoclonal antibodies. *J Pathol* **146**:197–204.
6. Schwartz, S.M. (1997) Perspectives series: Cell adhesion in vascular biology. Smooth muscle migration in atherosclerosis and restenosis. *J Clin Invest* **99**:2814–2816.
7. Malek, A.M., Alper, S.L., and Izumo, S. (1999) Hemodynamic shear stress and its role in atherosclerosis. *JAMA* **282**:2035–2042.
8. Rhoades, R. and Pflanzer, R. (1996) *Human Physiology*, 3rd edn. Orlando, FL: Saunders College Publishing.
9. Lyon, R.T., Runyon-Hass, A., Davis, H.R., Glagov, S., and Zarins, C.K. (1987) Protection from atherosclerotic lesion formation by reduction of artery wall motion. *J Vasc Surg* **5**:59–67.

10. Cunningham, K.S. and Gotlieb, A.I. (2005) The role of shear stress in the pathogenesis of atherosclerosis. *Lab Invest* **85**(1):9–23.
11. White C.R. and Frangos J.A. (2007) The shear stress of it all: The cell membrane and mechanochemical transduction. *Philos Trans R Soc Lond B Biol Sci* **362**(1484):1459–1467.
12. Kurpinski, K., Park, J., Thakar, R.G., and Li, S. (2006) Regulation of vascular smooth muscle cells and mesenchymal stem cells by mechanical strain. *Mol Cell Biomech* **3**(1):21–34.
13. Haga, J.H., Li, Y.S., and Chien, S. (2007) Molecular basis of the effects of mechanical stretch on vascular smooth muscle cells. *J Biomech* **40**(5):947–960.
14. Liu, B., Qu, M.J., Qin, K.R., Li, H., Li, Z.K., Shen, B.R., and Jiang, Z.L. (2008) Role of cyclic strain frequency in regulating the alignment of vascular smooth muscle cells in vitro. *Biophys J* **94**(4):1497–1507.
15. Li, Y.S., Haga, J.H., and Chien, S. (2005) Molecular basis of the effects of shear stress on vascular endothelial cells. *J Biomech* **38**(10):1949–1971.
16. Lehoux, S., Castier, Y., and Tedgui, A. (2006) Molecular mechanisms of the vascular responses to haemodynamic forces. *J Intern Med* **259**(4):381–392.
17. Rensen, S.S., Doevendans, P.A., and Van Eys, G.J. (2007) Regulation and characteristics of vascular smooth muscle cell phenotypic diversity. *Neth Heart J* **15**(3):100–108.
18. Villaschi, S., Nicosia, R.F., and Smith, M.R. (1994) Isolation of a morphologically and functionally distinct smooth muscle cell type from the intimal aspect of the normal rat aorta. Evidence for smooth muscle cell heterogeneity. *In Vitro Cell Dev Biol Anim* **30A**:589–595.
19. Huang, Y.L., Shi, G.Y., Jiang, M.J., Lee, H., Chou, Y.W., Wu, H.L., and Yang, H.Y. (2008) Epidermal growth factor up-regulates the expression of nestin through the Ras-Raf-ERK signaling axis in rat vascular smooth muscle cells. *Biochem Biophys Res Commun* **377**(2):361–366.
20. Orr, A.W., Hastings, N.E., Blackman, B.R., and Wamhoff, B.R. (2010) Complex regulation and function of the inflammatory smooth muscle cell phenotype in atherosclerosis. *J Vasc Res* **47**(2):168–180.
21. Owens, G.K. (1995) Regulation of differentiation of vascular smooth muscle cells. *Physiol Rev* **75**(3):487–517.
22. Li, S., Sims, S., Jiao, Y., Chow, L.H., and Pickering, J.G. (1999) Evidence from a novel human cell clone that adult vascular smooth muscle cells can convert reversibly between noncontractile and contractile phenotypes. *Circ Res* **85**(4):338–348.
23. Regan, C.P., Adam, P.J., Madsen, C.S., and Owens, G.K. (2000) Molecular mechanisms of decreased smooth muscle differentiation marker expression after vascular injury. *J Clin Invest* **106**(9):1139–1147.
24. Hipper, A. and Isenberg, G. (2000) Cyclic mechanical strain decreases the DNA synthesis of vascular smooth muscle cells. *Pflugers Arch* **440**(1):19–27.
25. Albinsson, S., Nordström, I., and Hellstrand, P. (2004) Stretch of the vascular wall induces smooth muscle differentiation by promoting actin polymerization. *J Biol Chem* **279**(33):34849–34855.
26. Li, W., Chen, Q., Mills, I., and Sumpio, B.E. (2003) Involvement of S6 kinase and p38 mitogen-activated protein kinase pathways in strain-induced alignment and proliferation of bovine aortic smooth muscle cells. *J Cell Physiol* **195**(2):202–209.
27. Riha, G.M., Lin, P.H., Lumsden, A.B., Yao, Q., and Chen, C. (2005) Roles of hemodynamic forces in vascular cell differentiation. *Ann Biomed Eng* **33**(6):772–779.
28. Cappadona, C., Redmond, E.M., Theodorakis, N.G., McKillop, I.H., Hendrickson, R., Chhabra, A., Sitzmann, J.V., and Cahill, P.A. (1999) Phenotype dictates the growth response of vascular smooth muscle cells to pulse pressure in vitro. *Exp Cell Res* **250**(1):174–186.
29. Vouyouka, A.G., Salib, S.S., Cala, S., Marsh, J.D., and Basson, M.D. (2003) Chronic high pressure potentiates the antiproliferative effect and abolishes contractile phenotypic changes caused by endothelial cells in cocultured smooth muscle cells. *J Surg Res* **110**(2):344–351.
30. Hishikawa, K., Nakaki, T., Marumo, T., Hayashi, M., Suzuki, H., Kato, R., and Saruta, T. (1994) Pressure promotes DNA synthesis in rat cultured vascular smooth muscle cells. *J Clin Invest* **93**:1975–1980.
31. Schwartz, S.M. (1984) Smooth muscle proliferation in hypertension: State-of-the-art lecture. *Hypertension* **6**:I56-I61.
32. Levy, B.I., Michel, J.B., Salzmann, J.L., Azizi, M., Poitevin, P., Safar, M., and Camilleri, J.P. (1988) Effects of chronic inhibition of converting enzyme on mechanical and structural properties of arteries in rat renovascular hypertension. *Circ Res* **63**:227–239.
33. Zanchi, A., Wiesel, P., Aubert, J.F., Brunner, H.R., and Hayoz, D. (1997) Time course changes of the mechanical properties of the carotid artery in renal hypertensive rats. *Hypertension* **29**(5):1199–1203.
34. Gariepy, J., Massonneau, M., Levenson, J., Heudes, D., and Simon, A. (1993) Evidence for in vivo carotid and femoral wall thickening in human hypertension. *Hypertension* **22**:111–118.

35. Hadrava, V., Tremblay, J., and Hamet, P. (1991) Intrinsic factors involved in vascular smooth muscle cell proliferation in hypertension. *Clin Invest Med* **14**:535–544.
36. Bardy, N., Karillon, G.J., Merval, R., Samuel, J.L., and Tedgui, A. (1995) Differential effects of pressure and flow on DNA and protein synthesis and on fibronectin expression by arteries in a novel organ culture system. *Circ Res* **77**:684–694.
37. Birukov, K.G., Lehoux, S., Birukova, A.A., Merval, R., Tkachuk, V.A., and Tedgui, A. (1997) Increased pressure induces sustained protein kinase C-independent herbimycin A-sensitive activation of extracellular signal-related kinase 1/2 in the rabbit aorta in organ culture. *Circ Res* **81**:895–903.
38. Birukov, K.G., Bardy, N., Lehoux, S., Merval, R., Shirinsky, V.P., and Tedgui, A. (1998) Intraluminal pressure is essential for the maintenance of smooth muscle caldesmon and filamin content in aortic organ culture. *Arterioscler Thromb Vasc Biol* **18**:922–927.
39. Dela Cruz, S. (2002) Effects of pressure on vascular smooth muscle cell function. MS thesis, Rensselaer Polytechnic Institute, Troy, NY.
40. Watase, M., Awolesi, M.A., Ricotta, J., and Sumpio, B.E. (1997) Effect of pressure on cultured smooth muscle cells. *Life Sci* **61**:987–996.
41. Vouyouka, A.G., Jiang, M.S., and Basson, M.D. (2004) Pressure alters endothelial effects upon vascular smooth muscle cells by decreasing smooth muscle cell proliferation and increasing smooth muscle cell apoptosis. *Surgery* **136**:282–290.
42. Tsuda, Y., Okazaki, M., Uezono, Y., Osajima, A., Kato, H., Okuda, H., Oishi, Y., Yashiro, A., and Nakashima, Y. (2002) Activation of extracellular signal-regulated kinases is essential for pressure-induced proliferation of vascular smooth muscle cells. *Eur J Pharmacol* **446**(1–3):15–24.
43. Wynne, B.M., Chiao, C.W., and Webb, R.C. (2009) Vascular smooth muscle cell signaling mechanisms for contraction to angiotensin II and endothelin-1. *J Am Soc Hypertens* **3**(2):84–95.
44. Xu, Q., Schett, G., Li, C., Hu, Y., and Wick, G. (2000) Mechanical stress-induced heat shock protein 70 expression in vascular smooth muscle cells is regulated by Rac and Ras small G proteins but not mitogen-activated protein kinases. *Circ Res* **86**(11):1122–1128.
45. Berk, B.C. (2001) Vascular smooth muscle growth: Autocrine growth mechanisms. *Physiol Rev* **81**(3):999–1030.

6 Mechanobiology of Heart Valves

Joshua D. Hutcheson, Michael P. Nilo,
and W. David Merryman

CONTENTS

6.1 INTRODUCTION TO HEART VALVES

6.1.1 HEART VALVE FUNCTION

The heart acts as a pump to ensure that blood is distributed appropriately throughout the body. This is accomplished through a coordinated contraction of the heart muscles causing the blood to be pumped through four cardiac chambers. During the beginning of each cardiac cycle—a period known as diastole—deoxygenated blood from the body fills the right atrium and newly oxygenated blood from the lungs fills the left atrium. The flow of blood then proceeds through the tricuspid valve from the right atrium into the right ventricle or through the mitral valve from the left atrium into the left ventricle. Ventricular contraction—or systole—is the main impetus for moving blood out of the heart. The contraction of the ventricles forces blood from the right ventricle into the pulmonary artery to be delivered to the lungs for oxygenation and delivery of blood to the systemic circulation from the left ventricle through the aortic valve. The cycle completes when blood returns to the heart during the next diastolic period. Within the heart, the four valves maintain unidirectional flow of blood (Figure 6.1) by the coordinated action of leaflets that open and close during each cardiac cycle. In fact, the leaflets within each valve will open and close over 3 billion times in an average lifetime.[1] Therefore, the leaflets must be able to withstand dynamic, cyclic stresses while maintaining the structural integrity that is crucial to their function. The interplay between the forces caused by the fluid mechanics of blood flow and the biomechanical properties of the valve leaflets make the

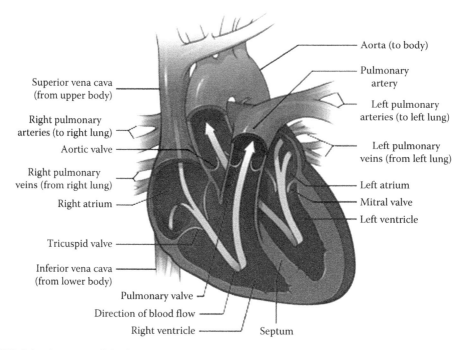

FIGURE 6.1 Anatomy of the heart. All of the major components of the heart are depicted including the locations of the four heart valves and the direction of blood flow. (Reprinted from the National Heart, Lung, and Blood Institute, a part of the National Institutes of Health and the U.S. Department of Health and Human Services.)

tissues that form the heart valve leaflets interesting subjects in mechanobiology. For introductory purposes, we will consider the two separate types of heart valves—the atrioventricular valves and the semilunar valves—and discuss the corresponding characteristics of each class.

The atrioventricular valves (i.e., the tricuspid valve and mitral valve on the right and left side of the heart, respectively) remain open during ventricular diastole to allow blood to move from the atria into the ventricles. As the ventricles begin to contract during systole, the valve leaflets snap shut to prevent blood from returning into the atria for the duration of systolic contraction. This continual cycling between opening and closing of the atrioventricular valves is controlled by both hemodynamic pressure and chordae tendineae that tether the valve leaflets to papillary muscles within the ventricle walls. The structure of the two atrioventricular valves is similar, with the main difference being the number of leaflets. The mitral valve has two leaflets (Figure 6.2), whereas the tricuspid valve has three leaflets. During diastole, the pressure in the atria is higher than that in the ventricles due to the influx of blood into the heart. This pressure difference causes the leaflets of the atrioventricular valves to extend outward into the ventricles; however, as the ventricles contract, the pressure gradient is reversed, causing the atrioventricular leaflets to be pushed back toward the atria. Tension from the chordae tendineae ensure that the leaflets come together to make a seal and do not prolapse—or protrude—back into the atria.[2] The interplay between the various components of these valves is more complicated when compared to the relatively simple semilunar valves directing blood flow out of the ventricles.

The valves that direct blood flow from the heart to the body and lungs are known as semilunar valves due to the crescent shape of the leaflets (Figure 6.3a, a view of the aortic valve from the perspective of the aorta). The pulmonary valve is situated between the right ventricle and pulmonary artery, and the aortic valve directs flow from the left ventricle to the aorta. These two valves differ from the atrioventricular valves in that they lack chordae tendineae and rely solely on the hemodynamic forces of blood flow to direct opening and closing of their leaflets. Ventricular contraction

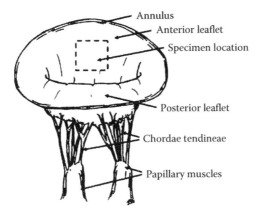

FIGURE 6.2 Anatomy of the mitral valve. All major portions of the tissue are shown. The dotted square indicates a portion of the anterior leaflet utilized in many of the studies discussed. (Reprinted from Grashow, J.S. et al., *Ann. Biomed. Eng.*, 34(2), 315, 2006. With permission.)

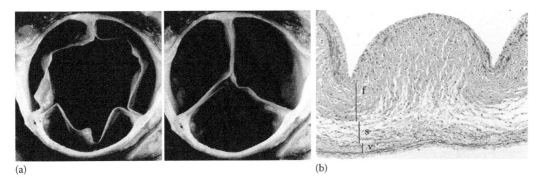

(a)   (b)

FIGURE 6.3 Anatomy of the aortic valve. (a) The aortic valve leaflets are pushed into the aorta during ventricular systole and coapt to form a seal during diastole. (b) A histological image of the layers of the aorta is shown. The ventricularis (v) is composed of elastin fibers. The spongiosa (s) provides a lubricating border between the ventricularis and the collagen-containing fibrosa (f). (Reprinted from Schoen, F.J. and Edwards, W.D., Valvular heart disease: General principles and stenosis, in *Cardiovascular Pathology*, Silver, M.D., Gotlieb, A.I., and Schoen, F.J. editors, Churchill Livingstone, New York, 402–442, 2001. With permission.)

during systole forces the leaflets of the semilunar valves open. During diastole, the leaflets coapt to prevent blood from flowing back into the ventricles. As we will discuss in much greater detail, the pressure of the blood on the leaflets of the closed semilunar valves during diastole introduces a high amount of stress on the tissues. This stress can lead to mechano-dependent signal transduction of pathologic responses at the cellular level within the leaflets that can greatly alter valve function.

The diminished ability of the valve to regulate blood flow can result in significantly altered cardiac function. Furthermore, heart valve disease (HVD; acronyms and abbreviations used in this chapter are shown in Table 6.1) is one of the leading causes of cardiovascular disease, especially among the elderly. Studies have shown that HVD increases in prevalence from 0.7% in people between 18 and 44 years of age to over 13% of people over 75 years of age. Moreover, early symptoms of aortic valve disease have been detected in 29% of patients over 65 years of age. These symptoms are associated with a 50% increase in cardiovascular-related morbidity and a similar increase in the risk of myocardial infarction.[3] Currently, the only effective long-term treatment for advanced HVD is replacement surgery, an invasive, high risk procedure for elderly patients that requires substantial recovery time; therefore, understanding of HVD etiology is an important step in developing new, less-invasive therapeutics or preventative strategies.

TABLE 6.1
Acronyms and Abbreviations

AV	Aortic valve
AVEC	Aortic valve endothelial cell
AVIC	Aortic valve interstitial cell
CAM	Cell adhesion molecule
CICP	Collagen type-1 c-terminal propeptide
EC	Endothelial cell
ECM	Extracellular matrix
GAG	Glucosaminoglycan
HSP47	Heat shock protein 47
HVD	Heart valve disease
LV	Left ventricle
MMP	Matrix metalloproteinase
MR	Mitral regurgitation
MV	Mitral valve
MVEC	Mitral valve endothelial cell
MVIC	Mitral valve interstitial cell
PG	Proteoglycan
PM	Papillary muscle
PVIC	Pulmonary valve interstitial cell
TGF-β1	Transforming growth gactor-β1
TVIC	Tricuspid valve interstitial cell
TVP	Transvalvular pressure
VEC	Valve endothelial cell
VHD	Valvular heart disease
VIC	Valve interstitial cell
αSMA	Smooth muscle α-actin

6.1.2 GLOBAL VIEW OF HEART VALVE MECHANOBIOLOGY

The importance of physiologic forces from normal cardiac function on cellular function has been extensively studied in recent years. In fact, the interrelationship between the mechanics and biological responses within the heart valves is evident from observing the differences between the biomechanical properties of the cells from the corresponding valves on each side of the heart. The pressures involved with systemic body circulation on the left side of the heart are dramatically higher than those of the pulmonary circulation on the right side of the heart; therefore, these pressure differences are transferred to the cells within the leaflets and produce dissimilarities that can be observed within cellular populations between the valves. The difference in mechanobiological response between the two sides was recently quantified by correlating transvalvular pressures (TVP) to heart valve interstitial cell (VIC) stiffness.[4] VICs are the main cellular component of the valve leaflets and are thought to play a crucial role in maintaining the mechanical integrity of the valve tissues by regulating extracellular matrix biosynthesis.[5,6] Isolated VICs from each of the four heart valves were analyzed for cellular stiffness using micropipette aspiration.[7,8] VICs isolated from the right-side valves—i.e., the tricuspid (TVIC) and pulmonary (PVIC) valves—displayed a significantly lower effective stiffness than the mitral (MVIC) and aortic (AVIC) valve cells isolated from the left-side valves (Figure 6.4). Stiffness data correlated well with the level of collagen synthesis by the VICs, indicating that the AVICs and MVICs are responsible for producing more robust ECM proteins, which yields the higher overall stiffness seen in the left-side valves.

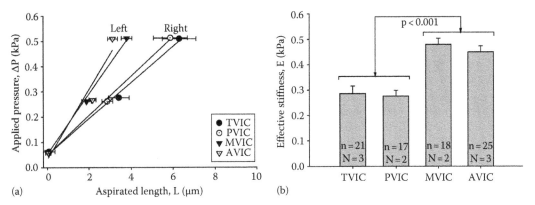

FIGURE 6.4 Comparison of interstitial cells from each side of the heart. The cells from the valves on the left side of the heart (MVICs and AVICs) exhibit a significantly higher stiffness than the cells from the valves on the right side of the heart (TVICs and PVICs). (a) Applied pressure vs. aspirated length during the micropipette aspiration procedure. (b) Calculated effective stiffness for each population in (a). (Reprinted from Merryman, W.D. et al., *Am. J. Physiol. Heart Circ. Physiol.*, 290(1), H224, 2006. With permission.)

All of the heart valves rely on their structural properties for correct function. This structural integrity is dictated by the material properties of the tissues comprising the valves. As mentioned above, the constant changes in TVP (due to contraction and relaxation of the cardiac chambers) place different stresses (i.e., force/area) across the heart and on different parts of the valves over the course of the cardiac cycle. The elevated pressures on the left-side of the heart lead to an increased importance on cellular mechanotransduction and also a heightened prevalence in disease for the aortic and mitral valves.[9] Therefore, we will continue to discuss the left-side valves only. However, the reader should note that many of the valve properties mentioned here are also applicable to the right-side heart valves. In the following sections, we will review current understanding of how valves respond to demanding physical forces present in the heart and how improper cellular response results in altered tissue architecture, which leads to valve pathologies that are of clinical relevance.

6.2 PHYSIOLOGY AND PATHOLOGY OF THE MITRAL VALVE

6.2.1 BIOMECHANICS AND STRUCTURE OF MITRAL VALVE

Before we can begin to discuss the response of the valvular cells and tissues to mechanical forces, we must first understand the nature of the forces as well as the importance of valvular cell and tissue structure in maintaining proper cardiac function. The mitral valve (MV) directs blood flow from the left atrium into the left ventricle by opening into the ventricles during diastole and closing to seal off the atria during systole. The mechanics of the valve play a very important role in the maintenance of the directionality of this flow. The MV consists of two leaflets (Figure 6.2), the anterior and posterior leaflets, connected to the boundary between the left atrium and the left ventricle known as the valve annulus.[10] During diastole, the leaflets remain passively open to allow blood to flow into the ventricle. Pressure changes from systolic contraction cause the leaflets to be pushed back toward the atrium. The valve is "closed" when apposition of the leaflets seals the gap between the atrium and the ventricle, preventing blood from flowing back into the atrium. Proper closure is ensured by the action of the chordae tendineae, which act as anchors, preventing the leaflets from collapsing back into the atrium during systole.[11] The high pressure on the leaflets and tendineae during the time of ventricular contraction confers the need for specialized mechanical properties to maintain proper valve function.[12]

The MV leaflets are composed of three layers: the ventricularis or fibrosa, the spongiosa, and the atrialis.[13] As indicated by the names, the ventricularis and atrialis portions of the leaflets face the left ventricle and left atrium, respectively. The spongiosa forms the barrier between these two regions and is composed of loosely packed collagen and proteoglycans, thought to serve as a lubricating layer within the leaflets. The atrialis is composed mostly of elastic fibers that allow the leaflets to expand during systole and retract during diastole, and the fibrosa is composed mainly of tightly packed collagen fibers that determine the structural integrity of the leaflets. The collagen fibers within the leaflets have been found to be circumferentially oriented.[14] This arrangement allows for the valve to deform in a unique orthogonal manner. During systole, the anterior and posterior leaflets must be allowed to stretch until they meet to seal the valve; therefore, the circumferential arrangement of the collagen fibers allows the leaflets to deform until the point at which these fibers become fully elongated—simultaneous to apposition of the MV leaflets—and further deformation of the valve is prevented.[15] Further, the tensile stresses on the leaflets during ventricular contraction result in significantly greater strain than those caused by the radially oriented shear stresses of blood flow and, therefore, the tensile stresses of the tissue dictate the circumferential alignment of the collagen fibers.

Pioneering studies have been done both in vitro and in vivo to quantify the strains of the MV leaflets during the cardiac cycle.[15–18] Strain is defined as the fractional change in length due to an applied load from an unloaded state. By studying the three-dimensional motion of a 4 × 4 mm central portion of the anterior MV leaflet, Sacks et al. found that the leaflet experiences high anisotropic strain rates of 500%–1000% during valve closure. After this initial rise, the strain remains constant as the collagen fibers become fully straightened when the TVP exceeds ~20 mm Hg, preventing further deformation of the leaflets, and thus, preventing regurgitation caused by MV prolapse.[15] The initial deformation is then symmetrically reversed during valvular opening. Figure 6.5 shows the stretch experienced by the anterior MV leaflet as a function of time. These results were corroborated with the use of a left heart simulator. Jimenez et al. found that the areal strains on both of the MV leaflets rapidly rise (approximately 75% change in the anterior leaflet corresponding to a marginally higher strain rate than that observed by Sacks et al.[15]) during normal MV closure, followed by a plateau period of no further deformation and complete reversal when the MV reopens.[19] The larger anterior leaflet also showed consistently higher strain than that of the smaller posterior leaflet.

As mentioned previously, the chordae tendineae anchor the leaflets to the papillary muscle (PM) on the ventricular side of the MV annulus. These chordae consist of two distinct structural layers: an inner layer of dense collagen and an outer layer of loose collagen and interwoven elastin fibers.[20–22] When the chordae are stretched, the collagen becomes aligned along the center axis (from the PM

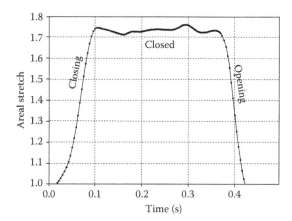

FIGURE 6.5   The areal stretch of the mitral valve. The mitral valve leaflets reach maximal stretch quickly during valve closing; whereupon, the straightened collagen fibers prevent further deformation. The stretch is reversed as the valve opens during diastole. (Adapted from Sacks, M.S. et al., *Ann. Biomed. Eng.*, 30(10), 1281, 2002.)

to the MV leaflets). This alignment gives the chordae the tensile strength necessary to prevent leaflet prolapse under high pressure from left ventricular (LV) contraction. The intermediate chordae are stretched farther than the marginal chordae, and therefore, must be capable of withstanding larger mechanical loads. Consequently, the intermediate chordae have been shown to contain more collagen to accommodate the tensile forces.[20] The total force distribution of the tendineae has been found to depend on the mechanical environment of the MV and the geometric relationships between MV components. The importance of the spatial orientations of the valve components to MV function will become more evident as we discuss MV pathologies in more detail below.

Further biomechanical properties of the MV can be ascertained by studying the material properties conferred by the collagen, proteoglycans (PG), and glycosaminoglycans (GAGs) that make up the valve tissues. Collagen is the main tensile-bearing component of many tissues. The PGs decorin and biglycan along with the GAG dermatan 4-sulfate regulate the formation of collagen fibers and, therefore, also play an important role in regions of increased tension.[23] As noted above, the chordae tendineae must bear a high tensile load that is distributed among the collagen fibers. The circumferential alignment of the collagen fibers in the MV leaflets allows the leaflets to deform enough to cause leaflet coaptation, whereupon the straightened collagen fibers prevent further leaflet deformation. Correspondingly, the regions with the highest tensile loads (e.g., the chordae tendineae) also contain an increased amount of decorin, biglycan, and 4-sulfated GAGs that serve to regulate the correct alignment of the collagen fibers.[24] The compressive regions of the MV also have specific structural components that are important in tissue function. For example, GAGs play a very important role in leaflet apposition. The free edges of the leaflets contain an abundance of the GAG hyaluronan, which resists compression by swelling in the presence of water.[25] This resistance to compression ensures further deformation is prevented at the closure region, where the edges of the leaflets meet to form a seal. All of these important biomechanical tissue components are produced and organized by specialized cells within the MV tissue.

6.2.2 Cellular Mechanobiology of Mitral Valves

The biomechanics discussed in the previous section depend heavily on the function of the cells that make up these tissues. These cells are directly responsible for maintenance of the extracellular matrix (ECM) contents (e.g., collagen, elastin, and GAG synthesis) that determine the material properties of the tissue.[25–29] The leaflets of the MV contain two types of cells: mitral valve endothelial cells (MVECs) and MVICs, which were mentioned earlier. Like all tissues of the vasculature, the MVECs provide the cellular boundary layer between the blood and the tissue; however, compared to MVICs, little research has been done on the importance of the MVECs to heart valve function, especially in the MV. Valve endothelial cell function will be discussed more thoroughly in the AV section. Presumably, many of the same endothelial functions in the AV discussed below are also applicable to the endothelial cells in the MV.

The VICs in all valves are a heterogeneous population of cells that are present in all three of the valve layers.[30,31] These cells have been extensively studied and are of special research interest due to their phenotypic plasticity. Under normal valvular conditions, these cells maintain a quiescent fibroblast-like phenotype; however, during times of development, disease, and/or valvular repair, these cells exhibit an activated myofibroblast phenotype.[1,5,32] Myofibroblast cells exhibit both fibroblast and smooth muscle cell characteristics, and once activated, the VICs display increased contractility and ECM synthesis. Many researchers in mechanotransduction have begun to focus on these cells as the main regulators of valve ECM maintenance due to the observation that VIC phenotype and function are highly dependent on their mechanical environment.

Recent work has focused specifically on the mechanobiology of MVICs and how these specialized cells regulate MV function.[25–27,29,33] In one study, MVICs were seeded into collagen gels and stretched or relaxed in 24 h increments for 1 week.[29] MVICs obtained from MV leaflets and chordae showed increased GAG synthesis after cyclic stretch compared to MVICs that were not stretched

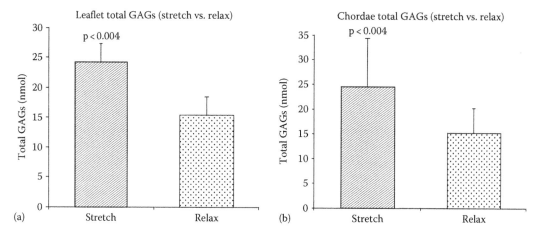

FIGURE 6.6   The effect of stretch on mitral GAG synthesis. (a) Total GAG synthesis is significantly increased in leaflets that were stretched versus those that did not receive mechanical stimulation. (b) This same trend was observed in the chordae of the mitral valves. (Reprinted from Gupta, V. et al., *Ann. Biomed. Eng.*, 36(7), 1092, 2008. With permission.)

in the gels (Figure 6.6). Furthermore, 4-sulfated GAGs were the most common GAG found to be secreted after the addition of the mechanical loads. As mentioned previously, GAGs are extremely important in tissues bearing large tensile burdens, such as the chordae tendineae. Therefore, one would expect to find more MVIC activity in the chordae. This hypothesis is supported by the observation that MVICs isolated from the chordae contain approximately 50% more smooth muscle α-actin (αSMA)—a marker of myofibroblast activation—than leaflet MVICs, indicating that the MVICs in the chordae are actively involved in maintaining the physiological structural integrity along the tendineae.[27] However, as discussed in the next section, many disease pathologies can arise from overactivity of these cells.

6.2.3   Mitral Valve Pathologies and Current Research

In Section 6.2.1, we have discussed the importance of mechanical properties to MV function. Alterations of structural integrity of the MV, is characteristic of most valve. Valve problems usually occur due to changes in the biomechanical responsiveness of the tissue, resulting in a reduced ability of the valve to maintain the directionality of blood flow. In the MV, this often occurs when the anterior and posterior leaflets do not meet at their free edges during systole. Instead, the stresses caused by LV contraction causes the leaflets to prolapse back into the left atrium resulting in the blood flowing along a reverse path, a condition known as mitral regurgitation (MR).[34] This malfunction results in blood being pushed back into the atrium, which increases the ventricular workload due to excess stress required to maintain the correct blood pressure in the systemic circulatory system (i.e., throughout the rest of the body). Remodeling of the LV to compensate for the hemodynamic changes caused by valve dysfunction can further exacerbate the problem by changing the orientation of the PM and valve annulus and causing the leaflets to be pulled farther apart.[35–38] Simultaneously, systemic and/or cardiac changes that lead to LV remodeling can cause valvular dysfunction. For example, MR has been observed in 11% of patients with coronary artery disease.[37] Even though the leaflets of these MV appear normal, apical displacement caused by changes in PM or LV geometry leads to insufficient leaflet coaptation.[38,39] Nielsen et al. modeled the cause of geometric-induced MR by quantifying changes in chordal force distribution; wherein, tethering forces that are imparted onto the leaflet by the chordae connection to the PM overcome the coapting forces required to allow apposition of the leaflets.[37]

Pathological changes in MV geometry and blood flow also impart non-physiological stresses on the cells within the MV, which can cause functional changes in the biomechanics of the tissue.[40,41] Cells responding to these irregular stresses have been shown to alter the properties and components of the ECM. One of the most common causes of MV prolapse is myxomatous degeneration, a disease thought to be of genetic origins.[42] The leaflets of myxomatous valves no longer exhibit the collagen-mediated stiffening required for appropriate apposition, as described above in the biomechanics section. In a study performed by Rabkin et al., the cellular and protein composition of 14 MVs removed from patients with myxomatous degeneration was compared to 11 MVs from autopsy controls.[43] Histological images showed clear morphological changes associated with myxomatous degeneration of the MV leaflets (Figure 6.7a). The fibrosa, while expanded in the diseased tissue, mainly consists of loose, scattered collagen fibers. In the myxomatous leaflet, the spongiosa is much larger with an increased amount of PGs and GAGs. These findings were corroborated by Grande-Allen et al. where the anterior portions of diseased MVs were found to contain 59% more GAGs and 15% more collagen than control MVs.[39] The atrialis layer of the diseased leaflets was fragmented with diminished elastin staining. Taken together, these morphological changes all contribute to the floppy leaflet performance often observed in myxomatous MV prolapse. Notably, the increased thickness of the fibrosa and spongiosa also corresponds to a large increase in cell density (Figure 6.7b), indicating an active cellular role in the ECM remodeling observed in the myxomatous MVs.

The role of MVECs in MV disease has received relatively little attention; however, researchers are beginning to fully appreciate the role that these cells play in potential HVD etiologies. Again, the MVECs establish the barrier between blood and the MV tissue. In other tissues, endothelial cells

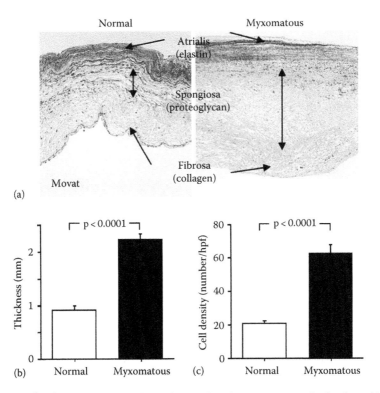

FIGURE 6.7 (See color insert.) Comparison of normal and myxomatous mitral valves. (a) Myxomatous valves show a decrease in elastin, an expanded spongiosa, and a fibrosa with more loosely connected collaged fibers. (b) Myxomatous valves are significantly thicker and (c) contain a higher cell density than normal valves. (Adapted from Rabkin, E. et al., *Circulation*, 104, 2525, 2001.)

have been long known to control local inflammatory responses by recruiting cells of the immune system to sites of tissue injury,[44-47] and these cells likely play a similar role in the MV tissues. Again, since most of the research into VEC function has been done using AVECs, we will reserve this discussion for the section on AV pathologies.

Due to their unique ability to remodel valvular tissue, VICs have received a lot of attention concerning their involvement in degenerative valve disease. In fact, Rabkin et al. observed that the cells in myxomatous MVs contained high amounts of αSMA indicative of activated, or myofibroblastic, MVICs. These cells were also found to highly express both matrix metalloproteinases (MMPs), enzymes involved in the degradation of ECM, and pro-collagen I mRNA compared to cells from control MVs, indicating a cellular-mediated remodeling of the ECM components in diseased MV leaflets.[43] Recently, MVICs have been implicated in modulating tissue changes that lead to chordae tendineae rupture—another common cause of MV prolapse and regurgitation. MVICs in the normal chordae have been found to secrete high levels of the anti-angiogenic factor tenomodulin; however, this factor is not observed in the vicinity of ruptured chordate.[48] Similar results were found by subjecting healthy chordae tendineae to a variety of stresses in vitro. In these studies, tenomodulin secretion was found to be dramatically reduced in tendineae that were exposed to pathologic mechanical stretching, hypoxia, or oxidative stress.[48] In vivo, suppression of this factor leads to increased angiogenesis, and consequently, ruptured tendineae tend to be highly vascularized.[49] Therefore, we speculate that increased vascularization of the tissue may allow for more nutrients and oxygen to reach the inner portions of the tendineae, which in turn could explain the increased cellularity, as measured by total DNA, observed within the diseased chordae. The increase in the number of cells has also been found to correspond to an increase in MMPs, which may lead to the overall weakening and subsequent rupture of the diseased chordae by breaking down the collagen required to maintain the tensile integrity of the tissue.[43]

6.3 PHYSIOLOGY AND PATHOLOGY OF THE AORTIC VALVE

6.3.1 BIOMECHANICS AND STRUCTURE OF AORTIC VALVE

As shown in Figure 6.1, the AV is situated between the LV and the aorta and functions to prevent regurgitation of blood into the LV that has been pumped to the aorta for distribution throughout the body. The AV accomplishes this task using three adjacent leaflets that open during systole (Figure 6.3a, left panel) and coapt (Figure 6.3a, right panel) during diastole. Unlike the MV, which has PM and tendineae to help control movement of the leaflets, the AV relies solely on hemodynamic forces for opening and closing. Therefore, the function of the AV is heavily reliant on the mechanical properties of the leaflets.[50]

Similar to the MV, each AV leaflet contains a layered ECM structure that is composed almost exclusively of collagen, elastin, and GAGs.[51] These ECM components are organized into three distinct layers: the ventricularis, the spongiosa, and the fibrosa (Figure 6.3b). In the AV, the elastin layer is located on the ventricular side of the leaflets, and the collagen-rich fibrosa is on the aortic side. The spongiosa is situated between these two layers and is composed mainly of GAGs that help direct collagen orientation during opening and closing of the valve.[52] This structural alignment of the ECM within the leaflets allows them to withstand repeated cardiac strains caused by pressure of blood backflow during diastole.

To completely coapt during diastole and withstand the force of backflow, the leaflet must expand extensively in the radial direction, and have an almost immediate stress response at low strain levels in the circumferential direction.[53] The collagen fibers, which provide the greatest amount of structural support, align in the circumferential direction and this amounts to a cyclic, repeated strain of 10%–20% in the circumferential direction with each diastolic cycle.[50] As with the MV during systole, the collagen in the AV quickly straightens during diastole, conferring structural rigidity to the leaflets. Leaflet deformation and straightening of the collagen causes the elastin of the ventricularis layer to stretch such that the three leaflets meet to seal the valve.[54,55] Once the diastolic pressure reaches 4 mm Hg, the collagen fibers completely align, and the valve closes completely at 20 mm Hg. Few leaflet changes are observed from 4 mm Hg to the peak diastolic pressure of 80 mm Hg.[52] During systolic opening of the valve, the TVP difference drops to zero causing the elongated elastin to recoil, which results in crimping of the collagen fibers and unloading of the AV leaflets.

6.3.2 CELLULAR MECHANOBIOLOGY OF AORTIC VALVES

In other parts of the vascular system, endothelial cells have been observed to play a crucial role in the transduction of mechanical signals such as changes in shear stress.[56] Because these cells are the first to experience any changes in outside forces that affect valve dynamics, they have received growing attention in their potential roles in regulating valve homeostasis. VECs are phenotypically distinct from other endothelia cells in the vasculature.[28] For instance, when subjected to the shear stresses of fluid flow, VECs align perpendicular to the flow[28]; whereas, endothelial cells from the aorta align parallel to the flow.[56,57] This indicates that the VECs may play a distinct role in regulating the biomechanical properties of the valve through specially evolved mechanisms. Furthermore, AVECs have been observed to exhibit genotypic heterogeneity along the surface of each leaflet. Simmons et al. found that 584 genes were differentially expressed between AVECs from the aortic side and the ventricular side.[58] These genotypic differences may be important in regulating the differing biomechanical properties on each side of the valve, i.e., the collagen of the fibrosa and the elastin of the ventricularis, or conversely, the gene expression of these cells may be influenced by the differing shear stresses seen on each side of the leaflet. We will further discuss the implications of the AVEC heterogeneity in the section on AV pathologies.

Some have suggested that AVECs may regulate valve properties through complex paracrine signaling to the AVICs.[58–62] Butcher and Nerem showed that AVECs can prevent activation of the AVICs when the two cell types are grown in coculture.[59] In this study, porcine AVICs were seeded into leaflet model type I collagen hydrogels and subjected to 20 dyn/cm^2 of steady shear for up to 96 h. The presence of the shear stress resulted in AVIC proliferation as well as an increase in αSMA and a decrease in GAGs—indicative of myofibroblast activation. When AVECs were introduced onto the surface of these gels in coculture with the AVICs, these trends were completely reversed and the AVICs exhibited a more quiescent fibroblastic phenotype. This suggests that the AVECs may play a role in maintaining physiological homeostasis in the AV, and a loss or dysfunction of these cells may play a role in the onset and/or progression of AV disease.

Though the AVECs appear to play a crucial role in valve tissue maintenance, the tissue structure and mechanical properties of the valve seem to be regulated largely by the AVICs, the most abundant cell type in the AV. As mentioned before, these cells exhibit a plastic phenotype that alternates between a fibroblast-like state and a myofibroblast state depending on the needs of the tissue. Within the AV leaflets, AVICs are aligned with the collagen fibers in the circumferential direction, and studies have shown that they experience similar levels of strain, indicating that macro, tissue level deformations (strains) effectively transduce to the cellular level.[63] To continually respond to and monitor tissue level changes, AVICs are spread throughout the leaflet where they actively control and remodel the ECM to maintain its structural integrity. During times of tissue remodeling, AVICs become activated from their quiescent phenotype and gain smooth muscle cell characteristics.[64,65] This phenotypic change results in increased cellular stiffness and contractility due to the formation of αSMA fibers within the AVICs. Once activated, these cells also produce an excess of extracellular collagen that increases total tissue stiffness.[66,67]

The detailed cellular signaling pathways associated with AVIC myofibroblast activation are poorly understood; however, TGF-β1 activation and receptor binding has been identified as a major contributor to myofibroblast activation.[68–70] TGF-β1 is a member of the transforming growth factor superfamily of cytokines. TGF-β1 is secreted from cells in an inactive, latent form and stored as part of a large complex in the ECM. Various extracellular cues can result in the activation of TGF-β1 during which the active subunit of the protein breaks away from the latent complex. This active form then binds to and activates the TGF-β serine/threonine kinase receptors. These receptors then function through a canonical signaling pathway to modulate a wide variety of cellular processes such as apoptosis, proliferation, and differentiation by signaling through a family of transcription factors known as Smads.[71]

Using porcine AV leaflets in a tension bioreactor, the affects of TGF-β1 and cyclic stretch on myofibroblast activation were studied in vitro.[69] The measured outputs were αSMA, heat shock protein 47 (HSP47), type I collagen C-terminal propeptide (CICP), and TGF-β1. CICP and HSP47 are both surrogates for collagen biosynthesis.[72–74] The baseline control for this study was tissue in static culture that was not treated with active TGF-β1 (Null). Tissues receiving 15% stretch for 2 weeks (Tension) showed a significant increase in myofibroblast activation and collagen synthesis over the Null group and compared to day 0 controls (Figure 6.8). The same trend was observed for tissue samples in the treated daily with 0.5 ng/mL active TGF-β1 for 2 weeks (TGF). Most interestingly, the combination of these two treatments (Tension + TGF) resulted in a very significant increase in myofibroblast activation and matrix remodeling than either independent treatment. This suggests a synergism between TGF-β1 signaling and mechanical signal transduction. Furthermore, the large increase in TGF-β1 within the tissues indicates a feed-forward mechanism in which the AVICs respond to the combination of the two stimuli by producing even more TGF-β1. In vivo, this result may translate into an autocrine/paracrine signaling mechanism by which myofibroblasts produce TGF-β1 to remain activated in times of pathologic strain and signal for the activation of other AVICs to aid in tissue remodeling. In normal valves, the AVICs remain active until repair is complete and AV homeostasis is restored.[64] As discussed below, AV sclerosis may be a result of the overactivity of the AVICs resulting in a loss in AV compliance.

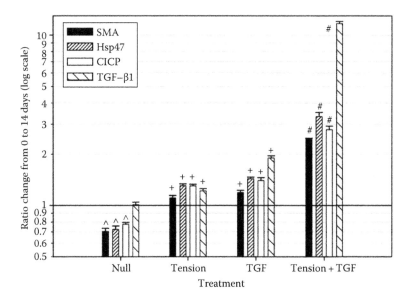

FIGURE 6.8 Effect of TGF-β1 and tension on AVICs. Both the presence of tension (Tension) and TGF-β1 (TGF) individually significantly increase markers of myofibroblast activation (SMA), ECM remodeling (Hsp47 and CICP), and the synthesis of native TGF-β1 (TGF-β1) over the nontreated controls (Null). A combined treatment of tension and TGF-β1 (Tension + TGF) showed a synergistic increase in all of the measured outcomes over either of the individual treatments. (Reprinted from Merryman, W.D. et al., *Cardiovasc. Pathol.*, 16, 268, 2007. With permission.)

6.3.3 AORTIC VALVE PATHOLOGIES AND CURRENT RESEARCH

AV disease can be classified as an inflammatory, congenital, or degenerative disease.[75] Until the mid-1980s, inflammatory valve pathologies such as those caused by rheumatic heart disease were much more common than degenerative diseases; however, with the eradication of such diseases as scarlet fever in the Western world, degenerative AV disease has become more prevalent in the United States, but inflammatory valve disease remains a major problem in developing countries.[42] Though a recent study has suggested that pathogen infections may also play a role in modulating the activity of AVICs and thus could lead to the development of AV disease, this is still an area of ongoing research.[76] Therefore, for continued discussion we will mainly focus on better-established AV pathologies with congenital and degenerative etiologies.

Congenital diseases are those of a genetic origin and are inherited at birth. The most frequent congenital valvular defect is the bicuspid AV (with a prevalence of approximately 1%).[77] This defect occurs due to the fusion of two of the three AV leaflets, leaving the individual with two functional leaflets. Early in life, this genetic malfunction does not cause problems in the majority of individuals; however, adults with this malformation are at much greater risk of developing degenerative valve pathologies.[52] The mechanisms through which this defect results in valve dysfunction is still largely unknown. Current theories associate the failing of the bicuspid valve to AV-mediated alterations of the aortic wall. The changes in aortic stress may then be transferred to the AV leaflets, which could exacerbate deleterious AV alterations—a theory explained in more detail below. High levels of MMPs have been observed in the aortic walls of patients with bicuspid AVs.[78] MMPs work by degrading ECM structural components and, in these cases, cause dilatation of the aortic wall. One study has also implicated changes in endothelial expression of nitric oxide in inducing alterations of the aortic wall.[79] Nitric oxide signaling causes smooth muscle cell relaxation, which also results in aortic dilatation. The link between the bicuspid leaflets and these signaling mechanisms are unclear. Regardless of the mechanism, however, the expanded aorta results in a higher volume of blood being forced back onto the leaflets during diastole. Accordingly, this increased pressure is transferred from the aorta to the cells within the AV leaflet, which mediate tissue remodeling.

For many years, the prevailing view among many researchers and clinicians was that AV degeneration was similar in nature to—if not identical to—the progression of atherosclerosis.[74] More recently, this view has been the subject of some controversy in the field. Both diseases involve the same risk factors and exhibit similar histological morphology; however, the AV leaflets lack smooth muscle cells, which play a significant role in atherosclerosis. Also, recent findings have shown that AV disease progression is more highly dependent on the interplay between the mechanical factors and inflammatory cytokines than the immunological processes that seem to mediate atherosclerosis. In fact, Robicsek et al. have observed that changes in the mechanical compliance of the aorta can lead to changes in AV function,[75] similar to the observations in the bicuspid AV studies. The reduced aortic compliance can lead to a significant stress overload on the AV leaflets, and thus result in increased matrix remodeling in the absence of other internal or external cues. Furthermore, this hyper-stressed diastolic state of the leaflets and resulting overactivity of the ECM remodeling mechanisms leads to a loss in the biomechanical integrity and functionality of the AV.

The most common degenerative AV pathology is known as AV sclerosis, which is characterized by thickening and stiffening of the valve leaflets. Changes in AV stiffness can lead to improper opening of the valve or insufficient coaptation of the AV leaflets to prevent backflow from the aorta to the LV.[80] The progression of this degenerative disease seems to occur over the course of years, and can lead to valve stenosis in the most serious cases. Stenosis occurs when the leaflets become so rigid that they are unable to coapt during diastole, resulting in partial occlusion of the valve. This pathology represents the most extreme case of AV degeneration and may be largely mediated by improper activity of the AVICs.[31,43,70,81]

Again, under normal physiological conditions AVICs exhibit a fibroblastic phenotype; however, the initiation of VHD has been found to be caused by the activation of AVICs to a myofibroblast phenotype.[5] As discussed above, AVICs can be activated in times of tissue development and repair. During these physiological circumstances, activated myofibroblasts remodel the ECM and return the tissue to homeostasis after which many of the myofibroblasts return to a quiescent state. The characteristic thickening observed in AV sclerosis occurs when these cells remain activated, creating an excess amount of collagen. This overactivation may occur due to a chronic physical change in the systemic circulation such as the stress transduction caused by the lost aortic compliance as observed by Robicsek et al.[75] (see above). The constant increased diastolic pressure on the tissue is transferred to the AVICs within the leaflets. As discussed before, this heightened mechanical stimulation may lead to AVIC activation and release of important signaling cytokines such as TGF-β1. The combination of a sustained pathological mechanical load and a constant increase in the production of TGF-β1 may lie at the root of the cellular cause of degenerative AV disease.[82]

VECs have also been shown to play a role in the etiology and progression of valve disease. In diseased or injured vascular tissues, endothelial cells direct the appropriate immune response by upregulating cell adhesion molecules (CAMs) that mediate platelet attachment and activation, as well as by secreting cytokines that are recognized by specialized inflammatory cells.[44–47] Similarly, AVECs have been observed to increase expression of CAMs (ICAM-1 and VCAM-1) and secretion of inflammatory cytokines such as IL-8, IL-1β and IFN-γ in response to oscillatory, turbulent fluid flows and pathological mechanical strains.[60] The exact mechanisms by which the AVECs may contribute to pathogenesis in heart valves remain unclear; however, these cells may mediate tissue changes through paracrine signaling to the AVICs within the valve. In support of this theory, IL-1β has also been found in the cells within the valvular interstitium and may be important in modulating the overall disease progression.[43]

It is also important to note that the decreased leaflet compliance that is associated with degenerative AV disease occurs due to an increase in type I collagen and the formation of calcium deposits on the aortic side of the valve leaflets.[83,84] This is intuitive because the mechanical load placed on the leaflets during diastole occurs due to a backflow in blood from the ascending aorta; therefore, this stress would be transferred to the AVICs on the aortic side of the valve. However, the AVECs may also play an important, though lesser known, function in the degenerative effects. The spatial heterogeneity of these cells may cause side-specific susceptibility to AV calcification.[58] The AVECs directly adjacent to the fibrosa have been found to express much lower levels of anti-calcification enzymes when compared to the AVECs on the ventricular side of the leaflets.[60] This suggests that the AVECs may also play an important role in sensing and responding to the various stresses on each side of the leaflets. Though the AVICs may be more directly involved in the tissue changes that are observed in AV degeneration, signaling from the AVECs may direct AVIC phenotype in vivo through the differential expression of signaling cytokines.

6.4 CONCLUDING REMARKS

The heart is the most dynamic organ in the human body. It relies on a coordinated action of chemical and electrical stimuli to produce the necessary mechanical responses needed to provide blood to the outermost regions of the body. Within the heart, the four valves work to direct blood flow and these heart valves rely heavily on their biomechanical properties to ensure proper function; therefore, the cells that compose these valves are extremely interesting case studies in the field of mechanobiology. Recent studies have begun to connect the effects of changes in the mechanical load on the valves to the biological response of the cells within the tissues. Throughout this chapter, we have focused on the important biomechanical features of the ECM of the valve tissues. We have also highlighted how the mechanical forces placed on the valve tissues can be transduced to the cellular level, which can lead to a feedback mechanism by which the cells actively work to remodel the tissues. In all valves, VECs likely play an important role in recruiting inflammatory cells to the surface of the leaflet and translating shear stress on the tissue into signaling cytokines that can modulate the activity of the VICs. The VICs themselves regulate tissue properties through secretion of ECM components in response to changes in tissue strain and signaling cytokines (Figure 6.9).

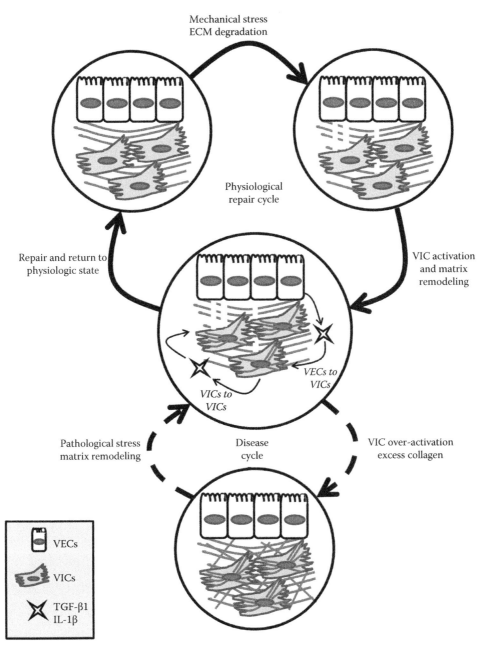

FIGURE 6.9 Hypothesized response of VECs and VICs to increased mechanical stress. The valve leaflets are composed of a layer of VECs with an underlying layer of VICs that maintain the ECM components. Heightened mechanical stresses may result in paracrine signaling from the VECs to the VICs and between different VICs. This causes the VICs to become activated, whereupon they begin the process of remodeling the ECM within the valve tissues. Under normal conditions, the valve tissues return to homeostasis after the ECM has been repaired (Physiological Repair Cycle); however, in disease state, the VICs remain constitutively active resulting in an accumulation in extracellular collagen (Disease Cycle). This disease cycle is likely a feed-forward mechanism as stiffer leaflets lead to greater stress applied to the VICs, which in turn leads to more ECM production and stiffer leaflets.

Under normal conditions, these processes would persist until valve remodeling is complete; whereupon, the valve would return to its physiologic state. However, the overactivity of these signaling mechanisms has been shown to affect a variety of HVDs. A better understanding of this complex interplay between mechanical stimulation as the signaling mechanisms utilized by the various cells that compose the valves may lead to the development of less invasive therapies and preventative techniques for heart valve disease.

ACKNOWLEDGMENTS

MPN was supported in part by an Ireland Tuition Scholarship from the University of Alabama at Birmingham. JDH was supported as an American Heart Association Pre-doctoral Fellow (10PRE4290020). WDM's research is supported by the American Heart Association (09GRNT2010125 and 0835496N), the Wallace H. Coulter Foundation (Early Career Award), and the National Institutes of Health (HL094707).

REFERENCES

1. Sacks, M.S. and A.P. Yoganathan, Heart valve function: A biomechanical perspective. *Philos Trans R Soc Lond B Biol Sci*, 2007, **362**(1484): 1369–1391.
2. Yoganathan, A.P., Z. He, and S. Casey Jones, Fluid mechanics of heart valves. *Annu Rev Biomed Eng*, 2004, **6**: 331–362.
3. Lloyd-Jones, D. et al., Heart disease and stroke statistics—2009 update: A report from the American Heart Association Statistics Committee and Stroke Statistics Subcommittee. *Circulation*, 2009, **119**(3): 480–486.
4. Merryman, W.D. et al., Correlation between heart valve interstitial cell stiffness and transvalvular pressure: Implications for collagen biosynthesis. *Am J Physiol Heart Circ Physiol*, 2006, **290**(1): H224–H231.
5. Rabkin-Aikawa, E. et al., Dynamic and reversible changes of interstitial cell phenotype during remodeling of cardiac valves. *J Heart Valve Dis*, 2004, **13**(5): 841–847.
6. Filip, D.A., A. Radu, and M. Simionescu, Interstitial cells of the heart valves possess characteristics similar to smooth muscle cells. *Circ Res*, 1986, **59**(3): 310–320.
7. Hochmuth, R.M., Micropipette aspiration of living cells. *J Biomech*, 2000, **33**(1): 15–22.
8. Sato, M. et al., Application of the micropipette technique to the measurement of cultured porcine aortic endothelial cell viscoelastic properties. *J Biomech Eng*, 1990, **112**(3): 263–268.
9. Banasik, J.L., *Pathophysiology*, 3rd edn., 2005, Philadelphia, PA: Elsevier Saunders.
10. Kunzelman, K.S., D.R. Einstein, and R.P. Cochran, Fluid-structure interaction models of the mitral valve: Function in normal and pathological states. *Philos Trans R Soc Lond B Biol Sci*, 2007, **362**(1484): 1393–1406.
11. Obadia, J.F. et al., Mitral subvalvular apparatus: Different functions of primary and secondary chordae. *Circulation*, 1997, **96**(9): 3124–3128.
12. Chen, L., F.C. Yin, and K. May-Newman, The structure and mechanical properties of the mitral valve leaflet-strut chordae transition zone. *J Biomech Eng*, 2004, **126**(2): 244–251.
13. Kunzelman, K.S. et al., Differential collagen distribution in the mitral valve and its influence on biomechanical behavior. *J Heart Valve Dis*, 1993, **2**(2): 236–244.
14. Liao, J. et al., The relation between collagen fibril kinematics and mechanical properties in the mitral valve anterior leaflet. *J Biomech Eng*, 2007, **129**(1): 78–87.
15. Sacks, M.S. et al., Surface strains in the anterior leaflet of the functioning mitral valve. *Ann Biomed Eng*, 2002, **30**(10): 1281–1290.
16. He, Z. et al., In vitro dynamic strain behavior of the mitral valve posterior leaflet. *J Biomech Eng*, 2005, **127**(3): 504–511.
17. Kroeze, W.K., D.J. Sheffler, and B.L. Roth, G-protein-coupled receptors at a glance. *J Cell Sci*, 2003, **116**(Pt 24): 4867–4869.
18. Sacks, M.S. et al., In-vivo dynamic deformation of the mitral valve anterior leaflet. *Ann Thorac Surg*, 2006, **82**(4): 1369–1377.
19. Jimenez, J.H. et al., Mechanics of the mitral valve: In vitro studies. *Conf Proc IEEE Eng Med Biol Soc*, 2004, **5**: 3727–3729.

20. Liao, J. and I. Vesely, Relationship between collagen fibrils, glycosaminoglycans, and stress relaxation in mitral valve chordae tendineae. *Ann Biomed Eng*, 2004, **32**(7): 977–983.
21. Millington-Sanders, C. et al., Structure of chordae tendineae in the left ventricle of the human heart. *J Anat*, 1998, **192**(Pt 4): 573–581.
22. Ritchie, J. et al., The material properties of the native porcine mitral valve chordae tendineae: An in vitro investigation. *J Biomech*, 2006, **39**(6): 1129–1135.
23. Schonherr, E. et al., Decorin-type I collagen interaction: Presence of separate core protein-binding domains. *J Biol Chem*, 1995, **270**(15): 8877–8883.
24. Grande-Allen, K.J. et al., Glycosaminoglycans and proteoglycans in normal mitral valve leaflets and chordae: Association with regions of tensile and compressive loading. *Glycobiology*, 2004, **14**(7): 621–633.
25. Gupta, V. et al., Synthesis of glycosaminoglycans in differently loaded regions of collagen gels seeded with valvular interstitial cells. *Tissue Eng*, 2007, **13**(1): 41–49.
26. Blevins, T.L. et al., Phenotypic characterization of isolated valvular interstitial cell subpopulations. *J Heart Valve Dis*, 2006, **15**(6): 815–822.
27. Blevins, T.L. et al., Mitral valvular interstitial cells demonstrate regional, adhesional, and synthetic heterogeneity. *Cells Tissues Organs*, 2008, **187**(2): 113–122.
28. Butcher, J.T. et al., Unique morphology and focal adhesion development of valvular endothelial cells in static and fluid flow environments. *Arterioscler Thromb Vasc Biol*, 2004, **24**(8): 1429–1434.
29. Gupta, V. et al., Reversible secretion of glycosaminoglycans and proteoglycans by cyclically stretched valvular cells in 3D culture. *Ann Biomed Eng*, 2008, **36**(7): 1092–1103.
30. Mulholland, D.L. and A.I. Gotlieb, Cell biology of valvular interstitial cells. *Can J Cardiol*, 1996, **12**(3): 231–236.
31. Taylor, P.M. et al., The cardiac valve interstitial cell. *Int J Biochem Cell Biol*, 2003, **35**(2): 113–118.
32. Rabkin, E. et al., Evolution of cell phenotype and extracellular matrix in tissue-engineered heart valves during in-vitro maturation and in-vivo remodeling. *J Heart Valve Dis*, 2002, **11**(3): 308–314; discussion 314.
33. Stephens, E.H. et al., Fibronectin-based isolation of valve interstitial cell subpopulations: Relevance to valve disease. *J Biomed Mater Res A*, 2010, **92**(1): 340–349.
34. Valentin Fuster, M. et al., *Hurst's The Heart*, 11th edn., 2004, New York, NY: McGraw-Hill Professional.
35. Jimenez, J.H. et al., Effects of annular size, transmitral pressure, and mitral flow rate on the edge-to-edge repair: An in vitro study. *Ann Thorac Surg*, 2006, **82**(4): 1362–1368.
36. Nielsen, S.L. et al., Imbalanced chordal force distribution causes acute ischemic mitral regurgitation: Mechanistic insights from chordae tendineae force measurements in pigs. *J Thorac Cardiovasc Surg*, 2005, **129**(3): 525–531.
37. Nielsen, S.L. et al., Chordal force distribution determines systolic mitral leaflet configuration and severity of functional mitral regurgitation. *J Am Coll Cardiol*, 1999, **33**(3): 843–853.
38. Nielsen, S.L. et al., Mechanism of incomplete mitral leaflet coaptation—Interaction of chordal restraint and changes in mitral leaflet coaptation geometry. Insight from in vitro validation of the premise of force equilibrium. *J Biomech Eng*, 2002, **124**(5): 596–608.
39. Grande-Allen, K.J. et al., Apparently normal mitral valves in patients with heart failure demonstrate biochemical and structural derangements: An extracellular matrix and echocardiographic study. *J Am Coll Cardiol*, 2005, **45**(1): 54–61.
40. Firstenberg, M.S. et al., Noninvasive estimation of transmitral pressure drop across the normal mitral valve in humans: Importance of convective and inertial forces during left ventricular filling. *J Am Coll Cardiol*, 2000, **36**(6): 1942–1949.
41. Stephens, E.H. et al., The effects of mitral regurgitation alone are sufficient for leaflet remodeling. *Circulation*, 2008, **118**(14 Suppl): S243–S249.
42. Otto, C.M., *Valvular Heart Disease*, 2004, Philadelphia, PA: Saunders.
43. Rabkin, E. et al., Activated interstitial myofibroblasts express catabolic enzymes and mediate matrix remodeling in myxomatous heart valves. *Circulation*, 2001, **104**(21): 2525–2532.
44. Barakat, A.I. and P.F. Davies, Mechanisms of shear stress transmission and transduction in endothelial cells. *Chest*, 1998, **114**(1 Suppl): 58S–63S.
45. Davies, P.F. and S.C. Tripathi, Mechanical stress mechanisms and the cell. An endothelial paradigm. *Circ Res*, 1993, **72**(2): 239–245.
46. Ingber, D.E., Mechanobiology and diseases of mechanotransduction. *Ann Med*, 2003, **35**(8): 564–577.
47. Jaalouk, D.E. and J. Lammerding, Mechanotransduction gone awry. *Nat Rev Mol Cell Biol*, 2009, **10**(1): 63–73.

48. Kimura, N. et al., Local tenomodulin absence, angiogenesis, and matrix metalloproteinase activation are associated with the rupture of the chordae tendineae cordis. *Circulation*, 2008, **118**(17): 1737–1747.
49. Soini, Y., T. Salo, and J. Satta, Angiogenesis is involved in the pathogenesis of nonrheumatic aortic valve stenosis. *Hum Pathol*, 2003, **34**(8): 756–763.
50. Thurbrikar, M., *The Aortic Valve*, 1990, Boca Raton, FL: CRC Press.
51. Schoen, F., Aortic valve structure-function correlations: Role of elastic fibers no longer a stretch of the imagination. *J Heart Valve Dis*, 1997, **6**: 1–6.
52. Schoen, F.J., Evolving concepts of cardiac valve dynamics: The continuum of development, functional structure, pathobiology, and tissue engineering. *Circulation*, 2008, **118**(18): 1864–1880.
53. Thubrikar, M., *The Aortic Valve*, 1990, Boca Raton, FL: CRC Press, 221.
54. Vesely, I. and R. Noseworthy, Micromechanics of the fibrosa and the ventricularis in aortic valve leaflets. *J Biomech*, 1992, **25**(1): 101–113.
55. Kershaw, J.D. et al., Specific regional and directional contractile responses of aortic cusp tissue. *J Heart Valve Dis*, 2004, **13**(5): 798–803.
56. Stamatas, G.N. and L.V. McIntire, Rapid flow-induced responses in endothelial cells. *Biotechnol Prog*, 2001, **17**(3): 383–402.
57. Cucina, A. et al., Shear stress induces changes in the morphology and cytoskeleton organisation of arterial endothelial cells. *Eur J Vasc Endovasc Surg*, 1995, **9**(1): 86–92.
58. Simmons, C.A. et al., Spatial heterogeneity of endothelial phenotypes correlates with side-specific vulnerability to calcification in normal porcine aortic valves. *Circ Res*, 2005, **96**(7): 792–799.
59. Butcher, J.T. and R.M. Nerem, Valvular endothelial cells regulate the phenotype of interstitial cells in co-culture: Effects of steady shear stress. *Tissue Eng*, 2006, **12**(4): 905–915.
60. Butcher, J.T. and R.M. Nerem, Valvular endothelial cells and the mechanoregulation of valvular pathology. *Philos Trans R Soc Lond B Biol Sci*, 2007, **362**(1484): 1445–1457.
61. El-Hamamsy, I. et al., Endothelium-dependent regulation of the mechanical properties of aortic valve cusps. *J Am Coll Cardiol*, 2009, **53**(16): 1448–1455.
62. Simmons, C.A., Aortic valve mechanics: An emerging role for the endothelium. *J Am Coll Cardiol*, 2009, **53**(16): 1456–1458.
63. Merryman, W.D., Mechanobiology of the aortic valve interstitial cell, doctoral dissertation Bioengineering, 2007, University of Pittsburgh: Pittsburgh, PA, p. 172.
64. Chester, A.H. and P.M. Taylor, Molecular and functional characteristics of heart-valve interstitial cells. *Philos Trans R Soc Lond B Biol Sci*, 2007, **362**(1484): 1437–1443.
65. Hinz, B. et al., The myofibroblast: One function, multiple origins. *Am J Pathol*, 2007, **170**(6): 1807–1816.
66. Desmouliere, A., Factors influencing myofibroblast differentiation during wound healing and fibrosis. *Cell Biol Int*, 1995, **19**(5): 471–476.
67. Desmouliere, A. et al., Transforming growth factor-beta 1 induces alpha-smooth muscle actin expression in granulation tissue myofibroblasts and in quiescent and growing cultured fibroblasts. *J Cell Biol*, 1993, **122**(1): 103–111.
68. Grinnell, F. and C.H. Ho, Transforming growth factor beta stimulates fibroblast-collagen matrix contraction by different mechanisms in mechanically loaded and unloaded matrices. *Exp Cell Res*, 2002, **273**(2): 248–255.
69. Merryman, W.D. et al., Synergistic effects of cyclic tension and transforming growth factor-beta1 on the aortic valve myofibroblast. *Cardiovasc Pathol*, 2007, **16**(5): 268–276.
70. Walker, G.A. et al., Valvular myofibroblast activation by transforming growth factor-beta: Implications for pathological extracellular matrix remodeling in heart valve disease. *Circ Res*, 2004, **95**(3): 253–260.
71. Howe, P.H., Transforming growth factor beta, in *The Cytokine Handbook*, A.W. Thomson and M.T. Lotze, editors, 2003, London, U.K.: Elsevier Science. pp. 1119–1141.
72. Rocnik, E.F. et al., Functional linkage between the endoplasmic reticulum protein Hsp47 and procollagen expression in human vascular smooth muscle cells. *J Biol Chem*, 2002, **277**(41): 38571–38578.
73. Tasab, M., M.R. Batten, and N.J. Bulleid, Hsp47: A molecular chaperone that interacts with and stabilizes correctly-folded procollagen. *Embo J*, 2000, **19**(10): 2204–2211.
74. Sauk, J.J., N. Nikitakis, and H. Siavash, *Hsp47:* A novel collagen binding serpin chaperone, autoantigen and therapeutic target. *Front Biosci*, 2005, **10**: 107–118.
75. Robicsek, F., M.J. Thubrikar, and A.A. Fokin, Cause of degenerative disease of the trileaflet aortic valve: Review of subject and presentation of a new theory. *Ann Thorac Surg*, 2002, **73**(4): 1346–1354.
76. Shun, C.T. et al., Activation of human valve interstitial cells by a viridians streptococci modulin induces chemotaxis of mononuclear cells. *J Infect Dis*, 2009, **199**(10): 1488–1496.

77. Fedak, P.W. et al., Bicuspid aortic valve disease: Recent insights in pathophysiology and treatment. *Expert Rev Cardiovasc Ther*, 2005, **3**(2): 295–308.

78. Fedak, P.W. et al., Vascular matrix remodeling in patients with bicuspid aortic valve malformations: Implications for aortic dilatation. *J Thorac Cardiovasc Surg*, 2003, **126**(3): 797–806.

79. Aicher, D. et al., Endothelial nitric oxide synthase in bicuspid aortic valve disease. *Ann Thorac Surg*, 2007, **83**(4): 1290–1294.

80. Schoen, F.J., Cardiac valves and valvular pathology: Update on function, disease, repair, and replacement. *Cardiovasc Pathol*, 2005, **14**(4): 189–194.

81. Berk, B.C., K. Fujiwara, and S. Lehoux, ECM remodeling in hypertensive heart disease. *J Clin Invest*, 2007, **117**(3): 568–575.

82. Merryman, W.D., Insights into (the interstitium of) degenerative aortic valve disease. *J Am Coll Cardiol*, 2008, **51**(14): 1415.

83. Yip, C.Y. et al., Calcification by valve interstitial cells is regulated by the stiffness of the extracellular matrix. *Arterioscler Thromb Vasc Biol*, 2009, **29**(6): 936–942.

84. Stewart, B.F. et al., Clinical factors associated with calcific aortic valve disease. Cardiovascular Health Study. *J Am Coll Cardiol*, 1997, **29**(3): 630–634.

7 Mechanobiology of Cardiac Fibroblasts

Peter A. Galie and Jan P. Stegemann

CONTENTS

7.1 INTRODUCTION

The healthy myocardium contains cardiomyocytes, mast cells, fibroblasts, and other vascular cell types. Cardiac fibroblasts constitute the largest cell population in the heart and therefore understanding how these cells function is essential for studying heart disease. Recent research has focused on how cardiac fibroblasts affect a broad range of cardiac pathologies: arrhythmias, dilated cardiomyopathy, cardiac hypertrophy, and other systolic- and diastolic-related diseases. In order to understand how cardiac fibroblasts contribute to these various disease states, it is necessary to learn how these cells respond to changes in biochemical and mechanical stimuli. Both have important impact on cell function, and this chapter focuses primarily on the effect of the mechanical environment. In addition, it addresses relevant biochemical factors, since the mechanical and biochemical environments are intimately linked. The process by which cells transduce mechanical signals into a biological response has been termed mechanobiology. The mechanobiology of cardiac fibroblasts has been the subject of extensive research in recent years, the results of which have had direct implications for the treatment of heart disease. This chapter provides an overview of the mechanical forces present in the myocardium, the cardiac fibroblast phenotype, and presents key recent results in the area of cardiac fibroblast mechanobiology.

The mechanical environment in the myocardium is constantly in flux due to the cyclical function of the organ. The cells of the myocardium are exposed to a range of stresses and strains that originate from the contraction of the tissue, shearing between lamella of cardiomyocytes and interstitial fluid flow through the myocardium. The range of mechanical stress can be drastically affected by pathology, which alters not only the mechanical environment but also the structural organization of the myocardial tissue. Changes in tissue structure, whether by increasing the stiffness or compliance of the extracellular matrix, have direct influence on the stresses and strains experienced by cardiac cells. Diseases that are associated with extensive remodeling and fibrosis provide a prime example. Cells in a fibrotic environment are exposed to significantly higher levels of stress due to larger levels of cross-linked collagen in the extracellular matrix. These changes in mechanical environment may perpetuate the problem,since cardiac fibroblasts are the cells that are primarily responsible for scar formation in fibrosis. As this example demonstrates, understanding the mechanobiology of cardiac fibroblasts has direct implications for the study of heart disease.

During the last decade, a variety of studies have provided an insight into how cardiac fibroblasts transduce mechanical stimuli into biological responses. Such responses typically include increased expression of extracellular matrix (ECM) proteins, release of growth factors and cytokines, and alignment of actin and intermediate filaments in the cytoskeleton. The biology of cardiac fibroblasts is complicated by the phenotypic fluidity of these cells. Though a healthy myocardium largely comprises cardiac fibroblasts, some environmental conditions can promote the differentiation of the cells to myofibroblasts. This contractile phenotype is characterized by increased collagen deposition and prominent stress fibers in the cytoskeleton. The protein smooth muscle alpha-actin (SMA), a component of the contractile apparatus, is often used as a marker for this transition. However, it should be noted that there is a spectrum of phenotypes between fibroblasts and myofibroblasts. It is now established that the phenotype of cardiac fibroblasts is strongly influenced by both biochemical and mechanical factors. Changing both the type and the magnitude of mechanical stress applied to cardiac fibroblasts has been shown to have direct effects on this phenotypic transition. This dependence of cell function on the surrounding mechanical environment has direct implications for heart disease and provides further impetus for studying the mechanobiology of these cells.

7.2 MECHANICAL ENVIRONMENT IN THE MYOCARDIUM

Cardiac fibroblasts are exposed to a complex mechanical environment that includes tensile, compressive, and shear stresses. These stresses are exerted by the contraction of surrounding cardiomyocytes, interaction with the microstructure of the extracellular environment, as well as interstitial fluid flow within the myocardium. Due to the anisotropy and dynamic nature of the heart, these stresses vary both spatially and temporally and are coupled with one another. Figure 7.1 illustrates this unique stress state. Due to the complexity of the mechanical milieu in the heart, computational and analytical modeling is an important tool in studying both the cell and tissue behavior of the

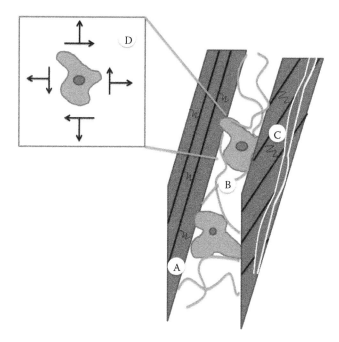

FIGURE 7.1 Mechanical environment in the myocardium: (A) laminar sheet, (B) connecting fiber between lamina, (C) intercalating disks and capillary, and (D) normal and shear stresses exerted on resident cardiac fibroblasts.

myocardium. Experimental approaches to measuring strains in the heart also have been developed,[1] though such experiments are technically challenging.

Nonetheless, a substantial amount of work has studied the complex structure of the myocardium, which is one of the primary determinants of the stress state. The structure of mammalian heart tissue is largely anisotropic, yet highly ordered. The wall consists of repeating laminar sheets, each approximately four myocytes thick.[2] The myocytes within each lamina are connected through intercalated disks, which contain both gap junctions and desmosomes to link adjacent cells electrically and mechanically. These branching muscle fibers are surrounded by fibrillar ECM proteins like collagen I and III, as well as fibroblasts and the endothelial cells of capillaries. The myocytes branch between laminae, but these connections are less frequent. This affects the mechanics of the heart in addition to how the wall is depolarized. The existence of discrete laminae causes significant shear to adjacent layers and causes the conduction velocity to be two to three times greater along the plane of the laminae compared to the transverse plane.[2] Collagen fibers also connect adjacent layers, though these fibers are often thin and uncoiled, allowing for shear strain.[2–4]

The overall architecture of the laminae is also ordered. The myocardium exhibits subtle differences between fiber orientation within the sheets and the orientation of the sheets themselves with respect to an orthogonal coordinate system. Fiber angles vary from −60° at the endocardium to +60° at the epicardium, creating a helical pattern to the radial direction of the wall.[3] However, the sheet angles of the lamina are restricted to either −45° or +45° and alternate depending upon the location of the sheet in the myocardium.[3] Because the sheet orientations are alternating, the laminae slide along one another in different directions even though the magnitude of the shear deformation is uniform across the wall.[3] This leads to thickening of the wall in systole. Interestingly, research has indicated the orientation of these sheets aligns with planes of maximum shear strain.[1,4] During systole the heart produces ejection fractions of 60%–70%, which is accompanied by radial thickening of the heart wall of around 40%. However, the maximum contraction of an individual myocyte is limited to approximately 15%.[3] Unfortunately, there has not yet been a study to determine the approximate strain on a cardiac fibroblast. The discrepancy in dimensional changes is thought to be a result of the shear exerted by slip between laminae in the myocardium, which contributes to the orientation of the cells within the sheets.[1–3]

Fibroblasts in the myocardium are also exposed to substantial fluid stresses from interstitial fluid flow.[5–8] Because of the large deformation of the heart wall during the cardiac cycle, the interstitial fluid undergoes substantial flux and consequently exerts a pressure and shear force on cells within the myocardium. The fluid balance within the myocardium is controlled by the perfusion of the coronary artery and corresponding microcirculation as well as by lymphatic drainage that transports the products of cell metabolism.[9] The factors that control fluid balance are strongly dependent on one another. For example, the pressure of the interstitial fluid is a regulator of the rate of perfusion. This effect is manifested during systole, when perfusion in the coronary capillaries is impeded due to the high pressure existing in the contracting myocardium. In addition, the myocardium separates two different bodies of fluid: the endocardium surrounds the blood-filled lumen, and the pericardial fluid surrounds the epicardium. Consequently, there is a pressure gradient in the radial direction of the myocardium that may influence interstitial fluid flow.[10]

The cardiac mechanical environment is strongly affected by, and may be a contributing factor to, the damaging effects of various pathologies. Many of the pathologies affecting the heart, including acute infarction, dilated cardiomyopathy, and others, are associated with changes in the extracellular matrix. These conditions affect the interaction between the sheets of cardiomyocytes and consequently the stresses exerted on resident cells like fibroblasts. In the case of acute myocardial infarction, the heart progresses through ischemic, necrotic, fibrotic, and remodeling phases that are each characterized by different mechanical properties.[11] The necrotic phase is characterized by edema, which affects transport of the interstitial fluid within the myocardium. During the fibrotic phase, the mechanics are influenced by excess collagen, the cross-linking of which changes the mechanical environment during remodeling.[11] In a canine model of infarction, researchers found that the twisting motion of the heart during systole was significantly reduced by fibrosis, indicating a substantial change in cardiac mechanics.[12] Other studies have found that the wall stress increases with the size of a remodeling left ventricle, indicating that excess collagen increases the stress on cells in the myocardium.[13] Cardiac fibroblasts play a central role in remodeling processes, and the cell phenotype may be influenced by the changes associated with fibrosis. As mentioned, these cells can transition into a myofibroblast phenotype that exhibits pro-fibrotic cell functions including augmented matrix production and contractile intracellular mechanisms. The relationship between increased stress from fibrosis and changes in fibroblast phenotype has important implications for cardiac pathologies.

Recent studies have focused on computational modeling of the mechanical environment experienced by cells in the myocardial wall. However, similarly to the experimental approach, modeling is made difficult by the complex microstructure of the tissue. Hence, validation of these computational models is also a substantial obstacle, though some empirical data is available.[1,4] Moreover, the tissue undergoes finite, time-dependent deformation that depends upon interaction between the fluid mechanics of blood in the lumen as well as interstitial flow within the tissue and the structural mechanics of the heart wall. Nonetheless, useful models have been constructed for elucidating the stresses applied to cells in the heart wall. For example, one recent model utilizes an arbitrary Lagrangian–Eulerian reference frame to model the fluid-structure interaction present in the heart. Using finite element analysis, this study estimated the stress in the left ventricle wall, and confirmed the thickening of the wall during systole.[14] In another study, a model was created to predict the planes of maximum shear strain in the heart laminae.[1] Although complex, such modeling is necessary to complement empirical studies to discern the exact mechanical environment surrounding cardiac fibroblasts in the heart.

7.3   CARDIAC FIBROBLASTS: A FLUID PHENOTYPE

Fibroblasts in the myocardium are similar in structure and function to fibroblasts in other parts of the body; they produce vital components of the extracellular matrix, and by doing so serve important roles in maintaining tissue structure. These cells also are able to take on a contractile, smooth muscle-like phenotype. In the skin, this transition from fibroblast to myofibroblast is a crucial step in wound healing. However, in the heart, this transition occurs in fibrosis and can lead to hypertrophy of the wall and eventual heart failure. In a healthy myocardium, the ratio of myofibroblasts to fibroblasts is very low. However, the distinction between these two phenotypes is not entirely clear. Though increased ECM production and expression of SMA are often utilized as markers of the transition, cardiac fibroblast phenotype comprises a continuous spectrum of states. The most common markers for the myofibroblast phenotype are illustrated in Figure 7.2. Studies have shown that mechanical stress can influence the location of the cell in this spectrum, and the pathways by which these cells transduce mechanical signals to biological changes have been extensively studied.

Directly measuring the cellular response to mechanical load is complicated by several factors that affect the phenotype of cardiac fibroblasts. For example, cardiac fibroblasts isolated from tissue typically begin to spontaneously differentiate to a myofibroblast phenotype. Studies have shown that this transition may be related to increased activation of protein kinase A through the G protein coupled receptor–adenylyl cyclase–cAMP pathway.[15] Moreover, the cells appear to be sensitive to the concentration of serum used in the culture medium. Serum deprivation has been shown to not only affect the expression of SMA,[15] but also the viability and ECM production of cardiac fibroblasts.[16,17] Another recent study observed that the response of cardiac fibroblasts to cyclic mechanical loading depended upon the serum concentration of the culture media.[18] The data indicated that the cell response to cyclic loading was negligible when cultured in 1% FBS, but became significant in the presence of 10% serum. In this case, the cellular response was characterized by the amount of procollagen produced by the cells, as well as labeled thymidine incorporation as a measure of cell proliferation.

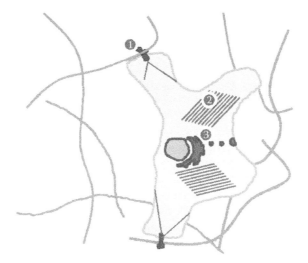

FIGURE 7.2 Markers for the myofibroblast transition: (1) augmented contractility, (2) expression of alpha-SMA, and (3) increased production of extracellular proteins.

As these results indicate, understanding the relationship between mechanical transduction and the action of biochemical factors is vital to understanding the phenotype response of cardiac fibroblasts. Serum contains a host of growth factors that affect the cardiac fibroblast, so modifying serum concentrations does not indicate specifically what factors influence phenotype. Nonetheless, other studies have investigated how several growth factors, especially those associated with fibrosis, affect the response of cardiac fibroblasts to mechanical loading. For example, angiotensin II is an important peptide in heart disease, since elevated levels are associated with fibrosis and heart failure.[19–23] Studies have shown that not only are angiotensin II and other components of the RAAS system present in the myocardium, but also that static 20% strain initially reduces the expression of angiotensinogen, a key RAS substrate.[24] Hence, these results indicate that static stretch diminishes the presence of the myofibroblast phenotype, since angiotensin II is an important factor in this fibroblast–myofibroblast transition. Moreover, it provides a clear example by which mechanical stimuli in the myocardium influences the biochemical environment.

There has been extensive research on the effect of different biochemical factors on cardiac fibroblast phenotype. Factors as diverse as glucose, osteopontin and angiotensin II have been studied for their effects on the fibroblast–myofibroblast transition.[25–30] Another important biochemical factor that affects the phenotype of cardiac fibroblasts is the oxygen concentration of the environment surrounding the cells. The change in oxygen concentration has been speculated as the impetus for the transition from fibroblasts to myofibroblasts in culture, since oxygen concentrations in surrounding tissue (~14%) are significantly less than the atmospheric conditions encountered once isolated (~21%). Moreover, hypoxic conditions can reduce oxygen concentrations to as low as 3%.[31] Research by Roy et al. has shown that cardiac fibroblasts respond to surrounding hyperoxic conditions through a specific signaling pathway involving induced p21, cyclin D1, cyclin D2, cyclin G1, Fos-related antigen-2, and TGF-beta.[31] This finding is nonintuitive because it indicates the myofibroblast transition is initiated by both ischemia and isolation to oxygen concentrations higher than normal tissue levels. The authors of this study speculate that cardiac fibroblasts are able to reset their normoxic oxygen concentration to much lower levels in hypoxic conditions, so that a return to the 14% concentration is sensed not as a return to normoxia, but rather a hyperoxic condition. Hence, as the heart is reperfused, this triggers the myofibroblast transition, in a manner similar to cells isolated and plated in atmospheric oxygen levels. Further research by the group indicated that blocking this p21 associated pathway attenuates the expression of SMA, effectively maintaining a fibroblast phenotype even in the presence of hyperoxic conditions.[32] Others have postulated that reactive oxygen species also play a central role in producing the myofibroblast phenotype. Cucoranu et al. found evidence that NADPH oxidase 4 is significantly upregulated in cardiac fibroblasts stimulated with TGF-beta, a growth factor known for its pro-fibrotic properties. Using si-RNA, the results suggest that production of reactive oxygen species by NADPH oxidase 4 is required for the fibroblast to myofibroblast transition.[33]

The importance of TGF-beta in the production of fibrosis in the heart as well as other tissues has been well documented; Refs.[34–38] provide a cursory snapshot of the many studies conducted on this topic. In general, the growth factor works in various tissues to increase matrix production, stimulate the proliferation of myofibroblasts, and initiate other processes associated with fibrosis. Recent research has investigated its effect on cardiac fibroblasts in both three-dimensional (3D) and two-dimensional (2D) environments. A study by Drobic et al. indicated that TGF-beta and cardiotrophin-1 may have opposing effects on the contraction of collagen type I hydrogels. Cardiac fibroblasts stimulated by TGF-beta augmented contraction of the gels compared to controls, with and without the presence of cardiotrophin-1.[39] This response is consistent with the ability of TGF-beta to promote the contractile myofibroblast phenotype, and shows that the effect of the growth factor is resilient to the presence of opposing factors. The augmented gel contraction has been repeated in another study, which showed that the extent of contraction was TGF-beta dose dependent.[40] This study also showed that TGF-beta stimulated an increase in protein production by the cells. This is consistent with a myofibroblast phenotype, which is characterized by higher contractility and expression of stress fibers. This study found that the DNA content of cells was unchanged by TGF-beta[40]; however, another study utilized TUNEL staining to show that myofibroblasts stimulated by TGF-beta have increased DNA fragmentation and an inhibition of telomerase activity that halts cell proliferation, though the cells are non-apoptotic.[41] As mentioned previously, angiotensin II is also an important fibrotic factor in cardiac fibroblasts. A recent study by Chen et al. investigated a relationship between this growth factor and the action of TGF-beta. The results of this study showed that a specific TGF-beta receptor can modulate the response of cells to angiotensin II.[42] Hence, the biochemical environment that stimulates the transition from fibroblasts to myofibroblasts is complex—relying on multiple, related factors in addition to mechanical stimuli. These effects must be understood in order to evaluate the cell's response to mechanical loading.

An important component of both the biochemical and mechanical environments in tissue is the ECM, which also directly affects the fibroblast/myofibroblast phenotype. Studies have shown that specific elements of the ECM facilitate a transition to myofibroblasts. Though complex, the orientation of cardiac fibroblasts in the surrounding ECM is ordered and well-defined.[43] The interaction between fibroblasts and the surrounding ECM is potentially as important as the relationship between fibroblasts and myocytes.[44] A study by Naugle et al. showed that one component of the ECM, collagen type VI, is over-expressed in vivo after myocardial infarction. This protein attenuated cell proliferation in the presence of angiotensin II compared to collagens type I and III.[45] The authors hypothesized that collagen VI produced this effect by stimulating a transition of fibroblasts to myofibroblasts, which are relatively non-proliferative. Not only does the matrix environment affect the cell biology, but the inverse is also true. Increased matrix production by the myofibroblast phenotype alters the structural and mechanical properties of the matrix. A recent study found that not only is the quantity of matrix production augmented in hypertensive heart failure, but the quality of this production is also affected. An imbalance in matrix metalloproteinases and tissue inhibitors of these enzymes were found in these tissues, indicating that pathology alters the pattern of collagen deposition.[46]

Several in vivo studies utilizing animal models have been conducted to further investigate the cardiac fibroblast to myofibroblast transition.[47–49] As discussed in this chapter, the mechanical environment in the heart is complex and varies both spatially and temporally. The added complexity of the interaction between biochemical and mechanical stimuli adds to the inherent variability of studies analyzing cell response to mechanical loading. The following section focuses on in vitro studies that aimed to isolate the response of cardiac fibroblasts to specific, well-defined mechanical and fluidic stimuli.

7.4 CARDIAC FIBROBLAST RESPONSE TO NORMAL AND SHEAR STRESSES

Cells in the myocardium experience a combination of normal and shear stresses from mechanical (solid and fluid) stimuli. As with many biological systems, a feedback loop exists in the cellular response to loading. Mechanical stress affects the cell phenotype, which in turn alters the surrounding ECM, modifying the applied mechanical stress. This feedback is amplified in pathologies like fibrosis; pressure overload stimulates a transition of fibroblasts to myofibroblasts, these cells deposit greater amounts of ECM, and the ECM enlarges and stiffens. This change in the ECM further increases the stresses exerted on the cells. A schematic of this interaction is diagramed in Figure 7.3.

Understanding how cardiac fibroblasts respond to loading can affect the development of treatments for these diseases. Furthermore, understanding how to prevent this positive feedback can enable physicians to impede or prevent heart failure. This section addresses key results from these recent studies which involve in vitro mechanical testing of isolated cardiac fibroblasts, with an emphasis on studies that elucidate how cardiac fibroblasts respond to stress, and how these responses are relevant for the treatment of cardiac pathologies.

Methods of applying mechanical stimuli using in vitro environments are necessarily limited; an in vitro test setup can never fully represent the actual in vivo condition. Nonetheless, research findings from such experiments provide useful insight into the mechanobiology of cardiac fibroblasts. An important consideration when analyzing these results is the conditions used for the in vitro testing. Cell responses to identical mechanical stimuli can change depending on whether the loading is applied in 2D or 3D culture systems, whether the cells are derived from cell lines or isolated directly, and whether the rats used for isolations are neonatal or adult. As discussed previously in this chapter, the culture conditions also have considerable effects on the biology of cardiac fibroblasts. Thus, when studies are discussed in this section, the culture method and testing configurations used in the experiments will be addressed along with the results.

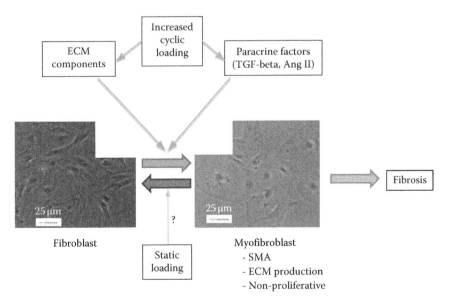

FIGURE 7.3 The effect of mechanical loading on the cardiac fibroblast phenotype.

Early studies on the mechanobiology of cardiac fibroblasts utilized various flexure units able to apply fixed biaxial strain to cells attached to a 2D substrate. A study by Lee et al. applied 10%–20% strain to cardiac fibroblasts isolated from adult rats, which were cultured in 10% serum media. The cells were plated on a 2D silicone substrate coated with fibronectin. Two-dimensional biaxial strain of these substrates was associated with early activation of G proteins in the cardiac fibroblasts, and the rate of activation was dependent upon the strain rate utilized to stimulate the sample.[50] Although this result does not point to a specific signaling pathway, it demonstrates the ability of the cardiac fibroblasts to transduce mechanical stimuli into specific biological responses. In a related study by Lee using similar mechanical stimulation, the authors found that static, biaxial strain increased levels of TGF-beta produced by the cells,[51] suggesting a shift toward the myofibroblast state. In contrast Wang et al. used magnetic beads coated with collagen to apply static stresses of $0.65\,pN/\mu m^2$ and measured attenuated expression of SMA by the cardiac fibroblasts,[52] suggestive of the fibroblast phenotype. In both studies, cells were isolated from adult rats and cultured in 10% serum. Wang theorized that the transition from fibroblast to myofibroblast occurs during pressure overload because of angiotensin II released in response to hypertension, not from the augmented loading itself. In a related study by the Lee group, cyclic stress produced significant amounts of IGF-1, another pro-fibrotic growth factor, while static stress did not have this effect.[53] This result suggests that static stretch does cause an anti-fibrotic effect. Based on these studies, static stretch can influence the fibroblast phenotype; however, the effects can vary depending on the method of loading and the type of ECM used for cell adhesion.

Several recent studies have investigated the effect of ECM composition on the cellular response of cardiac fibroblasts to mechanical loading. For example, one study applied 2D static strain to cardiac fibroblasts isolated from neonatal rats and cultured in 15% serum.[54] The cells were plated onto the loading substrate using the ECM component laminin. Although the authors did not study protein expression of the cells, their results did indicate that cytoskeletal arrangement was significantly affected by loading. In a related study, authors from the same group found that mechanical stretch increased activation of MAP kinase when cells were plated on collagen and fibronectin compared to cells plated on aligned collagen and laminin.[55] The results also indicated a difference in collagen type I deposition depending on the ECM component utilized. Although the authors of the study did not directly evaluate whether the static stretch promoted a transition to myofibroblasts, they did find that strain reduced proliferation rates, a characteristic of myofibroblasts. These results underscore the importance of the ECM on cardiac fibroblast response to mechanical loading. In another recent study it was shown that cyclic loading affects the type of ECM produced by the cells. Two-dimensional cyclic strain was applied to cardiac fibroblasts isolated from adult rats cultured in varying serum conditions. The authors found that both serum concentration and magnitude of strain affected the amount of collagen type I and III expressed by the cells.[56]

Recent research has also investigated the extent to which mechanical loading affects the release of paracrine factors from cardiac fibroblasts. Factors like TGF-beta and angiotensin II have potent effects on the microenvironment surrounding cardiac fibroblasts, which includes cardiomyocytes. The interaction between fibroblasts and cardiomyocytes has implications for fibroblast mechanobiology. Research has shown that myofibroblasts can form gap junctions with surrounding cells, and that these junctions influence the mechanical stresses experienced by the cells.[57] The role of IGF-1 secreted by cyclically strained cardiac fibroblasts as a potent paracrine factor for cardiac hypertrophy was investigated by Lee et al.[53] Research has also shown that 2D strain does not cause the release of apoptotic factors by cardiac fibroblasts.[58] These studies show that mechanical loading not only affects fibroblast physiology, but may also induce the release of certain factors that directly affect neighboring cardiomyocytes. This relationship may also play an important role in the progression of cardiac pathologies like fibrosis.

Several studies have begun to focus on specific signaling pathways associated with cardiac fibroblast mechano-transduction. In a study utilizing magnetic beads to apply strain to cells, authors investigated the role of a Rho-Rho kinase pathway in overexpressing SMA in stretched cardiac fibroblasts.[59] Other studies have found that mechanical stretch is involved with specific changes in gene transcription, and activation of specific kinases.[60–63] Moreover, other studies have elucidated the role of ion exchange and ion channels in the response of cardiac fibroblasts to mechanical loading.[64,65] This body of work contributes to our understanding of the specific mechanisms by which cardiac fibroblasts transduce mechanical stimuli into biological changes. However, most of this work was performed in 2D culture systems that do not fully recreate the 3D nature of the loading experienced by cells in vivo.

There also have been several studies focused on the effects of 3D mechanical loading on cardiac fibroblasts. Interestingly, the change from 2D to 3D does not necessarily cause differences in results. A study analyzing the response of cardiac fibroblasts to static stretch in a 3D environment supported the results of Wang's study.[66] The authors found that static stretch reverted myofibroblasts to fibroblasts, reducing the expression of SMA and attenuating collagen and fibronectin production, two hallmarks of the myofibroblast phenotype. Hence, both 2D and 3D static stress may block the transition to myofibroblast phenotypes. Other studies have examined how cardiac fibroblasts compact 3D hydrogels, suggesting that the morphology of the cells depends on whether the gels are free to compact or restrained.[67] Another method of recreating the 3D environment is to use trabecula isolated directly from heart tissue. One study utilized this method and found that cyclic loading increased the size of the trabecula compared to control, which can be interpreted as a model of cardiac hypertrophy.[68] However, the specific action of the fibroblasts in this testing is unknown.

The majority of studies investigating cardiac fibroblast mechanobiology focus on normal stress, but shear stress is also a crucial component of the mechanical environment in the myocardium. The laminae of the myocardium experience significant sliding, which generates shear stress on the cells contained within these layers. Moreover, interstitial fluid flow also imparts shear stress on the cells in the myocardium. In conditions of ischemia following myocardial infarction, the cells experience an absence of perfusion and the interstitial flow that may affect the cellular responses of cardiac fibroblasts and myocytes. Shear stress has been shown to affect fibroblasts isolated from skin[69] through the release of TGF-beta. To date, a similar study has not been conducted on cardiac fibroblasts. Nonetheless, researchers have constructed bioreactors to study the effects of perfusion on tissue engineered cardiac constructs.[70] In the future, insight into how shear stress affects cardiac fibroblasts will be invaluable for understanding the mechanobiology of these cells.

7.5 CONCLUSION

This chapter provides an overview of the broad field of research investigating cardiac fibroblast mechanobiology. Understanding how these cells respond to mechanical loads provides crucial insight into how pathologies like fibrosis initiate and affect heart function, as well as how better therapies can be developed to both prevent and treat these diseases. The study of cardiac fibroblast mechanobiology is complicated by the fluid nature of the phenotype of these cells. Even small changes in culture conditions can have important effects on the cellular responses to mechanical loading. For example, we have discussed how changes in the serum concentration of the growth media can accelerate the transition to a myofibroblast phenotype. Such phenotypic shifts caused by the biochemical environment complicate the study of how cells are affected by mechanical forces. A firm understanding of the factors which affect the fibroblast/myofibroblasts transition in cardiac cells is necessary to study cell mechanobiology.

This chapter also highlights the importance of loading protocols and configurations on cellular response. It has been shown that cells respond differently depending on what type of surrounding ECM is present, whether the loading is in a 2D or 3D system, and whether the applied load is static or cyclic. Therefore, comparing results between studies is difficult, since even though the general loading scheme may be similar, inconsistency in other factors can lead to very different results between studies. Studying cell responses in 2D configurations has provided important information about how cardiac fibroblasts sense, transduce, and react to mechanical forces. However, it is becoming increasingly clear that the cell response may be significantly different in 3D systems that more closely mimic the cellular environment in tissues. For these reasons, despite the thorough and robust data generated from research on this topic, there is no unified theory explaining how cardiac fibroblasts generate biological responses when strained mechanically. In the future, further research on the response of these cells to mechanical stimuli in 3D environments can be expected to provide insight into how they behave in the in vivo microenvironment. Such research is important to understand a variety of disease states, including cardiac fibrosis, in which changes in mechanical environment have been implicated in causing undesirable changes in cell function.

REFERENCES

1. LeGrice I.J., Takayama Y., and Covell J.W. Transverse shear along myocardial cleavage planes provides a mechanism for normal systolic wall thickening. *Circ Res* 77:182–193, 1995.
2. LeGrice I.J. et al. Laminar structure of the heart: Ventricular myocyte arrangement and connective tissue architecture in the dog. *Am J Physiol* 269:H571–H582, 1995 (*Heart Circ Physiol* 38).
3. Harrington K. et al. Direct measurement of transmural laminar architecture in the anterolateral wall of the ovine left ventricle: New implications for wall thickening mechanics. *Am J Physiol Heart Circ Physiol* 288:1324–1330, 2005.
4. Arts T., Costa K.D., Covell J.W., and McCulloch A.D. Relating myocardial laminar architecture to shear strain and muscle fiber orientation. *Am J Physiol Heart Circ Physiol* 280:H2222–H2229, 2001.
5. Lorenzen-Schmidt I. et al. Chronotropic response of cultured neonatal rat ventricular myocytes to short-term fluid shear. *Cell Biochem Biophys* 46(2):113–122, 2006.
6. Dvir T. et al. Activation of the ERK1/2 cascade via pulsatile interstitial fluid flow promotes cardiac tissue assembly. *Tissue Eng* 13(9):2185–2193, 2007.
7. Sinay R. et al. Intramyocardial pressure: Interaction of myocardial fluid pressure and fiber stress. *Am J Physiol* 257:H357–H364, 1989 (*Heart Circ Physiol* 26).
8. Ashikaga T. et al. Changes in regional myocardial volume during the cardiac cycle: Implications for transmural blood flow and cardiac structure. *Am J Physiol Heart Circ Physiol* 295:610–618, 2008.
9. Mehlhom U., Geissler H., Laine G., and Allen S. Myocardial fluid balance. *Eur J Cardiothorac Surg* 20:1220–1230, 2001.
10. Lee D. et al. Magnetic resonance versus radionuclide pharmacological stress perfusion imaging for flow-limiting stenoses of varying severity. *Circulation* 110:58–65, 2004.
11. Holmes J., Borg T., and Covell J. Structure and mechanics of healing myocardial infarcts. *Annu Rev Biomed Eng* 7:223–253, 2005.
12. Wang J. et al. Left ventricular twist mechanics in a canine model of reversible congestive heart failure: A pilot study. *J Am Soc Echocardiogr* 22:95–98, 2009.
13. Aikawa Y. et al. Regional wall stress predicts ventricular remodeling after anteroseptal myocardial infarction in the healing and early afterload reducing trial (HEART): An echocardiography based structural analysis. *Am Heart J* 141:234–42, 2001.
14. Watanabe H. et al. Multiphysics simulation of left ventricular filling dynamics using fluid-structure interaction finite element method. *Biophys J* 87:2074–2085, 2004.
15. Swaney J. et al. Inhibition of cardiac myofibroblast formation and collagen synthesis by activation and overexpression of adenylyl cyclase. *Proc Nat Acad Sci* 102(2):437–442, 2005.
16. Leicht M. et al. Serum depletion induces cell loss of rat cardiac fibroblasts and increased expression of extracellular matrix proteins in surviving cells. *Cardiovasc Res* 52:429–437, 2001.
17. Leicht M. et al. Mechanism of cell death of rat cardiac fibroblasts induced by serum depletion. *Mol Cell Biochem* 251:119–126, 2003.
18. Butt R.P. and Bishop J.E. Mechanical load enhances the stimulatory effect of serum growth factors on cardiac fibroblast procollagen synthesis. *J Mol Cell Cardiol* 29:1141–1151, 1997.
19. Fujita K. et al. Adiponectin protects against angiotensin II–induced cardiac fibrosis through activation of PPAR-alpha. *Arterioscler Thromb Vasc Biol* 28:863–870, 2008.
20. Olson E. et al. Angiotensin II–induced extracellular signal–regulated kinase 1/2 activation is mediated by protein kinase C and intracellular calcium in adult rat cardiac fibroblasts. *Hypertension* 51:704–711, 2008.
21. Yokoyama T. et al. Angiotensin II and mechanical stretch induce production of tumor necrosis factor in cardiac fibroblasts. *Am J Physiol Heart Circ Physiol* 276:1968–1976, 1999.

22. Kawano H. et al. Angiotensin II enhances integrin and alpha-actinin expression in adult rat cardiac fibroblasts. *Hypertension* 35(Pt 2):273–279, 2000.

23. Thibault G. et al. Upregulation of a8b1-integrin in cardiac fibroblast by angiotensin II and transforming growth factor-b1. *Am J Physiol Cell Physiol* 281:C1457–C1467, 2001.

24. Lal H. et al. Stretch-induced regulation of angiotensinogen gene expression in cardiac myocytes and fibroblasts: Opposing roles of JNK1/2 and p38α MAP kinases. *J Mol Cell Cardiol* 45:770–778, 2008.

25. Tanaka T. et al. Marked elevation of brain natriuretic peptide levels in pericardial fluid is closely associated with left ventricular dysfunction. *J Am Coll Cardiol* 31(2):399–403, 1998.

26. Tsuruda T. et al. Brain natriuretic peptide is produced in cardiac fibroblasts and induces matrix metalloproteinases. *Circ Res* 91:1127–1134, 2002.

27. Tamura N. et al. Cardiac fibrosis in mice lacking brain natriuretic peptide. *Proc Nat Acad Sci* 97(8): 4239–4244, 2000.

28. Pho M. et al. Cofilin is a marker of myofibroblast differentiation in cells from porcine aortic cardiac valves. *Am J Physiol Heart Circ Physiol* 294:H1767–H1778, 2008.

29. Zhang X. et al. Effects of elevated glucose levels on interactions of cardiac fibroblasts with the extracellular matrix. *In Vitro Cell Dev Biol Anim* 43:297–305, 2007.

30. Lenga Y. et al. Osteopontin expression is required for myofibroblast differentiation. *Circ Res* 102: 319–327, 2008.

31. Roy S. et al. Oxygen sensing by primary cardiac fibroblasts: A key role of p21 waf1/cip1/sdi1. *Circ Res* 92:264–271, 2003.

32. Roy S. et al. P21waf1/cip1/sdi1 as a central regulator of inducible smooth muscle actin expression and differentiation of cardiac fibroblasts to myofibroblasts. *Mol Biol Cell* 18:4837–4846, 2007.

33. Cucoranu I. et al. NAD(P)H oxidase 4 mediates transforming growth factor-beta1–induced differentiation of cardiac fibroblasts into myofibroblasts. *Circ Res* 97:900–907, 2005.

34. Landeen L. et al. Sphingosine-1-phosphate receptor expression in cardiac fibroblasts is modulated by in vitro culture conditions. *Am J Physiol Heart Circ Physiol* 292:H2698–H2711, 2007.

35. Brown R. et al. Enhanced fibroblast contraction of 3D collagen lattices and integrin expression by TGF-beta1 and -beta3: Mechanoregulatory growth factors? *Exp Cell Res* 274:310–322, 2002.

36. Liu X. et al. Smad3 mediates the TGF-beta-induced contraction of Type I collagen gels by mouse embryo fibroblasts. *Cell Motil Cytoskeleton* 54:248–253, 2003.

37. Wipff P.J. et al. Myofi broblast contraction activates latent TGF-β1 from the extracellular matrix. *J Cell Biol* 179(6):1311–1323, 2007.

38. Merryman W.D. et al. Differences in tissue-remodeling potential of aortic and pulmonary heart valve interstitial cells. *Tissue Eng* 13, (9):2281–2289, 2007.

39. Drobic V. et al. Differential and combined effects of cardiotrophin-1 and TGF-beta1 on cardiac myofibroblast proliferation and contraction. *Am J Physiol Heart Circ Physiol* 293:H1053–H1064, 2007.

40. Lijnen P. et al. Transforming growth factor-beta1-mediated collagen gel contraction by cardiac fibroblasts. *J Renin Angiotensin Aldosterone Syst* 4:113, 2003.

41. Petrov V. et al. TGF-β1-induced cardiac myofibroblasts are nonproliferating functional cells carrying DNA damages. *Exp Cell Res* 314:1480–1494, 2008.

42. Chen K. et al. Transforming growth factor beta receptor endoglin is expressed in cardiac fibroblasts and modulates profibrogenic actions of angiotensin II. *Circ Res* 95:1167–1173, 2004.

43. Goldsmith E. et al. Organization of fibroblasts in the heart. *Dev Dyn* 230:787–794, 2004.

44. Banerjee I. et al. Dynamic interactions between myocytes, fibroblasts, and extracellular matrix. *Ann NY Acad Sci* 1080:76–84, 2006.

45. Naugle J. et al. Type VI collagen induces cardiac myofibroblast differentiation: Implications for postinfarction remodeling. *Am J Physiol Heart Circ Physiol* 290:H323–H330, 2006.

46. Lopez B. et al. Alterations in the pattern of collagen deposition may contribute to the deterioration of systolic function in hypertensive patients with heart failure. *J Am CollCardiol* 48(1):97–98, 2006.

47. Flack E.C. et al. Alterations in cultured myocardial fibroblast function following the development of left ventricular failure. *J MolCell Cardiol* 40 :474–483, 2006.

48. Sun M. et al. Tumor necrosis factor-alpha mediates cardiac remodeling and ventricular dysfunction after pressure overload state. *Circulation* 115:1398–1407, 2007.

49. Rosenkranz S. et al. Alterations of beta-adrenergic signaling and cardiac hypertrophy in transgenic mice overexpressing TGF-beta1. *Am J Physiol Heart Circ Physiol* 283:H1253–H1262, 2002.

50. Gudi S. et al. Equibiaxial strain and strain rate stimulate early activation of G proteins in cardiac fibroblasts. *Am J Physiol* 274:C1424–C1428, 1998 (*Cell Physiol* 43).

51. Lee A. et al. Differential responses of adult cardiac fibroblasts to in vitro biaxial strain patterns. *J Mol Cell Cardiol* 31:1833–1843, 1999.
52. Wang J. et al. Force regulates smooth muscle actin in cardiac fibroblasts. *Am J Physiol Heart Circ Physiol* 279:H2776–H2785, 2000.
53. Hu B.S. et al. An analysis of the effects of stretch on IGF-I secretion from rat ventricular fibroblasts. *Am J Physiol Heart Circ Physiol* 293:H677–H683, 2007.
54. Fuseler J.W. et al. Fractal and image analysis of morphological changes in the actin cytoskeleton of neonatal cardiac fibroblasts in response to mechanical stretch. *Microsc Microanal* 13:133–143, 2007.
55. Atance J. et al. Influence of the extracellular matrix on the regulation of cardiac fibroblast behavior by mechanical stretch. *J Cell Physiol* 200:377–386, 2004.
56. Husse B. et al. Cyclical mechanical stretch modulates expression of collagen I and collagen III by PKC and tyrosine kinase in cardiac fibroblasts. *Am J Physiol Regul Integr Comp Physiol* 293:R1898–R1907, 2007.
57. Follonier L. et al. Myofibroblast communication is controlled by intercellular mechanical coupling. *J Cell Sci* 121:3305–3316, 2008.
58. Persoon-Rothert M. et al. Mechanical overload-induced apoptosis: A study in cultured neonatal ventricular myocytes and fibroblasts. *Mol Cell Biochem* 241:115–124, 2002.
59. Zhao X.H. et al. Force activates smooth muscle-actin promoter activity through the Rho signaling pathway. *J Cell Sci* 120:1801–1809, 2007.
60. Liang F. et al. Integrin dependence of brain natriuretic peptide gene promoter activation by mechanical strain. *J Biol Chem* 275(27):20355–20360, 2000.
61. Prante C. et al. Transforming growth factor beta1-regulated xylosyltransferase I activity in human cardiac fibroblasts and its impact for myocardial remodeling. *J Biol Chem* 282(36):26441–26449, 2007.
62. Wang J. et al. Mechanical force activates eIF-2a phospho-kinases in fibroblast. *Biochem Biophys Res Commun* 330:123–130, 2005.
63. Lindahl G.E. et al. Activation of fibroblast procollagen _1(I) transcription by mechanical strain is transforming growth factor-beta-dependent and involves increased binding of CCAAT-binding factor (CBF/NF-Y) at the proximal promoter. *J Biol Chem* 277(8):6153–6161, 2002.
64. Raizman J.E. et al. The participation of the Na+–Ca2+ exchanger in primary cardiac myofibroblast migration, contraction, and proliferation. *J Cell Physiol* 213:540–551, 2007.
65. Chilton L. et al. K_currents regulate the resting membrane potential, proliferation, and contractile responses in ventricular fibroblasts and myofibroblasts. *Am J Physiol Heart Circ Physiol* 288: H2931–H2939, 2005.
66. Poobalarahi F. et al. Cardiac myofibroblasts differentiated in 3D culture exhibit distinct changes in collagen I production, processing, and matrix deposition. *Am J Physiol Heart Circ Physiol* 291: H2924–H2932, 2006.
67. Baxter S., Morales M., and Goldsmith E. Adaptive changes in cardiac fibroblast morphology and collagen organization as a result of mechanical environment. *Cell Biochem Biophys* 51:33–44, 2008.
68. Bupha-Intr T., Holmes J., and Janssen P. Induction of hypertrophy in vitro by mechanical loading in adult rabbit myocardium. *Am J Physiol Heart Circ Physiol* 293:H3759–H3767, 2007.
69. Ahamed J. et al. In vitro and in vivo evidence for shear-induced activation of latent transforming growth factor-beta1. *Blood* 112:3650–3660, 2008.
70. Brown M., Iyer R., and Radisic M. Pulsatile perfusion bioreactor for cardiac tissue engineering. *Biotechnol. Prog* 24:907–920, 2008.

8 Mechanobiological Evidence for the Control of Neutrophil Activity by Fluid Shear Stress

Hainsworth Y. Shin, Xiaoyan Zhang, Aya Makino, and Geert W. Schmid-Schönbein

CONTENTS

8.1 INTRODUCTION: CONTROL OF NEUTROPHIL ACTIVITY

The central topic of this chapter is the regulation of polymorphonuclear leukocyte (particularly, neutrophil) activity by circulatory hemodynamics-derived mechanical stresses. Neutrophil activation in the microcirculation plays a critical role in the initiation and control of inflammation (definition: a cascade of biological processes by multiple immune and tissue-specific cell types that promote an adaptive response of host tissues to noxious insult) (Medzhitov, 2008). In this respect, neutrophils play an essential role in the acute-stage defense against pathogens (e.g., microorganisms, foreign bodies, and inorganic materials) as well as in the repair and management of tissue damage due to injury. As a consequence of their capacity to express potent antimicrobial and tissue degradative agents during early inflammatory processes, cellular mechanisms must exist to ensure tight regulation of the destructive potential of the polymorphonuclear leukocytes that, if unchecked, may lead to damage to host tissues. Thus, turning off neutrophil inflammatory processes during the resolution stages of wound healing and infection is just as critical as turning on these activities at the time of infection or tissue injury.

Moreover, because of their destructive potential as well as hemorheological issues (especially in the microcirculation), maintaining neutrophils (as well as other leukocytes) in an inactive state while circulating in blood under physiologic (i.e., noninflamed) situations is critical as well. Although neutrophil inactivity in the circulation is predicated on the mere absence of inflammatory stimulators (e.g., products of tissue damage, pathogenic particles, cytokines, and bacteria), cellular and molecular mechanisms do exist to proactively ensure that these cells remain in an inactivated state under physiological conditions (Rinaldo and Basford, 1987). In addition to biochemical mediators (e.g., endothelial secretions) that inhibit inflammatory processes, substantial evidence shows that responses of neutrophils to fluid shear stresses are essential for normal capillary perfusion under physiological conditions in addition to traditional leukocyte functions associated with cell adhesion in postcapillary venules, migration across the endothelium, and phagocytosis during inflammation.

Under pathological conditions, infection or products of tissue damage irreversibly commit neutrophils to an activated phenotype. Circulating neutrophils become stimulated to express a preprogrammed cascade of adhesion molecules, which facilitate their recruitment out of the bloodstream and onto the blood vessel lumen. In these situations, cellular activation above a threshold level of stimulation overrides the ability of fluid flow to deactivate neutrophils (Makino et al., 2007). Moreover, while blood flow itself is a barrier to cell adhesion and migration, the hemorheological properties of blood (i.e., its cellular constituents) facilitate neutrophil–vessel wall interactions by mechanically transporting the cells to the endothelium. In fact, the persistence of vascular wall–neutrophil interactions, despite the presence of flow-induced shearing forces that serve to "wash away" the cells, points to fluid mechanical stimuli as bioactive facilitators of cell–cell interactions between neutrophils and endothelial cells or with other leukocytes (Simon and Goldsmith, 2002). In addition to this, fluid stresses influence the functions of activated leukocytes, after their transit out of the bloodstream; for example, the case of fluid flow and pressure stimulation of macrophages (Sakamoto and Cabrera, 2003; Shiratsuchi and Basson, 2004, 2005, 2007; Shiratsuchi et al., 2009).

Neutrophil mechanobiology, therefore, encompasses a broad field of study ranging from cellular/molecular mechanics experimentation (e.g., mechanical properties of the leukocyte and its components) to research regarding the cell biological responses to fluid stress. Whereas the assessment of mechanical properties provides information regarding the ability of the neutrophil to passively deform in the microvasculature as well as to detect perturbations in their local mechanoenvironment, investigations on the biological outcomes associated with mechanical force stimulation give insight into the sensitivity of neutrophil biological processes to the surrounding blood flow environment. In this chapter, we restrict the focus to the mechanobiological regulation of neutrophil activity under physiological flow conditions, a major aspect that maintains circulatory homeostasis. We, however, recognize that the mechanobiology of the activated neutrophils, particularly as it relates to their recruitment out of the bloodstream, also contributes to an understanding of the influence of the fluid mechanics of blood flow on the overall regulation of neutrophil behavior; information regarding this can be found in extensive review articles provided by the published literature (Smith, 2000; Ley, 2002; Simon and Green, 2005; Zhu et al., 2005; Astrof et al., 2006).

8.2 BASIC CONCEPTS: THE GRANULOCYTES

This section focuses on key basic concepts of behavior of the neutrophils relevant to the current understanding of how fluid flow exposure under physiological conditions regulates their activity. For a complete discussion of traditional polymorphonuclear leukocyte physiology, the reader is referred to any number of textbook sources (Alberts et al., 1983; Guyton and Hall, 1996; Edwards, 2005). The family of cells responsible for immune processes associated with the body's defense against infection and tissue damage include the nonnucleated platelets, as well as populations of either polymorphonucleated or mononuclear leukocytes (Table 8.1). The leukocytes vary in size, function, and nuclear morphology, all of which impact their transit through the circulation in one way or another. In total, the leukocytes represent 20% of the cellular content of blood. The relative abundance of the neutrophils in blood as compared to the other nucleated white cell types lends a clue to the important role of these cells in the acute inflammatory response.

8.2.1 POLYMORPHONUCLEAR LEUKOCYTES

The polymorphonuclear leukocytes are granule-enriched phagocytes that enter the bloodstream from the bone marrow after undergoing a maturation process from resident myeloid precursor cells. In the bloodstream, the polymorphonucleated leukocytes (consisting primarily of the neutrophils with small percentages of basophiles and eosinophiles) make up the major component of nucleated white blood cells with the remaining percentages of cells being of the lymphocytic and monocytic lineages.

Neutrophils floating in blood patrol the circulation in a mechanically passive (Mazzoni and Schmid-Schönbein, 1996), freely suspended (i.e., floating), and relatively quiescent (inactive) state. A consequence of their lack of activity is their ability to freely deform and pass through the 5–8 μm

---

TABLE 8.1
Cell Types in the Blood

Cell Type	Diameter (μm)	Blood Cell Density (Cells/L)	Function(s)	Special Note
Red blood cells	6–8	5×10^{12}	• O_2 transport	• Anuclear
Platelets	1–4	3×10^{11}	• Coagulation/blood clot formation	• Microparticles
Polymorphonuclear granulocytes				• Multilobed nucleus
				• Granulated
Neutrophils	9–16	5×10^9	• Phagocytose pathogens	
Eosinophils	9–15	2×10^8	• Phagocytose parasites • Modulate allergic responses	
Basophils	10–15	4×10^7	• Release histamine	
Mononuclear cells				• Agranulated
Monocytes	10–30	4×10^8	• Precursor to tissue macrophages • Phagocytose pathogens and damaged or senescent cells	• Kidney-shaped nucleus
Lymphocytes	6–10	4×10^9	• Antibodies • Acquired immunity • T-cells, B-cells, natural killer cells	• Rounded nucleus

Sources: Alberts, B.D. et al., *Molecular Biology of the Cell*, Garland Publishing, Inc., New York, 1983; Guyton, A.C. and Hall, J.E., *Textbook of Medical Physiology*, W. B. Saunders, Philadelphia, PA, 1996; Edwards, S.W., *Biochemistry and Physiology of the Neutrophil*, Cambridge University Press, New York, 2005.

diameter capillaries of the microvasculature (Mazzoni and Schmid-Schönbein, 1996). Functionally, neutrophils are critical regulators of immune responses to pathogens or to trauma associated with (1) detecting/combating infection, (2) initiating acute inflammatory processes, and (3) controlling wound healing. The amount of time that neutrophils spend in free circulation is on the order of 4–8 h by which time they either (1) begin to undergo apoptosis (programmed cell death), or (2) upon activation by agonists, transit to the interstitium (e.g., lymphatics, organ tissue matrices) where they reside and/or patrol for 4–5 days while fighting infection (Guyton and Hall, 1996). The phenotype of a neutrophil, at a given instant in time, is therefore based on its level of activity including whether it is in an adherent or suspended state.

A critical aspect of the neutrophils is their acute and selective sensitivity to inflammatory stimuli and the rapidity with which they transition from an inactivated to an activated phenotype. In fact, this conversion occurs on the order of milliseconds and is related to the level of their stimulation that is best exemplified by the reports indicating that only 10–100 bacterial or chemokine peptides (Simon and Goldsmith, 2002) entering their local milieu are necessary to elicit cellular activation and downstream functional responses. Thus, low levels of activators can potentially lead to a potent acute neutrophil response. It is, therefore, intuitive that such a level of sensitivity to stimuli is balanced by precise control of neutrophil activity levels; this is, in fact, the case.

Upon activation, neutrophils undergo rapid changes in their biochemical status (e.g., membrane surface expression of proteins) as well as mechanical properties (e.g., stiffness), which together facilitate their recruitment out of the bloodstream (Schmid-Schönbein, 2006). After reaching their target tissue sites via chemotaxis down a concentration gradient of signaling factors (e.g., cytokines), these cells attack and degrade pathogens by engulfing foreign materials into cytosolic phagosomes that fuse with lysosomes containing protein degrading and bactericidal agents (e.g., reactive oxygen species) that destroy the phagocytosed particles. In fact, neutrophils are armed with a potent array of biochemical agents (e.g., proteases, phosphatases, cytokines, other inflammatory mediators) with which they fight infection within the controlled and confined regions of the local injury site (Edwards, 2005), but which may also be toxic to the host if released (or "leaked") into the environment in amounts that exceed the requirements for a given wound response scenario. In light of their enormous responsibilities and powerful arsenal of weapons to destroy and degrade foreign materials, it is not surprising that cellular mechanisms, involving both biological and mechanical signal transduction, exist to tightly regulate neutrophil activation.

8.2.2 NEUTROPHIL ACTIVATION AND RECRUITMENT DURING ACUTE INFLAMMATION

Neutrophil recruitment out of the blood flow during acute inflammation is a multistep cascade of events involving selectins, integrins, and other adhesion molecules (Figure 8.1). Initially, neutrophils entering postcapillary venules from a single-file arrangement with other red cells in the capillaries are forced to marginate from the centerline of the flow field and to make close contact with the endothelium by the deformable, smaller, and faster flowing erythrocytes (Schmid-Schönbein et al., 1980; Ley, 1996). This margination process (Figure 8.2), which is also observed at the confluent venular bifurcations (Bagge and Karlsson, 1980) and promoted by erythrocyte aggregation in large microvessels (Goldsmith and Spain, 1984; Nobis et al., 1985; Pearson and Lipowsky, 2004), is an important step in leukocyte recruitment to the vessel wall. Bagge et al. (1983) reported, from in vitro studies, that not a single activated neutrophil (or other leukocyte) will attach to inflamed endothelium in the absence of red cells. There, however, is actually no reported inflammatory deficiency caused by a lack of leukocyte margination (Ley, 1996). Despite this, red cell-induced margination of leukocytes to the vessel wall is an important step in acute inflammation.

Leukocytes near the endothelium due to margination are captured out of the blood flow and roll on the endothelium via rapid association and dissociation of bonds between selectins (e.g., L-selectin,

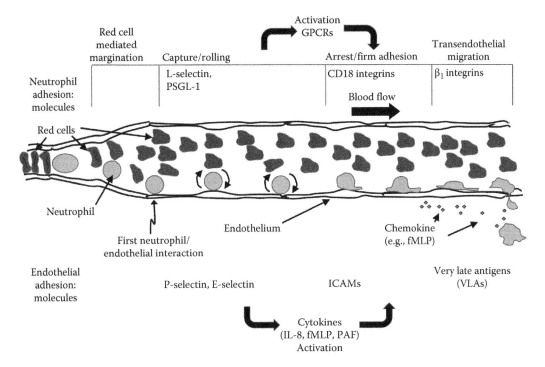

FIGURE 8.1 Schematic representation of leukocyte recruitment during acute inflammation. At vascular geometries associated with transitional blood flow, red cells induce margination of leukocytes to the blood vessel endothelium (its first contact). During acute inflammation, formation and breakage of bonds between selectins and their ligands (PSGL-1) mediate leukocyte capture and rolling on the endothelium after margination. A period of rolling interactions between the endothelium and leukocytes maintains close proximity between the two cells. Upon activation by cytokines, rolling cells arrest on the endothelium via CD18 integrin–ligand (ICAM-1) bonds and migrate along the vessel wall until they eventually spread via β_1 integrins and transmigrate across the endothelium. The leukocyte then migrates in the tissues to the site of trauma/infection.

E-selectin) and their ligands (e.g., "P-selectin glycoprotein ligand-1" or PSGL-1) (Tozeren et al., 1982; Chen and Springer, 1999; Simon and Green, 2005) for the purpose of prolonging their residence times on at the vessel wall. Upon inflammatory stimulation, the rolling neutrophils get activated through the ligation of G-protein coupled receptors (GPCRs) with chemokines or chemoattractants released by inflamed endothelial cells (Zarbock and Ley, 2009). This leads to upregulated activity of β_2 integrins (i.e., CD18) expressed on leukocytes and deployment of additional CD18 receptors (Laudanna et al., 2002; Kinashi and Katagiri, 2004; Luo et al., 2007; Zarbock and Ley, 2009); engagement of these adhesion receptors occurs cooperatively with or independently of the selectins. Subsequently, binding of the CD18 integrins with their ligands (i.e., members of the immunoglobulin superfamily such as intercellular adhesion molecule 1 [ICAM1], vascular cell-adhesion molecule 1 [VCAM1], etc.) not only leads to firm adhesion, but also stimulates downstream signaling pathways that regulate intraluminal crawling, proliferation, and phagocytosis (Radi et al., 2001), as well as respiratory burst and degranulation (Ley, 1996; Luo et al., 2007).

During the later stages of inflammation, leukocytes emigrate from the intravascular compartment to the target tissue via either paracellular or transcellular route, which, in general, is mediated by β_1 integrins. Upregulation of β_1 integrins (β subunit for integrins of the very late antigen lineage) marks the transition from an early adherent neutrophil migrating on the microvascular wall to a more adhered cell that migrates across the endothelium and basement membrane into the underlying tissues (Sixt et al., 2001; Luster et al., 2005).

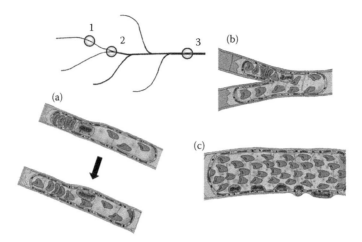

FIGURE 8.2 Examples of sites of red cell-induced margination of the neutrophil. Changes in vascular geometries induce regions of transitional blood flow. The number labels in the microvascular circuit diagram (top left corner) refer to enlarged views of margination events at various example sites in the microcirculation. (a) For instance, at sites of flow expansion (indicated as 1 on vascular tree diagram) where leukocytes (represented as nucleated cells) and red cells (demarcated by discoid or bullet-shaped appearance) transition from flowing in a single file configuration to a more dispersed arrangement in a larger-diameter venule, acceleration of the faster-moving red cells displace the leukocyte to the wall; this leads to the first contact made between the leukocyte and endothelium. (b) Similar margination of leukocytes occurs at branch points (2 on vascular tree diagram) where capillaries feed into the post-capillary venules. (c) In the larger veins (3 on vascular tree diagram), stochastic interactions between red cells and the leukocytes play a critical role in initiating and maintaining leukocyte interaction with the endothelium during recruitment stages of acute inflammation. Fluid flow is left to right in all panels. (Adapted from Schmid-Schönbein, G.W., 2006.)

8.3 FLUID MECHANICAL ENVIRONMENT OF THE NEUTROPHIL

In the circulation, the neutrophils (and other cardiovascular cell types) experience a diverse range of mechanical stresses in terms of types, magnitudes, and distributions. The source of the fluid mechanical stresses imposed on leukocytes (including the neutrophils) is the flow of blood, which imposes on the cell membrane a nonuniform and unsteady arrangement of tangential (shear) and normal (pressure) fluid stress components as well as tensile stresses in the extracellular matrix substrate for adherent cells. These stresses are generated due to the pumping of the heart, which is responsible for propelling blood through the vasculature.

In the macrocirculation, blood pressure, flow, and wall stretch are pulsatile as a result of the cyclical pumping of the heart. Heart rates in humans during resting conditions are on average 60–80 beats/min (frequency of 1–1.2 Hz); physical activity and pathological conditions are associated with acute and/or chronic elevations in these values up to 200 beats/min (Guyton and Hall, 1996). In the microcirculation (e.g., capillaries, postcapillary venules), as well as in most of the venous system, blood flow and pressure are nearly steady since pulsation pressure amplitudes of the heart are reduced by the viscous energy dissipation in the microvasculature due to the dramatic increase in total cross-sectional area of the microcirculation to fluid transport as well as by the compliance of the blood vessels that become more thin-walled with lower content of collagen, elastin, and smooth muscle toward the microcirculation(Guyton and Hall, 1996). Pulsatile flow resulting from heart and skeletal muscle activity, however, has been documented for the microvasculature and in veins, particularly in the lower limbs, where venous valves ensure forward transport of blood volume back to the heart (Intaglietta et al., 1971; Lipowsky et al., 1978; Guyton and Hall, 1996; Sherwood, 2001).

Neutrophils, both adherent and nonadherent, in the circulation are, therefore, exposed to time variant and invariant fluid shear stresses and fluid pressures (Figure 8.3). Additionally, activated

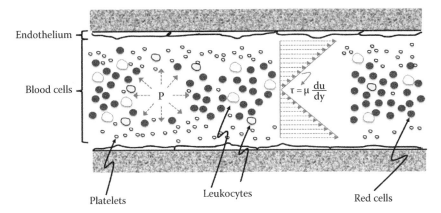

FIGURE 8.3 Schematic representation of the fluid stresses experienced by the blood-borne leukocytes. The platelets, leukocytes, and red cells are exposed to a linear fluid shear stress (τ) field due to blood movement (modeled as flow through a cylindrical tube). These cells also experience fluid pressures (P) resulting from containment of blood volume within the vasculature, the pumping of the heart, and blood flow dynamics.

neutrophils adhered to and migrating on the endothelium are exposed to pulsatile tensile stresses and stretch. The present chapter focuses on the role of the normal (i.e., pressure) and tangential (i.e., shear) stress components of fluid flow on the regulation of neutrophil activity.

8.3.1 FLUID PRESSURE

At any point in the circulation, the actual value of the fluid pressure in a vessel is derived from: (1) the ambient pressure (i.e., atmospheric pressure near the right atrium), (2) the hydrostatic pressure (i.e., pressure due to gravity and to the mass density of blood), (3) the dynamic pressure (i.e., pressure due to kinetic energy), and (4) the potential energy pressure (i.e., pulse pressure due to the pumping of the heart). Notably, in the circulation, the viscous pressure drop is a key determinant of the local pressure. At different sites in the vascular tree, blood pressures vary depending on location (e.g., systemic vs. pulmonary circuit, macro- vs. micro-circulation Table 8.2), time of measurement, level of physical activity, and pathological condition of the subject. Case in point, cyclic blood pressures in the pulmonary artery are 25–30/10–15 mmHg (systolic/diastolic pressure) but average approximately 121/79 mmHg (systolic pressure/diastolic pressure) in the aorta, as measured in the brachial artery (Schettini et al., 1999). In humans, these pressure levels change throughout the course of the day; for example, at night (typically associated with reduced physical activity), brachial pressures average approximately 107/64 mmHg (Schettini et al., 1999). Under pathological conditions (such as hypertension), brachial blood pressures in human subjects are either equal to or greater than approximately 140/83 mmHg (Toikka et al., 2000).

The pulsatile nature of blood flow is mostly dampened in the microcirculation with pressure values in the range of 0–30 mmHg (Guyton and Hall, 1996; Samet and Lelkes, 1999). At the beginning of the venous system, blood pressures are in the 10–12 mmHg range but fall to 0 mmHg at the right atrium (Attinger, 1963; Guyton and Hall, 1996; Samet and Lelkes, 1999). However, in some segments of the venous circulation (e.g., the lower extremities), blood pressures are also influenced by blood pooling and skeletal muscle action (contraction and relaxation cycles) to promote blood return back to the central circulation where it is recycled to the body. Under these circumstances, blood is pumped past sets of valves, with the effect that during quiet standing, venous pressure progressively rises to 20–30 mmHg (Attinger, 1963; Guyton and Hall, 1996; Sherwood, 2001). Cycles of muscle contraction and relaxation (e.g., during walking) lead to pulsatile compression of the veins

TABLE 8.2
Hemodynamic Parameters of the Human Circulation

Vessel Type	Blood Pressure (mmHg)	Mean Shear Stress (dyn/cm²)
Aorta	120/100[a,b,c]	4.5[c,d]
Pulmonary artery	25–30/10–15[a,b,c]	2.7[c]
Small arteries	*33[c]*	up to 32.0[c]
Arterioles	*≈30[c]*	*47.1[f]*
Capillaries	0–30[a,b,c]	*≈30[f]*
Venules	10–12[a]	*29.0[f]*
Small veins	*15[c]*	up to 11.0[c]
Large vein	0–2[b,g]	3.0[c]

Note: Italics indicate measurements taken in cat microvasculature.

[a] Attinger (1963).
[b] Guyton and Hall (1996).
[c] Samet and Lelkes (1999).
[d] Moore et al. (1994).
[e] Lipowsky et al. (1978).
[f] Intaglietta et al. (1971).
[g] Sherwood (2001).

that transports blood past the valves and in the direction of the heart; this reduces venous pressures in the leg restarting the whole process. Finally, based on measurements in the cat microcirculation (which are expected to be similar to those of humans), mean pressures in the arterioles and venules are ≈30 and ≈15 mmHg, respectively (Lipowsky et al., 1978). Blood-borne leukocytes are therefore exposed to a wide range of pressure regimes that vary in magnitudes (from 0 to ≈10^3 mmHg), pulsatility, and frequency.

8.3.2 Fluid Shear Stress

Compared to the normal stress components due to blood flow, shear stress magnitudes are considerably lower, typically in the range between 1 and 10 dyn/cm². One gets an idea of how low these shear stress magnitudes are compared to vascular pressures when considering that 1 mmHg = 1330 dyn/cm² or that a column of water 1 cm high exerts a pressure equivalent to 981 dyn/cm² on the container walls. Interestingly, shear stress levels on the blood vessel wall (i.e., wall shear stress) never vary by more than 1 order of magnitude over the entire circulation (Table 8.2) (Lipowsky et al., 1978; Kamiya et al., 1988).

Due to the nonuniform (typically parabolic) radial velocity gradient, fluid shear stress levels in the vasculature vary within a given vessel lumen. For a Newtonian fluid (e.g., plasma), the radial gradient in shear stress in a flow channel with cylindrical cross section (such as a blood vessel) and parabolic velocity profile depends on the location along the radius (r). With respect to the radial direction, fluid shear stress (τ) is linearly proportional to the velocity gradient ($\partial u/\partial r$) by the coefficient of viscosity (μ), as formulated in Equation 8.1:

$$\tau = \mu \left(\frac{\partial u}{\partial r} \right) r \tag{8.1}$$

The fluid shear stresses in the vasculature are, therefore, determined by the velocity gradients (i.e., shear rates) that characterize blood flow. For cylindrical Newtonian fluid flow, shear stress magnitudes are minimal (with values close to zero) at the center of a vessel and gradually increase to a maximum at the endothelium on the vessel wall (typically referred to as the wall shear stress). Because the white blood cells (as well as red cells) circulating in the vessels are freely suspended in the flowing blood, they are immersed in this shear stress gradient (Table 8.2). As such, suspended cells in the blood flow may rotate and experience time-variant shear stresses on their surface if they are flowing off the centerline; the time dependence is a function of their radial position in the flow field. Moreover, stochastic interactions with other flowing cells also affect the magnitude and time dependency of shear stresses that they experience.

In the blood vessel lumen, there are two major features that cause local surface-associated variations in the shear stress magnitudes at or near the wall: (1) the nonuniform thickness of the endothelium, and (2) the time-dependent nature of blood flow. The endothelium, although relatively flat in most vessels, still exhibits microscale variations in thickness, which may impact blood flow, particularly in microvessels such as the capillaries. For instance, regions of the cell membrane overlying the nuclei of the endothelial cells protrude into the bloodstream and are exposed to higher shear stress magnitudes compared to those membrane regions at the junction between cells where shear stresses are close to zero (Barbee et al., 1995). In addition, evidence is reported for endothelial cell projections (i.e., pseudopods) that are extended into the lumen of the capillaries and influence blood flow; these projections are proposed to serve a role in the regulation of microvascular perfusion (Hueck et al., 2008). Finally, the time dependence of the wall shear stress results from the pulsatile variations of the blood velocity occurring with each heartbeat, and the passage of blood cells that leads to variations in shear stresses at higher frequencies due to the stochastic nature of cell–cell interactions that affect blood movement.

Interestingly, the wall shear stress magnitude (estimated assuming a steady cylindrical flow), in general, is used as an indicator of fluid flow conditions in the circulation, particularly as it relates to the impact on the vascular cell components (e.g., the endothelium, adherent leukocytes). Mean wall shear stresses in the large arteries and veins (e.g., pulmonary artery, aorta, vena cava) range from approximately 2.7 to 4.5 dyn/cm² and can reach as high as 32 dyn/cm² in the small arteries and 11 dyn/cm² in the small veins (Samet and Lelkes, 1999). Moreover, Moore et al. (1994) determined that mean wall shear stresses in the abdominal aorta are approximately 1.3 dyn/cm² with exact values at a given instance in time during the pulsatile flow cycle ranging from −4.1 to 8.4 dyn/cm². The wall shear stresses, however, become near steady in the microcirculation and in the veins (with some exception such as those in the lower limbs) due to the capacitance of downstream blood vessels.

As with blood pressure, wall shear stresses fluctuate as the physical activity level of subject changes; for example, wall shear stresses in the human infrarenal aorta at rest (1.3 ± 0.8 dyn/cm²) can increase by a factor of 4 during lower limb exercise (5.2 ± 1.3 dyn/cm²) (Taylor et al., 2002). In addition, regions of disturbed flow (such as at arterial bifurcations where blood flow is high and flow contraction/expansion greatly alters the shear stress distributions on the vascular endothelium) are characterized by high wall shear gradients (in time and space) on the blood vessel lumen (DePaola et al., 1992). Adherent and/or migrating leukocytes on the vascular wall are, therefore, exposed to wall shear stresses that are governed by the existing flow conditions (e.g., associated with physical activity levels) and geometry of the blood vessel.

Based on estimates in the cat microcirculation, shear stresses in the arterioles, capillaries, and venules are expected to reach approximately 47, 30, and 29 dyn/cm², respectively (Intaglietta et al., 1971). In the capillaries, the situation is unique. Leukocytes travel in a single-file configuration and deform to the size of the vessel creating a thin viscous "lubrication layer" between the cell surface and the vessel wall (Figure 8.4). An important consequence of this lubrication fluid layer is that it prevents contact between the leukocyte and the endothelium lining the capillary walls. Moreover, shear stresses and shear rates are greatly enhanced in the plasma lubrication layer

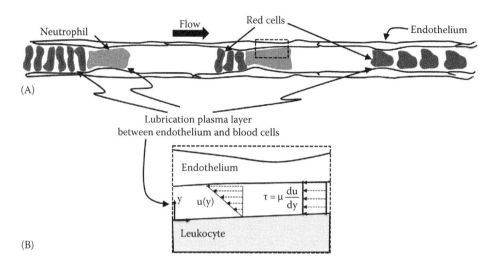

FIGURE 8.4 Single-file configuration of blood cells flowing through capillaries. (A) Leukocytes and red cells adopt a single-file arrangement when flowing through the capillaries. Since capillary diameters are typically smaller than the diameters of both red cells (8–10 μm) and nucleated leukocytes (10–15 μm), the cells deform. A lubrication film of plasma with a thickness much lower than the diameter of the capillaries prevents contact between the flowing blood cells and the endothelium on the blood vessel lumen. (B) This slipstream, however, imposes membrane fluid shear stresses governed by the thickness of the film layer. The shear stresses generated within these regions are modeled by Poiseuille approximation.

between the leukocytes and endothelial membrane in the capillaries (Tozeren et al., 1982); shear stresses generated in the lubrication layer have been computed to be on the order of 10–50 dyn/cm² (Secomb and Skalak, 1982; Sugihara-Seki and Schmid-Schönbein, 2003) resulting from the thin gap between the cell surface and the vessel lumen. Such shear stress values are based on a fluid flow geometry resembling two parallel surfaces with one surface (e.g., the leukocyte surface) moving laterally with respect to the other (e.g., endothelium); this is modeled by classic Couette flow approximation.

In summary, as with the case for circulatory pressures, blood-borne leukocytes are subjected to a range of shear stresses that vary in magnitudes, pulsatility, and frequency. These variations depend not only on the location within the vasculature, but also the leukocyte position in the flow field. Moreover, the magnitude of the shear stress applied to the leukocyte surface also depends on other factors at the microscale (on the order of the cell diameter).

8.3.3 Distribution of Fluid Mechanical Stresses over the Neutrophil Surface

In general, cells exhibit a high sensitivity to macroscale fluid (shear and normal) stress magnitudes in the circulation. In the vasculature, this is evident from a wide variety of responses by leukocytes and vascular endothelium to estimated mechanical stress values that mimic (typically at or near the wall of the vessels; that is, the fluid flow boundary layer) the high pressure/flow environment of the macrocirculation as well as to the considerably lower magnitudes of flow-related stresses in the microcirculation. To further illustrate this point, even osteocytes normally found in the dense mineralized bone matrix respond to fluid mechanical stresses that mimic the low shear/pressure magnitudes predicted to occur in the interstitial fluid environment of bone (Klein-Nulend et al., 2005; Robling et al., 2006; Rubin et al., 2006).

Given the evidence substantiating high cellular sensitivity to fluid normal and shear stresses, it is conceivable that variations in the magnitude of these stresses along the cell surface could impact neutrophil physiology in a spatially dependent fashion. As such, the need to characterize the complex distribution of fluid shear and normal stresses on the cell is evident. In fact, there are several factors that influence the fluid stress distribution on the surface of a neutrophil in the circulation: (1) location in the circulation, (2) positioning within a flow field (e.g., within a blood vessel), (3) geometric topography of the cell surface, and (4) the presence of neighboring cells or similar-sized particles (Schmid-Schönbein, 2006; Makino et al., 2007).

As previously mentioned, the normal and shear stresses applied to a cell vary whether in the arterial, venous or microvascular systems. In addition to the dependence on the velocity distribution within the blood vessel lumen (as discussed in Section 8.3.3), neutrophils located off the centerline will rotate due to the velocity gradient imposed on opposing sides of the cell (Sugihara-Seki and Schmid-Schönbein, 2003). A single point on the membrane as it rotates with the cell will, therefore, experience a time-variant shear stress (Figure 8.5).

Variations in shear stress on the cell membrane will also be influenced by the presence of neighboring cells. The effects of neighboring cells on the fluid stresses generated on the neutrophil surface will result primarily from the contributions of nearby red blood cells that influence the local flow field and that may collide with nearby leukocytes. The neutrophils near or at the wall of the blood vessel experience the highest shear stress magnitudes. In fact, as a neutrophil makes its first contact with the wall during rolling membrane interactions with the endothelium, the shear stress dramatically increases as the cell decelerates due to the formation of adhesive bonds and generation of a fluid drag on the membrane. As the cell activates and spreads, these drag force-associated shear stresses stay relatively constant (Su and Schmid-Schönbein, 2008). Additionally, a shear stress gradient is created between the apical and basal regions of the cell surface; shear stresses at the apical cell surface are at a maximum compared to those regions on the basal side where the cell interacts with the vascular endothelium through integrin and other cell–cell adhesion molecules (Su and Schmid-Schönbein, 2008).

Finally, the fine surface topography of a neutrophil also influences the spatial distribution of fluid shear stresses on the plasma membrane, resulting from the surrounding macroscale fluid flow conditions. As previously mentioned, the neutrophil surface consists of membrane folds, which provide the cell with the ability to undergo large deformations (approximately 2.6-fold increases in surface area), thereby facilitating transit through the small of the microcirculation. These membrane folds contribute to high spatial gradients of fluid shear stress on the cell membrane. The tips

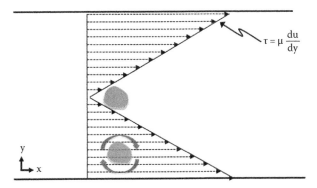

$$\tau = \mu \frac{du}{dy}$$

y

x

FIGURE 8.5 Schematic representation depicting induction of cell rotation for a leukocyte (modeled as a circular solid) flowing in blood. Given blood flow is modeled as Poiseuille flow in a cylindrical tube; leukocytes immersed in the blood and located off the centerline will experience different magnitudes of fluid shear stresses on opposing sides resulting in the induction of rotation. Due to this rotation, leukocytes experience time-variant membrane shear stresses on their surface that act in opposition to the viscous forces imposed by fluid flow. On the other hand, leukocytes flowing at the centerline experience minimal rolling. For the flow field depicted, the shear stress (τ) is linearly proportional to the velocity gradient with respect to the y-direction (du/dy) by the viscosity (μ).

of the membrane folds experience the highest shear stress magnitudes, while the valleys between neighboring folds are associated with low to near-zero shear stresses (Su and Schmid-Schönbein, 2008); this sheltering effect can serve as a means to control the activation of mechanosensitive receptors by fluid shear stimulation while at the same time maintain its permissiveness to soluble ligand interactions.

The existence of a protein surface layer (called the glycocalyx) on the cell surface that is proposed to facilitate ligand–receptor interactions leading to downstream molecular signaling points to another geometric surface feature that may influence the magnitudes of fluid shear stresses directly experienced by the neutrophil membrane or its constituent integral proteins (Tarbell and Ebong, 2008). The existence of cellular mechanisms (Mulivor and Lipowsky, 2002, 2004) associated with cleavage of the glycocalyx may thus not only affect neutrophil biochemical signaling but also expose putative mechanoreceptors on the cell surface to threshold fluid shear stress magnitudes resulting in a mechanotransduction response. In fact, heparinase-modification of the glycocalyx (enriched in heparan sulfate proteoglycans) on the cell surface of neutrophils has been shown to enhance neutrophil shear stress responses (Coughlin and Schmid-Schönbein, 2004), pointing to the presence of flow-sensitive transmembrane proteins that link the extracellular mechanoenvironment with intracellular mechanotransducing pathways. The nature of the mechanotransduction responses of neutrophils is the subject of the following sections.

8.4 LEUKOCYTE RESPONSES TO FLUID PRESSURE

To our knowledge, there is no reported evidence that hydrostatic or hydrodynamic fluid pressures, of magnitudes representative of those throughout the circulation, influence neutrophil functions. In fact, based on morphological parameters, neutrophils are insensitive to pressures or normal stresses up to $1000 \, dyn/cm^2$ (Moazzam et al., 1997), which are approximately 2–3 orders of magnitude higher in value compared to the shear stress levels present in the circulation. Although these values are well below those present in the circulation ($1000 \, dyn/cm^2 = 0.75 \, mmHg$), they are likely in the range of the dynamic pressure component (related to the kinetic energy associated with fluid flow), which contributes to the numerical value of the total pressure at a location in the vasculature (see Section 8.3.1). The major contribution to the total pressure is induced by the pumping actions of the heart, which are up to approximately 2–3 orders of magnitudes higher. It should be pointed out, however, that there is a lack of any evidence regarding pressure stimulation as a modulator of neutrophil signaling or other functions independent of morphological changes.

The possibility that pressures, in the range of in vivo values, modulate the functions of neutrophils is substantiated by evidence reported in the literature, particularly for endothelial cells as well as monocytes and the related macrophages. Endothelial cell functions associated with capillary/blood vessel formation (e.g., proliferation, morphology/migration, expression of angiogenic molecules) (Acevedo et al., 1993; Kato et al., 1994; Sumpio et al., 1994; Vouyouka et al., 1998; Schwartz et al., 1999; Shin et al., 2002a,b, 2004), vasomotor activity (Hishikawa et al., 1992, 1995), and barrier function (Shin et al., 2003; Muller-Marschhausen et al., 2008) are modulated by hydrostatic and/or pulsatile pressures. As an indicator of how sensitive these cells are to pressure stimulation, it is reported that endothelial cells respond to pressures as low as $1.1 \, mmHg$ (Acevedo et al., 1993; Schwartz et al., 1999), although their responses require at least 24 h. The time frame of a neutrophil chemotactic response is likely considerably shorter considering their rapid activation and responses to biochemical stimuli as well as their short half-life in the circulation.

In the case of leukocytes other than neutrophils, blood-derived monocytes and macrophages exhibit sensitivities to hydrostatic and hydrodynamic pressure stimulation. Exposure to elevated static pressures of 40–130 mmHg above atmospheric in a Boyden chamber setup stimulated migration of U937 monocytes in a magnitude-dependent fashion that involved cytosolic calcium (Ca^{2+}) levels (Sakamoto et al., 2001). Moreover, primary human monocytes exposed to 20 mmHg static pressures for 2 h exhibited enhanced phagocytic activity (Shiratsuchi and Basson, 2005, 2007; Shiratsuchi et al., 2009). In fact, static (within the range of 40–130 mmHg) as well as pulsatile (up to 1000 mmHg) pressures have been shown to modulate phagocytosis as well as alter the expression of scavenger receptor A and various proinflammatory cytokines by monocytes/macrophages (Sakamoto et al., 2001). The pressure mechanotransducing capacity of the monocytes and macrophages is further evidenced by the observations of Shiratsuchi et al. (Shiratsuchi and Basson, 2004, 2005, 2007; Shiratsuchi et al., 2009) demonstrating putative involvement of p130cas cytosolic signaling as well as altered MAPKα, FAK, Akt2, and ERK activity in the enhanced phagocytic capacity of THP-1-derived macrophages exposed to pressure.

Taken together, the evidence presented in this section points to the role of fluid pressure in modulating macrophage phagocytosis. As such, these studies, although for different cell types, give credence to the possibility that pressure mechanosensitivity is a fine tune controller of cellular activity levels in leukocytes. At this time, additional research is needed to address the critical issue regarding the regulation of neutrophil activity by fluid mechanical pressure under physiological and pathological conditions.

8.5   MECHANOBIOLOGICAL CONTROL BY FLUID SHEAR STRESS

8.5.1   Pseudopod Retraction Model of Neutrophil Deactivation by Fluid Shear Stress

Whether floating in suspension or migrating on a surface, neutrophils sense and respond to fluid shear stress (Moazzam et al., 1997; Fukuda et al., 2000; Fukuda and Schmid-Schönbein, 2002). Overall but with some exceptions (Coughlin and Schmid-Schönbein, 2004; Coughlin et al., 2008), evidence accumulated to date primarily paints a picture whereby physiological levels of fluid shear stress (ranging from approximately 1 to 5 dyn/cm^2) minimize neutrophil activity levels as well as prevent neutrophil activation below a putative threshold level (Makino et al., 2007). The most obvious indicator of the negative effect of shear stress stimulation on cell activity is the retraction of existing pseudopodia by non-cytokine-stimulated human neutrophils migrating on glass under the influence of a nonuniform flow field imposed by a micropipette with a tip of diameter in the range of 4–8 µm (Figure 8.6). This situation models brief and spontaneous periods of ischemia followed by reperfusion occurring in the microvessels. During periods of no flow, neutrophils present in the regions where flow is interrupted sediment, attach, extend pseudopods, and migrate on the vascular endothelium. Upon reintroduction of fluid flow, these cells retract pseudopods and detach into the flow field. Such a scenario has been documented using intravital microscopy preparations of in vivo microvascular networks of rats (predominantly in the mesentery, but also other microvascular beds such as the spinotrapezius muscle) (Moazzam et al., 1997; Fukuda et al., 2000) and mice (mainly the cremaster muscle) (Makino et al., 2005). The ability of fluid shear stress to influence pseudopod activity, however, has been further confirmed for nonadherent, but activated, heterogeneous populations of leukocytes (Komai and Schmid-Schönbein, 2005). In this case, suspensions of leukocytes exposed to a uniform shear field in a cone-plate viscometer (5 dyn/cm^2) (Figure 8.7) exhibit reduced percentages of neutrophils that project pseudopods likely formed due to the in vitro handling conditions of experiments.

Pseudopod formation is the most obvious morphological feature visible under microscopic examination that demarcates cell activation; it is, in fact, a hallmark characteristic of an activated leukocyte associated with cell adhesion, migration, and phagocytosis during acute inflammation (Mazzoni and Schmid-Schönbein, 1996). During rapid migration resulting from cellular activation (e.g., f-Met-Leu-Phe [fMLP]), pseudopods serve as "cellular extensions" to provide a structural basis for neutrophil membrane extension, migration, and chemotaxis by serving as an extendable anchor to facilitate displacement of the cell body; the trailing edge then turns into a short stretched out membrane region, called a "uropod," that is retracted (Ehrengruber et al., 1996; Cicchetti et al., 2002). In the case of suspended cells, pseudopodia help to establish membrane attachment points

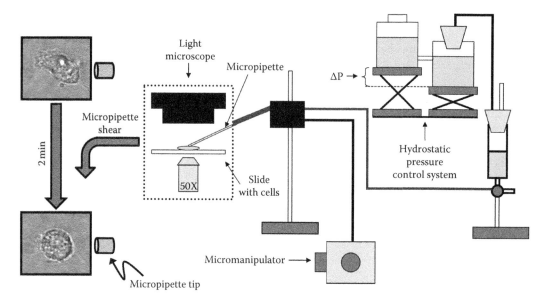

FIGURE 8.6 Deactivation of individual migrating neutrophils under flow stimulation. Single human neutrophils (shown in micrographs) were exposed to shear stress ($\tau \approx 2 \, dyn/cm^2$) using a micropipette flow system for 2 min. Flow was generated using an adjustable hydrostatic pressure head and micropipette tips (diameter $\approx$ 4–8 μm) were positioned $\approx$8 μm from the cell surface. During experiments, pseudopod activity was recorded in real time using a camera interfaced to an inverted light microscope. Neutrophils exhibited pseudopod retraction under fluid flow stimulation for 2 min.

that promote capture onto a flat substrate, for example, the vessel wall, or on a curved substrate, for example, phagocytic entrapment of particles.

At the same time, pseudopods contribute to a reduction in cell deformability due to their enriched content of polymerized cytoskeletal actin, and to an increase in geometric size resulting from unfolding of the neutrophil membrane that accommodates the added cell surface area for pseudopod projection As such, pseudopod extension by circulating leukocytes impairs their ability to transit the small vessels of the microvasculature (e.g., capillaries) (Worthen et al., 1989; Fukuda et al., 2004b), something that is possible for quiescent (inactivated) leukocytes that are rounded, freely floating, and mechanically passive. Quiescent neutrophils with dimensions of 12–15 μm undergo deformations in order to pass through the smaller capillaries with diameters of 5–10 μm. Another effect that pseudopods have on neutrophil behavior in the microcirculation is to influence stochastic interactions with red cells where margination plays a critical role in facilitating the first neutrophil–endothelial cell interactions (Helmke et al., 1998).

Exposure to shear magnitudes typically found in the macro- and micro-circulations is also associated with other attributes of neutrophil deactivation such as decreased cell surface levels of integrin adhesion receptors (i.e., CD18), depolymerization of the F-actin cytoskeleton related to reduced cell stiffness, cell detachment, and attenuated phagocytic activity (Moazzam et al., 1997; Fukuda et al., 2000). In fact, fluid shear stress exposure also enhances caspase 3-dependent neutrophil apoptosis (Shive et al., 2002), in line with the relatively short lifetime (18–24 h) of these cells when they are passively circulating under physiological conditions. These important results are consistent with the role of fluid shear stress as a biophysical stimulus that induces neutrophil inactivation below some threshold level of inflammatory activity.

It is important to point out, however, the existence of a subpopulation of cells within an ex vivo preparation of migrating leukocyte populations (or even suspended in blood) that either do not retract pseudopods or actually project pseudopods upon exposure to a shear flow (Coughlin and Schmid-Schönbein, 2004). Moreover, rounded neutrophils adhered to glass substrates exhibit

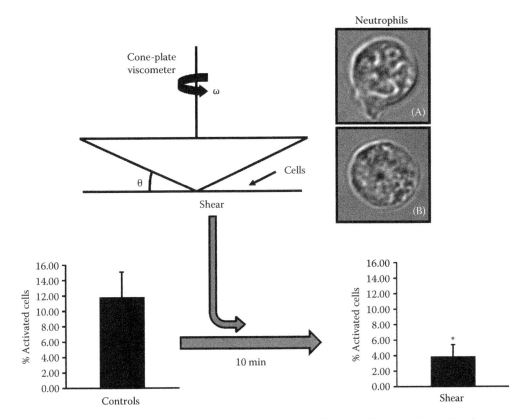

FIGURE 8.7 Deactivation of suspended (nonadherent) neutrophils under flow stimulation. Murine neutrophil populations in whole blood diluted 1:20 (v/v) in Hank's buffered saline were exposed to shear stress ($\tau \approx 5\,\text{dyn/cm}^2$) in a cone-plate viscometer for 10 min. Controls were unsheared cell populations. At the end of experiments, cells were fixed and microscopically examined for pseudopod activity (by microscopic observation). Cells with pseudopods (A) were considered activated while those without (B) were scored as inactivated. Bars are mean percentage of activated cells (expressing pseudopods) in each population tested $\pm$ SD; *p < 0.05 compared to controls using paired Student's t-test.

recoil and stiffening of their cell body when initially subjected to shear flow that occurs rapidly (within a minute) at flow onset and after (or possibly in response to) an initial passive displacement of the cell centroid in the flow direction (Coughlin et al., 2008). This recoil and stiffening is proposed to result from the recruitment of adhesion receptor–substrate interactions that promote neutrophil attachment and migration. Collectively, these data point to the sensitivity of the neutrophils to conditions in the extracellular milieu where, as yet, undetermined factors mitigate conversion of a neutrophil from a pseudopod-retracting to a pseudopod-projecting cell under shear stress exposure.

In fact, centrifugation, commonly used for purifying neutrophils from blood, attenuates and even reverses the pseudopod retraction responses of these cells to shear exposure (Fukuda and Schmid-Schönbein, 2002). This attenuating effect of centrifugation on the responsiveness of neutrophils to fluid shear is a testament to the sensitivity of their mechanotransducing abilities to environmental conditions and has proved to be an experimental impediment for studying the underlying cellular and molecular mechanotransduction mechanisms associated with neutrophil inactivation by fluid shear stress. Due to this technical limitation, the majority of experiments to assess neutrophil inactivation by fluid shear stress is conducted using either whole blood or buffy coat preparations. As such, experimental results can be quite variable and dependent on the composition and state of the blood at the time of its collection from human volunteers or animals.

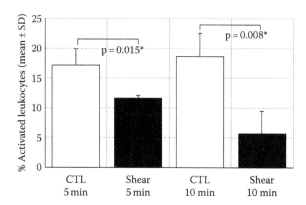

FIGURE 8.8 HL-60-derived neutrophils exhibit deactivation under flow stimulation. Pseudopod activity of HL-60 cells that had been differentiated into neutrophilic leukocytes were exposed to cone-plate flow ($\tau \approx 5\,\text{dyn/cm}^2$) for 5 or 10 min. Controls were cells under no-flow, but otherwise, similar experimental conditions. At the end of experiments, cells were fixed and microscopically examined for pseudopod activity (by observation). Bars are mean percentage of activated cells (expressing pseudopods) ± SD; p-values calculated with Student's t-test.

However, the fact that fluid shear stress, in addition to inducing pseudopod retraction by migrating neutrophils, correlates with reduced pseudopod activity of nonadherent populations of human neutrophils (Fukuda et al., 2000; Komai and Schmid-Schönbein, 2005), including those derived from HL-60 promyelocytes (a model neutrophilic cell line derived from the blood of a patient suffering from leukemia; Figure 8.8), substantiates that fluid flow stimulation negatively regulates cell activity levels.

8.5.2 SHEAR STRESS: A BIOPHYSICAL MODULATOR OF NEUTROPHIL BEHAVIOR

Fluid flow influences leukocytes in the vasculature by two interdependent modes of action: in the form of either (1) shear stress or (2) shear rate (i.e., the spatial velocity gradient). Whereas shear stress represents a physical force per unit surface area causing deformation of or changes in molecular configuration, shear rate determines macromolecular or intercellular transport in the local milieu of the cell, for example, associated with a paracrine/autocrine-based mechanotransduction mechanism whereby fluid flow facilitates transport of ligand to its receptors on the cell surface.

Shear rate appears to play a role in leukocyte–endothelial interactions in the postcapillary venules. For example, leukocyte tethering and rolling results from the formation of carbohydrate-selectin bonds between leukocytes and endothelial cells at the leading edge with simultaneous breakage of bonds at the trailing edge; these events are governed by rates of bond association (k_{on}) and dissociation (k_{off}). In terms of bond kinetics, shear rates influence bond association rates directly via regulating the transport of cell-associated receptors to counter-receptors on other cells (Finger et al., 1996; Chang and Hammer, 1999; Chen and Springer, 1999). Moreover, fluid shear rates may influence bond dissociation rates indirectly through shear stresses and thus result in tensile forces generated at the selectin–ligand bonds; tensile forces at the sites of selectin–ligand interaction may serve to either reduce dissociation rates to prolong bond lifetimes (catch bonds) or enhance dissociation rates to shorten bond lifetimes (slip bonds) (Dembo et al., 1988) depending on the magnitudes and loading rates of the stresses imposed by the surrounding fluid flow. Finally, shear rates also affect leukocyte residence times at the vascular endothelium; for example, higher shear rates are associated with reduced residence times for leukocytes at the endothelium and thus compromised efficiency of adhesion bond formation. Bond kinetics, bond mechanics, and cell transport, therefore, contribute to the existence of an optimal shear rate for selectin-mediated neutrophil rolling.

Chemokine-induced engagement of neutrophil CD18 integrins to their respective ligands on the endothelial surface during fluid flow, however, has been shown to depend on local fluid shear stresses. Specifically, fluid flow facilitates bond strengthening and cell capture via a putative mechanism whereby apical shear stresses on the neutrophil surface influence CD18 interactions at the basal surface in contact with the endothelium (Cinamon et al., 2001, 2004; Shulman et al., 2009). These results suggest bond formation due to shear-induced conformational shifts in adhesion receptors as they bind to their counter-receptors under fluid flow conditions. In the case of neutrophil–endothelial cell interactions during acute inflammation, it is, therefore apparent that the contributions of shear rate and shear stress are interdependent.

Typically, the dependence of a biological response on either fluid shear stress or shear rate can be discerned by taking advantage of the linear relationship between the two that is characterized by the coefficient of viscosity of a Newtonian fluid, for example, plasma or buffer. Thus, by changing the viscosity of a buffer solution while maintaining constant shear rate, one can determine the dependence of a biological response on either of these two fluid flow-related stimuli (i.e., shear rate vs. shear stress). This type of strategy has been utilized to confirm the dependence of nitric oxide (NO) release by endothelial cells (Bao et al., 1999) as well as by osteoblasts (McAllister and Frangos, 1999) exposed to fluid flow conditions on shear stress, and not shear rate. In the case of nonadherent neutrophils suspended in blood stimulated with low levels (10^{-5} M) of fMLP, fluid flow exposure reduces the percentage of activated cells in the sample solutions in a shear stress, and not shear rate, dependent fashion (Komai and Schmid-Schönbein, 2005). These results, therefore, provide evidence that fluid flow-induced neutrophil inactivation involves deformation-inducing stresses on the cell.

A key aspect of shear-induced neutrophil deactivation is that they occur in the absence of any passive cell deformation due to flow (Sugihara-Seki and Schmid-Schönbein, 2003), indicating the presence of a cell surface component(s) that initiates or modulates the cascade of biological events resulting from mechanical perturbations generated in the surrounding extracellular fluid milieu. Moreover, neutrophils retract pseudopods independently of the fluid shear stress distribution imposed on the cell surface by exposure to micropipette flow (Moazzam et al., 1997). This was further confirmed by Su and Schmid-Schönbein (2008) who used a computational fluid mechanics model in conjunction with three dimensional reconstructions of cultured leukocytes (using confocal microscopy) to assess the fluid stress distribution on a single HL-60-derived neutrophil retracting pseudopods under the influence of 2.2 dyn/cm^2 wall shear stress in a parallel plate flow chamber for 3 min. Interestingly, leukocytes retracted pseudopods in all regions independently of its orientation with respect to the flow field and the distribution of the fluid stress components (normal and shear) imposed by the macroscale fluid flow (Figure 8.9). Moreover, the shear stress component is the dominant fluid stress component experienced by a migrating neutrophil in a flow field. Taken together, these data substantiated the existence of a membrane surface-associated component sensitive to shear stress stimulation that initiates neutrophil mechanotransduction.

The actin cytoskeleton, a structure that regulates cell-signaling processes in addition to its role in governing cell shape and motile function, has also been implicated as a putative mechanotransducer. The actin cytoskeleton is a dynamic structure that is distributed throughout the cell cytoplasm and that, at any given instant in time, can be in different stages of fiber-bundle formation and crosslinking, both of which depend on the level of cell activity. In addition to contributing to the shape of a cell, the location of cytoskeletal fibers (including the actin bundles called stress fibers, Satcher and Dewey, 1996; Satcher et al., 1997) determines the number, location, and size of cell attachment points (focal adhesions) on the basal surfaces of adherent cells (e.g., endothelium, smooth muscle cells, fibroblasts, osteoblasts, migrating leukocytes). In the context of the surrounding mechanoenvironment, the resultant fluid force imparted by fluid shear stress on the surface of a cell that is exposed to the blood stream is balanced by the stresses at these focal adhesion

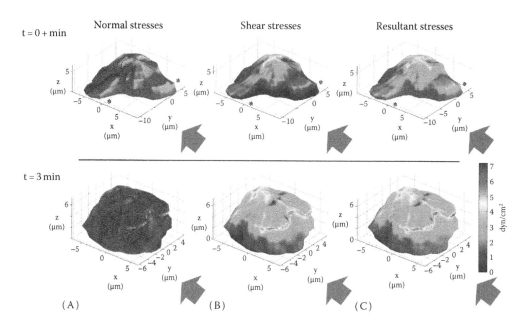

FIGURE 8.9 (See color insert.) Shear-induced retraction of neutrophil pseudopods is independent of their orientation in the flow field and the stress distribution on the cell surface. Three-dimensional confocal microscopic images of HL-60-derived neutrophils exposed to 2.2 dyn/cm² wall shear stress (direction indicated by block arrows) was acquired at t = 0+ and at t = 3 min after the onset of flow of a parallel plate geometry over the cell. Variations in the cell surface geometry leads to a heterogeneous stress distribution over the cell. Within the time frame of shear experiments, the cell retracted existing pseudopods (*). Three-dimensional reconstructions of the cell at the two time points were used for computational fluid mechanics estimations of the distributions of the normal (A) and shear stress components (B) as well as the resultant (the vector combination of the normal and shear) stress (C) on the cell surface. Note that the shear stress component imposed by fluid flow is the dominant stress experienced by the cell compared to the normal component. (Image courtesy of Dr. Susan Su.)

points. Changes in the fluid stresses at the top of cells can thus lead to perturbations in the stress state of the focal adhesions and subsequently alter cytoskeleton-related mechanobiological signaling activity (Ingber, 2008); this idea is elaborated by the so-called decentralized theory of cellular mechanotransduction.

The fact that nonadherent neutrophils freely suspended in blood also respond to shear stress, however, substantiates the involvement of mechanotransducing elements on the peripheral cell membrane that are independent of focal adhesion molecules. As such, the cell surface, particularly the plasma membrane, has been implicated in mechanotransduction due to its strategic location between the intra- and extra-cellular milieu as well as its rich content of receptors, signaling molecules, and second messengers (Seely et al., 2003; White and Frangos, 2007). The fact that the peripheral cell membrane is responsible for orchestrating pseudopod and uropod activities by influencing adhesion receptor function (such as CD18), transport of Ca²⁺ and other ions, and the actin cytoskeleton (Bodin and Welch, 2005) suggests that this structure mediates neutrophil shear responses.

In fact, the membrane itself may act as a mechanotransducer (Haidekker et al., 2000; Butler et al., 2001, 2002) via stress-induced changes in its fluidity. Membrane fluidity is known to regulate neutrophil phagocytosis (Berlin and Fera, 1977; Wiles et al., 1994) and chemotaxis by influencing chemokine-specific receptor number and affinity for ligands (Yuli et al., 1982; Tomonaga et al., 1983). The membrane may also transduce shear stress (Rizzo et al., 1998a,b; Ferraro et al., 2004) through lipid rafts, that are dynamic clusters of saturated lipids, cholesterol, and sphingolipids and that act as signaling platforms by providing spatial organization for select membrane proteins (Simons and Toomre, 2000). For example, lipid rafts regulate neutrophil signal transduction (chemokine-induced calcium flux, cell polarization, migration, integrin expression, and actin dynamics, etc.) and functions (shape change, motility, etc.) (Seely et al., 2003; Tuluc et al., 2003; Niggli et al., 2004) as well as CD18-mediated T-cell adhesion (Marwali et al., 2003) and neutrophil adherence (Solomkin et al., 2007). Thus, shear stress may modulate neutrophil activity via membrane mechanotransduction.

On the other hand, the concept that the membrane serves as a fluid stress sensor lacks the specificity that explains the diversity of cell type-specific responses to shear. The specificity associated with mechanotransduction, therefore, likely depends on the cell membrane's enriched content of a cell-specific variety of signaling molecules, some of which may be mechanosensitive. In fact, a multitude of cell transmembrane proteins including various GPCRs (Chachisvilis et al., 2006; Makino et al., 2006; Zhang et al., 2009), tyrosine kinase receptors (Chen et al., 1999; Shay-Salit et al., 2002; Jin et al., 2003; Lee and Koh, 2003; Milkiewicz et al., 2007), ion channels (Tarbell et al., 2005), and integrin-associated focal adhesions (Kamm and Kaazempur-Mofrad, 2004; Lee et al., 2007) as well as signaling proteins such as G-proteins (Gudi et al., 1996, 1998) have been implicated as fluid shear stress transducers for a variety of cells (e.g., endothelial cells, osteoblasts, neutrophils). Moreover, the cytoskeleton that links these surface receptors to the cytosolic machinery appears to play an integral role in this mechanotransduction process particularly since a majority of these transmembrane molecules are linked to this network of F-actin filaments and microtubules.

In the end, pseudopod retraction by migrating neutrophils in response to fluid shear stress points to two requirements that must be fulfilled: (1) depolymerization of the F-actin that provides structural integrity and serves as a signaling scaffold for promoting neutrophil motility, and (2) rapid disengagement of adhesion receptors that anchor the pseudopod to the underlying substrates. For suspended neutrophils, similar events are needed but, in this case, mechanisms must be in place to prevent the expression of adhesive proteins or interfere with adhesion molecule engagement with substrate counter-receptors, which can promote neutrophil attachment and migration.

8.5.3 THE ACTIN CYTOSKELETON, SMALL GTPASES, GPCRS, AND SHEAR-INDUCED PSEUDOPOD RETRACTION

Pseudopods are local cytoplasmic projections physically supported by polymerized cytoskeletal proteins such as F-actin. These structures are formed following a process whereby the cell surface projects membrane processes called lamellipodia (i.e., membrane ruffles; not to be confused with smaller membrane folds on a passive neutrophil surface) resulting from polymerization of monomeric cytoskeletal proteins (e.g., G-actin to F-actin fibers) into a filamentous network at the leading edge (Zhelev and Alteraifi, 2002). As these structures increase in size, cytoplasmic material flows. Hence, pseudopods projected by activated neutrophils are actin-enriched cellular projections (Figure 8.10) filled with cytoplasmic material and typically of a width that is on the same order of magnitude as the diameter of the neutrophil (Ehrengruber et al., 1996). In comparison, inactivated neutrophils (e.g., after shear stress exposure) do not extend pseudopods and exhibit F-actin localized to the cell cortex (Figure 8.10).

Rapid migration of neutrophils along the endothelium during early recruitment stages (i.e., capture/arrest) of acute inflammation has been extensively studied and reviewed in the literature (Cicchetti et al., 2002; Zhelev and Alteraifi, 2002; Niggli, 2003). In summary, neutrophil motility depends on swift formation and breakdown of cytoskeletal networks during cyclical lamellipod and pseudopod formation and eventual retraction of a uropod as the cell body moves. This leads to polarization of the cell with lamellipodia and/or pseudopodia at the leading edge; the uropod makes up the trailing edge of a motile cell. Notably, at any stage of lamellipod and pseudopod formation, these developing cellular projections may retract and either extend in a different direction or remain absent, thus, curtailing cell movement or phagocytic events. The speed with which a neutrophil

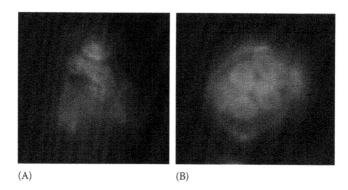

(A)   (B)

FIGURE 8.10 (See color insert.)   F-actin distribution in human neutrophils. Activated (A) and shear-inactivated (B) neutrophils were fixed, labeled with phalloidin (F-actin staining at the cell periphery) and propidium iodide (nuclear staining). *Note*: neutrophils are 10–15 μm in diameter.

moves is determined by cytosolic stores of monomeric G-actin that are transported toward the leading edge of the pseudopod where they incorporate and polymerize into the tips of filamentous F-actin fibers. Moreover, membrane unfolding allows the cell pseudopod to accommodate the increase in surface area over a growing pseudopod.

Pseudopods, in the process of serving as anchors for migration, are enriched with chemokine receptors at the leading edge, which provide directional cues toward sites of inflammation based on the presence of local chemokine gradients originating from the site of injury or infection. The integral role of pseudopods and rapid cytoskeletal remodeling in neutrophil motility, therefore, implicates regulatory mechanisms related with F-actin depolymerization in shear-induced pseudopod retraction responses. In fact, fluid shear stress-induced pseudopod retraction occurs in parallel with a decrease in F-actin content with significant reductions occurring within 1 min of the onset of flow. Shear-induced reductions in F-actin content also appear to be shear magnitude-dependent although the exact relationship between the two is still unclear. Shive et al. (2000) demonstrated that exposure to shear stresses between 2 and 7.5 dyn/cm^2 decreased F-actin content of neutrophils in a magnitude-dependent fashion while exposure to stresses above 7.5 dyn/cm^2 had no further effect. In contrast, Chen et al. (2004) showed that while low shear stresses reduced F-actin content, high shear stresses stimulated F-actin polymerization. Although such differences can be explained by experimental conditions, the fact that shear stress modulates F-actin assembly and organization substantiates the key role of cytoskeletal remodeling in the shear responses of neutrophils.

Processes associated with remodeling of the F-actin cytoskeleton in leukocytes are controlled predominantly by cytosolic signaling pathways of the Ras superfamily of small guanine triphosphate (GTP)-binding proteins, particularly the small GTP-binding phosphatases (GTPases; i.e., Rac1, Rac2, cdc42, and members of the Rho family (e.g., RhoA) as reviewed in the literature, Cicchetti et al., 2002; Niggli, 2003; Tybulewicz and Henderson, 2009). These small GTPases mediate the actions of chemokines on the leukocyte cytoskeleton through the classical GPCR signaling pathway. Principally, ligand-GPCR binding leads to recruitment of G-proteins (typically bound in an inactive state to the cell membrane) to the cytosolic receptor domain. This event leads to phosphoinositide-3 kinase (PI3K)-dependent cellular signaling that ultimately increases the activity of downstream effectors of cytoskeletal reorganization, that is, the small GTPases. More importantly, each individual small GTPase is responsible for specific cytoskeletal responses. Typically, activation of cdc42 is responsible for actin polymerization associated with the formation of thin finger-like filipodia, but also, has been shown to activate Rac. Rac, on the other hand, plays a critical role in cytoskeletal events associated with the formation of lamellipodia and pseudopodia while RhoA plays a prominent regulatory role in uropod retraction at the trailing edge of the migrating cell.

With this in mind, it is interesting to note that pseudopod retraction by HL-60-derived neutrophils under the influence of fluid shear stress was shown to be associated with decreased activity of Rac1 and Rac2 without any apparent effect on cdc42 activity (Makino et al., 2005). Moreover, shear-induced pseudopod retraction was independent of RhoA activity as well as its downstream target, myosin light chain kinase (MLCK) (Makino et al., 2005). Rather than stimulating the activity of molecules that coordinate retraction of cell processes (e.g., RhoA, MLCK), fluid shear stress, therefore, appears to either attenuate (e.g., possibly through stimulated release of an inhibitor) or interfere with the ability of neutrophils to form or sustain pseudopod projections via reduced cytosolic activity of key molecules (e.g., Rac1, Rac2) involved in actin polymerization. This finding was further substantiated by evidence showing (Makino et al., 2005) that neutrophils from Rac1- and Rac2-deficient mice exhibited an altered pseudopod retraction response to flow. Shear-induced pseudopod retraction, therefore, acts through signaling pathways associated with Rac GTPase-dependent regulation of cytoskeletal remodeling.

The reported effects of fluid shear stress on cytosolic Rac1 and Rac2 activity in conjunction with the role of fluid flow modulation of pseudopod activity (Makino et al., 2005) implicate upstream signal transduction pathways, particularly G-protein signaling associated with GPCRs that regulate neutrophil chemotaxis (e.g., fMLP receptor, platelet activating factor [PAF] receptor). Classically, downstream signaling initiated by stimulation of transmembrane GPCRs with cytokines is dependent on their conformation within the cell membrane which is, in turn, determinant of their cytoplasmic interactions with heterotrimeric G-proteins: G_α, G_β, and G_γ (as reviewed elsewhere Miyajima et al., 1992; Onishi et al., 1998). For the most part, selective activation of GPCRs by their respective cytokine ligands involves the inhibitory isoform of the G_α subunit, $G_{\alpha i}$, since it attenuates downstream adenylate cyclase-mediated production of the second messenger, cyclic adenosine monophosphate (cAMP); this is one of three possible G_α subunits that plays a role in GPCR signaling with the others being $G_{\alpha s}$ and $G_{\alpha q}$, which enhance downstream adenylate cyclase activity. The predominant involvement of $G_{\alpha i}$ in cytokine signaling is corroborated by the fact that the GPCRs participating in cytokine activation of neutrophil motility typically exhibit sensitivity to pertussis toxin, an inhibitor of $G_{\alpha i}$.

In this respect, fluid shear stress exposure modulates the activity of $G_{\alpha i}$ in HL-60-derived neutrophils; exposure to 5 dyn/cm^2 fluid shear stress rapidly (within 1 min) deactivated $G_{\alpha i}$ as indicated by a fluorescence resonance energy transfer (FRET) construct that reports conformational activity of this G-protein subunit (Makino et al., 2006). Indeed, shear exposure has also been shown to alter the activity of G-proteins in a variety of cells including endothelial cells (Gudi et al., 1996, 2003) and other organisms, for example, dino flagella (Chen et al., 2007). It is, in fact, proposed that fluid shear stress rapidly increases membrane fluidity, which removes restrictions on changes in the G-protein conformation (Gudi et al., 1998). Since the activity of G-proteins is highly dependent on their tertiary structure, these membrane fluidity-induced changes may be a mechanism of fluid flow mechanotransduction. It is not clear how G-protein mechanotransduction alone would account for differences in the responses of different cells (i.e., cell-type specificity) to fluid flow stimulation (e.g., endothelial cells vs. neutrophils).

The basis for cell-type specificity of shear stress responses likely results from the diverse surface expression patterns of transmembrane proteins, particularly receptors involved in providing the cell with information regarding its surrounding environment. The fact that shear stress evokes pseudopod retraction (Fukuda et al., 2000; Komai and Schmid-Schönbein, 2005) with actin depolymerization (Figure 8.11) by neutrophils even after activation with low levels of cytokines such as fMLP (which binds the formyl-peptide receptor [FPR]) substantiates the modulatory actions of fluid flow on receptor activity. On the other hand, stimulation of neutrophils with low levels of phorbol ester (PMA), which activates intracellular signaling cascades associated with protein kinase C (PKC)-dependent actin remodeling through a receptor-independent pathway (via target sites in the cell cytosol), is not affected by fluid shear stress exposure (Figure 8.11). Together, this evidence substantiated (1) the modulatory role of fluid flow on membrane-associated receptor activity and (2) the involvement of membrane-bound GPCRs, particularly FPR with its high constitutive activity in the neutrophil pseudopod retraction response to fluid shear stress. Reports (Fukuda et al., 2000)

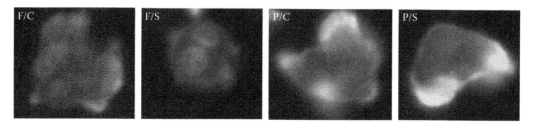

FIGURE 8.11 (See color insert.) Deactivation of neutrophils by fluid shear stress involves redistribution of cytoskeletal F-actin. Neutrophils that had been prestimulated with either 10 nM fMLP (F) or 10 nM PMA (P) were exposed to cone-plate flow ($\tau \approx 5 \, dyn/cm^2$) for 5 min (S). Controls (C) were cells under no-flow, but otherwise, similar experimental conditions prevailed. After experiments, cells were fixed, labeled with phalloidin (F-actin staining in pseudopods and cell periphery) and propidium iodide (nuclear staining). *Note*: neutrophils are 10–15 µm in diameter. F/C = FMLP under control; F/S = FMLP + Shear; P/C = PMA under control; P/S = PMA + Shear.

documenting shear-induced pseudopod retraction by cells stimulated with PAF further substantiate the relationship between GPCR activity and neutrophil shear responses.

It is important to note that the pseudopod retraction response under fluid shear stress is unique to the leukocytes (particularly, the neutrophils). Endothelial cells, and possibly osteoblasts, realign and elongate in response to fluid shear exposure while cardiac myocytes do not exhibit any detectable changes in cell shape (Makino et al., 2007). In fact, the swift membrane expansion and cytoskeletal remodeling required for rapid migration and phagocytosis is a key trait of neutrophils that must travel quickly to their target sites and engulf foreign particles. As such, transmembrane receptors that are involved in initiating these cellular processes, such as the cytokine receptors, are likely candidate mechanosensors. In the case of neutrophils, GPCRs were implicated due to the fact that shear-induced pseudopod retraction (1) was attenuated by monensin, which increases cytosolic Na^+ and interferes with GPCR constitutive activity and (2) exhibited pertusis toxin-sensitivity pointing to the involvement of a cytokine receptor (Makino et al., 2005).

Neutrophils express a diverse array of GPCRs such as those for FPR, complement 5a, leukotriene B4, interleukin (IL)-8, and PAF, all of which exhibit constitutive conformation-dependent activity; changes in conformation may result in either increased or decreased activity depending on how ligand binding-induced changes affect the tertiary structure. Evidence demonstrating that physiologically relevant magnitudes of mechanical stresses may be capable of physically altering the conformation of proteins (Kamm and Kaazempur-Mofrad, 2004; Mofrad et al., 2004; Lee et al., 2007) further suggests that transmembrane proteins such as GPCRs may serve not only as biochemical sensors but also as shear stress detectors. In fact, reports in the literature also provide evidence that fluid shear stress alters the conformation of other membrane-bound GPCRs in other cell types including the bradykinin B_2 receptor for endothelial cells and the type I parathyroid hormone for osteoblasts (Chachisvilis et al., 2006; Zhang et al., 2009). The fact that FPR is highly expressed in leukocytes (e.g., neutrophils, monocyte/macrophages), more so than other cytokine receptors, and exhibits a high level of constitutive activity provided the basis for Makino et al. (2006) to test the hypothesis that FPRs serve as a mechanosensors for pseudopod retraction.

Interestingly, while undifferentiated HL-60 promyelocytes do not express FPR and do not exhibit spontaneous pseudopod projections, induction of the neutrophil phenotype by treatment with dimethyl sulfoxide (DMSO) elicited high FPR expression and the ability to extend pseudopods; like blood-derived neutrophils, these cells also retract pseudopods in response to fluid shear stress (Makino et al., 2006). The critical piece of evidence supporting the role of FPR as a mechanosensory regulator of pseudopod retraction was the observation that transfection of undifferentiated HL-60 cells with an expression plasmid for FPR not only conferred expression of this receptor but imparted on these cells the ability to form pseudopods that retracted under the influence of fluid shear stress (Makino et al., 2006). Furthermore, transfection of differentiated HL-60-derived neutrophils with

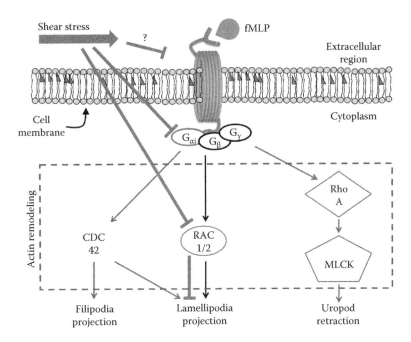

FIGURE 8.12 Schematic that summarizes regulation of actin remodeling during shear-induced pseudopod retraction by neutrophils. As demonstrated in this diagram, pseudopod retraction under the influence of fluid shear stress requires FPR receptor activity and involves attenuated activity of $G_{\alpha i}$ and Rac1/2, but not either CDC42, RhoA, or MLCK. Initiation of these signaling events is proposed to result from shear-induced reductions in FPR activity.

inhibitory small interfering RNA (siRNA) for FPR expression attenuated the effects of fluid shear on pseudopod retraction, despite the fact that the cells still projected pseudopods likely due to the presence of other cytokine-related GPCRs (Makino et al., 2006).

Taken together, the evidence accumulated, to date, point to FPR as a mechanosensor and an important regulator of pseudopod activity in neutrophils that depends on the surrounding flow environment (Figure 8.12). The fact that shear-induced pseudopod retraction involves reduced activity of $G_{\alpha i}$ and downstream Rac1 and Rac2 small GTPases, both of which are downstream of FPR, strongly suggests that fluid shear stress reduces the activity of the membrane-bound fMLP receptor. Putative modes of action of fluid shear stress on GPCRs include direct effects on protein conformation or enhancement of membrane fluidity that removes restrictions on the conformation of the receptors in the transmembrane domain, as well as an indirect effect via the force-multiplying action by the glycocalyx (Weinbaum et al., 2003). Another possibility is the shear-induced change in surface expression levels of FPR on the neutrophil surface. In fact, exposure of HL-60-derived neutrophils to $5\,dyn/cm^2$ wall shear stress for 10 min using a parallel plate flow chamber elicited redistribution of surface-associated FPRs (tagged with green-fluorescent protein) to a perinuclear compartment in the cytosol (Schmid-Schönbein and Su, 2010). These results point to the possibility that GPCR internalization under the influence of fluid shear stress stimulation leads to pseudopod retraction by counteracting the constitutive activity of FPRs serving to promote pseudopod extension.

8.5.4 INTEGRINS AND PROTEOLYTIC REGULATION OF SHEAR-INDUCED PSEUDOPOD RETRACTION

In addition to cytoskeletal remodeling, cyclical pseudopod projection and retraction depends on integrin receptors serving as anchorage points on the underlying substrate by which the cell propels itself (Anderson et al., 2003). In the case of fluid flow regulation of leukocyte activity under physiological conditions (i.e., below a threshold level of inflammatory stimulation), pseudopod retraction by neutrophils migrating on glass substrates under the influence of fluid shear stress depends on β_2, but not β_1 integrins (Marschel and Schmid-Schönbein, 2002). In addition to modulating interactions of β_2 integrins with their ligands (e.g., ICAM-1) during inflammation-related adhesion and migration (Simon and Goldsmith, 2002; Simon and Green, 2005), fluid shear also regulates integrin dynamics on the surfaces of neutrophils under conditions that mimic a low activation state specifically by eliciting redistribution of these receptors on the cell membrane from areas of maximal shear stress to regions where shear is minimal, such as at sites of cell–substrate interactions (Fukuda and Schmid-Schönbein, 2003). Moreover, exposure of neutrophils to fluid shear induces reductions in CD18 levels on the cell surface even in the presence of inflammatory mediators such as PAF and fMLP (Fukuda et al., 2000).

Since CD18 is critical for neutrophil migration during acute inflammatory processes and plays a major role in regulating the strength of attachment of neutrophils to the vascular wall, rapid reductions in CD18 due to shear exposure likely impact the ability of the cells to withstand mechanical stresses due to fluid flow and maintain adhesive attachments during migration (Fukuda and Schmid-Schönbein, 2003). Furthermore, a balance may exist between CD18-mediated pseudopod formation and retraction that depends on the extent of cell activation (i.e., local concentrations of inflammatory mediators). Thus, shear stress-induced reductions in CD18 surface levels as a result of fluid shear exposures appear to serve an anti-inflammatory role for both neutrophils in suspension as well as when migrating on the endothelium depending upon the surrounding biochemical environment. The combined actions of fluid shear stress in regulating neutrophil pseudopod activity and cell adhesion (through CD18) are, therefore, consistent with the hypothesis that neutrophil mechanotransduction under the action of fluid shear stress serves as a control mechanism to ensure neutrophils remain in a nonadhesive state during passage through the small vessels of the microcirculation.

The underlying mechanism for shear-induced reductions in CD18 surface levels appears to result from a fluid shear-dependent proteolytic process that elicits cleavage of the extracellular ligand-binding domain of full-length CD18 present on the surfaces of migrating as well as suspended neutrophils under fluid shear. Generally, proteolysis is one mechanism by which a cell modulates the levels of a wide variety of surface-associated receptors. For example, L-selectin (involved in rolling interactions with endothelium) is cleaved off the cell membrane by matrix metalloproteinases (Walcheck et al., 1996) marking the transition of an activated and rolling neutrophil into an adhesive phenotype (Hafezi-Moghadam et al., 2001). Tumor necrosis factor-α (TNF-α) (Lopez et al., 1998) induces serine protease-mediated cleavage (Carney et al., 1998) of CD43, an anti-adhesive mucin-like molecule on the neutrophil surface. Proteolysis is also involved in cleavage of β_2 integrins during neutrophil migration. For instance, β_2 integrin undergoes cleavage of the intracellular domain by calpain to promote detachment of the cell uropod (or trailing region) during leukocyte migration (Pfaff et al., 1999).

Shear-induced truncation of CD18 integrins, however, differs from calpain-mediated integrin cleavage in that fluid flow-dependent proteolysis involves lysosomal proteases that exert enzymatic activity on the extracellular domain. This was based on several key pieces of evidence reported by Fukuda and Schmid-Schönbein (2003) and Shin et al. (2008) as follows: (1) only broad-spectrum cysteine protease inhibitors (e.g., leupeptin, E64) as well as a cell-impermeable protease inhibitor with activity against cathepsin B and L (i.e., CA074) abolish shear-induced β_2 integrin cleavage and (2) chemical treatment of neutrophils with exogenous cathepsin B in the absence of shear exposure results in reduced surface binding of MEM48 antibodies to total β_2 integrins in a fashion similar to that of shear stress stimulation. Moreover, cleavage of CD18 in the extracellular domain implies cathepsin B is either released from lysosomes to the extracellular milieu or transported to the cell surface. Elastase (Salcedo et al., 1997) and calpain (Pontremoli et al., 1989) localize to the membranes of neutrophils upon activation with either chemokines (e.g., fMLP, PAF) or PMA. Cathepsin G and elastase are present on neutrophils during TNF-α-induced transmigration (Hermant et al.,

2003). Moreover, cathepsin B associates with surfaces of lymphocytes after T-cell receptor-induced degranulation (Balaji et al., 2002).

In addition to release of cathepsin B, recent evidence points to another requirement for this mechanism, namely conformational extension of the ectodomain of CD18 integrins on the neutrophil membrane due to shear stress exposure (Shin et al., 2008). Neutrophil recruitment results predominantly from ligand-induced activation of surface receptors (e.g., chemokine receptors, Fc receptors) followed by activation (into an open-extended conformation that exposes ligand binding sites) and deployment (Arnaout, 1990) of CD18 integrins from secondary granules that promote cell capture onto the vessel wall (Radi et al., 2001; Simon and Goldsmith, 2002). In this way, CD18-mediated adhesive interactions withstand physical stresses imposed by flow. The levels and structure of CD18 on the surface, therefore, modulate neutrophil migration (Hughes et al., 1992). Another consequence of this conformational change in CD18 induced by shear stress may be to increase its susceptibility to cathepsin B proteolysis.

The underlying cellular mechanism governing transport of cathepsin B and other lysosomal proteases from the cytosol to the extracellular milieu under the influence of fluid shear stress, however, remains unresolved. One possibility is that shear stress leads to degranulation due to its depolymerizing effects on the F-actin cytoskeletal matrix. The fact that the F-actin cytoskeleton rapidly depolymerizes upon exposure of neutrophils to fluid shear stress may result in increased random Brownian motion of cytoplasmic granules, an event that has been observed during experiments (Moazzam et al., 1997). This increase in granular motion is consistent with the possibility for enhanced degranulation due to increased collision and possible fusion of cytosolic lysosomes with the plasma membrane. But several attempts to demonstrate the degranulation process with confocal microscopy did not yield a detectable signal under the influence of the short shear periods. This result is not surprising. Neutrophils are in the circulation for just a few hours during which time they are constantly exposed to fluid shear stress. Moreover, shear-induced degranulation is likely a slow process, involving more than the 5 min duration of shear stress stimulation applied to neutrophils in our studies. We therefore expect that, within this time frame, neutrophils under shear released only a small percentage of granules (out of a total of about 3000–5000/cell) (Schmid-Schönbein and Chien, 1989).

Collectively, the current evidence regarding the role of shear stress in regulating CD18 surface expression and localization paints a picture whereby fluid shear serves to keep suspended neutrophils in a round, passive, nonadherent state by downregulating CD18. Furthermore, cathepsin B plays a critical role in the pseudopod retraction process as demonstrated by the lack of this response by leukocytes from cathepsin B-deficient mice in conjunction with the results from ex vivo studies (Fukuda et al., unpublished observations). CD18 proteolysis, therefore, relies on cathepsin B transport to the extracellular mileu where it may cleave shear-induced conformationally extended CD18 receptors important for rapid neutrophil migration (Figure 8.13). By doing so, fluid shear stress serves as a negative regulator of neutrophil motility. In fact, this agrees with recent evidence demonstrating Macrophage-1 antigen (Mac-1) (Luster et al., 2005) and Lymphocyte function-associate antigen (LFA-1) (Evans et al., 2006) shedding by macrophages as putative mechanisms to detach integrins from their counter-receptors or prevent further migration at sites of inflammation.

8.5.5 SECOND MESSENGERS: NITRIC OXIDE, CYCLIC GUANIDINE MONOPHOSPHATE, AND SUPEROXIDE

Reactive nitrogen species (e.g., nitric oxide or NO) are multifunctional regulators of acute inflammation serving not only as an antimicrobial agent in the environment of a phagosome, but also as a second messenger signaling molecule that influences downstream leukocyte functions (e.g., chemotaxis, phagocytosis) (Laroux et al., 2000; Guzik et al., 2003; Korhonen et al., 2005; Fialkow et al., 2007). NO from exogenous (from outside the cell or from NO donors) as well as endogenous sources (such as membrane-associated NO synthases) exerts inhibitory actions on

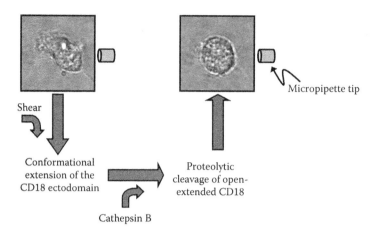

FIGURE 8.13 Schematic for regulation of CD18 surface levels during shear-induced pseudopod retraction by neutrophils. As demonstrated in this diagram, pseudopod retraction under the influence of fluid shear stress (imposed by micropipette flow) requires cathepsin B-mediated cleavage of CD18 integrins. CD18 cleavage by cathepsin B, in turn, requires shear-induced conformational extension of the CD18 ectodomain (i.e., the extracellular ligand-binding region of the receptor). Cleavage of CD18 also occurs for suspended, nonadherent neutrophils in cone-plate shear flow.

neutrophil recruitment out of the microvasculature during acute-stage inflammatory processes with reported attenuated leukocyte rolling, adhesion, and migration on vascular endothelium (Kubes et al., 1991; Dal Secco et al., 2006).

The involvement of NO in shear-induced pseudopod retraction is proposed to exist due to its influence on intracellular levels of the second messenger, cyclic guanidine monophosphate (cGMP). Both NO donors and cGMP analogs enhance neutrophil pseudopod retraction responses to fluid shear stress while blocking cGMP activity (downstream of NO) abolishes this shear response both in an in vitro and in vivo setting (Fukuda et al., 2000). Interestingly, inhibition of NO synthase ((NOS), an enzyme involved in NO generation during conversion of L-arginine to L-cirtulline) by N^G-methyl-L-arginine (L-MNA) or N^G-nitro-L-arginine methyl ester (L-NAME) treatment of neutrophils has no effect on shear-induced pseudopod retraction in an in vitro setting, although both NO donors and cGMP analogs counteract the blocking effects of fMLP and PAF on this response (Fukuda et al., 2000). Taken together, these observations are in line with the involvement of exogenous NO in stimulating increases in intracellular cGMP levels (Elferink and VanUffelen, 1996; Dal Secco et al., 2006; Fialkow et al., 2007).

It is possible, however, that the influence of NO on the shear stress response (i.e., pseudopod retraction) in migrating neutrophils results from the reported ability of NO to act as a scavenger of superoxide (O_2^-) (Fialkow et al., 2007) and, in this way, mediate cell migration processes (Gaboury et al., 1993; Kubes et al., 1993). In fact, Komai and Schmid-Schönbein (2005) reported that red blood cells play an important role in the cell deactivating effects of fluid shear stress on nonadherent neutrophils in whole blood stimulated with fMLP (which is known to induce O_2^- formation); the dependence of the neutrophil shear response on blood hematocrit was based primarily on the ability of the red blood cell membranes to scavenge O_2^-. In fact, this is consistent with additional data demonstrating the ability of exogenous superoxide dismutase (SOD; a scavenger of oxygen free radicals such as O_2^-) to enhance the shear responses of human neutrophils (Komai and Schmid-Schönbein, 2005) that had been stimulated with fMLP activator. Although it remains to be determined, these data point to the possibility that fluid shear stress is associated with the generation of large amounts of reactive oxygen species (e.g., O_2^-) and that this must be countered by the scavenging abilities of the red cells and, in areas of the circulation of reduced hematocrit (e.g., capillaries), by the NO-generating capabilities of the neutrophils.

The exact mechanism that links increased cGMP and NO generation to pseudopod retraction by adherent neutrophils in response to fluid shear, however, is currently unclear and is likely a complex process especially in light of literature reports (Elferink and VanUffelen, 1996) documenting stimulatory as well as attenuating effects of these two molecules on neutrophil migration. In fact, fMLP has been reported to enhance cGMP levels (Wyatt et al., 1993) and NO (Sodhi and Biswas, 2002). This may seem contradictory to the results of Fukuda et al. (2000) since cGMP analogs enhanced pseudopod retraction in untreated neutrophils exposed to shear stress and counteracted the blocking effects of fMLP. One can postulate, however, that above a threshold concentration (above $10\,\mu M$), cGMP can possibly inhibit neutrophil migration; however, at concentrations between 5 and $10\,\mu M$, it enhances fMLP-induced neutrophil migration (Elferink and de Koster, 1993). It is, therefore, quite possible that shear-induced pseudopod retraction may result from generation of NO and/or cGMP to a level above a threshold concentration.

Another interesting point is that cGMP and NO signaling link GPCR-related signaling, actin depolymerization, and CD18 cleavage processes associated with pseudopod retraction by neutrophils in response to fluid flow (Makino et al., 2007). Both cGMP and NO have been linked to changes in the F-actin and microtubule components of the cytoskeleton (Wyatt et al., 1991; Clancy et al., 1995; Boran and Garcia, 2007). Additionally, cGMP and NO production are linked to both fMLP and PAF simulation (Wyatt et al., 1993; Catalan et al., 1996; Sodhi and Biswas, 2002; Boran and Garcia, 2007). Moreover, surface expression of CD18 integrins is elevated in response to NOS inhibition and reduced in response to incubation of cells with NO donors (Lefer and Lefer, 1996; Sato et al., 1999). Interestingly, cGMP and shear stress separately induce reductions in CD18 surface expression, but cGMP analog does not enhance this response. Therefore, there may exist more than one mechanism that works independently but in a coordinated fashion to promote cytoskeletal disassembly and to induce a substrate debinding process required for retraction of pseudopods by neutrophils exposed to fluid shear stress.

The current evidence in the literature, therefore, depicts a picture where early signaling events resulting from GPCR deactivation are linked to pseudopod retraction through a pathway involving NO production and enhancement of cGMP levels.

8.6 NEUTROPHIL MECHANOBIOLOGY IN CARDIOVASCULAR PHYSIOLOGY AND DISEASE

8.6.1 Fluid Flow-Dependent Leukocyte Deactivation in the In Vivo Blood Vessel

In in vivo blood vessel, vascular endothelial cells produce NO through a process involving various isoforms of NOS such as endothelial NOS (eNOS) and inducible NOS (iNOS) (Laroux et al., 2000). These NOS isoforms have been shown to influence leukocyte rolling, adhesion, and migration (Kubes et al., 1991; Lefer et al., 1999). Moreover, NOS isoforms themselves have been shown to play roles in shear-induced pseudopod retraction (Fukuda et al., 2000). Specifically, leukocytes in the microvessels of the rat mesentery that had been treated with NOS inhibitors as well as those from genetically-engineered eNOS-deficient mice lack the pseudopod retraction response to shear stress in the absence or presence of PAF (Fukuda et al., 2000). Therefore, NO from exogenous sources such as due to the release by the blood vessel endothelium may serve to promote leukocyte inactivity under physiological flow conditions.

Interestingly, this mechanotransduction mechanism involves cGMP signaling that is downstream of NO stimulation. Evidence for this was provided by the observation that cGMP analogs rescue the pseudopod retraction responses of neutrophils in the presence of inhibitors of NO production, such as L-NMA (Fukuda et al., 2000). It should be pointed out that shear stress stimulation also induces production and release of NO by endothelial cells (Davies, 1995) that may enhance the neutrophil shear response by further elevating cGMP levels in attached cells. As such, fluid shear stress modulates the activities of multiple cell types in the vascular environment in such a fashion to prevent immune cell activation in the absence of inflammatory mediators.

8.6.2 CHRONIC INFLAMMATORY PHENOTYPE AND ITS PUTATIVE ROLE IN IMPAIRED LEUKOCYTE MECHANOTRANSDUCTION OF SHEAR STRESS

The importance of cell-deactivating effects of fluid flow stimulation on leukocytes is evidenced by the impairment of shear-induced pseudopod retraction after treating neutrophils with above-threshold concentrations of soluble agonists, such as fMLP (at $>10^{-8}$ M) or PAF (at $>10^{-7}$ M), which commit these cells to an activated phenotype (Moazzam et al., 1997; Fukuda et al., 2000). This can impact blood perfusion in the small vessels (particularly, the capillaries) since activation of leukocytes in the blood leads to their entrapment in the microcirculation due to an increase in membrane adhesion as well as pseudopod formation with enhanced cytoplasmic stiffness (Worthen et al., 1989; Helmke et al., 1998). Elevated leukocyte activity in the microcirculation may, therefore, play a role in capillary rarefaction leading to downstream local tissue ischemia and tissue damage. In contrast, neutrophils treated with below threshold concentrations of these activators exhibit shear-induced pseudopod retraction.

The attenuating effects of threshold cytokine stimulation levels on shear-induced neutrophil deactivation also imply that in vivo conditions involving elevated levels of leukocyte activation may link impaired fluid flow-related mechanotransduction and downstream cardiovascular pathologies. As such, agonists that stimulate leukocytes and trigger inflammatory processes are balanced by negative control of cell activity in the absence of trauma or at later tissue repair stages. Failure to regulate cell activation promotes chronic inflammation, an underlying trait (Mazzoni and Schmid-Schönbein, 1996; Schmid-Schönbein, 2006) for a variety of cardiovascular pathologies (e.g., atherosclerosis, myocardial infarction, ischemia-reperfusion injury, glucose intolerance).

In fact, chronic inflammation (a form of tissue repair with frustrated inflammatory resolution), associated with elevated systemic levels of leukocyte activation in the blood, has been recognized as a common denominator for a variety of diseases ranging from heart disorders to Alzheimer's, diabetes, and cancer (Gorman et al., 2004). For the case of cardiovascular disease, chronic inflammation due to dysregulated vascular cell activity has been implicated in various diseases resulting from such conditions as hypertension, hypercholesterolemia, and diabetes (Esch and Stefano, 2002). In this respect, the established role of circulatory hemodynamics in controlling the activity of the blood-borne leukocytes points to the mechanotransducing ability of these cells as critical for promoting circulatory homeostasis.

8.6.3 NEUTROPHIL MECHANOTRANSDUCTION OF FLUID SHEAR STRESS AND CARDIOVASCULAR DISEASES

The primary evidence supporting the importance of the neutrophil shear response in circulatory homeostasis derives from research investigating links between hypertension and elevated leukocyte activation. This work relied heavily on a rat model of essential hypertension: the spontaneously hypertensive rat (SHR) that has a genetic predisposition to develop high blood pressures in the systemic circulation. Experimental strategies for their use as an animal model to elucidate the underlying causes of spontaneous hypertension typically involve comparison of their behavior to either wild-type Wistar rat, the background genetic strain for the SHR, or the Wistar-Kyoto rat (WKY), the background genetic strain from which the SHR was bred.

A key feature of the blood of SHRs is the elevated numbers of circulating neutrophils which, at the same time, display a suppressed expression of adhesion molecules involved in neutrophil rolling (i.e., P-selectin) and adhesion (i.e., CD18) as well as an activated phenotype based on the presence of pseudopod extensions (Arndt et al., 1993; Suzuki et al., 1994; Suematsu et al., 2002). However, the increased prevalence of neutrophils with pseudopods is not associated with increased adhesion to microvascular endothelium (Makino et al., 2007). The increased number of circulating neutrophils and their spontaneous pseudopod formation, however, do appear to contribute to an elevation in peripheral vascular resistance (Fukuda et al., 2004b). One possibility is that these circulating activated neutrophils release vasoactive substances that constrict the small arteries and arterioles leading to an elevation in peripheral vascular resistance; this has been documented for the case of atherosclerosis (Faraci et al., 1991; Mügge et al., 1992; Kaul et al., 1994).

Extensive evidence, however, also points to the possibility of a hemorheological effect of leukocyte activation on microvascular resistance (Eppihimer and Lipowsky, 1996; Mazzoni and Schmid-Schönbein, 1996). Specifically, pseudopod formation causes a significant reduction in the already reduced velocity of white cells in the microcirculation, compared to the microvascular velocity of red cells in the absence of white cells. This difference in velocity between the two circulating cell types leads to an enhancement of hemodynamic resistance in microvasculature that is due to an enhanced disturbance of their motion within narrow vessels (Helmke et al., 1998). In turn, disturbances in red cell motion, particularly in the microcirculation, enhance peripheral resistance (Fukuda et al., 2004b) and consequently increase blood pressures in upstream arteriolar/artery segments (Figure 8.14). The key evidence for the involvement of fluid flow mechanotransduction in microvascular abnormalities associated with hypertension was the observation that neutrophils from SHRs lacked a pseudopod retraction in response to fluid shear stress; in fact, these cells extended cellular projections upon fluid shear exposure (Fukuda et al., 2004b).

The underlying mechanism associated with the blockade and possible reversal of the pseudopod retraction response to fluid shear was revealed to involve the dependence of blood pressure elevation in SHRs on levels of glucocorticoid-related steroid hormones circulating in the plasma as well as the increased density of glucocorticoid receptors on the neutrophil surface; removal of the glucocorticoid-producing adrenal gland (i.e., adrenalectomy) in SHRs reduces blood pressures down to levels observed in WKY rats (Sutanto et al., 1992; DeLano and Schmid-Schönbein, 2004). Moreover, neutrophils from Wistar (wild-type counterparts to the SHR) or WKY rats extend pseudopods upon fluid shear exposure in the presence of the synthetic steroid, dexamethasone (Fukuda et al., 2004a). Interestingly, dexamethasone exhibits an inhibitory effect on pseudopod formation in neutrophils maintained in the absence of shear stress stimulation pointing to the influence of the fluid flow environment on neutrophil behavior under physiological and pathological conditions.

The critical biological players that contribute to the switch in neutrophil phenotype from a cell that retracts to one that extends pseudopods under flow stimulation are still not known. It is, however,

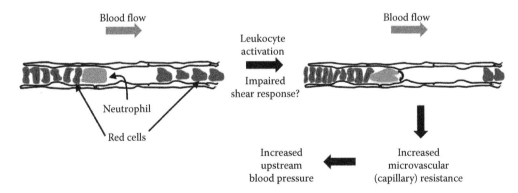

FIGURE 8.14 Relationship between leukocyte activity and microvascular resistance. Fluid shear stress maintains leukocytes in a passive state capable of conforming to the geometry of the capillaries. At equilibrium, passive neutrophils flow with minimal disturbance of the red cell stream. Upon activation and/or projection of pseudopods (due to impaired shear responses), the neutrophil rotates dramatically and interferes with blood movement. This is evident by the disturbed red cell flow upstream and an increase in the cell-free zone downstream of the leukocyte. This disturbed blood flow pattern elevates peripheral (microvascular) resistance and thus upstream blood pressures.

believed to involve shear-dependent modifications in the behavior of membrane-bound ion channels that regulate levels of intracellular calcium (i.e., $[Ca^{2+}]_i$). Neutrophils that retract pseudopods in response to fluid shear stress display parallel decreases in $[Ca^{2+}]_i$ (Fukuda and Schmid-Schönbein, 2002). In contrast, cells, such as dexamethasone-treated leukocytes (Fukuda et al., 2004a) or cells from SHR (Fukuda et al., 2004b) that project pseudopods under flow exhibit increased $[Ca^{2+}]_i$. In fact, $[Ca^{2+}]_i$ is likely a critical determinant of pseudopod activity in dexamethasone-treated leukocytes since either outward-directing calcium channel blockers or analogs of Ca^{2+}-dependent cAMP second messenger signaling interfere with shear-induced pseudopod projection (Fukuda et al., 2000, 2004a). This is further substantiated by evidence showing these same calcium channel blockers as well as either cAMP or NO/cGMP, both of which reduce $[Ca^{2+}]_i$, enhance the neutrophil pseudopod retraction response to fluid shear stress.

Fluid shear stress is, therefore, a regulator of neutrophil activity, i.e., a control mechanism that, in the absence of agonists, prevents neutrophil recruitment under physiological flows. Although extensive mechanobiological research has revealed a role for altered hemodynamics or complex flow patterns in the circulation (i.e., vascular bifurcations, bends/curves, tortuous vessels) as a causal factor for the progression of vascular diseases (e.g., atherosclerosis), the onset of these pathologies is proposed to be based on a dysregulated or inflamed vascular cell phenotype including the endothelial cells (Libby, 2002; Cunningham and Gotlieb, 2005) and leukocytes. The origin of the dysregulated phenotype for these cells, however, remains largely unknown, although recent evidence in the SHR suggests that proteolytic receptor cleavage may be involved in the pathophysiological behavior of these cells (DeLano and Schmid-Schönbein, 2008). In this regard, Chen et al. (2010) demonstrated a link between reduced FPR levels on the cell surface, resulting from MMP-9-related proteolysis, and an impaired shear stress response in leukocytes from SHR.

Therefore, physiological factors that alter the phenotype of neutrophils and other leukocytes dramatically influence their ability to sense the surrounding flow environment (i.e., mechanosensitivity) leading to the development of pathological behavior, including immune suppression. This is possibly due to activation-dependent reorganization and even cleavage of the membrane and/or its constituent proteins (i.e., cytokine receptors, ion channels, adhesion molecules, etc.) that serve as mechanosensors. In the case of hypertension, the attenuation or reversal of fluid shear stress responses of neutrophils results in elevated blood pressures. Whether this results from a biophysical effect on the ability of neutrophils to pass through the capillaries, their effect on arteriolar vasoconstriction, or the destructive effects of their secretions on the host vasculature, remains an issue for clarification.

In the end, the possibility that impairment of neutrophil mechanotransduction of fluid flow stimuli leads to pathological situations implicates a wide range of cardiovascular (and noncardiovascular) diseases that correlate with chronic inflammation as an underlying component. In addition to hypertension, conditions such as hypercholesterolemia and diabetes, which are associated with chronic inflammation, are also associated with elevated microvascular leukocyte activity (Lechi et al., 1984; Tomida et al., 2001). The critical issue is to determine if chronic inflammation precedes or results from an impairment of mechanotransduction mechanisms that link the extracellular environment to intracellular signaling machinery of the leukocyte. Further work is, therefore, currently underway to fully define the role of fluid mechanics in the physiological regulation of leukocyte activity (particularly, that of the neutrophil) in order to gain a better understanding of their role in the pathogenesis of cardiovascular disease.

ACKNOWLEDGMENTS

This work was supported in part by PHS HL-10881 (to G. W. Schmid-Schönbein) and F32 HL083740-01 (to H. Y. Shin) from the National Institutes of Health, the University of Kentucky Research Support Fund (to H. Y. Shin), and 09BGIA2250309 (to H. Y. Shin) from the American Heart Association.

REFERENCES

Alberts, B., D. Bray et al. (1983). *Molecular Biology of the Cell*. New York, Garland Publishing, Inc.

Acevedo, A. D., S. S. Bowser et al. (1993). Morphological and proliferative responses of endothelial cells to hydrostatic pressure: Role of fibroblast growth factor. *J Cell Physiol* 157(3): 603–614.

Anderson, S. I., B. Behrendt et al. (2003). Linked regulation of motility and integrin function in activated migrating neutrophils revealed by interference in remodelling of the cytoskeleton. *Cell Motil Cytoskeleton* 54(2): 135–146.

Arnaout, M. A. (1990). Structure and function of the leukocyte adhesion molecules CD11/CD18. *Blood* 75(5): 1037–1050.

Arndt, H., C. W. Smith et al. (1993). Leukocyte–endothelial cell adhesion in spontaneously hypertensive and normotensive rats. *Hypertension* 21(5): 667–673.

Astrof, N. S., A. Salas et al. (2006). Importance of force linkage in mechanochemistry of adhesion receptors. *Biochemistry* 45(50): 15020–15028.

Attinger, E. O. (1963). *Pulsatile Blood Flow*. New York, McGraw-Hill Book Company.

Bagge, U. and R. Karlsson (1980). Maintenance of white blood cell margination at the passage through small venular junctions. *Microvasc Res* 20(1): 92–95.

Bagge, U., A. Blixt et al. (1983). The initiation of post-capillary margination of leukocytes: Studies in vitro on the influence of erythrocyte concentration and flow velocity. *Int J Microcirc Clin Exp* 2(3): 215–227.

Balaji, K. N., N. Schaschke et al. (2002). Surface cathepsin B protects cytotoxic lymphocytes from self-destruction after degranulation. *J Exp Med* 196(4): 493–503.

Bao, X., C. Lu et al. (1999). Temporal gradient in shear but not steady shear stress induces PDGF-A and MCP-1 expression in endothelial cells: Role of NO, NF kappa B, and egr-1. *Arterioscler Thromb Vasc Biol* 19(4): 996–1003.

Barbee, K. A., T. Mundel et al. (1995). Subcellular distribution of shear stress at the surface of flow-aligned and nonaligned endothelial monolayers. *Am J Physiol* 268(4 Pt 2): H1765–H1772.

Berlin, R. D. and J. P. Fera (1977). Changes in membrane microviscosity associated with phagocytosis: Effects of colchicine. *Proc Natl Acad Sci U S A* 74(3): 1072–1076.

Bodin, S. and M. D. Welch (2005). Plasma membrane organization is essential for balancing competing pseudopod- and uropod-promoting signals during neutrophil polarization and migration. *Mol Biol Cell* 16(12): 5773–5783.

Boran, M. S. and A. Garcia (2007). The cyclic GMP-protein kinase G pathway regulates cytoskeleton dynamics and motility in astrocytes. *J Neurochem* 102(1): 216–230.

Butler, P. J., G. Norwich et al. (2001). Shear stress induces a time- and position-dependent increase in endothelial cell membrane fluidity. *Am J Physiol Cell Physiol* 280(4): C962–C969.

Butler, P. J., T. C. Tsou et al. (2002). Rate sensitivity of shear-induced changes in the lateral diffusion of endothelial cell membrane lipids: A role for membrane perturbation in shear-induced MAPK activation. *FASEB J* 16(2): 216–218.

Carney, D. F., M. A. Jagels et al. (1998). Effect of serine proteinase inhibitors on neutrophil function: Alpha-1-proteinase inhibitor, antichymotrypsin, and a recombinant hybrid mutant of antichymotrypsin (LEX032) modulate neutrophil adhesion interactions. *J Leukoc Biol* 63(1): 75–82.

Catalan, R. E., A. M. Martinez et al. (1996). Nitric oxide mediates the PAF-stimulated cyclic GMP production in hippocampal slices. *Biochem Biophys Res Commun* 226(1): 27–31.

Chachisvilis, M., Y. L. Zhang et al. (2006). G protein-coupled receptors sense fluid shear stress in endothelial cells. *Proc Natl Acad Sci U S A* 103(42): 15463–15468.

Chang, K. C. and D. A. Hammer (1999). The forward rate of binding of surface-tethered reactants: Effect of relative motion between two surfaces. *Biophysical Journal* 76(3): 1280–1292.

Chen, A. Y. et al. (2010). Receptor cleavage reduces the fluid shear response in neutrophils of the spontaneously hypertensive rat. *Am. J. of Physiol – Cell Physiol* (in press).

Chen, S. and T. A. Springer (1999). An automatic braking system that stabilizes leukocyte rolling by an increase in selectin bond number with shear. *J Cell Biol* 144(1): 185–200.

Chen, K. D., Y. S. Li et al. (1999). Mechanotransduction in response to shear stress. Roles of receptor tyrosine kinases, integrins, and Shc. *J Biol Chem* 274(26): 18393–18400.

Chen, H. Q., W. Tian et al. (2004). Effect of steady and oscillatory shear stress on F-actin content and distribution in neutrophils. *Biorheology* 41(5): 655–664.

Chen, A. K., M. I. Latz et al. (2007). Evidence for the role of G-proteins in flow stimulation of dinoflagellate bioluminescence. *Am J Physiol Regul Integr Comp Physiol* 292(5): R2020–R2027.

Cicchetti, G., P. G. Allen et al. (2002). Chemotactic signaling pathways in neutrophils: From receptor to actin assembly. *Crit Rev Oral Biol Med* 13(3): 220–228.

Cinamon, G., V. Shinder et al. (2001). Shear forces promote lymphocyte migration across vascular endothelium bearing apical chemokines. *Nat Immunol* 2(6): 515–522.

Cinamon, G., V. Shinder et al. (2004). Chemoattractant signals and beta 2 integrin occupancy at apical endothelial contacts combine with shear stress signals to promote transendothelial neutrophil migration. *J Immunol* 173(12): 7282–7291.

Clancy, R., J. Leszczynska et al. (1995). Nitric oxide stimulates ADP ribosylation of actin in association with the inhibition of actin polymerization in human neutrophils. *J Leukoc Biol* 58(2): 196–202.

Coughlin, M. F. and G. W. Schmid-Schönbein (2004). Pseudopod projection and cell spreading of passive leukocytes in response to fluid shear stress. *Biophys J* 87(3): 2035–2042.

Coughlin, M. F., D. D. Sohn et al. (2008). Recoil and stiffening by adherent leukocytes in response to fluid shear. *Biophys J* 94(3): 1046–1051.

Cunningham, K. S. and A. I. Gotlieb (2005). The role of shear stress in the pathogenesis of atherosclerosis. *Lab Invest* 85(1): 9–23.

Dal Secco, D., A. P. Moreira et al. (2006). Nitric oxide inhibits neutrophil migration by a mechanism dependent on ICAM-1: Role of soluble guanylate cyclase. *Nitric Oxide* 15(1): 77–86.

Davies, P. F. (1995). Flow-mediated endothelial mechanotransduction. *Physiol Rev* 75(3): 519–560.

DeLano, F. A. and G. W. Schmid-Schönbein (2004). Enhancement of glucocorticoid and mineralocorticoid receptor density in the microcirculation of the spontaneously hypertensive rat. *Microcirculation* 11(1): 69–78.

DeLano, F. A. and G. W. Schmid-Schönbein (2008). Proteinase activity and receptor cleavage: Mechanism for insulin resistance in the spontaneously hypertensive rat. *Hypertension* 52(2): 415–423.

Dembo, M., D. C. Torney et al. (1988). The reaction-limited kinetics of membrane-to-surface adhesion and detachment. *Proc R Soc Lond B Biol Sci* 234(1274): 55–83.

DePaola, N., M. A. Gimbrone et al. (1992). Vascular endothelium responds to fluid shear stress gradients. *Arterioscler Thromb* 12(11): 1254–1257.

Edwards, S. W. (2005). *Biochemistry and Physiology of the Neutrophil*. New York, Cambridge University Press.

Ehrengruber, M. U., D. A. Deranleau et al. (1996). Shape oscillations of human neutrophil leukocytes: Characterization and relationship to cell motility. *J Exp Biol* 199(Pt 4): 741–747.

Elferink, J. G. and B. M. de Koster (1993). Modulation of neutrophil migration by captopril. *Naunyn Schmiedebergs Arch Pharmacol* 347(5): 562–567.

Elferink, J. G. and B. E. VanUffelen (1996). The role of cyclic nucleotides in neutrophil migration. *Gen Pharmacol* 27(2): 387–393.

Eppihimer, M. J. and H. H. Lipowsky (1996). Effects of leukocyte-capillary plugging on the resistance to flow in the microvasculature of cremaster muscle for normal and activated leukocytes. *Microvasc Res* 51(2): 187–201.

Esch, T. and G. Stefano (2002). Proinflammation: A common denominator or initiator of different pathophysiological disease processes. *Med Sci Monit* 8(5): HY1–HY9.

Evans, B. J., A. McDowall et al. (2006). Shedding of lymphocyte function-associated antigen-1 (LFA-1) in a human inflammatory response. *Blood* 107(9): 3593–3599.

Faraci, F. M., A. G. Lopez et al. (1991). Effect of atherosclerosis on cerebral vascular responses to activation of leukocytes and platelets in monkeys. *Stroke* 22(6): 790–796.

Ferraro, J. T., M. Daneshmand et al. (2004). Depletion of plasma membrane cholesterol dampens hydrostatic pressure and shear stress-induced mechanotransduction pathways in osteoblast cultures. *Am J Physiol Cell Physiol* 286(4): C831–C839.

Fialkow, L., Y. Wang et al. (2007). Reactive oxygen and nitrogen species as signaling molecules regulating neutrophil function. *Free Radic Biol Med* 42(2): 153–156.

Finger, E. B., K. D. Puri et al. (1996). Adhesion through L-selectin requires a threshold hydrodynamic shear. *Nature* 379(6562): 266–269.

Fukuda, S. and G. W. Schmid-Schönbein (2002). Centrifugation attenuates the fluid shear response of circulating leukocytes. *J Leukoc Biol* 72(1): 133–139.

Fukuda, S. and G. W. Schmid-Schönbein (2003). Regulation of CD18 expression on neutrophils in response to fluid shear stress. *Proc Natl Acad Sci U S A* 100(23): 13152–13157.

Fukuda, S., T. Yasu et al. (2000). Mechanisms for regulation of fluid shear stress response in circulating leukocytes. *Circ Res* 86(1): E13–E18.

Fukuda, S., H. Mitsuoka et al. (2004a). Leukocyte fluid shear response in the presence of glucocorticoid. *J Leukoc Biol* 75(4): 664–670.

Fukuda, S., T. Yasu et al. (2004b). Contribution of fluid shear response in leukocytes to hemodynamic resistance in the spontaneously hypertensive rat. *Circ Res* 95(1): 100–108.

Gaboury, J., R. C. Woodman et al. (1993). Nitric oxide prevents leukocyte adherence: Role of superoxide. *Am J Physiol* 265(3 Pt 2): H862–H867.

Goldsmith, H. L. and S. Spain (1984). Margination of leukocytes in blood flow through small tubes. *Microvasc Res* 27(2): 204–222.

Gorman, C., A. Park et al. (2004). The fires within. *Time* 163(8): 30–46.

Gudi, S. R., C. B. Clark et al. (1996). Fluid flow rapidly activates G proteins in human endothelial cells. Involvement of G proteins in mechanochemical signal transduction. *Circ Res* 79(4): 834–839.

Gudi, S., J. P. Nolan et al. (1998). Modulation of GTPase activity of G proteins by fluid shear stress and phospholipid composition. *Proc Natl Acad Sci U S A* 95(5): 2515–2519.

Gudi, S., I. Huvar et al. (2003). Rapid activation of Ras by fluid flow is mediated by Galpha(q) and Gbetagamma subunits of heterotrimeric G proteins in human endothelial cells. *Arterioscler Thromb Vasc Biol* 23(6): 994–1000.

Guyton, A. C. and J. E. Hall (1996). *Textbook of Medical Physiology*. Philadelphia, PA, W. B. Saunders.

Guzik, T. J., R. Korbut et al. (2003). Nitric oxide and superoxide in inflammation and immune regulation. *J Physiol Pharmacol* 54(4): 469–487.

Hafezi-Moghadam, A., K. L. Thomas et al. (2001). L-selectin shedding regulates leukocyte recruitment. *J Exp Med* 193(7): 863–872.

Haidekker, M. A., N. L'Heureux et al. (2000). Fluid shear stress increases membrane fluidity in endothelial cells: A study with DCVJ fluorescence. *Am J Physiol Heart Circ Physiol* 278(4): H1401–H1406.

Helmke, B. P., M. Sugihara-Seki et al. (1998). A mechanism for erythrocyte-mediated elevation of apparent viscosity by leukocytes in vivo without adhesion to the endothelium. *Biorheology* 35(6): 437–448.

Hermant, B., S. Bibert et al. (2003). Identification of proteases involved in the proteolysis of vascular endothelium cadherin during neutrophil transmigration. *J Biol Chem* 278(16): 14002–14012.

Hishikawa, K., T. Nakaki et al. (1992). Transmural pressure inhibits nitric oxide release from human endothelial cells. *Eur J Pharmacol* 215(2–3): 329–331.

Hishikawa, K., T. Nakaki et al. (1995). Pressure enhances endothelin-1 release from cultured human endothelial cells. *Hypertension* 25(3): 449–452.

Hueck, I. S., K. Rossiter et al. (2008). Fluid shear attenuates endothelial pseudopodia formation into the capillary lumen. *Microcirculation* 15(6): 531–542.

Hughes, B. J., J. C. Hollers et al. (1992). Recruitment of CD11b/CD18 to the neutrophil surface and adherence-dependent cell locomotion. *J Clin Invest* 90(5): 1687–1696.

Ingber, D. E. (2008). Tensegrity and mechanotransduction. *J Bodyw Mov Ther* 12(3): 198–200.

Intaglietta, M., D. R. Richardson et al. (1971). Blood pressure, flow, and elastic properties in microvessels of cat omentum. *Am J Physiol* 221(3): 922–928.

Jin, Z. G., H. Ueba et al. (2003). Ligand-independent activation of vascular endothelial growth factor receptor 2 by fluid shear stress regulates activation of endothelial nitric oxide synthase. *Circ Res* 93(4): 354–363.

Kamiya, A., J. Ando et al. (1988). Roles of fluid shear stress in physiological regulation of vascular structure and function. *Biorheology* 25(1–2): 271–278.

Kamm, R. D. and M. R. Kaazempur-Mofrad (2004). On the molecular basis for mechanotransduction. *Mech Chem Biosyst* 1(3): 201–209.

Kato, S., Y. Sasaguri et al. (1994). Ambient pressure stimulates immortalized human aortic endothelial cells to increase DNA synthesis and matrix metalloproteinase 1 (tissue collagenase) production. *Virchows Arch* 425(4): 385–390.

Kaul, S., R. C. Padgett et al. (1994). Role of platelets and leukocytes in modulation of vascular tone. *Ann N Y Acad Sci* 714: 122–135.

Kinashi, T. and K. Katagiri (2004). Regulation of lymphocyte adhesion and migration by the small GTPase Rap1 and its effector molecule, RAPL. *Immunol Lett* 93(1): 1–5.

Klein-Nulend, J., R. G. Bacabac et al. (2005). Mechanobiology of bone tissue. *Pathol Biol (Paris)* 53(10): 576–580.

Komai, Y. and G. W. Schmid-Schönbein (2005). De-activation of neutrophils in suspension by fluid shear stress: A requirement for erythrocytes. *Ann Biomed Eng* 33(10): 1375–1386.

Korhonen, R., A. Lahti et al. (2005). Nitric oxide production and signaling in inflammation. *Curr Drug Targets Inflamm Allergy* 4(4): 471–479.

Kubes, P., M. Suzuki et al. (1991). Nitric oxide: An endogenous modulator of leukocyte adhesion. *Proc Natl Acad Sci U S A* 88(11): 4651–4655.

Kubes, P., S. Kanwar et al. (1993). Nitric oxide synthesis inhibition induces leukocyte adhesion via superoxide and mast cells. *FASEB J* 7(13): 1293–1299.

Laroux, F. S., D. J. Lefer et al. (2000). Role of nitric oxide in the regulation of acute and chronic inflammation. *Antioxid Redox Signal* 2(3): 391–396.

Laudanna, C., J. Y. Kim et al. (2002). Rapid leukocyte integrin activation by chemokines. *Immunol Rev* 186: 37–46.

Lechi, C., M. Zatti et al. (1984). Increased leukocyte aggregation in patients with hypercholesterolaemia. *Clin Chim Acta* 144(1): 11–16.

Lee, H. J. and G. Y. Koh (2003). Shear stress activates Tie2 receptor tyrosine kinase in human endothelial cells. *Biochem Biophys Res Commun* 304(2): 399–404.

Lee, S. E., R. D. Kamm et al. (2007). Force-induced activation of talin and its possible role in focal adhesion mechanotransduction. *J Biomech* 40(9): 2096–2106.

Lefer, A. M. and D. J. Lefer (1996). The role of nitric oxide and cell adhesion molecules on the microcirculation in ischaemia-reperfusion. *Cardiovasc Res* 32(4): 743–751.

Lefer, D. J., S. P. Jones et al. (1999). Leukocyte–endothelial cell interactions in nitric oxide synthase-deficient mice. *Am J Physiol* 276(6 Pt 2): H1943–H1950.

Ley, K. (1996). Molecular mechanisms of leukocyte recruitment in the inflammatory process. *Cardiovasc Res* 32(4): 733–742.

Ley, K. (2002). Integration of inflammatory signals by rolling neutrophils. *Immunol Rev* 186: 8–18.

Libby, P. (2002). Inflammation in atherosclerosis. *Nature* 420(6917): 868–874.

Lipowsky, H. H., S. Kovalcheck et al. (1978). The distribution of blood rheological parameters in the microvasculature of cat mesentery. *Circ Res* 43(5): 738–749.

Lopez, S., S. Seveau et al. (1998). CD43 (sialophorin, leukosialin) shedding is an initial event during neutrophil migration, which could be closely related to the spreading of adherent cells. *Cell Adhes Commun* 5(2): 151–160.

Luo, B. H., C. V. Carman et al. (2007). Structural basis of integrin regulation and signaling. *Annu Rev Immunol* 25: 619–647.

Luster, A. D., R. Alon et al. (2005). Immune cell migration in inflammation: Present and future therapeutic targets. *Nat Immunol* 6(12): 1182–1190.

Makino, A., M. Glogauer et al. (2005). Control of neutrophil pseudopods by fluid shear: Role of Rho family GTPases. *Am J Physiol Cell Physiol* 288(4): C863–C871.

Makino, A., E. R. Prossnitz et al. (2006). G protein-coupled receptors serve as mechanosensors for fluid shear stress in neutrophils. *Am J Physiol Cell Physiol* 290(6): C1633–C1639.

Makino, A., H. Y. Shin et al. (2007). Mechanotransduction in leukocyte activation: A review. *Biorheology* 44(4): 221–249.

Marschel, P. and G. W. Schmid-Schönbein (2002). Control of fluid shear response in circulating leukocytes by integrins. *Ann Biomed Eng* 30(3): 333–343.

Marwali, M. R., J. Rey-Ladino et al. (2003). Membrane cholesterol regulates LFA-1 function and lipid raft heterogeneity. *Blood* 102(1): 215–222.

Mazzoni, M. C. and G. W. Schmid-Schönbein (1996). Mechanisms and consequences of cell activation in the microcirculation. *Cardiovasc Res* 32(4): 709–719.

McAllister, T. N. and J. A. Frangos (1999). Steady and transient fluid shear stress stimulate NO release in osteoblasts through distinct biochemical pathways. *J Bone Miner Res* 14(6): 930–936.

Medzhitov, R. (2008). Origin and physiological roles of inflammation. *Nature* 454(7203): 428–435.

Milkiewicz, M., J. L. Doyle et al. (2007). HIF-1alpha and HIF-2alpha play a central role in stretch-induced but not shear-stress-induced angiogenesis in rat skeletal muscle. *J Physiol* 583(Pt 2): 753–766.

Miyajima, A., T. Kitamura et al. (1992). Cytokine receptors and signal transduction. *Annu Rev Immunol* 10: 295–331.

Moazzam, F., F. A. DeLano et al. (1997). The leukocyte response to fluid stress. *Proc Natl Acad Sci U S A* 94(10): 5338–5343.

Mofrad, M. R., J. Golji et al. (2004). Force-induced unfolding of the focal adhesion targeting domain and the influence of paxillin binding. *Mech Chem Biosyst* 1(4): 253–265.

Moore, J. E., C. Xu et al. (1994). Fluid wall shear stress measurements in a model of the human abdominal aorta: Oscillatory behavior and relationship to atherosclerosis. *Atherosclerosis* 110(2): 225–240.

Mügge, A., D. D. Heistad et al. (1992). Activation of leukocytes with complement C5a is associated with prostanoid-dependent constriction of large arteries in atherosclerotic monkeys in vivo. *Atherosclerosis* 95(2–3): 211–222.

Mulivor, A. W. and H. H. Lipowsky (2002). Role of glycocalyx in leukocyte–endothelial cell adhesion. *Am J Physiol Heart Circ Physiol* 283(4): H1282–H1291.

Mulivor, A. W. and H. H. Lipowsky (2004). Inflammation- and ischemia-induced shedding of venular glycocalyx. *Am J Physiol Heart Circ Physiol* 286(5): H1672–H1680.

Muller-Marschhausen, K., J. Waschke et al. (2008). Physiological hydrostatic pressure protects endothelial monolayer integrity. *Am J Physiol Cell Physiol* 294(1): C324–C332.

Niggli, V. (2003). Signaling to migration in neutrophils: Importance of localized pathways. *Int J Biochem Cell Biol* 35(12): 1619–1638.

Niggli, V., A. V. Meszaros et al. (2004). Impact of cholesterol depletion on shape changes, actin reorganization, and signal transduction in neutrophil-like HL-60 cells. *Exp Cell Res* 296(2): 358–368.

Nobis, U., A. R. Pries et al. (1985). Radial distribution of white cells during blood flow in small tubes. *Microvasc Res* 29(3): 295–304.

Onishi, M., T. Nosaka et al. (1998). Cytokine receptors: Structures and signal transduction. *Int Rev Immunol* 16(5–6): 617–634.

Pearson, M. J. and H. H. Lipowsky (2004). Effect of fibrinogen on leukocyte margination and adhesion in postcapillary venules. *Microcirculation* 11(3): 295–306.

Pfaff, M., X. Du et al. (1999). Calpain cleavage of integrin beta cytoplasmic domains. *FEBS Lett* 460(1): 17–22.

Pontremoli, S., E. Melloni et al. (1989). Activation of neutrophil calpain following its translocation to the plasma membrane induced by phorbol ester or fMet-Leu-Phe. *Biochem Biophys Res Commun* 160(2): 737–743.

Radi, Z. A., M. E. Kehrli, Jr. et al. (2001). Cell adhesion molecules, leukocyte trafficking, and strategies to reduce leukocyte infiltration. *J Vet Intern Med* 15(6): 516–529.

Rinaldo, J. E. and R. E. Basford (1987). Neutrophil-endothelial interactions: Modulation of neutrophil activation responses by endothelial cells. *Tissue Cell* 19(5): 99–606.

Rizzo, V., D. P. McIntosh et al. (1998a). In situ flow activates endothelial nitric oxide synthase in luminal caveolae of endothelium with rapid caveolin dissociation and calmodulin association. *J Biol Chem* 273(52): 34724–34729.

Rizzo, V., A. Sung et al. (1998b). Rapid mechanotransduction in situ at the luminal cell surface of vascular endothelium and its caveolae. *J Biol Chem* 273(41): 26323–26329.

Robling, A. G., A. B. Castillo et al. (2006). Biomechanical and molecular regulation of bone remodeling. *Annu Rev Biomed Eng* 8: 455–498.

Rubin, J., C. Rubin et al. (2006). Molecular pathways mediating mechanical signaling in bone. *Gene* 367: 1–16.

Sakamoto, H., M. Aikawa et al. (2001). Biomechanical strain induces class a scavenger receptor expression in human monocyte/macrophages and THP-1 cells: A potential mechanism of increased atherosclerosis in hypertension. *Circulation* 104(1): 109–114.

Sakamoto, T. and P. A. Cabrera (2003). Immunohistochemical observations on cellular response in unilocular hydatid lesions and lymph nodes of cattle. *Acta Trop* 85(2): 271–279.

Salcedo, R., K. Wasserman et al. (1997). Endogenous fibronectin of blood polymorphonuclear leukocytes: Stimulus-induced secretion and proteolysis by cell surface-bound elastase. *Exp Cell Res* 233(1): 33–40.

Samet, M. M. and P. L. Lelkes (1999). The hemodynamic environment of endothelium *in vivo* and its simulation *in vitro*. In: *Mechanical Forces and the Endothelium*, P. L. Lelkes (ed.). Amsterdam, the Netherlands, Harwood Academic Publishers, pp. 1–32.

Satcher, R. L. Jr. and C. F. Dewey Jr. (1996). Theoretical estimates of mechanical properties of the endothelial cell cytoskeleton. *Biophys J* 71(1): 109–118.

Satcher, R., C. F. Dewey Jr. et al. (1997). Mechanical remodeling of the endothelial surface and actin cytoskeleton induced by fluid flow. *Microcirculation* 4(4): 439–453.

Sato, Y., K. R. Walley et al. (1999). Nitric oxide reduces the sequestration of polymorphonuclear leukocytes in lung by changing deformability and CD18 expression. *Am J Respir Crit Care Med* 159(5 Pt 1): 1469–1476.

Schettini, C., M. Bianchi et al. (1999). Ambulatory blood pressure: Normality and comparison with other measurements. Hypertension Working Group. *Hypertension* 34(4 Pt 2): 818–825.

Schmid-Schönbein, G. W. (2006). Analysis of inflammation. *Annu Rev Biomed Eng* 8: 93–131.

Schmid-Schönbein, G. W. and S. Chien (1989). Morphometry of human leukocyte granules. *Biorheology* 26(2): 331–343.

Schmid-Schönbein, G. W. and S. S. Su (2010). Internalization of formyl peptide receptor in leukocytes subject to fluid stresses. *Cell Mol Bioeng.* 3(1):20–29.

Schmid-Schönbein, G. W., S. Usami et al. (1980). The interaction of leukocytes and erythrocytes in capillary and postcapillary vessels. *Microvasc Res* 19(1): 45–70.

Schwartz, E. A., R. Bizios et al. (1999). Exposure of human vascular endothelial cells to sustained hydrostatic pressure stimulates proliferation. Involvement of the alphaV integrins. *Circ Res* 84(3): 315–322.

Secomb, T. W. and R. Skalak (1982). A two-dimensional model for capillary flow of an asymmetric cell. *Microvasc Res* 24(2): 194–203.

Seely, A. J., J. L. Pascual et al. (2003). Science review: Cell membrane expression (connectivity) regulates neutrophil delivery, function and clearance. *Crit Care* 7(4): 291–307.

Shay-Salit, A., M. Shushy et al. (2002). VEGF receptor 2 and the adherens junction as a mechanical transducer in vascular endothelial cells. *Proc Natl Acad Sci U S A* 99(14): 9462–9467.

Sherwood, L. (2001). *Human Physiology: From Cells to Systems*. Pacific Grove, CA, Brooks/Cole.

Shin, H. Y., M. E. Gerritsen et al. (2002a). Regulation of endothelial cell proliferation and apoptosis by cyclic pressure. *Ann Biomed Eng* 30(3): 297–304.

Shin, H. Y., M. L. Smith et al. (2002b). VEGF-C mediates cyclic pressure-induced endothelial cell proliferation. *Physiol Genomics* 11(3): 245–251.

Shin, H. Y., R. Bizios et al. (2003). Cyclic pressure modulates endothelial barrier function. *Endothelium* 10(3): 179–187.

Shin, H. Y., E. A. Schwartz et al. (2004). Receptor-mediated basic fibroblast growth factor signaling regulates cyclic pressure-induced human endothelial cell proliferation. *Endothelium* 11(5–6): 285–291.

Shin, H. Y., S. I. Simon et al. (2008). Fluid shear-induced activation and cleavage of CD18 during pseudopod retraction by human neutrophils. *J Cell Physiol* 214(2): 528–536.

Shiratsuchi, H. and M. D. Basson (2004). Extracellular pressure stimulates macrophage phagocytosis by inhibiting a pathway involving FAK and ERK. *Am J Physiol Cell Physiol* 286(6): C1358–C1366.

Shiratsuchi, H. and M. D. Basson (2005). Activation of p38 MAPKalpha by extracellular pressure mediates the stimulation of macrophage phagocytosis by pressure. *Am J Physiol Cell Physiol* 288(5): C1083–C1093.

Shiratsuchi, H. and M. D. Basson (2007). Akt2, but not Akt1 or Akt3 mediates pressure-stimulated serum-opsonized latex bead phagocytosis through activating mTOR and p70 S6 kinase. *J Cell Biochem* 102(2): 353–367.

Shiratsuchi, H., Y. Kouatli et al. (2009). Propofol inhibits pressure-stimulated macrophage phagocytosis via the GABAA receptor and dysregulation of p130cas phosphorylation. *Am J Physiol Cell Physiol* 296(6): C1400–C1410.

Shive, M. S., M. L. Salloum et al. (2000). Shear stress-induced apoptosis of adherent neutrophils: A mechanism for persistence of cardiovascular device infections. *Proc Natl Acad Sci U S A* 97(12): 6710–6715.

Shive, M. S., W. G. Brodbeck et al. (2002). Activation of caspase 3 during shear stress-induced neutrophil apoptosis on biomaterials. *J Biomed Mater Res* 62(2): 163–168.

Shulman, Z., V. Shinder et al. (2009). Lymphocyte crawling and transendothelial migration require chemokine triggering of high-affinity LFA-1 integrin. *Immunity* 30(3): 384–396.

Simon, S. I. and H. L. Goldsmith (2002). Leukocyte adhesion dynamics in shear flow. *Ann Biomed Eng* 30(3): 315–332.

Simon, S. I. and C. E. Green (2005). Molecular mechanics and dynamics of leukocyte recruitment during inflammation. *Annu Rev Biomed Eng* 7: 151–185.

Simons, K. and D. Toomre (2000). Lipid rafts and signal transduction. *Nat Rev Mol Cell Biol* 1(1): 31–39.

Sixt, M., R. Hallmann et al. (2001). Cell adhesion and migration properties of beta 2-integrin negative polymorphonuclear granulocytes on defined extracellular matrix molecules. Relevance for leukocyte extravasation. *J Biol Chem* 276(22): 18878–18887.

Smith, C. W. (2000). Possible steps involved in the transition to stationary adhesion of rolling neutrophils: A brief review. *Microcirculation* 7(6 Pt 1): 385–394.

Sodhi, A. and S. K. Biswas (2002). fMLP-induced in vitro nitric oxide production and its regulation in murine peritoneal macrophages. *J Leukoc Biol* 71(2): 262–270.

Solomkin, J. S., C. T. Robinson et al. (2007). Alterations in membrane cholesterol cause mobilization of lipid rafts from specific granules and prime human neutrophils for enhanced adherence-dependent oxidant production. *Shock* 28(3): 334–338.

Su, S. S. and G. W. Schmid-Schönbein (2008). Fluid stresses on the membrane of migrating leukocytes. *Ann Biomed Eng* 36(2): 298–307.

Suematsu, M., H. Suzuki et al. (2002). The inflammatory aspect of the microcirculation in hypertension: Oxidative stress, leukocytes/endothelial interaction, apoptosis. *Microcirculation* 9(4): 259–276.

Sugihara-Seki, M. and G. W. Schmid-Schönbein (2003). The fluid shear stress distribution on the membrane of leukocytes in the microcirculation. *J Biomech Eng* 125(5): 628–638.

Sumpio, B. E., M. D. Widmann et al. (1994). Increased ambient pressure stimulates proliferation and morphologic changes in cultured endothelial cells. *J Cell Physiol* 158(1): 133–139.

Sutanto, W., M. S. Oitzl et al. (1992). Corticosteroid receptor plasticity in the central nervous system of various rat models. *Endocr Regul* 26(3): 111–118.

Suzuki, H., G. W. Schmid-Schönbein et al. (1994). Impaired leukocyte–endothelial cell interaction in spontaneously hypertensive rats. *Hypertension* 24(6): 719–727.

Tarbell, J. M. and E. E. Ebong (2008). The endothelial glycocalyx: A mechano-sensor and -transducer. *Sci Signal* 1(40): pt8.

Tarbell, J. M., S. Weinbaum et al. (2005). Cellular fluid mechanics and mechanotransduction. *Ann Biomed Eng* 33(12): 1719–1723.

Taylor, C. A., C. P. Cheng et al. (2002). In vivo quantification of blood flow and wall shear stress in the human abdominal aorta during lower limb exercise. *Ann Biomed Eng* 30(3): 402–408.

Toikka, J. O., H. Laine et al. (2000). Increased arterial intima-media thickness and in vivo LDL oxidation in young men with borderline hypertension. *Hypertension* 36(6): 929–933.

Tomida, K., K. Tamai et al. (2001). Hypercholesterolemia induces leukocyte entrapment in the retinal microcirculation of rats. *Curr Eye Res* 23(1): 38–43.

Tomonaga, A., M. Hirota et al. (1983). Effect of membrane fluidizers on the number and affinity of chemotactic factor receptors on human polymorphonuclear leukocytes. *Microbiol Immunol* 27(11): 961–972.

Tozeren, A., R. Skalak et al. (1982). Viscoelastic behavior of erythrocyte membrane. *Biophys J* 39(1): 23–32.

Tuluc, F., J. Meshki et al. (2003). Membrane lipid microdomains differentially regulate intracellular signaling events in human neutrophils. *Int Immunopharmacol* 3(13–14): 1775–1790.

Tybulewicz, V. L. and R. B. Henderson (2009). Rho family GTPases and their regulators in lymphocytes. *Nat Rev Immunol* 9(9): 630–644.

Vouyouka, A. G., R. J. Powell et al. (1998). Ambient pulsatile pressure modulates endothelial cell proliferation. *J Mol Cell Cardiol* 30(3): 609–615.

Walcheck, B., J. Kahn et al. (1996). Neutrophil rolling altered by inhibition of L-selectin shedding in vitro. *Nature* 380(6576): 720–723.

Weinbaum, S., X. Zhang et al. (2003). Mechanotransduction and flow across the endothelial glycocalyx. *Proc Natl Acad Sci U S A* 100(13): 7988–7995.

White, C. R. and J. A. Frangos (2007). The shear stress of it all: The cell membrane and mechanochemical transduction. *Philos Trans R Soc Lond B Biol Sci* 362(1484): 1459–1467.

Wiles, M. E., J. A. Dykens et al. (1994). Regulation of polymorphonuclear leukocyte membrane fluidity: Effect of cytoskeletal modification. *J Leukoc Biol* 56(2): 192–199.

Worthen, G. S., B. Schwab, III et al. (1989). Mechanics of stimulated neutrophils: Cell stiffening induces retention in capillaries. *Science* 245(4914): 183–186.

Wyatt, T. A., T. M. Lincoln et al. (1991). Vimentin is transiently co-localized with and phosphorylated by cyclic GMP-dependent protein kinase in formyl-peptide-stimulated neutrophils. *J Biol Chem* 266(31): 21274–21280.

Wyatt, T. A., T. M. Lincoln et al. (1993). Regulation of human neutrophil degranulation by LY-83583 and L-arginine: Role of cGMP-dependent protein kinase. *Am J Physiol* 265(1 Pt 1): C201–C211.

Yuli, I., A. Tomonaga et al. (1982). Chemoattractant receptor functions in human polymorphonuclear leukocytes are divergently altered by membrane fluidizers. *Proc Natl Acad Sci U S A* 79(19): 5906–5910.

Zarbock, A. and K. Ley (2009). New insights into leukocyte recruitment by intravital microscopy. *Curr Top Microbiol Immunol* 334: 129–152.

Zhang, Y. L., J. A. Frangos et al. (2009). Mechanical stimulus alters conformation of type 1 parathyroid hormone receptor in bone cells. *Am J Physiol Cell Physiol* 296(6): C1391–C1399.

Zhelev, D. V. and A. Alteraifi (2002). Signaling in the motility responses of the human neutrophil. *Ann Biomed Eng* 30(3): 356–370.

Zhu, C., J. Lou et al. (2005). Catch bonds: Physical models, structural bases, biological function and rheological relevance. *Biorheology* 42(6): 443–462.

Section II (Part 2)

Literature Review of Mechanobiology Research Findings and Theories

Musculoskeletal Systems

9 Skeletal Mechanobiology

Alesha B. Castillo and Christopher R. Jacobs

CONTENTS

9.1 INTRODUCTION

Skeletal tissue is a reservoir of calcium that is important for proper muscle and nerve function, and houses the marrow that supports the hematopoietic and mesenchymal stem cell niche. The skeleton also provides the framework of the body, supports organs and tissues against gravitational force, provides levers through which muscles act to aid in locomotion, and protects internal organs against blunt force trauma. For a seemingly inert tissue, bone has the ability to remodel, reorganize, and regenerate in response to mechanical loading and injury. Mechanical force is an important regulator of biology and plays a critical role in tissue development and maintenance.[1] Indeed, mechanical loading is a potent regulator of bone turnover, and the structural success of the skeleton is due in large part to the capacity of bones to respond to the prevailing mechanical environment by optimizing its shape and size to meet physical demands.[2] Mechanisms by which mechanical loading regulates bone remodeling and adaptation involve first, the sensing of an extracellular physical signal, and second, the conversion of this signal into a biochemical response, a process known as mechanotransduction. The purpose of this chapter is to systematically review the state of knowledge with respect to skeletal mechanobiology. After a brief outline of basic bone physiology, we review the research on the ability of bone cells to sense and to respond to physical signals and the nature of the physical signals experienced at a cellular level. Finally, we consider potential molecular mechanisms responsible for cellular mechanosensing and some recently uncovered novel mechanisms.

9.2 BONE MODELING AND REMODELING

Bone is shaped and renewed by highly specialized cells through the processes of modeling and remodeling. These processes are carried out through the coordinated and tightly controlled actions of osteoclasts, the bone-resorbing cells, osteoblasts, the bone-forming cells, and osteocytes, which are embedded deep in the mineralized matrix (Figure 9.1). Osteoclasts are multinucleated giant cells that originate from the fusion of mononuclear precursors of hematopoietic origin.[3] Osteoblasts originate from bone marrow stroma-derived adherent cells or colony forming unit cells, which can also form fat, muscle, cartilage, and fibrous tissue.[4] A fraction of bone-forming osteoblasts are embedded in bone during matrix formation and become terminally differentiated osteocytes, the cell that is likely one of the primary mechanosensors in bone.[5,6]

Modeling and remodeling require an elegant orchestration of spatial and temporal cues that regulate cell migration, proliferation, differentiation, and apoptosis. Modeling involves the *uncoupled* action of osteoclasts and osteoblasts on distinct bone surfaces, and is typically observed during growth and adaptation of skeletal shape and size. Remodeling involves *coupled* osteoblast and osteoclast activity where bone formation follows bone resorption, both temporally and spatially. Remodeling is the primary mechanism by which bone renewal occurs and its molecular regulation is discussed in more detail below. Both modeling and remodeling can occur in four distinct regions, including the periosteal, endosteal, trabecular, and intracortical envelopes, and is initiated by various stimuli including mechanical microenvironment,[7] matrix microdamage,[8] humoral mediators,[9] and hypoxia.[10]

9.2.1 RESORPTION

The regulation of osteoclast recruitment and differentiation during remodeling is carried out by a family of tumor necrosis factor (TNF) receptor (TNFR)/TNF-like proteins including the receptor activator of nuclear factor-κB (NF-κB) (RANK), RANK ligand (RANKL), and osteoprotegerin (OPG).[11] Osteoblasts and stromal cells enhance bone resorption through the expression of RANKL, a membrane-bound ligand, which binds to RANK on osteoclast precursors[12] (Figure 9.1a). Binding activates the expression of osteoclast-specific markers including tartrate-resistant acid phosphatase, cathepsin K (CATK), the calcitonin receptor, and the β3-integrin, and is necessary for osteoclast maturation.[12] Mature osteoclasts form a sealing zone and ruffled border atop bone surfaces, and create an acidic environment by releasing H^+ ions, CATK, and metalloproteinases, all which serve to degrade bone matrix.[13] Osteoblasts can also impede bone resorption through the expression of OPG, a soluble decoy receptor that competitively binds RANKL. Thus, the relative expression of OPG and RANKL in the local environment regulates bone resorption.

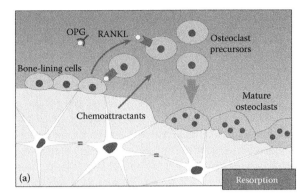

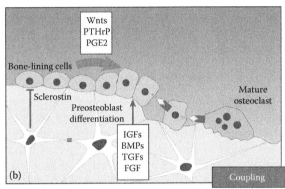

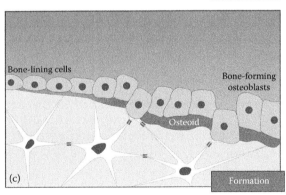

FIGURE 9.1 (See color insert.) Bone remodeling is carried out by teams of osteoblasts and osteoclasts. (a) Osteoclast precursors (green) are recruited to bone surfaces via chemoattractants released (straight red arrow) from bone tissue (gray) into the marrow space (blue). RANKL (yellow) is expressed on the surface of osteo-blasts and osteoblast progenitors, shown here as bone lining cells (pink). RANKL is also expressed in a soluble form. RANKL binds RANK on osteoclast precursors and initiates monocyte fusion and differentiation (blue arrow) into multinucleated mature osteoclasts. Osteoblasts and osteoblast progenitors also express osteoprote-gerin (OPG), which acts as a decoy receptor for RANKL and blocks osteoclast differentiation. Mature osteo-clasts adhere to bone surfaces and resorb matrix. (b) Coupling action of osteoclast and osteoblast activity occurs at bone surfaces where resorption of matrix by osteoclasts releases osteogenic growth factors (e.g., IGFs, BMPs, TGFs, and FGFs) thereby enhancing osteoblast differentiation. Wnts, PTHrP, and PGE2 released from neigh-boring osteoblastic cells (blue arrow) also enhance osteogenic activity. Cell–cell signaling (blue receptor/yellow ligand) can also play a role in osteoblast differentiation. Sclerostin, which is produced by osteocytes (white) embedded within the matrix, can negatively control this process by inhibiting osteoblast differentiation. (c) Mature cuboidal osteoblasts form new matrix or osteoid on previously resorbed bone surfaces. A fraction of the osteoblasts become embedded within the matrix, a process that is partially controlled by direct communication with osteocytes via gap junctions (red double dash).

9.2.2 COUPLING

Mechanisms by which osteoprogenitors are coupled to osteoclast activity during remodeling are less clear. Proposed mechanisms are reviewed by Martin et al.[14] and include (1) the release of homing and growth factors (e.g., TGF-β, IGF-I, FGF, and BMPs) during matrix resorption, (2) cell–cell communication via gap junctions and through cell–cell bidirectional signaling molecules, (3) an osteoclast-derived signal that activates osteoblast activity, and (4) an osteocyte-derived signal that activates osteoblast activity (Figure 9.1b). Recent data suggest that the coupling of osteoclast and osteoblast activity on bone surfaces is facilitated by the formation of bone remodeling compartments (BRCs) at the bone surface, wherein osteoclasts and osteoblasts are separated from the bone marrow by a layer of bone-lining cells.[15,16] This "canopy" is in direct physical contact with bone marrow capillaries, which may be a source of immature osteoclasts and osteoprogenitors.[17] Interestingly, disruption of the BRC appears to enhance resorption and prevent formation,[15] suggesting that a defined and intimate microenvironment is ideal for normal remodeling.

9.2.3 FORMATION

Once recruited to newly resorbed sites, osteoprogenitors differentiate into mature bone-forming osteoblasts (Figure 9.1c) via activation of several osteogenic signaling pathways. Of these, Wnt signaling is critical at all stages of skeletal maintenance.[18] Wnts are secreted glycoproteins that bind a receptor complex comprised of a seven-pass transmembrane protein, frizzled (Fz), and a single pass transmembrane protein of the low-density lipoprotein (LDL) receptor-related protein (LRP) family.[19] Canonical Wnt signaling involves several proteins, including dishevelled (Dsh), axin, adenomatous polyposis coli (APC), glycogen synthase kinase (GSK)-3β, and β-catenin. β-catenin is a cytoplasmic phosphoprotein, which, in the absence of Wnt signaling, is targeted for degradation through phosphorylation by GSK-3β. However, upon Wnt binding, Dsh is phosphorylated leading to the phosphorylation and inactivation of GSK-3β, allowing β-catenin to accumulate in the cytoplasm. Subsequently, β-catenin translocates to the nucleus where it interacts with the T-cell and lymphoid enhancer (TCF-LEF) transcription factors to affect gene transcription.[19] Target genes include Runx2[20] and Osterix,[21] both of which are osteoblast-specific transcription factors critical in osteoblast differentiation, proliferation, activity, and apoptosis.[22] Wnt signaling is inhibited by several proteins including sclerostin, which is encoded by the gene sclerosteosis (SOST),[23] dickkopf1 (Dkk1),[24] secreted frizzled-related protein 1 (sFRP1),[25] and Wise.[26] Thus, osteoblast differentiation is regulated by the spatial and temporal expression of Wnt-signaling modulators.

Osteocytes, terminally differentiated bone cells located throughout the mineralized matrix, are formed from osteoblasts that become embedded during bone formation. Osteocytes are stellate cells that communicate with one another at the bone surface via gap junctions to regulate bone metabolism. Osteoblasts and osteocytes exhibit different expression profiles,[27,28] and markers specific for osteocytes include E11/gp38, a glycoprotein important in the formation of dendritic processes,[29] and sclerostin, a glycoprotein that is a potent inhibitor of osteoblast function.[30] Additional osteocyte-derived factors that have been shown to regulate skeletal metabolism are the dentin matrix protein 1 (DMP-1),[31] the fibroblast growth factor 23 (FGF23),[32] Phex,[33] and the heparin-binding growth-associated molecule (HB-GAM).[34]

9.3 MECHANICAL ADAPTATION OF THE SKELETON

9.3.1 MECHANICALLY INDUCED BONE FORMATION

Bone adapts to its mechanical environment by adjusting its shape and size to withstand applied forces.[35,36] Loading activates progenitors on cortical and trabecular bone surfaces where they

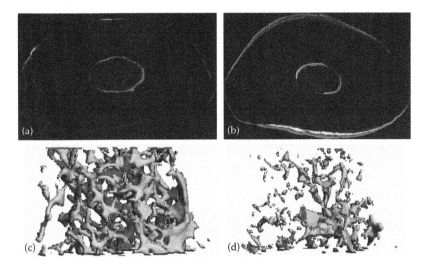

FIGURE 9.2 (See color insert.) Mechanical adaptation of bone. Mechanical loading and disuse result in new bone formation and bone loss, respectively. Cyclic compressive loading of the mouse ulna results in significant bone formation on the periosteal and endosteal surfaces at midshaft (b) compared to the nonloaded contralateral limb (a). Fluorescent bone labels (green calcein, red alizarin) administered 7 days apart are incorporated into the newly forming matrix and allow the visualization of newly formed bone. Hind limb suspension in mice results in significant trabecular bone loss in the distal femur (d) as compared to normally ambulating controls (c) depicted by a reduction in trabecular number and thickness and an increase in spacing between individual trabeculae.

divide and differentiate into bone-forming osteoblasts.[37] Additionally, there is mounting evidence that mechanical loading mobilizes and recruits mesenchymal stem cells to the osteoblastic lineage.[37,38] Mechanical loading data from humans and animals show that bone is preferentially added where strains are the greatest[35,39] so as to reduce strains to some preset value, also referred to as the "mechanostat."[40] A striking example of skeletal functional optimization of bone is the significantly greater bone mass observed in the playing versus nonplaying arm of racquetball athlete.[41] Likewise, studies using the rodent ulna loading model clearly show that bone forms principally on the medial and lateral periosteal surfaces where measured strains are greatest (Figure 9.2a and b),[42] resulting in a bone that is stronger in bending and fatigue.[43,44] Although the precise molecular mechanisms that govern skeletal mechanical adaptation are not clear, Turner[45] has described three general rules to which bones adhere: (1) a dynamic mechanical stimulus is required for bone adaptation;[46] (2) bone cells become desensitized to continued loading and resensitized with rest;[47,48] and (3) bone cells adapt to the current loading environment so that subsequent changes in loading are perceived.[49]

9.3.2 Mechanical Fatigue

Applied repetitive forces and mechanical overload of the skeleton can result in matrix damage in the formation of microscopic cracks[50] (Figure 9.2). Damage leads to reduced bone stiffness and strength,[51] and is a potent activator of bone remodeling.[8,52] Schaffler and colleagues[53] showed that load-induced microcracks initiated intracortical remodeling in the ulna of a rat, an animal that does not ordinarily display large amounts of cortical remodeling. In addition, Burr et al.[8] showed that new sites of remodeling were four to six times more likely to be spatially associated with microcracks than by chance alone, indicating a targeted versus random remodeling process. Recent data suggest that matrix cracking results in the disruption of osteocyte lacunae and interstitial fluid flow,[54–56] which in turn can lead to hypoxic and hyponutrient conditions and eventual osteocyte apoptosis.[53]

Apoptotic osteocytes are believed to then signal and recruit cells locally including bone-resorbing osteoclasts.[57] In this regard, targeted remodeling is thought to be the primary mechanism by which damage is repaired in the absence of full fracture.

9.3.3 MECHANICAL DISUSE

Mechanical disuse due to immobilization, microgravity, and paralysis leads to dramatic bone loss.[58] Significant losses in areal bone mineral density (aBMD) in volunteers undergoing 17 weeks of bed rest were observed at several sites including the lumbar spine (−3.9%), femoral neck (−3.6%), and calcaneus (−10.4%).[59] Similar rates of bone loss occur in microgravity conditions.[60] Jaworski and colleagues showed that long-term immobilization of forelimb of dogs resulted in significant bone loss.[61] Rodents subjected to hindlimb suspension exhibit severe bone loss (Figure 9.2c and d), inhibition of bone formation and mineralization,[62] a reduction in osteoblast number,[63,64] and a decrease in bone strength.[65] Interestingly, bed rest subjects, cosmonauts, and hindlimb suspended animals exhibit increased cranial bone mass,[66,67] an observation that has been linked to a fluid pressure shift from the lower extremities to the upper body and skull.[68] These data suggest an important link between interstitial fluid and bone metabolism.

As is the case of damage-induced remodeling, it is proposed that mechanical unloading leads to reduced interstitial fluid flow, hypoxia, and osteocyte apoptosis, which then serves as a beacon for osteoclast recruitment.[69–71] The nature of this signaling is unknown, although there appears to be several feasible candidates. Bidwell and colleagues[72] have shown that the high mobility group box 1 (HMBG1), a chromatin-associated nuclear protein and endogenous inflammatory trigger, is released from apoptotic osteocytes and may increase the RANKL/OPG ratio in the marrow. Another possible contender is osteopontin,[73] a glycosylated protein that is upregulated in hypoxic osteocytes, which is chemoattractive for osteoclasts,[74] and is required for disuse-induced bone loss in the hindlimb suspension mouse model.[75] More recently, Lin et al.[76] showed that SOST$^{-/-}$ mice did not lose bone in response to hindlimb suspension, which appeared to be regulated via Wnt signaling, suggesting that SOST expression may play a critical role in osteoclast recruitment.

How does the skeleton sense and respond to changes in the mechanical environment and matrix microdamage? Mechanical loading results in a number of changes in the physical environment of bone cells including the creation of extracellular streaming potentials, mechanical strain, pressure gradients, and fluid flow-induced shear stress (Figure 9.3a). Of these, interstitial fluid flow appears to be one of the primary physical stimuli acting on bone cells.[77] Qin et al.[78] provided compelling evidence for this showing that short daily bouts (60 mmHg, 20 Hz, 10 min/day) of increased intramedullary pressure, in the absence of mechanical strain, prevented disuse bone loss and led to new cortical bone formation in isolated turkey ulnae. These results suggest that interstitial fluid flow is indeed a critical stimulus for mechanical adaptation. Potential mechanoreceptors in bone include stretch-activated ion channels, primary cilia, G-protein-coupled receptors, integrins and adhesion-associated proteins, and cytoskeletal proteins. These receptors serve to transduce physical signals into a biochemical response by activating downstream signaling pathways and gene transcription (Figure 9.3b). In addition, mechanically induced gene products can provide autoregulatory feedback by binding membrane-bound receptors.

9.4 INTERSTITIAL FLUID FLOW IN BONE

Interstitial fluid is found between cells and blood vessels and is comprised of oxygen, chemical messengers, nutrients, and cellular waste.[79] It bathes all tissues including the mineralized matrix of bone, where it flows through the lacunocanalicular system.[80] Lacunocanalicular fluid flow is the sum of a steady wash-through flow due to circulatory-induced intramedullary pressure (IMP) and load-induced flow.[77] Blood supply to long bones is provided by a single nutrient artery that progressively bifurcates into arterioles and then into capillaries located in Haversian systems. IMP is pulsatile in

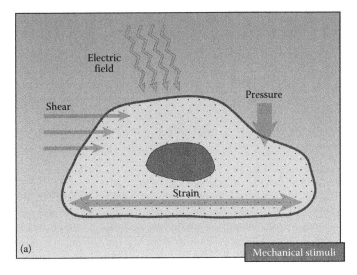

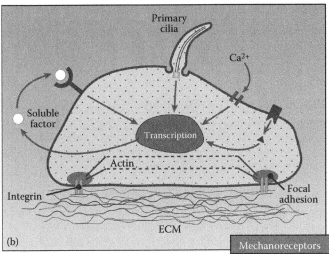

FIGURE 9.3 Mechanical stimuli and cellular mechanoreceptors. (a) Mechanical stimuli resulting from physical force applied directly to tissue in vivo can result in deformation or strain of the cell body, fluid flow shear stress, electric fields, and pressure. (b) Potential mechanoreceptors that may sense physical stimuli include primary cilia, ion channels, transmembrane receptors (rectangle), integrins, and focal adhesion signaling proteins. Activation of various receptors can ultimately result in changes in gene expression through numerous signaling pathways. Additionally, a release of soluble factors (white circles) in response to loading can bind surface receptors and provide autoregulatory feedback.

nature coincident with arterial blood pressure and respiration, and is higher than that found in most other tissues, resulting in a hydrostatic pressure drop across the bone cortex.[81] Fluid is driven from Haversian capillaries, through a layer of bone-lining cells and into the lacunocanalicular space.[82] Once in the lacunocanalicular space, fluid moves in a radial pattern from the endosteal surface and drains at the periosteal surface through lymphatics found in the connective tissues surrounding the periosteum.[83] When bone is compressed, fluid is displaced from regions of compression to regions of tension. Osteocytes are ideally situated to sense fluid flow in bone[84] as they are dispersed throughout the bone matrix (Figure 9.4) and can communicate with one another and with cells at the bone surface through gap junctions. Osteocytes are housed in individual lacunae (Figure 9.5) and depend on canalicular fluid flow to supply nutrients, remove cellular waste, and facilitate the

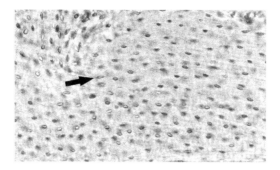

FIGURE 9.4 Osteocytes are dispersed throughout the bone matrix. Cortical bone cross section showing osteocytes within their lacunae (black arrow) dispersed throughout the bone matrix. Osteocytes communicate with one another and with cells on the bone surface through their processes via gap junctions. Their number and location within the bony matrix makes them likely master regulators of bone metabolism in response to mechanical and biochemical stimuli. Image width ~300 μm.

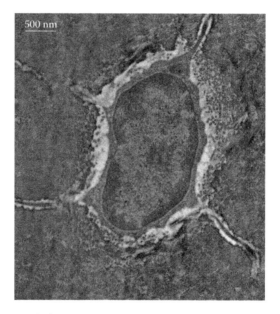

FIGURE 9.5 Single osteocyte in its lacunar space. The dense cortical bone contains osteocytes housed in individual lacunae dispersed throughout its matrix. Osteocytes are tethered to the extracellular matrix through focal adhesions composed primarily of integrins, and osteocyte dendritic processes penetrate the canalicular space allowing communication between osteocytes and bone lining cells via gap junctions. Osteocytes are surrounded by pericellular fluid within their lacunae, and fluid oscillates in the interstitial space (white) between the cell body and the lacunar walls. (From McNamara, L.M. et al., *Anat. Rec. (Hoboken)*, 292, 357, 2009. Copyright 2009, Wiley-Liss, Inc. Reprinted with permission of John Wiley & Sons, Inc.)

active transport of bone-reactive humoral substances including the parathyroid hormone (PTH), calcitonin, and estradiol, all of which regulate mineral metabolism and homeostasis.

9.4.1 TRACER STUDIES

Using in vivo and ex vivo tracer studies in rodents, Knothe Tate et al. have shown that the lacuno-canalicular network can accommodate molecules up to 40–70 kDa or ~7 nm in size by diffusion alone.[85] Larger molecules were not able to penetrate the cortical bone in mature animals suggesting that mechanical loading is necessary for the diffusion of molecules larger than 10 nm.[86] Indeed, they later showed that mechanical loading enhanced intracortical fluid flow[87] in a site-specific manner.[88] In a separate study, Dillaman et al.[89] showed that unloading reduced the perfusion of HRP (4–5 nm), a finding that was corroborated by Frangos and colleagues who found that interstitial fluid flow in the cortical long bone was directly related to IMP with a net fluid movement from the marrow and endosteal surface to the periosteal surface.[71] Qin et al. showed that short daily bouts of a sinusoidal fluid pressure signal (60 mmHg, 20 Hz), in the absence of strain, prevented disuse bone loss and led to new bone formation in isolated turkey ulnae,[78] thereby providing compelling evidence that fluid flow alone was a key regulator of load-induced adaptation as mentioned previously.

9.4.2 Theoretical Models

While in vivo tracing studies allow the observation of fluid movement after the fact, it is still very difficult to make detailed mechanical measurements due to limited access to the pericellular space. Conventional strain measurements and fluid velocimetry are nearly impossible. Thus, theoretical models of the osteocyte and osteocyte lacuna are very valuable. The initial demonstration that loading of bone tissue produces fluid motion through the canalicular space was described by Piekarski and Monroe.[80] They established a theoretical basis for a transport mechanism to exist between the vascular system and the osteocyte, whereby dynamic loading of the bone tissue results in a dynamic pericellular fluid flow. Little was added over the following decade until Kufahl et al.[90] further developed the relationship between tissue loading and flow and speculated that this was an important mechanism in the regulation of bone remodeling. Dillaman and colleagues first created theoretical models of interstitial fluid flow in bone and described the resulting shear stress acting on bone cells.[89,91] These studies focused on circulation driven flow and its impact on chemical exchange between the interstitial and vascular fluid spaces and specifically avoided the consideration of load-induced deformation of the mineralized tissue. McCarthy and Yang[92] modeled a similar effect and found that diffusion alone may be insufficient to supply nutrients to cells deep in the matrix. The transport of molecular tracers through the lacunar canalicular network has been modeled by[93] providing a quantitative demonstration that load-induced pericellular flows do indeed occur in bone. Weinbaum and Cowin[6,94–96] introduced more detailed analytical models that included the effects of the glycocalyx proteoglycans occupying the noncellular regions of the lacunar canalicular network. They specifically addressed the paradox of how tissue strain levels that occur with habitual bone loading appear to be insufficient to stimulate bone cells if directly applied in vitro.[97,98] Alternatives to the fluid-drag strain amplification theory include strain amplification of the solid matrix itself due to the void inclusion of the osteocyte lacuna. This has been predicted to result in strains as large as 3%.[99] However, these strains are highly localized within the lacuna and are at the low end of the range of strains that osteocytes have been shown to respond to.

This strain amplification effect due to fluid drag through the glycocalyx was extended to account for the mechanical behavior of the cytoplasmic actin in the osteocyte process and to make initial estimates of cellular deformations.[96] You et al.[100] reported that highly localized contacts exist between the osteocyte process and the surrounding mineralized matrix. These contacts were modeled by Han et al.[101] who found that the fluid drag force on the process would translate into highly local forces across transmembrane proteins. Although it is not known what specific proteins occur at these contacts, the forces were found to be of sufficient magnitude to regulate integrins. This multiscale approach demonstrates how tissue-level loading due to habitual activities of daily living translate into molecular forces with potential to modulate protein confirmation and to initiate intracellular signaling.

In addition to analytical models, numerical approaches have also been employed. For example, Steck et al.[102] utilized the finite element method to simulate fluid flow through the lacunar canalicular system as it relates to mechanosensing with more geometric complexity. Gururaja et al.[103] used a two-dimensional lumped parameter model validated by comparing predicted and measured bone tissue energy loss under dynamic loading and found that the relaxation time for flow through the canalicular system is orders of magnitude longer than the vascular Haversian system. Recently, Goulet et al.[104] utilized a multiscale continuum poroelastic approach that incorporated diffusive and convective fluid transport and osteocyte metabolism. Transport, pressure, and shear parameters were shown to be consistent with prior experimental and theoretical models.

9.4.3 STEADY VERSUS OSCILLATORY FLUID FLOW

Cyclic loading due to ordinary activities, such as ambulation, forces fluid from regions of high compressive strains and returns when loading is removed, thus exposing cells to an oscillating fluid flow.[105] Oscillatory flow profiles activate specific biochemical signaling pathways in bone cells with important distinctions from those activated by unidirectional steady flow.[96,105–107] For example, intracellular calcium mobilization was observed in both the response to steady as well as oscillatory flow.[107,108] However, in the case of steady flow, the source of this calcium increase was found to be from both intracellular stored calcium and extracellular calcium influx through stretch-activated membrane channels. In contrast, the oscillatory flow resulted in the release of calcium from the endoplasmic reticulum via IP3-sensitive calcium channels only with no involvement of stretch-activated channels. Similarly, medium-duration (1 h) exposure to steady flow produces a dramatic reorganization of the actin cytoskeleton including densification and stress fiber formation,[108] whereas oscillatory flow does not,[105,109] although longer oscillatory flow times (5 h) do eventually lead to actin filament reorganization.[109] The differences in the response to oscillatory versus nonoscillatory flow are likely due to large cellular viscoelastic deformations expected from a unidirectional steady flow that may activate additional pathways.[110]

9.4.4 FLUID FLOW IN MARROW-FILLED AND OSTEONAL SPACES

Less is understood about the flow of fluid in marrow-filled spaces and within in-filling osteons, both of which house osteoprogenitors and preosteoclasts. IMP and IMP-induced fluid flow are possible mechanical signals that regulate the milieu of cells found in the marrow. Dickerson et al.[111] showed that whole body vibration, a known anabolic stimulus for vertebral trabecular bone,[112] resulted in fluid flow shear stress (1–20 dyn/cm^2) at trabecular surfaces in the marrow similar in magnitude to previously reported estimates of what occurs in vivo.[95] It is also interesting that models predict that high frequency mechanical loads induce higher levels of fluid flow than the same level of mechanical loading applied at a low frequency.[104] This result, combined with experimental in vivo strain gage studies showing that high frequency (up to 20 Hz) loading does occur physiologically,[113] suggests the possibility that an important physical signal, both in the marrow and canalicular space, is high-frequency oscillatory fluid flow. In addition, due to the enclosed space within in-filling osteons, fluid shear acting over these cells is potentially larger than in the intramedullary space. In fact, low-amplitude high-frequency signals elicit an osteogenic response in bone cells,[114] and are anabolic for trabecular, but not cortical, bone.[115,116] In addition, preliminary studies by Judex and colleagues suggest that vibration enhances bone regeneration,[117] but further studies are needed to determine precisely how vibration affects bone cell function. In conclusion, load-induced fluid flow in the cortical bone and in the marrow certainly occurs, and cannot be dismissed as a potent regulator of bone metabolism in vivo.

9.5 SIGNAL PROPAGATION IN BONE

9.5.1 CELL–CELL COMMUNICATION

The responsiveness of osteocytes, osteoblasts, and stromal cells to fluid flow in vitro has been established, as evidenced by increases in calcium signaling, G-protein activation, ERK phosphorylation, nitric oxide production, and prostaglandin-2 (PGE-2) release,[107,118–124] all of which are important early mechanically induced signaling events in bone cells. Mechanisms by which mechanical and biochemical signals are integrated and communicated in vivo among osteocytes deep within the bone matrix and cells at bone surfaces is unclear, but gap junctions appear to play a prominent role. Gap junctions are membrane-spanning channels, which allow the passage of small molecules (<1 kD), such as calcium ions, inositol phosphates, and cyclic nucleotides from one cell to another. Each gap junction is comprised of two hexameres termed connexons, which in turn, are comprised of six subunits termed connexins (Cx). The most widely expressed of these include Cx26 and Cx32. Morphological and ultrastructural studies reveal that gap junctions directly connect adjoining osteocytes and osteoblasts[125] and are the key couplers in the osteocyte syncytium. Additionally, bone cells are functionally coupled both in vivo and in vitro.[126,127] Thus, gap junctions are well suited to communicate and integrate the fluid flow response of the osteocyte–osteoblast network. Indeed, recent evidence suggests that gap junctions may play a role in the response of cellular networks to diverse extracellular stimuli, both chemical and biophysical. Xia and Ferrier[128] showed that the mechanical perturbation of a single osteoblast results in Ca^{2+} propagation, presumably via gap junctions. Others have demonstrated that gap junctional intercellular communication (GJIC) contributes to bone cell responsiveness to electromagnetic fields[129] and that osteoblastic cells expressing antisense cDNA to Cx43 are functionally uncoupled and dramatically less responsive to the parathyroid hormone than well-coupled osteoblastic cells.[130] In addition, fluid flow has been shown to result in the opening of gap junctions[131] as well as increasing Cx43 phosphorylation[118] in osteoblasts. Further, gap junctions have been shown to be involved in fluid flow-induced ATP release[132] and prostaglandin release[133] in osteoblasts. Finally, the fluid flow-induced expression of osteopontin and osteocalcin, two important regulatory genes in osteoblasts, was significantly attenuated with the gap junction blocker, 18β-glycyrrhetinic acid.[134] These results establish GJIC as an important mechanism by which osteocytes and osteoblasts communicate in their response to mechanically induced fluid flow.

9.5.2 Long-Range Communication

Less is known about how osteocytes and osteoblasts achieve long-range communication with cells presumably outside of the syncytium (e.g., multipotent stromal cells and pre- and mature osteoclasts). Evidence suggests that both osteoblast progenitors and osteoclasts are mobilized and recruited to bone surfaces in response to mechanical loading,[37] which may represent a direct effect of mechanical stimuli on cellular function or a secondary effect through the release of soluble molecules that serve as chemoattractants.

9.6 MECHANOSENSITIVE RECEPTORS IN BONE

Mechanoreceptors are proteins present on the cell surface that can detect a physical stimulus and initiate an intracellular biochemical response via conformational changes or the activation of signaling molecules. It is likely that a combination of signaling events and a convergence of pathways ultimately control the mechanical adaptation of the skeleton. The first described mechanoreceptor is a stretch-activated ion channel that regulates calcium entrance into the cell.[135] Since then, several putative mechanoreceptors have been described in bone cells and in other cell types.

9.6.1 Ion Channels

Three classes of pressure-sensitive ion channels in human osteoblast-like cells were described by Davidson et al.[136] The channel exhibiting the highest conductance was found to be a voltage-gated and calcium-dependent potassium channel, which later was shown to be sensitive to membrane stretch.[137] Cyclic stretch has the ability to alter ion channel characteristics in osteoblast-like cells as evidenced by increased numbers of open channels and a reduced strain threshold compared to nonstrained controls.[138] Ion channels are also responsive to fluid flow. Fluid flow induced a rapid increase in intercellular calcium signaling, which was inhibited by pertussis toxin,[139] an inhibitor of G-proteins known to regulate calcium flux. L-Type voltage sensitive calcium channels (LVCC) were shown to mediate mechanically induced bone formation in vivo.[140] Ion channels are also activated by treatment with PTH,[141] a potent regular of serum calcium. In fact, PTH treatment enhances fluid flow-induced calcium signaling through activation of mechanosensitive and voltage sensitive calcium channels.[142] It has been shown that membrane strains of 800% were needed to open 50% of the channels in primary rat osteoblasts.[143] The precise mechanism by which mechanical stimuli activate ion channels is unclear; however, cellular localization of Annexin V, a calcium-dependent binding protein, can be regulated by fluid flow[144] suggesting that flow alters the spatial and temporal activity of ion channels, which may indeed also affect channel function.

Recent attention has focused on the role of the transient receptor potential (TRP) cation channel superfamily members in bone cell signaling. TRPV4, 5, and 6 are highly selective for calcium,[145] and TRPV5 null mice exhibit hypercalciuria and decreased cortical and trabecular thickness in long bones.[146] TRPV4 has been shown to localize to primary cilia,[147] which are known mechanosensors in kidney cells.[148] Recent data show that TRPV4 can be activated by mechanical stress,[149] and gene ablation of TRPV4 in mice resulted in increased trabecular bone mass due to reductions in osteoclast number and activity.[150] The recent observation that TRPV4 is also expressed in human and murine osteoblast-like cells[151] make it a possible candidate for fluid flow-induced calcium signaling in osteoblasts.

Another area of intense study is the role of polycystin-1 and -2 in skeletal mechanosensation. Polycystin-1 is an 11-pass transmembrane protein with a large extracellular domain, whereas polycystin-2 is a calcium channel that belongs to the TRP ion channel superfamily.[152] Polycystins are expressed in osteoblasts and osteocytes, and, like TRPV4, have been shown to localize to primary cilia.[122,153] Gene ablation of Pkd1, the gene encoding polycystin-1, resulted in the abnormal development of the axial skeleton and long bones[154,155] suggesting a role for polycystins in skeletal maintenance.

9.6.2 PRIMARY CILIA

The primary cilium is a solitary, nonmotile organelle that projects from the surface of almost all nondividing eukaryotic cells. It is comprised of a central axoneme made of nine outer doublet microtubules and is surrounded by a plasma membrane that is contiguous with the plasma membrane of the cell body.[156] Axonemal assembly and maintenance is carried out by intraflagellar transport (IFT) of large protein complexes up and down the cilium. Movement from base to tip is accomplished by kinesin-2 motors (Kif3a/Kif3b/KAP complex) and involves IFT88/Polaris, and movement from tip to base is regulated by cytoplasmic dynein 2 and involves IFT139/THM1. Regulatory mechanisms of IFT are not well understood, but mitogen-activated protein kinases (MAPK) and IFT172 have been identified as key players. Defects in primary cilia lead to several ciliopathies including polycystic kidney disease, polydactyly, and retinitis pigmentosa (for a review see Veland et al.[157]). Recently, it has been shown that the primary cilium has a prominent role as a mechanosensor in kidney cells.[148,158] Flow-induced bending of the primary cilium in renal epithelial cells increases ATP release, intracellular calcium, and opens the intermediate-conductance K+ channels.[159] With several signaling receptors localized to the primary cilium (Hh, Wnt, PDGFRα, integrin, TauT[160]), it is considered a nexus of mechanical (Figure 9.6a) and biochemical (Figure 9.6b) signal integration that is required for normal cell function.[121,161–163] Downstream signaling pathways include p53, JNK, JAK/STAT, and mTOR.[164,165]

Primary cilia in osteocytes and osteoblasts were first described in the early 1970s.[166,167] Recently, Quarles and colleagues[153] showed that an overexpression of polycystin-1 in MC3T3 osteoblasts resulted in increased Runx2 expression while primary osteoblasts from the Pkd1[+/m1Bei] mouse, which has an inactivating missense mutation of polycystin-1, showed reductions in Runx2 expression and osteogenic markers. They also showed that polycystin-1 colocalizes with polycystin-2 in MC3T3 osteoblast-like cells and regulates the Runx2-11 expression through a calcium-dependent mechanism in osteoblasts.[168] Additionally, Pkd1[+/m1Bei] mice exhibited significant reductions in femoral BMD and bone formation rates, most likely resulting from reduced osteogenic gene expression.

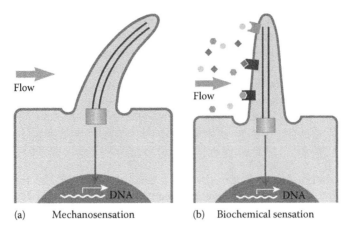

(a)   Mechanosensation   (b)   Biochemical sensation

FIGURE 9.6 The primary cilium is a signaling nexus. (a) Fluid flow across bone cells results in the bending of the cilium and downstream intracellular signaling via unknown mechanisms. (b) Several biochemical receptors have been proposed to localize to the primary cilium including integrins, Smo, Inversin, PDGFRα, and TauT, which bind ECM molecules, Hh, Wnts, RTK, and calcium, respectively. Thus, the primary cilium is an important structural complex where both mechanical and biochemical signaling are integrated to regulate cellular function.

The primary cilium as a mechanosensor in bone cells has recently been investigated. Inhibition of cilia formation in MC3T3 osteoblasts by IFT88/Polaris knockdown and chloral hydrate treatment resulted in reductions in osteopontin expression, the OPG/RANKL ratio, and PGE-2 release in response to flow.[122] In addition, the rapid decrease in cAMP in response to flow was inhibited in osteocytes treated with siRNA against Polaris.[121] Finally, ongoing work shows that the conditional ablation of Kif3a in osteoblasts and osteocytes results in attenuated load-induced cortical bone formation (unpublished data). Taken together, these data suggest that the primary cilium is indeed a mechanosensor in bone. Further work is needed to determine the precise molecular mechanisms governing ciliary mechanosensing in the skeleton.

9.6.3 G-Protein-Coupled Receptors

Guanine nucleotide-binding (G)-protein-coupled receptors (GPCRs) are the largest family of membrane receptors and are activated by biochemical and mechanical stimuli.[169,170] They play a critical role in a variety of cellular processes including proliferation, migration, and chemotaxis. GPCRs contain α, β, and γ subunits, and four main classes of Gα proteins (G_s, $G_{i/o}$, $G_{q/11}$, and $G_{12/13}$) engage unique downstream targets when activated. G_s activates cAMP, protein kinase A (PKA), and mitogen-activated protein kinase (MAPK) signaling—G_i activates Src and PI3K signaling, $G_{q/11}$ activates phospholipase C (PLC)/ protein kinase C (PKC) signaling, and $G_{12/13}$ can activate Rho GTP binding proteins. Mice, with an osteoblast-specific ablation of G_s[171] and G_i[172] in osteoblasts, exhibited significantly reduced cortical and trabecular bone volume, whereas an overexpression of G_s in osteoblasts results in a significant increase in bone mass.[173] Frangos and colleagues recently showed that osteoblasts expose to fluid flow undergo a conformational change in PTH1R, a G-protein-coupled receptor, in a ligand-independent manner.[174] They have also shown that a flow-induced activation of G-proteins independent of the membrane receptor is possible. Additional data are needed to further explore the role of G-protein-coupled receptors bone cell mechanotransduction.

9.6.4 Integrins and Adhesion-Associated Proteins

Integrins are transmembrane cell adhesion heterodimers comprised of α and β subunits that bind directly to the extracellular matrix (ECM) at anchoring hubs known as focal adhesions. The cytoplasmic tail binds talin and paxillin, both of which link directly to the actin cytoskeleton through various signaling molecules, including focal adhesion kinase (FAK). Integrin signaling is bidirectional in that extracellular domain-ECM binding activates intracellular signaling molecules (outside-in signaling) while conformational changes in the cytoplasmic tail activates the integrin head for ligand binding.[175] Osteoblasts express α2-α6, α8, αv, β1, β3, and β5.[176,177] Of these, the β1 subunit appears to play an important functional role in osteoblasts.[178] Mice expressing an osteoblast-specific dominant negative form of β1 integrin exhibited reduced bone mass and increased cortical porosity due to a defect in bone formation and increased osteoclast resorption. Expression of β1, β3, and CD44 has been reported in osteocytes.[84] In osteocytes, β1 is predominantly expressed in the cell body, whereas β3 is expressed along the osteocyte processes and appears to tether processes to the canalicular wall at focal adhesions,[84] which would be situated directly in the path of the canalicular fluid flow.[179] Mechanical stimulation activates integrins[180] and downstream signaling including MAPK,[181] tyrosine phosphorylation, intracellular calcium release,[182,183] c-fos,[184] and translocation of FAK from the cytosol to the cytoskeleton during the creation of focal adhesions,[180,185] all of which are important osteogenic signals. In addition, β1 expression is upregulated in osteoblasts in response to fluid flow[186] suggesting that β1 binding may be a primary response to physical forces in bone cells. Indeed, mice with osteocyte-specific gene ablation of β1 integrin exhibited significantly lower load-induced bone formation than wild type littermates,[187] suggesting that β1 signaling is an important component of flow sensing in osteocytes.

As integrins do not possess intrinsic catalytic activity, ECM-intracellular signal transduction is carried out by activation of downstream signaling molecules comprising focal adhesions. FAK, a nonreceptor tyrosine kinase,[188] is one of the first molecules recruited to focal adhesions upon integrin binding[189] and has been shown to be important in cell migration, proliferation, and survival. FAK has also been implicated in cellular mechanotransduction,[190,191] including bone cells.[192–194] The activation of FAK results in the autophosphorylation of tyrosine 397 creating a high-affinity binding site for the Src-homology 2 (SH2) domain[195] of the Src family protein tyrosine kinases. FAK and SH2 binding activates MAPK signaling[196] though interaction with c-src, Grb2, and the small GTPase Ras.[197,198] Fluid flow leads to FAK phosphorylation[199] and MAPK activation in bone and endothelial cells.[200,201] FAK phosphorylation has also been linked to NFκB activation[202] as well as calcium release via large conductance calcium channels.[203] Disruption of FAK in osteoblasts leads to decreases in the fluid flow-induced ERK phosphorylation, expression of c-fos and Cox-2, as well as PGE-2 release,[194] all of which are important signaling events in osteoblast function. Interestingly, preliminary data suggest that osteocyte-specific gene ablation of FAK does not affect load-induced cortical bone formation in vivo (unpublished data). Thus, FAK appears to play a lesser role in the sensing of and response to fluid flow in osteocytes in vivo, and a more prominent role in downstream bone formation events, such as osteoblast differentiation and recruitment. Indeed, the disruption of FAK in osteoblasts abolishes the response of bone marrow cells to mechanical stimuli in a tibial injury model[204] indicating a potential role for FAK in osteoprogenitor recruitment and homing. Recent data suggest that proline-rich tyrosine kinase 2 (Pyk2) may compensate for loss of FAK in various cell types including endothelial cells[205] and fibroblasts.[206] However, Pyk2 expression was not enhanced in FAK[−/−] osteoblasts,[194] and it is unclear whether Pyk2 has a compensatory effect on load-induced bone formation in vivo. In fact, Pyk2 null mice exhibit increased bone mass and bone formation suggesting that Pyk2 normally represses osteoblast differentiation.[207]

Additional adhesion-associated proteins that may regulate mechanosensing in bone are talin and paxillin. Talin binds the integrin cytoplasmic β tail[208] and alters integrin conformation thereby enhancing the integrin-binding affinity and receptor activation.[209] Talin also binds FAK and actin, and is a key linking protein between the ECM and the actin cytoskeleton.[210,211] Paxillin binds the cytoplasmic tail of α4 and contains several signaling motifs including phosphorylation sites, phosphatases, and regulators of the Rho family of small GTPases.[212] Paxillin has been shown to align with the direction of fluid flow in osteocytes,[213] suggesting that it is responsive to mechanical stimuli, but additional data are needed to understand the role of these adhesion molecules in mechanosensing.

9.6.5 Cytoskeletal Proteins and Tensegrity

While focal adhesions are potent signaling structures in their own right, their connection to the cytoskeleton may be just as important for mechanosensing. The cytoskeleton is composed primarily of three cytoskeletal filaments (actin microfilaments, intermediate filaments, and microtubules), and the dynamic interplay of these filaments with focal adhesions regulates cell shape and structure. Tensegrity (tensional integrity), a concept proposed by Ingber,[214] describes how the structural organization of the cell is modified by its physical environment. The tensegrity model predicts that the cytoskeleton serves to connect the ECM to the cell nucleus via focal adhesions making the cell "hard-wired" to respond to mechanical stress. That is, the cytoskeleton actively generates tensile forces via contractile microfilaments that are balanced by other structural elements, namely integrins and focal adhesions, to mechanically stabilize cell shape. Mechanical stress results in a perturbation of the entire prestressed network and triggers a series of signaling events that leads to further integrin activation and recruitment of additional focal adhesions.

The role of the cytoskeleton in mechanotransduction has been studied in several cell types.[215–217] Force applied directly to microbead-bound integrins resulted in cytoskeletal filament rearrangement and nuclear distortion indicating force transfer to the nucleus is actin filament-mediated.[218] Studies in osteoblasts showed that the upregulation of Cox-2 and c-fos in response to 1 h of steady flow involves the rearrangement of the actin cytoskeleton[219] (Figure 9.7). Conversely, oscillatory fluid flow in the same cell type did not lead to actin filament reorganization, and disruption of actin

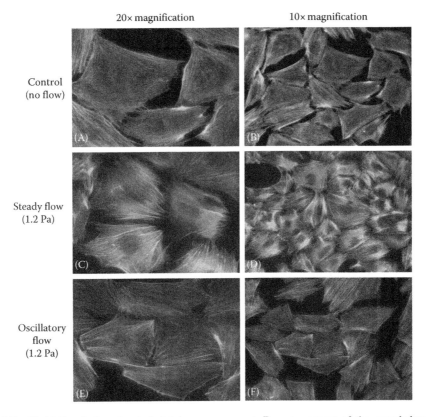

20× magnification   10× magnification

Control (no flow)

Steady flow (1.2 Pa)

Oscillatory flow (1.2 Pa)

FIGURE 9.7 Fluid flow-induced cytoskeletal rearrangement. Rearrangement of the cytoskeletal protein, F-actin, in MC3T3-E1 osteoblasts in response to 1 h of no flow (A and B), unidirectional steady flow (C and D), and oscillatory flow (E and F). The greatest response is observed in unidirectional steady flow. (From Malone, A.M. et al., *Am. J. Physiol. Cell Physiol.*, 292, C1832, 2007. Copyright 2007. Reprinted with permission of American Physiological Society.)

filaments did not affect flow-induced intracellular calcium and PGE-2 release.[123] The pharmaco-logical disruption of actin microfilaments, microtubules, or intermediate filaments did not affect laminar flow-induced increases in PGE-2 release, and of the three, only disruption of intermedi-ate filaments reduced flow-induced Cox-2 expression.[220] Microtubule disruption attenuated flow-induced proliferation,[221] collagen I, and matrix metalloproteinases 1 and 3,[222] whereas disruption of microfilaments had no effect. Finally, the disarrangement of the actin cytoskeleton in a multipotent cell line abrogates the flow-induced increase in Runx2 expression suggesting that the cytoskeleton plays a key role in flow-induced osteogenic differentiation.[120]

9.7 MECHANOSENSITIVE SIGNALING PATHWAYS IN BONE

There are numerous downstream signaling pathways initiated by mechanical stimuli in bone cells, including early (e.g., ERK phosphorylation, calcium signaling) and late (e.g., Cox-2 expression, PGE-2 release, gene transcription) signaling events. Here, we briefly review some of the most well-defined early signaling evens, as they descend directly from mechanoreceptor activation and likely precede pathway cross-talk.

9.7.1 CALCIUM SIGNALING

Intracellular calcium signaling regulates numerous basic cellular processes including proliferation, differentiation, and cellular motility.[223] In bone cells, calcium is an early second messenger, and an extracellular influx and intracellular release is rapidly activated by strain, pressure, and fluid flow.[105,106,224–226] The source of calcium appears to be intracellular stores following the activation of phospholipase C and IP3 signaling, and blocking calcium release attenuates the mechanically induced upregulation of the osteogenic gene expression in vitro[107,108,227] and abrogates load-induced bone formation in vivo.[140] Downstream signaling in osteoblasts and osteocytes includes PKA, MAPK, c-fox, NFKβ translocation to the nucleus, and Cox-2 expression,[228,229] all of which are important for cell survival. Alternatively, PGE-2 release, a common response to mechanical stimu-lation in osteoblasts, appears to be independent of calcium signaling.[230]

9.7.2 G-PROTEIN-RELATED SIGNALING

The α subunit of G-proteins exist as monomers known as Rho and Ras GTPases, which are active when bound to GTP.[231] Flow-activated G-protein recruit and activate phospholipase C (PLC), which hydrolyzes phophatidylinositol 4,5-bisphophonate and generates the second messengers DAG and IP3.[232] IP3 then induces calcium release from intracellular stores, which acts as a second messen-ger in bone cells.[233] Flow-activated G-protein also activate PKA, which phosphorylates the cAMP response element (CRE)-binding protein (CREB), a transcription factor that can bind the promoter of Cox-2.[234] Mechanical stimuli have been shown to activate G-proteins in vascular endothelial[235,236] and multipotent stromal[120] cells. Frangos and colleagues showed that G-protein activation leads to an increase in PGE-2 release, whereas G-protein inhibitors block the response.[237] They also showed that G-protein activation was initiated independent of a protein receptor,[238] which suggests that changes in membrane structure alone can activate G-proteins. Finally, flow-induced osteogenic dif-ferentiation as measured by Runx2 was significantly reduced in multipotent stromal cells treated with a RhoA inhibitor,[120] suggesting that G-proteins plays an important role in both early and late events in osteoblastic signaling.

9.7.3 MAPK

Mitogen-activated protein kinases (MAPKs) are serine/threonine protein kinases that play a role in cell differentiation, proliferation, and apoptosis via downstream regulators including ERK1/2, p38, and JNK/SAPK signaling.[239,240] Many groups have reported ERK1/2 activation in bone cells in response to fluid flow,[241–244] an event that appears to be calcium- and ATP-independent.[229] Flow-induced ERK phosphorylation has been shown to modulate RANKL expression,[245] proliferation,[199] nitric oxide synthase expression,[246] and MMP-13 expression.[244] In addition, recent data show that ERK1/2 is important in flow-induced osteogenic differentiation in multipotent stromal cells.[247]

9.7.4 NITRIC OXIDE

Nitric oxide (NO) is a second messenger that regulates many important biological processes including vascular tone, neuronal communication, and immune function.[248] NO binds the enzyme guanylyl cyclase, which then generates the small intracellular signaling molecule cyclic GMP (cGMP). NO is synthesized from arginine catalyzed by NO synthase (NOS). There are three isoforms of NOS: neuronal (nNOS), endothelial (eNOS), and inducible (iNOS). nNOS is constitutively expressed in nerve and muscle cells. eNOS is found in endothelial cells and generates a rapid and transient increase in NO in a calcium-dependent manner. iNOS is found in macrophages of the immune system and the effects are longer lasting with greater amounts of NO produced. iNOS is expressed in cytokine-stimulated osteoblasts and macrophages, but not osteoclasts.[249] eNOS is expressed in stromal cells, osteoblasts, osteocytes, and osteoclasts,[250,251] and nNOS mRNA and protein have been detected in whole bone and bone cells.[249,252]

eNOS-deficient mice exhibit reduced bone formation during growth resulting from defects in osteoblast function.[253] In vivo mechanical loading data has been shown to increase eNOS[251] and iNOS[254] activity in bone cells, and appears to be essential for load-induced bone formation.[255] The generation of NO in bone cells in response to mechanical stimulation in vitro has been reported by several groups,[186,246,256,257] and has been linked to signaling events including prostaglandin release,[246] MAPK signaling,[124] and cytoskeletal rearrangement.[258] Recent data suggest that NO production and calcium signaling is regulated by the ciliary proteins polycystin-1[162] and -2.[259]

9.8 PERSPECTIVE

The ability of nonspecialized, nonexcitable cells to sense and respond to mechanical stimulation is central to proper physiologic function in a surprisingly wide range of cell types, including endothelial cells, liver, lung and kidney epithelial cells, chondrocytes, and bone cells. Cellular mechanosensation is critical in diseases responsible for enormous human suffering including atherosclerosis, osteoarthritis, cancer, and osteoporosis. Despite the virtually ubiquitous nature of mechanics in biology and its importance in understanding disease, mechanobiology is a nascent field. In bone biology, the role of mechanics in the regulation of bone tissue is well accepted and characterized; however, how mechanical tissue loading is sensed by bone cells is an open question.

The first issue that arises from consideration of this question is what pericellular physical signal bone cells sense. A number of loading-induced cell-level physical signals have been considered including direct cellular deformation, electromagnetic fields, fluid flow, and fatigue damage of the bone tissue. The problem is compounded since the cellular responses elicited by these stimuli are often similar or overlapping when applied to cells in culture. Fluid flow through the lacunar-canalicular network is currently a strong candidate signal since flow in the absence of other stimuli has been shown to increase bone formation in vivo, and fatigue damage, which is known to activate remodeling, has been shown to interrupt flow in vivo. Of course, this is not a settled question and it is important to keep an open mind about novel possibilities, such as direct cellular vibration.

Numerous cellular components are suggested to play a role including membrane ion channels, the cytoskeleton, integrins, adhesion-associated proteins, pericellular matrix proteins, and most recently, primary cilia. Much of the data originating from cell culture experiments are only suggestive. Increasingly, corroborative in vivo data have been obtained using tissue-specific gene deletions in mice. The ultimate application of this knowledge is to develop novel pharmacologic interventions targeting specific pathways to induce bone formation and reduce bone loss in humans suffering from debilitating bone conditions and diseases. Since downstream signaling rapidly becomes complex and often redundant, targeting the initial transduction event is much more likely to yield the specific desired effects. Thus, future research in this area is a compelling priority.

ACKNOWLEDGMENT

The authors thank Kris Morrow (VA Palo Alto Medical Illustrations) for creating the illustrations. Funding sources include VA Career Development Award A6842-W (ABC), NIH AR45989 (CRJ), NIH AR5456 (CRJ) and New York Stem Cell Grant N089-210 (CRJ).

REFERENCES

1. Ingber D. 2003. Mechanobiology and diseases of mechanotransduction. *Ann Med* **35**(8):564–577.
2. Robling A, Castillo A, Turner C. 2006. Biomechanical and molecular regulation of bone remodeling. *Annu Rev Biomed Eng* **8**:455–498.
3. Athanasou NA, Bliss E, Gatter KC, Heryet A, Woods CG, McGee JO. 1985. An immunohistological study of giant-cell tumour of bone: Evidence for an osteoclast origin of the giant cells. *J Pathol* **147**(3):153–158.
4. Friedenstein AJ, Petrakova KV, Kurolesova AI, Frolova GP. 1968. Heterotopic of bone marrow. Analysis of precursor cells for osteogenic and hematopoietic tissues. *Transplantation* **6**(2):230–247.
5. Cowin SC, Moss-Salentijn L, Moss ML. 1991. Candidates for the mechanosensory system in bone. *J Biomech Eng* **113**(2):191–197.
6. Cowin SC, Weinbaum S, Zeng Y. 1995. A case for bone canaliculi as the anatomical site of strain generated potentials. *J Biomech* **28**(11):1281–1297.
7. Lanyon LE, Rubin CT. 1984. Static vs dynamic loads as an influence on bone remodelling. *J Biomech* **17**(12):897–905.
8. Burr DB, Martin RB, Schaffler MB, Radin EL. 1985. Bone remodeling in response to in vivo fatigue microdamage. *J Biomech* **18**(3):189–200.
9. Compston JE. 2001. Sex steroids and bone. *Physiol Rev* **81**(1):419–447.
10. Dodd JS, Raleigh JA, Gross TS. 1999. Osteocyte hypoxia: A novel mechanotransduction pathway. *Am J Physiol* **277**(3 Pt 1):C598–C602.
11. Boyce BF, Xing L. 2008. Functions of RANKL/RANK/OPG in bone modeling and remodeling. *Arch Biochem Biophys* **473**(2):139–146.
12. Lacey DL, Timms E, Tan HL, Kelley MJ, Dunstan CR, Burgess T, Elliott R et al. 1998. Osteoprotegerin ligand is a cytokine that regulates osteoclast differentiation and activation. *Cell* **93**(2):165–176.
13. Teitelbaum SL. 2007. Osteoclasts: What do they do and how do they do it? *Am J Pathol* **170**(2):427–435.
14. Martin T, Gooi J, Sims N. 2009. Molecular mechanisms in coupling of bone formation to resorption. *Crit Rev Eukaryot Gene Expr* **19**(1):73–88.
15. Andersen TL, Sondergaard TE, Skorzynska KE, Dagnaes-Hansen F, Plesner TL, Hauge EM, Plesner T, Delaisse JM. 2009. A physical mechanism for coupling bone resorption and formation in adult human bone. *Am J Pathol* **174**(1):239–247.
16. Hauge EM, Qvesel D, Eriksen EF, Mosekilde L, Melsen F. 2001. Cancellous bone remodeling occurs in specialized compartments lined by cells expressing osteoblastic markers. *J Bone Miner Res* **16**(9):1575–1582.
17. Matsumoto T, Kuroda R, Mifune Y, Kawamoto A, Shoji T, Miwa M, Asahara T, Kurosaka M. 2008. Circulating endothelial/skeletal progenitor cells for bone regeneration and healing. *Bone* **43**(3):434–439.
18. Leucht P, Minear S, Ten Berge D, Nusse R, Helms J. 2008. Translating insights from development into regenerative medicine: The function of Wnts in bone biology. *Semin Cell Dev Biol* **19**(5):434–443.
19. Gordon MD, Nusse R. 2006. Wnt signaling: Multiple pathways, multiple receptors, and multiple transcription factors. *J Biol Chem* **281**(32):22429–22433.
20. Gaur T, Lengner CJ, Hovhannisyan H, Bhat RA, Bodine PV, Komm BS, Javed A et al. 2005. Canonical WNT signaling promotes osteogenesis by directly stimulating Runx2 gene expression. *J Biol Chem* **280**(39):33132–33140.
21. Zhang C, Cho K, Huang Y, Lyons JP, Zhou X, Sinha K, McCrea PD, de Crombrugghe B. 2008. Inhibition of Wnt signaling by the osteoblast-specific transcription factor Osterix. *Proc Natl Acad Sci U S A* **105**(19):6936–6941.

22. Karsenty G. 2008. Transcriptional control of skeletogenesis. *Annu Rev Genomics Hum Genet* **9**:183–196.
23. Li X, Zhang Y, Kang H, Liu W, Liu P, Zhang J, Harris SE, Wu D. 2005. Sclerostin binds to LRP5/6 and antagonizes canonical Wnt signaling. *J Biol Chem* **280**(20):19883–19887.
24. Glinka A, Wu W, Delius H, Monaghan AP, Blumenstock C, Niehrs C. 1998. Dickkopf-1 is a member of a new family of secreted proteins and functions in head induction. *Nature* **391**(6665):357–362.
25. Bhanot P, Brink M, Samos CH, Hsieh JC, Wang Y, Macke JP, Andrew D, Nathans J, Nusse R. 1996. A new member of the frizzled family from Drosophila functions as a Wingless receptor. *Nature* **382**(6588):225–230.
26. Itasaki N, Jones CM, Mercurio S, Rowe A, Domingos PM, Smith JC, Krumlauf R. 2003. Wise, a context-dependent activator and inhibitor of Wnt signalling. *Development* **130**(18):4295–4305.
27. Yang W, Harris MA, Heinrich JG, Guo D, Bonewald LF, Harris SE. 2009. Gene expression signatures of a fibroblastoid preosteoblast and cuboidal osteoblast cell model compared to the MLO-Y4 osteocyte cell model. *Bone* **44**(1):32–45.
28. Paic F, Igwe JC, Nori R, Kronenberg MS, Franceschetti T, Harrington P, Kuo L, Shin DG, Rowe DW, Harris SE, Kalajzic I. 2009. Identification of differentially expressed genes between osteoblasts and osteocytes. *Bone* **45**(4):682–692.
29. Zhang K, Barragan-Adjemian C, Ye L, Kotha S, Dallas M, Lu Y, Zhao S, Harris M, Harris SE, Feng JQ, Bonewald LF. 2006. E11/gp38 selective expression in osteocytes: Regulation by mechanical strain and role in dendrite elongation. *Mol Cell Biol* **26**(12):4539–4552.
30. Winkler DG, Sutherland MK, Geoghegan JC, Yu C, Hayes T, Skonier JE, Shpektor D et al. 2003. Osteocyte control of bone formation via sclerostin, a novel BMP antagonist. *EMBO J* **22**(23):6267–6276.
31. Toyosawa S, Shintani S, Fujiwara T, Ooshima T, Sato A, Ijuhin N, Komori T. 2001. Dentin matrix protein 1 is predominantly expressed in chicken and rat osteocytes but not in osteoblasts. *J Bone Miner Res* **16**(11):2017–2026.
32. Liu S, Zhou J, Tang W, Jiang X, Rowe DW, Quarles LD. 2006. Pathogenic role of Fgf23 in Hyp mice. *Am J Physiol Endocrinol Metab* **291**(1):E38–E49.
33. Ruchon AF, Tenenhouse HS, Marcinkiewicz M, Siegfried G, Aubin JE, DesGroseillers L, Crine P, Boileau G. 2000. Developmental expression and tissue distribution of Phex protein: Effect of the Hyp mutation and relationship to bone markers. *J Bone Miner Res* **15**(8):1440–1450.
34. Imai S, Heino TJ, Hienola A, Kurata K, Buki K, Matsusue Y, Vaananen HK, Rauvala H. 2009. Osteocyte-derived HB-GAM (pleiotrophin) is associated with bone formation and mechanical loading. *Bone* **44**(5):785–794.
35. Robling AG, Hinant FM, Burr DB, Turner CH. 2002. Improved bone structure and strength after long-term mechanical loading is greatest if loading is separated into short bouts. *J Bone Miner Res* **17**(8):1545–1554.
36. Rubin CT, Lanyon LE. 1984. Regulation of bone formation by applied dynamic loads. *J Bone Joint Surg Am* **66**(3):397–402.
37. Turner CH, Owan I, Alvey T, Hulman J, Hock JM. 1998. Recruitment and proliferative responses of osteoblasts after mechanical loading in vivo determined using sustained-release bromodeoxyuridine. *Bone* **22**(5):463–469.
38. Carter DR, Beaupré GS, Giori NJ, Helms JA. 1998. Mechanobiology of skeletal regeneration. *Clin Orthop Relat Res* (355 Suppl):S41–S55.
39. Bass SL, Saxon L, Daly RM, Turner CH, Robling AG, Seeman E, Stuckey S. 2002. The effect of mechanical loading on the size and shape of bone in pre-, peri-, and postpubertal girls: A study in tennis players. *J Bone Miner Res* **17**(12):2274–2280.
40. Frost HM. 1987. Bone. "mass" and the "mechanostat": A proposal. *Anat Rec* **219**(1):1–9.
41. Ducher G, Daly R, Bass S. 2009. The effects of repetitive loading on bone mass and geometry in young male tennis players: A quantitative study using magnetic resonance imaging. *J Bone Miner Res* **24**(10):1686–1692.
42. Hsieh YF, Robling AG, Ambrosius WT, Burr DB, Turner CH. 2001. Mechanical loading of diaphyseal bone in vivo: The strain threshold for an osteogenic response varies with location. *J Bone Miner Res* **16**(12):2291–2297.
43. Warden SJ, Hurst JA, Sanders MS, Turner CH, Burr DB, Li J. 2005. Bone adaptation to a mechanical loading program significantly increases skeletal fatigue resistance. *J Bone Miner Res* **20**(5):809–816.
44. Warden SJ, Fuchs RK, Castillo AB, Nelson IR, Turner CH. 2007. Exercise when young provides lifelong benefits to bone structure and strength. *J Bone Miner Res* **22**(2):251–259.
45. Turner CH. 1998. Three rules for bone adaptation to mechanical stimuli. *Bone* **23**(5):399–407.

46. Hert J, Liskova M, Landa J. 1971. Reaction of bone to mechanical stimuli. 1. Continuous and intermittent loading of tibia in rabbit. *Folia Morphol* (*Praha*) 19(3):290–300.

47. Robling AG, Burr DB, Turner CH. 2000. Partitioning a daily mechanical stimulus into discrete loading bouts improves the osteogenic response to loading. *J Bone Miner Res* 15(8):1596–1602.

48. Srinivasan S, Weimer DA, Agans SC, Bain SD, Gross TS. 2002. Low-magnitude mechanical loading becomes osteogenic when rest is inserted between each load cycle. *J Bone Miner Res* 17(9):1613–1620.

49. Schriefer JL, Warden SJ, Saxon LK, Robling AG, Turner CH. 2005. Cellular accommodation and the response of bone to mechanical loading. *J Biomech* 38(9):1838–1845.

50. Frost HM. 1960. Presence of microscopic cracks in vivo in bone. *Henry Ford Hosp Med J* 8:25–35.

51. Carter DR, Hayes WC. 1977. Compact bone fatigue damage—I. Residual strength and stiffness. *J Biomech* 10(5–6):325–337.

52. Mori S, Burr DB. 1993. Increased intracortical remodeling following fatigue damage. *Bone* 14(2):103–109.

53. Verborgt O, Gibson GJ, Schaffler MB. 2000. Loss of osteocyte integrity in association with microdamage and bone remodeling after fatigue in vivo. *J Bone Miner Res* 15(1):60–67.

54. Hazenberg JG, Freeley M, Foran E, Lee TC, Taylor D. 2006. Microdamage: A cell transducing mechanism based on ruptured osteocyte processes. *J Biomech* 39(11):2096–2103.

55. Muir P, Sample S, Barrett JG, McCarthy J, Vanderby R, Markel M, Prokuski LJ, Kalscheur V. 2007. Effect of fatigue loading and associated matrix microdamage on bone blood flow and interstitial fluid flow. *Bone* 40(4):948–956.

56. Tami AE, Nasser P, Verborgt O, Schaffler MB, Knothe Tate ML. 2002. The role of interstitial fluid flow in the remodeling response to fatigue loading. *J Bone Miner Res* 17(11):2030–2037.

57. Cardoso L, Herman BC, Verborgt O, Laudier D, Majeska RJ, Schaffler MB. 2009. Osteocyte apoptosis controls activation of intracortical resorption in response to bone fatigue. *J Bone Miner Res* 24(4):597–605.

58. LeBlanc AD, Spector ER, Evans HJ, Sibonga JD. 2007. Skeletal responses to space flight and the bed rest analog: A review. *J Musculoskelet Neuronal Interact* 7(1):33–47.

59. Leblanc AD, Schneider VS, Evans HJ, Engelbretson DA, Krebs JM. 1990. Bone mineral loss and recovery after 17 weeks of bed rest. *J Bone Miner Res* 5(8):843–850.

60. Vico L, Collet P, Guignandon A, Lafage–Proust MH, Thomas T, Rehaillia M, Alexandre C. 2000. Effects of long-term microgravity exposure on cancellous and cortical weight-bearing bones of cosmonauts. *Lancet* 355(9215):1607–1611.

61. Uhthoff HK, Jaworski ZF. 1978. Bone loss in response to long-term immobilisation. *J Bone Joint Surg Br* 60-B(3):420–429.

62. Dehority W, Halloran BP, Bikle DD, Curren T, Kostenuik PJ, Wronski TJ, Shen Y, Rabkin B, Bouraoui A, Morey-Holton E. 1999. Bone and hormonal changes induced by skeletal unloading in the mature male rat. *Am J Physiol* 276(1 Pt 1):E62–E69.

63. Barou O, Palle S, Vico L, Alexandre C, Lafage-Proust MH. 1998. Hindlimb unloading in rat decreases pre-osteoblast proliferation assessed in vivo with BrdU incorporation. *Am J Physiol* 274(1 Pt 1):E108–E114.

64. Bikle DD, Halloran BP, Cone CM, Globus RK, Morey-Holton E. 1987. The effects of simulated weightlessness on bone maturation. *Endocrinology* 120(2):678–684.

65. Abram AC, Keller TS, Spengler DM. 1988. The effects of simulated weightlessness on bone biomechanical and biochemical properties in the maturing rat. *J Biomech* 21(9):755–767.

66. Hargens AR, Mortensen WW, Gershuni DH, Crenshaw AG, Lieber RL, Akeson WH. 1984. Long-term measurement of muscle function in the dog hindlimb using a new apparatus. *J Orthop Res* 1(3):284–291.

67. Roer RD, Dillaman RM. 1990. Bone growth and calcium balance during simulated weightlessness in the rat. *J Appl Physiol* 68(1):13–20.

68. Hargens AR. 1983. Fluid shifts in vascular and extravascular spaces during and after simulated weightlessness. *Med Sci Sports Exerc* 15(5):421–427.

69. Aguirre JI, Plotkin LI, Stewart SA, Weinstein RS, Parfitt AM, Manolagas SC, Bellido T. 2006. Osteocyte apoptosis is induced by weightlessness in mice and precedes osteoclast recruitment and bone loss. *J Bone Miner Res* 21(4):605–615.

70. Gross TS, Akeno N, Clemens TL, Komarova S, Srinivasan S, Weimer DA, Mayorov S. 2001. Selected contribution: Osteocytes upregulate HIF-1alpha in response to acute disuse and oxygen deprivation. *J Appl Physiol* 90(6):2514–2519.

71. Stevens HY, Meays DR, Frangos JA. 2006. Pressure gradients and transport in the murine femur upon hindlimb suspension. *Bone* 39(3):565–572.

72. Yang J, Shah R, Robling AG, Templeton E, Yang H, Tracey KJ, Bidwell JP. 2008. HMGB1 is a bone-active cytokine. *J Cell Physiol* 214(3):730–739.

73. Gross TS, King KA, Rabaia N, Pathare P, Srinivasan S. 2005. Upregulation of osteopontin by osteocytes deprived of mechanical loading or oxygen. *J Bone Miner Res* **20**(2):250–256.

74. Suzuki K, Zhu B, Rittling SR, Denhardt DT, Goldberg HA, McCulloch CA, Sodek J. 2002. Colocalization of intracellular osteopontin with CD44 is associated with migration, cell fusion, and resorption in osteoclasts. *J Bone Miner Res* **17**(8):1486–1497.

75. Ishijima M, Tsuji K, Rittling SR, Yamashita T, Kurosawa H, Denhardt DT, Nifuji A, Noda M. 2002. Resistance to unloading-induced three-dimensional bone loss in osteopontin-deficient mice. *J Bone Miner Res* **17**(4):661–667.

76. Lin C, Jiang X, Dai Z, Guo X, Weng T, Wang J, Li Y, Feng G, Gao X, He L. 2009. Sclerostin mediates bone response to mechanical unloading via antagonizing Wnt/beta-catenin signaling. *J Bone Miner Res* **24**(10):1651–1661.

77. Hillsley MV, Frangos JA. 1994. Bone tissue engineering: The role of interstitial fluid flow. *Biotechnol Bioeng* **43**(7):573–581.

78. Qin YX, Kaplan T, Saldanha A, Rubin C. 2003. Fluid pressure gradients, arising from oscillations in intramedullary pressure, is correlated with the formation of bone and inhibition of intracortical porosity. *J Biomech* **36**(10):1427–1437.

79. Aukland K, Reed RK. 1993. Interstitial-lymphatic mechanisms in the control of extracellular fluid volume. *Physiol Rev* **73**(1):1–78.

80. Piekarski K, Munro M. 1977. Transport mechanism operating between blood supply and osteocytes in long bones. *Nature* **269**(5623):80–82.

81. Wilkes CH, Visscher MB. 1975. Some physiological aspects of bone marrow pressure. *J Bone Joint Surg Am* **57**(1):49–57.

82. Cooper RR, Milgram JW, Robinson RA. 1966. Morphology of the osteon. An electron microscopic study. *J Bone Joint Surg Am* **48**(7):1239–1271.

83. Edwards JR, Williams K, Kindblom LG, Meis-Kindblom JM, Hogendoorn PC, Hughes D, Forsyth RG, Jackson D, Athanasou NA. 2008. Lymphatics and bone. *Hum Pathol* **39**(1):49–55.

84. McNamara LM, Majeska RJ, Weinbaum S, Friedrich V, Schaffler MB. 2009. Attachment of osteocyte cell processes to the bone matrix. *Anat Rec (Hoboken)* **292**(3):355–363.

85. Knothe Tate ML, Niederer P, Knothe U. 1998. In vivo tracer transport through the lacunocanalicular system of rat bone in an environment devoid of mechanical loading. *Bone* **22**(2):107–117.

86. Montgomery RJ, Sutker BD, Bronk JT, Smith SR, Kelly PJ. 1988. Interstitial fluid flow in cortical bone. *Microvasc Res* **35**(3):295–307.

87. Knothe Tate ML, Knothe U. 2000. An ex vivo model to study transport processes and fluid flow in loaded bone. *J Biomech* **33**(2):247–254.

88. Knothe Tate ML, Steck R, Forwood MR, Niederer P. 2000. In vivo demonstration of load-induced fluid flow in the rat tibia and its potential implications for processes associated with functional adaptation. *J Exp Biol* **203**(Pt 18):2737–2745.

89. Dillaman RM, Roer RD, Gay DM. 1991. Fluid movement in bone: Theoretical and empirical. *J Biomech* **24**(Suppl 1):163–177.

90. Kufahl RH, Saha S. 1990. A theoretical model for stress-generated fluid flow in the caniculi-lacunae network in bone tissue. *J Biomech* **23**(2):171–180.

91. Keanini RG, Roer RD, Dillaman RM. 1995. A theoretical model of circulatory interstitial fluid flow and species transport within porous cortical bone. *J Biomech* **28**(8):901–914.

92. McCarthy ID, Yang L 1992 A distributed model of exchange processes within the osteon. *J Biomech* **25**(4):441–450.

93. Wang L, Cowin SC, Weinbaum S, Fritton SP. 2000. Modeling tracer transport in an osteon under cyclic loading. *Ann Biomed Eng* **28**(10):1200–1209.

94. Cowin SC. 2002. Mechanosensation and fluid transport in living bone. *J Musculoskelet Neuronal Interact* **2**(3):256–260.

95. Weinbaum S, Cowin SC, Zeng Y. 1994. A model for the excitation of osteocytes by mechanical loading-induced bone fluid shear stresses. *J Biomech* **27**(3):339–360.

96. You L, Cowin SC, Schaffler MB, Weinbaum S. 2001. A model for strain amplification in the actin cytoskeleton of osteocytes due to fluid drag on pericellular matrix. *J Biomech* **34**(11):1375–1386.

97. Owan I, Burr DB, Turner CH, Qiu J, Tu Y, Onyia JE, Duncan RL. 1997. Mechanotransduction in bone: Osteoblasts are more responsive to fluid forces than mechanical strain. *Am J Physiol* **273**(3 Pt 1):C810–C815.

98. Smalt R, Mitchell FT, Howard RL, Chambers TJ. 1997. Mechanotransduction in bone cells: Induction of nitric oxide and prostaglandin synthesis by fluid shear stress, but not by mechanical strain. *Adv Exp Med Biol* **433**:311–314.

99. Nicolella DP, Lankford J. 2002. Microstructural strain near osteocyte lacuna in cortical bone in vitro. *J Musculoskelet Neuronal Interact* **2**(3):261–263.

100. You LD, Weinbaum S, Cowin SC, Schaffler MB. 2004. Ultrastructure of the osteocyte process and its pericellular matrix. *Anat Rec A Discov Mol Cell Evol Biol* **278**(2):505–513.

101. Han Y, Cowin S, Schaffler MB, Weinbaum S. 2004. Mechanotransduction and strain amplification in osteocyte cell processes. *Proc Natl Acad Sci U S A* **101**(47):16689–16694.

102. Steck R, Niederer P, Knothe Tate ML. 2003. A finite element analysis for the prediction of load-induced fluid flow and mechanochemical transduction in bone. *J Theor Biol* **220**(2):249–259.

103. Gururaja S, Kim HJ, Swan CC, Brand RA, Lakes RS. 2005. Modeling deformation-induced fluid flow in cortical bone's canalicular-lacunar system. *Ann Biomed Eng* **33**(1):7–25.

104. Goulet GC, Coombe D, Martinuzzi RJ, Zernicke RF. 2009. Poroelastic evaluation of fluid movement through the lacunocanalicular system. *Ann Biomed Eng* **37**(7):1390–1402.

105. Jacobs CR, Yellowley CE, Davis BR, Zhou Z, Cimbala JM, Donahue HJ. 1998. Differential effect of steady versus oscillating flow on bone cells. *J Biomech* **31**(11):969–976.

106. You J, Yellowley CE, Donahue HJ, Zhang Y, Chen Q, Jacobs CR. 2000. Substrate deformation levels associated with routine physical activity are less stimulatory to bone cells relative to loading-induced oscillatory fluid flow. *Journal Biomech Eng* **122**(4):387–393.

107. You J, Reilly GC, Zhen X, Yellowley CE, Chen Q, Donahue HJ, Jacobs CR. 2001. Osteopontin gene regulation by oscillatory fluid flow via intracellular calcium mobilization and activation of mitogen-activated protein kinase in MC3T3-E1 osteoblasts. *J Biol Chem* **276**(16):13365–13371.

108. Chen NX, Ryder KD, Pavalko FM, Turner CH, Burr DB, Qiu J, Duncan RL. 2000. Ca(2+) regulates fluid shear-induced cytoskeletal reorganization and gene expression in osteoblasts. *Am J Physiol Cell Physiol* **278**(5):C989–C997.

109. Ponik SM, Triplett JW, Pavalko FM. 2007. Osteoblasts and osteocytes respond differently to oscillatory and unidirectional fluid flow profiles. *J Cell Biochem* **100**(3):794–807.

110. Kwon RY, Jacobs CR. 2007. Time-dependent deformations in bone cells exposed to fluid flow in vitro: Investigating the role of cellular deformation in fluid flow-induced signaling. *J Biomech* **40**(14):3162–3168.

111. Dickerson DA, Sander EA, Nauman EA. 2008. Modeling the mechanical consequences of vibratory loading in the vertebral body: Microscale effects. *Biomech Model Mechanobiol* **7**(3):191–202.

112. Gilsanz V, Wren TA, Sanchez M, Dorey F, Judex S, Rubin C. 2006. Low-level, high-frequency mechanical signals enhance musculoskeletal development of young women with low BMD. *J Bone Miner Res* **21**(9):1464–1474.

113. Fritton SP, McLeod KJ, Rubin CT. 2000. Quantifying the strain history of bone: Spatial uniformity and self-similarity of low-magnitude strains. *J Biomech* **33**(3):317–325.

114. Tanaka SM, Li J, Duncan RL, Yokota H, Burr DB, Turner CH. 2003. Effects of broad frequency vibration on cultured osteoblasts. *J Biomech* **36**(1):73–80.

115. Garman R, Gaudette G, Donahue LR, Rubin C, Judex S. 2007. Low-level accelerations applied in the absence of weight bearing can enhance trabecular bone formation. *J Orthop Res* **25**(6):732–740.

116. Rubin C, Turner AS, Mallinckrodt C, Jerome C, McLeod K, Bain S. 2002. Mechanical strain, induced noninvasively in the high-frequency domain, is anabolic to cancellous bone, but not cortical bone. *Bone* **30**(3):445–452.

117. Hwang SJ, Lublinsky S, Seo YK, Kim IS, Judex S. 2009. Extremely small-magnitude accelerations enhance bone regeneration: A preliminary study. *Clin Orthop Relat Res* **467**(4):1083–1091.

118. Alford AI, Jacobs CR, Donahue HJ. 2003. Oscillating fluid flow regulates gap junction communication in osteocytic MLO-Y4 cells by an ERK1/2 MAP kinase-dependent mechanism small star, filled. *Bone* **33**(1):64–70.

119. Arnsdorf E, Tummala P, Jacobs CR. 2009. Non-canonical Wnt signaling and N-cadherin related beta-catenin signaling play a role in mechanically induced osteogenic cell fate. *PLoS One* **4**(4):e5388.

120. Arnsdorf E, Tummala P, Kwon RY, Jacobs CR. 2009. Mechanically induced osteogenic differentiation—The role of RhoA, ROCKII and cytoskeletal dynamics. *J Cell Sci* **122**(Pt 4):546–553.

121. Kwon RY, Temiyasathit S, Tummala P, Quah CC, Jacobs CR. 2009. Adenylyl cyclase 6 mediates primary cilia-regulated decreases in cAMP in bone cells exposed to dynamic fluid flow. *30th American Society for Bone and Mineral Research Annual Meeting*, Montréal, Québec, Canada.

122. Malone AM, Anderson C, Tummala P, Kwon RY, Johnston TR, Stearns T, Jacobs CR. 2007. Primary cilia mediate mechanosensing in bone cells by a calcium-independent mechanism. *Proc Natl Acad Sci U S A* **104**(33):13325–13330.

123. Malone AM, Batra NN, Shivaram G, Kwon RY, You L, Kim CH, Rodriguez J, Jair K, Jacobs CR. 2007. The role of actin cytoskeleton in oscillatory fluid flow-induced signaling in MC3T3-E1 osteoblasts. *Am J Physiol Cell Physiol* **292**(5):C1830–C1836.

124. Tan SD, Bakker AD, Semeins CM, Kuijpers-Jagtman AM, Klein-Nulend J. 2008. Inhibition of osteocyte apoptosis by fluid flow is mediated by nitric oxide. *Biochem Biophys Res Commun* **369**(4):1150–1154.

125. Doty SB. 1981. Morphological evidence of gap junctions between bone cells. *Calcif Tissue Int* **33**(5):509–512.

126. Civitelli R. 2008. Cell-cell communication in the osteoblast/osteocyte lineage. *Arch Biochem Biophys* **473**(2):188–192.

127. Schirrmacher K, Schmitz I, Winterhager E, Traub O, Brummer F, Jones D, Bingmann D. 1992. Characterization of gap junctions between osteoblast-like cells in culture. *Calcif Tissue Int* **51**(4):285–290.

128. Xia SL, Ferrier J. 1992. Propagation of a calcium pulse between osteoblastic cells. *Biochem Biophys Res Commun* **186**(3):1212–1219.

129. Vander Molen MA, Donahue HJ, Rubin CT, McLeod KJ. 2000. Osteoblastic networks with deficient coupling: Differential effects of magnetic and electric field exposure. *Bone* **27**(2):227–231.

130. Vander Molen MA, Rubin CT, McLeod KJ, McCauley LK, Donahue HJ. 1996. Gap junctional intercellular communication contributes to hormonal responsiveness in osteoblastic networks. *J Biol Chem* **271**(21):12165–12171.

131. Cheng B, Zhao S, Luo J, Sprague E, Bonewald LF, Jiang JX. 2001. Expression of functional gap junctions and regulation by fluid flow in osteocyte-like MLO-Y4 cells. *J Bone Miner Res* **16**(2):249–259.

132. Genetos DC, Kephart CJ, Zhang Y, Yellowley CE, Donahue HJ. 2007. Oscillating fluid flow activation of gap junction hemichannels induces ATP release from MLO-Y4 osteocytes. *J Cell Physiol* **212**(1):207–214.

133. Saunders MM, You J, Zhou Z, Li Z, Yellowley CE, Kunze EL, Jacobs CR, Donahue HJ. 2003. Fluid flow-induced prostaglandin E2 response of osteoblastic ROS 17/2.8 cells is gap junction-mediated and independent of cytosolic calcium. *Bone* **32**(4):350–356.

134. Jekir MG, Donahue HJ. 2009. Gap junctions and osteoblast-like cell gene expression in response to fluid flow. *J Biomech Eng* **131**(1):011005.

135. Guharay F, Sachs F. 1984. Stretch-activated single ion channel currents in tissue-cultured embryonic chick skeletal muscle. *J Physiol* **352**:685–701.

136. Davidson RM, Tatakis DW, Auerbach AL. 1990. Multiple forms of mechanosensitive ion channels in osteoblast-like cells. *Pflugers Arch* **416**(6):646–651.

137. Davidson RM. 1993. Membrane stretch activates a high-conductance K^+ channel in G292 osteoblastic-like cells. *J Membr Biol* **131**(1):81–92.

138. Duncan RL, Hruska KA. 1994. Chronic, intermittent loading alters mechanosensitive channel characteristics in osteoblast-like cells. *Am J Physiol* **267**(6 Pt 2):F909–F916.

139. McDonald F, Somasundaram B, McCann TJ, Mason WT, Meikle MC. 1996. Calcium waves in fluid flow stimulated osteoblasts are G protein mediated. *Arch Biochem Biophys* **326**(1):31–38.

140. Li J, Duncan RL, Burr DB, Turner CH. 2002. L-type calcium channels mediate mechanically induced bone formation in vivo. *J Bone Miner Res* **17**(10):1795–1800.

141. Duncan RL, Hruska KA, Misler S. 1992. Parathyroid hormone activation of stretch-activated cation channels in osteosarcoma cells (UMR-106.01). *FEBS Lett* **307**(2):219–223.

142. Ryder KD, Duncan RL. 2001. Parathyroid hormone enhances fluid shear-induced $[Ca^{2+}]i$ signaling in osteoblastic cells through activation of mechanosensitive and voltage-sensitive Ca^{2+} channels. *J Bone Miner Res* **16**(2):240–248.

143. Charras GT, Williams BA, Sims SM, Horton MA. 2004. Estimating the sensitivity of mechanosensitive ion channels to membrane strain and tension. *Biophys J* **87**(4):2870–2884.

144. Haut Donahue TL, Genetos DC, Jacobs CR, Donahue HJ, Yellowley CE. 2004. Annexin V disruption impairs mechanically induced calcium signaling in osteoblastic cells. *Bone* **35**(3):656–663.

145. Pedersen SF, Owsianik G, Nilius B. 2005. TRP channels: An overview. *Cell Calcium* **38**(3–4):233–252.

146. Hoenderop JG, van Leeuwen JP, van der Eerden BC, Kersten FF, van der Kemp AW, Merillat AM et al. 2003. Renal Ca^{2+} wasting, hyperabsorption, and reduced bone thickness in mice lacking TRPV5. *J Clin Invest* **112**(12):1906–1914.

147. Gradilone SA, Masyuk AI, Splinter PL, Banales JM, Huang BQ, Tietz PS, Masyuk TV, Larusso NF. 2007. Cholangiocyte cilia express TRPV4 and detect changes in luminal tonicity inducing bicarbonate secretion. *Proc Natl Acad Sci U S A* **104**(48):19138–19143.

148. Nauli SM, Alenghat FJ, Luo Y, Williams E, Vassilev P, Li X, Elia AE et al. 2003. Polycystins 1 and 2 mediate mechanosensation in the primary cilium of kidney cells. *Nat Genet* **33**(2):129–137.
149. Mochizuki T, Sokabe T, Araki I, Fujishita K, Shibasaki K, Uchida K, Naruse K, Koizumi S, Takeda M, Tominaga M. 2009. The TRPV4 cation channel mediates stretch-evoked Ca^{2+} influx and ATP release in primary urothelial cell cultures. *J Biol Chem* **284**(32):21257–21264.
150. Masuyama R, Vriens J, Voets T, Karashima Y, Owsianik G, Vennekens R, Lieben L et al. 2008. TRPV4-mediated calcium influx regulates terminal differentiation of osteoclasts. *Cell Metab* **8**(3):257–265.
151. Abed E, Labelle D, Martineau C, Loghin A, Moreau R. 2009. Expression of transient receptor potential (TRP) channels in human and murine osteoblast-like cells. *Mol Membr Biol* **26**(3):146–158.
152. Delmas P. 2004. Polycystins: From mechanosensation to gene regulation. *Cell* **118**(2):145–148.
153. Xiao Z, Zhang S, Mahlios J, Zhou G, Magenheimer BS, Guo D, Dallas SL et al. 2006. Cilia-like structures and polycystin-1 in osteoblasts/osteocytes and associated abnormalities in skeletogenesis and Runx2 expression. *J Biol Chem* **281**(41):30884–30895.
154. Boulter C, Mulroy S, Webb S, Fleming S, Brindle K, Sandford R. 2001. Cardiovascular, skeletal, and renal defects in mice with a targeted disruption of the Pkd1 gene. *Proc Natl Acad Sci U S A* **98**(21):12174–12179.
155. Lu W, Shen X, Pavlova A, Lakkis M, Ward CJ, Pritchard L, Harris PC, Genest DR, Perez-Atayde AR, Zhou J. 2001. Comparison of Pkd1-targeted mutants reveals that loss of polycystin-1 causes cystogenesis and bone defects. *Hum Mol Genet* **10**(21):2385–2396.
156. Satir P, Christensen ST. 2007. Overview of structure and function of mammalian cilia. *Annu Rev Physiol* **69**:377–400.
157. Veland IR, Awan A, Pedersen LB, Yoder BK, Christensen ST. 2009. Primary cilia and signaling pathways in mammalian development, health and disease. *Nephron Physiol* **111**(3):39–53.
158. Resnick A, Hopfer U. 2008. Mechanical stimulation of primary cilia. *Front Biosci* **13**:1665–1680.
159. Praetorius HA, Spring KR. 2001 Bending the MDCK cell primary cilium increases intracellular calcium. *J Membr Biol* **184**(1):71–79.
160. Christensen ST, Pedersen LB, Schneider L, Satir P. 2007. Sensory cilia and integration of signal transduction in human health and disease. *Traffic* **8**(2):97–109.
161. Anderson CT, Castillo AB, Brugmann SA, Helms JA, Jacobs CR, Stearns T. 2008. Primary cilia: Cellular sensors for the skeleton. *Anat Rec (Hoboken)* **291**(9):1074–1078.
162. Nauli SM, Kawanabe Y, Kaminski JJ, Pearce WJ, Ingber DE, Zhou J. 2008. Endothelial cilia are fluid shear sensors that regulate calcium signaling and nitric oxide production through polycystin-1. *Circulation* **117**(9):1161–1171.
163. Schwartz EA, Leonard ML, Bizios R, Bowser SS. 1997. Analysis and modeling of the primary cilium bending response to fluid shear. *Am J Physiol* **272**(1 Pt 2):F132–F138.
164. Bhunia AK, Piontek K, Boletta A, Liu L, Qian F, Xu PN, Germino FJ, Germino GG. 2002. PKD1 induces p21(waf1) and regulation of the cell cycle via direct activation of the JAK-STAT signaling pathway in a process requiring PKD2. Cell **109**(2):157–168.
165. Shillingford JM, Murcia NS, Larson CH, Low SH, Hedgepeth R, Brown N, Flask CA et al. 2006. The mTOR pathway is regulated by polycystin-1, and its inhibition reverses renal cystogenesis in polycystic kidney disease. *Proc Natl Acad Sci U S A* **103**(14):5466–5471.
166. Matthews JL, Martin JH. 1971. Intracellular transport of calcium and its relationship to homeostasis and mineralization. An electron microscope study. *Am J Med* **50**(5):589–597.
167. Tonna EA, Lampen NM. 1972 Electron microscopy of aging skeletal cells. I. Centrioles and solitary cilia. *J Gerontol* **27**(3):316–324.
168. Xiao Z, Zhang S, Magenheimer BS, Luo J, Quarles LD. 2008. Polycystin-1 regulates skeletogenesis through stimulation of the osteoblast-specific transcription factor RUNX2-II. *J Biol Chem* **283**(18):12624–12634.
169. Chachisvilis M, Zhang YL, Frangos JA. 2006. G protein-coupled receptors sense fluid shear stress in endothelial cells. *Proc Natl Acad Sci U S A* **103**(42):15463–15468.
170. Makino A, Prossnitz ER, Bunemann M, Wang JM, Yao W, Schmid-Schonbein GW. 2006. G protein-coupled receptors serve as mechanosensors for fluid shear stress in neutrophils. *Am J Physiol Cell Physiol* **290**(6):C1633–C1639.
171. Sakamoto A, Chen M, Nakamura T, Xie T, Karsenty G, Weinstein LS. 2005. Deficiency of the G-protein alpha-subunit G(s)alpha in osteoblasts leads to differential effects on trabecular and cortical bone. *J Biol Chem* **280**(22):21369–21375.

172. Peng J, Bencsik M, Louie A, Lu W, Millard S, Nguyen P, Burghardt A et al. 2008. Conditional expression of a Gi-coupled receptor in osteoblasts results in trabecular osteopenia. *Endocrinology* **149**(3):1329–1337.

173. Hsiao EC, Boudignon BM, Chang WC, Bencsik M, Peng J, Nguyen TD, Manalac C, Halloran BP, Conklin BR, Nissenson RA. 2008. Osteoblast expression of an engineered Gs-coupled receptor dramatically increases bone mass. *Proc Natl Acad Sci U S A* **105**(4):1209–1214.

174. Zhang YL, Frangos JA, Chachisvilis M. 2009. Mechanical stimulus alters conformation of type 1 parathyroid hormone receptor in bone cells. *Am J Physiol Cell Physiol* **296**(6):C1391–C1399.

175. Hynes RO. 2002. Integrins: Bidirectional, allosteric signaling machines. *Cell* **110**(6):673–687.

176. Grzesik WJ, Robey PG. 1994. Bone matrix RGD glycoproteins: Immunolocalization and interaction with human primary osteoblastic bone cells in vitro. *J Bone Miner Res* **9**(4):487–496.

177. Sinha RK, Tuan RS. 1996. Regulation of human osteoblast integrin expression by orthopedic implant materials. *Bone* **18**(5):451–457.

178. Zimmerman D, Jin F, Leboy P, Hardy S, Damsky C. 2000. Impaired bone formation in transgenic mice resulting from altered integrin function in osteoblasts. *Dev Biol* **220**(1):2–15.

179. Wang Y, McNamara LM, Schaffler MB, Weinbaum S. 2007. A model for the role of integrins in flow induced mechanotransduction in osteocytes. *Proc Natl Acad Sci U S A* **104**(40):15941–15946.

180. Weyts FA, Li YS, van Leeuwen J, Weinans H, Chien S. 2002. ERK activation and alpha v beta 3 integrin signaling through Shc recruitment in response to mechanical stimulation in human osteoblasts. *J Cell Biochem* **87**(1):85–92.

181. Ishida T, Peterson TE, Kovach NL, Berk BC. 1996. MAP kinase activation by flow in endothelial cells. Role of beta 1 integrins and tyrosine kinases. *Circ Res* **79**(2):310–316.

182. Miyauchi A, Gotoh M, Kamioka H, Notoya K, Sekiya H, Takagi Y, Yoshimoto Y et al. 2006. AlphaVbeta3 integrin ligands enhance volume-sensitive calcium influx in mechanically stretched osteocytes. *J Bone Miner Metab* **24**(6):498–504.

183. Pommerenke H, Schreiber E, Durr F, Nebe B, Hahnel C, Moller W, Rychly J. 1996. Stimulation of integrin receptors using a magnetic drag force device induces an intracellular free calcium response. *Eur J Cell Biol* **70**(2):157–164.

184. Peake MA, Cooling LM, Magnay JL, Thomas PB, El Haj AJ. 2000. Selected contribution: Regulatory pathways involved in mechanical induction of c-fos gene expression in bone cells. *J Appl Physiol* **89**(6):2498–2507.

185. Pommerenke H, Schmidt C, Durr F, Nebe B, Luthen F, Muller P, Rychly J. 2002. The mode of mechanical integrin stressing controls intracellular signaling in osteoblasts. *J Bone Miner Res* **17**(4):603–611.

186. Kapur S, Baylink DJ, Lau KH. 2003. Fluid flow shear stress stimulates human osteoblast proliferation and differentiation through multiple interacting and competing signal transduction pathways. *Bone* **32**(3):241–251.

187. Litzenberger JB, Tang WJ, Castillo AB, Jacobs CR. 2009. Deletion of beta-1 integrins from cortical astrocytes reduces load-induced bone formation. *Cell Mol Bioeng* **2**(3):416–424.

188. Schlaepfer DD, Hauck CR, Sieg DJ. 1999. Signaling through focal adhesion kinase. *Prog Biophys Mol Biol* **71**(3–4):435–478.

189. Schwartz MA, DeSimone DW. 2008. Cell adhesion receptors in mechanotransduction. *Curr Opin Cell Biol* **20**(5):551–556.

190. Lee HS, Millward-Sadler SJ, Wright MO, Nuki G, Salter DM. 2000. Integrin and mechanosensitive ion channel-dependent tyrosine phosphorylation of focal adhesion proteins and beta-catenin in human articular chondrocytes after mechanical stimulation. *J Bone Miner Res* **15**(8):1501–1509.

191. Orr AW, Murphy-Ullrich JE. 2004. Regulation of endothelial cell function BY FAK and PYK2. *Front Biosci* **9**:1254–1266.

192. Ponik SM, Pavalko FM. 2004. Formation of focal adhesions on fibronectin promotes fluid shear stress induction of COX-2 and PGE2 release in MC3T3-E1 osteoblasts. *J Appl Physiol* **97**(1):135–142.

193. Wozniak M, Fausto A, Carron CP, Meyer DM, Hruska KA. 2000. Mechanically strained cells of the osteoblast lineage organize their extracellular matrix through unique sites of alphavbeta3-integrin expression. *J Bone Miner Res* **15**(9):1731–1745.

194. Young SR, Gerard-O'Riley R, Kim JB, Pavalko FM. 2009. Focal adhesion kinase is important for fluid shear stress-induced mechanotransduction in osteoblasts. *J Bone Miner Res* **24**(3):411–424.

195. Toutant M, Studler JM, Burgaya F, Costa A, Ezan P, Gelman M, Girault JA. 2000. Autophosphorylation of Tyr397 and its phosphorylation by Src-family kinases are altered in focal-adhesion-kinase neuronal isoforms. *Biochem J* **348**(Pt 1):119–128.

196. MacKenna DA, Dolfi F, Vuori K, Ruoslahti E. 1998. Extracellular signal-regulated kinase and c-Jun NH2-terminal kinase activation by mechanical stretch is integrin-dependent and matrix-specific in rat cardiac fibroblasts. *J Clin Invest* **101**(2):301–310.

197. Schlaepfer DD, Hanks SK, Hunter T, van der Geer P. 1994. Integrin-mediated signal transduction linked to Ras pathway by GRB2 binding to focal adhesion kinase. *Nature* **372**(6508):786–791.

198. Schlaepfer DD, Hunter T. 1996. Evidence for in vivo phosphorylation of the Grb2 SH2-domain binding site on focal adhesion kinase by Src-family protein-tyrosine kinases. *Mol Cell Biol* **16**(10):5623–5633.

199. Boutahar N, Guignandon A, Vico L, Lafage-Proust MH. 2004. Mechanical strain on osteoblasts activates autophosphorylation of focal adhesion kinase and proline-rich tyrosine kinase 2 tyrosine sites involved in ERK activation. *J Biol Chem* **279**(29):30588–30599.

200. Berk BC, Corson MA, Peterson TE, Tseng H. 1995. Protein kinases as mediators of fluid shear stress stimulated signal transduction in endothelial cells: A hypothesis for calcium-dependent and calcium-independent events activated by flow. *J Biomech* **28**(12):1439–1450.

201. Li S, Kim M, Hu YL, Jalali S, Schlaepfer DD, Hunter T, Chien S, Shyy JY. 1997. Fluid shear stress activation of focal adhesion kinase. Linking to mitogen-activated protein kinases. *J Biol Chem* **272**(48):30455–30462.

202. Petzold T, Orr AW, Hahn C, Jhaveri K, Parsons JT, Schwartz MA. 2009. Focal adhesion kinase modulates activation of NF-{kappa}B by flow in endothelial cells. *Am J Physiol Cell Physiol* **297**(4):C814–C822.

203. Rezzonico R, Cayatte C, Bourget-Ponzio I, Romey G, Belhacene N, Loubat A, Rocchi S et al. 2003. Focal adhesion kinase pp125FAK interacts with the large conductance calcium-activated hSlo potassium channel in human osteoblasts: Potential role in mechanotransduction. *J Bone Miner Res* **18**(10):1863–1871.

204. Leucht P, Kim JB, Currey JA, Brunski J, Helms JA. 2007. FAK-Mediated mechanotransduction in skeletal regeneration. *PLoS ONE* **2**(4):e390.

205. Sieg DJ, Hauck CR, Schlaepfer DD. 1999. Required role of focal adhesion kinase (FAK) for integrin-stimulated cell migration. *J Cell Sci* **112** (Pt 16):2677–2691.

206. Weis SM, Lim ST, Lutu-Fuga KM, Barnes LA, Chen XL, Gothert JR, Shen TL, Guan JL, Schlaepfer DD, Cheresh DA. 2008. Compensatory role for Pyk2 during angiogenesis in adult mice lacking endothelial cell FAK. *J Cell Biol* **181**(1):43–50.

207. Buckbinder L, Crawford DT, Qi H, Ke HZ, Olson LM, Long KR, Bonnette PC et al. 2007. Proline-rich tyrosine kinase 2 regulates osteoprogenitor cells and bone formation, and offers an anabolic treatment approach for osteoporosis. *Proc Natl Acad Sci U S A* **104**(25):10619–10624.

208. Horwitz A, Duggan K, Buck C, Beckerle MC, Burridge K. 1986. Interaction of plasma membrane fibronectin receptor with talin—A transmembrane linkage. *Nature* **320**(6062):531–533.

209. Vinogradova O, Velyvis A, Velyviene A, Hu B, Haas T, Plow E, Qin J. 2002. A structural mechanism of integrin alpha(IIb)beta(3) "inside-out" activation as regulated by its cytoplasmic face. *Cell* **110**(5):587–597.

210. Critchley DR, Gingras AR. 2008. Talin at a glance. *J Cell Sci* **121**(Pt 9):1345–1347.

211. Schaller MD, Otey CA, Hildebrand JD, Parsons JT. 1995. Focal adhesion kinase and paxillin bind to peptides mimicking beta integrin cytoplasmic domains. *J Cell Biol* **130**(5):1181–1187.

212. Harburger DS, Calderwood DA. 2009. Integrin signalling at a glance. *J Cell Sci* **122**(Pt 2):159–163.

213. Vatsa A, Semeins CM, Smit TH, Klein-Nulend J. 2008. Paxillin localisation in osteocytes—Is it determined by the direction of loading? *Biochem Biophys Res Commun* **377**(4):1019–1024.

214. Ingber DE 1997 Tensegrity: The architectural basis of cellular mechanotransduction. *Annu Rev Physiol* **59**:575–599.

215. Alenghat FJ, Nauli SM, Kolb R, Zhou J, Ingber DE. 2004. Global cytoskeletal control of mechanotransduction in kidney epithelial cells. *Exp Cell Res* **301**(1):23–30.

216. Pritchard S, Guilak F. 2004. The role of F-actin in hypo-osmotically induced cell volume change and calcium signaling in anulus fibrosus cells. *Ann Biomed Eng* **32**(1):103–111.

217. Zhang L, Tran N, Chen HQ, Wang X. 2008. Cyclic stretching promotes collagen synthesis and affects F-actin distribution in rat mesenchymal stem cells. *Biomed Mater Eng* **18**(4–5):205–210.

218. Maniotis AJ, Chen CS, Ingber DE. 1997. Demonstration of mechanical connections between integrins, cytoskeletal filaments, and nucleoplasm that stabilize nuclear structure. *Proc Natl Acad Sci U S A* **94**(3):849–854.

219. Pavalko FM, Chen NX, Turner CH, Burr DB, Atkinson S, Hsieh YF, Qiu J, Duncan RL. 1998. Fluid shear-induced mechanical signaling in MC3T3-E1 osteoblasts requires cytoskeleton-integrin interactions. *Am J Physiol* **275**(6 Pt 1):C1591–C601.

220. Norvell SM, Ponik SM, Bowen DK, Gerard R, Pavalko FM. 2004. Fluid shear stress induction of COX-2 protein and prostaglandin release in cultured MC3T3-E1 osteoblasts does not require intact microfilaments or microtubules. *J Appl Physiol* **96**(3):957–966.

221. Rosenberg N. 2003. The role of the cytoskeleton in mechanotransduction in human osteoblast-like cells. *Hum Exp Toxicol* **22**(5):271–274.

222. Myers KA, Rattner JB, Shrive NG, Hart DA. 2007. Osteoblast-like cells and fluid flow: Cytoskeleton-dependent shear sensitivity. *Biochem Biophys Res Commun* **364**(2):214–219.

223. Berridge MJ, Lipp P, Bootman MD. 2000. The versatility and universality of calcium signalling. *Nat Rev Mol Cell Biol* **1**(1):11–21.

224. Donahue SW, Donahue HJ, Jacobs CR. 2003. Osteoblastic cells have refractory periods for fluid-flow-induced intracellular calcium oscillations for short bouts of flow and display multiple low-magnitude oscillations during long-term flow. *J Biomech* **36**(1):35–43.

225. Hung CT, Allen FD, Pollack SR, Brighton CT. 1996. Intracellular Ca^{2+} stores and extracellular Ca^{2+} are required in the real-time Ca^{2+} response of bone cells experiencing fluid flow. *J Biomech* **29**(11):1411–1417.

226. Wiltink A, Nijweide PJ, Scheenen WJ, Ypey DL, Van Duijn B. 1995. Cell membrane stretch in osteoclasts triggers a self-reinforcing Ca^{2+} entry pathway. *Pflugers Arch* **429**(5):663–671.

227. Batra NN, Li YJ, Yellowley CE, You L, Malone AM, Kim CH, Jacobs CR. 2005. Effects of short-term recovery periods on fluid-induced signaling in osteoblastic cells. *J Biomech* **38**(9):1909–1917.

228. Chen NX, Geist DJ, Genetos DC, Pavalko FM, Duncan RL. 2003. Fluid shear-induced NFkappaB translocation in osteoblasts is mediated by intracellular calcium release. *Bone* **33**(3):399–410.

229. Liu D, Genetos DC, Shao Y, Geist DJ, Li J, Ke HZ, Turner CH, Duncan RL. 2008. Activation of extracellular-signal regulated kinase (ERK1/2) by fluid shear is Ca(2+)- and ATP-dependent in MC3T3-E1 osteoblasts. *Bone* **42**(4):644–652.

230. Saunders MM, You J, Trosko JE, Yamasaki H, Li Z, Donahue HJ, Jacobs CR. 2001. Gap junctions and fluid flow response in MC3T3-E1 cells. *Am J Physiol Cell Physiol* **281**(6):C1917–C1925.

231. Cotton M, Claing A. 2009. G protein-coupled receptors stimulation and the control of cell migration. *Cell Signal* **21**(7):1045–1053.

232. Reich KM, Frangos JA. 1991. Effect of flow on prostaglandin E2 and inositol trisphosphate levels in osteoblasts. *Am J Physiol* **261**(3 Pt 1):C428–C432.

233. Hung CT, Pollack SR, Reilly TM, Brighton CT. 1995. Real-time calcium response of cultured bone cells to fluid flow. *Clin Orthop Relat Res* (313):256–269.

234. Ogasawara A, Arakawa T, Kaneda T, Takuma T, Sato T, Kaneko H, Kumegawa M, Hakeda Y. 2001. Fluid shear stress-induced cyclooxygenase-2 expression is mediated by C/EBP beta, cAMP-response element-binding protein, and AP-1 in osteoblastic MC3T3-E1 cells. *J Biol Chem* **276**(10):7048–7054.

235. Clark CB, McKnight NL, Frangos JA. 2002. Strain and strain rate activation of G proteins in human endothelial cells. *Biochem Biophys Res Commun* **299**(2):258–262.

236. Gudi SR, Lee AA, Clark CB, Frangos JA. 1998. Equibiaxial strain and strain rate stimulate early activation of G proteins in cardiac fibroblasts. *Am J Physiol* **274**(5 Pt 1):C1424–C1428.

237. Reich KM, McAllister TN, Gudi S, Frangos JA. 1997. Activation of G proteins mediates flow-induced prostaglandin E2 production in osteoblasts. *Endocrinology* **138**(3):1014–1018.

238. Gudi S, Nolan JP, Frangos JA. 1998. Modulation of GTPase activity of G proteins by fluid shear stress and phospholipid composition. *Proc Natl Acad Sci U S A* **95**(5):2515–2519.

239. Bonni A, Brunet A, West AE, Datta SR, Takasu MA, Greenberg ME. 1999. Cell survival promoted by the Ras-MAPK signaling pathway by transcription-dependent and -independent mechanisms. *Science* **286**(5443):1358–1362.

240. Pearson G, Robinson F, Beers Gibson T, Xu BE, Karandikar M, Berman K, Cobb MH. 2001. Mitogen-activated protein (MAP) kinase pathways: Regulation and physiological functions. *Endocr Rev* **22**(2):153–183.

241. Jessop HL, Rawlinson SC, Pitsillides AA, Lanyon LE. 2002. Mechanical strain and fluid movement both activate extracellular regulated kinase (ERK) in osteoblast-like cells but via different signaling pathways. *Bone* **31**(1):186–194.

242. Kapur S, Chen ST, Baylink DJ, Lau KH. 2004. Extracellular signal-regulated kinase-1 and -2 are both essential for the shear stress-induced human osteoblast proliferation. *Bone* **35**(2):525–534.

243. Plotkin LI, Mathov I, Aguirre JI, Parfitt AM, Manolagas SC, Bellido T. 2005. Mechanical stimulation prevents osteocyte apoptosis: requirement of integrins, Src kinases, and ERKs. *Am J Physiol Cell Physiol* **289**(3):C633–C643.

244. Yang CM, Chien CS, Yao CC, Hsiao LD, Huang YC, Wu CB. 2004. Mechanical strain induces collagenase-3 (MMP-13) expression in MC3T3-E1 osteoblastic cells. *J Biol Chem* **279**(21):22158–22165.

245. Rubin J, Murphy TC, Fan X, Goldschmidt M, Taylor WR. 2002. Activation of extracellular signal-regulated kinase is involved in mechanical strain inhibition of RANKL expression in bone stromal cells. *J Bone Miner Res* **17**(8):1452–1460.
246. Rubin J, Murphy TC, Zhu L, Roy E, Nanes MS, Fan X. 2003. Mechanical strain differentially regulates endothelial nitric-oxide synthase and receptor activator of nuclear kappa B ligand expression via ERK1/2 MAPK. *J Biol Chem* **278**(36):34018–34025.
247. Ward DF, Jr., Salasznyk RM, Klees RF, Backiel J, Agius P, Bennett K, Boskey A, Plopper GE. 2007. Mechanical strain enhances extracellular matrix-induced gene focusing and promotes osteogenic differentiation of human mesenchymal stem cells through an extracellular-related kinase-dependent pathway. *Stem Cells Dev* **16**(3):467–480.
248. Bryan NS, Bian K, Murad F. 2009. Discovery of the nitric oxide signaling pathway and targets for drug development. *Front Biosci* **14**:1–18.
249. Helfrich MH, Evans DE, Grabowski PS, Pollock JS, Ohshima H, Ralston SH. 1997. Expression of nitric oxide synthase isoforms in bone and bone cell cultures. *J Bone Miner Res* **12**(7):1108–1115.
250. Caballero-Alias AM, Loveridge N, Lyon A, Das-Gupta V, Pitsillides A, Reeve J. 2004. NOS isoforms in adult human osteocytes: multiple pathways of NO regulation? *Calcif Tissue Int* **75**(1):78–84.
251. Zaman G, Pitsillides AA, Rawlinson SC, Suswillo RF, Mosley JR, Cheng MZ, Platts LA, Hukkanen M, Polak JM, Lanyon LE. 1999. Mechanical strain stimulates nitric oxide production by rapid activation of endothelial nitric oxide synthase in osteocytes. *J Bone Miner Res* **14**(7):1123–1131.
252. Caballero-Alias AM, Loveridge N, Pitsillides A, Parker M, Kaptoge S, Lyon A, Reeve J. 2005. Osteocytic expression of constitutive NO synthase isoforms in the femoral neck cortex: A case-control study of intracapsular hip fracture. *J Bone Miner Res* **20**(2):268–273.
253. Aguirre J, Buttery L, O'Shaughnessy M, Afzal F, Fernandez de Marticorena I, Hukkanen M, Huang P, MacIntyre I, Polak J. 2001. Endothelial nitric oxide synthase gene-deficient mice demonstrate marked retardation in postnatal bone formation, reduced bone volume, and defects in osteoblast maturation and activity. *Am J Pathol* **158**(1):247–257.
254. Basso N, Heersche JN. 2006. Effects of hind limb unloading and reloading on nitric oxide synthase expression and apoptosis of osteocytes and chondrocytes. *Bone* **39**(4):807–814.
255. Turner CH, Takano Y, Owan I, Murrell GA. 1996. Nitric oxide inhibitor L-NAME suppresses mechanically induced bone formation in rats. *Am J Physiol* **270**(4 Pt 1):E634–E639.
256. Bacabac RG, Smit TH, Mullender MG, Dijcks SJ, Van Loon JJ, Klein-Nulend J. 2004. Nitric oxide production by bone cells is fluid shear stress rate dependent. *Biochem Biophys Res Commun* **315**(4):823–829.
257. Klein-Nulend J, Helfrich MH, Sterck JG, MacPherson H, Joldersma M, Ralston SH, Semeins CM, Burger EH. 1998. Nitric oxide response to shear stress by human bone cell cultures is endothelial nitric oxide synthase dependent. *Biochem Biophys Res Commun* **250**(1):108–114.
258. McGarry JG, Klein-Nulend J, Mullender MG, Prendergast PJ. 2005. A comparison of strain and fluid shear stress in stimulating bone cell responses—A computational and experimental study. *FASEB J* **19**(3):482–484.
259. AbouAlaiwi WA, Takahashi M, Mell BR, Jones TJ, Ratnam S, Kolb RJ, Nauli SM. 2009. Ciliary polycystin-2 is a mechanosensitive calcium channel involved in nitric oxide signaling cascades. *Circ Res* **104**(7):860–869.

10 Mechanical Control of Bone Remodeling

Natasha Case and Janet Rubin

CONTENTS

10.1 INTRODUCTION

Bone tissue has the capacity to adapt to its functional environment through alterations in bone remodeling orchestrated by the cells of the skeleton—the osteoblasts, osteocytes, and osteoclasts. To fulfill this capacity, those cells responsible for bone remodeling both perceive and respond to the mechanical environment. In this way, mechanical forces exert regulatory control over skeletal morphology, and ultimately produce a skeleton that can withstand daily functional use. Understanding how the cells of the skeleton convert mechanical signals into a programmed biological response has been the goal of our laboratory.

This chapter will discuss mechanical influences on osteoclastogenesis, osteoblastogenesis, and mesenchymal stem cell (MSC) lineage selection. Data will be considered from both in vivo and in vitro studies. Mechanical regulation of osteocytes, the other major cellular contributor to bone remodeling, was discussed in Chapter 9.

10.2 BIOMECHANICAL FACTORS IN BONE

The capacity of bone to remodel to meet functional structural demands was recognized by Wolff in 1892 as the "law of bone transformation."[1] Studies in humans[2] and animals[3,4] have shown that applied loads are associated with changes in bone density as well as skeletal macro- and microstructure.[5,6] Bone's sensitivity to site-specific mechanical demands is illustrated by the example of elite tennis players, in which the dominant humerae are 35% thicker than the arm that simply throws the ball into the air.[7,8] Exercise studies have demonstrated that certain parameters of loading must be exceeded to elucidate changes[9] and that the degree of the response is determined not only by the magnitude of the load,[10] but also by the rate,[11] cycle number,[12] and frequency of the load.[13,14] In the controlled environment of the laboratory, loading regimes have been designed to cause significant skeletal remodeling in rodents,[15,16] sheep,[17] and turkeys.[18] Distinct parameters of the applied loads have been correlated to changes in bone morphology[16,17,19] as well as changes in gene and protein expression.[20–22]

During weight-bearing activities, skeletal loading generates deformation in bone tissue, pressure changes in the intramedullary cavity and within the cortices, transient pressure waves, shear forces through cannaliculi, and dynamic electric fields as fluid flows past charged bone crystals. As such, the microenvironment of cells within skeletal tissue is rich in mechanical information. In vitro studies have shown that the multiple cell populations present in bone, including osteoprogenitor cells, osteoblasts, and osteocytes, undergo intracellular changes in response to various modes of mechanical stimulation, including substrate deformation, fluid flow, and hydrostatic pressure.[23–26] So far, isolating a specific component of the physical milieu that regulates skeletal morphology has been difficult—no single parameter of the mechanical environment reliably predicts bone remodeling in all naturally observed or experimentally created conditions.[27]

At the level of small volumes of tissue, all loads and bending moments resolve into strain. While peak strain magnitudes measured in the load-bearing regions of the skeleton of adult species range from 2000 to 3500 microstrain,[28] the strain levels experienced by cells within the hard tissue environment of bone are unclear. By virtue of adherence to components of the extracellular matrix, the architecture of the plasma membrane, and the organization of intracellular compartments, cells have a complex and heterogeneous strain distribution. When the cell is then deformed, discrete regions of the cell are subject to differential strain patterns. Mechanical strain is typically applied in vitro to bone cells by deforming the substrate on which the cells are attached. It has been demonstrated that bone cells respond to various types of dynamic strains, including uniaxial strain,[29] equibiaxial strain,[25] and strain resulting from four-point bending.[30]

Bone volume includes a significant fraction of unbound interstitial fluid surrounding all cells found in the tissue. Motion of interstitial fluid flow is driven by the pressure differential of the circulatory system[31,32] and secondly by externally applied mechanical loading.[33] When bone is loaded, fluid is forced out of regions of high compressive strain and returns when the load is removed, hence generating an oscillatory flow pattern. Theoretical models predict that bone cells are exposed to fluid-induced shear stress on the order of 0.8–3.0 Pa in vivo[34] (up to 30 dyn/cm²). In vitro studies have proven that bone cells respond directly to fluid flow, whether steady, pulsatile, or oscillatory.[35–37] Unidirectional and oscillatory fluid flow result in different but contextually similar responses, dependent on the cell studied.[38]

Since the complex loading environment of the skeleton results in a diverse range of mechanical forces in the bone tissue, it is difficult to tease apart responses dependent on one or another biophysical factor. Furthermore, a cell cannot experience any force in the absence of others because loading the skeleton generates all of these biophysical signals concurrently. It is not clear if variable mechanical factors truly elicit differential responses, as suggested by some in vitro studies,[39,40] or whether these are differences elicited by force level or rate of force application. To date, there is insufficient evidence to support that bone cells ignore any one physical signal during bone adaptation. Rather, in vitro application of forces to cells seems to stimulate many of the same signal-transduction pathways, all perhaps contributing to an integrated alteration in gene expression. Similarities, where present, in the cellular responses to diverse physical signals might suggest that a common set of molecular events can be induced by each of these mechanical forces or that different signaling pathways can be integrated into a sum total response.

In relating experimental data to dynamic changes in the viable skeleton, it must be emphasized that in vitro systems are models only. Cells plated in monolayer in a tissue culture vehicle or within a three-dimensional scaffold do not experience force as they would within an organ. As well, the times studied in tissue culture are short compared to a remodeling cycle in bone where, in humans, resorption occurs over ~8 days, and formation requires nearly 3 months. Thus, in vitro systems, while critical to understanding the molecules from whence the response is derived, can simply not replicate the temporal, spatial, and contextual experience of a cell in the living skeleton.

10.3 MECHANICAL REGULATION OF OSTEOCLASTOGENESIS

Bone tissue undergoes remodeling throughout life to adapt to changing mechanical demands. The remodeling process is tightly regulated by a balance between the number of bone-producing osteoblasts and bone-resorbing osteoclasts. The local mechanical environment is an important regulator of skeletal remodeling. In response to diminished physical activity, the skeleton seeks to adapt to the altered mechanical environment by initiating bone remodeling. Thus, in patients confined to bed rest or with paraplegic limbs, loss of physical loading initiates recruitment of osteoclasts to the unloaded site,[41] resulting in local bone resorption and loss of bone mass.[42] Microgravity conditions further illustrate the importance of normal physiologic loading for maintaining bone mass. Astronauts lose bone mineral in the lower skeleton at a rate approaching 1.6% per month,[43] a substantial amount when one considers that a similar amount is lost per year in the postmenopausal woman. These examples of exuberant bone resorption illustrate that mechanical factors contribute a tonic inhibition to osteoclastic bone resorption, and that a reduction in skeletal loading incites a wave of osteoclast activity.

Osteoclasts originate from hematopoeitic precursors in the bone marrow, and have life spans of hardly more than a week in which they can generate resorptive lacunae on hard tissue. A primary mechanism for modulating local osteoclast activity is thus through controlling the formation of new osteoclasts. This requires recruitment of granulocyte macrophage colony-forming units (GM-CFU) into the macrophage lineage, from whence osteoclasts diverge when presented with appropriate signals within skeletal tissue. Recruitment is subject to regulation by mechanical factors.

Our laboratory has shown that biophysical factors generated during loading, including electrical fields, pressure, and mechanical deformation, inhibit osteoclastogenesis in vitro.[44–46] In these experiments, primary murine marrow cultures were grown in the presence of $1\alpha,25$-Dihydroxyvitamin D3 to stimulate osteoclast formation. Mechanical loading was applied continuously and osteoclasts, identified as multinuclear, tartrate-resistant acid phosphatase positive cells, were counted after 1 week in culture. Loading was effective when dosed during the early stages of osteoclast recruitment, essentially during the initial entry of monoblastic precursors into the osteoclast lineage.[46] We found further that the mechanical inhibitory effect was not transferable via a soluble factor secreted into the medium, suggesting that either direct effects on osteoclast precursors, or on a fixed molecule necessary to osteoclast recruitment, was critical to inhibition. This fixed factor, as will be described below, was identified as Receptor activator of nuclear factor kappaB ligand (RANKL), a molecule that is displayed on the surface of bone stromal cells and osteoblasts and is necessary for osteoclastogenesis.[47]

Our first description of biophysical inhibition of osteoclastogenesis was to show that dynamic electric fields limited osteoclast formation. Skeletal loading generates dynamic electric fields within bone in the form of streaming potentials, as interstitial fluid flows past charged molecules trapped within the mineral phase of bone. To assess the effect of electric fields on osteoclast formation in vitro, a solenoid was used to generate uniform time-varying electrical fields similar in magnitude and frequency to those induced by dynamic functional activity. Continuous application of these extremely low-intensity, low-frequency sinusoidal electric fields inhibited osteoclastogenesis in primary marrow cultures by 25%.[44] This in vitro result supports the in vivo finding that dynamic electric fields inhibit bone loss in an experimental disuse model.[48]

We also examined the effect of hydrostatic pressure. Intramedullary pressure increases during skeletal loading, and anabolic oscillatory pressure alone can regulate bone mass.[49] Cultures exposed to a 30% increase in normal hydrostatic pressure, an increase similar to that arising in the marrow cavity during functional activity, showed an inhibition of osteoclast recruitment by 35%.[45] Hydrostatic pressure also decreased expression of osteoblast/stromal cell macrophage colony stimulating factor (MCSF). MCSF is necessary not only for entry into the macrophage lineage, but also for proliferation of cells within this lineage.[50] As it became clear that MCSF, while necessary for osteoclastogenesis, was not sufficient, and it was furthermore a soluble factor,[51] we needed to explore farther to find the specific factor by which mechanical input was able to limit osteoclast formation.

Cyclic mechanical strain, applied by controlled deformation of the flexible substrate on which the cells were grown, was particularly robust at repressing in vitro osteoclast formation. Uniform equibiaxial strain applied at a peak magnitude of 1.8% resulted in a 52% reduction in osteoclast number, and importantly, this low-magnitude strain was applied in the absence of shear stress.[46] To further explore how mechanical strain influenced osteoclast formation, experiments were conducted to examine whether the cellular response to strain was dependent upon the period of time during which strain was applied. In these experiments, mechanical strain was applied throughout the culture period (days 2–7), during the early culture period (days 2–4) or during the late culture period (days 5–7). Straining during the early culture period significantly reduced osteoclast formation, similar to the effect caused by straining throughout. In contrast, the number of osteoclasts formed in response to straining during the late culture period, when osteoclast precursors fuse into the mature multinuclear cell, was not different from the number formed in the unstrained control cultures. These data concurred with that from application of hydrostatic pressure, indicating that there was a fixed period during early entry and proliferation of osteoclast progenitors that was susceptible to mechanical control.

In initial strain experiments, nonuniform strain was applied to marrow cultures plated onto flexible membranes and osteoclast formation was inhibited across the membrane by 35%.[46] To evaluate whether the inhibitory effect of strain was influenced by strain magnitude, the membrane was divided into equal areas of distinct strain magnitudes. Analysis of osteoclast formation revealed that higher tensile strains were associated with greater degrees of inhibition (Figure 10.1). This localized response, taken together with the observation that conditioned medium from strained cultures was not able to transfer an inhibitory effect to unstrained cells, suggested that mechanical repression of osteoclastogenesis involved a signal captured on the surface of attached cells, and that this signal was sensitive to strain magnitude.

Three components are necessary for osteoclast recruitment: osteoclast progenitors, MCSF, and RANKL. (Note, RANKL was initially referred to by many names, including osteoclast differentiation factor [ODF], tumor necrosis factor-related activation-induced cytokine [TRANCE], osteoprotegerin [OPG] ligand [OPGL], and tumor necrosis factor ligand superfamily member 11 [TNFSF11].) As we have detailed above, MCSF is necessary for the expansion of osteoclast progenitors.[52] RANKL, which is specifically required to induce monoblasts to enter the osteoclast lineage, was found to be the necessary partner for MCSF.[53] MCSF and RANKL are expressed as both soluble and membrane-bound cytokines and elicit their actions by binding to receptors on the

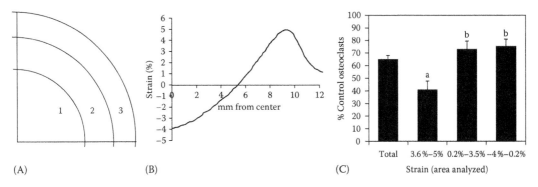

FIGURE 10.1 Osteoclast inhibition is site-specifically proportional to strain magnitude. Nonuniform strain was applied to primary marrow cultures plated onto flexible membranes. The membrane was divided into equal areas of distinct strain magnitudes for analysis (A and B). Fewer osteoclasts were formed in the outer region, where tensile strains were the highest, than in either the middle or central region (C). (From Rubin, J. et al., *J. Orthop. Res.*, 17, 639, 1999. With permission.)

surface of osteoclast precursors. Both are expressed on the surface of bone stromal cells, which are cells that form a minor population in primary marrow cultures, as well as stromal cells that differentiate into osteoblasts, and confer the ability to mediate ostoclastogenesis in the marrow compartment through cell–cell interactions. As noted, in nonuniformly stretched dishes, there was a variable reduction of osteoclast formation dependent on strain magnitude. The adherent stromal cells were exposed to the variable strain magnitude, and were the sole source of both MCSF and RANKL to the dish. This strongly suggested that this cell population was subject to mechanical control, and that mechanical strain effectively regulated local cellular expression of molecules determining osteoclastogenesis.

Thus, with data suggesting that mechanical factors regulated a membrane-bound molecule critical to osteoclast formation, effects of cyclic mechanical strain on expression of RANKL were evaluated. In primary marrow cultures, strain applied for 3 days reduced RANKL mRNA to 59% of the level in unstrained cultures.[25] The association between mechanical downregulation of RANKL expression and inhibition of osteoclast formation by strain was strengthened by experiments in which marrow cultures were treated with soluble RANKL and the effect of strain on osteoclastogenesis was negated.

Primary stromal cell cultures were prepared from bone marrow to clarify that strain directly influenced RANKL expression in this cell population. After 3 days of strain, RANKL mRNA in primary stromal cells was reduced to 60% of the level in unstrained cultures.[25] In contrast, expression of the membrane-bound form of MCSF was unchanged by strain application. Further experiments showed that RANKL mRNA was decreased within 24 h of strain initiation and that a 6-h period of strain was sufficient to inhibit RANKL expression by the following day.[54] RANKL was also found to be highly sensitive to mechanical strain (Figure 10.2), with strain magnitudes as low as 0.25% causing downregulation of expression.[55]

More recently, strain inhibition of RANKL has been demonstrated in preosteoblast cell lines.[23,56] RANKL expression in cells of the osteoblast lineage may be regulated by multiple forms of mechanical loading, as others have shown that shear has similar effects to reduce RANKL.[57] In the absence of mechanical load, RANKL expression was increased in osteoblasts exposed to simulated microgravity conditions.[58] Importantly, RANKL downregulation accompanied bone formation in loaded animals,[4] supporting the in vitro data on mechanical regulation of RANKL.

OPG is a soluble decoy receptor for RANKL and is produced by cells of the osteoblast lineage. Factors that increase OPG levels may inhibit osteoclastogenesis by disruption of RANKL signaling.

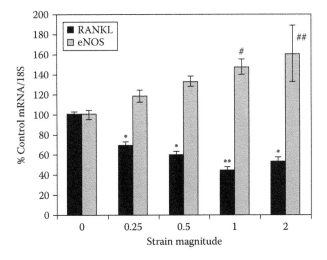

FIGURE 10.2 Strain magnitude dose-dependently regulates RANKL and eNOS gene expression. Strain inhibited RANKL expression and stimulated eNOS expression in primary murine stromal cells, and these responses were strain magnitude dependent. (From Rubin, J. et al., *J. Biol. Chem.*, 278, 34018, 2003.)

Recent evidence suggests that mechanical loading regulates OPG expression. Cyclic strain increased OPG synthesis in osteoblastic cells,[59,60] while oscillatory fluid flow upregulated OPG expression in osteoprogenitor cells.[57] Thus, regulation of both RANKL and OPG levels in bone cells could contribute to inhibition of osteoclastogenesis by mechanical input.

Nitric oxide (NO) is another potential factor mediating the effects of mechanical strain on osteoclastogenesis. This intracellular messenger is synthesized by NO synthase (NOS) from L-arginine. NO's effects on bone are complex,[61] but studies overall concur that NO can slow down bone remodeling and bone loss in animals and humans.[62,63] Consistent with in vivo findings, NO modulates osteoclast functions in vitro. Osteoclastic resorption was potentiated in the presence of NOS inhibition.[62] Furthermore, treatment with NO donors decreased RANKL expression and increased OPG expression in bone stromal cells,[64] suggesting that NO indirectly regulates osteoclastogenesis.

Our laboratory demonstrated that NOS was enhanced in response to strain application in primary marrow stromal cells.[55] NO production did not increase within the first hour after beginning strain, in contrast to findings that NO production was rapidly increased by both strain and shear in other bone cell types.[36,40,65–67] Rather, the level of NO doubled following 24 h of continuous strain. This increase required endogenous NOS activity, as treatment with the NOS inhibitor L-NAME blocked the effect of strain on NO production. Endogenous synthesis of NO was the result of enhanced expression of endothelial NOS (eNOS), the most prominent NOS isoform in bone cells[68]; strain increased eNOS mRNA by nearly twofold compared to unstrained cultures (Figure 10.2), and elevation of eNOS protein by strain was confirmed. Importantly, strain inhibition of RANKL expression was prevented by treatment with NOS inhibitors,[69] confirming a role for NO in mechanical repression of RANKL.

In summary, mechanical factors generated by loading in the skeleton are capable of inhibiting osteoclast formation. At the very least, this is via regulation of RANKL expression by bone cells, and may involve regulation of local levels of NO and MCSF. The sum total of mechanical regulation of these molecules is to control entry of precursors into the osteoclast lineage, although direct effects on osteoclast function have not been ruled out.

10.4   INTRACELLULAR SIGNALING BY WHICH STRAIN LIMITS RANKL GENE EXPRESSION

To expand our knowledge about mechanotransduction in bone cells, our laboratory focused on identifying those signaling pathways involved in the inhibition of RANKL expression by mechanical strain. Since information was lacking regarding mechanically activated signaling pathways in bone cells, we considered transduction cascades that were activated by mechanical stimuli in other cell types. Mitogen-activated protein kinases (MAPKs) had been identified as transducers of effects of both mechanical shear and strain in cardiovascular cells. Fluid shear applied to endothelial cells caused activation of three members of the MAP kinase family—extracellular signal-regulated kinases 1 and 2 (ERK1/2),[70] c-Jun N-terminal kinase (JNK),[71] and Big MAP kinase (BMK-1 or ERK5).[72] Applying strain to vascular smooth muscle cells rapidly activated both ERK1/2 and JNK.[73] Thus, studying strain effects on MAPKs in bone cells seemed a good starting point.

Application of low-magnitude strain caused a rapid and transient activation of ERK1/2 in primary marrow stromal cells. ERK1/2 phosphorylation was maximally increased by 5 min after strain initiation, with a return toward baseline by 60 min.[54] To assess the level of strain necessary to activate ERK1/2, strain magnitudes ranging from 0.5% to 8.5% were applied for 5 min. Application of strain at 0.5%, a level within the physiologic range of mineralized tissue in the skeleton,[74,75] strongly increased ERK1/2 phosphorylation. Increasing the strain magnitude did not further increase the level of phosphorylated ERK1/2.

To assess whether ERK1/2 signaling contributed to mechanical repression of RANKL, the MEK inhibitor PD98059 was used to disrupt strain activation of ERK1/2. Mechanical inhibition

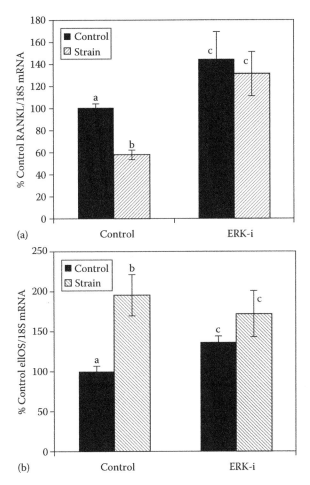

FIGURE 10.3 Strain stimulation of ERK1/2 MAP kinase is required for downstream gene regulation. The effect of strain to decrease RANKL (a) and increase eNOS (b) expression in bone stromal cells was completely blocked by the presence of an ERK1/2 inhibitor (ERK-i). (Adapted from Rubin, J. et al., *J. Biol. Chem.*, 278, 34018, 2003.)

of RANKL expression was blocked by treatment with PD98059 (Figure 10.3), as was strain induction of eNOS.[55] To further explore the link between ERK1/2 activation and RANKL inhibition, cultures were infected with a constitutively active MEK adenovirus (Ad.caMEK) to re-create the putative strain signal. As expected, Ad.caMEK caused activation of ERK1/2, and, as well, inhibited RANKL expression to 56% of the level displayed by cells treated with an empty adenovirus. eNOS expression was increased fourfold by treatment with Ad.caMEK. Together, these data implicate ERK1/2 in strain regulation of both RANKL and eNOS expression by osteoprogenitor cells.

Similar to ERK1/2 activation, strain induced activation of JNK that was rapid and maximal at a strain magnitude of 0.5%.[54] Although JNK activation was sustained, in contrast to the transient activation of ERK1/2, treatment of cultures with a JNK inhibitor did not block strain effects on RANKL or eNOS.[55] Thus, JNK activation was not required for these downstream effects, suggesting that the requirement for ERK1/2 was specific.

Numerous in vitro studies have now shown that mechanical strain and fluid shear stress both activate ERK1/2 in bone cells.[76–78] While the role of ERK1/2 in bone cell function is far from simple, with both positive and negative results on osteogenesis reported,[79,80] mechanical activation of

ERK1/2 appears to support anabolic effects on bone. We have shown that ERK1/2 is also involved in mechanical repression of RANKL in a preosteoblast cell line.[23] In addition to RANKL, ERK1/2 activation by mechanical loading in bone cells has been linked to changes in expression of osteopontin[81] and type I collagen,[82] as well as cell proliferation,[83,84] factors which are associated with enhanced bone formation.

In the signal cascade leading to ERK1/2 activation, the most proximal action activates a guanine exchange factor, which places a GTP on the Ras 21-kD GTPase. In the GTP-bound state, the GTPase associates with an MEKK, in this case Raf1, which phosphorylates MEK1/2, which then phosphorylates ERK1/2. To trace the mechanical signal proximally from ERK1/2 activation, our laboratory concentrated on the Ras-GTPase located in the membrane where deformation occurs. Using immobilized Raf beads to capture GTP-bound Ras, we showed that Ras was activated within 2 min of strain initiation in primary marrow stromal cells.[85] Mechanical activation of Ras was isoform-specific. The Ras subfamily includes the classical Ras proteins H-, K-, and N-Ras.[86] Strain strongly activated H-Ras, and to a lesser degree, activated K-Ras. N-Ras was not activated by strain.

To determine whether strain activation of Ras contributed to RANKL inhibition, RNAi technology was used to specifically silence H-Ras or K-Ras expression in marrow stromal cells.[85] Silencing H-Ras completely prevented mechanical repression of RANKL. In contrast, strain inhibition of RANKL was preserved in cultures where K-Ras was silenced. These results were reproduced in the CIMC-4 preosteoblast cell line, again implicating strain activation of H-Ras in specifically inhibiting RANKL expression.

Pharmacologic inhibition of H-Ras was used to complement the RNAi approach. As targeting of H-Ras to the plasma membrane requires palmitoylation, we used a specific farnesyl transferase inhibitor (FTI277) to disrupt H-Ras function. Treatment of cultures with FTI277 blocked strain activation of H-Ras and prevented mechanical inhibition of RANKL expression. Together, these data confirm the involvement of H-Ras in mechanical repression of RANKL transcription.

Control of RANKL transcription has proved to be exceedingly complex. Gene transcription enhancers and promoters are found thousands of bases proximal to the transcription start site, and the RANKL gene is silenced in most cell lineages.[87,88] The mechanism by which mechanical activation of ERK1/2 confers inhibition of these distant enhancers in bone cells has not yet been solved, and remains a critical question in the physiological regulation of bone remodeling. It is likely that other aspects of loading as well modulate the transcription of RANKL: as will be discussed below, loading of bone cells activates β-catenin signaling. β-Catenin has been implicated in decreasing RANKL promoter activity.[89]

10.5 MECHANICAL REGULATION OF OSTEOBLASTOGENESIS

It has been conclusively demonstrated that mechanical loading stimulates the appearance of osteoblasts and consequent bone formation (for reviews, see Skerry[90] and Rubin et al.[91]). A single exposure of load in rat tibia is associated with the appearance of periosteal osteoblasts,[92] and the degree of bone formation in response to loading is positively correlated with strain gradients.[18] The loading parameters that induce bone formation span from high magnitude[19] to very low magnitude signals,[16,93] and include signals dosed over very short times.[10] On the complementary side, unloading decreases bone formation, as evidenced by individuals subject to spinal cord injury[94] or astronauts experiencing microgravity.[95]

Osteoblasts, as well as osteoprogenitor cells, respond directly to in vitro mechanical loading by increasing cellular activity, including proliferation[24,83,96,97] and expression of osteogenic genes.[23,81,98,99] Mechanical input also increases release of bioactive factors, such as NO and prostaglandins. In vivo loading experiments have identified NO as critical to mechanical stimulation of bone formation.[100,101] NO production is increased in vitro by fluid flow and strain in osteoblasts,[36,40,66] an effect due in part to increased eNOS expression.[55,65] Prostaglandin levels are increased during in vivo loading.[102] A variety of in vitro mechanical signals, including strain,[30] pressure,[103] and fluid

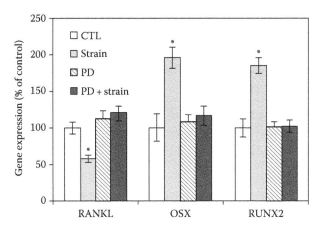

FIGURE 10.4 Mechanical loading increases osteogenic genes. Strain applied overnight to preosteoblast cells increased Runx2 and osterix (OSX) gene expression, while decreasing RANKL expression. These effects were completely abolished by treatment with the ERK1/2 MAP kinase inhibitor PD98059. (From Fan, X. et al., *J. Cell Physiol.*, 207, 454, 2006.)

flow,[104] markedly upregulate prostaglandin production in osteoblasts through mechanical regulation of an enzyme that generates prostaglandins in bone, cyclo-oxygenase-2 (COX-2). Expression of COX-2 is increased in response to fluid flow[105,106] and strain.[107,108]

As loading is anabolic,[12] genes that are associated with bone formation should be subject to mechanoregulation. Indeed, osteoblasts increase production of bone matrix proteins,[81,82,98,109–111] including type I collagen, osteopontin, and osteocalcin, in response to mechanical stimulation. Mechanical signals also upregulate expression of transcription factors that support osteoblast differentiation and function,[99,105,112–115] including Runx2, osterix, c-fos, and FosB. Our laboratory showed that expression of Runx2 and osterix is increased by strain (Figure 10.4) in an ERK1/2 MAP kinase-dependent fashion, linked to mechanical activation of H-Ras, and in parallel with strain inhibition of RANKL.[23] Studies in vivo where hindlimb unloading prevented mechanical use of back paws complement cell studies by showing that unloading is associated with a downregulation of osterix and Runx2 in bone,[116] confirming that mechanoregulation allows bone adaptation over a spectrum of loading conditions.

Mechanical input activates ERK1/2 MAPK in osteoblasts,[76–78] as discussed in the previous section. ERK1/2 activation by mechanical loading has been linked to proliferation[83,84,117] and to regulation of multiple genes in osteoblasts.[81,82,99,106] While many signaling molecules act proximally to ERK1/2, NOS has been implicated in mechanical activation of ERK1/2 by fluid flow.[117,118] Activation of JNK and p38 MAPKs by fluid flow has also been reported, with JNK implicated in flow regulation of type I collagen[82] and p38 implicated in flow regulation of osteopontin.[81]

Osteoblasts respond to fluid flow, whether steady or dynamic, with a rapid increase in intracellular calcium,[37,119] an effect also reported in response to strain.[110,120] Influx of calcium through mechanosensitive channels and release of calcium from intracellular stores, downstream of IP3 signaling, may contribute to this effect.[121] Mechanically regulated changes in gene expression,[81,112,122] as well as ERK1/2 activation by loading,[76,99] were disrupted when calcium signaling was blocked. Thus, mechanical effects at the membrane involve multiple signaling molecules which result in, at the very least, ERK1/2 activation with consequent changes in gene expression that regulate both bone formation and resorption.

Much progress has been made in understanding how cells of the osteoblast lineage respond to various types of mechanical input. A future challenge will be determining how mechanical stimuli from different sources (i.e., fluid flow, substrate deformation, pressure) are integrated into a unified cellular response. Furthermore, how bone cells sense mechanical stimuli is mostly unknown, in particular regarding what early events occur at the plasma membrane to transmit the external mechanical signal into the cell. Ion channels,[123] integrins,[124] and, recently, primary cilium[125] have been proposed as candidate mechanosensors in bone cells.

10.6 FOCUS ON MECHANICAL ACTIVATION OF β-CATENIN IN OSTEOBLASTS

An important role for canonical Wnt signaling in promoting bone anabolism via effects on both formation and resorption has been confirmed,[126] leading to studies which show that mechanical factors can activate downstream Wnt signals. Early evidence suggesting the importance of Wnt signaling to bone formation was based upon observations that gain- or loss-of-function mutations in LRP5, the Wnt co-receptor, were associated with high or low bone mass, respectively, in humans and mice.[127–129] Downstream of Wnt ligand binding to its transmembrane receptors, activation of β-catenin in the cytoplasm is critical to the cell response.[130] Targeted deletion of β-catenin has been found to disrupt both skeletal development and postnatal bone acquisition,[131,132] establishing the importance of β-catenin signaling in osteoblast differentiation and function.

The cytoplasmic level of β-catenin is regulated by a destruction complex consisting of adenomatous polyposis coli, axin, and glycogen synthase kinase 3β (GSK3β). Phosphorylation by cyclin-dependent kinase and GSK3β targets β-catenin for degradation by the ubiquitin/proteosome pathway, limiting levels of cytoplasmic β-catenin under resting conditions. In canonical Wnt signaling, binding of Wnt ligand to its transmembrane receptors leads to disruption of the β-catenin destruction complex, which allows accumulation of stabilized β-catenin and subsequent nuclear translocation. In the nucleus, β-catenin interacts with T cell factor/lymphoid enhancer-binding factor (TCF/LEF) transcription factors to promote changes in expression of its target genes.

More recently, mechanoregulation of this pathway in bone cells has been explored. Mechanical loading upregulates Wnt/β-catenin target genes in vivo[108] and in vitro.[96] This effect is enhanced in LRP5 gain-of-function transgenic mice, where Wnt signaling is constitutively activated.[108] β-Catenin translocates into the nucleus after loading of osteoblasts,[133,134] both in response to substrate deformation and fluid flow. Importantly, mice with a loss-of-function mutation in LRP5 respond poorly to local bone loading,[135] suggesting that mechanical activation of this pathway contributes to downstream loading effects in bone.

Our laboratory used a preosteoblastic cell line to study the effect of dynamic mechanical strain via substrate deformation on β-catenin signaling.[107] We focused on elucidating the proximal signals that contribute to the β-catenin response by mechanical loading. Consistent with previous reports, we found that uniform equibiaxial strain applied at a peak magnitude of 2% caused a rapid, transient accumulation of β-catenin in the cytoplasm and its translocation to the nucleus. Fifteen minutes after initiating the strain regimen, active β-catenin was strongly increased in the nuclei of strained cells measured by confocal microscopy. The increase in active nuclear β-catenin reached plateau by 60 min. Active β-catenin is recognized by an antibody that binds to β-catenin species dephosphorylated on serine 37 and threonine 41.[136] Lysates made from cells that were subjected to strain showed increases in both active and total β-catenin protein within 60 min after beginning strain (Figure 10.5a).

Consistent with the rapid activation and translocation of β-catenin, we measured a transient increase of the Wnt/β-catenin target gene WISP1 in response to strain, with a peak response 4 h after strain initiation (Figure 10.5b). COX-2, known to be upregulated by mechanical loading in osteoblasts[105,108] and a downstream target of β-catenin in skeletal lineage cells,[134,137] was also increased by loading. COX-2 expression responded robustly to mechanical strain, with a peak response at 1 h. To investigate whether the COX-2 product, PGE2, was critical to subsequent upregulation of WISP1, the COX inhibitor indomethacin was added. Indomethacin treatment did not disrupt the strain-induced increase in WISP1 expression.

The increase in active (dephosphorylated) β-catenin suggested that GSK3β might be a target of strain. Phosphorylation of GSK3β at Ser9, which inactivates GSK3β, was maximally increased

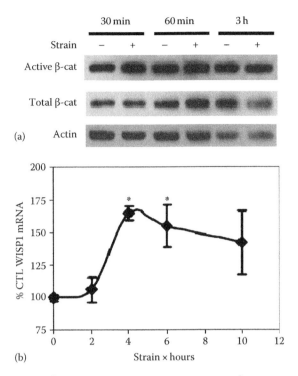

FIGURE 10.5 Strain activates β-catenin signaling. Active and total β-catenin levels (a) were increased within 60 min of strain initiation in preosteoblast cells. This was followed by induction of the β-catenin target gene WISP1 (b). (Adapted from Case, N. et al., *J. Biol. Chem.*, 283, 29196, 2008. With permission.)

60 min after beginning strain exposure. Activation of Akt, which phosphorylates and inactivates GSK3β, was also studied. Akt phosphorylation at Ser473 was strongly induced at 15 min of strain, with a return to near basal levels by 60 min. Akt activation was not dependent upon PI3-kinase activity, as PI3-kinase inhibition with LY294002 did not block Akt phosphorylation in strained cells. In sum, these data suggested that the increase of β-catenin was temporally related to the activation of Akt and subsequent inactivation of GSK3β.

Strain-induced activation of Akt in vascular cells has been shown to require the presence of caveolin-1,[138] and as such, caveolin-1 has been proposed as a key element of the mechanoresponse. Interestingly, although siRNA targeting caveolin-1 increased cytoplasmic β-catenin levels, silencing caveolin-1 did not disrupt strain activation of Akt. Furthermore, strain-induced inactivation of GSK3β and induction of WISP1 were actually enhanced when caveolin-1 was knocked down. Together, these data indicated that caveolin-1 was not required for Akt/GSK3β/β-catenin signaling in bone cells, but rather might provide tonic inhibition to strain-induced signals.

The cytoplasmic pool of β-catenin "accessible" for activation by strain could influence the distal response to strain, as perhaps suggested by experiments where mechanical stimulation of bone formation was increased in LRP5 gain-of-function mice.[108] We used lithium chloride to inactivate GSK3β and increase cytoplasmic β-catenin levels. The strain-induced increase in the β-catenin responder WISP1 was significantly greater in cultures pretreated with lithium chloride, similar to the effect seen with caveolin-1 silencing. Together these studies show that the size of the accessible pool of β-catenin influences the magnitude of the mechanical response.

The time course of β-catenin activation and translocation to the nucleus was not consistent with an autocrine loop requiring Wnt signaling. To prove that the mechanical effect bypassed Wnt signaling, cells were strained after treating with the Wnt inhibitor, DKK-1. β-Catenin activation and translocation to the nucleus after initiation of mechanical strain was not diminished in the presence of DKK-1. Furthermore, strain regulation of Akt and GSK3β did not require Wnt binding and, finally, DKK-1 did not disrupt strain upregulation of WISP1 and COX-2. These data suggest that paracrine Wnt signaling is not necessary to the machinery by which strain rapidly activates β-catenin in osteoblasts.

In summary, data now converge to show that β-catenin signaling is activated by mechanical input in osteoblasts. Linking mechanical activation of β-catenin to downstream cellular responses, an area of active investigation, will further increase our understanding of mechanotransduction in bone.

10.7 MECHANICAL REGULATION OF MESENCHYMAL STEM CELL LINEAGE SELECTION

Osteoblasts and adipocytes, along with chondrocytes and myocytes, originate from a common multipotent precursor, the mesenchymal stem cell (MSC).[139,140] Within the bone marrow microenvironment, the pools of osteoblastic and adipocytic progenitors appear to be subject to reciprocal control. For example, there is an inverse relation between bone marrow adiposity and the amount of bone in the axial and appendicular skeleton of young adults,[141] while in aging individuals trabecular bone is essentially replaced by fat tissue.[142] Expanding this relationship, a new study has shown that in young males, total body fat mass is negatively associated with cortical bone mass.[143] An increase in marrow fat, as well as a reduction in osteoprogenitor cell number, accompanies estrogen deficiency; postmenopausal women have twice as much fat in the marrow compared to premenopausal women, and women with low bone density have more bone marrow fat than women with normal bone density.[144] Conversely, when the Wnt co-receptor LRP5 is constitutively activated, causing an increase in bone mass, there is a consequent decrease in bone marrow fat.[145] The marrow mesenchymal progenitor population is also influenced by inactivity or unloading. Immobilization leads to a near doubling of marrow fat within 15 weeks of bed rest.[146] Space flight similarly results in bone loss and increased marrow fat.[146,147] These data from mice and men suggest that mechanical loading may influence bone formation and remodeling through regulating MSC lineage allocation.

In vivo studies support that mechanical factors regulate marrow MSC fate. Exposure to extremely low-magnitude mechanical signals increases MSC proliferation, with a biasing of the MSC population toward osteogenesis.[148] Mice subjected to a climbing regimen for 4 weeks have more osteoblasts and fewer adipocytes in the marrow cavity.[149] Running rats similarly decreases marrow fat.[150] In vitro mechanical loading of MSCs promotes proliferation[24,151,152] and enhances osteogenic differentiation.[153–156] The sensitivity of MSCs to mechanical input is further demonstrated by an inhibition of osteoblast recruitment during hindlimb unloading of rats, which results in a decrease of 76% in the number of osteoprogenitor cells in the marrow of non-weight-bearing bones within 14 days.[157] Hindlimb unloading also leads to increased marrow fat,[158] and, interestingly, increases the potential for adipogenesis in rat ex vivo marrow cultures.[159] Similarly, removal of mechanical load during microgravity simulation decreases osteoblastogenesis while increasing adipogenesis.[160]

Studies implicate the transcription factor PPARγ in the regulation of MSC lineage selection through its effects to promote adipogenesis. Wnt/β-catenin signaling and mechanical strain both suppress PPARγ and stimulate osteoblastogenesis.[150,161] Haploinsufficiency of PPARγ is also associated with reduced adipogenesis and augmented osteoblastogenesis.[162] In contrast, increased PPARγ activity enhances the proteasomal degradation of β-catenin, with the sum effect being a reduction in osteogenesis.[163] A decrease in β-catenin levels parallels the rise in PPARγ during adipocyte differentiation.[164] Disruption of β-catenin degradation by inhibitors of GSK3β limits adipogenesis,[165,166] suggesting that the reduction in β-catenin is permissive for differentiation. The ability of β-catenin to counteract an adipogenic stimulus not only involves repression of PPARγ,[167] but also blockade of downstream events initiated by PPARγ.[164] Thus, competing actions of PPARγ and β-catenin regulate MSC differentiation between the adipocyte and osteoblast lineages. Given that mechanical loading activates β-catenin signaling in osteoblasts,[107,133] β-catenin is a candidate for modulation by mechanical input to influence MSC fate.

Our laboratory has recently proved that mechanical signals participate in directly regulating lineage selection of mesenchymal precursors.[166] Our approach involved using a daily mechanical input to counteract an adipogenic stimulus. First, we showed that exposure to a strongly adipogenic medium caused differentiation of C3H10T1/2 mesenchymal progenitor cells, as delineated by expression of PPARγ and adiponectin and accumulation of intracellular lipid. A daily strain regimen (3600 cycles/day at 10 cpm, 2%) was applied beginning concurrent to the change to adipogenic medium. In cultures where strain was delivered, lipid accumulation was significantly reduced (Figure 10.6a). Mechanical strain further inhibited expression of PPARγ and adiponectin mRNA by up to 35% and 50%, respectively, after 5 days (Figure 10.6b), an effect confirmed at the protein level. The efficacy of the mechanical influence relied upon both repetitive and early exposure. A single day of loading was insufficient to restrict emergence of the adipocytic phenotype. Similarly, when loading was delayed for 2 days after exposure to adipogenic medium, neither PPARγ nor adiponectin protein levels were influenced by strain.

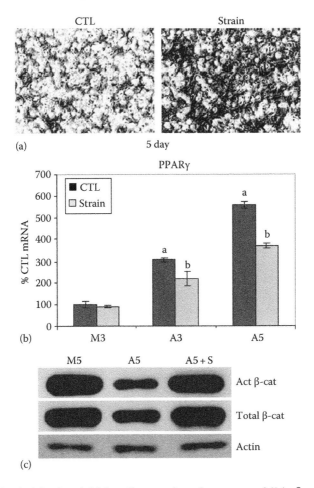

FIGURE 10.6 Mechanical loading inhibits adipogenesis and prevents a fall in β-catenin. Application of daily strain to C3H10T1/2 mesenchymal progenitor cells in adipogenic (A) medium prevented accumulation of cytoplasmic triglyceride droplets (a). Compared to cells in multipotential (M) medium, PPARγ expression was increased in response to adipogenic stimuli and this was reduced by daily loading (b). β-Catenin levels dropped in adipogenic medium, but were preserved by daily loading (c). (Adapted from Sen, B. et al., *Endocrinology*, 149, 6065, 2008. With permission.)

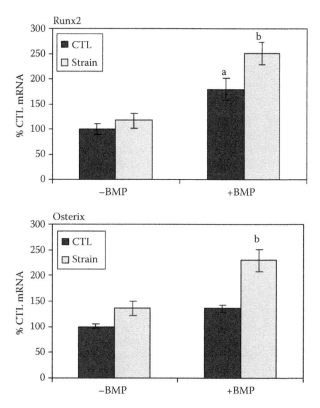

FIGURE 10.7 Strain preserves mesenchymal progenitor cells in a multipotential state. After 5 days in adipogenic medium, cells were exposed to BMP-2, an osteogenic stimulus. BMP-2 treatment for 2 days increased expression of the osteogenic markers Runx2 and osterix more strongly in cells that received daily loading. (From Sen, B. et al., *Endocrinology*, 149, 6065, 2008. With permission.)

Accompanying the adipogenic transformation in MSCs, we measured a progressive decrease in both active and total β-catenin (Figure 10.6c). Mechanical signals completely prevented the decrease in β-catenin, while simultaneously limiting expression of both PPARγ and adiponectin protein. Strain also induced β-catenin nuclear translocation, and confirmation of a significant downstream response to mechanical activation of β-catenin was measured by increases in WISP1 and cyclin D1 mRNA, both β-catenin responsive genes. The ability of strain to preserve β-catenin levels required daily exposure to mechanical signals. If cells in adipogenic medium were "exercised" with strain for 5 days and then cultured without strain for three further days, β-catenin levels fell by day 8 and adiponectin rose. Thus, continuous daily loading was necessary to support the β-catenin signal and suppress adipogenesis.

To confirm that mechanical preservation of β-catenin mediated the effects of daily strain, inhibitors of GSK3β were used to mimic this effect. Addition of either lithium chloride or SB415286, a specific GSK3β inhibitor, prevented the rise of PPARγ and adiponectin in adipogenic medium, consistent with a strain effect that relies on β-catenin signaling. As well, both inhibitors induced expression of WISP1 and cyclin D1. Furthermore, we found that mechanical stimulation increased the phosphorylation of Akt at Ser473, indicating increased activation, as well as phosphorylation of GSK3β at Ser9. Akt is known to directly phosphorylate and thereby inactivate GSK3 isoforms. Together, these results suggested that mechanical strain activated the β-catenin pathway through alteration of GSK3β phosphorylation via Akt.

As repetitive mechanical loading is able to prevent adipogenesis of C3H10T1/2 cells, we next asked whether these MSCs would more readily enter an osteoblastic lineage, i.e., whether mechanical input tipped the scale from fat to bone. In a strongly adipogenic environment, daily loading did not stimulate osteogenesis. To determine if loading preserved the ability of MSCs to respond to a strong osteogenic stimulus, we added BMP-2 and evaluated expression of the osteogenic markers Runx2 and osterix. Cells grown under adipogenic conditions responded to BMP-2 stimulation with an increase in Runx2 expression, but osterix was unchanged. In cultures which were exercised daily, not only was there an enhanced Runx2 response to BMP-2, but osterix expression more than doubled (Figure 10.7). As such, mechanical strain was able to enhance the potential of MSCs to enter the osteoblast lineage despite exposure to adipogenic conditions.

Our data suggest that adipogenesis can be inhibited when β-catenin levels are supported by mechanical input, allowing MSCs to remain available for alternate cell lineage allocation. As such, one critical contribution of a daily input of exercise may be to preserve MSCs in a multipotential state, suppressing the emergence of adipocytes from the MSC pool by stimulating a durable β-catenin signal. While there is very little downside to exercise, many barriers prevent the population from achieving an adequate exercise regimen, including difficulties due to aging and obesity. There is no question that reproducing and enhancing the effects of exercise on the skeleton through drugs or devices would be of great benefit to society.

10.8 CONCLUSION

The role of mechanical signals in the achievement and maintenance of bone quantity and quality is clear. As we have discussed, normal physiologic loading is both anabolic, stimulating cells of the osteoblast lineage, and anticatabolic, inhibiting osteoclast formation. Understanding the signals which transmit and translate mechanical load into a tissue response is critical for understanding the bone pathophysiology underlying skeletal fragility and aging, where these signals are either mis-sensed or mis-directed.

The fact that mechanical signals are both "normal" and "effective" should invite their use for therapeutics. We predict that the mechanical factors and mechanoresponses which generate and preserve an effective structure in the intact skeleton can be translated into therapeutic applications to promote bone repair following injury, such as in fracture healing or osseointegration. Not only does exercise build bone,[9,168] but engineered mechanical signals have already been shown to be useful for improving skeletal fitness.[93,169] Finally, as we have discussed in the last section of this chapter, recent advances indicate that the precursors of bone cells should be targets for therapeutics as well. As such, we can envision that mechanical signals will be of use to bias mesenchymal stem cells away from becoming adipocytes, and into the osteogenic lineage to build bone.

REFERENCES

1. Wolff J. 1892. *Das gesetz der transformation der knochen.* Berlin, Germany: Verlag August Hirschwald.
2. Bennell KL, Malcolm SA, Khan KM, Thomas SA, Reid SJ, Brukner PD, Ebeling PR, Wark JD. 1997. Bone mass and bone turnover in power athletes, endurance athletes, and controls: A 12-month longitudinal study. *Bone* **20**(5):477–484.
3. Judex S, Gross TS, Zernicke RF. 1997. Strain gradients correlate with sites of exercise-induced bone-forming surfaces in the adult skeleton. *J Bone Miner Res* **12**(10):1737–1745.
4. Rubin C, Xu G, Judex S. 2001. The anabolic activity of bone tissue, suppressed by disuse, is normalized by brief exposure to extremely low-magnitude mechanical stimuli. *FASEB J* **15**(12):2225–2229.
5. Haapasalo H, Kontulainen S, Sievanen H, Kannus P, Jarvinen M, Vuori I. 2000. Exercise-induced bone gain is due to enlargement in bone size without a change in volumetric bone density: A peripheral quantitative computed tomography study of the upper arms of male tennis players. *Bone* **27**(3):351–357.
6. Haapasalo H, Sievanen H, Kannus P, Heinonen A, Oja P, Vuori I. 1996. Dimensions and estimated mechanical characteristics of the humerus after long-term tennis loading. *J Bone Miner Res* **11**(6):864–872.
7. Jones HH, Priest JD, Hayes WC, Tichenor CC, Nagel DA. 1977. Humeral hypertrophy in response to exercise. *J Bone Joint Surg Am* **59**(2):204–208.

8. Kontulainen S, Sievanen H, Kannus P, Pasanen M, Vuori I. 2003. Effect of long-term impact-loading on mass, size, and estimated strength of humerus and radius of female racquet-sports players: A peripheral quantitative computed tomography study between young and old starters and controls. *J Bone Miner Res* **18**(2):352–359.

9. MacKelvie KJ, Khan KM, Petit MA, Janssen PA, McKay HA. 2003. A school-based exercise intervention elicits substantial bone health benefits: A 2-year randomized controlled trial in girls. *Pediatrics* **112**(6 Pt 1):e447.

10. Rubin C, Lanyon L. 1985. Regulation of bone mass by mechanical strain magnitude. *Calcif Tissue Int* **37**:411–417.

11. O'Connor JA, Lanyon LE, MacFie H. 1982. The influence of strain rate on adaptive bone remodelling. *J Biomech* **15**(10):767–781.

12. Rubin C, Lanyon L. 1984. Regulation of bone formation by applied dynamic loads. *J Bone Joint Surg* **66A**:397–402.

13. Rubin CT, McLeod KJ. 1994. Promotion of bony ingrowth by frequency-specific, low-amplitude mechanical strain. *Clin Orthop Relat Res* (298):165–174.

14. Qin YX, Rubin CT, McLeod KJ. 1998. Nonlinear dependence of loading intensity and cycle number in the maintenance of bone mass and morphology. *J Orthop Res* **16**(4):482–489.

15. Gross TS, Srinivasan S, Liu CC, Clemens TL, Bain SD. 2002. Noninvasive loading of the murine tibia: An in vivo model for the study of mechanotransduction. *J Bone Miner Res* **17**(3):493–501.

16. Srinivasan S, Weimer DA, Agans SC, Bain SD, Gross TS. 2002. Low-magnitude mechanical loading becomes osteogenic when rest is inserted between each load cycle. *J Bone Miner Res* **17**(9):1613–1620.

17. Rubin C, Turner AS, Bain S, Mallinckrodt C, McLeod K. 2001. Anabolism. Low mechanical signals strengthen long bones. *Nature* **412**(6847):603–604.

18. Gross TS, Edwards JL, McLeod KJ, Rubin CT. 1997. Strain gradients correlate with sites of periosteal bone formation. *J Bone Miner Res* **12**(6):982–988.

19. Rubin C, Gross T, Qin YX, Fritton S, Guilak F, McLeod K. 1996. Differentiation of the bone-tissue remodeling response to axial and torsional loading in the turkey ulna. *J Bone Joint Surg Am* **78**(10):1523–1533.

20. Sun YQ, McLeod KJ, Rubin CT. 1995. Mechanically induced periosteal bone formation is paralleled by the upregulation of collagen type one mRNA in osteocytes as measured by in situ reverse transcript-polymerase chain reaction. *Calcif Tissue Int* **57**(6):456–462.

21. Lee K, Jessop H, Suswillo R, Zaman G, Lanyon L. 2003. Endocrinology: Bone adaptation requires oestrogen receptor-alpha. *Nature* **424**(6947):389.

22. Judex S, Zhong N, Squire ME, Ye K, Donahue LR, Hadjiargyrou M, Rubin CT. 2005. Mechanical modulation of molecular signals which regulate anabolic and catabolic activity in bone tissue. *J Cell Biochem* **94**(5):982–994.

23. Fan X, Rahnert JA, Murphy TC, Nanes MS, Greenfield EM, Rubin J. 2006. Response to mechanical strain in an immortalized pre-osteoblast cell is dependent on ERK1/2. *J Cell Physiol* **207**(2):454–460.

24. Li YJ, Batra NN, You L, Meier SC, Coe IA, Yellowley CE, Jacobs CR. 2004. Oscillatory fluid flow affects human marrow stromal cell proliferation and differentiation. *J Orthop Res* **22**(6):1283–1289.

25. Rubin J, Murphy T, Nanes MS, Fan X. 2000. Mechanical strain inhibits expression of osteoclast differentiation factor by murine stromal cells. *Am J Physiol Cell Physiol* **278**(6):C1126–C1132.

26. Klein-Nulend J, van der Plas A, Semeins CM, Ajubi NE, Frangos JA, Nijweide PJ, Burger EH. 1995. Sensitivity of osteocytes to biomechanical stress in vitro. *FASEB J* **9**(5):441–445.

27. Brown TD, Pedersen DR, Gray ML, Brand RA, Rubin CT. 1990. Toward an identification of mechanical parameters initiating periosteal remodeling: A combined experimental and analytic approach. *J Biomech* **23**(9):893–905.

28. Rubin CT, Lanyon LE. 1984. Dynamic strain similarity in vertebrates: An alternative to allometric limb bone scaling. *J Theor Biol* **107**(2):321–327.

29. Kaspar D, Seidl W, Neidlinger-Wilke C, Ignatius A, Claes L. 2000. Dynamic cell stretching increases human osteoblast proliferation and CICP synthesis but decreases osteocalcin synthesis and alkaline phosphatase activity. *J Biomech* **33**(1):45–51.

30. Zaman G, Suswillo RF, Cheng MZ, Tavares IA, Lanyon LE. 1997. Early responses to dynamic strain change and prostaglandins in bone-derived cells in culture. *J Bone Miner Res* **12**(5):769–777.

31. Dillaman RM, Roer RD, Gay DM. 1991. Fluid movement in bone: Theoretical and empirical. *J Biomech* **24**(Suppl 1):163–177.

32. Keanini RG, Roer RD, Dillaman RM. 1995. A theoretical model of circulatory interstitial fluid flow and species transport within porous cortical bone. *J Biomech* **28**(8):901–914.

33. Piekarski K, Munro M. 1977. Transport mechanism operating between blood supply and osteocytes in long bones. *Nature* **269**(5623):80–82.
34. Weinbaum S, Cowin SC, Zeng Y. 1994. A model for the excitation of osteocytes by mechanical loading-induced bone fluid shear stresses. *J Biomech* **27**(3):339–360.
35. Smalt R, Mitchell FT, Howard RL, Chambers TJ. 1997. Mechanotransduction in bone cells: Induction of nitric oxide and prostaglandin synthesis by fluid shear stress, but not by mechanical strain. *Adv Exp Med Biol* **433**:311–314.
36. McAllister TN, Frangos JA. 1999. Steady and transient fluid shear stress stimulate NO release in osteoblasts through distinct biochemical pathways. *J Bone Miner Res* **14**(6):930–936.
37. Jacobs CR, Yellowley CE, Davis BR, Zhou Z, Cimbala JM, Donahue HJ. 1998. Differential effect of steady versus oscillating flow on bone cells. *J Biomech* **31**(11):969–976.
38. Ponik SM, Triplett JW, Pavalko FM. 2007. Osteoblasts and osteocytes respond differently to oscillatory and unidirectional fluid flow profiles. *J Cell Biochem* **100**(3):794–807.
39. You J, Yellowley CE, Donahue HJ, Zhang Y, Chen Q, Jacobs CR. 2000. Substrate deformation levels associated with routine physical activity are less stimulatory to bone cells relative to loading-induced oscillatory fluid flow. *J Biomech Eng* **122**(4):387–393.
40. Smalt R, Mitchell F, Howard R, Chambers T. 1997. Induction of NO and prostaglandin E2 in osteoblasts by wall-shear stress but not mechanical strain. *Am J Phys* **273**:E751–E758.
41. Bain SD, Rubin CT. 1990. Metabolic modulation of disuse osteopenia: Endocrine-dependent site specificity of bone remodeling. *J Bone Miner Res* **5**(10):1069–1075.
42. Leblanc AD, Schneider VS, Evans HJ, Engelbretson DA, Krebs JM. 1990. Bone mineral loss and recovery after 17 weeks of bed rest. *J Bone Miner Res* **5**(8):843–850.
43. LeBlanc A, Schneider V, Shackelford L et al. 1996. Bone mineral and lean tissue loss after long duration spaceflight. *J Bone Miner Res* **11**(S1):567.
44. Rubin J, McLeod KJ, Titus L, Nanes MS, Catherwood BD, Rubin CT. 1996. Formation of osteoclast-like cells is suppressed by low frequency, low intensity electric fields. *J Orthop Res* **14**(1):7–15.
45. Rubin J, Biskobing D, Fan X, Rubin C, McLeod K, Taylor WR. 1997. Pressure regulates osteoclast formation and MCSF expression in marrow culture. *J Cell Physiol* **170**(1):81–87.
46. Rubin J, Fan X, Biskobing D, Taylor W, Rubin C. 1999. Osteoclastogenesis is repressed by mechanical strain in an in vitro model. *J Orthop Res* **17**(5):639–645.
47. Lacey DL, Timms E, Tan HL et al. 1998. Osteoprotegerin ligand is a cytokine that regulates osteoclast differentiation and activation. *Cell* **93**(2):165–176.
48. Rubin CT, McLeod KJ, Lanyon LE. 1989. Prevention of osteoporosis by pulsed electromagnetic fields. *J Bone Joint Surg Am* **71**(3):411–417.
49. Qin YX, Kaplan T, Saldanha A, Rubin C. 2003. Fluid pressure gradients, arising from oscillations in intramedullary pressure, is correlated with the formation of bone and inhibition of intracortical porosity. *J Biomech* **36**(10):1427–1437.
50. Sherr CJ. 1990. Colony-stimulating factor-1 receptor. *Blood* **75**(1):1–12.
51. Fan X, Fan D, Gewant H, Royce C, Nanes M, Rubin J. 2001. Increasing membrane-bound MCSF does not enhance OPGL-driven osteoclastogenesis from marrow cells. *Am J Physiol Endocrinol Metab* **280**:E103–E111.
52. Perkins S, Kling S. 1995. Local concentrations of MCSF mediate osteoclastic differentiation. *Am J Physiol* **269**:E1024–E1030.
53. Yasuda H, Shima N, Nakagawa N et al. 1998. Osteoclast differentiation factor is a ligand for osteoprotegerin/osteoclastogenesis-inhibitory factor and is identical to TRANCE/RANKL. *Proc Natl Acad Sci U S A* **95**(7):3597–3602.
54. Rubin J, Murphy T, Fan X, Goldschmidt M, Taylor W. 2002. Activation of extracellular signal-regulated kinase is involved in mechanical strain inhibition of RANKL expression in bone stromal cells. *J Bone Miner Res* **17**:1452–1460.
55. Rubin J, Murphy TC, Zhu L, Roy E, Nanes MS, Fan X. 2003. Mechanical strain differentially regulates endothelial nitric-oxide synthase and receptor activator of nuclear kappa B ligand expression via ERK1/2 MAPK. *J Biol Chem* **278**(36):34018–34025.
56. Tang L, Lin Z, Li YM. 2006. Effects of different magnitudes of mechanical strain on osteoblasts in vitro. *Biochem Biophys Res Commun* **344**(1):122–128.
57. Kim CH, You L, Yellowley CE, Jacobs CR. 2006. Oscillatory fluid flow-induced shear stress decreases osteoclastogenesis through RANKL and OPG signaling. *Bone* **39**(5):1043–1047.
58. Rucci N, Rufo A, Alamanou M, Teti A. 2007. Modeled microgravity stimulates osteoclastogenesis and bone resorption by increasing osteoblast RANKL/OPG ratio. *J Cell Biochem* **100**(2):464–473.

59. Kusumi A, Sakaki H, Kusumi T, Oda M, Narita K, Nakagawa H, Kubota K, Satoh H, Kimura H. 2005. Regulation of synthesis of osteoprotegerin and soluble receptor activator of nuclear factor-kappaB ligand in normal human osteoblasts via the p38 mitogen-activated protein kinase pathway by the application of cyclic tensile strain. *J Bone Miner Metab* **23**(5):373–381.

60. Saunders MM, Taylor AF, Du C, Zhou Z, Pellegrini VD, Jr., Donahue HJ. 2006. Mechanical stimulation effects on functional end effectors in osteoblastic MG-63 cells. *J Biomech* **39**(8):1419–1427.

61. Ralston SH. 1997. The Michael Mason prize essay. 1997. Nitric oxide and bone: What a gas! *Br J Rheumatol* **36**(8):831–838.

62. Kasten TP, Collin-Osdoby P, Patel N, Osdoby P, Krukowski M, Misko TP, Settle SL, Currie MG, Nickols GA. 1994. Potentiation of osteoclast bone-resorption activity by inhibition of nitric oxide synthase. *Proc Natl Acad Sci U S A* **91**(9):3569–3573.

63. Jamal SA, Browner WS, Bauer DC, Cummings SR. 1998. Intermittent use of nitrates increases bone mineral density: The study of osteoporotic fractures. *J Bone Miner Res* **13**(11):1755–1759.

64. Fan X, Roy E, Zhu L, Murphy T, Ackert-Bicknell C, Hart C, Rosen C, Nanes M, Rubin J. 2004. Nitric oxide regulates receptor activator of nuclear factor-kappaB ligand and osteoprotegerin expression in bone marrow stromal cells. *Endocrinology* **145**(2):1–9.

65. Klein-Nulend J, Helfrich MH, Sterck JG, MacPherson H, Joldersma M, Ralston SH, Semeins CM, Burger EH. 1998. Nitric oxide response to shear stress by human bone cell cultures is endothelial nitric oxide synthase dependent. *Biochem Biophys Res Commun* **250**(1):108–114.

66. Pitsillides AA, Rawlinson SC, Suswillo RF, Bourrin S, Zaman G, Lanyon LE. 1995. Mechanical strain-induced NO production by bone cells: A possible role in adaptive bone (re)modeling? *FASEB J* **9**(15):1614–1622.

67. Zaman G, Pitsillides AA, Rawlinson SC, Suswillo RF, Mosley JR, Cheng MZ, Platts LA, Hukkanen M, Polak JM, Lanyon LE. 1999. Mechanical strain stimulates nitric oxide production by rapid activation of endothelial nitric oxide synthase in osteocytes. *J Bone Miner Res* **14**(7):1123–1131.

68. MacPherson H, Noble BS, Ralston SH. 1999. Expression and functional role of nitric oxide synthase isoforms in human osteoblast-like cells. *Bone* **24**(3):179–185.

69. Rahnert J, Fan X, Case N, Murphy TC, Grassi F, Sen B, Rubin J. 2008. The role of nitric oxide in the mechanical repression of RANKL in bone stromal cells. *Bone* **43**(1):48–54.

70. Schwachtgen J, Houston P, Campbell C, Sukhatme V, Braddock M. 1998. Fluid shear stress activation of egr-1 transcription in cultured human endothelial and epithelial cells is mediated via the extracellular signal-related kinase 1/2 mitogen-activated protein kinase pathway. *J Clin Invest* **101**(11):2540–2549.

71. Chen CS, Ingber DE. 1999. Tensegrity and mechanoregulation: From skeleton to cytoskeleton. *Osteoarth Cartil* **7**(1):81–94.

72. Yan C, Takahashi M, Okuda M, Lee JD, Berk BC. 1999. Fluid shear stress stimulates big mitogen-activated protein kinase 1 (BMK1) activity in endothelial cells. Dependence on tyrosine kinases and intracellular calcium. *J Biol Chem* **274**(1):143–150.

73. Li C, Hu Y, Mayr M, Xy Q. 1999. Cyclic strain stress-induced MAP kinase phosphatase 1 expression in vascular smooth muscle cells is regulated by Ras/Rac-MAPK pathways. *J Biol Chem* **274**(36):25273–25380.

74. Rubin C, Rubin J. 1999. Biomechanics of bone. In: Favus M (ed.) *Primer on the Metabolic Bone Diseases and Disorders of Mineral Metabolism*, Vol. Chapter 5. Philadelphia, PA: Lippincott Williams & Wilkins, pp 39–44.

75. Cowin SC, Weinbaum S. 1998. Strain amplification in the bone mechanosensory system. *Am J Med Sci* **316**(3):184–188.

76. Jessop HL, Rawlinson SC, Pitsillides AA, Lanyon LE. 2002. Mechanical strain and fluid movement both activate extracellular regulated kinase (ERK) in osteoblast-like cells but via different signaling pathways. *Bone* **31**(1):186–194.

77. Weyts FA, Li YS, van Leeuwen J, Weinans H, Chien S. 2002. ERK activation and alpha v beta 3 integrin signaling through Shc recruitment in response to mechanical stimulation in human osteoblasts. *J Cell Biochem* **87**(1):85–92.

78. Yang CM, Chien CS, Yao CC, Hsiao LD, Huang YC, Wu CB. 2004. Mechanical strain induces collagenase-3 (MMP-13) expression in MC3T3-E1 osteoblastic cells. *J Biol Chem* **279**(21):22158–255165.

79. Lu X, Gilbert L, He X, Rubin J, Nanes MS. 2006. Transcriptional regulation of the osterix (Osx, Sp7) promoter by tumor necrosis factor identifies disparate effects of mitogen-activated protein kinase and NF kappa B pathways. *J Biol Chem* **281**(10):6297–6306.

80. Schindeler A, Little DG. 2006. Ras-MAPK signaling in osteogenic differentiation: Friend or foe? *J Bone Miner Res* **21**(9):1331–1338.

81. You J, Reilly GC, Zhen X, Yellowley CE, Chen Q, Donahue HJ, Jacobs CR. 2001. Osteopontin gene regulation by oscillatory fluid flow via intracellular calcium mobilization and activation of mitogen-activated protein kinase in MC3T3-E1 osteoblasts. *J Biol Chem* **276**(16):13365–13371.

82. Wu CC, Li YS, Haga JH, Wang N, Lian IY, Su FC, Usami S, Chien S. 2006. Roles of MAP kinases in the regulation of bone matrix gene expressions in human osteoblasts by oscillatory fluid flow. *J Cell Biochem* **98**(3):632–641.

83. Boutahar N, Guignandon A, Vico L, Lafage-Proust MH. 2004. Mechanical strain on osteoblasts activates autophosphorylation of focal adhesion kinase and proline-rich tyrosine kinase 2 tyrosine sites involved in ERK activation. *J Biol Chem* **279**(29):30588–30599.

84. Jiang GL, White CR, Stevens HY, Frangos JA. 2002. Temporal gradients in shear stimulate osteo-blastic proliferation via ERK1/2 and retinoblastoma protein. *Am J Physiol Endocrinol Metab* **283**(2):E383–E389.

85. Rubin J, Murphy TC, Rahnert J, Song H, Nanes MS, Greenfield EM, Jo H, Fan X. 2006. Mechanical inhibition of RANKL expression is regulated by H-Ras-GTPase. *J Biol Chem* **281**(3):1412–1418.

86. Ehrhardt A, Ehrhardt GR, Guo X, Schrader JW. 2002. Ras and relatives—Job sharing and networking keep an old family together. *Exp Hematol* **30**(10):1089–1106.

87. Kim S, Yamazaki M, Zella LA, Shevde NK, Pike JW. 2006. Activation of receptor activator of NF-{kappa} B ligand gene expression by 1,25-dihydroxyvitamin D3 is mediated through multiple long-range enhanc-ers. *Mol Cell Biol* **26**(17):6469–6486.

88. Fan X, Roy EM, Murphy TC, Nanes MS, Kim S, Pike JW, Rubin J. 2004. Regulation of RANKL promoter activity is associated with histone remodeling in murine bone stromal cells. *J Cell Biochem* **93**(4):807–818.

89. Spencer GJ, Utting JC, Etheridge SL, Arnett TR, Genever PG. 2006. Wnt signalling in osteoblasts regu-lates expression of the receptor activator of NFkappaB ligand and inhibits osteoclastogenesis in vitro. *J Cell Sci* **119**(Pt 7):1283–1296.

90. Skerry TM. 2008. The response of bone to mechanical loading and disuse: Fundamental principles and influences on osteoblast/osteocyte homeostasis. *Arch Biochem Biophys* **473**(2):117–123.

91. Rubin J, Rubin C, Jacobs CR. 2006. Molecular pathways mediating mechanical signaling in bone. *Gene* **367**:1–16.

92. Boppart MD, Kimmel DB, Yee JA, Cullen DM. 1998. Time course of osteoblast appearance after in vivo mechanical loading. *Bone* **23**(5):409–415.

93. Rubin C, Recker R, Cullen D, Ryaby J, McCabe J, McLeod K. 2004. Prevention of postmenopausal bone loss by a low-magnitude, high-frequency mechanical stimuli: A clinical trial assessing compliance, efficacy, and safety. *J Bone Miner Res* **19**(3):343–351.

94. Uebelhart D, Demiaux-Domenech B, Roth M, Chantraine A. 1995. Bone metabolism in spinal cord injured individuals and in others who have prolonged immobilisation. A review. *Paraplegia* **33**(11):669–673.

95. Carmeliet G, Vico L, Bouillon R. 2001. Space flight: A challenge for normal bone homeostasis. *Crit Rev Eukaryot Gene Expr* **11**(1–3):131–144.

96. Lau KH, Kapur S, Kesavan C, Baylink DJ. 2006. Up-regulation of the Wnt, estrogen receptor, insulin-like growth factor-I, and bone morphogenetic protein pathways in C57BL/6J osteoblasts as opposed to C3H/HeJ osteoblasts in part contributes to the differential anabolic response to fluid shear. *J Biol Chem* **281**(14):9576–9588.

97. Zhuang H, Wang W, Tahernia AD, Levitz CL, Luchetti WT, Brighton CT. 1996. Mechanical strain-induced proliferation of osteoblastic cells parallels increased TGF-beta 1 mRNA. *Biochem Biophys Res Commun* **229**(2):449–453.

98. Harter L, Hruska K, Duncan R. 1995. Human osteoblast-like cells respond to mechanical strain with increased bone matrix protein production independent of hormonal regulation. *Endocrinology* **136**:528–535.

99. Inoue D, Kido S, Matsumoto T. 2004. Transcriptional induction of fosB/delta fosB gene by mechanical stress in osteoblasts. *J Biol Chem* **279**:49795–49803.

100. Fox SW, Chambers TJ, Chow JW. 1996. Nitric oxide is an early mediator of the increase in bone forma-tion by mechanical stimulation. *Am J Physiol* **270**(6 Pt 1):E955–E960.

101. Turner CH, Takano Y, Owan I, Murrell GA. 1996. Nitric oxide inhibitor L-NAME suppresses mechani-cally induced bone formation in rats. *Am J Physiol* **270**(4 Pt 1):E634–E639.

102. Rawlinson SC, el-Haj AJ, Minter SL, Tavares IA, Bennett A, Lanyon LE. 1991. Loading-related increases in prostaglandin production in cores of adult canine cancellous bone in vitro: A role for prostacyclin in adaptive bone remodeling? *J Bone Miner Res* **6**(12):1345–1351.

103. Ozawa H, Imamura K, Abe E, Takahashi N, Hiraide T, Shibasaki Y, Fukuhara T, Suda T. 1990. Effect of a continuously applied compressive pressure on mouse osteoblast-like cells (MC3T3-E1) in vitro. *J Cell Physiol* **142**(1):177–185.
104. Klein-Nulend J, Burger EH, Semeins CM, Raisz LG, Pilbeam CC. 1997. Pulsating fluid flow stimulates prostaglandin release and inducible prostaglandin G/H synthase mRNA expression in primary mouse bone cells. *J Bone Miner Res* **12**(1):45–51.
105. Pavalko FM, Chen NX, Turner CH, Burr DB, Atkinson S, Hsieh YF, Qiu J, Duncan RL. 1998. Fluid shear-induced mechanical signaling in MC3T3-E1 osteoblasts requires cytoskeleton-integrin interactions. *Am J Physiol* **275**(6 Pt 1):C1591–C1601.
106. Wadhwa S, Godwin SL, Peterson DR, Epstein MA, Raisz LG, Pilbeam CC. 2002. Fluid flow induction of cyclo-oxygenase 2 gene expression in osteoblasts is dependent on an extracellular signal-regulated kinase signaling pathway. *J Bone Miner Res* **17**(2):266–274.
107. Case N, Ma M, Sen B, Xie Z, Gross TS, Rubin J. 2008. Beta-catenin levels influence rapid mechanical responses in osteoblasts. *J Biol Chem* **283**(43):29196–29205.
108. Robinson JA, Chatterjee-Kishore M, Yaworsky PJ et al. 2006. WNT/beta-catenin signaling is a normal physiological response to mechanical loading in bone. *J Biol Chem* **281**(42):31720–31728.
109. Klein-Nulend J, Roelofsen J, Semeins CM, Bronckers AL, Burger EH. 1997. Mechanical stimulation of osteopontin mRNA expression and synthesis in bone cell cultures. *J Cell Physiol* **170**(2):174–181.
110. Walker LM, Publicover SJ, Preston MR, Said Ahmed MA, El Haj AJ. 2000. Calcium-channel activation and matrix protein upregulation in bone cells in response to mechanical strain. *J Cell Biochem* **79**(4):648–661.
111. Liu X, Zhang X, Luo ZP. 2005. Strain-related collagen gene expression in human osteoblast-like cells. *Cell Tissue Res* **322**(2):331–334.
112. Peake MA, Cooling LM, Magnay JL, Thomas PB, El Haj AJ. 2000. Selected contribution: Regulatory pathways involved in mechanical induction of c-fos gene expression in bone cells. *J Appl Physiol* **89**(6):2498–2507.
113. Granet C, Vico AG, Alexandre C, Lafage-Proust MH. 2002. MAP and Src kinases control the induction of AP-1 members in response to changes in mechanical environment in osteoblastic cells. *Cell Signal* **14**(8):679–688.
114. Koike M, Shimokawa H, Kanno Z, Ohya K, Soma K. 2005. Effects of mechanical strain on proliferation and differentiation of bone marrow stromal cell line ST2. *J Bone Miner Metab* **23**(3):219–225.
115. Kanno T, Takahashi T, Tsujisawa T, Ariyoshi W, Nishihara T. 2007. Mechanical stress-mediated Runx2 activation is dependent on Ras/ERK1/2 MAPK signaling in osteoblasts. *J Cell Biochem* **101**(5):1266–1277.
116. Salingcarnboriboon R, Tsuji K, Komori T, Nakashima K, Ezura Y, Noda M. 2006. Runx2 is a target of mechanical unloading to alter osteoblastic activity and bone formation in vivo. *Endocrinology* **147**(5):2296–2305.
117. Kapur S, Baylink DJ, William Lau KH. 2003. Fluid flow shear stress stimulates human osteoblast proliferation and differentiation through multiple interacting and competing signal transduction pathways. *Bone* **32**(3):241–51.
118. Rangaswami H, Marathe N, Zhuang S, Chen Y, Yeh JC, Frangos JA, Boss GR, Pilz RB. 2009. Type II cGMP-dependent protein kinase mediates osteoblast mechanotransduction. *J Biol Chem* **284**(22):14796–14808.
119. Hung CT, Pollack SR, Reilly TM, Brighton CT. 1995. Real-time calcium response of cultured bone cells to fluid flow. *Clin Orthop Relat Res* (313):256–269.
120. Danciu TE, Adam RM, Naruse K, Freeman MR, Hauschka PV. 2003. Calcium regulates the PI3K-Akt pathway in stretched osteoblasts. *FEBS Lett* **536**(1–3):193–197.
121. Hung CT, Allen FD, Pollack SR, Brighton CT. 1996. Intracellular Ca^{2+} stores and extracellular Ca^{2+} are required in the real-time Ca^{2+} response of bone cells experiencing fluid flow. *J Biomech* **29**(11):1411–1417.
122. Chen NX, Ryder KD, Pavalko FM, Turner CH, Burr DB, Qiu J, Duncan RL. 2000. Ca(2+) regulates fluid shear-induced cytoskeletal reorganization and gene expression in osteoblasts. *Am J Physiol Cell Physiol* **278**(5):C989–C997.
123. Li J, Duncan RL, Burr DB, Turner CH. 2002. L-type calcium channels mediate mechanically induced bone formation in vivo. *J Bone Miner Res* **17**(10):1795–1800.
124. Wozniak M, Fausto A, Carron CP, Meyer DM, Hruska KA. 2000. Mechanically strained cells of the osteoblast lineage organize their extracellular matrix through unique sites of alphavbeta3-integrin expression. *J Bone Miner Res* **15**(9):1731–1745.
125. Anderson CT, Castillo AB, Brugmann SA, Helms JA, Jacobs CR, Stearns T. 2008. Primary cilia: Cellular sensors for the skeleton. *Anat Rec (Hoboken)* **291**(9):1074–1078.

126. Krishnan V, Bryant HU, Macdougald OA. 2006. Regulation of bone mass by Wnt signaling. *J Clin Invest* **116**(5):1202–1209.

127. Little RD, Carulli JP, Del Mastro RG et al. 2002. A mutation in the LDL receptor-related protein 5 gene results in the autosomal dominant high-bone-mass trait. *Am J Hum Genet* **70**(1):11–19.

128. Gong Y, Slee RB, Fukai N et al. 2001. LDL receptor-related protein 5 (LRP5) affects bone accrual and eye development. *Cell* **107**(4):513–523.

129. Babij P, Zhao W, Small C et al. 2003. High bone mass in mice expressing a mutant LRP5 gene. *J Bone Miner Res* **18**(6):960–974.

130. Westendorf JJ, Kahler RA, Schroeder TM. 2004. Wnt signaling in osteoblasts and bone diseases. *Gene* **341**:19–39.

131. Rodda SJ, McMahon AP. 2006. Distinct roles for Hedgehog and canonical Wnt signaling in specification, differentiation and maintenance of osteoblast progenitors. *Development* **133**(16):3231–3244.

132. Holmen SL, Zylstra CR, Mukherjee A, Sigler RE, Faugere MC, Bouxsein ML, Deng L, Clemens TL, Williams BO. 2005. Essential role of beta-catenin in postnatal bone acquisition. *J Biol Chem* **280**(22):21162–21168.

133. Armstrong VJ, Muzylak M, Sunters A, Zaman G, Saxon LK, Price JS, Lanyon LE. 2007. Wnt/beta-catenin signaling is a component of osteoblastic bone cells' early responses to load-bearing, and requires estrogen receptor alpha. *J Biol Chem* **282**(28):20715–20727.

134. Norvell SM, Alvarez M, Bidwell JP, Pavalko FM. 2004. Fluid shear stress induces beta-catenin signaling in osteoblasts. *Calcif Tissue Int* **75**(5):396–404.

135. Sawakami K, Robling AG, Ai M et al. 2006. The Wnt co-receptor LRP5 is essential for skeletal mechanotransduction but not for the anabolic bone response to parathyroid hormone treatment. *J Biol Chem* **281**(33):23698–711.

136. van Noort M, Meeldijk J, van der Zee R, Destree O, Clevers H. 2002. Wnt signaling controls the phosphorylation status of beta-catenin. *J Biol Chem* **277**(20):17901–17905.

137. Kim SJ, Im DS, Kim SH, Ryu JH, Hwang SG, Seong JK, Chun CH, Chun JS. 2002. Beta-catenin regulates expression of cyclooxygenase-2 in articular chondrocytes. *Biochem Biophys Res Commun* **296**(1):221–226.

138. Sedding DG, Hermsen J, Seay U, Eickelberg O, Kummer W, Schwencke C, Strasser RH, Tillmanns H, Braun-Dullaeus RC. 2005. Caveolin-1 facilitates mechanosensitive protein kinase B (Akt) signaling in vitro and in vivo. *Circ Res* **96**(6):635–642.

139. Beresford JN. 1989. Osteogenic stem cells and the stromal system of bone and marrow. *Clin Orthop Relat Res* **240**:270–280.

140. Pittenger MF, Mackay AM, Beck SC, Jaiswal RK, Douglas R, Mosca JD, Moorman MA, Simonetti DW, Craig S, Marshak DR. 1999. Multilineage potential of adult human mesenchymal stem cells. *Science* **284**(5411):143–147.

141. Di Iorgi N, Rosol M, Mittelman SD, Gilsanz V. 2008. Reciprocal relation between marrow adiposity and the amount of bone in the axial and appendicular skeleton of young adults. *J Clin Endocrinol Metab* **93**(6):2281–2286.

142. Meunier P, Courpron P, Edouard C, Bernard J, Bringuier J, Vignon G. 1973. Physiological senile involution and pathological rarefaction of bone. Quantitative and comparative histological data. *Clin Endocrinol Metab* **2**(2):239–256.

143. Taes YE, Lapauw B, Vanbillemont G, Bogaert V, De Bacquer D, Zmierczak H, Goemaere S, Kaufman JM. 2009. Fat mass is negatively associated with cortical bone size in young healthy male siblings. *J Clin Endocrinol Metab* **94**(7):2325–2331.

144. Yeung DK, Griffith JF, Antonio GE, Lee FK, Woo J, Leung PC. 2005. Osteoporosis is associated with increased marrow fat content and decreased marrow fat unsaturation: A proton MR spectroscopy study. *J Magn Reson Imaging* **22**(2):279–285.

145. Qiu W, Andersen TE, Bollerslev J, Mandrup S, Abdallah BM, Kassem M. 2007. Patients with high bone mass phenotype exhibit enhanced osteoblast differentiation and inhibition of adipogenesis of human mesenchymal stem cells. *J Bone Miner Res* **22**(11):1720–1731.

146. Minaire P, Edouard C, Arlot M, Meunier PJ. 1984. Marrow changes in paraplegic patients. *Calcif Tissue Int* **36**(3):338–340.

147. Wronski TJ, Morey-Holton E, Jee WS. 1981. Skeletal alterations in rats during space flight. *Adv Space Res* **1**(14):135–140.

148. Luu YK, Capilla E, Rosen CJ, Gilsanz V, Pessin JE, Judex S, Rubin CT. 2009. Mechanical stimulation of mesenchymal stem cell proliferation and differentiation promotes osteogenesis while preventing dietary-induced obesity. *J Bone Miner Res* **24**(1):50–61.

149. Menuki K, Mori T, Sakai A, Sakuma M, Okimoto N, Shimizu Y, Kunugita N, Nakamura T. 2008. Climbing exercise enhances osteoblast differentiation and inhibits adipogenic differentiation with high expression of PTH/PTHrP receptor in bone marrow cells. *Bone* **43**(3):613–620.

150. David V, Martin A, Lafage-Proust MH, Malaval L, Peyroche S, Jones DB, Vico L, Guignandon A. 2007. Mechanical loading down-regulates peroxisome proliferator-activated receptor gamma in bone marrow stromal cells and favors osteoblastogenesis at the expense of adipogenesis. *Endocrinology* **148**(5):2553–2562.

151. Riddle RC, Taylor AF, Genetos DC, Donahue HJ. 2006. MAP kinase and calcium signaling mediate fluid flow-induced human mesenchymal stem cell proliferation. *Am J Physiol Cell Physiol* **290**(3):C776–C784.

152. Song G, Ju Y, Soyama H, Ohashi T, Sato M. 2007. Regulation of cyclic longitudinal mechanical stretch on proliferation of human bone marrow mesenchymal stem cells. *Mol Cell Biomech* **4**(4):201–210.

153. Holtorf HL, Jansen JA, Mikos AG. 2005. Flow perfusion culture induces the osteoblastic differentiation of marrow stroma cell-scaffold constructs in the absence of dexamethasone. *J Biomed Mater Res A* **72**(3):326–334.

154. Ward DF Jr., Salasznyk RM, Klees RF, Backiel J, Agius P, Bennett K, Boskey A, Plopper GE. 2007. Mechanical strain enhances extracellular matrix-induced gene focusing and promotes osteogenic differentiation of human mesenchymal stem cells through an extracellular-related kinase-dependent pathway. *Stem Cells Dev* **16**(3):467–480.

155. Friedl G, Schmidt H, Rehak I, Kostner G, Schauenstein K, Windhager R. 2007. Undifferentiated human mesenchymal stem cells (hMSCs) are highly sensitive to mechanical strain: Transcriptionally controlled early osteo-chondrogenic response in vitro. *Osteoarth Cartil* **15**(11):1293–1300.

156. Haasper C, Jagodzinski M, Drescher M, Meller R, Wehmeier M, Krettek C, Hesse E. 2008. Cyclic strain induces FosB and initiates osteogenic differentiation of mesenchymal cells. *Exp Toxicol Pathol* **59**(6):355–363.

157. Basso N, Bellows CG, Heersche JN. 2005. Effect of simulated weightlessness on osteoprogenitor cell number and proliferation in young and adult rats. *Bone* **36**(1):173–183.

158. Ahdjoudj S, Lasmoles F, Holy X, Zerath E, Marie PJ. 2002. Transforming growth factor beta2 inhibits adipocyte differentiation induced by skeletal unloading in rat bone marrow stroma. *J Bone Miner Res* **17**(4):668–677.

159. Pan Z, Yang J, Guo C, Shi D, Shen D, Zheng Q, Chen R, Xu Y, Xi Y, Wang J. 2008. Effects of hindlimb unloading on ex vivo growth and osteogenic/adipogenic potentials of bone marrow-derived mesenchymal stem cells in rats. *Stem Cells Dev* **17**(4):795–804.

160. Zayzafoon M, Gathings WE, McDonald JM. 2004. Modeled microgravity inhibits osteogenic differentiation of human mesenchymal stem cells and increases adipogenesis. *Endocrinology* **145**(5):2421–2432.

161. Kang S, Bennett CN, Gerin I, Rapp LA, Hankenson KD, Macdougald OA. 2007. Wnt signaling stimulates osteoblastogenesis of mesenchymal precursors by suppressing CCAAT/enhancer-binding protein alpha and peroxisome proliferator-activated receptor gamma. *J Biol Chem* **282**(19):14515–14524.

162. Akune T, Ohba S, Kamekura S et al. 2004. PPARgamma insufficiency enhances osteogenesis through osteoblast formation from bone marrow progenitors. *J Clin Invest* **113**(6):846–855.

163. Liu J, Wang H, Zuo Y, Farmer SR. 2006. Functional interaction between peroxisome proliferator-activated receptor gamma and beta-catenin. *Mol Cell Biol* **26**(15):5827–5837.

164. Liu J, Farmer SR. 2004. Regulating the balance between peroxisome proliferator-activated receptor gamma and beta-catenin signaling during adipogenesis. A glycogen synthase kinase 3beta phosphorylation-defective mutant of beta-catenin inhibits expression of a subset of adipogenic genes. *J Biol Chem* **279**(43):45020–45027.

165. Bennett CN, Ross SE, Longo KA, Bajnok L, Hemati N, Johnson KW, Harrison SD, MacDougald OA. 2002. Regulation of Wnt signaling during adipogenesis. *J Biol Chem* **277**(34):30998–31004.

166. Sen B, Xie Z, Case N, Ma M, Rubin C, Rubin J. 2008. Mechanical strain inhibits adipogenesis in mesenchymal stem cells by stimulating a durable beta-catenin signal. *Endocrinology* **149**(12):6065–6075.

167. Ross SE, Hemati N, Longo KA, Bennett CN, Lucas PC, Erickson RL, MacDougald OA. 2000. Inhibition of adipogenesis by Wnt signaling. *Science* **289**(5481):950–953.

168. Friedlander AL, Genant HK, Sadowsky S, Byl NN, Gluer CC. 1995. A two-year program of aerobics and weight training enhances bone mineral density of young women. *J Bone Miner Res* **10**(4):574–585.

169. Ward K, Alsop C, Caulton J, Rubin C, Adams J, Mughal Z. 2004. Low magnitude mechanical loading is osteogenic in children with disabling conditions. *J Bone Miner Res* **19**(3):360–369.

11 Cartilage Mechanobiology

Hai Yao, Yongren Wu, and Xin L. Lu

CONTENTS

11.1 INTRODUCTION

Until the 1990s, research in cartilage biomechanics mainly focused on tissues and cell properties in order to identify the bioclinical problems linked to the mechanical properties of cells and tissues or to develop cartilage replacements. Advances in molecular biology and new knowledge of cellular biology over the last two decades have given access to a more physiological approach to study the effects of physical forces on cells and tissues. Indeed, all cells and tissues in the body are constantly exposed to physical forces and these can influence the biological behaviors of cells, including gene expression, phenotype, paracrine or autocrine factor secretion, and metabolism. These mechanically induced cellular alterations may constitute major factors affecting the physiological and pathological conditions of the organism. This new approach is known as mechanobiology because it requires the use of cellular/molecular biology methods to identify the various steps or stages by which changes occur in the cells and tissues as a result of the permanently applied mechanical forces. This new avenue of research, crucial for understanding tissue remodeling phenomena or pathological processes like osteoarthritis (OA), or to develop new strategies for tissue engineering/regeneration, requires understanding mechanisms of mechanotransduction through which physical stimuli are transformed into cellular responses.

Cartilage is a soft connective tissue that overlies the articulating bony ends in diarthrodial joints. It provides the joint with essential biomechanical functions such as wear resistance, load bearing, and shock absorption for eight decades or more. Cartilage is composed of a single cell type, chondrocytes, and a highly specialized extracellular matrix (ECM). Physiologically, articular cartilage is an isolated tissue because it is devoid of blood vessels, lymphatic supplies, and nerves [1]. The cells inside the tissue, which are usually termed as chondrocytes, are responsible for the synthesis, assembly, and organization of the ECM to a limited extent [2,3]. Thus, the ECM of articular cartilage is solely maintained by chondrocyte metabolism. The activities of chondrocytes are dramatically regulated and controlled by the physical stimuli surrounding the cells. The relative simple structure of cartilage and the prevalence of cartilage disease make it one of the most understood and engineered tissue, and a model tissue for mechanotransduction research.

The development and maintenance of cartilage structure and mechanical characteristics are tied directly to the effect of mechanical loading on the biology of cartilage cells and ECM. Cartilage mechanobiology research aims to gain sufficient knowledge about how chondrocytes convert physical stimuli into cellular responses, which can ultimately guide the maintenance of healthy cartilage and treatment of related disease. In this chapter, we present a review of cartilage mechanobiology research, including in vitro and in vivo experiments on the role of mechanical loading in cartilage/chondrocyte biology. First, a brief introduction is presented to illustrate the composition and function of articular cartilage. Second, the effects of physical factors on articular cartilage remodeling in both in vivo and in vitro studies are explained. These studies present us with a better understanding of chondrocyte function and its interactions with the surrounding native environment. Finally, a review of mechanotransduction and intracellular signaling pathways in chondrocytes at cellular and molecular level is presented.

11.2 COMPOSITION AND FUNCTION OF ARTICULAR CARTILAGE

Articular cartilage is a layer of low friction, load-bearing, soft tissue that overlies the articulating bony surfaces in diarthrodial joints. Under normal physiological conditions, articular cartilage provides a nearly frictionless surface for the transmission and distribution of joint load, exhibiting little to no wear over decades of use (Figure 11.1A) [4,5]. This remarkable function of cartilage is granted by its unique composition and the microstructure of the fluid-filled ECM or by the multiphasic nature of articular cartilage (Figure 11.1B). In engineering terms, the tissue is a porous viscoelastic material consisting of two principal phases: a fluid phase primarily composed of water with dissolved solutes and mobile ions in it, and a solid phase composed of a densely woven, strong, collagen (mainly type II) fibrillar network enmeshed with proteoglycan (PG) macromolecules [1,6–8]. In biomechanical models used to quantitatively describe the cartilage osmotic swelling, the dissolved electrolytes (Na^+, Ca^{2+}, Cl^-, etc.) within the interstitial free water are often treated as a separate third phase [9–12]. Indeed, each phase of the tissue contributes significantly to its known mechanical and physicochemical properties.

Collagen is the most abundant organic component of cartilage, which comprises about 10%–20% of the wet weight (~65% dry weight). Articular cartilage contains primarily type II collagen, with smaller amounts of other types of collagens such as types VI, IX, and XI. The diameter of collagen fibers can reach up to 300 nm [13,14]. Large collagen fibers are stabilized by covalent intermolecular cross-links, which further confer the fibers with extremely high tensile strength [15]. The density of collagen fibers, fiber diameter, and orientation determine the tensile and shear stiffness of the ECM [16–18]. They also function to restrain the swelling pressure introduced by the PGs and further determine the water amount in the tissue. The collagen in articular cartilage is inhomogeneously distributed, and the structure varies with depth, giving the tissue a three-layered structure (Figure 11.1C).

Another major component of the cartilage solid matrix is the large aggregating PG (5%–10% of wet weight, molecular weight of approximately 200 million daltons) [8]. A PG unit generally

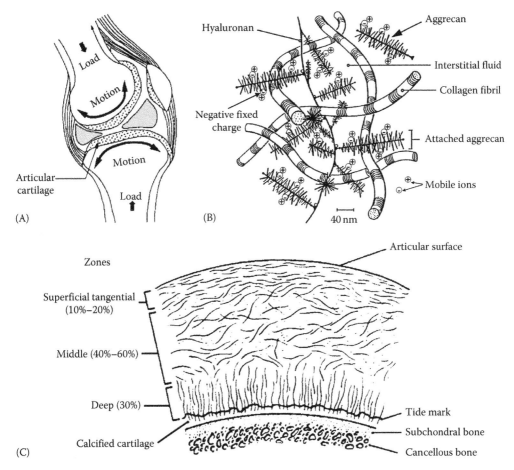

FIGURE 11.1 (A) The structure of a typical diarthrodial joint. Each articulating bone end within the joint is lined with a thin layer of hydrated soft tissue, i.e., the articular cartilage. Diarthrodial joints are subjected to an enormous range of loading conditions, particularly with slow cyclical conditions and high loads. The synovial fluid, articular cartilage, and supporting bone form a "closed" biomechanical system that provides the smooth, nearly frictionless bearing system of the body. (B) A schematic of collagen–PG matrix in cartilage, which consists of three major phases: a charged solid phase largely composed of collagen and PG aggregates, a fluid phase primarily composed of water with dissolved solutes, and an ion phase that is required to neutralize the charges fixed to the solid matrix. (C) Layered structure of cartilage collagen network indicating three distinct regions: superficial, middle, and deep zones. (Adapted from Mow, V.C. et al., *Biomaterials*, 13, 67, 1992. With permission.)

consists of a long protein core to which one or more glycosaminoglycan (GAG) chains are covalently attached [19,20]. The major PG in articular cartilage is aggrecan, and distinct small PGs are also found [21–23]. PGs vary in length, molecular weight, and composition among different types of cartilage. For example, the PG size decreases with age and fluctuates with OA stages. Thus, PG content is often used as a biomarker for OA diagnosis [24,25]. Keratan sulfate and chondroitin sulfate are the two major GAGs found in cartilage. The sulfate and carboxyl groups on these macromolecules become negatively charged in electrolyte solution or in vivo [26,27]. Because these charges are fixed on the GAG chains, they are generally termed as fixed charges. This charged nature of PGs contributes significantly to the mechanical ability of articular cartilage to sustain compressive loadings.

The fixed charge density (FCD) in cartilage ranges from 0.05 to 0.3 mEq/g wet tissue at physiological pH [1,28]. Since each negative charge requires a counter-ion to maintain electrical neutrality, the total ion concentration inside the tissue is higher than that in the external bathing solution. This imbalance of ion concentration introduces a pressure variance between inside and outside of the tissue, which further generates the famous Donnan osmotic pressure in the ECM [28,29]. The osmotic pressure inside the tissue is sustained by the tensile strength of collagen fibers and contributes to the compressive stiffness of articular cartilage [9]. Therefore, even without external mechanical loading the collagen network exists in a constantly stretched state due to the osmotic pressure.

Water is the most abundant component in articular cartilage (68%–85% wet weight) [7,30,31]. A portion of the water (~30%) is contained in the intrafibrillar "pore" space of collagen and is not available for transport under mechanical loading. This part of water is also generally believed to be unavailable for PG solving [30,32]. The remaining portion of the water (~70%) within the tissue may freely move under loading and be forced out of the tissue by compression [6,33]. The volume fraction of free water is modulated by both the osmotic pressure and the tensile stiffness of collagen network [34,35]. Movement of this water through the ECM can generate immense frictional drag between fluid and solid phases, which is the dominant mechanism responsible for the viscoelastic behavior of cartilage [6]. The frictional force causes fluid pressurization in the tissue, and the fluid pressure can support up to 99% of the external compressive loading [36,37]. Fluid exudation during compression also contributes to the fluid film required for lubrication during diarthrodial joint movement [38,39]. Moreover, the flow of the free water is important for the transport of the nutrients and waste products between cartilage solid matrix and the surrounding synovial fluid, which can significantly affect the metabolism of chondrocytes [40].

Chondrocytes are sparsely distributed cells in adult articular cartilage and account for 1%–10% of the tissue volume [41]. Mechanical properties of the chondrocyte itself barely contribute to the overall mechanical behavior of cartilage [42]. However, these cells are responsible for the synthesis and repair of the organic components of cartilage by manufacturing and organizing the collagen and PG network.

In articular cartilage, important molecular interactions exist between the networks formed by the collagens and PG aggregates [18,43]. These interactions, along with forces acting within the tissue (e.g., swelling pressure) and those applied on the tissue (i.e., external loading), can affect chondrocyte metabolism, collagen fibrillogenesis, and collagen network organization [27]. From a material standpoint, the ECM of cartilage is a fiber-reinforced composite solid consisting of a dense stable network of collagen fibers embedded with a high concentration of PG gel. According to the varying structure of collagen along tissue depth, articular cartilage can be separated into three different zones (Figure 11.1C) [44,45]. The orientation of collagen fibers is parallel to the joint surface in the superficial zone [46,47] and perpendicular to the subchondral bone in the deep zone [48,49]. In the three different zones formed by collagen fibers, PG content, water volume fraction, and cell morphology all vary significantly. This inhomogeneous composition underlies the complexity of mechanical behaviors and physicochemical fields inside the physically loaded cartilage.

Various constitutive models have been used to describe articular cartilage. Generalized Hooke's law, the simplest 3D stress–strain law possible, was adopted to model the tissue as a linear isotropic elastic material in early biomechanics research [50,51]. Later, viscoelastic models composed of springs and dashpots were proposed for cartilage to account for its creep and stress relaxation behaviors [52–54]. Today, the most successful theories for cartilage biomechanics are the mixture theories based on poroelasticity, i.e., biphasic and triphasic theories. The biphasic theory models the soft hydrated tissues as composite materials consisting of two continuum and immiscible phases: solid phase and fluid phase [6]. Besides Young's modulus and Poisson's ratio, hydraulic permeability of ECM is the third critical parameter to determine the compressive viscoelastic behavior of the tissue. The relative movement between the two phases and resultant frictional drag were described by Darcy's law. To account for the Donnan osmotic pressure, ion transport, and other electrokinetic-related effects, the same research group further developed the triphasic theory, which separates ions from the fluid phase as a third phase [9,55]. To accommodate these models and determine the

TABLE 11.1
Mechanical Properties of Patellar Groove Cartilage for Various Species

	Human	Bovine	Canine	Monkey	Rabbit
Aggregate modulus (MPa)	0.53	0.47	0.55	0.52	0.51
Poisson's ratio	0.00	0.25	0.09	0.20	0.21
Permeability (10^{-15} m^4/N s)	2.17	1.42	0.93	4.74	3.84

Source: Athanasiou, K.A. et al., *J. Orthop. Res.*, 9, 330, 1991. With permission.

intrinsic mechanical parameters of cartilage, both destructive and nondestructive mechanical testing methods were widely employed in cartilage-related research. The former includes confined compression (used to determine aggregate modulus) [56,57], unconfined compression (Young's modulus and Poisson's ratio) [58,59], and pure shear by a rheometer (shear modulus) [60]. Indentation is the most popular nondestructive testing method for most in vivo and in situ studies and small animal models [50,61,62]. The biphasic and triphasic indentation solutions can simultaneously determine the aggregate modulus, Poisson's ratio, permeability, and FCD of cartilage without removal from the subchondral bone [63–65]. The average mechanical properties for the femoral groove cartilage of normal human, canine, monkey, and rabbit are provided in Table 11.1 [64].

In summary, articular cartilage is a unique, charged, hydrated soft tissue with multiple phases. The ability of articular cartilage to support physical loading depends on its biomechanical properties, which are in turn related to its own biochemical composition and architectural arrangements. Mechanical loadings produce complex physical–chemical environment changes around chondrocytes. For example, oscillatory and intermittent loading on cartilage produce dynamic fluid flow, which further produce convective transport and redistribution of the interstitial fluid phase, nutrients, molecular messengers, and electrolyte ions. Therefore, it is necessary to fully understand the biomechanical behaviors of cartilage in order to quantitatively describe the spatial–temporal characteristics of load-induced mechano-electrochemical signals surrounding the chondrocytes.

11.3 EFFECTS OF PHYSICAL FORCES: TISSUE LEVEL STUDIES

Chondrocytes are the predominant cell type in articular cartilage, which are responsible for the formation, maintenance, and repair of ECM in this avascular tissue. The composition and structure of articular cartilage are controlled by a balance of anabolic and catabolic activities of the chondrocyte population. As most cell types in musculoskeletal system, the metabolism of chondrocytes is controlled not only by biological factors (e.g., growth factors, cytokines, and hormones) but also by physical factors such as mechanical loading. Clinical observations and animal studies have shown that chondrocytes can perceive and respond to a wide range of signals generated by normal physical activities [66]. As stated in the previous section, dynamic compression of cartilage results in physical deformation of chondrocytes, hydrostatic pressure, fluid flow, and streaming potentials and currents surrounding the cells [67,68]. Furthermore, deformation of the ECM also leads to physicochemical environment changes within the tissue for water content, FCD, nutrients and ion concentration, electrical potential, and swelling pressure. All of these mechanical, chemical, and electrical signals may affect the synthesis, assembly, and degradation of PGs, collagens, and other matrix molecules [9,69–71]. This mechanotransduction mechanism allows cartilage to alter its structure and composition to meet the physical demands of the body [72]. To understand the cellular transduction mechanisms that govern chondrocyte responses to mechanical stimuli, numerous efforts, including both in vivo and in vitro studies, have been undertaken over the last two decades, which are briefly reviewed in the following sections. The findings will eventually aid future studies of cell-based repair and cartilage tissue engineering.

11.3.1 In Vivo Cartilage Response to Physical Loading

Chondrocytes have been known to respond to physical stimuli and have limited ability to remodel the ECM of cartilage under the prevailing stress–strain [66]. By using models of disuse and immobilization, overuse, impact loading, and joint instability, recent clinical observations and in vivo animal studies have revealed to what extent this remodeling can occur [73–75]. It has also been proven that a certain amount of physiological loading from daily activities is necessary for the maintenance of the ECM and its proper function. Deviations from normal loading patterns could be a significant source of cartilage tissue degeneration [66,76,77].

Decreased loading or immobilization of a joint can cause profound decreases in PG synthesis and softening of the tissue, although the PG content can be gradually restored once the joint is remobilized [78–80]. In contrast, aggrecan concentration is often higher in areas of habitually loaded cartilage and can be further increased by in vivo dynamic loading of a joint [81,82]. Due to the higher FCD and osmotic pressure in ECM, it was found that load-bearing regions in joints are often thicker and stronger than neighboring less load-bearing surfaces in the same joint [83]. Studies of animals with amputated limbs demonstrate that the loss of PG appears to be more sensitive to diminished mechanical loading rather than to the lack of motion. The cartilage tissue in joints above the amputation, which experiences diminished loading but continues to be subjected to motion, significantly loses its volume by decreasing in thickness [84]. The volume of the chondrocytes in exercised rabbits were found to be larger than those in sedentary control units [78]. Impact loading or strenuous exercise loading, however, on a normal joint can cause both immediate and progressive damage to the cartilage [85]. The focal lesions of OA in the knee and hip joints always occur together with regions of peak stress, suggesting that excessive mechanical forces are involved in OA pathogenesis [86]. Disruption of other functional capsular soft tissue, such as ligaments and menisci, can also induce degenerative changes in joints. It is well known that surgically induced destabilization produces altered patterns of load distribution on the articular surface and further degradation of the cartilage, which serves as a widely adopted animal model for in vivo OA etiology study [87].

The in vivo observations demonstrated that a critical level of joint loading is necessary to generate effective physical signals to the chondrocytes in order to maintain the composition, stiffness, and mechanical function of cartilage ECM. The alteration of joint loading results in altered biological activity of chondrocytes. Thus, the chondrocyte population is important in sustaining the correlation between joint loading and functioning of articular cartilage. While in vivo studies reveal important characteristics of chondrocyte mechanotransduction, they also indicate more unknowns in the chondrocyte–matrix relationship. Further techniques and studies are required to determine the nature of chondrocyte metabolic responses to physical signals, especially, the precise relationships between isolated factors of mechanical loading and specific physical signaling mechanisms.

11.3.2 In Vitro Culture of Explants and Chondrocytes

Since in vivo mechanical loading produces a complex combination of electrical, chemical, and mechanical signals in ECM, it is difficult to identify and quantify which factors are the key mediators for chondrocyte responses. Models such as 3D cell culture and ex vivo explants have been widely used to quantitatively correlate mechanical loading parameters with metabolic activities of chondrocytes, such as gene expression or biosynthesis. Most studies were designed to investigate the net response of cultured structure, almost always with ^{35}S-sulfate or ^{3}H-proline incorporation, under a variety of external loading conditions, with the focus on whether signals and changes in the physical environment would promote or inhibit ECM synthesis. Two main types of mechanical loading patterns employed in in vitro studies are prolonged static and cyclic or intermittent loadings. In addition, alternative mechanical loading protocols, including hydrostatic pressure and electrical fields, have also been used as regulatory effects on chondrocyte metabolism. To apply different physical signals, a variety of incubator-housed loading instruments were developed for mechanobiology and tissue engineering studies using 3D agarose and alginate gel culture systems.

11.3.2.1 Static Compression

Static loading on animal and human cartilage explants is catabolic and results in reduced matrix production and increased protease activities. It causes a dose-dependent decrease in the biosynthesis of PGs, collagen, and other proteins, which can occur in a few hours [88–94]. Static compression also increases the production of various matrix metalloproteinase (MMPs). Studies on bovine cartilage explants showed up-regulated catabolic gene expressions of MMPs-3, -9, -13, aggrecanase-1, and the matrix protease regulator cyclooxygenase-2 [95]. To better understand the mechanisms that may mediate the inhibition of ECM, it is necessary to appreciate the changes in physical, chemical, and mechanical environments induced by static compression. Static mechanical loading initiates an accumulation of interstitial hydrostatic pressure and fluid flow within the cartilage [7,36,58]. This pressure and fluid flow soon diminishes after the ECM reaches an equilibrium state. The persistent changes are essentially related to the loss of water, ECM consolidation, subsequent increase of FCD, and cell volume change [1,9,89,96]. Since transient changes diminish shortly after loading, it is believed that the effects of persistent changes, such as osmotic pressure, dominate over transient changes. The inhibition of ECM biosynthesis under static compression is predominantly associated with ECM consolidation and dramatically changed osmotic effects. Similarly, inhibition has been observed during mechanical compression of cartilage explants and incubation of cartilage plugs in media containing osmolarity active solutes such as polyethylene glycol [96–98]. Static compression also inhibited biosynthesis of cartilage chondrocytes seeded in 3D agarose gel but only after prolonged culture in which a PG-rich ECM had formed [99,100]. These results further confirmed that the biosynthesis decrease under static compression is related to the associated tissue consolidation and variation of osmotic pressure.

11.3.2.2 Dynamic Compression

Dynamic compression on cartilage explants and chondrocyte-seeded agarose constructs can dramatically increase the production of ECM proteins by chondrocytes. The extent of effects is associated with the characteristics of the loading profiles, such as amplitude and frequency, as well as the developmental stage of the explants [88,91,94,99,101–106]. Results indicate that small- and medium-amplitude (<15%) dynamic compression stimulates PG synthesis and increases proliferation of chondrocytes [107–109]. Dynamic loading is shown to up-regulate expression of anabolic genes such as ACAN, COL2A1, and tissue inhibitor of MMP-3 (TIMP3) [95], while down-regulating specific genes of the MMP families [95,110–112]. Cyclic compression on cartilage explants increases the expression of cartilage oligomeric protein (COMP) [113] and influences the orientation and organization of collagen fibers in the superficial zone [114]. More importantly, dynamic compression at low amplitude can be anti-inflammatory [115–117]. However, large-amplitude dynamic compression has been found to be traumatic [104]. Amplitudes of 25%–50% are catabolic and can substantially down-regulate expression of ACAN and COL2A1 and up-regulate MMP-3, -9,-13, and ADAMTS4 expression, which can induce severe ECM destruction [112,118]. The dynamic loading frequency is of great importance for chondrocyte activities. High-frequency (0.01–1 Hz) compression increased the incorporation rate of ^{35}S-sulfate and ^{3}H-proline into cartilage by 20%–40% [99], while low-frequency (~0.001Hz) loading had little effect when coupled with low amplitudes but increased the incorporation of ^{35}S-sulfate and ^{3}H-proline at medium-loading amplitudes [91].

Small-amplitude dynamic oscillations have been extensively employed in the last two decades in identifying the specific roles of each individual physical, chemical, and electrical stimuli on chondrocyte metabolism. This loading profile best mimics the daily in vivo loading situations, for example, walking. While static compression results in dramatic cell volume and chemical environment change, small-amplitude dynamic compression barely disturbs pericellular ion concentrations. Both static and cyclic compression generate fluid flow, streaming potentials, convective transport of ions, and shear stress. These responses are transient in static compression models but continuous in dynamic models. The quantitative description of physical signals in dynamic loading models has not been well characterized due to the complexity of this multiphasic system. For example, under the same magnitude of applied stress, different loading frequencies will result in different magnitudes of tissue strain. Dynamic loadings that are "load controlled" must be interpreted separately from those that are "deformation controlled."

11.3.2.3 Tensile Stretch

Two-dimensional cell culture systems are often used to examine the response of chondrocytes to tensile stretch. Significant increases in aggrecan synthesis were observed after 24 h of tensile stretch on collagen matrix generated by growth plate chondrocytes [119]. Tensile strains of 10% on a supportive elastin membrane seeded with chick sternal chondrocytes resulted in twofold to threefold increases in aggrecan synthesis [120]. Similar with compressive stimuli, chondrocytes perceive tensile stretch in a magnitude-dependent manner. At low and medium magnitudes, dynamic tensile stretch acts as a potent anti-inflammatory signal and inhibits interleukin-1β (IL-1β), tumor necrosis factor-α (TNF-α), and lipopolysaccharide-induced proinflammatory gene transcription. Tensile stretch of high magnitude acts as a traumatic signal and induces production of proinflammatory mediators such as NOS2A, COX2, MMPS, and NO [121]. It should be noted that the relationship between substrate strain and cellular strain may be complex. Even though the strain in the substrate can be precisely controlled in above studies, the forces transformed to the cell and final cell body deformation remain unclear.

11.3.2.4 Hydrostatic Pressure

The effects of hydrostatic pressure on cartilage biosynthesis have been clearly identified because physiological levels of hydrostatic pressure on explants or cell cultures can be easily achieved without inducing confounding physical factors such as fluid flow, FCD variation, or large ECM deformation [122]. Explants of articular cartilage exposed to dynamic hydrostatic pressure at a physiologically relevant level increase the synthesis of PGs [123,124], while excessive loading has the opposite effect and may cause non-physiological changes in the Golgi and cytoskeleton [125]. In cell culture studies, exposing chondrocytes to cyclic hydrostatic pressure at physiological magnitudes increases sulfate incorporation and the expression of ECM proteins and reduces MMP levels [126,127]. Once hydrostatic pressure reaches non-physiological levels, heat shock protein, interleukins, and TNF-β can be induced [128–130]. It is important to note that sustained high pressures are less physiologically relevant under in vivo situation since the porous feature of ECM can quickly dissipate the hydrostatic pressure with fluid flow.

11.3.2.5 Electric Field

Several studies have investigated the role of electric fields on chondrocyte activity and suggest that mechanically induced electrical fields may be a mechanism for the regulation of aggrecan biosynthesis [120,131–133]. Extremely low-level electric current applied to chick sternal chondrocytes induced similar changes in aggrecan synthesis rates as those induced by dynamic stretching of substrate [120]. In bovine cartilage explants, current densities up to $1\,mA/cm^2$ at frequencies of 1–$10\,Hz$ showed little effect [134], whereas higher amplitude current densities of 10–$30\,mA/cm^2$ at frequencies of 10–$1000\,Hz$ stimulated aggrecan synthesis [132]. The observed results in these studies were not related to heating accompanied with application of electric field since the temperature of the medium was not altered and there was no production of characteristic heat shock proteins. Thus, pulsed electromagnetic fields stimulate aggrecan synthesis in cartilage explants and are predicted to induce electric currents of a variety of amplitudes and frequencies within cartilage.

Results from tremendous in vivo and in vitro studies provide significant evidence that changes in physical environment can dramatically modulate the metabolic response of chondrocytes, whether the cell is within native tissues or artificial constructs. Further understanding of the effects of physical forces on chondrocyte activities awaits a more thorough investigation of mechanotransduction at cellular and molecular levels.

11.4 CHONDROCYTE MECHANOBIOLOGY: CELLULAR AND MOLECULAR LEVEL STUDIES

From the evidence presented thus far, it is clear that mechanical loading has a strong influence on the metabolic activity of chondrocytes. However, the precise sequence of events and mechanisms involved in regulating the synthesis and breakdown of ECM components is not well understood. The mechanical, electrical, and chemical environments surrounding the chondrocytes are coupled together and depend on the specific loading configuration to which cartilage or the cultured construct is exposed. Therefore, it is difficult to isolate the effects of each individual stimulus on chondrocyte activity and to determine the intracellular pathway in which the load-induced changes to the matrix are signaled to the cell. In recent years, a large quantity of cartilage biomechanics research has been focused on the physiological consequences of applied physical forces at the cellular and molecular levels. This opens a new avenue for "cartilage mechanobiology" research. This young field strives to better understand the etiology or pathological conditions of OA, which can aid the development of novel therapeutic treatments using tissue engineering concepts such as multifunctional engineered cartilage tissue. The main question in chondrocyte mechanobiology is how the chondrocytes sense physical stimuli and convert them into corresponding biochemical and physiological responses.

As introduced in previous sections, the mechano-electrochemical environment surrounding chondrocytes is complicated for both in vivo or in vitro situations [135]. Thus, it is essential to obtain a precise quantitative description of the physical signals on cells within the ECM in order to understand the biological response of tissue to physical stimuli. Figure 11.2 illustrates how the mechanical forces on the tissue level are coupled with the physical signals at cellular level, and how the tissue remodeling (growth or degeneration) changes the physical signals by altering the ECM mechanical properties [135,136]. As pointed out by Grodzinsky et al. [71], "These physical stimuli and the resulting cellular responses should be studied at the molecular, cellular and tissue levels for fully understanding the feedback between applied macroscopic forces, ECM molecular structure, and the resulting macrocontinuum tissue material properties—a feedback process that is orchestrated by cells in vivo."

11.4.1 MECHANICALLY INDUCED PHYSICOCHEMICAL SIGNALS

Mechanocoupling involves the conversion of applied physical force at tissue level into detectable physical signals at cellular level. Under physiological condition, the chondrocyte population is exposed to changes in a mechanical, physicochemical, and electrical environment of ECM, including spatial–temporal variations of stress, strain, fluid flow, fluid pressure, osmotic pressure, FCD, pH, electrical field, and solute transport within the ECM (Figure 11.2). At the final steady state of static loading condition (no fluid movement), there is a change (usually an increase) in the FCD, the concentration of positive counterions (Na^+, H^+, Ca^{2+}), and the osmotic pressure due to the dilatation of ECM. This change may also inhibit aggrecan synthesis. Dynamic loading conditions, however, may promote the metabolic activities of chondrocytes through temporal variations of hydrostatic pressure, fluid flow, streaming potentials, or oscillations in cell shape. During excessive compression or impact loading, high levels of strain or strain rate can induce ECM disruption, tissue swelling, increased diffusion, and loss of ECM macromolecules through abrupt fluid convection. All of these physical phenomena (cell deformation, osmotic pressure, hydrostatic pressure, fluid flow, shear

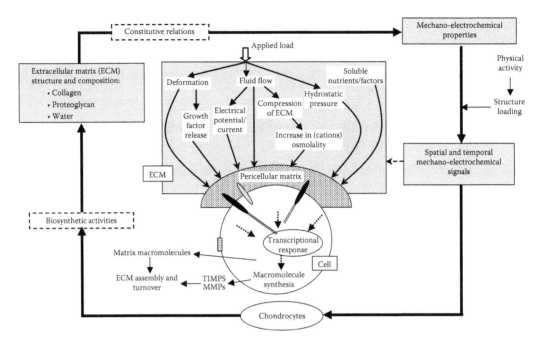

FIGURE 11.2 Illustration of the inter-relationships between ECM composition and structure, mechano-electrochemical signals, and cell biosynthetic activities in cartilage. Mechanical loading induces various extracellular signals (e.g., deformation, pressure, electrical, as well as fluid, solute (e.g., nutrient), and ion flow fields). These signals influence activities of the chondrocytes by acting independently or together. (Adapted from Mow, V.C. et al., *Osteoarth, Carti,* 7, 41, 1999. With permission.)

stress, and streaming potential) induced by mechanical loading at various levels can activate different signaling pathways in chondrocytes and can further result in distinct biochemical responses.

The physicochemical signals across the ECM induced by mechanical loading can be quantitatively determined by using appropriate theoretical models for the mechano-electrochemical behaviors of cartilage. These models provide an essential framework for correlating the spatial–temporal distributions of physical stimuli surrounding cells with external loading at tissue level. The unconfined compression test is a common testing configuration in cartilage biomechanics and cartilage mechanobiology studies [137]. The Young's modulus and Poisson's ratio of cylindrical cartilage tissue can be easily determined from the load-displacement data. The homogeneous strain distribution across the tissue thickness and the fluid convection through the sidewall makes it a perfect setup to investigate biosynthetic responses of cartilage explants to mechanical loading [58,71,91,92,97,99,105]. The mechanical signals within a cartilage explant under unconfined compression have been simulated with the biphasic theory [6,58,138]. The chemical and electrical signals within the cartilage sample were later analyzed using the electromechanical theory or the mechano-electrochemical theory [71,137]. More recently, a specialized model for charged hydrated soft tissue containing uncharged solutes (such as glucose and uncharged growth factor) was developed based on the general mechano-electrochemical mixture theory [9,139]. This unified model can predict the mechanical, chemical, and electrical signals within the tissue under dynamic unconfined compression using a finite element solution [40]. The effects of permeable loading platen, loading profiles, and FCD on the physical signals and transport of water, ions, and uncharged solutes were investigated (Figure 11.3). Such quantitative prediction provides a means to understand the physicochemical environmental changes in cartilage induced by various loading conditions and further facilitates the study on the role of specific physical factors in modulating chondrocyte proliferation, differentiation, and metabolic activities.

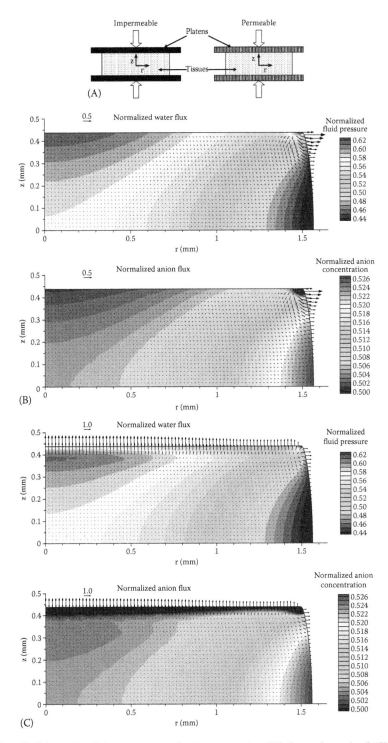

FIGURE 11.3 (A) Schematic of dynamic unconfined compression (5% dynamic strain, 0.1Hz) on cartilage explants with impermeable platens or porous platens. (B) The distribution of physical signals with the impermeable platen. (C) The distribution of physical signals with the permeable platen. Numerical analyses show that a permeable platen will increase the rate of dynamic fluxes of fluid, ion, and uncharged solute in the region near the permeable platen, but not the fluid pressure and electrical potential in the central region of the tissue. (From Yao, H. and Gu W.Y. *Ann. Biomed. Eng.*, 32, 380, 2004. With permission.)

11.4.2 Mechanotransduction in Chondrocytes

Mechanical loading on cartilage results in complex physicochemical environmental changes within the tissue that include ECM and cell deformation, hydrostatic pressure gradients, fluid flow, altered matrix water content, and changes in osmotic pressure, ion concentrations, and FCD [140]. To transfer these signals into corresponding chondrocyte activities and gene expressions, all potential physical signals must be transduced across the plasma membrane into an intracellular biochemical signal. The chondrocyte interacts with its pericellular environment through numerous membrane-bound receptors, adhesion molecules, and ion channels. Although mechanisms of this mechanotransduction process in chondrocytes are not fully understood, it is well known that mechanosensitive ion channels [141] and integrins [142] must be involved in the recognition of physicochemical signals.

11.4.2.1 Ion Channels

Due to the negatively charged nature of cartilage ECM, interstitial fluid and ion fluxes accompanied with applied mechanical loadings are associated with electrokinetic phenomena such as streaming potential and streaming current [9,10,28,137,143–146]. Since the streaming potential is always coupled with fluid flow, there is only indirect evidence supporting its involvement in elevating biosynthetic activity of chondrocytes [91,120]. In addition to flow-mediated electrokinetic effects, several studies proved that low-level electric fields can directly influence chondrocyte biosynthesis activities [120,131,132,147–149]. The candidate mechanisms through which the electric fields interact with chondrocytes include gating of voltage-dependent ion channels, hyper- or hypo-polarization of the membrane [136]. In monolayer cell culture experiments, cyclical loading (0.33 Hz, 20 min) on human articular chondrocytes causes membrane hyperpolarization [150]. The hyperpolarization is a direct result of the activation of slow conductance Ca^{2+}-sensitive K^+ channels (SK). It has also been shown that continuous loading may activate Na^+ ion channel, cause Na^+ influx, and depolarize the cell membrane [151].

The deformation of the plasma membrane can activate the stretch-activated ion channels (SAC) [141,152–154]. By turning on or off these channels, cellular deformation and volume change may directly regulate specific ion transport pathways, which will conceivably affect membrane potential and other second messenger activities [153,155,156]. After blocking SAC with 10 mM gadolinium, it has been shown that mechanical loading can no longer induce chondrocyte membrane hyperpolarization [150]. Results from various cell culture systems also showed that SAC can affect tyrosine phosphorylation of focal adhesion kinase (FAK) and paxillin [157]. Aggrecan and MMP-3 gene expression [126], cell proliferation [158], and chondrocyte phenotype [159] are all correlated with the activities of SAC. Physical signals can directly activate L-type calcium channels together with some other ion channels, though the mechanisms remain unclear. Under dynamic compression, extracellular calcium influx facilitated by L-type calcium channel can affect the protein synthesis [150,160]. The treatment of nifedipine, a calcium channel blocker, inhibits cell proliferation and matrix protein mRNA expression [158]. In response to the early rapid fluctuation of calcium concentration, the activation of multiple cell signaling pathways (e.g., phospholipase C [PLC], calmodulin, tyrosine protein kinase, and protein kinase C [PKC]) leads to later and long-term responses in chondrocytes and is ultimately involved in tissue remodeling.

11.4.2.2 Integrins

Chondrocytes can respond to the deformation of the ECM by sensing the composition and microstructure changes in local tissue through plasma-membrane-binding proteins, which are receptors to specific macromolecules. These receptors are particularly important to mechanotransduction, as studies have shown that chondrocytes cultured in a 3D solid matrix respond differently to mechanical loading after a pericellular matrix is constructed [99,161]. Direct interactions between chondrocytes and its ECM include integrin binding to collagens and fibronectin [162,163], anchorin CII (annexin V) binding to collagens [164,165], and CD44 binding to hyaluronan [166]. Changes in the local concentrations or conformations of these matrix proteins due to the consolidation of ECM can also significantly affect the kinetics of ligand–receptor binding and ultimately influence cell activities through intracellular second messenger and other downstream signaling pathways.

Integrins are receptors regulating the attachment between the ECM and cytoskeleton and also function as mechanoreceptors to perform signal transduction. The receptor has an extracellular ligand-binding site and intracellular tail, which interacts with actin cytoskeleton and intracellular molecules. It plays an important role in mechanotransduction in various cell types [167]. Extracellular mechanical signals could be transduced to an intracellular biochemical response due to the specific location and molecular structure of integrins [141,168]. Integrins consist of 17α and 8β subunits, which can result in atleast 22 distinct receptors for ECM proteins. α5β1 integrin is one of the major mechanoreceptors in chondrocytes through which chondrocytes can attach to fibronectin. It has been shown that α5β1 integrin plays a role in chondrocyte proliferation and adhesion [169]. Under dynamic compression, PG synthesis and chondrocyte proliferation were enhanced by α5β1 through transforming growth factor-β3 (TGF-β3)-dependent pathway [170]. Under mechanical stimulation, interaction between α5β1 integrin and integrins, which associate with protein CD47, causes elevation of aggrecan mRNA, tyrosine phosphorylation, and membrane hyperpolarization [171]. Specific integrin subunit blocking antibodies or RGD peptides blocker can inhibit the flux of K^+ through chondrocyte cell membrane following mechanical stimulation [172], as well as tyrosine phosphorylation of paxillin and FAK and activation of PKC [173]. RGD peptides were also found to be able to decrease cell proliferation, PG production, and nitric oxide release of chondrocytes under dynamic loading [116]. α2β1 integrin, the receptor for type II collagen, also plays a role in chondrocyte mechanotransduction [174]. β1 integrins are correlated with the up-regulation of COMP in chondrocyte, as well as the release of IL-4 induced by pressurization on cells. In bovine cartilage explants and alginate/chondrocyte constructs, gene expression of COMP under cyclical compression was inhibited by blocking β1 integrin [161]. Both α5β1 and α2β1 integrins act through mitogen-activated protein kinase (MAPK) pathway [175]. Nuclear factor-kB (NF-kB) and PLC may also be triggered by integrin activation.

11.4.2.3 Soluble Mediators

Mechanical loading can induce the release of soluble mediators from chondrocytes into extracellular environment that can bind to and activate membrane receptors. More importantly, the interstitial fluid flow generated by mechanical loadings significantly enhances the transport of these messenger molecules, such as growth-factor-binding protein complexes, cytokines, and enzymes, between the entire chondrocyte population. These mediators could bind to corresponding receptors in an autocrine or paracrine manner and initiate downstream activation of MAPK, PKC, and NF-kB pathways. For example, basic fibroblast growth factor (bFGF), which attaches to the heparin sulfate PG perlecan [176], can be released from the ECM and bind to FGF-receptor (FGFR) under mechanical loading [177,178], and can activate MAPK resulting in the increase of MMP and TIMP gene expression [178]. Many other growth factors are involved in mechanotransduction by regulating integrin expression and function [179–181], such as insulin-like growth factor-1 (IGF-1) and TGF-β. α3/α5 integrin expressions can be stimulated by IGF-1 and TGF-β and consequently enhance the adhesion of chondrocytes to ECM [180]. In addition, IGF-1 and TGF-β are able to increase the protein and PG synthesis in chondrocytes without coupling with any mechanical stimulus. Dynamic compression can of course enhance the effect of these growth factors by increasing the transport and cell exposure to these molecules [182].

Although chondrocytes can respond to various cytokines, IL-4 was shown to be critically important in mechanotransduction of normal chondrocytes [183]. This pleiotropic cytokine exerts its biological actions by binding to a heterodimeric receptor complex, the IL-4 receptor (IL-4R), present on the cell surface. There are two types of IL-4 receptors, type I (IL-4Rα and γc subunits) and type II (IL-4Rα and IL-13Rα1 subunits). Type II receptor appears to play a major role in IL-4 receptor signal pathway [184]. IL-4 is secreted as an autocrine signal, which leads to receptor dimerization and induces cell membrane hyperpolarization. It plays an essential role in mechanically stimulated electrophysiological responses and the up-regulation of aggrecan mRNA [126,183]. MMP-13 and cathepsin B inductions by mechanical compression, as well as cyclical tensile loading induced IL-1β, were found to be inhibited by IL-4 activities [185]. Several studies have shown that anabolic cytokines and growth factors enhance mechanically induced production of ECM molecules in cell culture and in tissue-engineered cartilage constructs [186,187]. Thus, mechanical stimuli and soluble mediators may activate similar intracellular signal cascades to induce anabolic or catabolic responses, which together might be antagonistic, additive, or synergistic.

11.4.2.4 Other Potential Mechanisms

The effect of shear and hydrostatic pressure on cartilage biology may involve the activation of distinct pathways that are not yet understood. Up-regulation of catabolic factors by shear stress may involve the MAPK and NF-kB pathways and possibly calcium signaling pathway [188–190]. Similarly, the response of chondrocytes to hydrostatic pressure most likely involves the interaction of many complex pathways. Salter et al. demonstrated the interactions between integrin signaling, SAC, and the autocrine/paracrine release of a soluble factor (IL-4) when chondrocytes were exposed to pressure [126,183]. Recently, Myers et al. suggested that hydrostatic pressure affects cytoskeletal polymerization [191]. The balance between free monomers and cytoskeletal polymers is shifted by alternations in hydrostatic pressure, which could initiate a cellular response by releasing and/or activating cytoskeleton-associated proteins.

11.4.3 Intracellular Pathways and Molecular Mechanisms

Mechanical loading on cartilage or chondrocytes directly leads to an intracellular cell signaling cascade, which results in the production of various molecules involved in chondrocyte viability and the maintenance of the ECM (Figure 11.4). Several second messenger systems, such as cAMP, IP_3, and calcium, are involved in intracellular signaling induced by physical stimuli. Voltage-gated or mechanical-gated membrane ion channels modulate the concentration of intracellular messengers, and the activation of G-protein pathway leads to elevation of cytosolic cAMP or IP_3 amount. Transmembrane signaling may occur by a direct link between ECM molecules and intracellular organelles through integrins. Additionally, enzymes producing reactive oxygen species, such as nitric oxide synthase (NOS), may be directly activated by mechanical stress or indirectly activated through second messengers such as calcium [192,193].

Mechanical loading can cause robust fluctuation of intracellular calcium concentration by activating PLC and IP_3 pathway [172,194] through IL-4 [183]. As one of the earliest intracellular responses in chondrocytes under mechanical stimulation, calcium signaling can initiate or regulate the secretion of growth factors and cytokines. The released autocrine or paracrine factors may bind to transmembrane receptors, such as G-protein coupled receptors (GPCR), to further initiate MAPK, PKC, and NF-kB pathways. Additionally, it can activate calmodulin kinase, resulting in the activation of transcription factor AP-1. It has been shown that aggrecan gene expression and electrophysiological response were suppressed under cyclic loading (0.33 Hz, 20 min) when the calmodulin pathway was inhibited [172]. Cyclic AMP (cAMP), another important second messenger for PG synthesis, can also cross-talk with calcium signaling [119]. It has been shown that gene expressions of bovine cartilage explants were regulated by cAMP through PKA pathway under static compression [95].

GPCRs are transmembrane receptors sensitive to extracellular molecules and consequently activate intracellular mechanotransduction pathways. G-proteins are associated with various intracellular signal cascades depending on the subclass. Gs subtype could enhance synthesis of cAMP and activation of PKA. Gq and Go subtypes could active PLC pathway, causing intracellular calcium release from storage sites. Evidence also suggests that the GPCRs may interact with integrins to regulate mechanotransduction process [195].

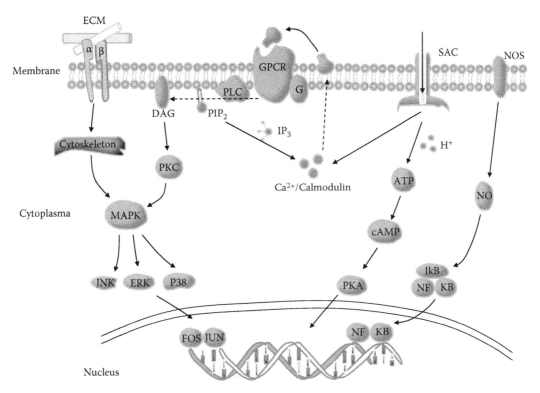

FIGURE 11.4 Potential mechanotransduction signaling pathways by which extracellular physicochemical signals induced by mechanical loading may be transduced into an intracellular signal. Stretch-activated ion channels allow rapid rise in intracellular calcium that can trigger secretion of regulatory molecules like interleukins. ECM components can bind to integrins causing their activation and downstream activation of the MAPK pathway. Binding of ligands to GPCRs or protein tyrosine receptors can lead to a release of intracellular calcium from intracellular stores via the IP$_3$ pathways or can activate the PKC pathways via DAG. Mechanical signals can up-regulate the activity of enzymes of reactive oxygen species (e.g., nitric oxide synthase). Nitric oxide, in turn, activates NF-kB pathways. All pathways lead to synthesis and/or activation of transcription factors, then gene expression.

It has been long conjectured that activation of gene expression by mechanical signals is dependent on the mechano-activated transcription factors. SOX-9 was found to be one of the mechano-active gene transcription factors. SOX-9 is a master gene that regulates a variety of matrix proteins such as collagen II [196], aggrecan [197], and link protein [198]. The effect of SOX-9 on matrix protein gene expressions depends on mechanical loading profiles and cell sources. Cyclic compression on alginate/chondrocyte constructs does not promote SOX-9 expression [199]. In contrast, repeated mechanical loading on mandibular condylar cartilage activates SOX-9 expression and condylar cartilage growth [200]. Also, genes that have similar responses to mechanical stimulation are thought to have common mechano-activated elements such as serum responsive elements (SRE), calcium responsive elements (CRE), and other mechano-responsive or inhibitory elements.

11.5 SUMMARY

The function of cartilage is predominantly mechanical. The tissue is subjected to a wide range of mechanical loading associated with the daily activities of diarthrodial joints. When cartilage is under mechanical loading, chondrocytes are exposed to all sorts of physical signals, in a manner dependent on the magnitudes, frequencies, and durations of the applied load. Mechanical signals are transmitted to chondrocytes via its interaction with the ECM, fluid flow, and fluid flow induced transport. These physical stimuli can dramatically regulate the chondrocyte behaviors in the synthesis, assembly, and degradation of ECM proteins. Chondrocytes integrate these various signals and react with appropriate responses, though the detailed mechanisms are not well understood. The combination of theoretical modeling and experimental studies in recent years starts to reveal the cellular and molecular mechanisms according to tissue level observations of cartilage mechanobiology. Exciting research on mechanical receptor signaling cascades as well as gene expression, translation, and posttranslational modification also leads us into new directions and poses new questions. This progress may lead to effective pharmacological and physical treatments for cartilage disease. For instance, a more thorough understanding of chondrocyte response to mechanical stress and other physical regulators within native and artificial ECM/scaffold would improve the ability to form functional tissue replacement for cartilage repair.

ACKNOWLEDGMENT

The authors acknowledge grant support from the National Institutes of Health and the AO Foundation.

REFERENCES

1. Maroudas A. (1979) Physicochemical properties of articular cartilage. In: Freeman M.A.R, editor, *Adult Articular Cartilage*. Kent, U.K.: Pitman Medical, pp. 215–290.
2. Aydelotte M. B., Schumacher B. L., and Kuettner K. E. (1992) Heterogeneity of articular chondrocytes. In: Kuettner K. E., editor, *Articular Cartilage and Osteoarthritis*. New York: Raven Press, pp. 237–249.
3. Lee J. H., Kisiday J., and Grodzinsky A. J. (2003) Tissue-engineered versus native cartilage: Linkage between cellular mechano-transduction and biomechanical properties. *Novartis Foundation Symposium* **249**, 52–64.
4. Mow V. C., Gu W. Y., and Chen F. H. (2005) Structure and function of articular cartilage and meniscus. In: Mow V. C., Huiskes R., editors, *Basic Orthopaedic Biomechanics and Mechano-Biology*, 3rd edn. Philadelphia, PA: Lippincott Williams & Wilkins, pp. 181–258.
5. Mow V. C., Ateshian G. A., and Ratcliffe A. (1992) Anatomic form and biomechanical properties of articular cartilage of the knee joint. In: Finerman G. A. M. and Noyes F. R., editors, *Biology and Biomechanics of the Traumatized Synovial Joint: The Knee as a Model: AAOS Symposium*, American Academy of Orthopaedic Surgeons, Rosemont, IL.
6. Mow V. C., Kuei S. C., Lai W. M., and Armstrong C. G. (1980) Biphasic creep and stress relaxation of articular cartilage in compression? Theory and experiments. *Journal of Biomechanical Engineering* **102**, 73–84.
7. Mow V. C., Holmes M. H., and Lai W. M. (1984) Fluid transport and mechanical properties of articular cartilage: A review. *Journal of Biomechanics* **17**, 377–394.
8. Mow V. C., Ratcliffe A., and Poole A. R. (1992) Cartilage and diarthrodial joints as paradigms for hierarchical materials and structures. *Biomaterials* **13**, 67–97.
9. Lai W. M., Hou J. S., and Mow V. C. (1991) A triphasic theory for the swelling and deformation behaviors of articular cartilage. *Journal of Biomechanical Engineering* **113**, 245–258.
10. Lai W. M., Mow V. C., Sun D. N., and Ateshian G. A. (2000) On the electric potentials inside a charged soft hydrated biological tissue: Streaming potential versus diffusion potential. *Journal of Biomechanical Engineering* **122**, 336–346.
11. Mow V. C., Ateshian G. A., Lai W. M., and Gu W. Y. (1998) Effects of fixed charges on the stress-relaxation behavior of hydrated soft tissues in a confined compression problem. *International Journal of Solids and Structure* **35**, 4945–4962.
12. Narmoneva D. A., Wang J. Y., and Setton L. A. (2001) A noncontacting method for material property determination for articular cartilage from osmotic loading. *Biophysical Journal* **81**, 3066–3076.
13. Van der Rest M., Garrone R. (1991) Collagen family of proteins. *Federation of American Societies for Experimental Biology of Journal* **5**, 2814–2823.
14. Thomas J. T., Ayad S., and Grant M. E. (1994) Cartilage collagens: Strategies for the study of their organisation and expression in the extracellular matrix. *Annals of the Rheumatic Diseases* **53**, 488–496.

15. Chapman J. A. and Hulmes D. J. (1984) Electron microscopy of the collagen fibril. In: Ruggeri A, Motta PM, editors, *Ultrastructure of the Connective Tissue Matrix*. Boston, MA: Martinus Nijhoff, pp. 1–33.
16. Akizuki S., Mow V. C., Muller F., Pita J. C., Howell D. S., and Manicourt D. H. (1987) Tensile properties of human knee joint cartilage: II. Correlations between weight bearing and tissue pathology and the kinetics of swelling. *Journal of Orthopaedic Research* **5**, 173–186.
17. Kempson G. E. (1975) Mechanical properties of articular cartilage and their relationship to matrix degradation and age. *Annals of the Rheumatic Diseases* **34**, 111–113.
18. Schmidt M. B., Mow V. C., Chun L. E., and Eyre D. R. (1990) Effects of proteoglycan extraction on the tensile behavior of articular cartilage. *Journal of Orthopaedic Research* **8**, 353–363.
19. Fosang A. J. and Hardingham T. E. (1996) Matrix proteoglycans. In: Comper W.D., editor, *Extracellular Matrix*. Amsterdam, the Netherlands: Harwood Academic Publishers.
20. Ratcliffe A. and Mow V. C. (1996) Structure and function of articular cartilage. In: Comper W.D., editor, *Extracellular Matrix*. Melbourne, Australia: Harwood Academic Publishers, pp. 234–302.
21. Hascall V. C. and Hascall G. K. (1981) Proteoglycans. In: Hay E. D., editor, *Cell Biology of Extracellular Matrix*. New York: Plenum Press, pp. 39–63.
22. Poole A. R. (1986) Proteoglycans in health and disease: Structure and functions. *Biochemical Journal* **236**, 1–14.
23. Hardingham T. E. (1981) Proteoglycans: Their structure, interactions and molecular organization in cartilage. *Biochemical Society Transaction* **9**, 489–497.
24. Allen R. G., Burstein D., and Gray M. L. (1999) Monitoring glycosaminoglycan replenishment in cartilage explants with gadolinium-enhanced magnetic resonance imaging. *Journal of Orthopaedic Research* **17**, 430–436.
25. Bashir A., Gray M. L., Hartke J., and Burstein D. (1999) Nondestructive imaging of human cartilage glycosaminoglycan concentration by MRI. *Magnetic Resonance in Medicine* **41**, 857–865.
26. Hardingham T. E., Perkins S. J., and Muir H. (1983) Molecular conformations in proteoglycan aggregation. *Biochemical Society Transactions* **11**, 128–130.
27. Muir H. (1983) Proteoglycans as organizers of the intercellular matrix. *Biochemical Society Transaction* **9**, 613–622.
28. Maroudas A. (1968) Physicochemical properties of cartilage in the light of ion exchange theory. *Biophysical Journal* **8**, 575–595.
29. Donnan F. G. (1924) The theory of membrane equilibria. *Chemical Reviews* **1**, 73–90.
30. Torzilli P. A. (1988) Water content and equilibrium water partition in immature cartilage. *Journal of Orthopaedic Research* **6**, 766–769.
31. Mankin H. J. and Thrasher A. Z. (1975) Water content and binding in normal and osteoarthritic human cartilage. *Journal of Bone and Joint Surgery* **64A**, 76–79.
32. Maroudas A. and Bannon C. (1981) Measurement of swelling pressure in cartilage and comparison with the osmotic pressure of constituent proteoglycans. *Biorheology* **18**, 619–632.
33. Edwards J. (1967) Physical characteristics of articular cartilage. *Proceedings of the Institution of Mechanical Engineering* **181–3J**, 16–24.
34. Katz E. P., Wachtel E. J., and Maroudas A. (1986) Extrafibrillar proteoglycans osmotically regulate the molecular packing of collagen in cartilage. *Biochimica et Biophysica Acta* **882**, 136–139.
35. Maroudas A. and Schneiderman R. (1987) "Free" and "exchangeable" or "trapped" and "non-exchangeable" water in cartilage. *Journal of Orthopaedic Research* **5**, 133–138.
36. Soltz M. A. and Ateshian G. A. (1998) Experimental verification and theoretical prediction of cartilage interstitial fluid pressurization at an impermeable contact interface in confined compression. *Journal of Biomechanics* **31**, 927–934.
37. Soltz M. A. and Ateshian G. A. (2000) Interstitial fluid pressurization during confined compression cyclical loading of articular cartilage. *Annals of Biomedical Engineering* **28**, 150–159.
38. McCutchen C. W. (1978) Lubrication of joints. In: Sokoloff L., editor, *The Joints and Synovial Fluid*. New York: Academic Press, pp. 437–483.
39. Ateshian G. A., Mow V. C. (2005) Friction, lubrication, and wear of articular cartilage and diarthrodial. In: Mow V. C. and Huiskes R., editors, *Basic Orthopaedic Biomechanics and Mechano-Biology*, 3rd edn. Philadelphia, PA: Lippincott Williams & Wilkins, pp. 447–494.
40. Yao H. and Gu W. Y. (2004) Physical signals and solute transport in cartilage under dynamic unconfined compression: Finite element analysis. *Annals of Biomedical Engineering* **32**, 380–390.
41. Stockwell R. A. (1979) *Biology of Cartilage Cells*. Cambridge, U.K.: Cambridge University Press.
42. Guilak F., Sato M., Stanford C. M., and Brand R. A. (2000) Cell mechanics. *Journal of Biomechanics* **33**, 1–2.

43. Kempson G. E. (1976) The effects of proteoglycan and collagen degradation on the mechanical properties of adult human articular cartilage. In: Burleigh P. M. C. and Poole A. R., editors, *Dynamics of Connective Tissue Macromolecules.* New York: American Elsevier, pp. 277–305.

44. Lane J. M. and Weiss C. (1975) Review of articular cartilage collagen research. *Arthritis and Rheumatism* **18**, 553–562.

45. Weiss C., Rosenberg L., and Helfet A. J. (1968) An ultrastructural study of normal young adult human articular cartilage. *Journal of Bone and Joint Surgery* **50A**, 663–674.

46. Roth V. and Mow V. C. (1980) The intrinsic tensile behavior of the matrix of bovine articular cartilage and its variation with age. *Journal of Bone and Joint Surgery American volume* **62**, 1102–1117.

47. Kempson G. E., Muir H., Pollard C., and Tuke M. (1973) The tensile properties of the cartilage of human femoral condyles related to the content of collagen and glycosaminoglycans. *Biochimica et Biophysica Acta* **297**, 456–472.

48. Broom N. D. and Poole C. A. (1983) Articular cartilage collagen and proteoglycans. Their functional interdependency. *Arthritis and Rheumatism* **26**, 1111–1119.

49. Clark J. M. (1985) The organization of collagen in cryofractured rabbit articular cartilage: A scanning electron microscopic study. *Journal of Orthopaedic Research* **3**, 17–29.

50. Sokoloff L. (1966) Elasticity of aging cartilage. *Federation Proceedings* **25**, 1089–1095.

51. Hayes W. C., Keer L. M., Herrmann G., and Mockros L. F. (1972) A mathematical analysis for indentation test of articular cartilage. *Journal of Biomechanics* **5**, 541–551.

52. Hayes W. C. and Mockros L. F. (1971) Viscoelastic properties of human articular cartilage. *Journal of Applied Physiology* **31**, 562–568.

53. Parsons J. R. and Black J. (1977) The viscoelastic shear behavior of normal rabbit articular cartilage. *Journal of Biomechanics* **10**, 21–30.

54. Woo S. L. Y., Simon B. R., Kuei S. C., and Akeson W. H. (1980) Quasi-linear viscoelastic properties of normal articular cartilage. *Journal of Biomechanical Engineering* **102**, 85–90.

55. Gu W. Y., Lai W. M., and Mow V. C. (1997) A triphasic analysis of negative osmotic flows through charged hydrated soft tissues. *Journal of Biomechanics* **30**, 71–78.

56. Ateshian G. A., Warden W. H., Kim J. J., Grelsamer R. P., and Mow V. C. (1997) Finite deformation biphasic material properties of bovine articular cartilage from confined compression experiments. *Journal of Biomechanics* **30**, 1157–1164.

57. Buschmann M. D., Soulhat J., Shirazi-Adl A., Jurvelin J. S., and Hunziker E. B. (1998) Confined compression of articular cartilage: Linearity in ramp and sinusoidal tests and the importance of interdigitation and incomplete confinement. *Journal of Biomechanics* **31**, 171–178.

58. Armstrong C. G., Lai W. M., and Mow V. C. (1984) An analysis of the unconfined compression of articular cartilage. *Journal of Biomechanical Engineering* **106**, 165–173.

59. Chen S. S., Falcovitz Y. H., Schneiderman R., Maroudas A., and Sah R. L. (2001) Depth-dependent compressive properties of normal aged human femoral head articular cartilage: Relationship to fixed charge density. *Osteoarthritis and Cartilage* **9**, 561–569.

60. Zhu W., Mow V. C., Koob T. J., and Eyre D. R. (1993) Viscoelastic shear properties of articular cartilage and the effects of glycosidase treatments. *Journal of Orthopaedic Research* **11**, 771–781.

61. Hori R. Y. and Mockros L. F. (1976) Indentation tests of human articular cartilage. *Journal of Biomechanics* **9**, 259–268.

62. Bae W. C., Temple M. M., Amiel D., Coutts R. D., Niederauer G. G., and Sah R. L. (2003) Indentation testing of human cartilage: Sensitivity to articular surface degeneration. *Arthritis and Rheumatism* **48**, 3382–3394.

63. Mow V. C., Gibbs M. C., Lai W. M., Zhu W. B., and Athanasiou K. A. (1989) Biphasic indentation of articular cartilage-II. A numerical algorithm and an experimental study. *Journal of Biomechanics* **22**, 853–861.

64. Athanasiou K. A., Rosenwasser M. P., Buckwalter J. A., Malinin T. I., and Mow V. C. (1991) Interspecies comparisons of in situ intrinsic mechanical properties of distal femoral cartilage. *Journal of Orthopaedic Research* **9**, 330–340.

65. Lu X. L., Miller C., Chen F. H., Guo X. E., and Mow V. C. (2007) The generalized triphasic correspondence principle for simultaneous determination of the mechanical properties and proteoglycan content of articular cartilage by indentation. *Journal of Biomechanics* **40**, 2434–2441.

66. Helminen H., Jurvelin J., Kiviranta I., Paukkonen K., Saamanen A. M., and Tammi M. (1987) Joint loading effects on articular cartilage: A historical review In: Helminen H.J., Kiviranta I., and Tammi M. et al., editors, *Joint Loading: Biology and Health of Articular Structures.* Bristol, U.K.: Wright and Sons, pp. 1–46.

67. Guilak F. (2000) The deformation behavior and viscoelastic properties of chondrocytes in articular cartilage. *Biorheology* **37**, 27–44.

68. Kim Y. J., Bonassar L. J., and Grodzinsky A. J. (1995) The role of cartilage streaming potential, fluid flow, and pressure in the stimulation of chondrocyte biosynthesis during dynamic compression. *Journal of Biomechanics* **28**, 1055–1066.

69. Frank E., Grodzinsky A., Phillips S., and Grimshaw P. (1990) Physiocochemical and bioelectrical determinants of cartilage material properties. In: Mow V. C., Ratcliffe A., and Woo S. L. Y., editors, *Biomechanics of Diarthrodial Joints*. New York: Springer Verlag, pp. 261–282.

70. Grodzinsky A. J. (1983) Electromechanical and physicochemical properties of connective tissue. *Critical Reviews in Biomedical Engineering* **9**, 133–199.

71. Grodzinsky A. J., Levenston M. E., Jin M., and Frank E. H. (2000) Cartilage tissue remodeling in response to mechanical forces. *Annual Review of Biomedical Engineering* **2**, 691–713.

72. Van Campen G. P. J. and Van de Stadt R. J. (1987) Cartilage and chondrocyte responses to mechanical loading in vitro. In: Helminen H. J., Kiviranta I., Tammi M., Saamanen A.-M., Paukkonen K., and Jurvelin J., editors, *Joint Loading: Biology and Health of Articular Structures*. Bristol, U.K.: Wright & Sons, pp. 112–125.

73. Moskowitz R. W. (1992) Experimental models of osteoarthritis. In: Moskowitz R. W., Howell D. S., Goldberg V. M. et al., editors, *Osteoarthritis: Diagnosis and Medical/Surgical Management*. Philadelphia, PA: WB Saunders, pp. 213–232.

74. Pritzker K. P. (1994) Animal models for osteoarthritis: Processes, problems and prospects. *Annals of the Rheumatic Diseases* **53**, 406–420.

75. Helminen H. J., Saamanen A. M., Jurvelin J., Kiviranta I., Parkkinen J. J., and Lammi M. J. et al. (1992) The effect of loading on articular cartilage. *Duodecim* **108**, 1097–1107.

76. Brandt K. D., Myers S. L., Burr D., and Albrecht M. (1991) Osteoarthritic changes in canine articular cartilage, Subchondral bone, and synovium fifty-four months after transection of the anterior cruciate ligament. *Arthritis and Rheumatism* **34**, 1560–1570.

77. Brandt K. D. (1991) Animal models: Insights into osteoarthritis (OA) provided by the cruciate-deficient dog. *British Journal of Rheumatology* **30**, 5–9.

78. Jurvelin J., Helminen H. J., Lauritsalo S., Kiviranta I., Saamanen A. M., Paukkonen K. et al. (1985) Influences of joint immobilization and running exercise on articular cartilage surfaces of young rabbits. A semiquantitative stereomicroscopic and scanning electron microscopic study. *Acta Anatomica* **122**, 62–68.

79. Palmoski M., Perricone E., and Brandt K. D. (1979) Development and reversal of a proteoglycan aggregation defect in normal canine knee cartilage after immobilization. *Arthritis and Rheumatism* **22**, 508–517.

80. Saamanen A. M., Tammi M., Kiviranta I., Jurvelin J., and Helminen H. J. (1987) Maturation of proteoglycan matrix in articular cartilage under increased and decreased joint loading. A study in young rabbits. *Connective Tissue Research* **16**, 163–175.

81. Kiviranta I., Tammi M., Jurvelin J., Saamanen A. M., and Helminen H. J. (1988) Moderate running exercise augments glycosaminoglycans and thickness of articular cartilage in the knee joint of young beagle dogs. *Journal of Orthopaedic Research* **6**, 188–195.

82. Eckstein F., Hudelmaier M., and Putz R. (2006) The effects of exercise on human articular cartilage. *Journal of Anatomy* **208**, 491–512.

83. Slowman S. D. and Brandt K. D. (1986) Composition and glycosaminoglycan metabolism of articular cartilage from habitually loaded and habitually unloaded sites. *Arthritis and Rheumatism* **29**, 88–94.

84. Helminen H. J., Kiviranta I., Saamanen A-M., Jurvelin J. S., Arokoski J., Oettmeier R. et al. (1992) Effect of motion and load on articular cartilage in animal models. In: Kuettner K. E., Schleyerbach R., Peyron J. G., and Hascall V. C., editors, *Articular Cartilage and Osteoarthritis*. New York: Raven Press, pp. 501–510.

85. Afoke A., Hutton W. C., and Byers P. D. (1990) Mechanical and electrical properties and their relevance to physiological processes: Pressure measurement in the human hip joint using Fujifilm. *Methods in Cartilage Research*. San Diego, CA: Academic Press, pp. 281–287.

86. Dieppe P. and Kirwan J. (1994) The localization of osteoarthritis. *British Journal of Rheumatology* **33**, 201–203.

87. Hoch D. H., Grodzinsky A. J., Koob T. J., Albert M. L., and Eyre D. R. (1983) Early changes in material properties of rabbit articular cartilage after meniscectomy. *Journal of Orthopaedic Research* **1**, 4–12.

88. Burton-Wurster N., Vernier-Singer M., Farquhar T., and Lust G. (1993) Effect of compressive loading and unloading on the synthesis of total protein, proteoglycan, and fibronectin by canine cartilage explants. *Journal of Orthopaedic Research* **11**, 717–729.

89. Gray M. L., Pizzanelli A. M., Grodzinsky A. J., and Lee R. C. (1988) Mechanical and physiochemical determinants of the chondrocyte biosynthetic response. *Journal of Orthopaedic Research* **6**, 777–792.

90. Guilak F., Meyer B. C., Ratcliffe A., and Mow V. C. (1994) The effects of matrix compression on proteoglycan metabolism in articular cartilage explants. *Osteoarthritis Cartilage* **2**, 91–101.

91. Kim Y. J., Sah R. L., Grodzinsky A. J., Plaas A. H., and Sandy J. D. (1994) Mechanical regulation of cartilage biosynthetic behavior: Physical stimuli. *Archives of Biochemistry and Biophysics* **311**, 1–12.

92. Sah R. L., Doong J. Y., Grodzinsky A. J., Plaas A. H., and Sandy J. D. (1991) Effects of compression on the loss of newly synthesized proteoglycans and proteins from cartilage explants. *Archives of Biochemistry and Biophysics* **286**, 20–29.

93. Sah R. L., Grodzinsky A. J., Plaas A. H., and Sandy J. D. (1990) Effects of tissue compression on the hyaluronate-binding properties of newly synthesized proteoglycans in cartilage explants. *The Biochemical Journal* **267**, 803–808.

94. Larsson T., Aspden R. M., and Heinegard D. (1991) Effects of mechanical load on cartilage matrix biosynthesis in vitro. *Matrix* **11**, 388–394.

95. Fitzgerald J. B., Jin M., Dean D., Wood D. J., Zheng M. H., and Grodzinsky A. J. (2004) Mechanical compression of cartilage explants induces multiple time-dependent gene expression patterns and involves intracellular calcium and cyclic AMP. *Journal of Biological Chemistry* **279**, 19502–19511.

96. Urban J. P. G. and Hall A. C. (1994) The effects of hydrostatic and osmotic pressures on chodrocyte metabolism. In: Mow V. C., Guilak F., Tran-Son-Tay R., and Hochmuth R. M., editors, *Cell Mechanics and Cellular Engineering*. New York: Springer-Verlag, pp. 398–419.

97. Schneiderman R., Keret D., and Maroudas A. (1986) Effects of mechanical and osmotic pressure on the rate of glycosaminoglycan synthesis in the human adult femoral head cartilage: An in vitro study. *Journal of Orthopaedic Research* **4**, 393–408.

98. Urban J. P., Hall A. C., and Gehl K. A. (1993) Regulation of matrix synthesis rates by the ionic and osmotic environment of articular chondrocytes. *Journal of Cellular Physiology* **154**, 262–270.

99. Buschmann M. D., Gluzband Y. A., Grodzinsky A. J., and Hunziker E. B. (1995) Mechanical compression modulates matrix biosynthesis in chondrocyte/agarose culture. *Journal of Cell Science* **108**, 1497–1508.

100. Ragan P. M., Chin V. I., Hung H. H., Masuda K., Thonar E. J., Arner E. C. et al. (2000) Chondrocyte extracellular matrix synthesis and turnover are influenced by static compression in a new alginate disk culture system. *Archives of Biochemistry and Biophysics* **383**, 256–264.

101. Korver T. H., Van de Stadt R. J., Kiljan E., Van Kampen G. P., and Van der Korst J. K. (1992) Effects of loading on the synthesis of proteoglycans in different layers of anatomically intact articular cartilage in vitro. *Journal of Rheumatology* **19**, 905–912.

102. Ostendorf R. H., Van de Stadt R. J., and Van Kampen G. P. (1994) Intermittent loading induces the expression of 3-B-3(-) epitope in cultured bovine articular cartilage. *Journal of Rheumatology* **21**, 287–292.

103. Palmoski M. J. and Brandt K. D. (1984) Effects of static and cyclic compressive loading on articular cartilage plugs in vitro. *Arthritis and Rheumatism* **27**, 675–681.

104. Parkkinen J., Lammi M. J., Helminen H. J., and Tammi M. (1992) Local stimulation of proteoglycan synthesis in articular cartilage explants by dynamic compression in vitro. *Journal of Orthopaedic Research* **10**, 610–620.

105. Sah R. L., Kim Y. J., Doong J. Y., Grodzinsky A. J., Plaas A. H., and Sandy J. D. (1989) Biosynthetic response of cartilage explants to dynamic compression. *Journal of Orthopaedic Research* **7**, 619–636.

106. Van Kampen G. P., Korver G. H., and Van de Stadt R. J. (1994) Modulation of proteoglycan composition in cultured anatomically intact joint cartilage by cyclic loads of various magnitudes. *International Journal of Tissue Reactions* **16**, 171–179.

107. Lee D. A. and Bader D. L. (1997) Compressive strains at physiological frequencies influence the metabolism of chondrocytes seeded in agarose. *Journal of Orthopaedic Research* **15**, 181–188.

108. Shelton J. C., Bader D. L., and Lee D. A. (2003) Mechanical conditioning influences the metabolic response of cell-seeded constructs. *Cells Tissues Organs* **175**, 140–150.

109. Mauck R. L., Soltz M. A., Wang C. C., Wong D. D., Chao P. H., Valhmu W. B. et al. (2000) Functional tissue engineering of articular cartilage through dynamic loading of chondrocyte-seeded agarose gels. *Journal of Biomechanical Engineering* **122**, 252–260.

110. Fitzgerald J. B., Jin M., and Grodzinsky A. J. (2006) Shear and compression differentially regulate clusters of functionally related temporal transcription patterns in cartilage tissue. *Journal of Biological Chemistry* **281**, 24095–24103.

111. Mio K., Saito S., Tomatsu T., and Toyama Y. (2005) Intermittent compressive strain may reduce aggrecanase expression in cartilage: A study of chondrocytes in agarose gel. *Clinical Orthopaedics and Related Research* **433**, 225–232.

112. Lee J. H., Fitzgerald J. B., Dimicco M. A., and Grodzinsky A. J. (2005) Mechanical injury of cartilage explants causes specific time-dependent changes in chondrocyte gene expression. *Arthritis and Rheumatism* **52**, 2386–2395.

113. Wong M., Siegrist M., and Cao X. (1999) Cyclic compression of articular cartilage explants is associated with progressive consolidation and altered expression pattern of extracellular matrix proteins. *Matrix Biology* **18**, 391–399.

114. Kiraly K., Hyttinen M. M., Parkkinen J. J., Arokoski J. A., Lapvetelainen T., Torronen K. et al. (1998) Articular cartilage collagen birefringence is altered concurrent with changes in proteoglycan synthesis during dynamic in vitro loading. *Anatomical Record* **251**, 28–36.

115. Murata M., Bonassar L. J., Wright M., Mankin H. J., and Towle C. A. (2003) A role for the interleukin-1 receptor in the pathway linking static mechanical compression to decreased proteoglycan synthesis in surface articular cartilage. *Archives of Biochemistry and Biophysics* **413**, 229–235.

116. Chowdhury T. T., Appleby R. N., Salter D. M., Bader D. A., and Lee D. A. (2006) Integrin-mediated mechanotransduction in IL-1 beta stimulated chondrocytes. *Biomechanics and Modeling in Mechanobiology* **5**, 192–201.

117. Chowdhury T. T., Bader D. L., and Lee D. A. (2001) Dynamic compression inhibits the synthesis of nitric oxide and PGE(2) by IL-1beta-stimulated chondrocytes cultured in agarose constructs. *Biochemical and Biophysical Research Communications* **285**, 1168–1174.

118. Griffin T. M. and Guilak F. (2005) The role of mechanical loading in the onset and progression of osteoarthritis. *Exercise and Sport Sciences Reviews* **33**, 195–200.

119. De Witt M. T., Handley C. J., Oakes B. W., and Lowther D. A. (1984) In vitro response of chondrocytes to mechanical loading: The effect of short term mechanical tension. *Connective Tissue Research* **12**, 97–109.

120. Lee R. C., Rich J. B., Kelley K. M., Weiman D. S., and Mathews M. B. (1982) A comparison of in vitro cellular responses to mechanical and electrical stimulation. *American Surgeon* **48**, 567–574.

121. Guilak F., Fermor B., Keefe F. J., Kraus V. B., Olson S. A., Pisetsky D. S., Setton L. A., and Weinberg J. B. (2004) The role of biomechanics and inflammation in cartilage injury and repair. *Clinical Orthopaedics and Related Research* **423**, 17–26.

122. Bachrach N. M., Mow V. C., and Guilak F. (1998) Incompressibility of the solid matrix of articular cartilage under high hydrostatic pressures. *Journal of Biomechanics* **31**, 445–451.

123. Hall A. C., Urban J. P. G., and Gehl K. A. (1991) The effects of hydrostatic pressure on matrix synthesis in articular cartilage. *Journal of Orthopaedic Research* **9**, 1–10.

124. Parkkinen J. J., Ikonen J., Lammi M. J., Laakkonen J., Tammi M., and Helminen H. J. (1993) Effects of cyclic hydrostatic pressure on proteoglycan synthesis in cultured chondrocytes and articular cartilage explants. *Archives of Biochemistry and Biophysics* **300**, 458–465.

125. Jortikka M. O., Parkkinen J. J., Inkinen R. I., Karner J., Jarvelainen H. T., Nelimarkka L. O. et al. (2000) The role of microtubules in the regulation of proteoglycan synthesis in chondrocytes under hydrostatic pressure. *Archives of Biochemistry and Biophysics* **374**, 172–180.

126. Millward-Sadler S. J., Wright M. O., Davies L. W., Nuki G., and Salter D. M. (2000) Mechanotransduction via integrins and interleukin-4 results in altered aggrecan and matrix metalloproteinase 3 gene expression in normal, but not osteoarthritic, human articular chondrocytes. *Arthritis and Rheumatism* **43**(9), 209–219.

127. Smith R. L., Rusk S. F., Ellison B. E., Wessells P., Tsuchiya K., Carter D. R. et al. (1996) In vitro stimulation of articular chondrocyte mRNA and extracellular matrix synthesis by hydrostatic pressure. *Journal of Orthopaedic Research* **14**, 53–60.

128. Takahashi K., Kubo T., Kobayashi K., Imanishi J., Takigawa M., Arai Y. et al. (1997) Hydrostatic pressure influences mRNA expression of transforming growth factor-beta 1 and heat shock protein 70 in chondrocyte-like cell line. *Journal of Orthopaedic Research* **15**, 150–158.

129. Sironen R., Elo M., Kaarniranta K., Helminen H. J., and Lammi M. J. (2000) Transcriptional activation in chondrocytes submitted to hydrostatic pressure. *Biorheology* **37**, 85–93.

130. Takahashi K., Kubo T., Arai Y., Kitajima I., Takigawa M., Imanishi J. et al. (1998) Hydrostatic pressure induces expression of interleukin 6 and tumour necrosis factor alpha mRNAs in a chondrocyte-like cell line. *Annals of the Rheumatic Diseases* **57**, 231–236.

131. Aaron R. K. and Ciombor D. M. (1993) Enhancement of extracellular matrix synthesis in cartilage explant cultures by exposure to an electric field. *Transactions of the Orthopaedic Research Society* **18**, 630.

132. MacGinitie L. A., Gluzband Y. A., and Grodzinsky A. J. (1994) Electric field stimulation can increase protein synthesis in articular cartilage explants. *Journal of Orthopaedic Research* **12**, 151–160.

133. Iannacone W. M., Pienkowski D., Pollack S. R., and Brighton C. T. (1988) Pulsing electromagnetic field stimulation of the in vitro growth plate. *Journal of Orthopaedic Research* **6**, 239–247.

134. Sah R. L. and Grodzinsky A. J. (1989) Biosynthetic response to mechanical and electrical forces: Calf articular cartilage in organ culture. In: Norton L. A., Burston C. J., editors, *Biology of Tooth Movement.* Boca Raton, FL: CRC Press, pp. 335–347.

135. Mow V. C., Wang C. C., and Hung C. T. (1999) The extracellular matrix, interstitial fluid and ions as a mechanical signal transducer in articular cartilage. *Osteoarthritis Cartilage* **7**, 41–58.

136. Urban J. P. (2000) Present perspectives on cartilage and chondrocyte mechanobiology. *Biorheology* **37**, 185–190.

137. Sun D. D., Guo X. E., Likhitpanichkul M., Lai W. M., and Mow V. C. (2004) The influence of the fixed negative charges on mechanical and electrical behaviors of articular cartilage under unconfined compression. *Journal of Biomechanical Engineering* **126**, 6–16.

138. Suh J. K. (1996) Dynamic unconfined compression of articular cartilage under a cyclic compressive load. *Biorheology* **33**, 289–304.

139. Gu W. Y., Lai W. M., and Mow V. C. (1998) A mixture theory for charged-hydrated soft tissues containing multi- electrolytes: Passive transport and swelling behaviors. *Journal of Biomechanical Engineering* **120**, 169–180.

140. Urban J. and Hall A. C. (1993) Adaptive responses of chondrocytes to changes in their physical environment. *Transactions of the Orthopaedic Research Society* **18**, 260.

141. Martinac B. (2004) Mechanosensitive ion channels: Molecules of mechanotransduction. *Journal of Cell Science* **117**, 2449–2460.

142. Ingber D. E. (1991) Integrins as mechanochemical transducers. *Current Opinions in Cell Biology* **3**, 841–848.

143. Bassett C. A. and Pawluk R. J. (1972) Electrical behavior of cartilage during loading. *Science* **178**, 982–983.

144. Frank E. H. and Grodzinsky A. J. (1987) Cartilage electromechanics—I. Electrokinetic transduction and the effects of electrolyte pH and ionic strength. *Journal of Biomechanics* **20**, 615–627.

145. Gu W. Y., Lai W. M., and Mow V. C. (1993) Transport of fluid and ions through a porous-permeable charged-hydrated tissue, and streaming potential data on normal bovine articular cartilage. *Journal of Biomechanics* **26**, 709–723.

146. Lotke P. A., Black J., and Richardson S. (1974) Electromechanical properties in human articular cartilage. *Journal of Bone and Joint Surgery American* **56**, 1040–1046.

147. Brighton C. T., Unger A. S., and Stambough J. L. (1984) In vitro growth of bovine articular cartilage chondrocytes in various capacitively coupled electrical fields. *Journal of Orthopaedic Research* **2**, 15–22.

148. Norton L. A., Rodan G. A., and Bourret L. A. (1977) Epiphyseal cartilage cAMP changes produced by electrical and mechanical perturbations. *Clinical Orthopaedics and Related Research*, 59–68.

149. Rodan G. A., Bourret L. A., and Norton L. A. (1978) DNA synthesis in cartilage cells is stimulated by oscillating electric fields. *Science* **199**, 690–692.

150. Wright M., Jobanputra P., Bavington C., Salter D. M., and Nuki G. (1996) Effects of intermittent pressure-induced strain on the electrophysiology of cultured human chondrocytes: Evidence for the presence of stretch-activated membrane ion channels. *Clinical Science (Lond)* **90**, 61–71.

151. Wright M. O., Stockwell R. A., and Nuki G. (1992) Response of plasma membrane to applied hydrostatic pressure in chondrocytes and fibroblasts. *Connective Tissue Research* **28**, 49–70.

152. Morris C. E. (1990) Mechanosensitive ion channels. *Journal of Membrane Biology* **113**, 93–107.

153. Sachs F. (1991) Mechanical transduction by membrane ion channels: A mini review. *Molecular and Cellular Biochemistry* **104**, 57–60.

154. Sokabe M., Sachs F., and Jing Z. Q. (1991) Quantitative video microscopy of patch clamped membranes stress, strain, capacitance, and stretch channel activation. *Biophysical Journal* **59**, 722–728.

155. Christensen O. (1987) Mediation of cell volume regulation by Ca^{2+} influx through stretch-activated channels. *Nature* **330**, 66–68.

156. Watson P. A. (1991) Function follows form: Generation of intracellular signals by cell deformations. *The FASEB Journal* **5**, 2013–2019.

157. Lee H. S., Millward-Sadler S. J., Wright M. O., Nuki G., and Salter D. M. (2000) Integrin and mechano-sensitive ion channel-dependent tyrosine phosphorylation of focal adhesion proteins and beta-catenin in human articular chondrocytes after mechanical stimulation. *Journal of Bone and Mineral and Research* **15**, 1501–1509.

158. Wu Q. Q. and Chen Q. (2000) Mechanoregulation of chondrocyte proliferation, maturation, and hypertrophy: Ion-channel dependent transduction of matrix deformation signals. *Experimental Cell Research* **256**, 383–391.

159. Perkins G. L., Derfoul A., Ast A., and Hall D. J. (2005) An inhibitor of the stretch-activated cation receptor exerts a potent effect on chondrocyte phenotype. *Differentiation* **73**, 199–211.

160. Mouw J. K., Imler S. M., and Levenston M. E. (2007) Ion-channel regulation of chondrocyte matrix synthesis in 3D culture under static and dynamic compression. *Biomechanics and Modeling in Mechanobiology* **6**, 33–41.

161. Giannoni P., Siegrist M., Hunziker E. B., and Wong M. (2003) The mechanosensitivity of cartilage oligomeric matrix protein (COMP). *Biorheology* **40**, 101–109.

162. Holmvall K., Camper L., Johansson S., Kimura J. H., and Lundgren-Akerlund E. (1995) Chondrocyte and chondrosarcoma cell integrins with affinity for collagen type II and their response to mechanical stress. *Experimental Cell Research* **221**, 496–503.

163. Loeser R. F. (2000) Chondrocyte integrin expression and function. *Biorheology* **37**, 109–116.

164. Reid D. L., Aydelotte M. B., and Mollenhauer J. (2000) Cell attachment, collagen binding, and receptor analysis on bovine articular chondrocytes. *Journal of Orthopaedic Research* **18**, 364–373.

165. Von der Mark K., Mollenhauer J., Pfaffle M., Van Menxel M., and Muller P. K. (1986) Role of anchorin CII in the interaction of chondrocytes with extracellular collagen. In: Kuettner K. E., Schleyerbach R., Hascall V. C., editors, *Articular Cartilage Biochemistry*. New York: Raven Press, pp. 125–141.

166. Knudson W. and Loeser R. F. (2002) CD44 and integrin matrix receptors participate in cartilage homeostasis. *Cellular and Molecular Life Sciences* **59**, 36–44.

167. Wang J. H. and Thampatty B. P. (2006) An introductory review of cell mechanobiology. *Biomechanics and Modeling in Mechanobiology* **5**, 1–16.

168. Hynes R. O. (1992) Integrins: Versatility, modulation, and signaling in cell adhesion. *Cell* **69**, 11–25.

169. Enomoto-Iwamoto M., Iwamoto M., Nakashima K., Mukudai Y., Boettiger D., Pacifici M. et al. (1997) Involvement of alpha5beta1 integrin in matrix interactions and proliferation of chondrocytes. *Journal of Bone and Mineral Research* **12**, 1124–1132.

170. Chowdhury T. T., Salter D. M., Bader D. L., and Lee D. A. (2004) Integrin-mediated mechanotransduction processes in TGFbeta-stimulated monolayer-expanded chondrocytes. *Biochemical and Biophysical Research Communication* **318**, 873–881.

171. Orazizadeh M., Lee H. S., Groenendijk B., Sadler S. J., Wright M. O., Lindberg F. P. et al. (2008) CD47 associates with alpha 5 integrin and regulates responses of human articular chondrocytes to mechanical stimulation in an in vitro model. *Arthritis Research and Therapy* **10**, R4.

172. Wright M. O., Nishida K., Bavington C., Godolphin J. L., Dunne E., Walmsley S. et al. (1997) Hyperpolarisation of cultured human chondrocytes following cyclical pressure-induced strain: Evidence of a role for alpha5beta1 integrin as a chondrocyte mechanoreceptor. *Journal of Orthopaedic Research* **15**, 742–747.

173. Lee H. S., Millward-Sadler S. J., Wright M. O., Nuki G., Al-Jamal R., and Salter D. M. (2002) Activation of Integrin-RACK1/PKCalpha signalling in human articular chondrocyte mechanotransduction. *Osteoarthritis Cartilage* **10**, 890–897.

174. Loeser R. F., Sadiev S., Tan L., and Goldring M. B. (2000) Integrin expression by primary and immortalized human chondrocytes: Evidence of a differential role for alpha1beta1 and alpha2beta1 integrins in mediating chondrocyte adhesion to types II and VI collagen. *Osteoarthritis Cartilage* **8**, 96–105.

175. Schaeffer H. J. and Weber M. J. (1999) Mitogen-activated protein kinases: Specific messages from ubiquitous messengers. *Molecular and Cellular Biology* **19**, 2435–2444.

176. Vincent T. L., McLean C. J., Full L. E., Peston D., and Saklatvala J. (2007) FGF-2 is bound to perlecan in the pericellular matrix of articular cartilage, where it acts as a chondrocyte mechanotransducer. *Osteoarthritis Cartilage* **15**, 752–763.

177. Vincent T., Hermansson M., Bolton M., Wait R., and Saklatvala J. (2002) Basic FGF mediates an immediate response of articular cartilage to mechanical injury. *Proceedings of the National Academy of Sciences of the United States of America* **99**, 8259–8264.

178. Vincent T. L., Hermansson M. A., Hansen U. N., Amis A. A., and Saklatvala J. (2004) Basic fibroblast growth factor mediates transduction of mechanical signals when articular cartilage is loaded. *Arthritis and Rheumatism* **50**, 526–533.

179. Jobanputra P., Lin H., Jenkins K., Bavington C., Brennan F. R., Nuki G. et al. (1996) Modulation of human chondrocyte integrins by inflammatory synovial fluid. *Arthritis and Rheumatism* **39**, 1430–1432.

180. Loeser R. F. (1997) Growth factor regulation of chondrocyte integrins. Differential effects of insulin-like growth factor 1 and transforming growth factor beta on alpha1beta1 integrin expression and chondrocyte adhesion to type VI collagen. *Arthritis and Rheumatism* **40**, 270–276.
181. Giancotti F. G. and Ruoslahti E. (1999) Integrin signaling. *Science* **285**, 1028–1032.
182. Bonassar L. J., Grodzinsky A. J., Srinivasan A., Davila S. G., and Trippel S. B. (2000) Mechanical and physicochemical regulation of the action of insulin-like growth factor-I on articular cartilage. *Archives of Biochemistry and Biophysics* **379**, 57–63.
183. Millward-Sadler S. J., Wright M. O., Lee H., Nishida K., Caldwell H., Nuki G. et al. (1999) Integrin-regulated secretion of interleukin 4: A novel pathway of mechanotransduction in human articular chondrocytes. *The Journal of Cell Biology* **145**, 183–189.
184. Millward-Sadler S. J., Khan N. S., Bracher M. G., Wright M. O., and Salter D. M. (2006) Roles for the interleukin-4 receptor and associated JAK/STAT proteins in human articular chondrocyte mechanotransduction. *Osteoarthritis Cartilage* **14**, 991–1001.
185. Doi H., Nishida K., Yorimitsu M., Komiyama T., Kadota Y., Tetsunaga T. et al. (2008) Interleukin-4 downregulates the cyclic tensile stress-induced matrix metalloproteinases-13 and cathepsin B expression by rat normal chondrocytes. *Acta Medica Okayama* **62**, 119–126.
186. Jin M., Emkey G. R., Siparsky P., Trippel S. B., and Grodzinsky A. J. (2003) Combined effects of dynamic tissue shear deformation and insulin-like growth factor I on chondrocyte biosynthesis in cartilage explants. *Archives of Biochemistry and Biophysics* **414**, 223–231.
187. Mauck R. L., Nicoll S. B., Seyhan S. L., Ateshian G. A., and Hung C. T. (2003) Synergistic action of growth factors and dynamic loading for articular cartilage tissue engineering. *Tissue Engineering* **9**, 597–611.
188. Hung C. T., Henshaw D. R., Wang C. C., Mauck R. L., Raia F., Palmer G. et al. (2000) Mitogen-activated protein kinase signaling in bovine articular chondrocytes in response to fluid flow does not require calcium mobilization. *Journal of Biomechanics* **33**, 73–80.
189. Jin G., Sah R. L., Li Y. S., Lotz M., Shyy J. Y., and Chien S. (2000) Biomechanical regulation of matrix metalloproteinase-9 in cultured chondrocytes. *Journal of Orthopaedic Research* **18**, 899–908.
190. Yellowley C. E., Jacobs C. R., and Donahue H. J. (1999) Mechanisms contributing to fluid-flow-induced Ca^{2+} mobilization in articular chondrocytes. *Journal of Cellular Physiology* **180**, 402–408.
191. Myers K. A., Rattner J. B., Shrive N. G., and Hart D. A. (2007) Hydrostatic pressure sensation in cells: Integration into the tensegrity model. *Biochemistry and Cell Biology* **85**, 543–551.
192. Fermor B., Weinberg J. B., Pisetsky D. S., Misukonis M. A., Fink C., and Guilak F. (2002) Induction of cyclooxygenase-2 by mechanical stress through a nitric oxide-regulated pathway. *Osteoarthritis Cartilage* **10**, 792–798.
193. Fermor B., Haribabu B., Weinberg J. B., Pisetsky D. S., and Guilak F. (2001) Mechanical stress and nitric oxide influence leukotriene production in cartilage. *Biochemical and Biophysical Research Communications* **285**, 806–810.
194. D'Andrea P., Calabrese A., Capozzi I., Grandolfo M., Tonon R., and Vittur F. (2000) Intercellular Ca^{2+} waves in mechanically stimulated articular chondrocytes. *Biorheology* **37**, 75–83.
195. Zhang M., Chen Y. J., Ono T., and Wang J. J. (2008) Crosstalk between integrin and G protein pathways involved in mechanotransduction in mandibular condylar chondrocytes under pressure. *Archives of Biochemistry and Biophysics* **474**, 102–108.
196. Bell D. M., Leung K. K., Wheatley S. C., Ng L. J., Zhou S., Ling K. W. et al. (1997) SOX9 directly regulates the type-II collagen gene. *Nature Genetics* **16**, 174–178.
197. Sekiya I., Tsuji K., Koopman P., Watanabe H., Yamada Y., Shinomiya K. et al. (2000) SOX9 enhances aggrecan gene promoter/enhancer activity and is up-regulated by retinoic acid in a cartilage-derived cell line, TC6. *The Journal of Biological Chemistry* **275**, 10738–10744.
198. Kou I. and Ikegawa S. (2004) SOX9-dependent and -independent transcriptional regulation of human cartilage link protein. *The Journal of Biological Chemistry* **279**, 50942–50948.
199. Wong M., Siegrist M., and Goodwin K. (2003) Cyclic tensile strain and cyclic hydrostatic pressure differentially regulate expression of hypertrophic markers in primary chondrocytes. *Bone* **33**, 685–693.
200. Ng A. F., Yang Y. O., Wong R. W., Hagg E. U., and Rabie A. B. (2006) Factors regulating condylar cartilage growth under repeated load application. *Frontiers in Bioscience* **11**, 949–954.

12 Cell Mechanobiology: The Forces Applied to Cells and Generated by Cells

Bin Li, Jeen-Shang Lin, and James H.-C. Wang

CONTENTS

12.1 INTRODUCTION

Many types of cells live in a mechanical environment and are sensitive to changes in mechanical forces. A few examples of these mechanical load-responsive cells include fibroblasts in skin, osteocytes in bones, chondrocytes in cartilage, and endothelial cells lining the blood vessels. These cells in vivo are subjected to tension, compression, shear stresses, hydrostatic pressure, or a combination of these forces (Figure 12.1). Mechanical forces on cells regulate a wide range of cellular events, including proliferation, differentiation, gene expression, and protein secretion.[13,23,35,52,67,72,128,132,135,161,175] As such, mechanical forces on cells have a profound effect on tissue homeostasis and pathophysiology. Therefore, in vitro model systems have been developed over the years to investigate cellular mechanobiological responses under well-controlled mechanical loading conditions.[21,68]

Besides external mechanical forces that act on cells, cells themselves generate internal mechanical forces, which are then transmitted to the ECM and are referred to as CTFs. CTFs are necessary for cells to migrate, maintain shape, organize ECM, probe physical environments, and generate mechanical signals. Hence, CTFs play a fundamental role in many biological processes including tissue homeostasis, wound healing, angiogenesis, and metastasis.[175] Quantitative analysis of CTFs, therefore, enables better understanding of these physiological and pathological events at the tissue and organ levels.

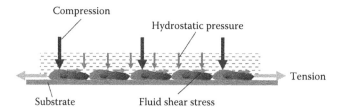

FIGURE 12.1 A schematic illustration of the mechanical forces acting on cells.

This chapter centers on two aspects of cell mechanobiology research. First, different techniques for applying external mechanical forces to cells are described. Because tensile and compressive forces are the major types of mechanical stresses that act on most musculoskeletal tissues, we focus on discussing those model systems that apply these two types of forces to cells. Second, the mechanisms of generation and transmission of CTFs as well as the techniques for CTF measurement are discussed, followed by examples of CTF applications. Finally, perspectives for future research and development in the area of cell mechanobiology are suggested.

12.2 MECHANICAL FORCES APPLIED TO CELLS

To simulate the loading conditions of cells in vivo, mechanical forces such as tensile and compressive forces are commonly applied to cells. Due to the relative simplicity and convenience in device fabrication and experimental setup, two-dimensional (2D) models that apply tensile or compressive loads to cells have been used in the majority of cell mechanobiological studies.

12.2.1 APPLICATION OF TENSILE FORCES TO CELLS

Many types of tissues are subjected to tensile stresses in vivo. Cells in these tissues are generally sensitive to tensile stresses. For example, the tensile stress-responsive cells include fibroblasts in skin, tendons, and ligaments[11,188]; smooth muscle cells (SMCs) in blood vessels, lungs, and intestines[5,57,87]; osteoblasts[80]; chondrocytes[66]; and mesenchymal stem cells (MSCs).[90,115,120] In the majority of tensile stress models, cells are grown on flexible substrates, which are predominantly 2D polyurethane or silicone polymer membranes[20,114,154,170] or elastic scaffolds or hydrogels.[76,81,138] The application of tensile, bending, or distending forces to the substrate results in deformation of the substrate, which in turn stretches cells that adhere to underlying substrate. The substrate can be stretched in two manners: uniaxially or biaxially (Figure 12.2). Under uniaxial stretching, the substrate is elongated along the stretching direction; however, it is compressed in the perpendicular direction due to the Poisson effect of the substrate material. Uniaxial stretching is appropriate for application of mechanical loading to cells from connective tissues including tendons and ligaments, as these cells are aligned with their long axis along the tendons or ligaments and are therefore subjected primarily to uniaxial stretching in vivo. On the other hand, with biaxial stretching the substrate is stretched at two orthogonal directions and hence can be stretched in all directions. This type of mechanical stretching is most suitable for cells that are randomly oriented in the ECM and are stretched in all directions in vivo, e.g., dermal fibroblasts. It should be noted that while stretching is applied to a substrate in the form of tensile force, the mechanical forces sensed by cells are measured by substrate strains, which are more easily defined than mechanical force on cells. However, the substrate strain is, in general, not the actual strain experienced by the cells; in fact, only a percentage of the substrate strain may actually be delivered to the cells. This is partially due to the differential adherence of individual cells in the population of cells—some cells may adhere to the matrix more strongly than others, subjecting them to varying strain levels.[33,142,152]

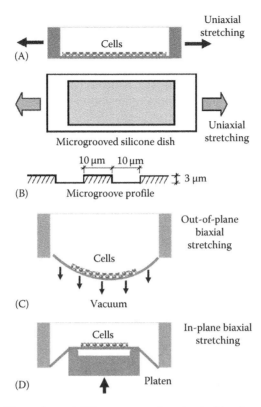

FIGURE 12.2 Schematic illustrations of different types of mechanical loading systems for applying tensile stresses to cells. (A) Uniaxial/longitudinal strain system. (B) A typical microgrooved silicone chamber. The microgrooves have a rectangular profile and align either in parallel or perpendicular to stretching direction (arrows). (Reprinted from Fig. 1 in Wang, J.H. et al., *J. Biomech.*, 37, 573, 2004. With permission.) (C) Out-of-plane biaxial strain system. (D) In-plane equibiaxial strain system.

12.2.1.1 Uniaxial Stretching Systems

Initially, uniaxial mechanical loading systems utilized a motorized plunger–linkage combination to stretch rectangular elastin or collagen ribbons that were fixed at their two ends.[96,97,185] While such systems make it possible to investigate cellular activities in response to mechanical loading, e.g., DNA and protein synthesis of rabbit aortic SMCs,[96,97] it has a few inherent limitations, including the structural irregularity and anisotropy of the substrate as well as the grip end effect, which largely results in nonuniformity in local strains and therefore heterogeneity in cellular responses. Later, an improved technique was developed in that a stepper motor was used to cyclically stretch a plastic base substrate, which was side sealed to form a chamber so that cells could grow inside.[113] The use of stepper motors is advantageous because it permits a broad range of strain inputs, which allows investigators to look into the effect of a wide range of mechanical loading levels.

To improve the substrate strain uniformity, the subsequent studies used rectangular culture wells/chambers that were molded from elastic polymers such as silicone elastomers[114,160,167] (Figure 12.2A). In such systems, strain distribution on the cell culture surface in the middle of a silicone chamber is nearly uniform, and the longitudinal strains along the stretching direction are markedly larger than transverse strains, which are caused by Poisson's effect of the substrate material. For example, when a culture well was stretched at a strain of 3%, the average longitudinal elongation was 2.84% ± 0.05%, while the transverse strain was 0.91% ± 0.03%.[114]

In most uniaxial stretching systems, substrates with smooth culture surfaces have been used. One problem with such stretching systems is that the orientation and organization of cells on these substrates are not controlled; consequently, cells during stretching tend to orient away from the stretching direction[38,39] and move towards a direction that has minimal substrate deformation.[167,170] This results in nonuniformity among individual cells in terms of their "force-bearing" level and duration and therefore leads to heterogeneity in cellular mechanobiological responses. Such a problem can be overcome by using substrates on which cells are guided to align in specific directions.[107,164,165,166,171] Microgrooved substrates are commonly used to align cells along the microgroove direction by taking advantage of cell contact guidance[165,177] (Figure 12.2B). Application of cyclic uniaxial stretching to cells on the microgrooved surface does not cause changes in cell alignment regardless of the cell's initial orientation with respect to stretching direction.[105,176] Because of these advantages, microgrooved substrates have been used in many mechanobiological studies for fibroblasts from connective tissues such as tendons and ligaments[75,173,176,177,182,183] and dermal tissues,[105] SMCs,[148] and stem cells.[89,90] An additional advantage of using the microgrooved substrate is that it can also control the organization of collagen matrix produced by cells in culture.[172] By using microgrooved substrates for mechanical stretching, it is possible to decouple the effects of strain anisotropy on different cellular activities. For instance, the strain anisotropy of SMCs likely determines their structural and functional adaptations to mechanical loading.[148]

Recently, uniaxial stretching has also been achieved by modification of a vacuum-driven Flexcell system, where a straight loading post is placed directly under a portion of the membrane such that its deformation is restricted to a single axis.[159] In addition, miniature stretching apparatuses, which can fit in the microscope stage, have also been developed to apply uniaxial stretching to cells.[169,187] These systems are especially useful when imaging cells and monitoring cell activities during mechanical stretching are desired.

In uniaxial stretching systems, the longitudinal strains can be created to range from 1% up to a few tens percentage, which are mostly suitable for mechanically stimulating fibroblasts from tendons, ligaments, and skin,[4,82,104,105,182] endothelial cells, cardiac fibroblasts, cardiomyocytes, and SMCs from cardiovascular tissues,[6,12,17,55] and (MSCs).[115,130,189] However, such relatively large strain levels are not physiologically amiable for bone cells, where much smaller strain levels ($\ll$1%) occur in vivo. In such cases, small strains can be achieved by means of substrate flexure in a four-point bending apparatus.[18,102,109,118,151,181] In such systems, cells are cultured on substrates including rubber, elastic plastic or alloy strips, or culture plates.[18,118,127,144,181] The substrates are bent by mechanical loads applied directly by motorized and usually computer-controlled machinery. The strain magnitude can be calculated from the displacement of substrate using elastic mechanics or measured directly by a strain gauge attached to the substrate. Depending on the side on which the substrate is loaded, either tensile or compressive stresses in such a four-point bending system can be applied to cells.

12.2.1.2 Biaxial Stretching Systems

The substrate strains realized by uniaxial stretching systems are basically anisotropic; in other words, the strain magnitude and type (tensile vs. compressive) change with respect to stretching direction. Cells in such systems reorient toward the direction of minimum substrate deformation.[171] Consequently, cellular responses in these in vitro systems are heterogeneous, and the gene and protein expressions measured only represent the "averaged" response of a population of cells to mechanical stretching. To overcome this problem, biaxial stretching systems have been developed, in which cells are subjected to the same type of strain in all directions. In addition, in many biaxial stretching systems, standard culture plates, from 6-well to 24-well ones, are used.[37,92,115,148,160] An additional advantage of such a system is its compact size and relatively simple actuating mechanisms, which facilitate live cell imaging as they are compatible with many types of microscopes and related on-stage incubation systems. These features enable investigators to perform cell stretching experiments more efficiently.

In biaxial stretching systems, circular elastic substrates (usually thin membranes) are used to culture cells. Depending on the manner in which mechanical stretching is applied, either anisotropic or isotropic strains are produced on the membrane. In the case of anisotropic strains, substrate strain varies with respect to stretching direction; conversely, isotropic strain does not depend on stretching direction.[91,137] The anisotropic strains are produced when two stretching magnitudes at two orthogonal directions are not equal, whereas the isotropic strains can be produced by so-called equi-biaxial stretching.[8,69,114,143]

Biaxial stretching systems are usually vacuum-driven stretching apparatuses[9,21] (Figure 12.2C). In such systems, elastic membrane-bottomed plastic culture plates are used. A matched vacuum manifold system provides suction power to stretch the membrane from underneath, resulting in biaxial distention of the membrane. Use of a vacuum as the driving force for generation of substrate strains has the advantage that many variables, including vacuum magnitude, frequency, wave pattern, and duty cycle, can be readily adjusted to fulfill the specific requirements of mechanobiological studies for different types of cells. The system has been modified in many aspects to improve its overall quality, uniformity, and adaptability.[21,53,54,123,124] This type of system has been used in many investigations, which address a broad spectrum of cellular events (morphology, phenotype, proliferation, differentiation, gene expression, cytokine or growth factor release, protein production, etc.) in response to various mechanical loading conditions or a combination of mechanical loading and biochemical intervention.[17,37,66,84,86,103,115,117,122,148,157,162]

Similar systems were also created using hydrostatic fluid pressure to press the membrane instead of vacuum.[19,46,180] While such devices could achieve strains as high as 72%,[46] they are relatively more complicated than their vacuum-driven counterparts in terms of assembly and calibration. Also, a common drawback of differential pressure-driven systems is that membrane surface strains are largely influenced by the complicated interaction between culture medium and its underlying substrate due to the relatively thin membrane used. Pretension of the tethered membrane substrate also introduces an additional complication to the strain profile.[21]

Besides the use of vacuum and hydrostatic pressure, another type of system utilizes a templated platen to deform the membrane substrate. Early versions of such systems were prototyped by embedding an orthodontic screw in acrylic resin molded to fit a Petriperm culture dish, which had a flexible plastic membrane in the bottom that could be distended by turning the screw.[22,60] While simple in design and in the ease of device assembly, one drawback of such a system is that the amount of stretching was not uniform throughout the dish. Alternatively, in another design, the bottom of the membrane was distended by pressing it against a convex surface (platen) instead of using a screw. Such an improvement markedly modified the strain pattern, which seemed to be uniform throughout the dish base if the platen was uniformly curved.[14,47,179] The average strain in these systems can be estimated from the arc length of spherical distention of the substrate. However, stretched Petriperm dish does not have ideal optical qualities, which often presents a problem in imaging.

A modified version of these systems used a prong to directly deform the substrate membrane to achieve biaxial strain instead of using platen to template the membrane.[160] Cells growing on the substrate were repetitively stretched and relaxed by vertical movement of a ball-ended prong. Since the activity-controlling unit was directly interfaced with the cell growth chamber via optical data links, such a system was capable of simulating many mechanical loading patterns that cells experience in vivo. The same idea was adopted in a number of similar biaxial stretching systems, which differed only in the geometrical features of the prong and the way of actuating.[2,108,140]

It should be noted that in these so-called out-of-plane biaxial stretching systems, whether driven by differential pressure or platen/prong displacement, large negative radial strains can arise near the tethered edge of the membrane. Consequently, the strains on the membrane substrate are heterogeneous and anisotropic.[21,137,179]

To produce an isotropic strain field on the membrane substrate, in-plane biaxial stretching systems were developed (Figure 12.2D). In such systems, the substrate is confined so that it can be biaxially distended within a plane without vertical deformation. One design uses a chamber with a low-friction O ring as a platen to upwardly indent the center of a peripherally tethered membrane.[26,69,137] Such an arrangement makes it possible to peripherally stretch the central region of membrane in the plane set by the edge of the O ring. The radial and circumferential strains have been shown to be equal using finite element analysis and through optical interference microscopy. The strain levels in such systems can be exactly controlled by the displacement of the platen. A similar device was created, which displaces the membrane downward from the top surface of the membrane instead.[91] In such a system, however, it is essential to keep the platen inert and sterile because it is in direct contact with culture medium. In addition, the movement of the platen causes fluid shear fluctuations, which alter the mechanical environment of the cells and thus may influence their mechanobiological responses to substrate strains.

To accommodate similar needs in a vacuum-driven system, a technique was developed using a low-friction flat surface circular platen that was attached to the center of a peripherally tethered membrane. Here, the vacuum pulls down only the outlying annular portion of the membrane and, as a result, peripherally distends the circular center portion of membrane on top of the platen. Note that in all such systems, frictionless sliding of substrate over the edge or on top of platen is assumed, which in reality is not true, resulting in the strain distribution falling slightly away from ideal homogeneity.

In an entirely different design, a system consisting of a cross-shaped elastic membrane and a movable apparatus has been developed.[116] The membrane is stretched in two perpendicular directions by a movable apparatus clamped at its four legs. While feasible in achieving strain homogeneity in such a system, it is complicated in mechanical design and requires delicate tuning to function properly; therefore, the system is utilized far less.

12.2.2 Application of Compressive Forces to Cells

Besides tensile stress, compressive forces are known to play an important role in tissue physiology including inducing new bone and cartilage formation.[29,83,121] To understand the underlying mechanisms of such mechano-regulation, the development of efficient in vitro systems for application of compressive stress is desirable. One way to apply compressive stresses to cells is through the use of a four-point bending system[118] (Figure 12.3A). Here, cells are cultured on the concave side of a substrate, which is compressed as a result of substrate bending. More often, compressive forces are applied to cells within a monolayer or embedded in a three-dimensional (3D) construct through

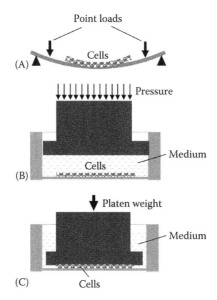

FIGURE 12.3 Schematic illustrations of different types of compressive stress models for applying compressive stresses to cells. (A) Four-point bending system. (B) Hydrostatic pressurization system. (C) Direct platen abutment system.

application of hydrostatic pressure (Figure 12.3B). Such a system typically consists of a pressure vessel and a hydraulic cylinder.[48] The magnitude and frequency of the hydrostatic pressure applied to the cells is controlled by an actuator, which powers the hydraulic cylinder. Because of the simplicity of the equipment, spatial homogeneity of stimulus, and ease of configuring different loading profiles, hydrostatic pressurization has been used in many studies.[3,58,59,70,93,111,133,146,163] However, such a model system is less physiological, as pure hydrostatic pressure on tissues does not likely exist in vivo. Another problem with the system is that the incubator gas pressures cause very high concentrations of O_2 and CO_2 in the medium, which require compensatory treatments.[71,119,150]

As an alternative approach to hydrostatic pressure, direct platen abutment is utilized to exert compressive stress (Figure 12.3C). Many studies used unconfined compression systems, in which constant or low-cycle intermittent loads were delivered by manually applying weights.[24,145] Typically, a thin plate was placed over a layer of confluent cells and a compressive load was applied to cells by placing weights on top of the plate.[110] Such a configuration has been successfully used to apply unconfined axial compression to bone, cartilage, endothelial cells, and tissue biopsies.[34,45] In particular, a low-strain regime (200–40,000 με) and a wide frequency range (a few tenths of 1 Hz up to several hundred Hz) can be readily achieved with this strategy, which is important for mechanical stimulation of bone and cartilage cells.[149] To apply unconfined uniaxial compression in a more precise and controllable way, pneumatically driven systems were developed, in which small cartilage explant discs were loaded by pneumatically driven or servo-controlled platen systems.[10,127,155] Precise loading control over three orders of magnitude (0.001–1.0 MPa) was achieved by using a high-sensitivity load.[10] On the other hand, confined compression systems were developed by placing cells between two pieces of porous platen, which were contained by a confining chamber[32,41] or by employing microscope stage-mounted devices suitable for rectangular specimens of agarose-embedded chondrocytes.[49] In general, the configuration of platen abutment systems is typically simple and straightforward. In addition to the fact that a wide range of specimen deformation is achievable in such systems, this type of mechanical loading regime is somewhat similar to the in vivo environment, especially for cartilage or osteochondral explants. Nonetheless, shortcomings exist, including anisotropic strain field due to the Poisson effect and strain heterogeneity associated with friction at the cell/platen interface.

One common problem in many model systems described above is that cells reside on 2D substrate without matrix surrounding them, which is less physiological as cells in vivo are present in a 3D matrix. Hence, it is desirable to place cells in the 3D environment in vitro to investigate the mechano-responses of cells in a way similar to their in vivo situation.[27,56] In fact, there has been plenty of evidence that differences largely exist between 2D and 3D culture conditions in terms of cell morphology, adhesion, cytoskeletal structures, stress generation, migration, gene expression, and protein production.[125] Fortunately, with slight modification most model systems can be utilized to apply mechanical loads to deformable 3D matrices embedded with cells. Such cell-matrix constructs can be mechanically loaded through various methods ranging from simple uniaxial stretching to multidimensional strain, compressive strains, and fluid shear stresses.

Two major categories of substrate materials are commonly used in embedding cells in 3D matrix. One is hydrogels formed from natural polymers such as collagen, gelatin, fibroin, chitosan, starch, and agar/agarose and synthetic polymers such as poly(N-isopropylacrylamide) (NIPAAm) and its copolymers, PEG-acrylates, and poly(vinyl alcohol) and its copolymers.[65] Examples of such applications include type I collagen sponges as autogenous tissue engineering constructs for rabbit MSCs for patellar tendon repair,[76] 3D rectangular collagen lattices seeded with myoblasts to mimic skeletal muscle,[30,31] and collagen rings seeded with SMCs to mimic blood vessels.[78,79] In situ gelling, in which injectable hydrogels solidify spontaneously under physiological conditions, fosters interest for future research.[158] After cells are embedded in gels, they can be subjected to various mechanical conditioning regimens within a certain range that the gels can bear without breaking. The other category is polymer scaffold matrices, most of which are made from biodegradable polymers, including polyurethanes, polyalkylene esters, polylactic acid and its copolymers, polyamide esters, polyvinyl esters, polyvinyl alcohol, and polyanhydrides. In the artificial constructs composed of cells and matrix materials, mechanical loading not only stimulates the cells inside to change their phenotypes and activities accordingly, but also to modulate the composition and mechanical properties of the constructs.[51]

12.3 MECHANICAL FORCES GENERATED BY CELLS

While external mechanical forces are applied to cells using in vitro model systems, cells also generate internal mechanical force or intracellular tension.[23,62] The intracellular tension is transmitted to their underlying substrate and becomes so-called CTF (Figure 12.4). CTFs function to direct ECM assembly,[95] control cell shape,[88,101,168] permit cell movement,[1,15,73,94,131] and maintain cellular tensional homeostasis.[44,61,136] In addition, CTFs can deform the ECM network and result in stress and strain in the network, which in turn modulate cellular functions, including DNA synthesis, cell differentiation, and ECM protein secretion.[62,156] As such, CTFs play a critical role in many fundamental biological processes including embryogenesis, angiogenesis, and wound healing. In addition, since the vast majority of mechanobiological investigations rely on cell–substrate adhesion to transmit substrate deformation and thereby to apply mechanical forces to cells, CTFs are essential in mediating the mechano-responses of cells.

12.3.1 GENERATION AND TRANSMISSION OF CTFs

In nonmuscle cells, the stress fibers, or bundles of actin filaments, form a semi-sarcomere structure. Inside this structure, the actomyosin cross-bridge, powered by ATP hydrolysis, generates the intracellular tension.[85,134] This tension is transmitted to the ECM via FAs that are located at both ends of the stress fiber and on the substrate or the ECM[7] and as such, CTFs are produced. The FAs consist of assemblies of ECM proteins, transmembrane receptors, and cytoplasmic structural and signaling proteins.

Besides the actin–myosin interaction, a second source for generating CTFs is actin polymerization, which drives forward protrusion of the leading edge of a migrating cell.[16] As CTF generation involves actin–myosin interaction, a number of molecules that are associated with actin and myosin

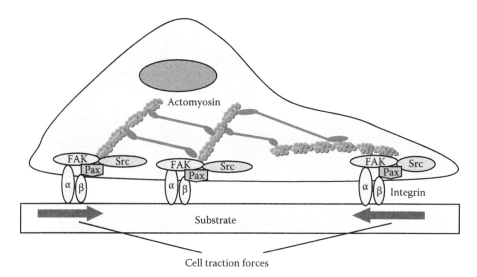

FIGURE 12.4 A schematic illustration of CTFs generated by a cell. CTFs are generated through the actomyosin interactions and act on the underlying substrate through focal adhesion proteins, such as integrins. (Adapted from Fig. 1 in Wang, J.H. and Lin, J.S., *Biomech. Model Mechanobiol.*, 6, 361, 2007. With permission.)

can regulate CTF. These include myosin light chain kinase (MLCK), MLC phosphatase,[36,64,141] and Rho, Rac, and Cdc42.[129]

12.3.2 CTF Measurement

CTFs can be determined using a number of qualitative and quantitative techniques. Using a thin silicone membrane, it has been qualitatively demonstrated that individual fibroblasts transmitted traction forces to their substrate because cells created wrinkles on the membrane.[63] Later, using the same silicone membrane approach, the traction forces generated during cytokinesis of individual cells were quantified using a calibration technique for wrinklable substrates.[25] Force applied to the wrinklable silicone membrane by the adherent cell was estimated by applying a flexible micro-needle of known stiffness to reverse the wrinkles. However, there is currently no mathematical solution available for accurately predicting the wrinkles caused by a complex, non-isotropic traction force field. As a result, absolute values of CTFs cannot be determined through this method. To improve measurement accuracy, fluorescent dot embedded elastic membranes were constructed for measuring CTFs.[7,40,74]

To measure CTFs quantitatively, a micro-machined device has been developed.[50] This device consists of an array of cantilever beams with calibrated stiffness. A cell is plated on the pads underlined with the cantilever beams. When the cell generates traction force and bends a cantilever beam, the extent of bending is recorded and the traction force is then determined through calculation. However, it can only determine the force in one direction and cannot determine the complex, non-isotropic traction force field within the whole cell spreading area. To overcome such a problem, the micropost force sensor array (MFSA) has been developed to measure CTFs.[43,100,147,190] In an MFSA, each micropost functions as an individual force-sensing unit and independently senses the CTF locally applied by the cell[100] (Figure 12.5). Specifically, when a cell adheres to the microposts, it exerts tensile forces at their tops and bends the microposts. The CTF can then be determined according to the deflection of microposts using beam theory.[153] Compared to micro-cantilevers which can only determine CTFs in a single direction,[50] the microposts in an array can detect CTFs in all directions. Therefore, MFSA technology is especially useful in mapping the forces during cell migration.[43] Moreover, such microposts could be used to apply localized mechanical forces to a cell by embedding magnetic cobalt nanowires inside them.[139] Similarly, taking advantage of the periodicity of the micropost array, an optical Moire-based CTF mapping technique was explored by acquiring the diffracted Moire fringe pattern of the array instead of tracking the displacements of individual microposts.[190] Due to the magnification effect of the Moire fringe pattern and having no need to track or visualize individual microposts, this method may potentially determine CTF at a higher resolution.

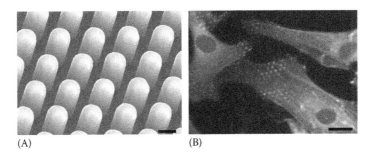

(A) (B)

FIGURE 12.5 (See color insert.) Micropost force sensor array. (A) SEM image of MFSA. Scale bar, 2 μm. (B) Fluorescence microscopy image of human skin fibroblasts cultured on a MFSA. TRITC-conjugated phalloidin (red), anti-vinculin monoclonal antibody and FITC-conjugated secondary antibody (green), and BisBenzimide H33258 (blue) were used to reveal actin filaments, focal contacts, and nuclei, respectively. Scale bar, 20 μm. (Adapted from Figs. 5 and 6 in Li, B. et al., *Cell Motil. Cytoskeleton*, 64, 509, 2007. With permission.)

One shortcoming of the MFSA technique is that it can only measure CTFs at prescribed discrete points. A further advancement in CTF measurement is the development of CTF microscopy (CTFM)[27,42,174] (Figure 12.6), which provides CTFs on the entire cell spreading area. In CTFM method, elastic hydrogels such as polyacrylamide gel (PAG) or gelatin gel are used as cell culture substrate. One advantage of using PAG substrate is that it is linearly elastic in response to a wide range of stress and is completely recoverable upon removal of stress. In addition, the stiffness of PAG can be easily adjusted from as low as 10 Pa up to as high as 40 kPa to cater towards specific requirements for measuring CTFs of certain cell types.[126,178,186] In a typical CTFM test, the hydrogel is embedded with fluorescent microbeads and undergo surface treatment with cross-linking reagents and coating with ECM proteins. Adherent cells spreading on the gel generate CTFs, deform the gel, and cause the embedded beads to displace along. By tracking the movement of beads using a pair of images, the "force-loaded" and the "null-force" images, CTFM reconstructs the CTF-induced displacement field on the surface of the gel. "Force-loaded" images are those taken while the adherent cells remain on the gels, and null-force" images after the cells have been removed. Two steps are involved in deriving CTF from this pair of images. The first step is to solve an image registration problem in which beads from two images are matched and thus the gel displacement field is derived. The second step is to solve an inverse problem in which the CTFs are computed that will give a best match to the corresponding displacement field. By dividing each image into a number of overlapped regions, cross-correlation functions provide a convenient way of matching each region from one image to the other. By further assuming that the gel is thick enough to behave like an elastic half space, the application of the analytical Boussinesq solution obtains the discrete loads as an estimate of the CTF. This can be achieved either by applying inverse Fourier transform, or through solving a general regularized inverse problem. The latter approach is more involved,[28] but allows for the incorporation of a priori information such as levels of errors. Recently, a new CTFM solution has been developed, in which effective pattern recognition algorithms and finite element method (FEM) were introduced to ensure reliable displacement calculation and enable simultaneous determination of CTFs of multiple cells.[184] In this solution, a feature registration scheme was devised to match individual beads from the image pair. Furthermore, the gel is modeled as a 3D object with its actual thickness in FEM using brick elements. The boundary conditions are that the bottom of the gel is fixed, the nodes outside the cell boundary are free from loading, and the surface nodes that are in contact with the cell are given prescribed displacement as interpolated from the extracted displacement field. By applying static condensation, FEM obtains CTF in the form of forces on the surface nodes that lie within the boundary of the cells.

12.3.3 Applications of CTFM Technology

One of the major functions of CTF is to enable cell locomotion. Technologies that measure CTFs, such as CTFM, have been used extensively and effectively to study cell migration and its underlying mechanisms. For example, a recent study sought to learn how CTFs permit cell migration using CTFM.[106] It was found that directed cell movement occurred when there was an asymmetrical distribution of CTF across the cell, i.e., CTFs being significantly higher at the rear than at the front. Furthermore, the speed of cell movement was dependent on how fast CTF asymmetry developed, not the absolute magnitude of CTF. This revealed that CTF asymmetry is critical for polarized cell movement—high CTFs at the rear promote retraction, whereas low CTFs at the front allow protrusion of the cell. Such an observation was further confirmed in another study by simultaneous recording of F-actin and CTFs, which demonstrated that CTFs were exerted under spot-like regions where F-actin accumulated.[74] These stress spots acted as scaffolds to transmit the propulsive forces at the leading edge generated by actin polymerization. Meanwhile, accumulation of filamentous myosin II and a subsequent increase in CTFs demonstrated that the source of retraction force was the accumulated myosin II.

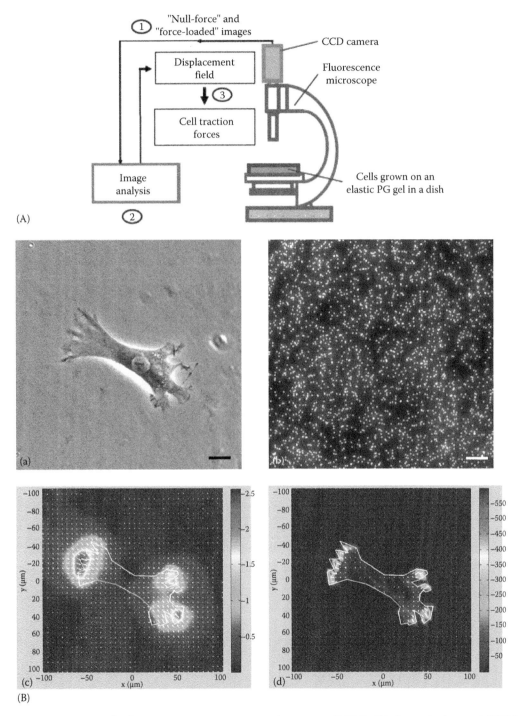

FIGURE 12.6 (See color insert.) Cell traction force microscopy (CTFM). (A) General scheme of CTFM. (Adapted from Fig. 2 in Wang and Thampatty, *Int. Rev Cell Mol. Biol.*, 271, 301, 2008. With permission.) (B) Application of CTFM to determine CTFs of a human patellar tendon fibroblast (HPTF): (a) HPTF on a PAG, (b) fluorescence image of fluorescent beads embedded in gel, (c) substrate displacement field, and (d) cell traction force field. Scale bars, 20 μm.

Moreover, because generation of CTFs involves the interaction between actin and myosin, the components that are involved in almost all cellular biological processes, any biological, biochemical, or biomechanical changes of cells will likely be reflected in CTF changes. For example, in tumorigenesis and metastasis, transformed cells have lower CTFs than non-transformed cells,[62,112] but CTFs increase in activated fibroblasts.[77] Therefore, CTF may serve as a useful "biophysical marker" to characterize specific cell phenotype. As a result, CTFM can be used to characterize phenotypic changes of cells. Results from CTFM will aid in understanding many fundamental biological processes, including embryogenesis, angiogenesis, and metastasis.

The integration of cell micropatterning and CTFM may provide new insights into fundamental understanding of certain biological problems. Here, an example is shown with respect to using CTFM technology to develop useful in vitro models for disease study. This model was developed by studying the contractility of micropatterned C2C12 skeletal muscle cells using CTFM.[99] In this model, several C2C12 cells were micropatterned on a rectangular adhesive island such that the cells assumed a shape similar to that of a typical myotube. During differentiation, these cells gradually fused into a myotube; meanwhile, their contractile forces, represented by CTFs, continually increased until the myotube reached maturation (Figure 12.7). Therefore, CTFM can directly

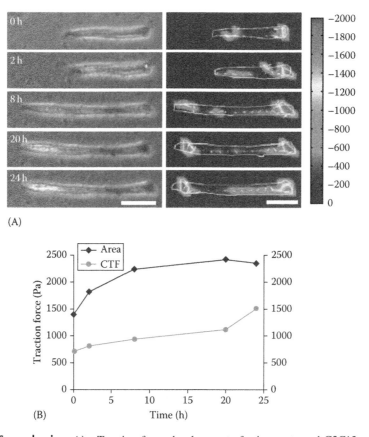

(A)

(B)

FIGURE 12.7 (See color insert.) Traction force development of micropatterned C2C12 cells during differentiation. (A) Phase contrast images (left panel) and corresponding traction force maps (right panel), respectively. The color bar represents traction force level. Note that prior to plating on PAG, cells were pre-differentiated for 2 days in differentiation medium. Note the phase contrast images were slightly out of focus due to the requirement of in situ fluorescence image acquisition. Scale bars, 50 μm. (B) Changes in cell spreading area and cell traction force during C2C12 differentiation between first and second days of culture. (Reprinted from Fig. 5 in Li, B. et al., *J. Biomech.*, 41, 3349, 2008. With permission.)

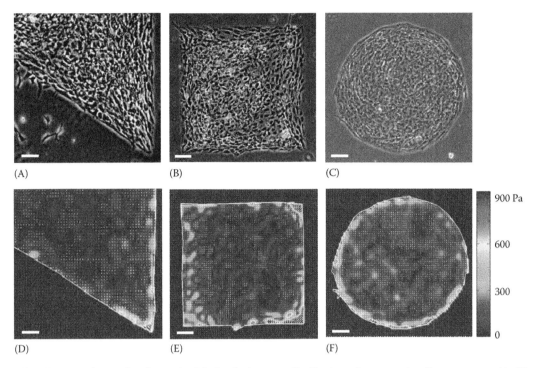

FIGURE 12.8 (See color insert.)   Mechanical stress distribution of patterned cell aggregates. (A–C) Micropatterned cell islands of distinctive shapes. (D–F) CTF distribution of cell islands determined by CTFM. Scale bars, 100 µm. (Adapted from Fig. 2 in Li, B. et al., *J. Biomech.*, 42, 1622, 2009. With permission.)

quantify the contractile forces of myotube, a precursor of myofiber which is the fundamental structural unit of muscle tissues and is responsible for muscle contraction. Such a model can potentially be used as a fast screening approach for therapies of Duchenne muscular dystrophy (DMD), a serious muscle atrophy disease.

Not only can CTFM be used for individual cells, it can also be applied to a population of cells, i.e., cell aggregate. In a recent study aiming to explore the exact mechanical signal that leads to local tissue pattern formation during tissue morphogenesis, fibroblasts were cultured on micropatterned adhesive islands and formed cell islands of specific shapes (Figure 12.8A-C). Using CTFM, it was found that the cells at the islands' perimeters produced large mechanical stresses (Figure 12.8D-F), which is consistent with the result of stress analysis by finite element method. The stress pattern closely resembled cell proliferation and differentiation patterns. However, when overall mechanical stress levels of cell islands were elevated by application of mechanical stretching, neither cell proliferation nor differentiation patterns followed the new mechanical stress pattern although both cell proliferation and differentiation increased. Thus, the results indicate that that mechanical stress magnitude, instead of stress gradient as previously proposed, determines the formation of specific cell proliferation and differentiation patterns.[98]

12.4   CONCLUDING REMARKS

Mechanical forces are known to regulate various cellular functions, including proliferation, differentiation, gene expression, and protein secretion. While appropriate mechanical loading is beneficial to cells, excessive mechanical loading leads to detrimental changes in cells. Thus, to better understand the mechanisms of tissue homeostasis and pathophysiology, it is highly desirable to determine how cells respond to mechanical forces in terms of their anabolic and catabolic responses. To this end, many in vitro model systems have been developed and they have helped generate much knowledge of cell mechanobiology. It is now clear that cell mechanobiological responses depend on mechanical loading conditions (e.g., mechanical force type, magnitude, frequency, and duration). In addition, surface chemistry and texture of cell culture substrates also influence cell mechanobiological responses in distinct ways. Therefore, one has to carefully consider various factors when designing experiments for specific cell types as well as interpreting experimental results.

Improvements in the understanding of cell mechanobiology have come with technological innovations. In order to further advance research in this area, many of the existing technologies need to be improved. First, the key to cellular mechanobiological studies is to determine how much mechanical force is applied to cells and what levels of mechanical stimulation cause anabolic or catabolic responses. Therefore, new techniques need to be developed to measure external mechanical forces in vivo so that in vitro studies can be performed under conditions similar to physiological situations. Second, due to the mechanical loading duration dependence of cell mechanobiological responses, technologies that permit real-time monitoring of gene and protein expression of the ECM in response to mechanical stimuli need to be developed and refined; these technologies will be useful for determining optimal mechanical loading parameters. Third, while current cell mechanobiological investigations have provided valuable insights in determining cell mechanobiological responses as well as mechanotransduction (or the mechanisms of how cells transduce mechanical loading signals into a cascade of biological events), most investigations use 2D substrates, instead of 3D matrices as in physiological conditions. Therefore, future research should focus on studying cell mechanobiology in a more physiological 3D context, which enables one to examine the integrated effect of an ECM scaffold, transportation of molecules in such matrices, and mechanical loading conditions on cells. This research endeavor requires new methodologies that permit manipulating cells and measuring cellular behaviors in 3D environments. Finally, as an essential biophysical and functional characterization method of cells, new improvements in CTF techniques also need to be developed to quantify CTFs in a 3D matrix, which is more physiological compared to 2D substrates currently being used.

REFERENCES

1. Ananthakrishnan, R. and Ehrlicher, A. 2007. The forces behind cell movement. *Int J Biol Sci.* 3:303–317.
2. Andersen, K.L. and Norton, L.A. 1991. A device for the application of known simulated orthodontic forces to human cells in vitro. *J Biomech.* 24:649–654.
3. Angele, P., Yoo, J.U., Smith, C., Mansour, J., Jepsen, K.J., Nerlich, M., and Johnstone, B. 2003. Cyclic hydrostatic pressure enhances the chondrogenic phenotype of human mesenchymal progenitor cells differentiated in vitro. *J Orthop Res.* 21:451–457.
4. Archambault, J., Tsuzaki, M., Herzog, W., and Banes, A.J. 2002. Stretch and interleukin-1beta induce matrix metalloproteinases in rabbit tendon cells in vitro. *J Orthop Res.* 20:36–39.
5. Asanuma, K., Magid, R., Johnson, C., Nerem, R.M., and Galis, Z.S. 2003. Uniaxial strain upregulates matrix-degrading enzymes produced by human vascular smooth muscle cells. *Am J Physiol Heart Circ Physiol.* 284:H1778–H1784.
6. Atance, J., Yost, M.J., and Carver, W. 2004. Influence of the extracellular matrix on the regulation of cardiac fibroblast behavior by mechanical stretch. *J Cell Physiol.* 200:377–386.
7. Balaban, N.Q., Schwarz, U.S., Riveline, D., Goichberg, P., Tzur, G., Sabanay, I., Mahalu, D., Safran, S., Bershadsky, A., Addadi, L., and Geiger, B. 2001. Force and focal adhesion assembly: A close relationship studied using elastic micropatterned substrates. *Nat Cell Biol.* 3:466–472.
8. Balestrini, J.L. and Billiar, K.L. 2006. Equibiaxial cyclic stretch stimulates fibroblasts to rapidly remodel fibrin. *J Biomech.* 39:2983–2990.
9. Banes, A.J., Gilbert, J., Taylor, D., and Monbureau, O. 1985. A new vacuum-operated stress-providing instrument that applies static or variable duration cyclic tension or compression to cells in vitro. *J Cell Sci.* 75:35–42.

10. Banes, A.J., Sanderson, M., Boitano, S., Hu, P., Brigman, B., Tzusaki, M., Fischer, T., and Lawrence, W.T. 1994. *Mechanical Load +/− Growth Factors Induce [Ca²⁺]ᵢ Release, Cyclin D₁ Expression and DNA Synthesis in Avian Tendon Cells.* Springer-Verlag, New York, pp. 210–232.
11. Barkhausen, T., Van Griensven, M., Zeichen, J., and Bosch, U. 2003. Modulation of cell functions of human tendon fibroblasts by different repetitive cyclic mechanical stress patterns. *Exp Toxicol Pathol.* 55:153–158.
12. Barron, V., Brougham, C., Coghlan, K., McLucas, E., O'Mahoney, D., Stenson-Cox, C., and McHugh, P.E. 2007. The effect of physiological cyclic stretch on the cell morphology, cell orientation and protein expression of endothelial cells. *J Mater Sci Mater Med.* 18:1973–1981.
13. Bartling, B., Tostlebe, H., Darmer, D., Holtz, J., Silber, R.E., and Morawietz, H. 2000. Shear stress-dependent expression of apoptosis-regulating genes in endothelial cells. *Biochem Biophys Res Commun.* 278:740–746.
14. Basdra, E.K., Papavassiliou, A.G., and Huber, L.A. 1995. Rab and Rho Gtpases are involved in specific response of periodontal-ligament fibroblasts to mechanical stretching. *Biochim Biophys Acta* 1268:209–213.
15. Beningo, K.A., Dembo, M., Kaverina, I., Small, J.V., and Wang, Y.L. 2001. Nascent focal adhesions are responsible for the generation of strong propulsive forces in migrating fibroblasts. *J Cell Biol.* 153:881–888.
16. Bereiter-Hahn, J. 2005. Mechanics of crawling cells. *Med Eng Phys.* 27:743–753.
17. Birukov, K.G., Shirinsky, V.P., Stepanova, O.V., Tkachuk, V.A., Hahn, A.W., Resink, T.J., and Smirnov, V.N. 1995. Stretch affects phenotype and proliferation of vascular smooth muscle cells. *Mol Cell Biochem.* 144:131–139.
18. Bottlang, M., Simnacher, M., Schmitt, H., Brand, R.A., and Claes, L. 1997. A cell strain system for small homogeneous strain applications. *Biomed Tech.* 42:305–309.
19. Brighton, C.T., Fisher, J.R.S., Levine, S.E., Corsetti, J.R., Reilly, T., Landsman, A.S., Williams, J.L., and Thibault, L.E. 1996. The biochemical pathway mediating the proliferative response of bone cells to a mechanical stimulus. *J Bone Joint Surg.* 78A:1337–1347.
20. Brighton, C.T., Sennett, B.J., Farmer, J.C., Iannotti, J.P., Hansen, C.A., Williams, J.L., and Williamson, J. 1992. The inositol phosphate pathway as a mediator in the proliferative response of rat calvarial bone cells to cyclical biaxial mechanical strain. *J Orthop Res.* 10:385–393.
21. Brown, T.D. 2000. Techniques for mechanical stimulation of cells in vitro: A review. *J Biomech.* 33:3–14.
22. Brunette, D.M. 1984. Mechanical stretching increases the number of epithelial cells synthesizing DNA in culture. *J Cell Sci.* 69:35–45.
23. Burridge, K. and Chrzanowska-Wodnicka, M. 1996. Focal adhesions, contractility, and signaling. *Annu Rev Cell Dev Biol.* 12:463–518.
24. Burton-Wurster, N., Vernier-Singer, M., Farquhar, T., and Lust, G. 1993. Effect of compressive loading and unloading on the synthesis of total protein, proteoglycan, and fibronectin by canine cartilage explants. *J Orthop Res.* 11:717–729.
25. Burton, K. and Taylor, D.L. 1997. Traction forces of cytokinesis measured with optically modified elastic substrata. *Nature.* 385:450–454.
26. Butcher, J.T., Barrett, B.C., and Nerem, R.M. 2006. Equibiaxial strain stimulates fibroblastic phenotype shift in smooth muscle cells in an engineered tissue model of the aortic wall. *Biomaterials.* 27:5252–5258.
27. Butler, D.L., Juncosa-Melvin, N., Boivin, G.P., Galloway, M.T., Shearn, J.T., Gooch, C., and Awad, H. 2008. Functional tissue engineering for tendon repair: A multidisciplinary strategy using mesenchymal stem cells, bioscaffolds, and mechanical stimulation. *J Orthop Res.* 26:1–9.
28. Butler, J.P., Tolic-Norrelykke, I.M., Fabry, B., and Fredberg, J.J. 2002. Traction fields, moments, and strain energy that cells exert on their surroundings. *Am J Physiol Cell Physiol.* 282:C595–C605.
29. Chambers, T.J., Evans, M., Gardner, T.N., Turner-Smith, A., and Chow, J.W. 1993. Induction of bone formation in rat tail vertebrae by mechanical loading. *Bone Miner.* 20:167–178.
30. Cheema, U., Brown, R., Mudera, V., Yang, S.Y., McGrouther, G., and Goldspink, G. 2005. Mechanical signals and IGF-I gene splicing in vitro in relation to development of skeletal muscle. *J Cell Physiol.* 202:67–75.
31. Cheema, U., Yang, S.Y., Mudera, V., Goldspink, G.G., and Brown, R.A. 2003. 3-D in vitro model of early skeletal muscle development. *Cell Motil Cytoskeleton.* 54:226–236.
32. Chen, A.C., and Sah, R.L. 1998. Effect of static compression on proteoglycan biosynthesis by chondrocytes transplanted to articular cartilage in vitro. *J Orthop Res.* 16:542–550.

33. Chen, S.J., Yuan, W., Mori, Y., Levenson, A., Trojanowska, M., and Varga, J. 1999. Stimulation of type I collagen transcription in human skin fibroblasts by TGF-beta involvement of Smad 3. *J Invest Dermatol*. 112:49–57.

34. Cheng, G.C., Libby, P., Grodzinsky, A.J., and Lee, R.T. 1996. Induction of DNA synthesis by a single transient mechanical stimulus of human vascular smooth muscle cells. Role of fibroblast growth factor-2. *Circulation*. 93:99–105.

35. Chien, S., Li, S., Shiu, Y.T., and Li, Y.S. 2005. Molecular basis of mechanical modulation of endothelial cell migration. *Front Biosci*. 10:1985–2000.

36. Chrzanowska-Wodnicka, M. and Burridge, K. 1996. Rho-stimulated contractility drives the formation of stress fibers and focal adhesions. *J Cell Biol*. 133:1403–1415.

37. Clarke, M.S. and Feeback, D.L. 1996. Mechanical load induces sarcoplasmic wounding and FGF release in differentiated human skeletal muscle cultures. *FASEB J*. 10:502–509.

38. Dartsch, P.C. and Hammerle, H. 1986. Orientation response of arterial smooth muscle cells to mechanical stimulation. *Eur J Cell Biol*. 41:339–346.

39. Dartsch, P.C., Hammerle, H., and Betz, E. 1986. Orientation of cultured arterial smooth muscle cells growing on cyclically stretched substrates. *Acta Anatomica*. 125:108–113.

40. Das, T., Maiti, T.K., and Chakraborty, S. 2008. Traction force microscopy on-chip: Shear deformation of fibroblast cells. *Lab Chip*. 8:1308–1318.

41. Davisson, T., Kunig, S., Chen, A., Sah, R., and Ratcliffe, A. 2002. Static and dynamic compression modulate matrix metabolism in tissue engineered cartilage. *J Orthop Res*. 20:842–848.

42. Dembo, M., Oliver, T., Ishihara, A., and Jacobson, K. 1996. Imaging the traction stresses exerted by locomoting cells with the elastic substratum method. *Biophys J*. 70:2008–2022.

43. Du Roure, O., Saez, A., Buguin, A., Austin, R.H., Chavrier, P., Silberzan, P., and Ladoux, B. 2005. Force mapping in epithelial cell migration. *Proc Natl Acad Sci U S A*. 102:2390–2395.

44. Eckes, B. and Krieg, T. 2004. Regulation of connective tissue homeostasis in the skin by mechanical forces. *Clin Exp Rheumatol*. 22:S73–S76.

45. El Haj, A.J., Minter, S.L., Rawlinson, S.C., Suswillo, R., and Lanyon, L.E. 1990. Cellular responses to mechanical loading in vitro. *J Bone Miner Res*. 5:923–932.

46. Ellis, E.F., Mckinney, J.S., Willoughby, K.A., Liang, S., and Povlishock, J.T. 1995. A new model for rapid stretch-induced injury of cells in culture: Characterization of the model using astrocytes. *J Neurotrauma*. 12:325–339.

47. Felix, J.A., Woodruff, M.L., and Dirksen, E.R. 1996. Stretch increases inositol 1,4,5-yriphosphate concentration in airway epithelial cell. *Am J Respir Cell Mol Biol*. 14:296–301.

48. Finger, A.R., Sargent, C.Y., Dulaney, K.O., Bernacki, S.H., and Loboa, E.G. 2007. Differential effects on messenger ribonucleic acid expression by bone marrow-derived human mesenchymal stem cells seeded in agarose constructs due to ramped and steady applications of cyclic hydrostatic pressure. *Tissue Eng*. 13:1151–1158.

49. Freeman, P.M., Natarajan, R.N., Kimura, J.H., and Andriacchi, T.P. 1994. Chondrocyte cells respond mechanically to compressive loads. *J Orthop Res*. 12:311–320.

50. Galbraith, C.G. and Sheetz, M.P. 1997. A micromachined device provides a new bend on fibroblast traction forces. *Proc Natl Acad Sci U S A*. 94:9114–9118.

51. Garvin, J., Qi, J., Maloney, M., and Banes, A.J. 2003. Novel system for engineering bioartificial tendons and application of mechanical load. *Tissue Eng*. 9:967–979.

52. Geiger, B. and Bershadsky, A. 2002. Exploring the neighborhood: Adhesion-coupled cell mechanosensors. *Cell*. 110:139–142.

53. Gilbert, J.A., Weinhold, P.S., Banes, A.J., Link, G.W., and Jones, G.L. 1994. Strain profiles for circular cell-culture plates containing flexible surfaces employed to mechanically deform cells in-vitro. *J Biomech*. 27:1169–1177.

54. Gilbert, J.L., Banes, A.J., Link, G.W., and Jones, G.L. 1990. Video analysis of membrane strain—An application in cell stretching. *Exp Tech*. 14:43–45.

55. Gopalan, S.M., Flaim, C., Bhatia, S.N., Hoshijima, M., Knoell, R., Chien, K.R., Omens, J.H., and McCulloch, A.D. 2003. Anisotropic stretch-induced hypertrophy in neonatal ventricular myocytes micropatterned on deformable elastomers. *Biotechnol Bioeng*. 81:578–587.

56. Guilak, F., Butler, D.L., and Goldstein, S.A. 2001. Functional tissue engineering: The role of biomechanics in articular cartilage repair. *Clin Orthop Relat Res*. 391:S295–S305.

57. Gutierrez, J.A. and Perr, H.A. 1999. Mechanical stretch modulates TGF-beta1 and alpha1(I) collagen expression in fetal human intestinal smooth muscle cells. *Am J Physiol*. 277:G1074–G1080.

58. Hall, A.C., Urban, J.P., and Gehl, K.A. 1991. The effects of hydrostatic pressure on matrix synthesis in articular cartilage. *J Orthop Res*. 9:1–10.

59. Hansen, U., Schunke, M., Domm, C., Ioannidis, N., Hassenpflug, J., Gehrke, T., and Kurz, B. 2001. Combination of reduced oxygen tension and intermittent hydrostatic pressure: A useful tool in articular cartilage tissue engineering. *J Biomech*. 34:941–949.

60. Harell, A., Dehel, S., and Binderman, I. 1977. Biochemical effect of mechanical stress on cultured bone cells. *Calc Tiss Res*. 22:202–207.

61. Harris, A.K. 1987. Cell motility and the problem of anatomical homeostasis. *J Cell Sci Suppl*. 8:121–140.

62. Harris, A.K., Stopak, D., and Wild, P. 1981. Fibroblast traction as a mechanism for collagen morphogenesis. *Nature*. 290:249–251.

63. Harris, A.K., Wild, P., and Stopak, D. 1980. Silicone rubber substrata: A new wrinkle in the study of cell locomotion. *Science*. 208:177–179.

64. Hartshorne, D.J., Ito, M., and Erdodi, F. 1998. Myosin light chain phosphatase: Subunit composition, interactions and regulation. *J Muscle Res Cell Motil*. 19:325–341.

65. Hoffman, A.S. 2002. Hydrogels for biomedical applications. *Adv Drug Deliv Rev*. 54:3–12.

66. Holmvall, K., Camper, L., Johansson, S., Kimura, J.H., and Lundgren-Akerlund, E. 1995. Chondrocyte and chondrosarcoma cell integrins with affinity for collagen type II and their response to mechanical stress. *Exp Cell Res*. 221:496–503.

67. Hsieh, M.H., and Nguyen, H.T. 2005. Molecular mechanism of apoptosis induced by mechanical forces. *Int Rev Cytol*. 245:45–90.

68. Huang, H., Kamm, R.D., and Lee, R.T. 2004. Cell mechanics and mechanotransduction: Pathways, probes, and physiology. *Am J Physiol Cell Physiol*. 287:C1–C11.

69. Hung, C.T. and Williams, J.L. 1994. A method for inducing equi-biaxial and uniform strains in elastomeric membranes used as cell substrates. *J Biomech*. 27:227–232.

70. Ikenoue, T., Trindade, M.C., Lee, M.S., Lin, E.Y., Schurman, D.J., Goodman, S.B., and Smith, R.L. 2003. Mechanoregulation of human articular chondrocyte aggrecan and type II collagen expression by intermittent hydrostatic pressure in vitro. *J Orthop Res*. 21:110–116.

71. Imamura, K., Ozawa, H., Hiraide, T., Takahashi, N., Shibasaki, Y., Fukuhara, T., and Suda, T. 1990. Continuously applied compressive pressure induces bone resorption by a mechanism involving prostaglandin E2 synthesis. *J Cell Physiol*. 144:222–228.

72. Ingber, D. 1991. Integrins as mechanochemical transducers. *Curr Opin Cell Biol*. 3:841–848.

73. Ingber, D.E. 2003. Mechanobiology and diseases of mechanotransduction. *Ann Med*. 35:564–577.

74. Iwadate, Y. and Yumura, S. 2008. Actin-based propulsive forces and myosin-II-based contractile forces in migrating Dictyostelium cells. *J Cell Sci*. 121:1314–1324.

75. Jones, B.F., Wall, M.E., Carroll, R.L., Washburn, S., and Banes, A.J. 2005. Ligament cells stretch-adapted on a microgrooved substrate increase intercellular communication in response to a mechanical stimulus. *J Biomech*. 38:1653–1664.

76. Juncosa-Melvin, N., Shearn, J.T., Boivin, G.P., Gooch, C., Galloway, M.T., West, J.R., Nirmalanandhan, V.S., Bradica, G., and Butler, D.L. 2006. Effects of mechanical stimulation on the biomechanics and histology of stem cell-collagen sponge constructs for rabbit patellar tendon repair. *Tissue Eng*. 12:2291–2300.

77. Kalluri, R. and Zeisberg, M. 2006. Fibroblasts in cancer. *Nat Rev Cancer*. 6:392–401.

78. Kanda, K. and Matsuda, T. 1994. Mechanical stress-induced orientation and ultrastructural change of smooth muscle cells cultured in three-dimensional collagen lattices. *Cell Transplant*. 3:481–492.

79. Kanda, K., Matsuda, T., and Oka, T. 1993. Mechanical stress induced cellular orientation and phenotypic modulation of 3-D cultured smooth muscle cells. *ASAIO J*. 39:M686–M690.

80. Kaspar, D., Seidl, W., Neidlinger-Wilke, C., Ignatius, A., and Claes, L. 2000. Dynamic cell stretching increases human osteoblast proliferation and CICP synthesis but decreases osteocalcin synthesis and alkaline phosphatase activity. *J Biomech*. 33:45–51.

81. Kessler, D., Dethlefsen, S., Haase, I., Plomann, M., Hirche, F., Krieg, T., and Eckes, B. 2001. Fibroblasts in mechanically stressed collagen lattices assume a "synthetic" phenotype. *J Biol Chem*. 276:36575–36585.

82. Kim, S.G., Akaike, T., Sasagawa, T., Atomi, Y., and Kurosawa, H. 2002. Gene expression of type I and type III collagen by mechanical stretch in anterior cruciate ligament cells. *Cell Struct Funct*. 27:139–144.

83. Klein-Nulend, J., Veldhuijzen, J.P., and Burger, E.H. 1986. Increased calcification of growth plate cartilage as a result of compressive force in vitro. *Arthritis Rheum*. 29:1002–1009.

84. Koike, M., Shimokawa, H., Kanno, Z., Ohya, K., and Soma, K. 2005. Effects of mechanical strain on proliferation and differentiation of bone marrow stromal cell line ST2. *J Bone Miner Metab*. 23:219–225.

85. Kolega, J., Janson, L.W., and Taylor, D.L. 1991. The role of solation-contraction coupling in regulating stress fiber dynamics in nonmuscle cells. *J Cell Biol.* 114:993–1003.
86. Ku, C.H., Johnson, P.H., Batten, P., Sarathchandra, P., Chambers, R.C., Taylor, P.M., Yacoub, M.H., and Chester, A.H. 2006. Collagen synthesis by mesenchymal stem cells and aortic valve interstitial cells in response to mechanical stretch. *Cardiovasc Res.* 71:548–556.
87. Kulik, T.J. and Alvarado, S.P. 1993. Effect of stretch on growth and collagen synthesis in cultured rat and lamb pulmonary arterial smooth muscle cells. *J Cell Physiol.* 157:615–624.
88. Kumar, S., Maxwell, I.Z., Heisterkamp, A., Polte, T.R., Lele, T.P., Salanga, M., Mazur, E., and Ingber, D.E. 2006. Viscoelastic retraction of single living stress fibers and its impact on cell shape, cytoskeletal organization, and extracellular matrix mechanics. *Biophys J.* 90:3762–3773.
89. Kurpinski, K., Chu, J., Hashi, C., and Li, S. 2006. Anisotropic mechanosensing by mesenchymal stem cells. *Proc Natl Acad Sci U S A.* 103:16095–16100.
90. Kurpinski, K. and Li, S. 2007. Mechanical stimulation of stem cells using cyclic uniaxial strain. *J Vis Exp.* 6:242.
91. Lee, A.A., Delhaas, T., Waldman, L.K., MacKenna, D.A., Villarreal, F.J., and McCulloch, A.D. 1996. An equibiaxial strain system for cultured cells. *Am J Physiol.* 271:C1400–C1408.
92. Lee, C.H., Shin, H.J., Cho, I.H., Kang, Y.M., Kim, I.A., Park, K.D., and Shin, J.W. 2005. Nanofiber alignment and direction of mechanical strain affect the ECM production of human ACL fibroblast. *Biomaterials.* 26:1261–1270.
93. Lee, D.A. and Bader, D.L. 1997. Compressive strains at physiological frequencies influence the metabolism of chondrocytes seeded in agarose. *J Orthop Res.* 15:181–188.
94. Lee, J., Leonard, M., Oliver, T., Ishihara, A., and Jacobson, K. 1994. Traction forces generated by locomoting keratocytes. *J Cell Biol.* 127:1957–1964.
95. Lemmon, C.A., Chen, C.S., and Romer, L.H. 2009. Cell traction forces direct fibronectin matrix assembly. *Biophys J.* 96:729–738.
96. Leung, D.Y., Glagov, S., and Mathews, M.B. 1976. Cyclic stretching stimulates synthesis of matrix components by arterial smooth muscle cells in vitro. *Science.* 191:475–477.
97. Leung, D.Y., Glagov, S., and Mathews, M.B. 1977. A new in vitro system for studying cell response to mechanical stimulation. Different effects of cyclic stretching and agitation on smooth muscle cell biosynthesis. *Exp Cell Res.* 109:285–298.
98. Li, B., Li, F., Puskar, K.M., and Wang, J.H. 2009. Spatial patterning of cell proliferation and differentiation depends on mechanical stress magnitude. *J Biomech.* 42:1622–1627.
99. Li, B., Lin, M., Tang, Y., Wang, B., and Wang, J.H. 2008. A novel functional assessment of the differentiation of micropatterned muscle cells. *J Biomech.* 41:3349–3353.
100. Li, B., Xie, L., Starr, Z.C., Yang, Z., Lin, J.S., and Wang, J.H. 2007. Development of micropost force sensor array with culture experiments for determination of cell traction forces. *Cell Motil Cytoskeleton.* 64:509–518.
101. Li, F., Li, B., Wang, Q.M., and Wang, J.H. 2008. Cell shape regulates collagen type I expression in human tendon fibroblasts. *Cell Motil Cytoskeleton.* 65:332–341.
102. Li, J., Chen, G.P., Zheng, L.L., Luo, S.J., and Zhao, Z.H. 2007. Osteoblast cytoskeletal modulation in response to compressive stress at physiological levels. *Mol Cell Biochem.* 304:45–52.
103. Li, Q., Muragaki, Y., Hatamura, I., Ueno, H., and Ooshima, A. 1998. Stretch-induced collagen synthesis in cultured smooth muscle cells from rabbit aortic media and a possible involvement of angiotensin II and transforming growth factor-beta. *J Vasc Res.* 35:93–103.
104. Li, Z., Yang, G., Khan, M., Stone, D., Woo, S.L., and Wang, J.H. 2004. Inflammatory response of human tendon fibroblasts to cyclic mechanical stretching. *Am J Sports Med.* 32:435–440.
105. Loesberg, W.A., Walboomers, X.F., van Loon, J.J., and Jansen, J.A. 2005. The effect of combined cyclic mechanical stretching and microgrooved surface topography on the behavior of fibroblasts. *J Biomed Mater Res A.* 75:723–732.
106. Lombardi, M.L., Knecht, D.A., Dembo, M., and Lee, J. 2007. Traction force microscopy in Dictyostelium reveals distinct roles for myosin II motor and actin-crosslinking activity in polarized cell movement. *J Cell Sci.* 120:1624–1634.
107. Mata, A., Boehm, C., Fleischman, A.J., Muschler, G., and Roy, S. 2002. Analysis of connective tissue progenitor cell behavior on polydimethylsiloxane smooth and channel micro-textures. *Biomed Microdevices.* 4:267–275.
108. Matsuo, T., Uchida, H., and Matsuo, N. 1996. Bovine and porcine trabecular cells produce prostaglandin F2 alpha in response to cyclic mechanical stretching. *Jpn J Ophthalmol.* 40:289–296.

109. Mauney, J.R., Sjostorm, S., Blumberg, J., Horan, R., O'Leary, J.P., Vunjak-Novakovic, G., Volloch, V., and Kaplan, D.L. 2004. Mechanical stimulation promotes osteogenic differentiation of human bone marrow stromal cells on 3-D partially demineralized bone scaffolds in vitro. *Calcif Tissue Int.* 74:458–468.

110. Mitsui, N., Suzuki, N., Koyama, Y., Yanagisawa, M., Otsuka, K., Shimizu, N., and Maeno, M. 2006. Effect of compressive force on the expression of MMPs, PAs, and their inhibitors in osteoblastic Saos-2 cells. *Life Sci.* 79:575–583.

111. Miyanishi, K., Trindade, M.C., Lindsey, D.P., Beaupre, G.S., Carter, D.R., Goodman, S.B., Schurman, D.J., and Smith, R.L. 2006. Effects of hydrostatic pressure and transforming growth factor-beta 3 on adult human mesenchymal stem cell chondrogenesis in vitro. *Tissue Eng.* 12:1419–1428.

112. Munevar, S., Wang, Y., and Dembo, M. 2001. Traction force microscopy of migrating normal and H-ras transformed 3T3 fibroblasts. *Biophys J.* 80:1744–1757.

113. Murray, D.W. and Rushton, N. 1990. The effect of strain on bone cell prostaglandin E2 release: A new experimental method. *Calcif Tissue Int.* 47:35–39.

114. Neidlinger-Wilke, C., Wilke, H.J., and Claes, L. 1994. Cyclic stretching of human osteoblasts affects proliferation and metabolism: A new experimental method and its application. *J Orthop Res.* 12:70–78.

115. Nieponice, A., Maul, T.M., Cumer, J.M., Soletti, L., and Vorp, D.A. 2007. Mechanical stimulation induces morphological and phenotypic changes in bone marrow-derived progenitor cells within a three-dimensional fibrin matrix. *J Biomed Mater Res A* 81:523–530.

116. Norton, L.A., Andersen, K.L., Arenholtbindslev, D., Andersen, L., and Melsen, B. 1995. A methodical study of shape changes in human oral cells perturbed by a simulated orthodontic strain in-vitro. *Arch Oral Biol.* 40:863–872.

117. O'Callaghan, C.J. and Williams, B. 2000. Mechanical strain-induced extracellular matrix production by human vascular smooth muscle cells: Role of TGF-beta(1). *Hypertension.* 36:319–324.

118. Owan, I., Burr, D.B., Turner, C.H., Qiu, J.Y., Tu, Y., Onyia, J.E., and Duncan, R.L. 1997. Mechanotransduction in bone: Osteoblasts are more responsive to fluid forces than mechanical strain. *Am J Physiol Cell Physiol.* 42:C810-C815.

119. Ozawa, H., Imamura, K., Abe, E., Takahashi, N., Hiraide, T., Shibasaki, Y., Fukuhara, T., and Suda, T. 1990. Effect of a continuously applied compressive pressure on mouse osteoblast-like cells (MC3T3-E1) in vitro. *J Cell Physiol.* 142:177–185.

120. Park, J.S., Chu, J.S.F., Cheng, C., Chen, F.Q., Chen, D., and Li, S. 2004. Differential effects of equiaxial and uniaxial strain on mesenchymal stem cells. *Biotechnol Bioeng.* 88:359–368.

121. Park, S.H., Sim, W.Y., Park, S.W., Yang, S.S., Choi, B.H., Park, S.R., Park, K., and Min, B.H. 2006. An electromagnetic compressive force by cell exciter stimulates chondrogenic differentiation of bone marrow-derived mesenchymal stem cells. *Tissue Eng.* 12:3107–3117.

122. Parsons, M., Kessler, E., Laurent, G.J., Brown, R.A., and Bishop, J.E. 1999. Mechanical load enhances procollagen processing in dermal fibroblasts by regulating levels of procollagen C-proteinase. *Exp Cell Res.* 252:319–331.

123. Pedersen, D.R., Bottlang, M., Brown, T.D., and Banes, A.J. 1993. Hyperelastic constitutive properties of polydimethyl siloxane cell culture membranes. *Adv Bioeng.* 26:607–609.

124. Pedersen, D.R., Brown, T.D., and Banes, A.J. 1992. Mechanical behavior of a new substratum for strain-mediated cell culture experiments. *Proc second N Amer Cong Biomech.* 355–356.

125. Pedersen, J.A. and Swartz, M.A. 2005. Mechanobiology in the third dimension. *Ann Biomed Eng.* 33:1469–1490.

126. Pelham, R.J. Jr. and Wang, Y. 1997. Cell locomotion and focal adhesions are regulated by substrate flexibility. *Proc Natl Acad Sci U S A.* 94:13661–13665.

127. Pitsillides, A.A., Rawlinson, S.C.F., Suswillo, R.F.L., Bourrin, S., Zaman, G., and Lanyon, L.E. 1995. Mechanical strain-induced NO production by bone cells: A possible role in adaptive bone (re)modeling? *FASEB J.* 9:1614–1622.

128. Pradhan, S. and Sumpio, B. 2004. Molecular and biological effects of hemodynamics on vascular cells. *Front Biosci.* 9:3276–3285.

129. Ridley, A.J. 2001. Rho GTPases and cell migration. *J Cell Sci.* 114:2713–2722.

130. Riha, G.M., Wang, X.W., Wang, H., Chai, H., Mu, H., Lin, P.H., Lumsden, A.B., Yao, Q.Z., and Chen, C.Y. 2007. Cyclic strain induces vascular smooth muscle cell differentiation from murine embryonic mesenchymal progenitor cells. *Surgery.* 141:394–402.

131. Rosel, D., Brabek, J., Tolde, O., Mierke, C.T., Zitterbart, D.P., Raupach, C., Bicanova, K., Kollmannsberger, P., Pankova, D., Vesely, P., Folk, P., and Fabry, B. 2008. Up-regulation of Rho/ROCK signaling in sarcoma cells drives invasion and increased generation of protrusive forces. *Mol Cancer Res.* 6:1410–1420.

132. Sadoshima, J. and Izumo, S. 1997. The cellular and molecular response of cardiac myocytes to mechanical stress. *Annu Rev Physiol*. 59:551–571.
133. Sah, R.L., Kim, Y.J., Doong, J.Y., Grodzinsky, A.J., Plaas, A.H., and Sandy, J.D. 1989. Biosynthetic response of cartilage explants to dynamic compression. *J Orthop Res*. 7:619–636.
134. Sanger, J.W., Sanger, J.M., and Jockusch, B.M. 1983. Differences in the stress fibers between fibroblasts and epithelial cells. *J Cell Biol*. 96:961–969.
135. Sarasa-Renedo, A. and Chiquet, M. 2005. Mechanical signals regulating extracellular matrix gene expression in fibroblasts. *Scand J Med Sci Sports*. 15:223–230.
136. Sawhney, R.K. and Howard, J. 2004. Molecular dissection of the fibroblast-traction machinery. *Cell Motil Cytoskeleton*. 58:175–185.
137. Schaffer, J.L., Rizen, M., Litalien, G.J., Benbrahim, A., Megerman, J., Gerstenfeld, L.C., and Gray, M.L. 1994. Device for the application of a dynamic biaxially uniform and isotropic strain to a flexible cell-culture membrane. *J Orthop Res*. 12:709–719.
138. Seliktar, D., Nerem, R.M., and Galis, Z.S. 2001. The role of matrix metalloproteinase-2 in the remodeling of cell-seeded vascular constructs subjected to cyclic strain. *Ann Biomed Eng*. 29:923–934.
139. Sniadecki, N.J., Lamb, C.M., Liu, Y., Chen, C.S., and Reich, D.H. 2008. Magnetic microposts for mechanical stimulation of biological cells: fabrication, characterization, and analysis. *Rev Sci Instrum*. 79:044302.
140. Soma, S., Matsumoto, S., and Takano-Yamamoto, T. 1997. Enhancement by conditioned medium of stretched calvarial bone cells of the osteoclast-like cell formation induced by parathyroid hormone in mouse bone marrow cultures. *Arch Oral Biol*. 42:205–211.
141. Somlyo, A.P. and Somlyo, A.V. 2000. Signal transduction by G-proteins, rho-kinase and protein phosphatase to smooth muscle and non-muscle myosin II. *J Physiol*. 522 Pt 2:177–185.
142. Song, G., Ju, Y., Shen, X., Luo, Q., Shi, Y., and Qin, J. 2007. Mechanical stretch promotes proliferation of rat bone marrow mesenchymal stem cells. *Colloids Surf B Biointerfaces*. 58:271–277.
143. Sotoudeh, M., Jalali, S., Usami, S., Shyy, J.Y., and Chien, S. 1998. A strain device imposing dynamic and uniform equi-biaxial strain to cultured cells. *Ann Biomed Eng*. 26:181–189.
144. Stanford, C.M., Welsch, F., Kastner, N., Thomas, G., Zaharias, R., Holtman, K., and Brand, R.A. 2000. Primary human bone cultures from older patients do not respond at continuum levels of in vivo strain magnitudes. *J Biomech*. 33:63–71.
145. Steinmeyer, J., Torzilli, P.A., Burton-Wurster, N., and Lust, G. 1993. A new pressure chamber to study the biosynthetic response of articular cartilage to mechanical loading. *Res Exp Med (Berl)*. 193:137–142.
146. Takano-Yamamoto, T., Soma, S., Nakagawa, K., Kobayashi, Y., Kawakami, M., and Sakuda, M. 1991. Comparison of the effects of hydrostatic compressive force on glycosaminoglycan synthesis and proliferation in rabbit chondrocytes from mandibular condylar cartilage, nasal septum, and spheno-occipital synchondrosis in vitro. *Am J Orthod Dentofacial Orthop*. 99:448–455.
147. Tan, J.L., Tien, J., Pirone, D.M., Gray, D.S., Bhadriraju, K., and Chen, C.S. 2003. Cells lying on a bed of microneedles: An approach to isolate mechanical force. *Proc Natl Acad Sci U S A*. 100:1484–1489.
148. Tan, W., Scott, D., Belchenko, D., Qi, H.J., and Xiao, L. 2008. Development and evaluation of microdevices for studying anisotropic biaxial cyclic stretch on cells. *Biomed Microdevices*. 10:869–882.
149. Tanaka, S.M. 1999. A new mechanical stimulator for cultured bone cells using piezoelectric actuator. *J Biomech*. 32:427–430.
150. Tanck, E., van Driel, W.D., Hagen, J.W., Burger, E.H., Blankevoort, L., and Huiskes, R. 1999. Why does intermittent hydrostatic pressure enhance the mineralization process in fetal cartilage? *J Biomech*. 32:153–161.
151. Tang, L.L., Wang, Y.L., Pan, J., and Cai, S.X. 2004. The effect of step-wise increased stretching on rat calvarial osteoblast collagen production. *J Biomech*. 37:157–161.
152. Thomas, W.A. and Yancey, J. 1988. Can retinal adhesion mechanisms determine cell-sorting patterns: A test of the differential adhesion hypothesis. *Development*. 103:37–48.
153. Timoshenko, S. and Woinowskey-Kreiger, S. 1959. *Theory of Plates and Shells*. McGraw-Hill, New York.
154. Tomei, A.A., Boschetti, F., Gervaso, F., and Swartz, M.A. 2008. 3D collagen cultures under well-defined dynamic strain: A novel strain device with a porous elastomeric support. *Biotechnol Bioeng*. 103:217–225.
155. Torzilli, P.A., Grigiene, R., Huang, C., Friedman, S.M., Doty, S.B., Boskey, A.L., and Lust, G. 1997. Characterization of cartilage metabolic response to static and dynamic stress using a mechanical explant test system. *J Biomech*. 30:1–9.

156. Tranquillo, R.T., Durrani, M.A., and Moon, A.G. 1992. Tissue engineering science—Consequences of cell traction force. *Cytotechnology*. 10:225–250.

157. Upchurch, G.R. Jr., Loscalzo, J., and Banes, A.J. 1997. Changes in the amplitude of cyclic load biphasically modulate endothelial cell DNA synthesis and division. *Vasc Med*. 2:19–24.

158. Van Tomme, S.R., Storm, G., and Hennink, W.E. 2008. In situ gelling hydrogels for pharmaceutical and biomedical applications. *Int J Pharm*. 355:1–18.

159. Vande Geest, J.P., Di Martino, E.S., and Vorp, D.A. 2004. An analysis of the complete strain field within Flexercell membranes. *J Biomech*. 37:1923–1928.

160. Vandenburgh, H.H. 1988. A computerized mechanical cell stimulator for tissue culture: Effects on skeletal muscle organogenesis. *In Vitro Cell Dev Biol*. 24:609–619.

161. Vandenburgh, H.H. 1992. Mechanical forces and their second messengers in stimulating cell growth in vitro. *Am J Physiol*. 262:R350–R355.

162. von Offenberg Sweeney, N., Cummins, P.M., Birney, Y.A., Cullen, J.P., Redmond, E.M., and Cahill, P.A. 2004. Cyclic strain-mediated regulation of endothelial matrix metalloproteinase-2 expression and activity. [see comment]. *Cardiovasc Res*. 63:625–634.

163. Wagner, D.R., Lindsey, D.P., Li, K.W., Tummala, P., Chandran, S.E., Smith, R.L., Longaker, M.T., Carter, D.R., and Beaupre, G.S. 2008. Hydrostatic pressure enhances chondrogenic differentiation of human bone marrow stromal cells in osteochondrogenic medium. *Ann Biomed Eng*. 36:813–820.

164. Walboomers, X.F., Ginsel, L.A., and Jansen, J.A. 2000. Early spreading events of fibroblasts on microgrooved substrates. *J Biomed Mater Res*. 51:529–534.

165. Walboomers, X.F. and Jansen, J.A. 2001. Cell and tissue behavior on micro-grooved surfaces. *Odontology*. 89:2–11.

166. Walboomers, X.F., Monaghan, W., Curtis, A.S., and Jansen, J.A. 1999. Attachment of fibroblasts on smooth and microgrooved polystyrene. *J Biomed Mater Res*. 46:212–220.

167. Wang, H., Ip, W., Boissy, R., and Grood, E.S. 1995. Cell orientation response to cyclically deformed substrates: experimental validation of a cell model. *J Biomech*. 28:1543–1552.

168. Wang, H.B., Dembo, M., and Wang, Y.L. 2000. Substrate flexibility regulates growth and apoptosis of normal but not transformed cells. *Am J Physiol Cell Physiol*. 279:C1345–C1350.

169. Wang, J.G., Miyazu, M., Xiang, P., Li, S.N., Sokabe, M., and Naruse, K. 2005. Stretch-induced cell proliferation is mediated by FAK-MAPK pathway. *Life Sci*. 76:2817–2825.

170. Wang, J.H., Goldschmidt-Clermont, P., Wille, J., and Yin, F.C. 2001. Specificity of endothelial cell reorientation in response to cyclic mechanical stretching. *J Biomech*. 34:1563–1572.

171. Wang, J.H. and Grood, E.S. 2000. The strain magnitude and contact guidance determine orientation response of fibroblasts to cyclic substrate strains. *Connect Tissue Res*. 41:29–36.

172. Wang, J.H., Jia, F., Gilbert, T.W., and Woo, S.L. 2003. Cell orientation determines the alignment of cell-produced collagenous matrix. *J Biomech*. 36:97–102.

173. Wang, J.H., Jia, F., Yang, G., Yang, S., Campbell, B.H., Stone, D., and Woo, S.L. 2003. Cyclic mechanical stretching of human tendon fibroblasts increases the production of prostaglandin E2 and levels of cyclooxygenase expression: A novel in vitro model study. *Connect Tissue Res*. 44:128–133.

174. Wang, J.H. and Lin, J.S. 2007. Cell traction force and measurement methods. *Biomech Model Mechanobiol*. 6:361–371.

175. Wang, J.H. and Thampatty, B.P. 2008. Mechanobiology of adult and stem cells. *Int Rev Cell Mol Biol*. 271:301–346.

176. Wang, J.H., Yang, G., and Li, Z. 2005. Controlling cell responses to cyclic mechanical stretching. *Ann Biomed Eng*. 33:337–342.

177. Wang, J.H., Yang, G., Li, Z., and Shen, W. 2004. Fibroblast responses to cyclic mechanical stretching depend on cell orientation to the stretching direction. *J Biomech*. 37:573–576.

178. Wang, Y.L. and Pelham, R.J. Jr. 1998. Preparation of a flexible, porous polyacrylamide substrate for mechanical studies of cultured cells. *Methods Enzymol*. 298:489–496.

179. Williams, J.L., Chen, J.H., and Belloli, D.M. 1992. Strain fields on cell stressing devices employing clamped circular elastic diaphragms as substrates. *J Biomech Eng*. 114:377–384.

180. Winston, F.K., Macarak, E.J., Gorfien, F., and Thibault, L.E. 1989. A system to reproduce and quantify the biomechanical environment of the cell. *J Appl Physiol*. 67:397–405.

181. Winter, L.C., Gilbert, J.A., Elder, S.H., and Bumgardner, J.D. 2002. A device for imposing cyclic strain to cells growing on implant alloys. *Ann Biomed Eng*. 30:1242–1250.

182. Yang, G., Crawford, R.C., and Wang, J.H. 2004. Proliferation and collagen production of human patellar tendon fibroblasts in response to cyclic uniaxial stretching in serum-free conditions. *J Biomech*. 37:1543–1550.

183. Yang, G., Im, H.J., and Wang, J.H. 2005. Repetitive mechanical stretching modulates IL-1beta induced COX-2, MMP-1 expression, and PGE2 production in human patellar tendon fibroblasts. *Gene.* 363:166–172.
184. Yang, Z., Lin, J.S., Chen, J., and Wang, J.H. 2006. Determining substrate displacement and cell traction fields—A new approach. *J Theor Biol.* 242:607–616.
185. Yeh, C.K. and Rodan, G.A. 1984. Tensile forces enhance prostaglandin E synthesis in osteoblastic cells grown on collagen ribbons. *Calcif Tissue Int.* 36 (Suppl 1):S67–S71.
186. Yeung, T., Georges, P.C., Flanagan, L.A., Marg, B., Ortiz, M., Funaki, M., Zahir, N., Ming, W., Weaver, V., and Janmey, P.A. 2005. Effects of substrate stiffness on cell morphology, cytoskeletal structure, and adhesion. *Cell Motil Cytoskeleton.* 60:24–34.
187. Yung, Y.C., Vandenburgh, H., and Mooney, D.J. 2009. Cellular strain assessment tool (CSAT): Precision-controlled cyclic uniaxial tensile loading. *J Biomech.* 42:178–182.
188. Zeichen, J., Van Griensven, M., and Bosch, U. 2000. The proliferative response of isolated human tendon fibroblasts to cyclic biaxial mechanical strain. *Am J Sports Med.* 28:888–892.
189. Zhang, L., Tran, N., Chen, H.Q., Kahn, C.J., Marchal, S., Groubatch, F., and Wang, X. 2008. Time-related changes in expression of collagen types I and III and of tenascin-C in rat bone mesenchymal stem cells under co-culture with ligament fibroblasts or uniaxial stretching. *Cell Tissue Res.* 332:101–109.
190. Zheng, X.Y. and Zhang, X. 2008. An optical Moire technique for cell traction force mapping. *J Micromech Microeng.* 18:125006.

Section II (Part 3)

Literature Review of Mechanobiology Research Findings and Theories

Other Organ Systems

13 Pulmonary Vascular Mechanobiology

Diana M. Tabima and Naomi C. Chesler

CONTENTS

13.1 INTRODUCTION

The pulmonary circulation is a low-pressure, high-flow (i.e., low-resistance) vascular network that carries deoxygenated blood into the lung microcirculation where gas exchange occurs. Like many vascular beds, it functions in a mechanically challenging environment and adapts and remodels in response to sustained changes in biomechanical loading. With each cardiac cycle, blood flow exerts shear, tensile, and compressive stresses on the walls of blood vessels. With each breath, air flow and subsequent alveolar pressure changes also exert shear, tensile, and compressive stresses on blood vessel walls. These forces determine patterns of blood flow, air flow, gas exchange, and cellular-level biological responses, including gene expression, protein expression, and enzyme activation.

Understanding the mechanobiology of the pulmonary vasculature requires integrating the effects of blood flow, air flow, and bloodborne and airborne stimuli on the biology of the vessel walls. Chronic physiological or pathological changes in the lung mechanical environment can cause arterial remodeling, which involves cell proliferation and apoptosis, and extracellular matrix (ECM) synthesis, degradation, reorganization, and restructuring. The links between the mechanical stimuli and their biological consequences are common signaling pathways that control physiological adaptation and pathological remodeling. A current challenge is to fully understand these signals and their targets and then to manipulate them in ways that redirect remodeling processes with the goal of preventing or reducing pathology.

In this chapter, we review the mechanobiology of the pulmonary vasculature, focusing first on the relevant anatomy and biology, including vascular cell types and component matrix proteins, and highlighting the ways in which cells and tissues in the pulmonary circulation are distinct from their counterparts in the systemic circulation. We then describe the complex and dynamic mechanical environment of the lung so that the mechanical forces acting on pulmonary vascular cells and tissue can be properly understood. Next, we review mechanosensors that are known or suspected to act in the pulmonary circulation. Finally, we illustrate the importance of pulmonary vascular mechanobiology to health and disease, using pulmonary hypertension as a case in point. Throughout, we attempt to highlight the exciting and important questions that remain to be answered in pulmonary vascular mechanobiology, including the scientific and clinical relevance of each.

13.2   LUNG VASCULATURE

Adequate supply of oxygen to tissues via blood flow is a requirement for survival. These oxygen demands change throughout growth and development and with physiological adaptation in adult life. The blood flow circuit that facilitates the exchange of oxygen and carbon dioxide in the alveolar spaces of the lung is the pulmonary circulation. The right ventricle ejects deoxygenated blood into the pulmonary arterial trunk, which then bifurcates into the left and right main (extralobar) pulmonary arteries. These branch into increasingly smaller arteries within the lung parenchyma (intralobar arteries) that ultimately feed the capillaries. The intralobar pulmonary arteries can be classified sequentially according to their position in the lung. The first generation intrapulmonary arteries are termed *segmental* arteries; they accompany the bronchi (large airways) and each supply approximately 20 bronchopulmonary segments in humans. The next smaller arteries are termed *intra-acinar* which accompany the terminal bronchioles that supply the airspaces of the lung or *acini*. The intra-acinar arteries branch into the microcirculation, which consists of precapillaries, capillaries and postcapillaries.[23] Anatomically and functionally, the pulmonary microcirculation is like two membranes or sheets, separating airspaces above and below from a central blood-filled region. Upon leaving the pulmonary microcirculation, blood is collected by the pulmonary venules, which anastamose into the pulmonary veins, which in turn terminate in the left atrium of the heart.

All vessels except the capillaries have three layers, which are the tunica intima, tunica media, and tunica adventitia. The innermost layer, the tunica intima, consists of a monolayer of endothelial cells and subendothelial layer of connective tissue. The tunica media is mainly composed of elastin, collagen, and vascular smooth muscle cells (vSMCs). The vSMCs are typically circumferentially oriented and arranged in layers separated by elastic fibers and an ECM rich in collagen. The outermost layer, the tunica adventitia, is composed of fibroblasts (10%) and connective tissue which consists mainly of the ECM proteins: collagen (63%), ground substances (25%) and elastin (2%).[45] Unlike in the tunica media where collagen fibers are oriented circumferentially, in the tunica adventitia rope-like bundles of collagen are oriented longitudinally.

The pulmonary arterial trunk, like the aorta, receives the entire cardiac output (CO). However, due to the relatively low resistance of the pulmonary circuit, the pressure in the pulmonary artery is about fivefold to sixfold lower than in the aorta. As a consequence, the large, elastic pulmonary arteries including the trunk and extralobar arteries are thinner walled, less muscular, more extensible, and greater in diameter than the large elastic arteries of the systemic circulation.[78] In humans, the extrapulmonary arteries measure more than 1 mm in outer diameter.[78] Whereas in the aorta, elastic lamina in the tunica media are continuous, parallel, and regularly arranged, in the pulmonary trunk these lamina tend to be interrupted and fragmented.[20] Like the aorta, the elastic pulmonary arteries are the main capacitance vessels of the lung vasculature.

The structure of the intralobar arteries varies according to location in the pulmonary vascular tree. As in the systemic circulation, the ratio of elastin to collagen falls with increasing distance from the heart, while the number of vSMCs per unit volume increases.[34] For any given caliber intralobar artery, the medial vSMC layer is thinner than that of a systemic artery, reflecting the different mechanical environments in these two major circulations.[4] The smallest intralobar arteries and arterioles have the highest percentage of vSMC per wall thickness and function to control blood flow distribution and pressure. These are termed the *resistance* arteries. In humans, the resistance arteries range in outer diameter from 100 to 1000 µm.[20]

The pulmonary capillaries, where gas exchange occurs, wrap over the alveolar surfaces, forming a thin interface between airspace and blood volume. In humans, pulmonary capillaries are defined as those vessels measuring less than 100 µm in outer diameter.[20] The structure of the pulmonary capillary network is unique. As noted above, it is typically conceptualized as a sheet of blood flow occasionally interrupted by posts that prevent collapse,[45] and membranes above and below that form the interface with airspaces through which gas exchange occurs. Pulmonary capillaries are smaller in diameter, shorter in length and form a much denser network than capillaries in the systemic circulation.[170] In addition to being the site for gas exchange, the pulmonary capillaries function as a blood filter and also a metabolysis site for some endogenous vasoactive agents.[81]

In all pulmonary blood vessels, ECM proteins, particularly collagen and elastin, largely determine *passive* mechanical behavior, i.e., the mechanical behavior in the absence of vSMC tone and activity. The presence of collagen provides tensile strength and prevents failure at high pressures,[45] whereas elastin bears load at lower pressures and provides elasticity during cyclic loading.[109] The presence of vSMCs may contribute to passive arterial viscoelasticity[8,35] but in the pulmonary circulation vSMCs are not typically load-bearing unless stimulated by an agonist. That is, even in a normal tone state, vSMCs have little influence on vascular mechanical properties because resting pulmonary vSMC tone is quite low in the absence of disease. When activated by endogenous or exogenous agents, however, pulmonary vSMCs can dramatically alter vascular function, which is then termed *active* mechanical behavior.[157]

Pulmonary vSMC activity is modulated by a series of endothelium-derived and circulating mediators.[7] Autonomic nerves, humoral factors and airway gasses can act as agonists to stimulate vSMC contraction. Also, lung diseases, such as chronic obstructive pulmonary disease,[6,67,175] cystic fibrosis[43] and obstructive sleep apnea[59] are associated with hypoxia, which is known to increase vSMC activity in pulmonary resistance arteries[70,101,136,161] but not larger, extralobar arteries.[102,157]

In order to understand the function–structure relationships in the lung vasculature, it is important to explore their different components, i.e., the pulmonary vascular cells and the components of ECM. In this section, we describe the principal elements of the pulmonary vasculature and some of their interactions.

13.2.1 Pulmonary Vascular Cells

Multiple different cell types, including endothelial cells, vSMC and fibroblasts, contribute to normal pulmonary vascular function and to the biological response to changes in the local mechanical environment. However, the basic cellular biology of these cells and their response to mechanical stimuli remain incompletely understood.

13.2.1.1 Endothelial Cells

The primary function of ECs is that of a barrier or interface between circulating blood and other tissues such as airways where gas exchange occurs. As in the systemic circulation, the pulmonary endothelium has long been considered a homogeneous cell layer with a rich diversity of functions. However, new evidence suggests that the endothelium is a heterogeneous cell layer in which more than three different subtypes of endothelial cells (EC) can be distinguished.[155] Some of the differences between these subpopulations are their permeability, proliferation ability, response to environmental factors, and response to mechanical forces. For example, pulmonary venules allow more macromolecular diffusion than pulmonary arterioles,[90] and microvascular ECs allow less diffusion than macrovascular ECs.[18] Indeed, lung microvascular ECs allow less macromolecular diffusion than ECs from either pulmonary arterioles or venules.[48]

A second important function of pulmonary ECs is to sense physical changes in the local environment and transform them into biochemical signals. In this way, ECs are natural cellular mechanotransducers. Elements of the EC membrane connected to the cytoskeleton are thought to sense mechanical stress created by flow and initiate the subsequent signaling responses that regulate vascular structure and function.[37] Due to their ability to detect flow (via shear stress) and pressure changes (via stretch), endothelial cells are considered mechanosensors for shear stress,[42] mechanical stretch and strain[3] and distinct patterns of flow.[17] EC mechanosensory machinery is even able to distinguish between steady and unsteady flow, which may enable them to participate in the modulation of blood flow waveform and pulsatile frequency.[5] Other EC functions include the control of cellular trafficking, the regulation of vasomotor tone, and the maintenance of homeostasis.[48,120]

Endothelial cells from different sites of the vascular tree differ in size, shape, nuclear orientation, and responsiveness to stimuli.[2] For example, pulmonary artery ECs are aligned in the direction of blood flow whereas microvascular ECs are not. Also, microvascular ECs grow faster in culture than their macrovascular counterparts[81] and exhibit distinctive calcium-mediated signal transduction mechanisms.[31] Some differences have been reported in gene expression profiles as well. For example, microvascular ECs express more cell adhesion molecules than pulmonary arterioles or venules[48] and increase their production of adhesion molecules, inflammatory cytokines, and upregulation expression of pro-proliferative genes in response to pulsatile shear stress than their macrovascular counterparts.[95] These data suggest that microvasculature ECs are more sensitive to pulsatile flow than microvascular ECs, which may be important in some disease processes.

Between pulmonary and systemic artery ECs we can find differences in (1) morphology, (2) protein expression, (3) gene expression, and (4) response to stimuli. With regard to morphology, for example, ECs in pulmonary arteries are larger and more rectangular than those lining the aorta.[80] Many differences in protein and gene expression exist, as one might expect given the different anatomy and biology of pulmonary and systemic arteries and their different mechanical environments. An important way in which pulmonary and systemic ECs respond differently to stimuli is their response to hypoxia. Given the critical role of the pulmonary vasculature in controlling blood oxygenation, a unique sensitivity to oxygen levels in ECs derived from the pulmonary vasculature is not unexpected. For example, with exposure to hypoxia, gene expression for ATP synthase, H^+ transport and glycolysis is reduced in pulmonary ECs but elevated in aortic ECs. Also, genes involved in lipid metabolism are elevated by hypoxia in pulmonary ECs but reduced in aortic ECs.[127] Hypoxia occurs clinically in many disease states, thus the pulmonary vascular response to hypoxia (inadequate oxygen supply), or its counterpart, hypoxemia (inadequate oxygen in the blood), is a critical current area of research.

Like their systemic counterparts, pulmonary ECs synthesize and express vasoactive mediators, such as endothelin-1 (ET-1) and endothelial-derived nitric oxide (EDNO), in response to changes in blood flow and pressure, oxygen tension, and circulating cytokines. These mediators modulate vSMC function, which is the focus of the next section.

13.2.1.2 Vascular Smooth Muscle Cells

The primary function of pulmonary vSMCs in mature animals is to control blood flow and pressure distribution via contraction and relaxation. SMC contraction is directly related to the concentration of calcium in the cytosol. An increase in the amount of Ca^{2+} results in activation of myosin light-chain kinase, which allows for cycling of cross bridges between actin and myosin resulting in SMC contraction.[149]

In healthy pulmonary arteries and veins, vSMCs have an extremely low rate of proliferation and low synthetic activity. Nevertheless, SMCs maintain plasticity throughout life and can exhibit diverse functions in response to changes in local environments.[124] Many different vSMC subpopulations with different origins and diverse functional capabilities exist in the pulmonary vasculature at various locations along the arterial tree.[44] The various subpopulations appear to respond differently to stimuli in terms of contractility and synthesis of ECM proteins.[153] Some of the stimuli known to affect SMC growth, differentiation, and synthetic activity are locally produced mitogens, hypoxia, and mechanical stress.

The degree of contractile response to hypoxia differentiates proximal pulmonary artery vSMCs from distal, resistance pulmonary arterial vSMCs. Madden et al.[102] showed that vSMCs isolated from conduit pulmonary arteries did not constrict in response to acute hypoxia, whereas those from resistance pulmonary arteries did. Some evidence suggests that preferential occurrence of oxygen-sensitive potassium channels in SMCs of resistance-size vessels is responsible for their heightened sensitivity to low oxygen.[4] The distribution of these potassium channels in arteries throughout the pulmonary vascular tree is consistent with the overall response to hypoxia.[172]

As found for pulmonary ECs and not surprisingly given the critical role of the pulmonary circulation in controlling blood oxygenation, the pulmonary vSMC response to hypoxia is significantly different than the systemic vSMC response. Via a poorly understood mechanism termed *hypoxic pulmonary vasoconstriction* (HPV; see Moudgil et al.[114] for a critical review of the literature), isolated pulmonary arteries without endothelium[181] and isolated vSMCs[116] contract in response to hypoxia, whereas vSMCs from most systemic arteries (e.g., cerebral and renal) relax.[68]

Another important function of the vSMCs is the secretion of various molecules, including ECM proteins that support cells in the medial layer and soluble substances such as growth factors.[4]

Although recent research has provided valuable insights into the effects of normal and pathological mechanical stimuli on vSMC biology, more work is needed to improve our understanding of how the mechanical signal is transmitted to the cell nucleus as well as how signal transmission is modulated by the magnitude and pattern of the stimulus. A more detailed understanding of vSMC mechanotransduction mechanisms will provide important future directions for preventative therapies related to the imbalances between proliferation and apoptosis, and ECM protein synthesis and degradation, found in vascular disease.

13.2.1.3 Fibroblasts

Until recently, adventitial fibroblasts, the principal cells in the adventitia, were thought to play a relatively minor role in pulmonary vascular homeostasis and pathology. However, new experimental evidence suggests that the adventitial compartment of arteries from both the pulmonary and systemic circulations regulates vessel function from the "outside-in." According to some authors, the adventitia acts as a master control center where key regulators of vessel wall function are monitored, stored, integrated, and released.[155] Supporting this idea, environmental changes such as hypoxia are known to activate adventitial fibroblasts, increase cell proliferation, and increase expression of contractile and ECM proteins in a manner that influences overall vascular tone and wall structure.[125]

More specifically, in response to hypoxia, fibroblasts differentiate into myofibroblasts, which are the principal producers of collagen and other ECM components including fibronectin, tenascin and elastin.[46,128] Pulmonary but not systemic artery fibroblasts proliferate in response to hypoxia via activation of p38 mitogen-activated protein kinase (p38MPAK),[9,113] the transcription of hypoxia-inducible factor-1 αHIF-1 α,[151] and the expression and activity of early growth response-1 (Egr-1).[51] There is evidence that changes in the behavior of adventitial fibroblasts may induce medial hypertrophy in response to chronic hypoxia, suggesting that adventitial remodeling is a precursor of pathological changes in the media and/or intima,[151] supporting the "outside-in" hypothesis.

Like ECs and SMCs, fibroblasts have the ability to rapidly respond to stimuli (e.g., hypoxia, shear stress, and pressure) and modulate their function. Fibroblasts are particularly responsive to a variety of stimuli because of their relative lack of differentiation, which confers capacity for contraction, migration, proliferation, differentiation, and synthesis of ECM proteins and cytokines. Their role in maintaining and adapting the ECM is particularly important to pulmonary vascular function because the ECM provides critical mechanical support and modulates biological signaling, as we discuss below.

13.2.2 EXTRACELLULAR MATRIX

The ECM contributes substantially to pulmonary vascular function in several ways. First, the ECM constitutes the scaffold for cells in the vessel wall and bears the mechanical forces levied on the vessel by pulsatile flow and pressure and vSMC contraction.[40] Second, the ECM biochemical and biophysical properties relay important information to the vascular cells, thus regulating migration, adhesion, proliferation, differentiation, and organization.[21,74] The ECM also provides nutrition to cells, regulates angiogenesis, and allows the formation of intracellular contacts.[21] The ways in which these functions are regulated depend on the mechanical properties of the tissue and the mechanical loading environment.[109] Finally, ECM molecules can store, mask, present or sequester growth factors, thereby modulating their effects remarkably.

The main ECM component in the tunica intima is collagen type IV in the basement membrane supporting ECs, which is produced by ECs. The ECM of the tunica media is mainly produced by vSMCs. Embedded in the network of microfibrils, every SMC in the tunica media is encapsulated by a basement membrane, which contains typical basement proteins, such as laminin, collagen IV, and persican.[154] The composition of ECM surrounding vSMCs determines in part their phenotype: contractile (differentiated) and secretory (less differentiated).[11] Typically, collagen types I and III surround contractile vSMCs in the tunica media. The adventitial ECM is characterized by longitudinally oriented fibrillar collagen bundles (mainly type I and III) as well as other matrix molecules, such as versican.[151]

The components of the ECM can be divided into four broad categories: collagen, glycosaminoglycans (GAGs) and proteoglycans, glycoproteins, and elastic fibers. Understanding the different natural configurations, material properties, and rates of turnover of these different components of the ECM is useful for understanding pulmonary vascular function.

13.2.2.1 Collagen

The principal structural element of the ECM is collagen. Collagen fibers play an important role in maintaining the biological and structural integrity of the pulmonary vasculature and provide tensile stiffness, elasticity, and strength. Collagens are a family of closely related but distinct ECM proteins and are composed of a triple helix of three polypeptide α chains, each having a Gly-X-Y repeating sequence. Although 13 different types of collagen are present in the vasculature, types I and III are predominant within the tunica media and adventitia of pulmonary arteries and assemble into cross-banded fibrils that provide tensile strength to the vessel wall. Collagens type IV, VI, and VIII are nonfibrillar collagens which form three-dimensional networks that serve as an anchoring substrate and help form a permeability barrier.[117]

Collagen type I usually consists of three coiled subunits: two α1 (I) chains and one α2 (I) chain forming fibrils of approximately 50 nm in diameter, collagen type II is composed of three identical α1 (II) chains forming fibrils less than 8 nm in diameter, and collagen type III fibrils are formed from three α1 (III) chains resulting in diverse fibril diameters between 30 and 130 nm.[50] While collagen I is the most abundant protein in the pulmonary vasculature, type IV collagen, a nonfibrillar collagen, is the most abundant protein in the basement membranes (up about 50%). Collagen IV comprises six highly homologous α-chains with three different domains: the amino-terminal domain rich in cysteine and lysine residues; the Gly-X-Y residues; and the long carboxyl-terminal non-collagenous domain. Collagen IV plays an important role in tissue organization, stability and differentiation.

Collagen fibers are crucial from the biomechanical point of view. These fibers are not elastically deformable, but bear the tensile forces of the pulmonary vessels and set the limits of its elastic expansibility. In unloaded arteries, collagen typically appears wavy or crimped histologically. This initial laxity allows some arterial stretch to occur before collagen fibers are loaded. At increasing levels of stretch above this engagement point, more and more collagen fibers become loaded and taut until no more extension is possible without fiber damage. In the single fiber configuration, estimates of collagen elastic modulus are ~10–100 MPa.[24,58]

In addition to the mechanical function, collagens have important biological functions. For example, polymerized collagens type I and IV surrounding SMC are known to promote a more quiescent and contractile SMC phenotype.[131] They also are involved in the regulation of endothelial cell behavior during angiogenesis[103] and the proliferation, migration, and phenotypic organization of EC.[16,87]

13.2.2.2 Glycosaminoglycans and Proteoglycans

GAGs and proteoglycans establish and maintain pulmonary vascular cell quiescence once morphogenesis is complete. When pulmonary vascular cells are in a proliferative state during different phases of vascular development and pathogenesis, production of GAGs and proteoglycans increases. A key function of proteoglycans and GAGs is to cushion the response to mechanical forces, protecting the cells against excessive or excessively rapid deformation.

GAGs interact with other ECM proteins, influencing the macromolecular organization and also controlling the phenotype of pulmonary cells such as ECs and SMCs. GAGs such as hyaluronic acid and heparin play an important role in regulating cell proliferation. For example, the interaction of hyaluronic acid with pericellular macromolecules leads to softening of the ECM, thereby facilitating cell shape changes that are required for cell division, migration or plasticity.[93] Heparin stimulates the growth of pulmonary vascular cells such as ECs, SMCs, and fibroblasts.[178] Another function of heparin is to stabilize the biological activity of fibroblast growth factor[144] through protection from proteolytic inactivation and conformational change. Finally, the secretion of heparin is responsible, at least in part, for the inhibition of SMC hypertrophy and hyperplasia.[32,162] As a result, heparin has been administered exogenously in patients with pulmonary hypertension with positive effects: it has been shown to reduce pulmonary arterial thickening in certain disease states.[57,162]

Proteoglycans are macromolecules composed of a protein core linked to one or more polysaccharides. They are important mechanically because their high fixed charge density imbibes water which regulates hydration and enables the arterial wall to resist compressive forces.[145] Combined with collagen and elastin, proteoglycans prevent plastic deformation of vessels subjected to pulsatile pressure and flow. Furthermore, proteoglycans help regulate vSMC migration, proliferation, and apoptosis as well as ECM protein synthesis and assembly.[10,138]

The synthesis of GAGs and proteoglycans are regulated by changes in the mechanical environment. For example, SMCs under mechanical strain increase synthesis of the proteoglycans versican, biglycan, and perlecan and decrease decorin synthesis.[94] Mechanical strain also causes changes in GAG and proteoglycan chemistry. In particular, it enhances the incorporation of sulfate groups, which in turn improves the cushioning properties of the wall.

The effects of GAGs on pulmonary vascular disease progression and vice versa represent a particularly important and active area of research.

13.2.2.3 Glycoproteins

Fibronectin is a glycoprotein found in the tunica media and adventitia that plays an important role as a cell adhesion substrate. It binds to other proteins including collagen and fibrin, making it a uniquely important "universal glue" of matrix proteins.[126] In the pulmonary circulation, fibronectin appears to play a critical role in facilitating proliferation of fibroblasts as well as in their differentiation into myofibroblasts.[39] Fibronectin has also been shown to induce vSMC proliferation and migration.[135] Deposition of fibronectin coincides with the expression and activity of matrix metalloproteinases (MMPs), which are molecules responsible for the degradation of ECM components including basement membrane collagen, interstitial collagen, and fibronectin itself.[126] In animal models of pulmonary hypertension increases in fibronectin have been found,[66] which likely facilitate smooth muscle cell migration and proliferation.[74]

Thrombospondin-1 is a matrix-associated adhesive glycoprotein secreted by pulmonary SMCs, ECs and fibroblasts. It is a glycoprotein dimer that binds collagens, fibrin, and proteoglycans via specific domains as well as vascular cells through specific integrins. Its expression in normal pulmonary vessels is limited[177] but it is upregulated when ECs in culture are exposed to cyclic stretch. The secretion of this glycoprotein is responsible for promoting SMC proliferation.[104,120,122]

Laminin is an important glycoprotein found in the subendothelial basement membrane. Laminin is expressed by ECs and deposited subendothelially throughout the vasculature.[60] The principal function of laminin is to create connections between cells and other ECM components such as collagen type IV via specific integrin receptors.[182] Laminin is the most abundant glycoprotein in EC and vSMC basement membranes. In conjunction with perlecan and type IV collagen, laminin helps regulate the growth of ECs and vSMCs. Laminin, like fibronectin, appears to increase in pulmonary arteries exposed to hypertension[66] and has been suggested to facilitate vSMC migration and proliferation in this disease state.[74]

Tenascin is a glycoprotein involved in lung and vascular morphogenesis. It is a matricellular proteins, which is a secreted glycoprotein, that interacts with other ECM constituents via multiple specific cell surface receptors, as well as growth factors, to modulate cell–matrix interactions. Tenascin-C, like fibronectin, is associated with fibroblast and SMC proliferation and has also been shown to contribute to fibroblast differentiation into a myofibroblast.[74] Deposition of tenacin-C also coincides with the expression and activity of MMPs.[49] The expression of tenascin-C in normal pulmonary vasculature is limited,[131] but in certain diseases states such as pulmonary hypertension tenascin-C is strongly expressed,[71,73] perhaps because of its sensitivity to circulating growth factors.[74,130]

13.2.2.4 Elastic Fibers

The insoluble elastin surrounded by a lattice of microfibrils in the vascular wall is more generally referred to as *elastic fibers*.[137] Microfibrils constitute less than 10% of the mature fiber and contain several glycoproteins and fibrillin.[140] The other 90% is composed exclusively of elastin, which is synthesized and secreted as tropoelastin by vSMCs. Tropoelastin is a non-glycosylated, highly hydrophobic, soluble monomer that is organized into a microfibrillar polymer that forms concentric rings around the arterial lumen.[137] Cross-linked elastin molecules also form concentric rings around the vessel lumen that alternate with a ring of vSMCs, which together provide the compliance that arteries need to absorb and transmit hemodynamic forces.[76]

Elastin is a major structural component of large elastic arteries and the principal source of vascular elasticity. Elastin molecules can extend up to 220% of their unstretched length without fracture or plastic deformation.[112] Their mechanical properties are caused by their unusual chemical composition rich in glycine, proline, and hydrophobic amino acids. Unlike collagen, elastin molecules are load-bearing even at very low strains. Thus, elastin contributes to the structural support of arteries at low pressures and up to the normal pressure range.[112,137] While collagens provide tensile strength, elastin provides the elasticity needed to accommodate the pulsatile nature of blood flow.[137] Elastic fibers also ensure uniform tension and distribution throughout the vessel walls and prevent dynamic creep.[146] The elastic modulus of elastin fibers and sheets has been estimated to be ~0.1–3 MPa.[24] Recent studies suggest that vascular elastin is anisotropic, meaning the modulus in the axial direction is different from the modulus in the circumferential direction.[91,166]

In addition to its structural role, elastin also regulates arterial morphogenesis and activity of pulmonary vascular cells. In the case of arterial morphogenesis, elastin instructs vSMCs to localize around the elastic fibers in an organized configuration and in a contractile, quiescent state.[76] The presence of elastin also inhibits growth of SMC more strongly than other proteins such as basement membrane proteins and collagen.

The contributions of vascular elastin to the development and progression of pulmonary hypertension have received significant attention[91,112,130,146]; however, much remains to be learned about the mechanical and regulatory roles of elastin in the arterial wall.

13.3 MECHANICAL ENVIRONMENT

The pulmonary vasculature is exposed to a variety of mechanical forces including circumferential stress and strain, longitudinal stress and strain, and shear stress and strain caused by cyclical changes in blood pressure and flow and volume during breathing (Figure 13.1). The two main mechanical stimuli that act on the non-capillary vessel walls are pressure, which acts throughout the vessel wall, and flow-induced shear stress, which acts at the luminal surface. In general, changes in blood pressure-induced circumferential wall stress control wall thickness, while changes in blood flow-induced wall shear stress control arterial caliber. Capillaries are exposed to pressure and flow-induced shear stress as well as extraluminal airway pressure and traction forces. In this section, we describe the mechanical environment of the lung vasculature in detail.

13.3.1 BLOOD PRESSURE AND VESSEL STRAIN

Blood pressure is the major determinant of vessel stretch both axially and circumferentially, which affects ECs, vSMCs and fibroblasts. Normal blood pressures in the pulmonary artery are 24 mmHg in systole and 9 mmHg in diastole, which are fivefold to sixfold lower than in the systemic circulation. The mean arterio-venous pressure difference in the lungs is 10 mmHg, which is about

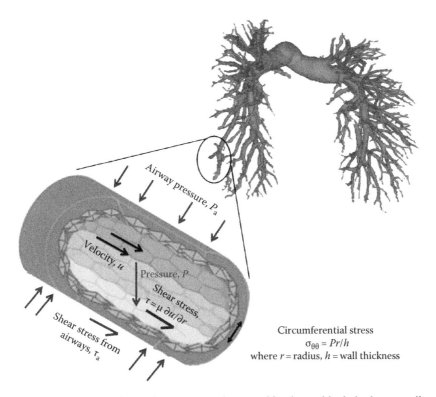

FIGURE 13.1 The mechanical forces that act on a pulmonary blood vessel include shear, tensile, and compressive stresses due to blood flow and pressure in the lumen and shear, tensile, and compressive stresses due to air flow in adjacent airways. These forces determine patterns of blood flow, air flow, gas exchange, and cellular-level biological responses, including gene expression, protein expression, and enzyme activation.

10-fold lower than the difference in the systemic circulation.[100] The low pressure in the pulmonary circulation prevents transudation of fluid from the blood spaces of the lung into the airspaces of the lung (i.e., pulmonary edema) and maintains a relatively low right ventricular workload. Current guidelines stipulate that the upper limit of normal adult mean pulmonary artery pressure (mPAP) is 25 mmHg and that pulmonary capillary pressures (PCWP) are 8–10 mmHg.[47]

Mean PAP can increase because of elevated left atrial pressures, increased CO or increased pulmonary vascular resistance (PVR = [mPAP-PCWP]/CO). Increased left atrial pressures are typically caused by left ventricular failure or mitral valve disease. Increases in CO can occur physiologically (e.g., with exercise) or pathologically (e.g., with atrial and ventricular septal defects). With exercise, the increase in mPAP is not as dramatic as one might expect given the increase in flow. This is because at rest many pulmonary capillaries and small arterioles are not required for adequate blood oxygenation and are thus not perfused. With exercise, these capillaries are recruited such that CO can increase without a dramatic increase in pressure. Once all capillaries are recruited, capillary and arteriolar dilation also occurs to permit further increases in blood flow without further increases in pressure.[100] Increases in PVR are caused by arteriolar narrowing or obstructions to flow upstream of the capillaries. Hypoxia increases PVR via constriction of pulmonary arterioles. Emboli increase PVR by obstructing large or small arteries. Other diseases can also increase PVR and thereby increase mPAP to cause pulmonary hypertension.[61]

For given upstream (pulmonary arterial) and downstream (left atrial) pressures, the distribution of pressure throughout the lung is related to the local resistance to blood flow in the arterial, capillary and venous compartments. Especially in the capillary region, respiration can alter intravascular pressure and the distribution of blood pressure throughout the lung.

As blood enters the pulmonary arteries from the right ventricle, it is highly pulsatile. The large, elastic arteries expand with systole to accommodate the ejected stroke volume; when the right ventricle enters diastole, the arteries recoil, propelling the blood forward. This property of the large, elastic arteries is known as *cushioning* because it reduces the pulse pressure. The determinants of cushioning are the mechanical properties of the elastic arterial walls, wave reflections from downstream sites, and the arterial pressure loading pattern caused by ventricular ejection.[119] In the systemic circulation, pulsatile perfusion has been shown to have beneficial effects on the microcirculation, metabolism, and organ functions.[85] Recent studies have recognized the clinical importance of the pulsatile component of blood pressure,[65,111] and central pulse pressure has been postulated as a good indicator of cardiovascular risk.[52] However, too much pulsatility may cause damage to the microcirculation either via overstretching or simply because the microcirculation is more sensitive to pulsatile flow than the macrocirculation,[95] as noted above.

While mean pressures are lower in the pulmonary circulation, pulsatility is inherently greater compared to the systemic circulation. The ratio of pulse pressure to mean pressure in the pulmonary artery is about one, whereas in the aorta, pulse pressure is about 40% of mean pressure.[110] It is not surprising, therefore, that the compliance of large pulmonary arteries—or their percent deformation for a given change in pressure—is much greater than the compliance of large systemic arteries. When the elastic arteries become stiffened, as occurs with aging or hypertension, this cushioning effect is impaired and highly pulsatile pressures in the microcirculation likely result.

The effects of gravity on blood pressure in the lung can be significant. The hydrostatic pressure gradient that results from upright posture leads to higher local blood pressures at the base of the lung and lower local blood pressures at the apex. As a consequence, at the apex, normal alveolar pressure exceeds pulmonary arterial pressure such that capillaries are collapsed and not perfused. This is typically termed "zone 1" conditions either anatomically or functionally.[176] In the mid-lung or zone 2, pulmonary arterial pressure exceeds pulmonary alveolar pressure, but the alveolar pressure is still higher than venous pressure; therefore, perfusion is dependent on airway pressure and not venous pressure. Because the resulting flow is insensitive to the true downstream (venous) pressure, the phenomenon is known as the waterfall effect. Finally, at the base of the lung or in zone 3 conditions, pulmonary arterial pressure is greater than venous pressure which is greater than airway pressures. In this case, flow is dependent on the difference between arterial and venous pressures.

The consequence of positive vascular transmural pressures is circumferential stretch or deformation of the pulmonary vessel wall. The degree of stretch that occurs is dependent on the transmural pressure and the mechanical properties of the vascular wall. That is, for a stiffer or thicker wall, a vessel will deform less in response to a given change in transmural pressure. If we assume that all vessels are cylindrical and thin walled, the circumferential force per unit length (the wall tension, T) is related to the transmural blood pressure (P) and the vessel mid-wall radius (r) by Laplace's law

$$T = P \cdot r \tag{13.1}$$

The relation between this tension and deformation of the vessel depends on the geometry and the elastic characteristics of the wall. In the case of the circumferential tension, each element of the wall supports part of this tension. This circumferential tension per unit of thickness or force per unit area is the circumferential stress ($\sigma_{\theta\theta}$) in the wall. For a thin-walled vessel of homogenous material, this is expressed as

$$\sigma_{\theta\theta-\text{thin wall}} = \frac{P \cdot r}{h} \tag{13.2}$$

where h is the thickness of the wall. When the vessel cannot be considered "thin" (i.e., wall thickness is greater than 1/10 of the vessel diameter), a thick-walled formulation can be used to describe the wall stress

$$\sigma_{\theta\theta-\text{thick wall}} = \left[\frac{\left(P_i R_i^2 - P_e R_e^2 \right)}{(R_e^2 - R_i^2)} \right] + \left[\frac{\left(R_i^2 R_e^2 (P_i - P_e) \right)}{R^2 (R_e^2 - R_i^2)} \right] \tag{13.3}$$

where
   P_i and P_e are the intra- and extravascular pressures
   R_i and R_e are the internal and external radii
   R is the radius at which the stress is calculated ($R_i \le R \le R_e$)

To obtain wall material properties, circumferential (or axial) stress–strain relations are required. The equations for strain in the circumferential direction is

$$\varepsilon_{\theta\theta} = \frac{2\pi R - 2\pi R_0}{2\pi R_0} = \frac{\Delta R}{R_0} \tag{13.4}$$

where R_0 is measured from a zero-stress state. However, even when no distending pressure acts, arteries are not in a zero stress state; *residual stresses* exist that alter the distribution of stress in the wall at loaded states. Residual stresses can be determined with an *opening angle test* in which an unloaded ring segment is cut radially.[45] The artery ring springs open and the cross section can be approximated as a circular sector with an opening angle α. Assuming that a single radial cut releases all residual stresses, the open configuration can be considered a zero-stress state and used as the reference configuration.

It's also important to note that this equation for strain (Equation 13.4) assumes *infinitesimal* or small deformations, which is not the case for arteries exposed to physiological pressures. Given the complexities of properly calculating stress and strain in arteries, and thus determining mechanical properties from the corresponding stress–strain relationships, it is often more convenient to calculate an approximate elastic modulus directly from pressure and diameter. For example, Bergel[12] proposed the following elastic modulus based on small changes in pressure and diameter:

$$E_{inc} = 1.5 D_i^2 D_e \frac{(\Delta P / \Delta D_e)}{(D_e^2 - D_i^2)} \tag{13.5}$$

Here, one must assume a thick-walled tube made of a linear elastic, isotropic, and incompressible (constant density) material. Hudetz[62] modified this expression assuming an orthotropic (properties depend on the direction of loading) cylindrical vessel having a nonlinear stress–strain relationship, which is more realistic.

$$H_{\theta\theta} = \frac{\Delta P_t}{\Delta D_e} \frac{2 D_i^2 D_e}{D_e^2 - D_i^2} + \frac{2 P_t D_e^2}{D_e^2 - D_i^2} \tag{13.6}$$

Finally, it is often convenient to discuss the actual forces required to increase arterial diameters in vivo and there is a long list of relationships that have been used to describe blood vessel behavior both in vivo and ex vivo. Table 13.1 summarizes some of the most often used parameters to describe blood vessel mechanical properties based on pressure-diameter measurements.

In vitro studies have shown that both the pattern and degree of mechanical stretch increase vascular SMC and EC proliferation, cause cell shape changes, and regulate the expression of several genes.[1,141] In addition, in vivo studies have demonstrated a direct relationship between the circumferential stress to which the pulmonary vascular wall is exposed and the structure of the wall itself. For example, SMC hypertrophy and collagen and elastin synthesis occur in response to increases in mechanical stress.[41,79,83,84] These are just a few examples of how increased pressure—by increasing either stretch or stress—alters pulmonary vascular biology. We discuss this topic in more depth in the final section of this chapter.

13.3.2 Pulmonary Blood Flow and Luminal Shear Stress

Mean pulmonary blood flow, which is equal to mean aortic blood flow, is approximately 5 L/min at rest and can increase to 20 L/min or higher with exercise in a healthy human. As blood enters the pulmonary arteries from the right ventricle, it is highly pulsatile. As the large arteries convert

TABLE 13.1
Summary of Structural Parameters of Elasticity for Pulmonary Vessels

Parameter [Units]	Expression	Explanation
Pressure-strain modulus (E_p) [dyn/cm^2]	$E_p = \dfrac{\Delta P \cdot R}{\Delta R}$	Change in transmural pressure times the initial radius divided by the change in radius. An increase in E_p could be caused by an increase in elastic modulus, wall thickening, or an increase in initial radius
Stiffness coefficient (β) [dimensionless]	$\beta = \dfrac{\ln\left(P_s / P_d\right)}{\left((D_s - D_d)/D_d\right)}$	Behavior of the vessel in the physiological range. P_s and P_d are the systolic and diastolic blood pressures, respectively, and D_s and D_d are the systolic and diastolic diameters, respectively
Volume distensibility coefficient (D_v [cm^2/dyn])	$D_v = \dfrac{\Delta V}{V \cdot \Delta P}$	Ratio of the change in volume induced by a change in pressure divided by the initial volume
Compliance (C) [cm^5/dyn])	$C = \dfrac{\Delta V}{\Delta P} = D_v \cdot V$	Change in volume for a given change in pressure

proximal pulsatile pressure to distal steady pressures, they also convert proximal pulsatile flows to distal steady flows.

In order to understand the distribution of blood flow from proximal to distal vessels in the pulmonary vasculature, we can use different theoretical explanations. One such theory was presented by Murray, who postulated that when a vessel immediately upstream from a branch point (parent vessel) branches into two (or more) daughter vessels, the cube of the radius of the parent artery is equal to the sum of the cubes of the radii of daughter vessels.[115] This "cube" law assumes a constant, steady flow of blood, a minimum pressure energy requirement for driving blood flow, and a minimum metabolic energy requirement for maintaining temperature. Two important consequences of the Cube Law are that shear stress is constant throughout the circulation and that the total cross-sectional area of the small vessels exceeds that of large vessels such that average velocities in small vessels are lower than those in large vessels.

Another theoretical explanation is based on the often fractal nature of biological and physiological systems. With this approach, the branching pattern of the vascular tree is assumed to be fractal and the pulmonary blood flow distribution is calculated accordingly, with equal resistance outlet boundary conditions. Fractal distributions have been shown to produce flow distribution and vessel dimensions that are similar to those seen in experimental animals.[55]

An important characteristic of the pulmonary vasculature is the extensive number of supernumerary vessels that branch off from the accompanying vessels, and that rapidly bifurcate into additional branches. The functional significance of these vessels remains unclear.

The regional distribution of blood in the lung is also determined by the effect of gravity, which creates hydrostatic pressure gradients as noted above, by the topology of the vascular trees, their mechanical properties, and by local vasocontrol.[7,63] These factors are species dependent and subject to the effects in changes of posture.[159] In order to study the contribution of these factors to the variation in the distribution of blood, different approaches have been developed. For example, perfusion distribution in human lungs and isolated perfused lungs preparation have been used as the basis for the generally accepted zone model (see above). In humans, imaging techniques such as positron emission tomography (PET), magnetic resonance (MR), and electron beams can be used to measure perfusion distribution. In the isolated perfused lung preparation, the lungs are continuously ventilated and the pulmonary circulation is typically cleared of blood by perfusion with a physiological synthetic medium. Then, blood flow and pressure distribution can be investigated with various manipulations.[56] Finally, computational studies using models such as fractal or anatomically based models have been performed.[54,55,88,158] All of these approaches have provided important insights into the physiological mechanisms of regional distribution of blood flow in the lung, which directly affect shear stress, an important mechanical stimulus.

As blood flows, it exerts a frictional shear stress on the endothelial surface in the direction of flow, and is influenced by changes in lung volume during respiration. Under steady flow conditions in a straight, cylindrical tube (for a Newtonian fluid), shear stress is directly proportional to the volumetric flow rate and viscosity, and inversely proportional to the inner radius cubed. Shear stress is primarily sensed by endothelial cells, which are located at the interface between the blood and the vessel wall. Shear stress is well known to affect EC morphology, structure, and function.[22]

Under physiologic conditions, the mean shear stress to which the pulmonary arterial endothelium surface is exposed is remarkably constant, close to 10–25 dyn/cm^2, in regions of uniform geometry and away from branch vessels.[38,129] In the case of small pulmonary arteries, shear stress is the major mechanical stimulus because circumferential stretch and stress, as a consequence of transmural pressure, are minimal.[95]

Shear stress may be altered in the pulmonary circulation by changes in (1) mean pulmonary blood flow or blood velocity distribution; (2) hematocrit, which affects viscosity; or (3) under- or overdevelopment of the pulmonary circulation that alters geometry and branching patterns.[29]

Shear stress plays a major role in modulating vascular function via different proteins including vasoactive components, growth factors and others. Shear stress induces gene expression of several endothelial cell-specific genes including EDNO, ET-1, and basic fibroblast growth factor (β-FGF).[29,106,167,168] EDNO is secreted by ECs in response to increases in shear stress and acts to dilate arteries via vSMC relaxation. Although the mechanisms involved are unclear, shear stress appears to activate endothelial nitric oxide synthase (eNOS) gene transcription via the protein kinase C system[167] and by Akt-mediated phosphorylation.[169] ET-1 production is increased and EDNO production is decreased by decreases in shear stress, resulting in vasoconstriction. Increases in shear stress also stimulate production of β-FGF, which in turn stimulates the migration of, and is mitogenic for, ECs, vSMCs, and fibroblasts.

Finally, synthesis of fibrous proteins, such as collagen I and elastin, is modulated by blood flow and shear stress.[51,96,158] Thus, it is a potent mechanical stimulus for changes in vascular biology.

In summary, the mechanical environment plays an important role in homeostasis of the pulmonary vasculature. In the case of sustained alterations of blood pressure and flow, the arterial wall appears to grow and remodel in order to restore circumferential stretch and stress and luminal shear stress toward homeostatic targets. NO appears to be a particularly important player in pulmonary vascular mechanobiology. Although there have been advances in the understanding of the relationships between hemodynamic forces and vascular biological responses, there are still many missing pieces. One important missing piece is an understanding of the cellular structures which are responsible for mechanotransduction and the ways in which these structures malfunction to cause disease.

13.4 PULMONARY VASCULAR MECHANOSENSORS

In the pulmonary vasculature, mechanical stimuli induce a variety of responses including activation of ion channels and production of vasoactive substances by EC, proliferation, apoptosis, and changes in gene expression in EC, vSMC and fibroblasts and SMC contraction and relaxation (Figure 13.2). Dysregulation of these mechanical responses contributes to some of the most important diseases of the pulmonary vasculature. Identification and characterization of the relevant mechanosensors, which are the structures capable of efficiently transmitting mechanical forces through linkages to the cytoskeleton and/or cell nucleus to elicit a biological response, is an active area of research. A variety of potential mechanosensors have been described, but the exact mechanisms by which cells sense and respond to mechanical forces remain largely unknown.[29,42] It has been postulated that nearly all mechanotransduction can be accounted for by changes in protein folding, which is the process by which proteins adopt the configuration that yields the lowest value of required energy. Since physical forces are able to modify the energy state of a protein, protein folding can be directly influenced by mechanical forces.[121] Furthermore, since most proposed mechanotransducers are made up of proteins, this mechanism could be universal. Some of the most well known mechanotransducers in the pulmonary vasculature are ion channels, integrins, G-proteins, and G-coupled protein receptors. In the case of potassium channels, phosphorylation of the channel is fundamental for changing states, which may involve a folded protein; for the integrins, protein rearrangement can be observed as filaments, microfilaments, and microtubules response to the mechanical forces; G-proteins and G-coupled protein receptors change their structure by phosphorylation, which again may stimulate protein folding.

Below we review the current state of knowledge regarding these mechanosensors and how they act in ECs and SMCs, which have received the most empirical attention. An improved understanding of the pulmonary vascular mechanotransduction mechanisms will surely lead to new insights into, and possibly provide novel therapeutic targets for, pulmonary vascular disease.

13.4.1 Ion Channels

Ion channels are rapidly responding elements that are present in the plasma membrane of vascular cells and thus are strategically located to respond to mechanical changes. Two different mechanosensitive ion channels have been described in pulmonary vascular cells: shear stress-activated

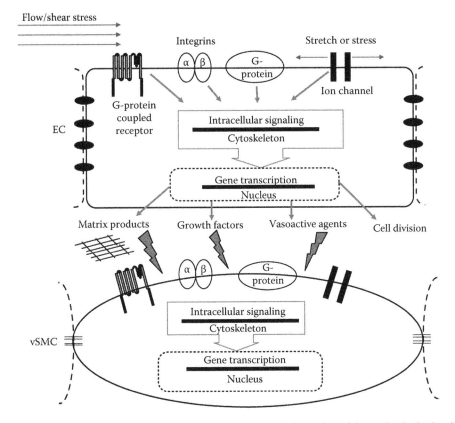

FIGURE 13.2 A conceptual illustration of the known pathways through which mechanical stimuli (flow-induced shear stress and cellular stretch or stress) evoke a biological response.

potassium channels and stretch-activated ion channels.[19] These channels have been shown to regulate different responses in EC, vSMC and fibroblasts.

Shear stress-activated potassium channels in EC have been proposed as cellular flow sensors and have been shown to regulate NO generation in response to shear stress[33] and the release of cGMP.[143] Potassium channels participate in the ability of ECs to sense both steady and oscillatory shear stress patterns.[27] These structures have an essential role in regulating overall EC responsiveness to flow. Studies using pharmacological blockers of potassium channels have shown that a number of important EC responses to flow are attenuated by channel inactivation including release of NO,[33] upregulation of transforming growth factor-beta, and downregulation of ET-1 protein expression.[19,27] By modulating flow-induced responses in ECs, these channels likely play an important role in regulating aspects of normal pulmonary vascular physiology including vasoregulation in response to acute changes in arterial blood flow and arterial wall remodeling in response to abnormal hemodynamics.[105,171]

The other type of ion channel present in EC is stretch-activated ion channels.[92] When a micro-pipette suction preparation is used, the degree of stretch has been found to be related to the electrical activity of the cell membrane and, more specifically, to the opening of transmembrane cation channels.[36]

In vSMCs, pressure-activated and stretch-activated ion channels exist.[25,26,86] Stretch-sensitive ion channels respond to cell deformation, not the action of forces on the cell surface.[13,26] Stretch-activated calcium and potassium channels have been identified in vSMCs.[82] Increased activity of ion channels in vSMCs has been correlated to pro-apoptotic signaling and a decrease in SMC contractility.[105] An important difference between potassium channels in pulmonary arterial SMC and systemic arterial SMC is their activation in acute hypoxia. In the case of pulmonary arteries, activity is reduced, which causes membrane depolarization, Ca^{2+} influx, increased cytosolic Ca^{2+}, and ultimately pulmonary vasoconstriction. Hypoxia-induced reduction of potassium channel activity may also contribute to SMC proliferation and inhibition of SMC apoptosis.[107] On the other hand, in SMC from systemic arteries (mesenteric and renal), activity is not reduced in response to hypoxia.[179]

13.4.2 INTEGRINS

When pulmonary arteries are circumferentially stretched by elevated blood pressure, transmission of the mechanical strain to cells occurs in part through integrins.[139] Integrins have been shown to play key roles in determining cell shape, levels of cytoskeletal tension, and other types of cell response to mechanical stimuli.[28,139]

ECs must be anchored to their substrate in order to sense and transduce signals related to shear stress associated with blood flow.[69] In the anchoring process, cells attach to a substrate via integrins. While there are many different types of integrins expressed in EC with different configurations— some are RGD-peptide sensitive in their binding to ECM ($\alpha_V\beta_3$, $\alpha_5\beta_1$, and $\alpha_3\beta_1$), while others are insensitive ($\alpha_2\beta_1$)—all transmembrane integrins are embedded in the cell membrane and have an extracellular and an intracellular domain. Shear stress sensed by the extracellular domain is thought to activate the intracellular domain, which interacts with cytoskeletal proteins that recruit signaling proteins to begin the biological response cascade.[77]

In vSMCs, integrins are responsible for the transduction of the mechanical deformation but are sensitive to signaling molecules generated by ECs and cell–cell interactions.[118,139] Integrins are also involved in the signaling pathways that lead to the production of endogenous vasoconstrictors of the prostanoid family including thromboxane A_2 and different types of prostanglandins ($F_{2\alpha}$, E_2, I_2).[139]

In all vascular cell types, stretch can enhance integrin adhesion and integrin-associated signaling, which can lead to diverse effects such as the release of cytokines, the activation of transcription factors, altered gene and protein expression, and cell division or death.[121] Finally, integrins play a crucial role in the interactions between ECM molecules and their response to mechanical stimuli.[29]

13.4.3 HETERODIMERIC G PROTEINS AND G-COUPLED PROTEIN RECEPTORS

Heterodimeric G proteins play a crucial role in the transduction of mechanical forces in ECs,[14] vSMC,[76] and fibroblasts.[113] Shear stress activates some G proteins, which may be linked to the activation of K^+ channels[27,36] and the increase of intracellular Ca^{2+}, both of which cause vSMC contraction. Another important function of this family of proteins is their participation in the upregulation of pulmonary EC permeability.[15]

G protein-coupled protein receptors are 7-protein transmembrane receptors that are also likely to be involved in mechanotransduction. Activation of these receptors may be by a direct physical contact between the receptor and the mechanical stimuli (flow or pressure) or secondary to mechanical-mediated binding of a known receptor. In the case of EC, the data supporting a role for G-coupled protein receptors can be identified for a number of shear stress-related responses; for example, the generation of different substances (i.e., IP_3, DAG, NO), which are associated with changes of intracellular calcium,[89] regulation of PDGF gene expression via a PKC-dependent mechanism, and the activation of potassium channels.[75] In the case of vSMC, G-coupled protein receptors are involved in the transduction of mechanical strain, which is related to changes in the regulation of stretch-sensitive genes.[26] The activation of these protein receptors occurs through a physical binding and induces a multistep series of transducing signals including elevation of intracellular calcium.[25]

Recently, cross-talk has been observed between signals transduction stimulated by integrins and G-protein-coupled receptors, suggesting multiple mechanisms by which different pathways may be activated in response to shear stress.[97] Experimental and modeling studies that help us understand the interactions between different vascular mechanotransducers in a given cell, between cells and between cells and matrix proteins are likely to be a particularly fruitful area of future research, especially as these studies help us better understand disease development and progression.

13.5 MECHANOBIOLOGY AND DISEASES

Pulmonary vascular mechanobiology is relevant to most if not all diseases of the pulmonary vasculature, either because abnormal mechanical forces trigger normal biological responses or because abnormal biological responses occur in reaction to normal mechanical forces. Examples of both can be found in the disease state of pulmonary hypertension, which is considered the most serious and potentially devastating chronic disorder of the pulmonary circulation. Recently, an expert consensus document was published on pulmonary hypertension detailing the classification and epidemiology, natural history and survival, pathology and pathogenesis, and strategies for diagnosis and treatment.[108] Here we review these concerns briefly and then focus on the known and speculated mechanobiological abnormalities that are relevant to pulmonary hypertension to illustrate the clinical relevance of pulmonary vascular mechanobiology.

Pulmonary hypertension is a family of diseases that includes pulmonary arterial hypertension (PAH), pulmonary hypertension secondary to left heart disease, pulmonary hypertension associated with lung diseases and/or hypoxemia, chronic thromboembolic disease, and pulmonary hypertension secondary to other "miscellaneous" diseases (e.g., sarcoidosis, lymphangiomatosis and histiocytosis X)[147] (Figure 13.3). PAH was previously classified as a rare disease but more recent data suggest the prevalence is 15 per million.[64] The prognosis is poor even with modern treatment options: the 1 year mortality rate is about 15%.[160] The cause of death is typically right ventricular failure and right ventricular function is the major determinant of outcomes in PAH.[164]

Like cancer and atherosclerosis, PAH is not caused by a single genetic defect or environmental insult. The pathology and pathogenesis are complex and multifaceted. Of the suspected initiating events in PAH, most are examples of abnormal biological responses to normal mechanical forces.

For example, in familial PAH (6%–10% of cases[64,133]) in which mutations in the transforming growth factor β (TGF-β) receptor pathway are inherited as autosomal dominant with incomplete penetrance and genetic anticipation, over-proliferation of EC and vSMC occur prior to any elevations in pressure or changes in flow.[156] Similarly, in idiopathic PAH (nearly 50% of cases[64] and a female/male ratio of 1.7:1[133]) mutations in the TGF-β receptor pathway have been found in up to 25% of

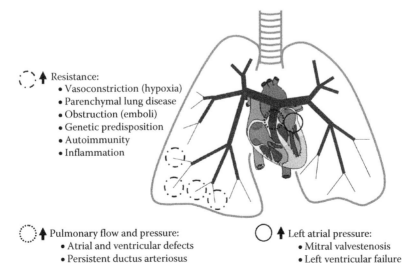

FIGURE 13.3 Schematic illustration of the known causes of pulmonary hypertension grouped by site of mechanical or biological abnormality.

the identified cases,[163] with consequent abnormalities in pulmonary vascular cell proliferation and apoptosis in the absence of detectable hemodynamic abnormalities.[163] Both of these diseases are characterized by an overabundance of vSMC in small, peripheral arteries such that the lumens are nearly occluded, the result of which is increased PVR and right ventricular overload.

An example of abnormal mechanical forces leading to normal, but not desirable, biological responses is PAH associated with congenital heart disease. In this situation, shunting from the left to the right side of the heart increases pulmonary blood flow and pressure. As a consequence, postnatal thinning of pulmonary arteries is impaired[54,79] and significant EC dysfunction develops.[150] Thereafter, there is an apparent progression from EC dysfunction to SMC dysfunction throughout the pulmonary vasculature that leads to heightened pulmonary vascular reactivity, which leads to intermittent arteriolar narrowing and further development of hypertension.[150] Importantly, if the defect is corrected in utero, no pulmonary hypertension develops, demonstrating that the pulmonary vascular cells are able to respond normally to normal mechanical forces but have a maladaptive response to abnormal mechanical forces.

Another example of abnormal mechanical forces leading to normal, but not desirable, biological responses is conduit artery stiffening in hypoxia-induced PAH. Chronic hypoxia, whether due to chronic obstructive pulmonary disease,[67] cystic fibrosis,[43] obstructive sleep apnea[59] or living at high altitude,[142] is known to cause persistent SMC contraction in pulmonary resistance arteries.[70] The subsequent increase in pressure throughout the pulmonary vasculature causes remodeling in the large conduit arteries, characterized by hypertrophy and hyperplasia of vSMCs, increased deposition of ECM proteins, including collagen and elastin, and proliferation of adventitial fibroblasts.[83,91,152] As a result of these changes, conduit arteries become stiffer, causing an increase in flow pulsatility, which can damage downstream peripheral arteries via overstretch and the upstream right ventricle via systolic overload. However, as with congenital heart disease, if the hypoxia-induced hypertension can be alleviated, the large and small arteries return to a normal state.[98,99,174]

Although there is no cure for PAH, treatments have improved significantly in the past decades, improving life quality and survival. The most effective therapy, which has demonstrated a 95% 5 year survival,[134] unfortunately works only in a very small subset of patients, variously estimated to be 12% to 55% of those with idiopathic PAH.[134,148,173] The therapy consists of utilizing calcium channel blockers, which cause sufficient vasodilation in patients who are "responders" to decrease mPAP by at least 10 mmHg to a mPAP of at most 40 mmHg with no change in CO.[108] In the majority of PAH patients, prostanoid therapies are the mainstay, which have both antiproliferative and vasodilatory effects and appear to somewhat compensate for inadequate prostacyclin synthase in patients with PAH.[108] Other approaches to treating PAH include blockade of the endothelin receptor pathway, which stimulates vasoconstriction and vSMC migration, and phosphodiesterase inhibitors, which may enhance or prolong the vasodilating and perhaps antiproliferative effects of cyclic nucleotides such as cGMP.[108]

Several curative strategies are in development, an example of which is adenoviral vector delivery of functional BMPR2 genes to the pulmonary vascular endothelium; in a rat model of hypoxic pulmonary hypertension, this strategy successfully reduced pulmonary arterial and right ventricular pressure and the pulmonary vascular SMC proliferative response to hypoxic pulmonary hypertension.[132] Inhibition of potassium channels, which may be defective in some PAH patients and which are responsible for the contraction of vSMC, has also shown promise.[179] Additionally, stem cell therapies in which regeneration of lung microvasculature is a strategy for reducing PVR in patients with advanced PAH[53] are being pursued. Finally, with increased empirical investigation we may learn that PAH is a malignant disease process much like cancer[164] that may be caused by multiple stimuli and, as such, might respond to multiple therapeutic strategies.

13.6 CONCLUSIONS

One of the big questions in the field of mechanobiology is how a mechanical signal is *initially* sensed in the pulmonary cells. Two different theories have been put forward—that there exists a direct, physical pathway (mechanotransduction) and that there exists an indirect pathway (mechanotransmission). In the direct pathway theory, transduction between the stimulus and the cell occurs via physical interactions between the transducer and either the cell membrane or other structures such as the cytoskeleton.[14,18,123] The indirect pathway theory postulates that although there is no physical connection between the transducer and the cell, mechanical forces initiated intercellular communication and the activation of the second type of cell is caused by signaling molecules from the first type of cell that are directly stimulated by flow or pressure.[22,30] There are several ways in which cells may communicate with each other including through the use of cytokines, growth factors, and other small soluble molecules. Cells can also directly transmit messages through cellular junctions or by their interactions with the ECM. In reality, both theories are not mutually exclusive; direct mechanical stimuli acting on surface protein and receptors may modify enzyme-substrate and agonist-receptor interactions at the same time that the surface concentration of the agonist is being influenced by active or passive transport (i.e., convective and diffusive).

Evidence shows that both theories can partially explain the response of the cell to the stimuli, but there is an urgent need for fully determining the pathways involved in the transmission of mechanical forces to the cell surfaces and the conversion to chemical or electrical responses, as well as understanding the nature and identity of most of the mechanosensors involved in the response.

Cloning some of the potential mechanosensors (e.g., ion channels, integrins, G-proteins) will allow researchers to better understand some of the mechanisms of activation and transduction. For example, expressing cloned mechanosensors in otherwise non-mechanosensitive cells may be an ideal platform in which the activity and mechanisms of action of these mechanosensors could be determined. In addition, the responsiveness of certain sensors, but not others, to particular characteristics of the stimuli such as steady vs. oscillatory pressure or flow, highlights the need for a better understanding of the impact of the signals on the transducers. Finally, the ways in which the different transducers interact with each other and between cells remain unknown. Elucidating these interactions promises to provide an integrated understanding of how the pulmonary vascular cells respond to flow and pressure under both normal and diseased conditions, which may yield important insights into the development, progression, and ideal treatments for pulmonary vascular disease.

REFERENCES

1. Acevedo, A. D., Bowser, S. S., Gerritsen, M. E., and Bizios, R. (1993). Morphological and proliferative responses of endothelial cells to hydrostatic pressure: Role of fibroblast growth factor. *Journal of Cellular Physiology, 157*(3), 603–614.
2. Aird, W. C. (2003). Endothelial cell heterogeneity. *Critical Care Medicine, 31*(4 Suppl), S221–S230.
3. Ali, M. H. and Schumacker, P. T. (2002). Endothelial responses to mechanical stress: Where is the mechanosensor? *Critical Care Medicine, 30*(5 Suppl), S198–S206.
4. Archer, S. L. (1996). Diversity of phenotype and function of vascular smooth muscle cells. *The Journal of Laboratory and Clinical Medicine, 127*(6), 524–529.
5. Barakat, A. and Lieu, D. (2003). Differential responsiveness of vascular endothelial cells to different types of fluid mechanical shear stress. *Cell Biochemistry and Biophysics, 38*(3), 323–343.
6. Barbera, J. A., Peinado, V. I., and Santos, S. (2003). Pulmonary hypertension in chronic obstructive pulmonary disease. *The European Respiratory Journal: Official Journal of the European Society for Clinical Respiratory Physiology, 21*(5), 892–905.
7. Barnes, P. J. and Liu, S. F. (1995). Regulation of pulmonary vascular tone. *Pharmacological Reviews, 47*(1), 87–131.
8. Barra, J. G., Armentano, R. L., Levenson, J., Fischer, E. I., Pichel, R. H., and Simon, A. (1993). Assessment of smooth muscle contribution to descending thoracic aortic elastic mechanics in conscious dogs. *Circulation Research, 73*(6), 1040–1050.
9. Belknap, J. K., Orton, E. C., Ensley, B., Tucker, A., and Stenmark, K. R. (1997). Hypoxia increases bromodeoxyuridine labeling indices in bovine neonatal pulmonary arteries. *American Journal of Respiratory Cell and Molecular Biology, 16*(4), 366–371.

10. Belknap, J. K., Weiser-Evans, M. C., Grieshaber, S. S., Majack, R. A., and Stenmark, K. R. (1999). Relationship between perlecan and tropoelastin gene expression and cell replication in the developing rat pulmonary vasculature. *American Journal of Respiratory Cell and Molecular Biology, 20*(1), 24–34.

11. Benitz, W. E., Kelley, R. T., Anderson, C. M., Lorant, D. E., and Bernfield, M. (1990). Endothelial heparan sulfate proteoglycan. I. inhibitory effects on smooth muscle cell proliferation. *American Journal of Respiratory Cell and Molecular Biology, 2*(1), 13–24.

12. Bergel, D. H. (1961). The static elastic properties of the arterial wall. *The Journal of Physiology, 156*(3), 445–457.

13. Bialecki, R. A., Kulik, T. J., and Colucci, W. S. (1992). Stretching increases calcium influx and efflux in cultured pulmonary arterial smooth muscle cells. *The American Journal of Physiology, 263*(5 Pt 1), L602–L606.

14. Birukov, K. G., Birukova, A. A., Dudek, S. M., Verin, A. D., Crow, M. T., Zhan, X. et al. (2002). Shear stress-mediated cytoskeletal remodeling and cortactin translocation in pulmonary endothelial cells. *American Journal of Respiratory Cell and Molecular Biology, 26*(4), 453–464.

15. Birukova, A. A., Arce, F. T., Moldobaeva, N., Dudek, S. M., Garcia, J. G., Lal, R. et al. (2009). Endothelial permeability is controlled by spatially defined cytoskeletal mechanics: Atomic force microscopy force mapping of pulmonary endothelial monolayer. *Nanomedicine: Nanotechnology, Biology, and Medicine, 5*(1), 30–41.

16. Bischoff, J. (1995). Approaches to studying cell adhesion molecules in angiogenesis. *Trends in Cell Biology, 5*(2), 69–74.

17. Blackman, B. R., Garcia-Cardena, G., and Gimbrone, M. A. Jr. (2002). A new in vitro model to evaluate differential responses of endothelial cells to simulated arterial shear stress waveforms. *Journal of Biomechanical Engineering, 124*(4), 397–407.

18. Blum, M. S., Toninelli, E., Anderson, J. M., Balda, M. S., Zhou, J., O'Donnell, L. et al. (1997). Cytoskeletal rearrangement mediates human microvascular endothelial tight junction modulation by cytokines. *The American Journal of Physiology, 273*(1 Pt 2), H286–H294.

19. Bonnet, S. and Archer, S. L. (2007). Potassium channel diversity in the pulmonary arteries and pulmonary veins: Implications for regulation of the pulmonary vasculature in health and during pulmonary hypertension. *Pharmacology & Therapeutics, 115*(1), 56–69.

20. Brenner, O. (1935). Pathology of the vessels of the pulmonary circulation. *Archives of Internal Medicine, 56*, 211–237.

21. Brooke, B. S., Karnik, S. K., and Li, D. Y. (2003). Extracellular matrix in vascular morphogenesis and disease: Structure versus signal. *Trends in Cell Biology, 13*(1), 51–56.

22. Bruder, J. L., Hsieh, T., Lerea, K. M., Olson, S. C., and Wu, J. M. (2001). Induced cytoskeletal changes in bovine pulmonary artery endothelial cells by resveratrol and the accompanying modified responses to arterial shear stress. *BMC Cell Biology, 2*, 1.

23. Bucher, U. and Reid, L. (1961). Development of the intrasegmental bronchial tree: The pattern of branching and development of cartilage at various stages of intra-uterine life. *Thorax, 16*, 207–218.

24. Chandran, K. B., Yoganathan, A. P., and Rittgers, S. E. (Eds.). (2007). *Biofluid Mechanics the Human Circulation*. Boca Raton, FL: CRC Press (Taylor & Francis Group).

25. Chao, W. and Olson, M. S. (1993). Platelet-activating factor: Receptors and signal transduction. *The Biochemical Journal, 292*(Pt 3), 617–629.

26. Chaqour, B., Howard, P. S., Richards, C. F., and Macarak, E. J. (1999). Mechanical stretch induces platelet-activating factor receptor gene expression through the NF-kappaB transcription factor. *Journal of Molecular and Cellular Cardiology, 31*(7), 1345–1355.

27. Chatterjee, S., Al-Mehdi, A. B., Levitan, I., Stevens, T., and Fisher, A. B. (2003). Shear stress increases expression of a KATP channel in rat and bovine pulmonary vascular endothelial cells. *American Journal of Physiology. Cell Physiology, 285*(4), C959–C967.

28. Chen, C. S., Mrksich, M., Huang, S., Whitesides, G. M., and Ingber, D. E. (1997). Geometric control of cell life and death. *Science, 276*(5317), 1425–1428.

29. Chen, K. D., Li, Y. S., Kim, M., Li, S., Yuan, S., Chien, S. et al. (1999). Mechanotransduction in response to shear stress. Roles of receptor tyrosine kinases, integrins, and shc. *The Journal of Biological Chemistry, 274*(26), 18393–18400.

30. Christman, B. W., McPherson, C. D., Newman, J. H., King, G. A., Bernard, G. R., Groves, B. M. et al. (1992). An imbalance between the excretion of thromboxane and prostacyclin metabolites in pulmonary hypertension. *The New England Journal of Medicine, 327*(2), 70–75.

31. Cioffi, D. L., Moore, T. M., Schaack, J., Creighton, J. R., Cooper, D. M., and Stevens, T. (2002). Dominant regulation of interendothelial cell gap formation by calcium-inhibited type 6 adenylyl cyclase. *The Journal of Cell Biology, 157*(7), 1267–1278.

32. Clowes, A. W. and Karnowsky, M. J. (1977). Suppression by heparin of smooth muscle cell proliferation in injured arteries. *Nature, 265*(5595), 625–626.

33. Cooke, J. P., Rossitch, E. Jr., Andon, N. A., Loscalzo, J., and Dzau, V. J. (1991). Flow activates an endothelial potassium channel to release an endogenous nitrovasodilator. *The Journal of Clinical Investigation, 88*(5), 1663–1671.

34. Cox, R. H. (1982). Comparison of mechanical and chemical properties of extra- and intralobar canine pulmonary arteries. *The American Journal of Physiology, 242*(2), H245–H253.

35. Cox, R. H. (1984). Viscoelastic properties of canine pulmonary arteries. *The American Journal of Physiology, 246*(1 Pt 2), H90–H96.

36. Davies, P. F. (1995). Flow-mediated endothelial mechanotransduction. *Physiological Reviews, 75*(3), 519–560.

37. Davies, P. F., Robotewskyj, A., Griem, M. L., Dull, R. O., and Polacek, D. C. (1992). Hemodynamic forces and vascular cell communication in arteries. *Archives of Pathology & Laboratory Medicine, 116*(12), 1301–1306.

38. Davies, P. F., Spaan, J. A., and Krams, R. (2005). Shear stress biology of the endothelium. *Annals of Biomedical Engineering, 33*(12), 1714–1718.

39. Desmouliere, A., Chaponnier, C., and Gabbiani, G. (2005). Tissue repair, contraction, and the myofibroblast. *Wound Repair and Regeneration: Official Publication of the Wound Healing Society [and] the European Tissue Repair Society, 13*(1), 7–12.

40. Faury, G. (2001). Function-structure relationship of elastic arteries in evolution: From microfibrils to elastin and elastic fibres. *Pathologie-Biologie, 49*(4), 310–325.

41. Feihl, F., Liaudet, L., Levy, B. I., and Waeber, B. (2008). Hypertension and microvascular remodelling. *Cardiovascular Research, 78*(2), 274–285.

42. Fisher, A. B., Chien, S., Barakat, A. I., and Nerem, R. M. (2001). Endothelial cellular response to altered shear stress. *American Journal of Physiology. Lung Cellular and Molecular Physiology, 281*(3), L529–L533.

43. Fraser, K. L., Tullis, D. E., Sasson, Z., Hyland, R. H., Thornley, K. S., and Hanly, P. J. (1999). Pulmonary hypertension and cardiac function in adult cystic fibrosis: Role of hypoxemia. *Chest, 115*(5), 1321–1328.

44. Frid, M. G., Dempsey, E. C., Durmowicz, A. G., and Stenmark, K. R. (1997). Smooth muscle cell heterogeneity in pulmonary and systemic vessels. Importance in vascular disease. *Arteriosclerosis, Thrombosis, and Vascular Biology, 17*(7), 1203–1209.

45. Fung, Y. C. (1993). *Biomechanics: Mechanical Properties of Living Tissues*, 2nd edn. Berlin, Germany: Springer.

46. Gabbiani, G. (2003). The myofibroblast in wound healing and fibrocontractive diseases. *The Journal of Pathology, 200*(4), 500–503.

47. Galie, N., Torbicki, A., Barst, R., Dartevelle, P., Haworth, S., Higenbottam, T. et al. (2004). Guidelines on diagnosis and treatment of pulmonary arterial hypertension. The Task Force on Diagnosis and Treatment of Pulmonary Arterial Hypertension of the European Society of Cardiology. *European Heart Journal, 25*(24), 2243–2278.

48. Gebb, S. and Stevens, T. (2004). On lung endothelial cell heterogeneity. *Microvascular Research, 68*(1), 1–12.

49. Gebb, S. A. and Jones, P. L. (2003). Hypoxia and lung branching morphogenesis. *Advances in Experimental Medicine and Biology, 543*, 117–125.

50. Gelse, K., Poschl, E., and Aigner, T. (2003, November 28). Collagens—Structure, function, and biosynthesis. *Advanced Drug Delivery Reviews, 55*(12), 1531–1546.

51. Gerasimovskaya, E. V., Ahmad, S., White, C. W., Jones, P. L., Carpenter, T. C., and Stenmark, K. R. (2002). Extracellular ATP is an autocrine/paracrine regulator of hypoxia-induced adventitial fibroblast growth. Signaling through extracellular signal-regulated kinase-1/2 and the Egr-1 transcription factor. *The Journal of Biological Chemistry, 277*(47), 44638–44650.

52. Ghiadoni, L., Bruno, R. M., Stea, F., Virdis, A., and Taddei, S. (2009). Central blood pressure, arterial stiffness, and wave reflection: New targets of treatment in essential hypertension. *Current Hypertension Reports, 11*(3), 190–196.

53. Ghofrani, H. A., Barst, R. J., Benza, R. L., Champion, H. C., Fagan, K. A., Grimminger, F. et al. (2009). Future perspectives for the treatment of pulmonary arterial hypertension. *Journal of the American College of Cardiology, 54*(1 Suppl), S108–S117.

54. Ghorishi, Z., Milstein, J. M., Poulain, F. R., Moon-Grady, A., Tacy, T., Bennett, S. H. et al. (2007). Shear stress paradigm for perinatal fractal arterial network remodeling in lambs with pulmonary hypertension and increased pulmonary blood flow. *American Journal of Physiology. Heart and Circulatory Physiology, 292*(6), H3006–H3018.

55. Glenny, R. W. (1998). Blood flow distribution in the lung. *Chest, 114*(1 Suppl), 8S–16S.

56. Hakim, T. S., Maarek, J. M., and Chang, H. K. (1989). Estimation of pulmonary capillary pressure in intact dog lungs using the arterial occlusion technique. *The American Review of Respiratory Disease, 140*(1), 217–224.

57. Hassoun, P. M., Thompson, B. T., and Hales, C. A. (1992). Partial reversal of hypoxic pulmonary hypertension by heparin. *The American Review of Respiratory Disease, 145*(1), 193–196.

58. Hayashi, K., Stergiopulos, N., Meister, J. J., Greenwald, S. E., Rachev, A. (2001). Techniques in the determination of the mechanical properties and constitutive laws of arterial wall. In: C. Leondes (Ed.), *Cardiovascular Techniques—Biomechanical Systems Techniques and Applications.* Boca Raton, FL: CRC Press.

59. Hiestand, D. and Phillips, B. (2008). The overlap syndrome: Chronic obstructive pulmonary disease and obstructive sleep apnea. *Critical Care Clinics, 24*(3), 551–563, vii.

60. Ho, M. S., Bose, K., Mokkapati, S., Nischt, R., and Smyth, N. (2008). Nidogens-extracellular matrix linker molecules. *Microscopy Research and Technique, 71*(5), 387–395.

61. Houweling, B. (2007). Regulation of pulmonary vascular tone in health and disease: Special emphasis on exercise and pulmonary hypertension after myocardial infarction (unpublished data).

62. Hudetz, A. G. (1979). Incremental elastic modulus for orthotropic incompressible arteries. *Journal of Biomechanics, 12*(9), 651–655.

63. Hughes, M. and West, J. B. (2008). Point: Gravity is the major factor determining the distribution of blood flow in the human lung. *Journal of Applied Physiology, 104*(5), 1531–1533.

64. Humbert, M., Sitbon, O., Chaouat, A., Bertocchi, M., Habib, G., Gressin, V. et al. (2006). Pulmonary arterial hypertension in France: Results from a national registry. *American Journal of Respiratory and Critical Care Medicine, 173*(9), 1023–1030.

65. Hunter, K. S., Lee, P. F., Lanning, C. J., Ivy, D. D., Kirby, K. S., Claussen, L. R. et al. (2008). Pulmonary vascular input impedance is a combined measure of pulmonary vascular resistance and stiffness and predicts clinical outcomes better than pulmonary vascular resistance alone in pediatric patients with pulmonary hypertension. *American Heart Journal, 155*(1), 166–174, doi:10.1016/j.ahj.2007.08.014.

66. Ilkiw, R., Todorovich-Hunter, L., Maruyama, K., Shin, J., and Rabinovitch, M. (1989). SC-39026, a serine elastase inhibitor, prevents muscularization of peripheral arteries, suggesting a mechanism of monocrotaline-induced pulmonary hypertension in rats. *Circulation Research, 64*(4), 814–825.

67. Incalzi, R. A., Fuso, L., De Rosa, M., Di Napoli, A., Basso, S., Pagliari, G. et al. (1999). Electrocardiographic signs of chronic cor pulmonale: A negative prognostic finding in chronic obstructive pulmonary disease. *Circulation, 99*(12), 1600–1605.

68. Isaacson, T. C., Hampl, V., Weir, E. K., Nelson, D. P., and Archer, S. L. (1994). Increased endothelium-derived NO in hypertensive pulmonary circulation of chronically hypoxic rats. *Journal of Applied Physiology, 76*(2), 933–940.

69. Ishida, T., Takahashi, M., Corson, M. A., and Berk, B. C. (1997). Fluid shear stress-mediated signal transduction: How do endothelial cells transduce mechanical force into biological responses? *Annals of the New York Academy of Sciences, 811*, 12–23; discussion 23–24.

70. Jeffery, T. K. and Wanstall, J. C. (2001). Comparison of pulmonary vascular function and structure in early and established hypoxic pulmonary hypertension in rats. *Canadian Journal of Physiology and Pharmacology, 79*(3), 227–237.

71. Jones, F. S., Meech, R., Edelman, D. B., Oakey, R. J., and Jones, P. L. (2001). Prx1 controls vascular smooth muscle cell proliferation and tenascin-C expression and is upregulated with Prx2 in pulmonary vascular disease. *Circulation Research, 89*(2), 131–138.

72. Jones, P. L., Chapados, R., Baldwin, H. S., Raff, G. W., Vitvitsky, E. V., Spray, T. L. et al. (2002). Altered hemodynamics controls matrix metalloproteinase activity and tenascin-C expression in neonatal pig lung. *American Journal of Physiology. Lung Cellular and Molecular Physiology, 282*(1), L26–L35.

73. Jones, P. L., Cowan, K. N., and Rabinovitch, M. (1997). Tenascin-C, proliferation and subendothelial fibronectin in progressive pulmonary vascular disease. *The American Journal of Pathology, 150*(4), 1349–1360.

74. Jones, P. L. and Rabinovitch, M. (1996). Tenascin-C is induced with progressive pulmonary vascular disease in rats and is functionally related to increased smooth muscle cell proliferation. *Circulation Research, 79*(6), 1131–1142.

75. Kamouchi, M., Van Den Bremt, K., Eggermont, J., Droogmans, G., and Nilius, B. (1997). Modulation of inwardly rectifying potassium channels in cultured bovine pulmonary artery endothelial cells. *The Journal of Physiology, 504*(Pt 3), 545–556.

76. Karnik, S. K., Brooke, B. S., Bayes-Genis, A., Sorensen, L., Wythe, J. D., Schwartz, R. S. et al. (2003). A critical role for elastin signaling in vascular morphogenesis and disease. *Development, 130*(2), 411–423.

77. Kawasaki, J., Davis, G. E., and Davis, M. J. (2004). Regulation of Ca^{2+}-dependent K^+ current by alphav-beta3 integrin engagement in vascular endothelium. *The Journal of Biological Chemistry, 279*(13), 12959–12966.

78. Kay, J. M. (1983). Comparative morphologic features of the pulmonary vasculature in mammals. *The American Review of Respiratory Disease, 128*(2 Pt 2), S53–S57.

79. Kelly, D. A., Hislop, A. A., Hall, S. M., and Haworth, S. G. (2005). Relationship between structural remodeling and reactivity in pulmonary resistance arteries from hypertensive piglets. *Pediatric Research, 58*(3), 525–530.

80. Kibria, G., Heath, D., Smith, P., and Biggar, R. (1980). Pulmonary endothelial pavement patterns. *Thorax, 35*(3), 186–191.

81. King, J., Hamil, T., Creighton, J., Wu, S., Bhat, P., McDonald, F. et al. (2004). Structural and functional characteristics of lung macro- and microvascular endothelial cell phenotypes. *Microvascular Research, 67*(2), 139–151.

82. Kirber, M. T., Ordway, R. W., Clapp, L. H., Walsh, J. V. Jr., and Singer, J. J. (1992). Both membrane stretch and fatty acids directly activate large conductance Ca(2+)-activated K^+ channels in vascular smooth muscle cells. *FEBS Letters, 297*(1–2), 24–28.

83. Kobs, R. W. and Chesler, N. C. (2006). The mechanobiology of pulmonary vascular remodeling in the congenital absence of eNOS. *Biomechanics and Modeling in Mechanobiology, 5*(4), 217–225, doi:10.1007/s10237-006-0018-1.

84. Kobs, R. W., Muvarak, N. E., Eickhoff, J. C., and Chesler, N. C. (2005). Linked mechanical and biological aspects of remodeling in mouse pulmonary arteries with hypoxia-induced hypertension. *American Journal of Physiology. Heart and Circulatory Physiology, 288*(3), H1209–H1217.

85. Kohjima, T. (1998). Comparative study of artificial circulation for the liver after cardiogenic shock: Pulsatile or nonpulsatile? *The Japanese Journal of Thoracic and Cardiovascular Surgery: Official Publication of the Japanese Association for Thoracic Surgery = Nihon Kyobu Geka Gakkai Zasshi, 46*(11), 1117–1125.

86. Kohler, R., Grundig, A., Brakemeier, S., Rothermund, L., Distler, A., Kreutz, R. et al. (2001). Regulation of pressure-activated channel in intact vascular endothelium of stroke-prone spontaneously hypertensive rats. *American Journal of Hypertension, 14*(7 Pt 1), 716–721.

87. Kolacna, L., Bakesova, J., Varga, F., Kostakova, E., Planka, L., Necas, A. et al. (2007). Biochemical and biophysical aspects of collagen nanostructure in the extracellular matrix. *Physiological Research/Academia Scientiarum Bohemoslovaca, 56*(Suppl 1), S51–S60.

88. Krenz, G. S. and Dawson, C. A. (2002). Vessel distensibility and flow distribution in vascular trees. *Journal of Mathematical Biology, 44*(4), 360–374.

89. Kuchan, M. J., Jo, H., and Frangos, J. A. (1994). Role of G proteins in shear stress-mediated nitric oxide production by endothelial cells. *The American Journal of Physiology, 267*(3 Pt 1), C753–C758.

90. Lamm, W. J., Luchtel, D., and Albert, R. K. (1988). Sites of leakage in three models of acute lung injury. *Journal of Applied Physiology, 64*(3), 1079–1083.

91. Lammers, S. R., Kao, P. H., Qi, H. J., Hunter, K., Lanning, C., Albietz, J. et al. (2008). Changes in the structure-function relationship of elastin and its impact on the proximal pulmonary arterial mechanics of hypertensive calves. *American Journal of Physiology. Heart and Circulatory Physiology, 295*(4), H1451–H1459.

92. Lansman, J. B., Hallam, T. J., and Rink, T. J. (1987). Single stretch-activated ion channels in vascular endothelial cells as mechanotransducers? *Nature, 325*(6107), 811–813.

93. Lee, J. Y. and Spicer, A. P. (2000). Hyaluronan: A multifunctional, megaDalton, stealth molecule. *Current Opinion in Cell Biology, 12*(5), 581–586.

94. Lee, R. T., Yamamoto, C., Feng, Y., Potter-Perigo, S., Briggs, W. H., Landschulz, K. T. et al. (2001). Mechanical strain induces specific changes in the synthesis and organization of proteoglycans by vascular smooth muscle cells. *The Journal of Biological Chemistry, 276*(17), 13847–13851.

95. Li, M., Scott, D. E., Shandas, R., Stenmark, K. R., and Tan, W. (2009). High pulsatility flow induces adhesion molecule and cytokine mRNA expression in distal pulmonary artery endothelial cells. *Annals of Biomedical Engineering, 37*(6), 1082–1092.

96. Li, X., Jin, H., Bin, G., Wang, L., Tang, C., and Du, J. (2009). Endogenous hydrogen sulfide regulates pulmonary artery collagen remodeling in rats with high pulmonary blood flow. *Experimental Biology and Medicine, 234*(5), 504–512.

97. Linseman, D. A., Benjamin, C. W., and Jones, D. A. (1995). Convergence of angiotensin II and platelet-derived growth factor receptor signaling cascades in vascular smooth muscle cells. *The Journal of Biological Chemistry, 270*(21), 12563–12568.

98. Liu, J. Q., Zelko, I. N., Erbynn, E. M., Sham, J. S., and Folz, R. J. (2006). Hypoxic pulmonary hypertension: Role of superoxide and NADPH oxidase (gp91phox). *American Journal of Physiology. Lung Cellular and Molecular Physiology, 290*(1), L2–L10.

99. Liu, M., Tanswell, A. K., and Post, M. (1999). Mechanical force-induced signal transduction in lung cells. *The American Journal of Physiology, 277*(4 Pt 1), L667–L683.

100. Lucas, C. L. (1984). Fluid mechanics of the pulmonary circulation. *Critical Reviews in Biomedical Engineering, 10*(4), 317–393.

101. MacLean, M. R., Sweeney, G., Baird, M., McCulloch, K. M., Houslay, M., and Morecroft, I. (1996). 5-hydroxytryptamine receptors mediating vasoconstriction in pulmonary arteries from control and pulmonary hypertensive rats. *British Journal of Pharmacology, 119*(5), 917–930.

102. Madden, J. A., Vadula, M. S., and Kurup, V. P. (1992). Effects of hypoxia and other vasoactive agents on pulmonary and cerebral artery smooth muscle cells. *The American Journal of Physiology, 263*(3 Pt 1), L384–93.

103. Madri, J. A. and Williams, S. K. (1983). Capillary endothelial cell cultures: Phenotypic modulation by matrix components. *The Journal of Cell Biology, 97*(1), 153–165.

104. Majack, R. A., Cook, S. C., and Bornstein, P. (1986). Control of smooth muscle cell growth by components of the extracellular matrix: Autocrine role for thrombospondin. *Proceedings of the National Academy of Sciences of the United States of America, 83*(23), 9050–9054.

105. Mandegar, M. and Yuan, J. X. (2002). Role of K+ channels in pulmonary hypertension. *Vascular Pharmacology, 38*(1), 25–33.

106. Mata-Greenwood, E., Grobe, A., Kumar, S., Noskina, Y., and Black, S. M. (2005). Cyclic stretch increases VEGF expression in pulmonary arterial smooth muscle cells via TGF-beta1 and reactive oxygen species: A requirement for NAD(P)H oxidase. *American Journal of Physiology. Lung Cellular and Molecular Physiology, 289*(2), L288–L289.

107. Mauban, J. R., Remillard, C. V., and Yuan, J. X. (2005). Hypoxic pulmonary vasoconstriction: Role of ion channels. *Journal of Applied Physiology, 98*(1), 415–420.

108. McLaughlin, V. V., Archer, S. L., Badesch, D. B., Barst, R. J., Farber, H. W., Lindner, J. R. et al. (2009). ACCF/AHA 2009 expert consensus document on pulmonary hypertension: A report of the American College of Cardiology Foundation Task Force on Expert Consensus Documents and the American Heart Association: Developed in collaboration with the American College of Chest Physicians, American Thoracic Society, Inc., and the Pulmonary Hypertension Association. *Circulation, 119*(16), 2250–2294.

109. Mecham, R. P. (2001). Overview of extracellular matrix. *Current Protocols in Cell Biology*, Chapter 10, Unit 10.1.

110. Milnor, W. R., Conti, C. R., Lewis, K. B., and O'Rourke, M. F. (1969). Pulmonary arterial pulse wave velocity and impedance in man. *Circulation Research, 25*(6), 637–649.

111. Mitchell, G. F. (2008). Clinical achievements of impedance analysis. *Medical & Biological Engineering & Computing, 47*(2), 153–163.

112. Mithieux, S. M. and Weiss, A. S. (2005). Elastin. *Advances in Protein Chemistry, 70*, 437–461.

113. Mortimer, H. J., Peacock, A. J., Kirk, A., and Welsh, D. J. (2007). p38 MAP kinase: Essential role in hypoxia-mediated human pulmonary artery fibroblast proliferation. *Pulmonary Pharmacology & Therapeutics, 20*(6), 718–725.

114. Moudgil, R., Michelakis, E. D., and Archer, S. L. (2005). Hypoxic pulmonary vasoconstriction. *Journal of Applied Physiology, 98*(1), 390–403.

115. Murray, C. D. (1926). The physiological principle of minimum work: I. The vascular system and the cost of blood volume. *Proceedings of the National Academy of Sciences of the United States of America, 12*(3), 207–214.

116. Murray, T. R., Chen, L., Marshall, B. E., and Macarak, E. J. (1990). Hypoxic contraction of cultured pulmonary vascular smooth muscle cells. *American Journal of Respiratory Cell and Molecular Biology, 3*(5), 457–465.

117. Myllyharju, J. and Kivirikko, K. I. (2001). Collagens and collagen-related diseases. *Annals of Medicine, 33*(1), 7–21.

118. Nakayama, K., Ueta, K., Tanaka, Y., Tanabe, Y., and Ishii, K. (1997). Stretch-induced contraction of rabbit isolated pulmonary artery and the involvement of endothelium-derived thromboxane A2. *British Journal of Pharmacology, 122*(2), 199–208.

119. Nichols, W. W. and O'Rourke, M. (Eds.). (1998). *McDonald's Blood Flow in Arteries: Theoretical, Experimental and Clinical Principles*, 4th edn. London: Arnold, New York: Oxford University Press.

120. Ochoa, C. D., Baker, H., Hasak, S., Matyal, R., Salam, A., Hales, C. A. et al. (2008). Cyclic stretch affects pulmonary endothelial cell control of pulmonary smooth muscle cell growth. *American Journal of Respiratory Cell and Molecular Biology, 39*(1), 105–112.

121. Orr, A. W., Helmke, B. P., Blackman, B. R., and Schwartz, M. A. (2006). Mechanisms of mechanotransduction. *Developmental Cell, 10*(1), 11–20.

122. O'Shea, K. S. and Dixit, V. M. (1988). Unique distribution of the extracellular matrix component thrombospondin in the developing mouse embryo. *The Journal of Cell Biology, 107*(6 Pt 2), 2737–2748.

123. Osol, G. (1995). Mechanotransduction by vascular smooth muscle. *Journal of Vascular Research, 32*(5), 275–292.

124. Owens, G. K. (1995). Regulation of differentiation of vascular smooth muscle cells. *Physiological Reviews, 75*(3), 487–517.

125. Pagano, P. J. and Gutterman, D. D. (2007). The adventitia: The outs and ins of vascular disease. *Cardiovascular Research, 75*(4), 636–639.

126. Pankov, R. and Yamada, K. M. (2002). Fibronectin at a glance. *Journal of Cell Science, 115*(Pt 20), 3861–3863.

127. Peters, D. G., Ning, W., Chu, T. J., Li, C. J., and Choi, A. M. (2006). Comparative SAGE analysis of the response to hypoxia in human pulmonary and aortic endothelial cells. *Physiological Genomics, 26*(2), 99–108.

128. Phan, S. H. (2002). The myofibroblast in pulmonary fibrosis. *Chest, 122*(6 Suppl), 286S–289S.

129. Pries, A. R. and Kuebler, W. M. (2006). Normal endothelium. *Handbook of Experimental Pharmacology, 176*(Pt 1), 1–40.

130. Rabinovitch, M. (1999). EVE and beyond, retro and prospective insights. *The American Journal of Physiology, 277*(1 Pt 1), L5–L12.

131. Raines, E. W. (2000). The extracellular matrix can regulate vascular cell migration, proliferation, and survival: Relationships to vascular disease. *International Journal of Experimental Pathology, 81*(3), 173–182.

132. Reynolds, A. M., Xia, W., Holmes, M. D., Hodge, S. J., Danilov, S., Curiel, D. T. et al. (2007). Bone morphogenetic protein type 2 receptor gene therapy attenuates hypoxic pulmonary hypertension. *American Journal of Physiology. Lung Cellular and Molecular Physiology, 292*(5), L1182–L1192.

133. Rich, S., Dantzker, D. R., Ayres, S. M., Bergofsky, E. H., Brundage, B. H., Detre, K. M. et al. (1987). Primary pulmonary hypertension. A national prospective study. *Annals of Internal Medicine, 107*(2), 216–223.

134. Rich, S., Kaufmann, E., and Levy, P. S. (1992). The effect of high doses of calcium-channel blockers on survival in primary pulmonary hypertension. *The New England Journal of Medicine, 327*(2), 76–81.

135. Roberts, A. B. and Sporn, M. B. (1989). Regulation of endothelial cell growth, architecture, and matrix synthesis by TGF-beta. *The American Review of Respiratory Disease, 140*(4), 1126–1128.

136. Robertson, T. P., Dipp, M., Ward, J. P., Aaronson, P. I., and Evans, A. M. (2000). Inhibition of sustained hypoxic vasoconstriction by Y-27632 in isolated intrapulmonary arteries and perfused lung of the rat. *British Journal of Pharmacology, 131*(1), 5–9.

137. Rosenbloom, J., Abrams, W. R., and Mecham, R. (1993). Extracellular matrix 4: The elastic fiber. *The FASEB Journal: Official Publication of the Federation of American Societies for Experimental Biology, 7*(13), 1208–1218.

138. Ruoslahti, E. (1989). Proteoglycans in cell regulation. *The Journal of Biological Chemistry, 264*(23), 13369–13372.

139. Saito, M., Tanabe, Y., Kudo, I., and Nakayama, K. (2003). Endothelium-derived prostaglandin H2 evokes the stretch-induced contraction of rabbit pulmonary artery. *European Journal of Pharmacology, 467*(1–3), 151–161.

140. Sakai, L. Y., Keene, D. R., and Engvall, E. (1986). Fibrillin, a new 350-kD glycoprotein, is a component of extracellular microfibrils. *The Journal of Cell Biology, 103*(6 Pt 1), 2499–2509.

141. Salwen, S. A., Szarowski, D. H., Turner, J. N., and Bizios, R. (1998). Three-dimensional changes of the cytoskeleton of vascular endothelial cells exposed to sustained hydrostatic pressure. *Medical & Biological Engineering & Computing, 36*(4), 520–527.

142. Sandoval, J., Beltran, U., Gomez, A., Lopez, R., Martinez, W., Vazquez, V. et al. (1985). Effect of chronic altitude hypoxia on the behavior of the respiratory center. Study on normal subjects living at the altitude of mexico city (2,240 meters) [Efecto de la hipoxia cronica de la altitud en el comportamiento del centro respiratorio. Estudio en sujetos normales residentes a la altitud de la ciudad de Mexico, D.F. (2,240 mts)]. *Archivos Del Instituto De Cardiologia De Mexico, 55*(5), 381–387.

143. Saqueton, C. B., Miller, R. B., Porter, V. A., Milla, C. E., and Cornfield, D. N. (1999). NO causes perinatal pulmonary vasodilation through K⁺-channel activation and intracellular Ca²⁺ release. *The American Journal of Physiology, 276*(6 Pt 1), L925–L932.

144. Schreiber, A. B., Kenney, J., Kowalski, W. J., Friesel, R., Mehlman, T., and Maciag, T. (1985). Interaction of endothelial cell growth factor with heparin: Characterization by receptor and antibody recognition. *Proceedings of the National Academy of Sciences of the United States of America, 82*(18), 6138–6142.

145. Scott, J. E. (1995). Extracellular matrix, supramolecular organisation and shape. *Journal of Anatomy, 187*(Pt 2), 259–269.

146. Shifren, A., Durmowicz, A. G., Knutsen, R. H., Faury, G., and Mecham, R. P. (2008). Elastin insufficiency predisposes to elevated pulmonary circulatory pressures through changes in elastic artery structure. *Journal of Applied Physiology, 105*(5), 1610–1619.

147. Simonneau, G., Galie, N., Rubin, L. J., Langleben, D., Seeger, W., Domenighetti, G. et al. (2004). Clinical classification of pulmonary hypertension. *Journal of the American College of Cardiology, 43*(12 Suppl S), 5S–12S.

148. Sitbon, O., Humbert, M., Jais, X., Ioos, V., Hamid, A. M., Provencher, S. et al. (2005). Long-term response to calcium channel blockers in idiopathic pulmonary arterial hypertension. *Circulation, 111*(23), 3105–3111.

149. Somlyo, A. P. and Somlyo, A. V. (1994). Signal transduction and regulation in smooth muscle. *Nature, 372*(6503), 231–236.

150. Steinhorn, R. H. and Fineman, J. R. (1999). The pathophysiology of pulmonary hypertension in congenital heart disease. *Artificial Organs, 23*(11), 970–974.

151. Stenmark, K. R., Davie, N., Frid, M., Gerasimovskaya, E., and Das, M. (2006). Role of the adventitia in pulmonary vascular remodeling. *Physiology, 21*, 134–145.

152. Stenmark, K. R., Fagan, K. A., and Frid, M. G. (2006). Hypoxia-induced pulmonary vascular remodeling: Cellular and molecular mechanisms. *Circulation Research, 99*(7), 675–691.

153. Stenmark, K. R. and Frid, M. G. (1998). Smooth muscle cell heterogeneity: Role of specific smooth muscle cell subpopulations in pulmonary vascular disease. *Chest, 114*(1 Suppl), 82S–90S.

154. Stenmark, K. R. and Mecham, R. P. (1997). Cellular and molecular mechanisms of pulmonary vascular remodeling. *Annual Review of Physiology, 59*, 89–144.

155. Stevens, T., Phan, S., Frid, M. G., Alvarez, D., Herzog, E., and Stenmark, K. R. (2008). Lung vascular cell heterogeneity: Endothelium, smooth muscle, and fibroblasts. *Proceedings of the American Thoracic Society, 5*(7), 783–791.

156. Sullivan, C. C., Du, L., Chu, D., Cho, A. J., Kido, M., Wolf, P. L. et al. (2003). Induction of pulmonary hypertension by an angiopoietin 1/TIE2/serotonin pathway. *Proceedings of the National Academy of Sciences of the United States of America, 100*(21), 12331–12336.

157. Tabima, D. M. and Chesler, N. C. (2010). The effects of vasoactivity and hypoxic pulmonary hypertension on extralobar pulmonary artery biomechanics. *Journal of Biomechanics, 43*(10), 1864–1869.

158. Tanaka, Y., Schuster, D. P., Davis, E. C., Patterson, G. A., and Botney, M. D. (1996). The role of vascular injury and hemodynamics in rat pulmonary artery remodeling. *The Journal of Clinical Investigation, 98*(2), 434–442.

159. Tawhai, M. H. and Burrowes, K. S. (2008). Modelling pulmonary blood flow. *Respiratory Physiology & Neurobiology, 163*(1–3), 150–157.

160. Thenappan, T., Shah, S. J., Rich, S., and Gomberg-Maitland, M. (2007). A USA-based registry for pulmonary arterial hypertension: 1982–2006. *The European Respiratory Journal: Official Journal of the European Society for Clinical Respiratory Physiology, 30*(6), 1103–1110.

161. Thomas, B. J. and Wanstall, J. C. (2003). Alterations in pulmonary vascular function in rats exposed to intermittent hypoxia. *European Journal of Pharmacology, 477*(2), 153–161.

162. Thompson, B. T., Spence, C. R., Janssens, S. P., Joseph, P. M., and Hales, C. A. (1994). Inhibition of hypoxic pulmonary hypertension by heparins of differing in vitro antiproliferative potency. *American Journal of Respiratory and Critical Care Medicine, 149*(6), 1512–1517.

163. Thomson, J. R., Machado, R. D., Pauciulo, M. W., Morgan, N. V., Humbert, M., Elliott, G. C. et al. (2000). Sporadic primary pulmonary hypertension is associated with germline mutations of the gene encoding BMPR-II, a receptor member of the TGF-beta family. *Journal of Medical Genetics, 37*(10), 741–745.

164. Tuder, R. M. and Voelkel, N. F. (2002). Angiogenesis and pulmonary hypertension: A unique process in a unique disease. *Antioxidants & Redox Signaling, 4*(5), 833–843.

165. Voelkel, N. F., Quaife, R. A., Leinwand, L. A., Barst, R. J., McGoon, M. D., Meldrum, D. R. et al. (2006). Right ventricular function and failure: Report of a national heart, lung, and blood institute working group on cellular and molecular mechanisms of right heart failure. *Circulation, 114*(17), 1883–1891.

166. Wagenseil, J. E., Nerurkar, N. L., Knutsen, R. H., Okamoto, R. J., Li, D. Y., and Mecham, R. P. (2005). Effects of elastin haploinsufficiency on the mechanical behavior of mouse arteries. *American Journal of Physiology, Heart and Circulatory Physiology, 289*(3), H1209–H1217.

167. Wedgwood, S., Bekker, J. M., and Black, S. M. (2001). Shear stress regulation of endothelial NOS in fetal pulmonary arterial endothelial cells involves PKC. *American Journal of Physiology. Lung Cellular and Molecular Physiology, 281*(2), L490–L498.

168. Wedgwood, S., Devol, J. M., Grobe, A., Benavidez, E., Azakie, A., Fineman, J. R. et al. (2007). Fibroblast growth factor-2 expression is altered in lambs with increased pulmonary blood flow and pulmonary hypertension. *Pediatric Research, 61*(1), 32–36.

169. Wedgwood, S., Mitchell, C. J., Fineman, J. R., and Black, S. M. (2003). Developmental differences in the shear stress-induced expression of endothelial NO synthase: Changing role of AP-1. *American Journal of Physiology. Lung Cellular and Molecular Physiology, 284*(4), L650–L662.

170. Weibel, E. R. (1979). Morphometry of the human lung: The state of the art after two decades. *Bulletin Europeen De Physiopathologie Respiratoire, 15*(5), 999–1013.

171. Weir, E. K. and Olschewski, A. (2006). Role of ion channels in acute and chronic responses of the pulmonary vasculature to hypoxia. *Cardiovascular Research, 71*(4), 630–641.

172. Weir, E. K., Reeve, H. L., Cornfield, D. N., Tristani-Firouzi, M., Peterson, D. A., and Archer, S. L. (1997). Diversity of response in vascular smooth muscle cells to changes in oxygen tension. *Kidney International, 51*(2), 462–466.

173. Weir, E. K., Rubin, L. J., Ayres, S. M., Bergofsky, E. H., Brundage, B. H., Detre, K. M. et al. (1989). The acute administration of vasodilators in primary pulmonary hypertension. Experience from the national institutes of health registry on primary pulmonary hypertension. *The American Review of Respiratory Disease, 140*(6), 1623–1630.

174. Weissmann, N., Winterhalder, S., Nollen, M., Voswinckel, R., Quanz, K., Ghofrani, H. A. et al. (2001). NO and reactive oxygen species are involved in biphasic hypoxic vasoconstriction of isolated rabbit lungs. *American Journal of Physiology. Lung Cellular and Molecular Physiology, 280*(4), L638–L645.

175. Weitzenblum, E. and Chaouat, A. (2004). Sleep and chronic obstructive pulmonary disease. *Sleep Medicine Reviews, 8*(4), 281–294.

176. West, J. B., Dollery, C. T., and Naimark, A. (1964). Distribution of blood flow in isolated lung; relation to vascular and alveolar pressures. *Journal of Applied Physiology, 19*, 713–724.

177. Wight, T. N., Raugi, G. J., Mumby, S. M., and Bornstein, P. (1985). Light microscopic immunolocation of thrombospondin in human tissues. *The Journal of Histochemistry and Cytochemistry: Official Journal of the Histochemistry Society, 33*(4), 295–302.

178. Yamashita, Y., Nakagomi, K., Takeda, T., Hasegawa, S., and Mitsui, Y. (1992). Effect of heparin on pulmonary fibroblasts and vascular cells. *Thorax, 47*(8), 634–639.

179. Yuan, J. X., Aldinger, A. M., Juhaszova, M., Wang, J., Conte, J. V. Jr., Gaine, S. P. et al. (1998). Dysfunctional voltage-gated K+ channels in pulmonary artery smooth muscle cells of patients with primary pulmonary hypertension. *Circulation, 98*(14), 1400–1406.

180. Yuan, X. J., Goldman, W. F., Tod, M. L., Rubin, L. J., and Blaustein, M. P. (1993). Hypoxia reduces potassium currents in cultured rat pulmonary but not mesenteric arterial myocytes. *The American Journal of Physiology, 264*(2 Pt 1), L116–L123.

181. Yuan, X. J., Tod, M. L., Rubin, L. J., and Blaustein, M. P. (1990). Contrasting effects of hypoxia on tension in rat pulmonary and mesenteric arteries. *The American Journal of Physiology, 259*(2 Pt 2), H281–H289.

182. Yurchenco, P. D. and Schittny, J. C. (1990). Molecular architecture of basement membranes. *The FASEB Journal: Official Publication of the Federation of American Societies for Experimental Biology, 4*(6), 1577–1590.

14 Lung Mechanobiology

Daniel J. Tschumperlin, Francis Boudreault, and Fei Liu

CONTENTS

14.1 INTRODUCTION

The central function of the lung is to facilitate the exchange of gases between the external environment and the body's circulation. Nearly every aspect of this function is inextricably linked with mechanics. Respiratory muscles, chiefly the diaphragm, generate a transpulmonary pressure gradient that inflates the lung by prompting gas to flow through the branched structure of the airways to the lung's terminal gas-exchange units, called alveoli. On relaxation of the inspiratory effort, the lung deflates passively due to the elastic recoil of the respiratory system. The expansion of the lung during inhalation brings atmospheric oxygen to the alveoli, where it diffuses across the blood–gas barrier into a dense network of capillaries. In exchange, carbon dioxide leaves the capillaries and is exhaled as the lung deflates. The fundamental design problem of the lung is how to organize and optimize the structure to enable rapid gas exchange.[1] The solution is elegant, but delicate: divide the blood in the lung's capillaries among millions of alveoli, each with walls less than 1 μm thick.[2] Air and blood can then be brought into close apposition to facilitate rapid equilibration.

The mechanical challenge in distributing gas and blood in suitable proportions is great: the lung's delicate structure must be compliant and elastic to allow lung inflation and deflation with minimal effort, yet stable to prevent collapse of the airway and alveoli. The stability of this arrangement depends crucially on pulmonary surfactant, a complex mixture of phospholipids and proteins synthesized by specialized cells within the epithelial lining of the alveoli.[3] Pulmonary surfactant lowers surface tension in alveoli, and possesses the unique characteristic that it drives surface tension toward zero during dynamic film compression, helping to stabilize alveoli during lung

deflation.[4] Both acute and chronic derangements of the lung's delicate microstructure and surfactant system ultimately lead to inadequate gas exchange and, in extreme cases, to respiratory failure.
Understanding the physical origin of lung mechanical function, and the mechanism of failure in
various disease states, is central to the study of respiratory physiology and medicine. In this chapter,
we focus on efforts to elucidate the cellular, molecular, and microstructural mechanisms underpin-
ning mechanobiological function of the lung in health and disease.

14.2   DEFINING THE MECHANICAL STATE OF THE LUNG

To understand the role that mechanics plays in stimulating and shaping aspects of lung biology,
a detailed understanding of the mechanical state of the tissue under physiological conditions is a
prerequisite. The complex and delicate microstructure of the lungs poses a significant challenge for
mechanical characterization, one that is further magnified by the limited access to the lung, encased
within the thoracic cavity, under physiological conditions. Traditionally, then, lung mechanics have
been assessed by measuring pressure, volume, and flow relationships at the entrance to the airways.
Respiratory mechanics can be partitioned into elastic and flow resistance components by measuring
pressure changes in the absence and presence of flow, and the relative contributions of the parenchyma and chest wall to respiratory elastance can be evaluated by measuring the pressure in the
interposing pleural space.[5] While these methods have served admirably to advance our understanding of respiratory mechanics, they treat the lung as a homogeneous material. This assumption works
well in healthy lungs, but it breaks down in most pathological conditions, including asthma, interstitial fibrosis, and acute lung injury, because injury and remodeling are heterogeneously distributed.
In answer to this challenge, numerous advances in methodology are under development; together,
they promise a new level of understanding by making possible visualization of the microstructure
of living lung tissue under dynamic physiological conditions.

At the macro end of the spatial scale, noninvasive-imaging modalities with enhanced resolution
are providing new information about regional tissue deformations and may make possible local
measurements of intact lung mechanical properties. One example, microfocal x-ray imaging, is providing unprecedented measurements of airway dimensions in intact lungs (Figure 14.1A), allowing
quantitative analysis of regional, axial, and circumferential variations in airway strains that occur
with changing lung volume.[6,7] Magnetic resonance elastography (MRE), while new to measurements of lung tissue, is well established as a noninvasive means to sample tissue mechanics in soft
organs such as the liver.[8] Application of MRE to the lung is complicated by the lung's air-filled
structure, but preliminary efforts to apply MRE to lungs inflated with hyperpolarized 3He supports
the feasibility of using this method to measure tissue mechanical properties within intact lungs.[9]
Both of these methods should provide new perspectives on regional changes in lung mechanical
properties when applied in the context of experimental models of lung injury and remodeling, and
could ultimately find use in clinical diagnosis and disease monitoring.

At the micro end of the spatial scale, recent efforts have brought the power of light microscopy
to the visualization of living lung microstructure. The development of intravital microscopy has
provided new opportunities to visualize events within subpleural alveoli and microvasculature of
the intact lung.[10] Exciting opportunities are now on the horizon for coupling intravital imaging
tools with live-cell indicators of cellular signaling events,[10,11] such as fluorescence resonance energy
transfer-based molecular sensors and intracellular molecular indicators. These approaches, once
combined, could provide a powerful platform for studying mechanotransduction under physiological conditions in situ.

More immediately, intravital imaging approaches are providing an increasingly sophisticated
understanding of the local deformations that occur at the microstructural level within the alveoli of
intact lungs, a great advance over the previous assumption of homogeneous behavior and extrapola-
tion from populations of thousands or millions of alveoli.[12,13] In situ measures of deformations are
critically needed to assess the mechanical stimuli that individual cells and cell types experience

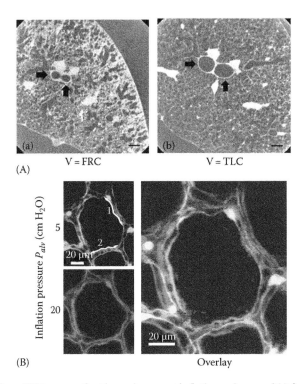

FIGURE 14.1 (A) Micro-CT images of rat lung airways at inflation volumes of (a) functional residual capacity (FRC) and (b) total lung capacity (TLC). Same-direction arrows indicate the same airways. Scale bar, 500 μm. (Adapted from Sera, T. et al., *J. Appl. Physiol.*, 96, 1665, 2004. With permission.) (B) A rat lung alveolus imaged with intravital microscopy at transpulmonary inflation pressures of 5 cm H_2O and 20 cm H_2O. Numbers in baseline image label two perimeter segments. An overlay of the images demonstrates inflation-induced alveolar expansion, which increased total alveolar perimeter length, L and alveolar diameter, D by 13% and 15%, respectively. (Adapted from Perlman, C.E. and Bhattacharya, J., *J. Appl. Physiol.*, 103, 1037, 2007. With permission.)

under typical loading conditions, and to design in vitro systems that faithfully recapitulate the mechanical environment for studies of cellular mechanotransduction. As a pertinent example, until recently, the best measure of strains experienced by lung epithelial cells had been obtained by quantitative morphometric analysis of electron micrographs from lung parenchyma fixed at various inflation volumes.[14,15] This measurement technique relied on averaging values across many cells and alveoli, and was vulnerable to artifacts introduced by fixation and processing of tissue to prepare for electron microscopy. A more recent study has employed intravital imaging to measure local intra-alveolar deformations from confocal imaging of subpleural alveoli within isolated rat lungs.[12] This study demonstrated striking heterogeneity in the behavior of alveolar wall segments (Figure 14.1B), with greater distention observed in wall segments associated with type I epithelial cells than in those segments associated with type II epithelial cells.[12] Strikingly, previous in vitro work demonstrated that type II cells are more vulnerable to stretch-induced cell injury and or death than are type I cells.[16] Together these findings suggest that the alveolus is constructed to protect type II cells from large distention, either through geometric or compositional effects on the underlying matrix, or through differences in the stiffness of the individual cell types themselves that populate the alveolus. Intriguingly, some support for the latter explanation comes from application of atomic force microscopy (AFM) to characterize the stiffness of isolated lung epithelial cells; the cytoplasm of type II cells is stiffer (~twofold median difference) than that of type I cells.[17] The results detailed above strongly suggest that while the lung is homogenous at the macroscale, at the microscale mechanical heterogeneity might play a central role in mechanical signaling in the pulmonary alveolus. Characterization of distribution of lung mechanical properties, particularly at sub-alveoli scale, should provide new insights into local heterogeneity in lung mechanics and its role in mechanotransduction in the lung.[17] AFM represents a more invasive new tool for characterizing the mechanical state of the lung, but one that offers spatial resolutions ranging from multimicron- to submicron-scale, making it a unique approach to characterize cell and tissue mechanical properties from the alveolar level down to the subcellular level. In a typical AFM force measurement, a silicon or Si_3N_4 cantilever with a borosilicate sphere tip several microns in diameter is pushed against a soft sample, with the displacement tracked and the force extrapolated from deflection of the calibrated cantilever. By collecting force–indentation curves over the sample surface, local viscoelastic parameters can be determined. If arrays of force–indentation curves are collected, microscale 2D maps of mechanical properties can be generated.[18] AFM has been widely used to map the structure and mechanical properties of cells,[19–22] including measurements of the local elasticity of migrating lung epithelial cells in a model of wound healing.[23] In this latter study, epithelial cell stiffness within $20 \mu m$ from the model wound edge was measured using AFM force mapping. Local stiffness measurement revealed a significant increase in cell stiffness between 10 and $15 \mu m$ from the wound edge suggesting cells near the wound may undergo localized changes in stiffness, which may provide signals for cell spreading and migration.[23]

AFM has also been used to characterize the local mechanical properties of intact soft tissues.[24–26] For example, Berry and coworkers isolated tissue sections of left ventricle from both infarcted and noninfarcted rat hearts and mapped myocardial elasticity using AFM. The authors measured the elasticity in the infarct and border region up to a range of $15 mm$, and compared changes in myocardial elasticity in the absence and presence of direct human mesenchymal stem cell (MSC) injection after infarction. Their AFM mechanical mapping suggested that MSC injection reduced scar stiffness and attenuated postinfarction remodeling. We have employed AFM microindentation to spatially map the stiffness of lung tissue slices harvested from mice with bleomycin-induced fibrosis (Figure 14.2A, F. Liu and D. Tschumperlin, unpublished data). Prior measurements of lung tissue stiffness at the macroscale indicated an approximate doubling or tripling of tissue stiffness with bleomycin-induced fibrosis.[27,28] In contrast, AFM microindentation reveals that tissue stiffening is highly localized, with some regions up to ~30-fold stiffer than the median observed in normal lung tissue.

Matrix stiffness has been shown by several groups to exert profound effects on fibroblast biology: for example, fibroblasts must exert traction against "stiff" substrates to undergo transition to a myofibroblast phenotype[29–32]; fibroblasts from various tissues undergo morphological changes with increases in substrate stiffness,[33,34] proliferate at increasing rates,[35] activate latent TGF-β in a stiffness-dependent fashion,[36] and even migrate up gradients of stiffness.[37] Given that fibroblast plays a prominent role in lung remodeling by undergoing proliferation to make up a larger fraction of parenchymal cells, and by undergoing differentiation to a contractile myofibroblast phenotype that also synthesizes matrix proteins such as collagen,[38,39] it is likely that matrix rigidity effects on lung cells are important in several pathologies of the lung, including asthmatic airway remodeling, fibrosis, emphysema, and pulmonary hypertension. So far, the role of matrix stiffness in the initiation and amplification of these lung diseases remains largely unexplored, though with local measurements of stiffness and deformation now in hand the stage is set for these investigations.

14.3 LUNG MECHANICS FROM MOLECULAR TO MACROSCALE

A major goal of mechanobiology is to understand how mechanical behaviors evident at the macroscale emerge from molecular, intermolecular, and cellular constituents. In the lung, this has been a longstanding focus, as mechanical dysfunction of the lung is associated with the major classes of restrictive and obstructive lung disease, but lack of molecular understanding has hampered targeted therapeutic interventions.[40,41]

Simplified cell-matrix model systems are one avenue currently being employed to identify stimuli that promote cell-mediated matrix remodeling, and to study the detailed changes in matrix

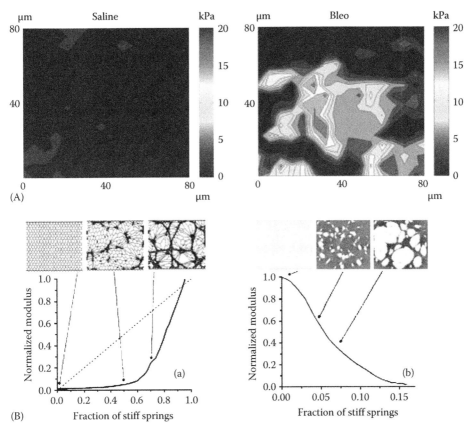

FIGURE 14.2 (A) Representative stiffness maps of normal and fibrotic (14 days post-bleomycin treatment) mouse lung parenchyma. Grayscale bar indicates shear modulus. AFM force–indentation profiles were acquired in a 16 × 16 sample grid separated by 5 μm spatially covering 80 × 80 μm area. Shear modulus at each point on the grid was calculated from fitting force–indentation data using a Hertz sphere model and resulting shear modulus data were plotted in a contour map (F. Liu and D. Tschumperlin, unpublished data). (B) Simulation of the progression of pulmonary fibrosis (a) and emphysema (b) based on percolation of sequential alveolar wall stiffening or rupture. (a) The curve shows the bulk modulus of the elastic network versus the fraction of springs randomly stiffened by a factor of 100. If all the spring constants were uniformly stiffened in a gradual manner from the baseline value of 1–100, the modulus would follow the dashed diagonal line. Top: Network configurations obtained when 0%, 50%, and 67% of the springs have been stiffened. (b) The curve shows the bulk modulus of the elastic network versus the fraction of springs cut on the basis of the amount of tension they carry. Top: Network configurations obtained at three points along this process. (Adapted from Bates, J.H. et al., *Am. J. Respir. Crit. Care Med.*, 176, 617, 2007. With permission.)

composition and organization that alter matrix mechanical behavior.[42–44] Analysis of second harmonic generation from multiphoton imaging of cell-matrix constructs[43,44] is being used to further explore linkages between structural and mechanical properties of collagen matrices.

An alternative approach to dissect molecular contributions to lung mechanics is selective suppression or digestion of individual matrix components. Mechanical testing of lung tissue strips treated to selectively remove matrix constituents is advancing our understanding of how collagen, elastin, and other matrix proteins contribute to tissue stability and deformability.[45–47] Similarly, lungs from mice genetically deficient for the proteoglycan decorin support a prominent role of this matrix protein in lung mechanics.[48] These studies are challenging the long-held assumption that lung elasticity can be simply partitioned into contributions from elastin at low volume, and collagen at high volume. Linking this new understanding of matrix mechanics to lung pathology, Suki and colleagues[47,49–52] have put forward a compelling hypothesis (Figure 14.2B) to explain progressive emphysema based on percolation of sequential alveolar wall rupture.[40,53,54] Together these approaches are building a multilevel hierarchical understanding of lung tissue mechanics, moving toward the ultimate goal of predictive power to understand how molecular perturbations alter lung micromechanics.

In parallel with advances in solid mechanics, ongoing investigations of biofluid mechanics in the pulmonary system[55] are providing an improved understanding of injury mechanisms associated with closure and opening of fragile lung tissue structures,[56,57] and generating multiscale models for investigation of fluid dynamics in airways and lung vasculature.[58] At the interface of solid and fluid surfaces, recent experimental work has demonstrated that the airway-lining layer of mucus, which is optimized for lung defense and clearance, is responsive to airway shear stresses,[59,60] and that pathological changes in mucus viscoelasticity can be targeted as a new therapeutic modality for patients with cystic fibrosis.[61,62]

The mechanical behavior of the lung's airways has been an area of intense and longstanding interest because flow resistance through the airways is highly responsive to the dimensions of airway lumen, and alterations in flow resistance are the prime pathophysiological endpoint in asthma. The dimensions of each airway lumen reflect a balance between forces which tend to narrow the airway (surface tension, elastic recoil) and forces which expand the airway (tethering of the airway to the parenchyma). Superimposed on this balance is dynamic regulation of airway smooth muscle contraction, which when activated can tilt the balance heavily in favor of airway narrowing. Over the past decade our understanding of the regulation of airway smooth muscle tone has undergone a revolution and it is now well accepted that the state of ASM is dynamically equilibrated, and thus sensitive to lung volume history and fluctuations in length.[63] The picture of ASM mechanics, and indeed cell mechanics in general,[64] has been thoroughly revised to account for the properties of cytoskeletal fluidization,[65] nanoscale remodeling,[66] and power law rheology.[64,67] However, fundamental questions remain regarding the relationship of such behaviors to the mechanisms of airways hyperresponsiveness, and whether the locus of hyperresponsiveness is within the smooth muscle itself, the remodeled airway, or some combination.[68–72]

14.4 MECHANOBIOLOGICAL RESPONSES TO LUNG STRETCH

The effects of stretch, as the central physical change required for lung inflation, continue to dominate lung mechanobiological investigation at organ and cellular scales. At the whole organ scale, the effects of stretch are manifested in three major settings: lung development, compensatory growth, and injury.

14.4.1 MECHANOBIOLOGY OF LUNG DEVELOPMENT

Physical forces stimulate lung growth at every stage of development, from the earliest branching of the lung buds to the continual expansion of the thoracic cavity throughout adolescence. In utero, the epithelium in the developing lung is secretory in a basal-to-apical direction, in contrast to its absorptive (apical-to-basal) role in mature lungs. The primitive epithelium in the developing lung expresses a basolateral bumetanide-sensitive NaKCl cotransporter and a lumenal chloride channel, perhaps CLC2,[73] to vectorially transport ions across the airway wall. Lumenal secretions across the epithelium flow through the developing airways and exit through the larynx and nasopharynx where the partial occlusion of the vocal cords acts as a one-way valve that generates back pressure to partially inflate the growing lungs. This tonic inflation is critical to lung development, as failure to inflate retards lung growth and maturation.[74] Tracheal occlusion increases lumenal expansion and accelerates branching and cellular maturation[75] (Figure 14.3). Experiments using organ cultures of primitive lung buds and perturbations targeting actin–myosin contractility demonstrate a critical role for local force generation in branching morphogenesis.[76,77] While the organ-level response

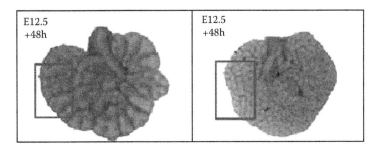

FIGURE 14.3 Mechanical strain enhances fetal lung growth. Murine fetal lung explants isolated on embryonic day 12.5 (E12.5) were incubated in DMEM F12 medium at 37°C and 5% CO_2. The trachea was left open (left) or occluded by cauterization (right) to increase lung intraluminal hydrostatic pressure. Tracheal occlusion resulted in accelerated lung development as evidenced by an increase in airways branching. After 48 h of this treatment, branching events in cauterized-occluded lungs occurred at approximately twice the control level. (Adapted from Unbekandt, M. et al., *Mech. Dev.*, 125, 314, 2008. With permission.)

to tracheal occlusion is well documented, the dissection of the cellular events that follow tracheal occlusion has only recently begun with the demonstration of rapid increases in proliferation of both epithelial and mesenchymal cells.[78] Further efforts directed at elucidating the molecular mechanisms coupling tracheal occlusion to accelerated airway branching point to the FGF10-FGFR2b signaling axis.[75] Similar to effects on airway branching, lung distention also plays a pivotal role in coordinating angiogenesis in the developing lung[79] and in the maturation of airway mesenchymal cells.[80]

In addition to the effects of tonic distention on lung development, cyclic and intermittent deformations also appear to be critical to lung development and maturation. Fetal breathing movements, caused by intermittent contractions of the diaphragm in the developing organisms, contribute to lung growth based on the observation that congenital diaphragmatic hernia results in smaller, less-mature lungs.[81–83] Within the developing airways, recent observations indicate that spontaneous peristaltic contractions (airway peristalsis) may contribute to lung growth and maturation of the fetal lung. The peristalsis of the embryonic airway is driven by spontaneous, regenerative, temperature-sensitive calcium waves propagating between airway smooth muscle cells via gap junctions. Calcium originates from extracellular entry via L-type voltage-gated channels and also intracellular stores.[84] Such spontaneous rhythmic contractions are the greatest in the fetus and absent in the adult but can be stimulated in adult as well with U46619, a thromboxane A2 analog. The amplitude and frequency of the stimulated phasic contractions are smaller, however, in the adult.[85]

14.4.2 Mechanobiology of Compensatory Lung Growth

Based on the essential role for mechanical forces in lung development, it is not surprising that mechanical forces also play a leading role in the compensatory growth of the lung following surgical resection.[86,87] This raises the possibility that harnessing the mechanical environment therapeutically could represent a powerful stimulus to treat hypoplastic lungs[88] or regenerate functional lung units lost to aging and disease.[89] Unfortunately, the robust compensatory growth seen in young animals is much weaker in adults,[86] and attempts to augment compensatory growth have met with mostly frustrating results.[90,91] Similarly, the compensatory growth responses are much weaker in larger, longer-lived mammals than in rodents, further clouding the potential for compensatory growth in human lungs.[92] The recent description of resident stem cells in a number of lung niches[93–95] and the possible connection between mechanical stimulation and stem cell activation[96] offer some hope that mechanobiological approaches may someday help to unlock the latent regenerative capacity in human lungs.

14.4.3 MECHANOBIOLOGY OF LUNG INJURY

Paradoxically, the lung response to stretch is not always beneficial. A vast body of literature demonstrates potentially pathological responses to lung stretch,[97] culminating in clinical trials demonstrating protective effects of reducing tidal volume during mechanical ventilation.[98,99] The trials that brought about alterations in clinical practice were the direct descendants of decades of research using experimental model systems[100–104] and represent a successful contribution of mechanobiology research (though preceding the usage of the term by many years) that translated into improvement in therapy. Current experimental efforts are exploring how even moderate volume ventilation, when superimposed on existing pulmonary conditions, can result in deleterious effects on lung function,[105–112] and these studies may help to further refine clinical management of ventilated patients.

Intense research efforts are directed at the molecular mechanisms that couple ventilatory mechanical stresses and strains into adverse physiological outcomes. TRPV4,[113] PI3K, Akt, and Src[114] have been implicated in capillary leakage, while hyaluronan fragmentation has been implicated in IL-8 upregulation,[115] and cellular stress failure demonstrated in lung over-distention injury.[116] Neutrophils play a prominent role in acute lung injury and their activation by injurious mechanical ventilation strategies continues to be an area of active investigation, with stretch-induced stiffening of neutrophils and signaling through c-Jun N-terminal kinase[117,118] both implicated in neutrophil margination and activation in the lung microcirculation. Mechanical ventilation can also generate deleterious oxidative stress, as evidenced by enhanced vascular permeability and inflammation in lungs of mechanically ventilated *Nrf2*-deficient mice, which lack a crucial transcription factor responsible for regulating expression of several antioxidant enzymes.[119] The observation that ventilation may result in profoundly different outcomes in neonatal versus adult populations has led to the development of comparative animal ventilation models, which should help to uncover the developmental basis for differential responses to lung distention.[120,121] While hypothesis-driven approaches provide for mechanistic dissection of ventilation-induced biochemical signaling, discovery-based approaches have also been implemented to survey the landscape of transcriptional events initiated by stretch. Microarray analyses highlight genomic scale events, transcriptional programs, and candidate genes evoked by ventilation in various model systems.[109,122–128] Ultimately, these approaches may allow elucidation of the mechanobiological pathways underlying lung stretch responses, and lead to the development of strategies to avoid or attenuate adverse responses to mechanical ventilation.

14.5 MECHANOBIOLOGICAL RESPONSES TO LUNG CELL STRETCH

In parallel with animal models of ventilation, cell culture approaches have been widely exploited to accelerate discovery and characterization of mechanical signaling events in lung cells. While limited in their capacity to recapitulate the complete in vivo environment and the complex intercellular interactions that are present in the lung, these models provide enhanced control of the mechanical and biochemical environments and offer ready opportunities to interrogate specific cellular and molecular events modulated by mechanical stimuli.

14.5.1 EPITHELIAL STRETCH RESPONSES

Many of the cellular mechanobiology studies relevant to the lung focus on the epithelium, which covers the vast surface area of the lung, and thus represents a potent source for mechanical regulation. Not surprisingly, studies of distal lung epithelium largely parallel whole lung studies with interrelated emphases on development and differentiation, growth responses, and injury mechanisms. In vitro studies of lung cells have identified a vast array of signaling responses to stretch ranging from activation of mechanosensitive receptor to generation of reactive oxygen species (ROS) and secretion of paracrine mediators.

14.5.2   Fetal Cells

The stimulatory effects of stretch on accelerated lung maturation and cellular differentiation are mirrored in cell culture studies where it has been demonstrated that stretch potently affects fetal type II epithelial differentiation through a variety of plasma membrane-associated proteins including receptor tyrosine kinases (RTK), integrins, and G-protein subunits. In the lung, the epidermal growth factor receptor (EGFR) has been shown to mediate stretch-induced alveolar type II cell differentiation.[129] Indeed, stretch-stimulated EGFR phosphorylation, Ras activation, and ERK1/2 phosphorylation were blocked by an EGFR inhibitor in those cells. Integrins, particularly subunits beta1, alpha6 and alpha3[130] and cAMP-PKA signaling,[131] also contribute to stretch-induced lung epithelial cell differentiation, while secreted products such as cyclooxygenase-derived prostanoids[132] and serotonin[133] appear to play key roles in lung epithelial cell responses to stretch.

Cyclic stretch dramatically stimulates proliferation of fetal-derived lung cells in vitro and the key signaling receptor for this process is the platelet-derived growth factor receptor.[134] Another putative mechanotransducer for lung growth is stretch-activated channels (SAC) permeable to calcium; transient elevations of cytosolic calcium in response to stretch were silenced by prior exposure to the lanthanide gadolinium, an SAC channel blocker, attenuating stretch-induced cell proliferation.[135] Calcium is a critical second messenger for strain-induced cell growth since inhibition of calcium mobilization prevents stretch-induced ERK activation, NF-KappaB nuclear translocation, and expression of early response genes such as Egr1, HSP70, IL-6, and MIP-2 in mechanically stimulated cells.[136] In addition, global microarray analysis of fetal type II epithelial cells exposed to stretch demonstrated enhanced expression of a variety of genes, including the amiloride-sensitive epithelial sodium channel gene (Scnn1a), suggesting a role for stretch in preparing the maturing epithelium for the transition from net secretion to absorption at birth.[137] Together these studies provide testable new hypotheses for critical regulatory nodes in mechanical regulation of lung development.

14.5.3   Mature Cells

In cells representative of those in the mature lung, stretch has been shown to exert potent effects on growth, structure, homeostasis, and differentiated function. Cyclic stretch drives the production of ROS in distal lung epithelial cells including superoxide anion O_2^- generated from the NADPH oxidase system and the mitochondria complex I (NADH dehydrogenase).[138] ROS exert opposing effects in lung epithelial cells, on the one hand they are necessary for stretch-induced cellular proliferation by an ERK/MAPK-dependent mechanism[139]; on the other hand, ROS production by mechanical stress has been associated with double-strand breakage DNA damage and subsequent cellular apoptosis.[140] That stretch-induced DNA damage can be completely prevented by exposing cells to FGF-10 prior to stretching hints to why diverging outcomes derive from stretch-generated ROS.[140] A gene macroarray analysis approach also revealed an antioxidant response in stretched pulmonary epithelial cells accompanied by phosphorylation of HSP-27, a downstream target of the oxidative stress pathway.[141] EGFR can play a role in this response by activation of *Nrf2*-dependent antioxidant response.[142]

MAPK activation by stretch is a key factor in growth regulation of epithelial cells. ERK1/2 can be activated by cyclic stretch in alveolar epithelial cells through a mechanism dependent on EGFR through transactivation of the receptor by shedding of HBEGF ligand as a result of stretch-induced GPCR activation.[143] A dystroglycan-dependent mechanotransduction response has also been observed in stretched alveolar cells[144] and contributes to activation of AMP kinase and ERK.[144,145] Activation of ERK also requires contributions from FAK and Src, emphasizing the complex interplay of regulatory signals governing cellular proliferation in response to mechanical stimulation.[146]

Stretch also leads to remodeling of the epithelial intermediate filament network[147] and actin cytoskeleton,[142] and calcium-dependent fusion of lamellar bodies with the cell membrane.[148] Lamellar bodies are the densely packed structures used by cells to store and deliver the phospholipid components of pulmonary surfactant. The stretch-induced fusion of lamellar bodies builds on pioneering work that demonstrated enhanced surfactant secretion caused by stretch in vitro[149] and lung inflation in situ.[150] Interestingly, further studies on cell stretch in vitro demonstrate that mixed cultures of type I and type II cells respond to stretch with greater phospholipid secretion than type II cells alone, implicating type I cells as primary mechanosensors, which stimulate secretion in neighboring type II cells though extracellular ATP signaling.[151]

14.5.4 FIBROBLAST STRETCH RESPONSES

Fibroblasts, a cell type that makes up roughly 30% of the cellular population in the normal lung, are also responsive to stretch. It has been long proposed that mechanical deformation stimulates human lung fibroblast proliferation by mediating the release of autocrine growth factors[152] or activation of phospholipid turnover.[153] The follow-up studies suggested the participation of signaling events downstream of PKC and pp60src.[154] However, another study reported that stretch of fibroblasts isolated during the canalicular stage of lung development inhibits cell cycle progression and activates apoptosis, emphasizing the important role of developmental context in determining cellular responses to mechanical stimulation.[155] All three major MAPKs, ERK1/2, p38, and SAP/JNK have also been shown to be potently but transiently activated in lung fibroblasts in response to stretch through a mechanism independent of RTKs and Ras.[156]

Mechanical stretch has a significant impact on the key fibroblast functions of ECM synthesis and secretion.[157] Stretch differentially regulates fibronectin production at posttranscriptional level.[158] Continuous cyclic stretch increases procollagen protein synthesis in human fetal lung fibroblast (IMR-90), which was associated with increased alpha1(I) procollagen mRNA.[159] Mechanical stretch also stimulates secretion of proteoglycans and glycosaminoglycans.[160] While the detailed signaling mechanisms and pathways are the subject of continuing investigations, current evidence suggests the increase in ECM molecule abundance is the result of increased synthesis and not decreasing activity of degrading enzymes.[161]

14.5.5 STRETCH AND CELL INJURY RESPONSES

As in the intact lung, not all effects of stretch are beneficial or benign. Cyclic stretch of lung epithelial cells causes an acidification response that promotes bacterial growth,[162] impairs the barrier function of cellular monolayers,[163] and leads to cell death and release of cytokines that play important roles in ventilator-induced lung injury.[164–166] Cell culture models also recapitulate the observation that moderate stretch alone is less provocative of cellular responses than is stretch in the presence of an infectious agent (LPS) or inflammatory cytokine (TNF-α).[167] Intriguingly, there is one example of using cell culture models to show that the epithelium can be treated to enhance its resistance to the negative effects of stretch, with compelling evidence that IL-10 has protective effects worthy of follow-on study in animal models.[168] Similarly, experimental and computational approaches have been employed to study stretch-induced increase in Na-K-ATPase pumping activity,[169] which could be a target to improve fluid clearance from airspaces in acute lung injury.

14.5.6 AIRWAY EPITHELIAL RESPONSES TO COMPRESSION

For the proximal airways, the dominant mechanical stress is not stretch, but rather the compressive stress experienced during bronchospasm. Substantial effort has been expended to elucidate the cellular and molecular responses of airway epithelial cells to this mechanical stimulus. Air-liquid interface cultures of primary bronchial epithelium faithfully recapitulate the differentiated character of the airway epithelium as found in vivo, and have been used to elucidate a unique EGFR-dependent mode of epithelial mechanotransduction.[170,171] The EGFR-dependent response is evoked by compressive stress shrinking the lateral intercellular space that separates neighboring

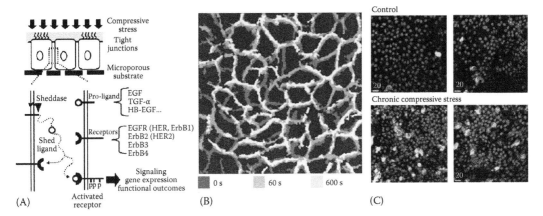

FIGURE 14.4 (A) Schematic of bronchial epithelial cells cultured at air–liquid interface on a microporous substrate. The lateral cellular surfaces express pro-ligands of the EGF family and their cognate EGFR receptors, forming a local autocrine circuit. (B) Compressive stress (apical to basal transcellular pressure gradient) shrinks the lateral intercellular space between neighboring bronchial epithelial cells, visualized by two-photon imaging of extracellular fluorescent dextran. Sequential images at baseline (0 s), 60, and 600 s after initiation of continuous compressive stress illustrate the gradual decline in intercellular gap distance. (C) Chronic intermittent exposure to compressive stress daily for 14 days enhances expression of a mucus secretory phenotype, visualized by immunofluorescent staining of MUC5AC (lighter shade of gray). Nuclear counterstain is shown in darker shade of gray.

epithelial cells (Figure 14.4A and B). The epithelium constitutively sheds EGF-family ligands into this intercellular space, and cognate receptors are localized on the lateral cell membranes. Hence, the mechanical stress-induced shrinkage of this space increases the local spatial concentration of ligands, enhancing receptor occupancy and downstream signaling through the EGFR. This mechanotransduction response has been further linked to positive feedback signal amplification through enhanced expression of EGF-family ligands, and also results in enhanced expression of the plasminogen family of enzymes that exert broad control over matrix remodeling pathways.[172,173] Application of chronic but intermittent compressive stress (Figure 14.4C), similar to episodic bronchospasm in asthma, enhances airway epithelial expression of mucin protein,[174] suggesting a causal link between mechanical perturbations in the airways and excessive mucus production that contributes to airway obstruction. Further development of this in vitro mechanotransduction system has led to the development of coculture models, which incorporate differentiated epithelium adjacent to cell-populated collagen matrices, allowing for cell-to-cell communication in configurations approximating the native airway wall.[175] These systems demonstrate an integrated tissue remodeling response to mechanical stress that resembles many of the features of asthmatic airway remodeling.[176] Together these model systems demonstrate that many features of airway remodeling, including matrix synthesis and metaplasia of mucus secretory cells, are responsive to changes in mechanical environment, suggesting that the mechanical stress associated with asthmatic airway narrowing may contribute to progressive detrimental changes in airway structure and function.[74]

14.6 OUTLOOK AND UNANSWERED QUESTIONS

This chapter has emphasized the diverse and profound effects that lung distention in vivo, and stretch in vitro exert on lung cells. When considered together, an unresolved paradox emerges: lung stretch is necessary for lung growth and regeneration, but also capable of causing or exacerbating injury in the context of mechanical ventilation. While it is tempting to dismiss this paradox as an issue of degree, with small distention promoting growth, and large distention causing injury, such a conclusion does not fit with two key lines of evidence. Moderate to large lung distention, in the absence of pre-existing injury or infection, can be quite well tolerated.[87] In contrast, even modest degrees of lung distention can cause or amplify injury when applied to lungs with local or systemic inflammation.[108,109] It thus appears that, to resolve this paradox, further intense study will be needed at the highly complex, and extraordinarily important intersection between the immune system, the lung's soluble milieu, and the cellular responses to stretch. Enticingly, the discovery of critical mediators or pathways that could shift cellular stretch responses from those associated with injury and inflammatory signaling toward those responses associated with growth and regeneration would have profound implications for both mechanical ventilation and lung regeneration.

Nascent efforts to engineer lung tissue[177–179] or develop scaffolds for lung tissue regeneration would appear natural beneficiaries of enhanced understanding of the key pathways that promote growth and regeneration responses to lung distention. The ongoing investigation of lung development and the differing responses of neonatal and mature lungs to mechanical perturbations should prove helpful in uncovering critical pathways selectively activated by stretch leading to lung growth.[120,121] As knowledge of progenitor cell function in the lung advances, the effects of the mechanical environment on these cells will likely become a natural area of focus. Key issues to be addressed include whether the mechanical environment is an important contributor in defining niches for progenitor cells, and whether changes in mechanical environment play any role in activation of progenitor cells or commitment of progenitor cells to various lung-specific lineages.

While stretch will continue to be a focal point for mechanobiological investigation in the lung, the role of "stiffness" appears poised for increased recognition. The lung parenchyma is typically highly compliant, and undergoes profound changes in stiffness either globally or locally in a variety of lung diseases, including pulmonary fibrosis, emphysema, and asthmatic airway remodeling. The role that matrix stiffness plays in initiating, amplifying, or prolonging these disease processes represents largely unexplored territory. Moreover, the effects of stretch on cell biological processes may need to be critically reevaluated with regards to stiffness—it is not at all clear that stretching cells on soft, physiologic stiffness substrates, as opposed to the much stiffer elastic substrates currently employed, will result in similar cell responses to stretch. To add to this complexity, it must be remembered that in the lung, the amount of tissue stretch is linked to the local and global tissue stiffness, thus as disease processes remodel the lung and change matrix stiffness, the tonic and cyclic levels of local distention will simultaneously change. The development of new in vitro models capable of simultaneously modulating stretch and stiffness will be needed to explore the intersection of these key aspects of the mechanical environment.[180]

Finally, the search for mechanisms of mechanotransduction in the lung has been quite successful over the past several years, but this success has led to an unanticipated question: what cellular processes and signaling pathways are unaffected by stretch (and perhaps variations in matrix stiffness)? While clearly there are cell-specific and stimulus-specific mechanotransduction responses, it is also increasingly clear that mechanical events can have wide-ranging and potent effects across a variety of pathways and cell types.[65] The emerging complexity of mechano-activated pathways, cellular responses, and contextual influences emphasize the growing need to apply integrative and systems-type analyses[181–183] to dissect critical mechanoregulated events in the lung and lung cells. Such approaches appear especially appealing when considering the diversity of responses that occur in response to lung distention, and the divergent physiological responses that can emerge when similar mechanical stimuli are applied in different contexts. The application of genomics, proteomics, and systems biology tools and approaches to lung mechanobiology research will be increasingly beneficial as we synthesize data from an ever-growing knowledge base and look toward a more integrated understanding of mechanobiological processes in lung health and disease.

ACKNOWLEDGMENTS

The authors would like to acknowledge grant support from the NIH and the Scleroderma Foundation.

REFERENCES

1. Fredberg, J. J. and Kamm, R. D. (2006) Stress transmission in the lung: Pathways from organ to molecule. *Annu Rev Physiol* **68**, 507–541.
2. Maina, J. N. and West, J. B. (2005) Thin and strong! The bioengineering dilemma in the structural and functional design of the blood–gas barrier. *Physiol Rev* **85**, 811–844.
3. Obladen, M. (2005) History of surfactant up to 1980. *Biol Neonate* **87**, 308–316.
4. Rugonyi, S., Biswas, S. C., and Hall, S. B. (2008) The biophysical function of pulmonary surfactant. *Respir Physiol Neurobiol* **163**, 244–255.
5. West, J. B. (2007) *Respiratory Physiology: The Essentials.* Lippincott Williams & Wilkins, Philadelphia, PA.
6. Sinclair, S. E., Molthen, R. C., Haworth, S. T., Dawson, C. A., and Waters, C. M. (2007) Airway strain during mechanical ventilation in an intact animal model. *Am J Respir Crit Care Med* **176**, 786–794.
7. Sera, T., Fujioka, H., Yokota, H., Makinouchi, A., Himeno, R., Schroter, R. C., and Tanishita, K. (2004) Localized compliance of small airways in excised rat lungs using microfocal X-ray computed tomography. *J Appl Physiol* **96**, 1665–1673.
8. Kruse, S. A., Smith, J. A., Lawrence, A. J., Dresner, M. A., Manduca, A., Greenleaf, J. F., and Ehman, R. L. (2000) Tissue characterization using magnetic resonance elastography: Preliminary results. *Phys Med Biol* **45**, 1579–1590.
9. McGee, K. P., Hubmayr, R. D., and Ehman, R. L. (2008) MR elastography of the lung with hyperpolarized 3He. *Magn Reson Med* **59**, 14–18.
10. Kuebler, W. M., Parthasarathi, K., Lindert, J., and Bhattacharya, J. (2007) Real-time lung microscopy. *J Appl Physiol* **102**, 1255–1264.
11. Sabouri-Ghomi, M., Wu, Y., Hahn, K., and Danuser, G. (2008) Visualizing and quantifying adhesive signals. *Curr Opin Cell Biol* **20**, 541–550.
12. Perlman, C. E. and Bhattacharya, J. (2007) Alveolar expansion imaged by optical sectioning microscopy. *J Appl Physiol* **103**, 1037–1044.
13. Popp, A., Wendel, M., Knels, L., Koch, T., and Koch, E. (2006) Imaging of the three-dimensional alveolar structure and the alveolar mechanics of a ventilated and perfused isolated rabbit lung with Fourier domain optical coherence tomography. *J Biomed Opt* **11**, 014015.
14. Tschumperlin, D. J. and Margulies, S. S. (1999) Alveolar epithelial surface area–volume relationship in isolated rat lungs. *J Appl Physiol* **86**, 2026–2033.
15. Bachofen, H., Schurch, S., Urbinelli, M., and Weibel, E. R. (1987) Relations among alveolar surface tension, surface area, volume, and recoil pressure. *J Appl Physiol* **62**, 1878–1887.
16. Tschumperlin, D. J. and Margulies, S. S. (1998) Equibiaxial deformation-induced injury of alveolar epithelial cells in vitro. *Am J Physiol* **275**, L1173–1183.
17. Azeloglu, E. U., Bhattacharya, J., and Costa, K. D. (2008) Atomic force microscope elastography reveals phenotypic differences in alveolar cell stiffness. *J Appl Physiol* **105**, 652–661.
18. A-Hassan, E., Heinz, W. F., Antonik, M. D., D'Costa, N. P., Nageswaran, S., Schoenenberger, C. A., and Hoh, J. H. (1998) Relative microelastic mapping of living cells by atomic force microscopy. *Biophys J* **74**, 1564–1578.
19. Gavara, N., Roca-Cusachs, P., Sunyer, R., Farre, R., and Navajas, D. (2008) Mapping cell-matrix stresses during stretch reveals inelastic reorganization of the cytoskeleton. *Biophys J* **95**, 464–471.
20. Kang, I., Panneerselvam, D., Panoskaltsis, V. P., Eppell, S. J., Marchant, R. E., and Doerschuk, C. M. (2008) Changes in the hyperelastic properties of endothelial cells induced by tumor necrosis factor-alpha. *Biophys J* **94**, 3273–3285.
21. Almqvist, N., Bhatia, R., Primbs, G., Desai, N., Banerjee, S., and Lal, R. (2004) Elasticity and adhesion force mapping reveals real-time clustering of growth factor receptors and associated changes in local cellular rheological properties. *Biophys J* **86**, 1753–1762.
22. Pesen, D. and Hoh, J. H. (2005) Micromechanical architecture of the endothelial cell cortex. *Biophys J* **88**, 670–679.
23. Wagh, A. A., Roan, E., Chapman, K. E., Desai, L. P., Rendon, D. A., Eckstein, E. C., and Waters, C. M. (2008) Localized elasticity measured in epithelial cells migrating at a wound edge using atomic force microscopy. *Am J Physiol Lung Cell Mol Physiol* **295**, L54–60.
24. Berry, M. F., Engler, A. J., Woo, Y. J., Pirolli, T. J., Bish, L. T., Jayasankar, V., Morine, K. J., Gardner, T. J., Discher, D. E., and Sweeney, H. L. (2006) Mesenchymal stem cell injection after myocardial infarction improves myocardial compliance. *Am J Physiol Heart Circ Physiol* **290**, H2196–H2203.

25. Engler, A. J., Griffin, M. A., Sen, S., Bonnemann, C. G., Sweeney, H. L., and Discher, D. E. (2004) Myotubes differentiate optimally on substrates with tissue-like stiffness: Pathological implications for soft or stiff microenvironments. *J Cell Biol* **166**, 877–887.

26. Engler, A. J., Rehfeldt, F., Sen, S., and Discher, D. E. (2007) Microtissue elasticity: Measurements by atomic force microscopy and its influence on cell differentiation. *Methods Cell Biol* **83**, 521–545.

27. Ebihara, T., Venkatesan, N., Tanaka, R., and Ludwig, M. S. (2000) Changes in extracellular matrix and tissue viscoelasticity in bleomycin-induced lung fibrosis. Temporal aspects. *Am J Respir Crit Care Med* **162**, 1569–1576.

28. Dolhnikoff, M., Mauad, T., and Ludwig, M. S. (1999) Extracellular matrix and oscillatory mechanics of rat lung parenchyma in bleomycin-induced fibrosis. *Am J Respir Crit Care Med* **160**, 1750–1757.

29. Arora, P. D., Narani, N., and McCulloch, C. A. (1999) The compliance of collagen gels regulates transforming growth factor-beta induction of alpha-smooth muscle actin in fibroblasts. *Am J Pathol* **154**, 871–882.

30. Goffin, J. M., Pittet, P., Csucs, G., Lussi, J. W., Meister, J. J., and Hinz, B. (2006) Focal adhesion size controls tension-dependent recruitment of alpha-smooth muscle actin to stress fibers. *J Cell Biol* **172**, 259–268.

31. Hinz, B. and Gabbiani, G. (2003) Mechanisms of force generation and transmission by myofibroblasts. *Curr Opin Biotechnol* **14**, 538–546.

32. Li, Z., Dranoff, J. A., Chan, E. P., Uemura, M., Sevigny, J., and Wells, R. G. (2007) Transforming growth factor-beta and substrate stiffness regulate portal fibroblast activation in culture. *Hepatology* **46**, 1246–1256.

33. Paszek, M. J., Zahir, N., Johnson, K. R., Lakins, J. N., Rozenberg, G. I., Gefen, A., Reinhart-King, C. A., Margulies, S. S., Dembo, M., Boettiger, D., Hammer, D. A., and Weaver, V. M. (2005) Tensional homeostasis and the malignant phenotype. *Cancer Cell* **8**, 241–254.

34. Yeung, T., Georges, P. C., Flanagan, L. A., Marg, B., Ortiz, M., Funaki, M., Zahir, N., Ming, W., Weaver, V., and Janmey, P. A. (2005) Effects of substrate stiffness on cell morphology, cytoskeletal structure, and adhesion. *Cell Motil Cytoskeleton* **60**, 24–34.

35. Wang, H. B., Dembo, M., and Wang, Y. L. (2000) Substrate flexibility regulates growth and apoptosis of normal but not transformed cells. *Am J Physiol Cell Physiol* **279**, C1345–1350.

36. Wipff, P. J., Rifkin, D. B., Meister, J. J., and Hinz, B. (2007) Myofibroblast contraction activates latent TGF-beta1 from the extracellular matrix. *J Cell Biol* **179**, 1311–1323.

37. Lo, C.-M., Wang, H.-B., Dembo, M., and Wang, Y.-l. (2000) Cell movement is guided by the rigidity of the substrate. *Biophys J* **79**, 144–152.

38. Phan, S. H. (2003) Fibroblast phenotypes in pulmonary fibrosis. *Am J Respir Cell Mol Biol* **29**, S87–92.

39. Zhang, K., Rekhter, M. D., Gordon, D., and Phan, S. H. (1994) Myofibroblasts and their role in lung collagen gene expression during pulmonary fibrosis. A combined immunohistochemical and in situ hybridization study. *Am J Pathol* **145**, 114–125.

40. Suki, B. and Bates, J. H. (2008) Extracellular matrix mechanics in lung parenchymal diseases. *Respir Physiol Neurobiol* **163**, 33–43.

41. Faffe, D. S. and Zin, W. A. (2009) Lung parenchymal mechanics in health and disease. *Physiol Rev* **89**, 759–775.

42. Leung, L. Y., Tian, D., Brangwynne, C. P., Weitz, D. A., and Tschumperlin, D. J. (2007) A new microrheometric approach reveals individual and cooperative roles for TGF-beta1 and IL-1beta in fibroblast-mediated stiffening of collagen gels. *FASEB J* **21**, 2064–2073.

43. Raub, C. B., Unruh, J., Suresh, V., Krasieva, T., Lindmo, T., Gratton, E., Tromberg, B. J., and George, S. C. (2008) Image correlation spectroscopy of multiphoton images correlates with collagen mechanical properties. *Biophys J* **94**, 2361–2373.

44. Raub, C. B., Suresh, V., Krasieva, T., Lyubovitsky, J., Mih, J. D., Putnam, A. J., Tromberg, B. J., and George, S. C. (2007) Noninvasive assessment of collagen gel microstructure and mechanics using multiphoton microscopy. *Biophys J* **92**, 2212–2222.

45. Cavalcante, F. S., Ito, S., Brewer, K., Sakai, H., Alencar, A. M., Almeida, M. P., Andrade, J. S., Jr., Majumdar, A., Ingenito, E. P., and Suki, B. (2005) Mechanical interactions between collagen and proteoglycans: Implications for the stability of lung tissue. *J Appl Physiol* **98**, 672–679.

46. Jesudason, R., Black, L., Majumdar, A., Stone, P., and Suki, B. (2007) Differential effects of static and cyclic stretching during elastase digestion on the mechanical properties of extracellular matrices. *J Appl Physiol* **103**, 803–811.

47. Kononov, S., Brewer, K., Sakai, H., Cavalcante, F. S., Sabayanagam, C. R., Ingenito, E. P., and Suki, B. (2001) Roles of mechanical forces and collagen failure in the development of elastase-induced emphysema. *Am J Respir Crit Care Med* **164**, 1920–1926.

48. Fust, A., LeBellego, F., Iozzo, R. V., Roughley, P. J., and Ludwig, M. S. (2005) Alterations in lung mechanics in decorin-deficient mice. *Am J Physiol Lung Cell Mol Physiol* **288**, L159–166.

49. Ito, S., Bartolak-Suki, E., Shipley, J. M., Parameswaran, H., Majumdar, A., and Suki, B. (2006) Early emphysema in the tight skin and pallid mice: roles of microfibril-associated glycoproteins, collagen, and mechanical forces. *Am J Respir Cell Mol Biol* **34**, 688–694.

50. Ito, S., Ingenito, E. P., Arold, S. P., Parameswaran, H., Tgavalekos, N. T., Lutchen, K. R., and Suki, B. (2004) Tissue heterogeneity in the mouse lung: Effects of elastase treatment. *J Appl Physiol* **97**, 204–212.

51. Ito, S., Ingenito, E. P., Brewer, K. K., Black, L. D., Parameswaran, H., Lutchen, K. R., and Suki, B. (2005) Mechanics, nonlinearity, and failure strength of lung tissue in a mouse model of emphysema: Possible role of collagen remodeling. *J Appl Physiol* **98**, 503–511.

52. Ritter, M. C., Jesudason, R., Majumdar, A., Stamenovic, D., Buczek-Thomas, J. A., Stone, P. J., Nugent, M. A., and Suki, B. (2009) A zipper network model of the failure mechanics of extracellular matrices. *Proc Natl Acad Sci U S A* **106**, 1081–1086.

53. Suki, B., Ito, S., Stamenovic, D., Lutchen, K. R., and Ingenito, E. P. (2005) Biomechanics of the lung parenchyma: Critical roles of collagen and mechanical forces. *J Appl Physiol* **98**, 1892–1899.

54. Bates, J. H., Davis, G. S., Majumdar, A., Butnor, K. J., and Suki, B. (2007) Linking parenchymal disease progression to changes in lung mechanical function by percolation. *Am J Respir Crit Care Med* **176**, 617–623.

55. Bertram, C. D. and Gaver, D. P. III (2005) Bio-fluid mechanics of the pulmonary system. *Ann Biomed Eng* **33**, 1681–1688.

56. Yalcin, H. C., Perry, S. F., and Ghadiali, S. N. (2007) Influence of airway diameter and cell confluence on epithelial cell injury in an in vitro model of airway reopening. *J Appl Physiol* **103**, 1796–1807.

57. Kay, S. S., Bilek, A. M., Dee, K. C., and Gaver, D. P. III. (2004) Pressure gradient, not exposure duration, determines the extent of epithelial cell damage in a model of pulmonary airway reopening. *J Appl Physiol* **97**, 269–276.

58. Tawhai, M. H. and Burrowes, K. S. (2008) Multi-scale models of the lung airways and vascular system. *Adv Exp Med Biol* **605**, 190–194.

59. Tarran, R., Button, B., and Boucher, R. C. (2006) Regulation of normal and cystic fibrosis airway surface liquid volume by phasic shear stress. *Annu Rev Physiol* **68**, 543–561.

60. Tarran, R., Button, B., Picher, M., Paradiso, A. M., Ribeiro, C. M., Lazarowski, E. R., Zhang, L., Collins, P. L., Pickles, R. J., Fredberg, J. J., and Boucher, R. C. (2005) Normal and cystic fibrosis airway surface liquid homeostasis. The effects of phasic shear stress and viral infections. *J Biol Chem* **280**, 35751–35759.

61. Donaldson, S. H., Bennett, W. D., Zeman, K. L., Knowles, M. R., Tarran, R., and Boucher, R. C. (2006) Mucus clearance and lung function in cystic fibrosis with hypertonic saline. *N Engl J Med* **354**, 241–250.

62. Tarran, R., Donaldson, S., and Boucher, R. C. (2007) Rationale for hypertonic saline therapy for cystic fibrosis lung disease. *Semin Respir Crit Care Med* **28**, 295–302.

63. An, S. S., Bai, T. R., Bates, J. H., Black, J. L., Brown, R. H., Brusasco, V., Chitano, P. et al. (2007) Airway smooth muscle dynamics: A common pathway of airway obstruction in asthma. *Eur Respir J* **29**, 834–860.

64. Hoffman, B. D., Massiera, G., Van Citters, K. M., and Crocker, J. C. (2006) The consensus mechanics of cultured mammalian cells. *Proc Natl Acad Sci U S A* **103**, 10259–10264.

65. Trepat, X., Deng, L., An, S. S., Navajas, D., Tschumperlin, D. J., Gerthoffer, W. T., Butler, J. P., and Fredberg, J. J. (2007) Universal physical responses to stretch in the living cell. *Nature* **447**, 592–595.

66. Bursac, P., Lenormand, G., Fabry, B., Oliver, M., Weitz, D. A., Viasnoff, V., Butler, J. P., and Fredberg, J. J. (2005) Cytoskeletal remodelling and slow dynamics in the living cell. *Nat Mater* **4**, 557–561.

67. Fabry, B., Maksym, G. N., Butler, J. P., Glogauer, M., Navajas, D., and Fredberg, J. J. (2001) Scaling the microrheology of living cells. *Phys Rev Lett* **87**, 148102.

68. Bates, J. H., and Lauzon, A. M. (2007) Parenchymal tethering, airway wall stiffness, and the dynamics of bronchoconstriction. *J Appl Physiol* **102**, 1912–1920.

69. Wagers, S., Lundblad, L. K., Ekman, M., Irvin, C. G., and Bates, J. H. (2004) The allergic mouse model of asthma: Normal smooth muscle in an abnormal lung? *J Appl Physiol* **96**, 2019–2027.

70. Wagers, S. S., Haverkamp, H. C., Bates, J. H., Norton, R. J., Thompson-Figueroa, J. A., Sullivan, M. J., and Irvin, C. G. (2007) Intrinsic and antigen-induced airway hyperresponsiveness are the result of diverse physiological mechanisms. *J Appl Physiol* **102**, 221–230.

71. McParland, B. E., Macklem, P. T., and Pare, P. D. (2003) Airway wall remodeling: Friend or foe? *J Appl Physiol* **95**, 426–434.

72. James, A. L. and Wenzel, S. (2007) Clinical relevance of airway remodelling in airway diseases. *Eur Respir J* **30**, 134–155.

73. Wilson, S. M., Olver, R. E., and Walters, D. V. (2007) Developmental regulation of lumenal lung fluid and electrolyte transport. *Respir Physiol Neurobiol* **159**, 247–255.
74. Tschumperlin, D. J. and Drazen, J. M. (2006) Chronic effects of mechanical force on airways. *Annu Rev Physiol* **68**, 563–583.
75. Unbekandt, M., del Moral, P. M., Sala, F. G., Bellusci, S., Warburton, D., and Fleury, V. (2008) Tracheal occlusion increases the rate of epithelial branching of embryonic mouse lung via the FGF10-FGFR2b-Sprouty2 pathway. *Mech Dev* **125**, 314–324.
76. Moore, K. A., Polte, T., Huang, S., Shi, B., Alsberg, E., Sunday, M. E., and Ingber, D. E. (2005) Control of basement membrane remodeling and epithelial branching morphogenesis in embryonic lung by Rho and cytoskeletal tension. *Dev Dyn* **232**, 268–281.
77. Moore, K. A., Huang, S., Kong, Y., Sunday, M. E., and Ingber, D. E. (2002) Control of embryonic lung branching morphogenesis by the Rho activator, cytotoxic necrotizing factor 1. *J Surg Res* **104**, 95–100.
78. Seaborn, T., St-Amand, J., Cloutier, M., Tremblay, M. G., Maltais, F., Dinel, S., Moulin, V., Khan, P. A., and Piedboeuf, B. (2008) Identification of cellular processes that are rapidly modulated in response to tracheal occlusion within mice lungs. *Pediatr Res* **63**, 124–130.
79. Cloutier, M., Maltais, F., and Piedboeuf, B. (2008) Increased distension stimulates distal capillary growth as well as expression of specific angiogenesis genes in fetal mouse lungs. *Exp Lung Res* **34**, 101–113.
80. Yang, Y., Beqaj, S., Kemp, P., Ariel, I., and Schuger, L. (2000) Stretch-induced alternative splicing of serum response factor promotes bronchial myogenesis and is defective in lung hypoplasia. *J Clin Invest* **106**, 1321–1330.
81. Inanlou, M. R., Baguma-Nibasheka, M., and Kablar, B. (2005) The role of fetal breathing-like movements in lung organogenesis. *Histol Histopathol* **20**, 1261–1266.
82. Joshi, S. and Kotecha, S. (2007) Lung growth and development. *Early Hum Dev* **83**, 789–794.
83. Harding, R. and Hooper, S. B. (1996) Regulation of lung expansion and lung growth before birth. *J Appl Physiol* **81**, 209–224.
84. Featherstone, N. C., Jesudason, E. C., Connell, M. G., Fernig, D. G., Wray, S., Losty, P. D., and Burdyga, T. V. (2005) Spontaneous propagating calcium waves underpin airway peristalsis in embryonic rat lung. *Am J Respir Cell Mol Biol* **33**, 153–160.
85. Parvez, O., Voss, A. M., de Kok, M., Roth-Kleiner, M., and Belik, J. (2006) Bronchial muscle peristaltic activity in the fetal rat. *Pediatr Res* **59**, 756–761.
86. (2004) Mechanisms and limits of induced postnatal lung growth. *Am J Respir Crit Care Med* **170**, 319–343.
87. Hsia, C. C. (2004) Signals and mechanisms of compensatory lung growth. *J Appl Physiol* **97**, 1992–1998.
88. Butter, A., Piedboeuf, B., Flageole, H., Meehan, B., and Laberge, J. M. (2005) Postnatal pulmonary distension for the treatment of pulmonary hypoplasia: Pilot study in the neonatal piglet model. *J Pediatr Surg* **40**, 826–831.
89. Fehrenbach, H., Voswinckel, R., Michl, V., Mehling, T., Fehrenbach, A., Seeger, W., and Nyengaard, J. R. (2008) Neoalveolarisation contributes to compensatory lung growth following pneumonectomy in mice. *Eur Respir J* **31**, 515–522.
90. Dane, D. M., Yan, X., Tamhane, R. M., Johnson, R. L. Jr., Estrera, A. S., Hogg, D. C., Hogg, R. T., and Hsia, C. C. (2004) Retinoic acid-induced alveolar cellular growth does not improve function after right pneumonectomy. *J Appl Physiol* **96**, 1090–1096.
91. Ravikumar, P., Yilmaz, C., Dane, D. M., Johnson, R. L. Jr., Estrera, A. S., and Hsia, C. C. (2007) Developmental signals do not further accentuate nonuniform postpneumonectomy compensatory lung growth. *J Appl Physiol* **102**, 1170–1177.
92. Hsia, C. C. (2004) Lessons from a canine model of compensatory lung growth. *Curr Top Dev Biol* **64**, 17–32.
93. Rawlins, E. L. and Hogan, B. L. (2006) Epithelial stem cells of the lung: privileged few or opportunities for many? *Development* **133**, 2455–2465.
94. Kim, C. F. (2007) Paving the road for lung stem cell biology: Bronchioalveolar stem cells and other putative distal lung stem cells. *Am J Physiol Lung Cell Mol Physiol* **293**, L1092–1098.
95. Stripp, B. R. and Shapiro, S. D. (2006) Stem cells in lung disease, repair, and the potential for therapeutic interventions: State-of-the-art and future challenges. *Am J Respir Cell Mol Biol* **34**, 517–518.
96. Nolen-Walston, R. D., Kim, C. F., Mazan, M. R., Ingenito, E. P., Gruntman, A. M., Tsai, L., Boston, R., Woolfenden, A. E., Jacks, T., and Hoffman, A. M. (2008) Cellular kinetics and modeling of bronchioalveolar stem cell response during lung regeneration. *Am J Physiol Lung Cell Mol Physiol* **294**, L1158–1165.
97. Oeckler, R. A. and Hubmayr, R. D. (2007) Ventilator-associated lung injury: A search for better therapeutic targets. *Eur Respir J* **30**, 1216–1226.

98. (2000) Ventilation with lower tidal volumes as compared with traditional tidal volumes for acute lung injury and the acute respiratory distress syndrome. The Acute Respiratory Distress Syndrome Network. *N Engl J Med* **342**, 1301–1308.

99. Meade, M. O., Cook, D. J., Guyatt, G. H., Slutsky, A. S., Arabi, Y. M., Cooper, D. J., Davies, A. R., Hand, L. E., Zhou, Q., Thabane, L., Austin, P., Lapinsky, S., Baxter, A., Russell, J., Skrobik, Y., Ronco, J. J., and Stewart, T. E. (2008) Ventilation strategy using low tidal volumes, recruitment maneuvers, and high positive end-expiratory pressure for acute lung injury and acute respiratory distress syndrome: A randomized controlled trial. *JAMA* **299**, 637–645.

100. Dreyfuss, D., Soler, P., and Saumon, G. (1995) Mechanical ventilation-induced pulmonary edema. Interaction with previous lung alterations. *Am J Respir Crit Care Med* **151**, 1568–1575.

101. Dreyfuss, D. and Saumon, G. (1993) Role of tidal volume, FRC, and end-inspiratory volume in the development of pulmonary edema following mechanical ventilation. *Am Rev Respir Dis* **148**, 1194–1203.

102. Webb, H. H. and Tierney, D. F. (1974) Experimental pulmonary edema due to intermittent positive pressure ventilation with high inflation pressures. Protection by positive end-expiratory pressure. *Am Rev Respir Dis* **110**, 556–565.

103. Tremblay, L. N. and Slutsky, A. S. (2006) Ventilator-induced lung injury: From the bench to the bedside. *Intensive Care Med* **32**, 24–33.

104. Tremblay, L., Valenza, F., Ribeiro, S. P., Li, J., and Slutsky, A. S. (1997) Injurious ventilatory strategies increase cytokines and c-fos m-RNA expression in an isolated rat lung model. *J Clin Invest* **99**, 944–952.

105. Levine, G. K., Deutschman, C. S., Helfaer, M. A., and Margulies, S. S. (2006) Sepsis-induced lung injury in rats increases alveolar epithelial vulnerability to stretch. *Crit Care Med* **34**, 1746–1751.

106. Tsuchida, S., Engelberts, D., Peltekova, V., Hopkins, N., Frndova, H., Babyn, P., McKerlie, C., Post, M., McLoughlin, P., and Kavanagh, B. P. (2006) Atelectasis causes alveolar injury in nonatelectatic lung regions. *Am J Respir Crit Care Med* **174**, 279–289.

107. Tsuchida, S., Engelberts, D., Roth, M., McKerlie, C., Post, M., and Kavanagh, B. P. (2005) Continuous positive airway pressure causes lung injury in a model of sepsis. *Am J Physiol Lung Cell Mol Physiol* **289**, L554–564.

108. Altemeier, W. A., Matute-Bello, G., Frevert, C. W., Kawata, Y., Kajikawa, O., Martin, T. R., and Glenny, R. W. (2004) Mechanical ventilation with moderate tidal volumes synergistically increases lung cytokine response to systemic endotoxin. *Am J Physiol Lung Cell Mol Physiol* **287**, L533–L542.

109. Altemeier, W. A., Matute-Bello, G., Gharib, S. A., Glenny, R. W., Martin, T. R., and Liles, W. C. (2005) Modulation of lipopolysaccharide-induced gene transcription and promotion of lung injury by mechanical ventilation. *J Immunol* **175**, 3369–3376.

110. Dhanireddy, S., Altemeier, W. A., Matute-Bello, G., O'Mahony, D. S., Glenny, R. W., Martin, T. R., and Liles, W. C. (2006) Mechanical ventilation induces inflammation, lung injury, and extra-pulmonary organ dysfunction in experimental pneumonia. *Lab Invest* **86**, 790–799.

111. O'Mahony, D. S., Liles, W. C., Altemeier, W. A., Dhanireddy, S., Frevert, C. W., Liggitt, D., Martin, T. R., and Matute-Bello, G. (2006) Mechanical ventilation interacts with endotoxemia to induce extrapulmonary organ dysfunction. *Crit Care* **10**, R136.

112. Bregeon, F., Delpierre, S., Chetaille, B., Kajikawa, O., Martin, T. R., Autillo-Touati, A., Jammes, Y., and Pugin, J. (2005) Mechanical ventilation affects lung function and cytokine production in an experimental model of endotoxemia. *Anesthesiology* **102**, 331–339.

113. Hamanaka, K., Jian, M. Y., Weber, D. S., Alvarez, D. F., Townsley, M. I., Al-Mehdi, A. B., King, J. A., Liedtke, W., and Parker, J. C. (2007) TRPV4 initiates the acute calcium-dependent permeability increase during ventilator-induced lung injury in isolated mouse lungs. *Am J Physiol Lung Cell Mol Physiol* **293**, L923–L932.

114. Miyahara, T., Hamanaka, K., Weber, D. S., Drake, D. A., Anghelescu, M., and Parker, J. C. (2007) Phosphoinositide 3-kinase, Src, and Akt modulate acute ventilation-induced vascular permeability increases in mouse lungs. *Am J Physiol Lung Cell Mol Physiol* **293**, L11–L21.

115. Mascarenhas, M. M., Day, R. M., Ochoa, C. D., Choi, W. I., Yu, L., Ouyang, B., Garg, H. G., Hales, C. A., and Quinn, D. A. (2004) Low molecular weight hyaluronan from stretched lung enhances interleukin-8 expression. *Am J Respir Cell Mol Biol* **30**, 51–60.

116. Vlahakis, N. E. and Hubmayr, R. D. (2005) Cellular stress failure in ventilator-injured lungs. *Am J Respir Crit Care Med* **171**, 1328–1342.

117. Choudhury, S., Wilson, M. R., Goddard, M. E., O'Dea, K. P., and Takata, M. (2004) Mechanisms of early pulmonary neutrophil sequestration in ventilator-induced lung injury in mice. *Am J Physiol Lung Cell Mol Physiol* **287**, L902–L910.

118. Li, L. F., Yu, L., and Quinn, D. A. (2004) Ventilation-induced neutrophil infiltration depends on c-Jun N-terminal kinase. *Am J Respir Crit Care Med* **169**, 518–524.

119. Papaiahgari, S., Yerrapureddy, A., Reddy, S. R., Reddy, N. M., Dodd, O. J., Crow, M. T., Grigoryev, D. N., Barnes, K., Tuder, R. M., Yamamoto, M., Kensler, T. W., Biswal, S., Mitzner, W., Hassoun, P. M., and Reddy, S. P. (2007) Genetic and pharmacologic evidence links oxidative stress to ventilator-induced lung injury in mice. *Am J Respir Crit Care Med* **176**, 1222–1235.

120. Copland, I. B., Martinez, F., Kavanagh, B. P., Engelberts, D., McKerlie, C., Belik, J., and Post, M. (2004) High tidal volume ventilation causes different inflammatory responses in newborn versus adult lung. *Am J Respir Crit Care Med* **169**, 739–748.

121. Kornecki, A., Tsuchida, S., Ondiveeran, H. K., Engelberts, D., Frndova, H., Tanswell, A. K., Post, M., McKerlie, C., Belik, J., Fox-Robichaud, A., and Kavanagh, B. P. (2005) Lung development and susceptibility to ventilator-induced lung injury. *Am J Respir Crit Care Med* **171**, 743–752.

122. dos Santos, C. C., Okutani, D., Hu, P., Han, B., Crimi, E., He, X., Keshavjee, S., Greenwood, C., Slutsky, A. S., Zhang, H., and Liu, M. (2008) Differential gene profiling in acute lung injury identifies injury-specific gene expression. *Crit Care Med* **36**, 855–865.

123. Ma, S. F., Grigoryev, D. N., Taylor, A. D., Nonas, S., Sammani, S., Ye, S. Q., and Garcia, J. G. (2005) Bioinformatic identification of novel early stress response genes in rodent models of lung injury. *Am J Physiol Lung Cell Mol Physiol* **289**, L468–L477.

124. Simon, B. A., Easley, R. B., Grigoryev, D. N., Ma, S. F., Ye, S. Q., Lavoie, T., Tuder, R. M., and Garcia, J. G. (2006) Microarray analysis of regional cellular responses to local mechanical stress in acute lung injury. *Am J Physiol Lung Cell Mol Physiol* **291**, L851–L861.

125. Grigoryev, D. N., Ma, S. F., Irizarry, R. A., Ye, S. Q., Quackenbush, J., and Garcia, J. G. (2004) Orthologous gene-expression profiling in multi-species models: search for candidate genes. *Genome Biol* **5**, R34.

126. Dolinay, T., Kaminski, N., Felgendreher, M., Kim, H. P., Reynolds, P., Watkins, S. C., Karp, D., Uhlig, S., and Choi, A. M. (2006) Gene expression profiling of target genes in ventilator-induced lung injury. *Physiol Genomics* **26**, 68–75.

127. Gharib, S. A., Liles, W. C., Matute-Bello, G., Glenny, R. W., Martin, T. R., and Altemeier, W. A. (2006) Computational identification of key biological modules and transcription factors in acute lung injury. *Am J Respir Crit Care Med* **173**, 653–658.

128. Wurfel, M. M. (2007) Microarray-based analysis of ventilator-induced lung injury. *Proc Am Thorac Soc* **4**, 77–84.

129. Sanchez-Esteban, J., Wang, Y., Gruppuso, P. A., and Rubin, L. P. (2004) Mechanical stretch induces fetal type II cell differentiation via an epidermal growth factor receptor-extracellular-regulated protein kinase signaling pathway. *Am J Respir Cell Mol Biol* **30**, 76–83.

130. Sanchez-Esteban, J., Wang, Y., Filardo, E. J., Rubin, L. P., and Ingber, D. E. (2006) Integrins beta1, alpha6, and alpha3 contribute to mechanical strain-induced differentiation of fetal lung type II epithelial cells via distinct mechanisms. *Am J Physiol Lung Cell Mol Physiol* **290**, L343–L350.

131. Wang, Y., Maciejewski, B. S., Lee, N., Silbert, O., McKnight, N. L., Frangos, J. A., and Sanchez-Esteban, J. (2006) Strain-induced fetal type II epithelial cell differentiation is mediated via cAMP-PKA-dependent signaling pathway. *Am J Physiol Lung Cell Mol Physiol* **291**, L820–L827.

132. Copland, I. B., Reynaud, D., Pace-Asciak, C., and Post, M. (2006) Mechanotransduction of stretch-induced prostanoid release by fetal lung epithelial cells. *Am J Physiol Lung Cell Mol Physiol* **291**, L487–L495.

133. Pan, J., Copland, I., Post, M., Yeger, H., and Cutz, E. (2006) Mechanical stretch-induced serotonin release from pulmonary neuroendocrine cells: implications for lung development. *Am J Physiol Lung Cell Mol Physiol* **290**, L185–L193.

134. Liu, M., Liu, J., Buch, S., Tanswell, A. K., and Post, M. (1995) Antisense oligonucleotides for PDGF-B and its receptor inhibit mechanical strain-induced fetal lung cell growth. *Am J Physiol* **269**, L178–L184.

135. Liu, M., Xu, J., Tanswell, A. K., and Post, M. (1994) Inhibition of mechanical strain-induced fetal rat lung cell proliferation by gadolinium, a stretch-activated channel blocker. *J Cell Physiol* **161**, 501–507.

136. Copland, I. B. and Post, M. (2007) Stretch-activated signaling pathways responsible for early response gene expression in fetal lung epithelial cells. *J Cell Physiol* **210**, 133–143.

137. Wang, Y., Maciejewski, B. S., Weissmann, G., Silbert, O., Han, H., and Sanchez-Esteban, J. (2006) DNA microarray reveals novel genes induced by mechanical forces in fetal lung type II epithelial cells. *Pediatr Res* **60**, 118–124.

138. Chapman, K. E., Sinclair, S. E., Zhuang, D., Hassid, A., Desai, L. P., and Waters, C. M. (2005) Cyclic mechanical strain increases reactive oxygen species production in pulmonary epithelial cells. *Am J Physiol Lung Cell Mol Physiol* **289**, L834–L841.

139. Chess, P. R., O'Reilly, M. A., Sachs, F., and Finkelstein, J. N. (2005) Reactive oxidant and p42/44 MAP kinase signaling is necessary for mechanical strain-induced proliferation in pulmonary epithelial cells. *J Appl Physiol* **99**, 1226–1232.

140. Upadhyay, D., Correa-Meyer, E., Sznajder, J. I., and Kamp, D. W. (2003) FGF-10 prevents mechanical stretch-induced alveolar epithelial cell DNA damage via MAPK activation. *Am J Physiol Lung Cell Mol Physiol* **284**, L350–L359.

141. Chess, P. R., O'Reilly, M. A., and Toia, L. (2004) Macroarray analysis reveals a strain-induced oxidant response in pulmonary epithelial cells. *Exp Lung Res* **30**, 739–753.

142. Papaiahgari, S., Yerrapureddy, A., Hassoun, P. M., Garcia, J. G., Birukov, K. G., and Reddy, S. P. (2007) EGFR-activated signaling and actin remodeling regulate cyclic stretch-induced NRF2-ARE activation. *Am J Respir Cell Mol Biol* **36**, 304–312.

143. Correa-Meyer, E., Pesce, L., Guerrero, C., and Sznajder, J. I. (2002) Cyclic stretch activates ERK1/2 via G proteins and EGFR in alveolar epithelial cells. *Am J Physiol Lung Cell Mol Physiol* **282**, L883–L891.

144. Jones, J. C., Lane, K., Hopkinson, S. B., Lecuona, E., Geiger, R. C., Dean, D. A., Correa-Meyer, E., Gonzales, M., Campbell, K., Sznajder, J. I., and Budinger, S. (2005) Laminin-6 assembles into multimolecular fibrillar complexes with perlecan and participates in mechanical-signal transduction via a dystroglycan-dependent, integrin-independent mechanism. *J Cell Sci* **118**, 2557–2566.

145. Budinger, G. R., Urich, D., DeBiase, P. J., Chiarella, S. E., Burgess, Z. O., Baker, C. M., Soberanes, S., Mutlu, G. M., and Jones, J. C. (2008) Stretch-induced activation of AMP kinase in the lung requires dystroglycan. *Am J Respir Cell Mol Biol* **39**, 666–672.

146. Chaturvedi, L. S., Marsh, H. M., and Basson, M. D. (2007) Src and focal adhesion kinase mediate mechanical strain-induced proliferation and ERK1/2 phosphorylation in human H441 pulmonary epithelial cells. *Am J Physiol Cell Physiol* **292**, C1701–C1713.

147. Felder, E., Siebenbrunner, M., Busch, T., Fois, G., Miklavc, P., Walther, P., and Dietl, P. (2008) Mechanical strain of alveolar type II cells in culture: changes in the transcellular cytokeratin network and adaptations. *Am J Physiol Lung Cell Mol Physiol* **295**, L849–L857.

148. Frick, M., Bertocchi, C., Jennings, P., Haller, T., Mair, N., Singer, W., Pfaller, W., Ritsch-Marte, M., and Dietl, P. (2004) Ca2+ entry is essential for cell strain-induced lamellar body fusion in isolated rat type II pneumocytes. *Am J Physiol Lung Cell Mol Physiol* **286**, L210–L220.

149. Wirtz, H. R. and Dobbs, L. G. (1990) Calcium mobilization and exocytosis after one mechanical stretch of lung epithelial cells. *Science* **250**, 1266–1269.

150. Massaro, G. D. and Massaro, D. (1983) Morphologic evidence that large inflations of the lung stimulate secretion of surfactant. *Am Rev Respir Dis* **127**, 235–236.

151. Patel, A. S., Reigada, D., Mitchell, C. H., Bates, S. R., Margulies, S. S., and Koval, M. (2005) Paracrine stimulation of surfactant secretion by extracellular ATP in response to mechanical deformation. *Am J Physiol Lung Cell Mol Physiol* **289**, L489–L496.

152. Bishop, J. E., Mitchell, J. J., Absher, P. M., Baldor, L., Geller, H. A., Woodcock-Mitchell, J., Hamblin, M. J., Vacek, P., and Low, R. B. (1993) Cyclic mechanical deformation stimulates human lung fibroblast proliferation and autocrine growth factor activity. *Am J Respir Cell Mol Biol* **9**, 126–133.

153. Liu, M., Xu, J., Liu, J., Kraw, M. E., Tanswell, A. K., and Post, M. (1995) Mechanical strain-enhanced fetal lung cell proliferation is mediated by phospholipase C and D and protein kinase C. *Am J Physiol* **268**, L729–L738.

154. Liu, M., Qin, Y., Liu, J., Tanswell, A. K., and Post, M. (1996) Mechanical strain induces pp60src activation and translocation to cytoskeleton in fetal rat lung cells. *J Biol Chem* **271**, 7066–7071.

155. Sanchez-Esteban, J., Wang, Y., Cicchiello, L. A., and Rubin, L. P. (2002) Cyclic mechanical stretch inhibits cell proliferation and induces apoptosis in fetal rat lung fibroblasts. *Am J Physiol Lung Cell Mol Physiol* **282**, L448–L456.

156. Boudreault, F. and Tschumperlin, D. J. (2001) Stretch-induced MAPK activation in lung fibroblasts is independent of receptor tyrosine kinases. *Am J Respir Cell Mol Biol.* **43**, 64–73.

157. Liu, M. and Post, M. (2000) Invited review: Mechanochemical signal transduction in the fetal lung. *J Appl Physiol* **89**, 2078–2084.

158. Mourgeon, E., Xu, J., Tanswell, A. K., Liu, M., and Post, M. (1999) Mechanical strain-induced posttranscriptional regulation of fibronectin production in fetal lung cells. *Am J Physiol* **277**, L142–L149.

159. Breen, E. C. (2000) Mechanical strain increases type I collagen expression in pulmonary fibroblasts in vitro. *J Appl Physiol* **88**, 203–209.

160. Xu, J., Liu, M., Liu, J., Caniggia, I., and Post, M. (1996) Mechanical strain induces constitutive and regulated secretion of glycosaminoglycans and proteoglycans in fetal lung cells. *J Cell Sci* **109** (Pt 6), 1605–1613.

161. Xu, J., Liu, M., and Post, M. (1999) Differential regulation of extracellular matrix molecules by mechanical strain of fetal lung cells. *Am J Physiol* **276**, L728–L735.

162. Pugin, J., Dunn-Siegrist, I., Dufour, J., Tissieres, P., Charles, P. E., and Comte, R. (2008) Cyclic stretch of human lung cells induces an acidification and promotes bacterial growth. *Am J Respir Cell Mol Biol* **38**, 362–370.

163. Cavanaugh, K. J., Cohen, T. S., and Margulies, S. S. (2006) Stretch increases alveolar epithelial permeability to uncharged micromolecules. *Am J Physiol Cell Physiol* **290**, C1179–C1188.

164. Hammerschmidt, S., Kuhn, H., Grasenack, T., Gessner, C., and Wirtz, H. (2004) Apoptosis and necrosis induced by cyclic mechanical stretching in alveolar type II cells. *Am J Respir Cell Mol Biol* **30**, 396–402.

165. Hammerschmidt, S., Kuhn, H., Sack, U., Schlenska, A., Gessner, C., Gillissen, A., and Wirtz, H. (2005) Mechanical stretch alters alveolar type II cell mediator release toward a proinflammatory pattern. *Am J Respir Cell Mol Biol* **33**, 203–210.

166. Hammerschmidt, S., Kuhn, H., Gessner, C., Seyfarth, H. J., and Wirtz, H. (2007) Stretch-induced alveolar type II cell apoptosis: Role of endogenous bradykinin and PI3K-Akt signaling. *Am J Respir Cell Mol Biol* **37**, 699–705.

167. dos Santos, C. C., Han, B., Andrade, C. F., Bai, X., Uhlig, S., Hubmayr, R., Tsang, M., Lodyga, M., Keshavjee, S., Slutsky, A. S., and Liu, M. (2004) DNA microarray analysis of gene expression in alveolar epithelial cells in response to TNFalpha, LPS, and cyclic stretch. *Physiol Genomics* **19**, 331–342.

168. Lee, H. S., Wang, Y., Maciejewski, B. S., Esho, K., Fulton, C., Sharma, S., and Sanchez-Esteban, J. (2008) Interleukin-10 protects cultured fetal rat type II epithelial cells from injury induced by mechanical stretch. *Am J Physiol Lung Cell Mol Physiol* **294**, L225–L232.

169. Fisher, J. L. and Margulies, S. S. (2007) Modeling the effect of stretch and plasma membrane tension on Na+-K+-ATPase activity in alveolar epithelial cells. *Am J Physiol Lung Cell Mol Physiol* **292**, L40–L53.

170. Tschumperlin, D. J., Dai, G., Maly, I. V., Kikuchi, T., Laiho, L. H., McVittie, A. K., Haley, K. J., Lilly, C. M., So, P. T., Lauffenburger, D. A., Kamm, R. D., and Drazen, J. M. (2004) Mechanotransduction through growth-factor shedding into the extracellular space. *Nature* **429**, 83–86.

171. Kojic, N., Kojic, M., and Tschumperlin, D. J. (2006) Computational modeling of extracellular mechanotransduction. *Biophys J* **90**, 4261–4270.

172. Chu, E. K., Foley, J. S., Cheng, J., Patel, A. S., Drazen, J. M., and Tschumperlin, D. J. (2005) Bronchial epithelial compression regulates epidermal growth factor receptor family ligand expression in an autocrine manner. *Am J Respir Cell Mol Biol* **32**, 373–380.

173. Chu, E. K., Cheng, J., Foley, J. S., Mecham, B. H., Owen, C. A., Haley, K. J., Mariani, T. J., Kohane, I. S., Tschumperlin, D. J., and Drazen, J. M. (2006) Induction of the plasminogen activator system by mechanical stimulation of human bronchial epithelial cells. *Am J Respir Cell Mol Biol* **35**, 628–638.

174. Park, J. A. and Tschumperlin, D. J. (2009) Chronic intermittent mechanical stress increases MUC5AC protein expression. *Am J Respir Cell Mol Biol.* **41**, 459–466.

175. Choe, M. M., Sporn, P. H., and Swartz, M. A. (2006) Extracellular matrix remodeling by dynamic strain in a three-dimensional tissue-engineered human airway wall model. *Am J Respir Cell Mol Biol* **35**, 306–313.

176. Swartz, M. A., Tschumperlin, D. J., Kamm, R. D., and Drazen, J. M. (2001) Mechanical stress is communicated between different cell types to elicit matrix remodeling. *Proc Natl Acad Sci U S A* **98**, 6180–6185.

177. Hoganson, D. M., Pryor, H. I. II, and Vacanti, J. P. (2008) Tissue engineering and organ structure: A vascularized approach to liver and lung. *Pediatr Res* **63**, 520–526.

178. Nichols, J. E. and Cortiella, J. (2008) Engineering of a complex organ: Progress toward development of a tissue-engineered lung. *Proc Am Thorac Soc* **5**, 723–730.

179. Zani, B. G., Kojima, K., Vacanti, C. A., and Edelman, E. R. (2008) Tissue-engineered endothelial and epithelial implants differentially and synergistically regulate airway repair. *Proc Natl Acad Sci U S A* **105**, 7046–7051.

180. Krishnan, R., Park, C. Y., Lin, Y. C., Mead, J., Jaspers, R. T., Trepat, X., Lenormand, G., Tambe, D., Smolensky, A. V., Knoll, A. H., Butler, J. P., and Fredberg, J. J. (2009) Reinforcement versus fluidization in cytoskeletal mechanoresponsiveness. *PLoS One* **4**, e5486.

181. Simpson, K. J., Selfors, L. M., Bui, J., Reynolds, A., Leake, D., Khvorova, A., and Brugge, J. S. (2008) Identification of genes that regulate epithelial cell migration using an siRNA screening approach. *Nat Cell Biol* **10**, 1027–1038.

182. Janes, K. A., Albeck, J. G., Gaudet, S., Sorger, P. K., Lauffenburger, D. A., and Yaffe, M. B. (2005) A systems model of signaling identifies a molecular basis set for cytokine-induced apoptosis. *Science* **310**, 1646–1653.

183. Miller-Jensen, K., Janes, K. A., Brugge, J. S., and Lauffenburger, D. A. (2007) Common effector processing mediates cell-specific responses to stimuli. *Nature* **448**, 604–608.

15 Mechanical Signaling in the Urinary Bladder

Aruna Ramachandran, Ramaswamy Krishnan, and Rosalyn M. Adam

CONTENTS

15.1 INTRODUCTION

All tissues in the human body are mechanically responsive; the sensing of physical pressure and stretch from the external environment modulates developmental or adaptive responses in cells from the earliest stages of embryogenesis to later stages of differentiated adult organs and tissues. Hollow organs such as the lungs, heart, vasculature, and the urinary bladder are particularly sensitive to mechanical stimuli since their primary function is to mediate an appropriate response to physical forces, whether regulating airway caliber during each breath or initiating voiding in response to mechanical pressure exerted on the bladder wall by stored urine. In this chapter, we review the current state of knowledge as it pertains to information transfer within the urinary bladder and explore the methods and approaches employed to study mechanotransduction, including their strengths and limitations. In addition, we consider how the signals that mediate mechanotransduction are subverted in disease. Finally, we review how a rigorous understanding of the mechanical and biochemical signaling webs that define the response of tissues to physical force may reveal novel therapeutic strategies for bladder dysfunction and diseases affecting other hollow organs.

15.1.1 Overview

The function of the urinary bladder is to store urine at low pressure and to expel it periodically under volitional control. These functions are achieved by cycles of smooth muscle relaxation and contraction during which the tissues of the bladder experience marked mechanical stimulation. The importance of mechanical signaling to bladder physiology is perhaps best appreciated by considering the deleterious health effects of conditions of the lower urinary tract characterized by excessive bladder wall distension. For instance, in response to chronic overdistension, as occurs following bladder outlet obstruction secondary to prostatic enlargement or to neural tube anomalies such as spina bifida, the smooth muscle compartment of the bladder wall undergoes marked fibroproliferative remodeling. While tissue remodeling is considered beneficial in the early stages, sustained remodeling leads to loss of the normal contractile function of smooth muscle resulting in incontinence and, in extreme cases, severe renal damage [1,2]. Although generally not life threatening in nature, nonmalignant urologic disease is associated with significant psychosocial distress and a marked decrease in quality of life. Given its incidence in the general population (>33 million office visits recorded annually) and estimated economic costs of over $16 billion annually [3], urologic diseases are clearly a significant healthcare burden.

Research from our laboratory and others has begun to delineate the cell physical responses, intracellular signaling cascades that are activated, and the gene expression changes that are evoked, in response to distension of bladder smooth muscle. In the following section, we discuss the diversity of techniques employed to study mechanotransduction in bladder smooth muscle.

15.2 HOW TO STUDY MECHANICAL SIGNALING IN BLADDER SMOOTH MUSCLE

Mechanotransduction is broadly defined as the process whereby physical forces are converted to biochemical signals within cells. Smooth muscle achieves its contractile function through the concerted shortening of single cells arranged in bundles or sheets. The coordinated activity is achieved through the functional coupling of cells via gap junctions. A diverse range of techniques has been employed to interrogate the process of mechanotransduction in bladder smooth muscle *in vitro*, ranging from analysis of single cells, as in patch clamp and traction force microscopy approaches to studies of the intact bladder in live organisms. In the following section, we describe *in vitro*, *ex vivo*, and *in vivo* approaches that have been applied to the investigation of mechanical signaling in bladder and other types of smooth muscle, and the types of information obtained from each. The advantages and limitations of different techniques are considered.

15.2.1 In Vitro Assays of Mechanical Signaling

15.2.1.1 Modified Patch-Clamp Single Cell Stretch

The behavior of smooth muscle–rich hollow organs such as the bladder relies on the coordinated contraction and relaxation of collections of smooth muscle cells (SMCs) arranged in sheets or bundles. However, valuable mechanistic information can be obtained from studies of SMC behavior at the single cell level. In two recent studies [4,5], a modified patch-clamp approach was used to apply stretch to single bladder smooth muscle cells (BSMCs), by using patch pipettes to form a high resistance seal with cells and subsequent attachment of the pipettes to manipulators. Movement of the pipettes along the long axis of the cells enabled stretch of defined parameters to be applied, following which biological responses could be assessed.

15.2.1.2 Traction Force Microscopy

Virtually every spread adherent cell exists in a state of mechanical tension. This tension—also referred to as "prestress"—is transmitted through the cell body to the extracellular microenvironment where it is manifest in the form of contractile or traction forces that the cell exerts upon its substrate [6–17].

Following the pioneering work of Harris et al. [10], Dembo and Wang demonstrated that traction forces within a cell could be mapped by studying the displacement field induced by the cell in the passive flexible substrate upon which it is adherent [9]. This novel technique, termed "traction force microscopy," was a technological breakthrough, but was extremely computationally intensive [18]. However, Fourier transform traction microscopy (FTTM) was subsequently developed [11,13] and is now extensively used to measure traction forces in a wide range of cell types [11–13,19–24]. An FTTM measurement proceeds as follows: cells are cultured upon deformable polyacrylamide gel substrates with embedded fluorescent nanobeads. Cell-exerted displacements are computed by comparing images of these nanobeads obtained during the experiment with a reference image obtained at the end of the experiment subsequent to detaching the cell from the substrate (Figure 15.1). The traction force field is computed from the displacement field using a mathematical approach called Fourier transform cytometry [11].

15.2.1.3 Cell Mapping Rheometry

In the approaches for traction force mapping described in Section 15.2.1.2, the isolated cell adheres to a deformable substrate that is purely passive. While such an approach suffices under some circumstances, it is inadequate for studying responses of cells from the heart, gut, lung, or bladder that routinely experience large transient stretch. Thus, in order to probe the effect of imposed stretch on cell traction forces, one of us recently developed a simple, quantitative, and highly precise dynamic traction force measurement approach called cell mapping rheometry (CMR) (Figure 15.2) [15]. In this approach, a cell is plated on a gel substrate and stretch is then applied using a punch-indentation system. When an annular punch indents the gel, the region in the gel center bulges and its surface

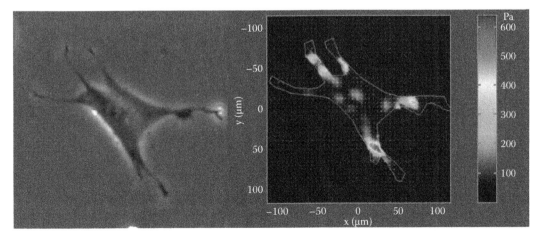

FIGURE 15.1 (See color insert.) Traction force microscopy. Traction forces exerted by a BSMC, analyzed using traction force microscopy. *Left*, Phase contrast image. *Right*, Corresponding traction force map.

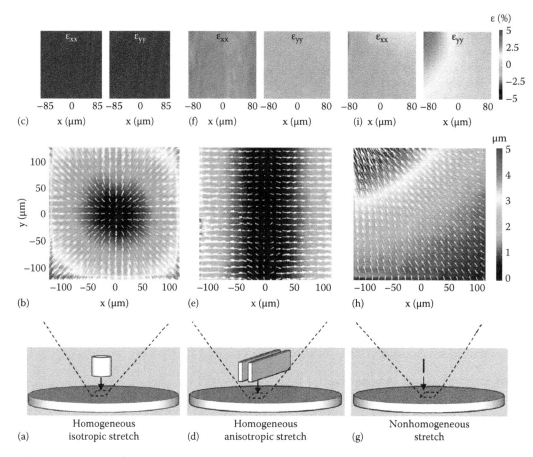

FIGURE 15.2 (See color insert.) Cell mapping rheometry. (a–c) An annular punch indenter imposes a homogeneous isotropic biaxial stretch within the central region of the indenter. (d–f) Two parallel plates impose a homogeneous, anisotropic, and uniaxial stretch. (g–i) A microneedle imposes a nonhomogeneous stretch. Localized gel displacements are indicated by arrows, and their magnitude by color. (From Krishnan, R. et al., *PLoS One*, 4, e5486, 2009.)

undergoes a strain that is termed biaxial and isotropic. When parallel plates indent the gel, however, the strain field is uniaxial and anisotropic. Depending upon the shape of the indenter, therefore, the cell adherent upon the gel surface can be subjected to a homogeneous stretch that is either biaxial and isotropic in the plane or uniaxial and anisotropic in the plane. If the punch is then lifted, the gel recoils elastically and the cell will have undergone one cycle of transient stretch-unstretch. After the stretch is applied using the punch-indentation system, traction force dynamics are measured using FTTM (described in Section 15.2.1.2).

The advantage of CMR is that it permits application of stretch that may be biaxial and isotropic in the plane or uniaxial and anisotropic. Nonhomogeneous stretch can be prescribed as well, and the applied stretch may be tensile or compressive, acute or chronic in nature. This allows us to study the changes in traction or contractile force response of SMC when subjected to a variety of stretch conditions that mimic physiological or pathological stimuli. Indeed, using CMR, it was found that a wide range of adherent cells respond to physiological levels of homogeneous stretch not by stiffening and increasing their traction forces above prestretch values as would be predicted [25–28], but by promptly and acutely decreasing their traction forces, followed by a slow recovery to baseline levels (Figure 15.3a) [15]. To understand the underlying mechanism, it is necessary to first consider an analogous but independent measure of cell tension called cell stiffness, as outlined below.

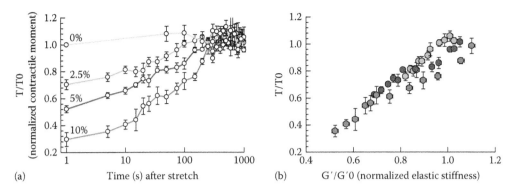

FIGURE 15.3 (See color insert.) In response to an acute homogeneous stretch, traction forces evolve in concert with cell stiffness. (a) Traction forces, represented by the ratio of the contractile moment T relative to the pre-stretch baseline value, T0, plotted as a function of time after stretch cessation. The greater was the applied stretch, the greater were the reductions in cell traction forces. Peak strains of: 0% (gray; $n=9$), 2.5% (green; $n=12$), 5.0% (blue; $n=11$) and 10.0% (red; $n=14$). (b) Normalized traction force data correlate linearly with normalized cell elastic stiffness (G'/G'0). (From Krishnan, R. et al., *PLoS One*, 4, e5486, 2009.)

15.2.1.4 Optical Magnetic Twisting Cytometry

Cell stiffness describes the extent to which a cell deforms when subjected to a external force. The external force is typically local such as a magnetic bead, a tweezer, a cantilever beam, or a microneedle pulling on specific focal adhesions on the cell surface [26,28–30]. A complementary approach, termed optical magnetic twisting cytometry (OMTC) allows the investigator to interrogate cell stiffness, another measure of cell tension [29]. Specifically, ferromagnetic beads (4.5 μm diameter) coated with a synthetic arginine-glycine-aspartic acid (RGD)-containing peptide are added to cells adherent upon a flexible membrane. The dishes are then incubated for a specified time to allow the beads to bind to integrin receptors on the cell surface and anchor tightly to the underlying cell cytoskeleton. The dishes are placed on a microscope stage within two pairs of coaxial magnetic coils. The beads are magnetized using a strong (1000 G) magnetic field pulse oriented in the horizontal plane and twisted using an oscillatory vertical magnetic field (5–75 G). This causes the beads to rotate and align with the twisting field and, as a result, experience a torque (Figure 15.4). The resulting bead motions deform structures deep in the cell interior. The cell stiffness is calculated from the ratio of the Fourier transform of this bead torque to the resulting bead lateral movement, as measured using a CCD camera, with appropriate mathematical transformations applied [31].

OMTC measurements are not limited to the static regime. They can be superimposed with stretch and extended to probe the cellular dynamic regime [32]. Shown in Figure 15.5 is how the stiffness

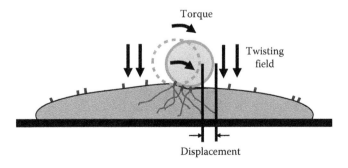

FIGURE 15.4 Schematic of OMTC technique.

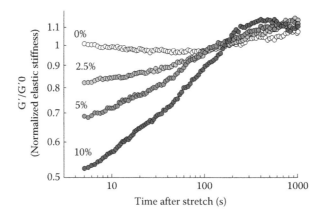

FIGURE 15.5 (See color insert.)   An acute homogeneous stretch fluidizes the airway SMC. Cell stiffness, represented by the ratio of the cell elastic stiffness G′ relative to the pre-stretch baseline value, G′0, reduces promptly and abruptly after stretch cessation and then recovers slowly. Peak strains of: 0% (gray; $n=9$), 2.5% (green; $n=12$), 5.0% (blue; $n=11$), and 10.0% (red; $n=14$). (From Trepat, X. et al. *Nature*, 447, 592, 2007.)

normalized to its pre-stretch baseline value changes in the human airway SMC following an acute transient stretch. When no stretch is applied, the stiffness is almost unchanged. However, when a transient stretch is applied, immediately after stretch cessation, the stiffness normalized to its pre-stretched value promptly and acutely decreases. These findings, taken together with other data, are strikingly similar to shear-induced fluidization, observed in soft inert materials [33,34]. Moreover, they highlight a novel paradigm for cellular mechanosensing. Cell responses to an acute stretch can be described through phenomenological laws from soft matter physics that are considerably less detailed and substantially more generic compared with specific molecular signaling cascades [15,32].

How do cell traction force dynamics described earlier using CMR compare with these fluidization responses? When dynamic cell stiffness changes are plotted not as a function time but as a function of dynamic traction force changes measured under similar experimental conditions, a strong linear correlation is observed (Figure 15.3b) [15].

TFM, CMR, and OMTC have not generally been applied to analysis of bladder smooth muscle, with much of the aforementioned data generated in airway smooth muscle, another mechanically responsive tissue. However, recent experiments in our laboratories have employed these techniques to probe fundamental aspects of BSMC mechanoresponsiveness at both the single-cell level and in bladder strips [25].

15.2.1.5   Membrane-Based Strain-Transducing Devices

A number of studies have employed devices of various configurations that comprise cells seeded in monolayer culture on a flexible-bottomed culture plate that is then subjected to mechanical stretch or strain by application of a vacuum to the underside of the membrane resulting in deformation (Figure 15.6). The best known of these devices is the FlexerCell Strain Unit developed by FlexCell Corporation, which has been in use for over 20 years. A major advantage of the FlexCell apparatus is the ability to precisely control the nature of the mechanical stimulus in terms of wave form, membrane configuration, frequency, and amplitude, thereby enabling appropriate modeling of different physiological or pathological environments experienced by SMC *in vivo*. In addition to the precise control and regulation afforded by the FlexCell apparatus, the flexible-bottomed culture plates may be coated with various extracellular matrix (ECM) components such as collagens and laminins, allowing investigation of the role of the ECM in modulating cellular responses to tensile stress. The use of the FlexCell to probe signal transduction in bladder smooth muscle is discussed in more detail in following sections.

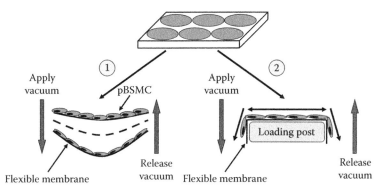

FIGURE 15.6 Application of mechanical stimuli using the FlexCell apparatus. Primary BSMC plated in flexible-bottomed 6-well plates can be exposed to mechanical stimulation by vacuum-mediated downward deformation of the membrane (1) or pulling of the membrane over a fixed loading post (2). Repetitive application and release of the vacuum confers cyclical stimulation on cells.

In an alternative approach, Haberstroh and colleagues developed a novel apparatus for the application of mechanical strain, similar to the FlexCell system, but with the added contribution of sustained hydrostatic pressure [35]. The rationale for development of this system was that it better approximated the physical stimuli, i.e., stretch and pressure experienced simultaneously by BSMC in the intact bladder. Using this approach, the authors were able to verify the stretch-stimulated increase in expression of heparin-binding EGF-like growth factor and alteration of the type I: type III collagen ratio previously described in the smooth muscle compartment of the distended intact bladder [36,37]. Although such membrane-based strategies have undoubtedly enhanced our understanding of the biochemical and molecular level changes in mechanically stimulated BSMC, they are limited in their ability to account for the three-dimensional (3D) architecture and appropriate tissue relationships encountered in the intact bladder wall.

15.2.2 EX VIVO ASSAYS OF MECHANICAL SIGNALING

15.2.2.1 Organ Bath Studies with Muscle Strips

The analysis of isolated muscle strips in organ bath assays is the classical approach for evaluating complex responses to agonists and antagonists in contractile tissue. In addition to assessment of alterations in contractility between normal and diseased muscle, the extent to which contractile responses are modulated by external stimuli, including putative therapeutic agents, can be investigated. One advantage of this approach is that it enables the behavior of muscle bundles working in concert to be measured. Briefly, strips isolated from bladder muscle are mounted either vertically or horizontally within an organ bath and perfused with physiological salt solution. Strips can be tested in various configurations depending on the purpose of the experiment. To measure isotonic contractions, one end of the muscle strip is connected to a load against which the strip can shorten. For evaluation of isometric contractions, the muscle strip is connected to an isometric force transducer designed to restrict muscle fiber shortening such that tension that develops in the stimulated muscle strip can be measured. To assess the impact of agents promoting relaxation, tissue strips are first pre-contracted with known contractile agonists such as cholinergic agents or electrical field stimulation and then exposed to relaxants. Alternatively, to compare relative contractility of muscle strips from normal and diseased muscle, strips are exposed to various inducers of contraction following equilibration, to ensure comparable baseline signals. The use of isolated muscle strips to investigate physiology of lower urinary tract smooth muscle has been extensively reviewed in a recent article [38] and additional experimental details are considered further here.

15.2.2.2 *Ex Vivo* Bladder Distension Models

A significant advance in our understanding of bladder smooth muscle mechanotransduction and mechanoresponsiveness occurred with the development of an *ex vivo* rodent model of bladder stretch injury in the laboratory of Dr. Darius Bägli [37]. Conceptually, the model is similar to the well-known Langendorff isolated perfused heart preparation, which has been in use for over a century [39]. Briefly, the bladder is hydrodistended transurethrally, followed by ligation of the ureters and bladder neck. Following incubation in culture medium for the desired time, distended tissues are then decompressed and harvested for assessment of histological, biochemical, and molecular endpoints. The advantages of this model include (a) the maintenance of native 3D architecture and tissue relationships; (b) the ability to isolate mechanical stimulation from other factors present *in vivo*, such as neural feedback and humoral influences; and (c) the tractability of the system to pharmacologic perturbation and genetic targeting through either RNA interference or application of the technique to transgenic rodent strains.

Two other *ex vivo* models have been described. In the murine whole bladder preparation developed by Zderic and colleagues [40], the bladder is excised along with a small segment of urethra to permit cannulation and subsequent perturbation with pharmacologic agents or electrical field stimulation in an organ bath. Although not used to evaluate mechanical signaling per se, this model enabled measurement of urodynamic parameters in response to known contractile agonists such as KCl or the cholinergic agent bethanechol and how these were modulated under conditions of hypoxia. Similarly, isolated bladders from guinea pigs were used to probe autonomous contractile activity and changes in intravesical pressure in the absence of neural input [41,42].

15.2.3 *In Vivo* Analysis of Mechanotransduction

The *in vitro* and *ex vivo* approaches mentioned in Sections 15.2.1 and 15.2.2 for analysis of mechanical signaling have many advantages including convenience, ease of manipulation and ease of perturbation with drugs and other modulators. However, because only a single cell type is represented, in the case of monolayer SMC culture, or because the tissue is isolated from its *in vivo* environment in the case of tissue strip and organ culture techniques, these methods do not model the totality of events occurring in the intact organism. To better approximate the response of the bladder to mechanical stimuli *in vivo*, many groups have created experimental bladder outlet obstruction in rodents. These procedures are designed to mimic the structural and/or functional obstruction that occurs in individuals afflicted with congenital anomalies such as posterior urethral valves in boys or that results from prostatic enlargement in the aging male.

15.2.3.1 Complete Bladder Outlet Obstruction

Complete occlusion of the bladder outlet is accomplished experimentally by ligating the urethra. As the animal produces urine, the bladder wall distends progressively beyond its natural limits, subjecting the tissues to marked mechanical stimulation. For reasons of animal welfare, complete occlusion models are limited in duration, but have proved useful in delineating early molecular responses to obstruction and the ensuing stretch of bladder tissues [36].

15.2.3.2 Partial Bladder Outlet Obstruction

Because of the limited duration of complete urethral ligation, many groups have employed partial bladder outlet obstruction (PBOO), in which the urethra is ligated over a catheter or other solid structure (e.g., syringe needle) placed parallel to the tissue. Subsequent removal of the catheter leaves a partially occluded urethral lumen, leading to increased urethral resistance and gradual but persistent elevation of intravesical pressure. Over time, the bladder wall undergoes fibroproliferative remodeling and a concomitant loss of contractility [43,44], consistent with that observed in human patients with urinary tract obstruction. Models of PBOO have been in use for over 25 years and have greatly enhanced our understanding of the processes, both at the tissue and molecular level, that contribute to remodeling and loss of normal contractile function.

15.2.3.3 Neurogenic Obstruction Models

In addition to the "mechanical" obstruction models described in Sections 15.2.3.1 and 15.2.3.2, which rely on physical occlusion of the urethral opening, several laboratories have used the strategy of perturbing bladder innervation to achieve functional, "neurogenic" obstruction. In humans afflicted with spinal cord injury (SCI) the bladder undergoes a spectrum of structural changes depending on the level of the spinal lesion. Damage to the thoracic spinal cord typically results in a small, low-volume bladder characterized by involuntary (i.e., spastic) contractions, whereas damage within the lumbosacral region often produces a large volume, flaccid bladder of low contractility. Thus, SCI and other conditions affecting innervation, such as spina bifida, elicit profound changes in the mechanical capabilities of the bladder. The most common experimental approach involves spinal cord transection between the eighth and tenth thoracic vertebrae, and leads to relatively rapid changes in bladder smooth muscle conformation and contractility. Consistent with observations in humans, Gloeckner et al. noted that the so-called inactive state tissue compliance of the SCI rat bladder exposed to biaxial stress was substantially greater than that of noninjured bladders [45]. A subsequent study demonstrated significant changes in the collagen and elastin content of bladders from spinal cord injured rats, consistent with the tissue remodeling predicted by altered compliance [46].

In addition to the use of trauma, i.e., spinal cord transection or crush injury, several groups have used chemical approaches to induce neural tube defects that lead to aberrant innervation and development of bladder dysfunction. Danzer et al. described a retinoic-acid-induced myelomeningocele (MMC) model in which administration of all-trans retinoic acid to maternal rats led to development of MMC in ~60% of offspring [47]. Moreover, despite similar morphology between control and MMC bladders, the contractile response of detrusor smooth muscle from MMC bladders to agonists such as KCl and bethanechol was diminished compared to controls [48]. Alterations in bladder smooth muscle mechanoresponsiveness were also demonstrated in a rodent model of hydrocephalus, generated by the injection of kaolin into the subarachnoid space [49]. In that study, rats subjected to experimental induction of hydrocephalus displayed decreased force generation in bladder muscle strips exposed to KCl, but showed an increased response to carbachol compared to controls. In addition, the relaxation response to adrenergic stimulation was diminished in the hydrocephalus group.

15.2.3.4 Bladder Hydrodistension

Finally, while not an experimental procedure, bladder hydrodistension is performed in patients as treatment for symptoms associated with interstitial cystitis (IC). IC is a debilitating condition of unknown etiology characterized by urinary frequency, urgency, and pelvic pain. Although the molecular events that underlie the development of IC are still poorly understood, the urine of patients with IC contains lower levels of heparin-binding EGF-like growth factor, a mitogen for urothelial cells, than the urine of patients without IC. Interestingly, Chai et al. showed that hydrodistension of the bladders of individuals with IC led to marked upregulation of urinary HB-EGF levels relative to that observed in controls [50]. The stretch-induced expression of HB-EGF observed in humans under these conditions is consistent with reports using cultured cells [51,52] and rodent models [36,53], that identified HB-EGF as a mechanically regulated gene in the bladder.

In summary, mechanosensitivity of bladder smooth muscle can be assessed using a wide variety of techniques, from the single cell level to the intact bladder in live organisms. Notably, experimental observations provide confidence that findings from reductionist approaches with cultured cells and tissue-based models are likely to be relevant to bladder function *in vivo*.

15.3 MECHANORESPONSIVE SIGNALING NETWORKS IN BLADDER SMOOTH MUSCLE

In the following section, we summarize findings obtained from the various experimental approaches above and how these have contributed to our understanding of mechanical signaling in bladder smooth muscle.

Studies in our laboratory and others using primary BSMC subjected to mechanical stretch using the FlexCell system have revealed stretch-induced signaling via multiple parallel kinase cascades, activation of select genes [53,54], and induction of a hyperplastic and hypertrophic response [51,52,55–57]. In particular, cyclic stretch relaxation of BSMC has been shown to activate the receptor tyrosine kinases (RTKs) epidermal growth factor receptor (EGFR)/ErbB1 [58] and ErbB2 [52], the mitogen-activated protein kinase Erk [59,60], the stress-activated kinases JNK/SAPK [52,61] and p38SAPK2 [52] and the phosphoinositide-3-kinase (PI3K)/Akt survival pathway [56]. In spite of the activation of several discrete signaling pathways following mechanical stimulation, the transcriptional response of stretched SMC has been shown to be highly selective. In a recent study from our laboratory, genome-scale analysis of gene expression in primary human BSMC exposed to cyclic stretch relaxation *in vitro* revealed less than 0.2% of the expressed genome in these cells to be mechanically responsive [53]. The gene found to be upregulated to the greatest extent in this study was that encoding heparin-binding EGF-like growth factor (HB-EGF), a member of the EGF-like growth factor family and activating ligand for the EGFR/ErbB1 and the related receptor ErbB4 [62,63]. Interestingly, a previous study from our program had provided the first demonstration that HB-EGF, a known SMC mitogen [62], was a stretch-sensitive gene in BSMC [51].

In a more recent study, Yang and colleagues performed expressing profiling on BSMC using stretch parameters similar to our study and observed differential gene activation profiles in BSMC exposed to mechanical strain for short (1 h) versus longer (24 h) durations [54]. A number of transcripts related to angiogenesis, ECM production and cell growth, signaling and inflammation were identified, several of which overlapped with candidates identified in our analysis [53]. Importantly, the findings from both microarray analyzes were confirmed independently using an *ex vivo* model of rodent bladder distension, establishing the usefulness of the FlexCell system in evaluating the molecular events that occur in mechanoresponsive tissues *in vivo*.

Other groups have employed alternative profiling approaches to probe mechanical signaling within the bladder. In an elegant approach, Fujita and colleagues used antibody array profiling to interrogate signaling events following creation of bladder outlet obstruction *in vivo* [64]. In that study, 64 out of 389 proteins were shown to change in bladder smooth muscle 7 days after partial urethral occlusion, and included transcription factors, regulators of cell cycle transit and apoptosis, mediators of adhesion, cytoskeletal proteins and components of the signal transduction apparatus. Of these, the transcription factor STAT3 was found to be markedly upregulated following obstruction, consistent with observations by Halachmi et al. who also implicated STAT3 as a mediator of mechanoresponsiveness in the intact bladder exposed to acute stretch injury [59]. In that study, the authors employed a commercial screening method for phosphorylated signaling proteins, and demonstrated convergence of growth factor–dependent and mechanical stimuli in regulation of STAT3 in BSMC.

A major finding from our microarray analysis was the identification of the AP-1 transcriptional complex as a potential mediator of the observed changes in gene expression. Using a novel informatics tool termed FIRED (Frequency-based Identification of Regulatory Elements in Differential gene expression) we observed an overrepresentation of AP-1 binding motifs in the promoters of stretch-responsive genes [53]. This finding was consistent with previous studies demonstrating a role for AP-1 in stretch-stimulated expression of the SMC mitogen HB-EGF in BSMC [65] as well as the demonstration of increased AP-1 activity in the intact bladder exposed to distension [66].

We have proceeded to confirm a role for distinct AP-1 dimers in transducing mechanical stimuli in BSMC. Consistent with the selective gene expression response of BSMC to cyclic stretch-relaxation, the same stimulus evokes a similarly specific pattern of AP-1 activation, with increased DNA binding activity of c-jun, c-Fos, and FosB, among the seven AP-1 subunits tested (A.R. and R.M.A, manuscript in preparation). This pattern of activation is in marked contrast to that observed in BSMC treated with platelet-derived growth factor (PDGF), in which all AP-1 subunits, with the exception of JunD and FosB showed increased binding to DNA. That FosB was not activated in response to PDGF suggests a potential role for FosB as a selective regulator of stretch-stimulated transcription. Studies are currently underway in our laboratory to test this hypothesis.

Another important observation from our expression profiling study was the demonstration that genes shown to be mechanically responsive were also sensitive to PDGF, a classical SMC mitogen. In an ongoing study, we have demonstrated that the PDGF receptor (PDGFRα/β) is activated by mechanical stimulation in the intact bladder exposed to acute overdistension [67]. Distension of rat bladders to a pressure of 40 cm H_2O led to a rapid (within 15 min) and transient increase in PDGFR phosphorylation. In addition, we observed increased Akt phosphorylation and increased nuclear localization of AP-1 subunits, consistent with our *in vitro* findings. Although we cannot rule out a role for ligand in PDGFR activation, the rapidity of receptor phosphorylation argues for a ligand-independent mode of regulation. The precise mechanisms involved in this process are still unknown but are likely to be mediated by mechanically sensitive elements including stretch-activated ion channels, integrins, or other components of focal adhesions.

Alterations in intracellular calcium are central to the process of smooth muscle contraction, such that the role of calcium in mechanoresponsiveness of bladder smooth muscle cannot be overlooked. Processes that alter the availability of cytosolic calcium, whether through influx from the extracellular environment or through release from intracellular stores, are central to BSMC contraction [68–71] and therefore mechanotransduction. Stretch of BSMC *in vitro* is known to promote both extracellular Ca^{2+} influx through stretch-activated channels in the plasma membrane [61] as well as release from intracellular stores through a ryanodine receptor–mediated process [4]. Ca^{2+} release from intracellular stores, in response to cholinergic receptor activation, initiates the contractile process through calcium-calmodulin-mediated activation of myosin light chain (MLC) kinase and subsequent phosphorylation of MLC (reviewed in [72]). As in smooth muscle from other organs, bladder smooth muscle displays the phenomenon of calcium sensitization, in that the degree of intracellular Ca^{2+} elevation does not correlate with the extent of contraction [73,74]. In particular, exposure to receptor agonists tends to evoke higher tension generation for a given intracellular Ca^{2+} level, compared to that observed with membrane depolarization. Takahashi and colleagues provided some of the first evidence for Ca^{2+} sensitization in human bladder smooth muscle, and demonstrated that the carbachol-induced increase in Ca^{2+} sensitivity was regulated by both ROCK- and PKC-dependent pathways [74]. In addition to its central role in regulating SM contraction, calcium also regulates the activity of a number of other ion channels and proteins that modulate BSMC contractility and growth. These include (a) the large-conductance calcium-activated potassium channel (BK) [75], the activity of which is enhanced with mechanical stimulation of bladder smooth muscle [76]; (b) the small-conductance calcium-activated potassium channels (SK1, SK2, and SK3), of which SK2 is essential in regulating BSM contractility [77]; (c) calcium-activated chloride channels (ClCa); and (d) the phosphatase calcineurin and its target NFATc. Interestingly, calcineurin expression was found to increase in bladder smooth muscle following PBOO and was implicated in the ensuing compensatory bladder hypertrophy and detrusor decompensation [78,79]. Moreover, the calcineurin effector NFATc3 has been reported to regulate BK channel activity and subsequent contractility in bladder smooth muscle *in vivo* [80].

Given the central role of Ca^{2+} in regulating contractility in bladder smooth muscle, it is not surprising that perturbations in bladder function are associated with aberrant calcium homeostasis. Several groups have observed changes in different components of the Ca^{2+}-sensing apparatus following experimental bladder outlet obstruction in animals [73,81–84], and in tissues from human patients with overactive bladder symptoms [85]. Interestingly, Lassmann and colleagues showed in an elegant report that genetic deletion of one allele of the sarco(endo)plasmic reticulum Ca^{2+}-ATPase isoform 2 (SERCA2) could confer protection against the deleterious hypertrophy that results from PBOO [86]. This was in marked contrast to the effect of SERCA2 deletion on the development of cardiac hypertrophy in response to aortic banding, in which mice heterozygous for SERCA2 showed greater development of hypertrophy, reduced cardiac function and increased death.

Consistent with a role for ROCK in Ca^{2+} sensitization as reported for smooth muscle in other organs, several recent studies have implicated signaling through the Rho-ROCK axis in bladder smooth muscle contractility and mechanotransduction [87–89]. The Rho-dependent kinase ROCK achieves its effects on smooth muscle contraction by phosphorylating and inactivating MLC phosphatase, thereby retaining MLC in its phosphorylated "contractile" state and promoting sustained contraction, without altering Ca^{2+} levels. Although not directly implicated in mechanotransduction in bladder SM, application of cyclic strain to various cell types activates RhoA and promotes ROCK-mediated assembly of actin filaments [90]. Thus, enhanced activity of the Rho-ROCK axis is implicated in regulation of contractile elements in cells and is likely to be a major mediator of mechanoresponsiveness in BSMC. In support of this idea, upregulation of Rho-ROCK activity has been demonstrated in isolated bladder muscle strips exposed to contractile agonists [87] and in the bladder wall following outlet obstruction [88,91,92].

15.4 THE ROLE OF THE EXTRACELLULAR MATRIX IN MECHANOTRANSDUCTION

The ECM is well established as an important component of mechanotransduction in intact tissues. Its contribution to mechanical signaling is perhaps best appreciated under conditions of aberrant mechanical signaling resulting from tissue fibrosis or other types of pathologic remodeling. Altered expression of collagens, and particularly a reduction in the collagen I:collagen III ratio has long been appreciated as contributing to loss of normal contractility in response to insult or injury within the lower urinary tract. However, the contribution of the ECM and the resultant changes in tissue architecture in pathological bladder wall remodeling has garnered increasing attention in recent years as a potential site of therapeutic intervention.

Using an acute bladder injury model together with assessment of primary BSMC, Bägli and colleagues implicated alterations in cell-ECM interactions as early events in bladder wall remodeling following bladder wall overdistension [93]. In particular, they demonstrated increased expression of the receptor for hyaluronic acid mediated cell motility (RHAMM) in BSMC following acute bladder injury, and showed that the addition of a function-blocking peptide to RHAMM was able to attenuate BSMC contraction *in vitro*. In subsequent analyzes, this group demonstrated the importance of specific ECM conformations in regulating BSMC behavior [94]. Specifically, the authors showed that exposure of BSMC to denatured collagen I, as is thought to occur following bladder wall overdistension, increased DNA synthesis severalfold compared to cells exposed to native collagen, and could be enhanced by further exposure to denatured collagen. This effect was shown to be mediated through the activation of the Erk-MAPK pathway and could be attenuated in the presence of the MEK inhibitor, PD98059. Signaling from the ECM to Erk-MAPK in response to bladder distension was shown to be regulated by matrix metalloproteinases, with pharmacologic targeting of MMP activity using a broad-spectrum inhibitor able to attenuate ECM-stimulated BSMC proliferation [60]. The ECM has also been shown to undergo dynamic and temporally regulated changes in response to stretch as evidenced by a reversal of the ratio of deposited collagen I and III isoforms at early versus late timepoints, with concomitant alterations in integrin expression in BSMC [95]. The production of ECM by BSMC is also affected by the frequency with which cells are stretched, with expression of type I and type III collagens displaying differential sensitivity to distension frequency [96].

Other ECM components have also been identified as important regulators of BSMC proliferation. The growth of BSMC under control or stretched conditions using the FlexCell system was shown to be greater when plated on laminin-coated silicone membranes than on collagen I-coated membranes [97]. A recent study by Wognum et al. described greater de novo production of elastin in the bladders of rats subjected to SCI that interacted with and contributed to deposition of a more highly supercoiled form of collagen [98]. These observations are consistent with previous observations by this group describing alterations in the viscoelastic properties of the bladder wall under normal conditions and following SCI [45,46,99]. Although the enhanced production of ECM proteins following SCI initially results in increased bladder compliance as it responds to the stress of overfilling, this adaptive response quickly becomes pathological since the underlying neural defect prevents full physiological recovery of bladder function.

Although alterations in production of ECM proteins following bladder wall distension have been well documented, the molecular events underlying changes in ECM protein expression have not been fully elucidated. The immediate early gene Cyr61/CCN1 encodes a secreted, cysteine-rich, heparin-binding protein and functions as an ECM-associated signaling molecule. Cyr61 was shown to be induced in BSMC following cyclic stretch-relaxation, in a protein kinase C-, ROCK-, and PI3K/Akt-dependent manner [100]. Cyr61 and another immediate early gene product, connective tissue growth factor (CTGF/CCN2), also identified as mechanosensitive in BSMC [101], were found to be induced in the detrusor smooth muscle following PBOO in rats, with expression sustained for the duration of obstruction [102]. Among their various diverse functions, Cyr61 and CTGF are key regulators of wound healing and regulate the expression of various matrix-associated, angiogenic, and ECM-remodeling proteins such as MMPs, TIMPs, VEGF, and integrins [103,104].

Interestingly, Cyr61 has been implicated as a mediator of dysregulated collagen production in certain cell types and tissues. In particular, the upregulation of Cyr61 in fibroblasts led to concomitant reduction of procollagen I expression and induction of MMP1, which promotes collagen degradation [104,105]. These observations may be consistent with the reduction in the collagen I:collagen III ratio observed following PBOO, a condition also associated with elevated Cyr61 levels [102]. Moreover, the upregulation of Cyr61 in fibroblasts led to the induction of AP-1, a transcription factor shown by us and others to be mechanoresponsive in bladder smooth muscle and to mediate expression of genes relevant to growth and remodeling in the bladder wall [53,65–67]. Finally, Cyr61 and CTGF have been shown to regulate angiogenesis and expression of angiogenic genes in various tissues, including bladder smooth muscle [54,106]. Alterations in bladder wall tension in response to mechanical stimulation promote tissue hypoxia [107], which is thought to stimulate feedback regulation of hypoxia-stimulated signaling pathways [108,109] and expression of angiogenic gene products [54]. Together, these observations suggest that the matricellular proteins Cyr61 and CTGF function to integrate mechanical and other environmental stimuli such as oxygen tension and inputs from the ECM and thereby regulate the fibroproliferative response of the bladder wall to injury.

15.5 SMOOTH MUSCLE-EPITHELIAL INTERACTIONS IN BLADDER MECHANOTRANSDUCTION

Although our research has focused on mechanical signaling in bladder smooth muscle, it is important to acknowledge the contribution of the urothelium lining the bladder lumen to organ mechanotransduction. In response to stretch, as occurs with bladder filling, the urothelium undergoes profound changes in membrane structure through processes of exocytosis and endocytosis, which lead to an increase in urothelial surface area [110]. Similar to smooth muscle, mechanotransduction in urothelial cells is mediated, at least in part, by the activation of stretch-sensitive ion channels [111].

Recent studies have shed light on the role of the urothelium in modulating bladder contractile function [112,113]. A study by Ikeda and Kanai [114] highlighted the contribution of the urothelium and lamina propria in mediating aberrant contractile activity in bladders from spinal-cord transected (SCT) rats [114]. Using novel optical mapping techniques, the mucosal surface of whole bladder sheets was assessed for intracellular Ca^{2+} levels and the frequency and amplitude of detrusor contractions under resting and stretched conditions. Bladders from normal animals showed an intrinsic pattern of disorganized low-amplitude, high-frequency contractions that was not significantly altered by stretch. In contrast, the SCT animals displayed a more regular low-frequency, high-amplitude signal with a further increase in the amplitude of contractions when subjected to stretch. Significantly, the changes in intrinsic contractile function observed in the SCT bladders under resting and stretched conditions were abolished by the removal of the mucosa from the bladder sheets. This suggests that changes in intrinsic contractile activity in pathological states are significantly dependent on signals from the urothelium, in contrast to predominantly detrusor-signaling-mediated contractions in the normal state.

The urothelium releases neurotransmitters such as NO, ATP, and Ach, in response to mechanical stimulation, that play critical roles in mechanosensory responses of the bladder [112]. Although the urothelium lacks voltage-gated Ca^{2+} channels and is not electrically excitable, it has other mechano-sensitive channels such as TRPs (transient receptor potential channels) and Deg/ENaC (degenerin/epithelial (Na+) channel). Increased expression of ENaC isoforms has been observed in the mucosa of overactive human bladder following BOO [115]. The same group went on to show, in rodents, that blockage of ENaC by amiloride significantly increased the intercontraction interval (the interval between two successive contraction events resulting in voiding) by interfering with stretch-evoked ATP release from the urothelium [116]. The requirement of the urothelial TRPV4 ion channel in maintaining normal voiding function has recently been established by Gevaert et al., who generated a TRPV4 knockout mouse that was found to be incontinent [117]. The authors present data suggesting, again, that this was due to a reduction in stretch-induced ATP release from the bladder epithelium.

Together, these studies highlight a critical role for the urothelium in both sensing and mediating responses to the mechanical stimulation that occurs with bladder filling and emptying, emphasizing the importance of paracrine interactions between the epithelial and smooth muscle tissue compartments in the response of the bladder wall to mechanical stimuli.

15.6 MECHANOTRANSDUCTION IN DISEASE AND POTENTIAL STRATEGIES FOR THERAPEUTIC INTERVENTION

The preceding sections have summarized some of the signaling pathways and molecular events that underlie mechanotransduction within the bladder wall. In addition to improving our understanding of the fundamental principles that govern bladder function, this knowledge has begun to provide information on potential therapeutic targets for a range of bladder diseases characterized by aberrant contractility. Existing treatments for bladder dysfunction associated with loss of normal contractility typically focus on the pharmacologic inhibition of muscarinic-receptor-mediated contraction with anticholinergic agents [118]. However, anticholinergic treatment is poorly tolerated in certain patients and has limited efficacy in others [119]. Thus, new therapeutic approaches are urgently required. Unlike other hollow organs, the bladder is considered to be a "privileged" site for introduction of drugs. The intravesical administration of therapeutic agents avoids the complications often observed with systemic therapy, and may enable drugs to be delivered to target tissues at higher concentrations than could be achieved with systemic administration. In the following section, we discuss ways in which our understanding of the effectors of smooth muscle mechanotransduction could be exploited for treatment of bladder dysfunction, with a particular emphasis on targeting kinase signaling (Figure 15.7).

15.6.1 RECEPTOR AND NON-RECEPTOR KINASE INHIBITION

A number of kinase-targeted drugs are currently under development or in clinical use for treatment of proliferative diseases such as cancer, as summarized in Table 15.1. These include the ErbB2/HER2 monoclonal antibody, Herceptin; the EGFR tyrosine kinase inhibitor Iressa (gefitinib), the PDGFR/c-Kit/Bcr-Abl kinase inhibitor, Gleevec (imatinib mesylate) and the VEGFR kinase inhibitor Sutent (sunitinib). The development of these agents grew out of extensive understanding of the signaling networks underlying tumor cell survival and proliferation, and the identification of specific molecular targets. A number of recent studies now suggest a role for such kinase inhibitors in nonmalignant conditions that display aberrant tissue growth. Gleevec was found to attenuate the spontaneous bladder contractions characteristic of detrusor overactivity in both human and

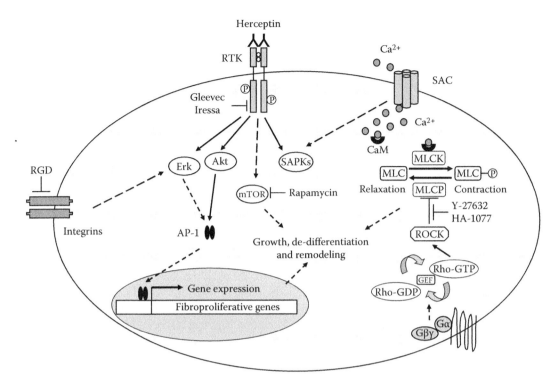

FIGURE 15.7 Targeting of signaling in BSMCs with selective inhibitors. The figure highlights drugs that target selected signaling mediators implicated in the process of BSMC mechanotransduction. RTK-mediated signaling is targeted by antibody-based therapy (e.g., Herceptin), or by inhibition of intrinsic receptor kinase activity (e.g., Gleevec, Iressa) attenuating the ErbB2, c-kit/PDGFR, and EGFR, respectively. Signaling through integrins can be blocked with RGD peptides. Rapamycin and its analogs inhibit signaling downstream of mTOR. The pyridine derivative Y-27632 and the isoquinoline derivative HA-1077 inhibit ROCK activity, thereby promoting smooth muscle relaxation. SAC, stretch-activated channels; CaM, calmodulin; MLCK, myosin light chain kinase; MLCP, myosin light chain phosphatase; GEF, guanine nucleotide exchange factor.

TABLE 15.1
The Table Indicates Clinically Approved and Investigational Agents Used in Treatment of Neoplastic or Cardiovascular Conditions that may also be Suitable for Treatment of Aberrant Smooth Muscle Growth Resulting from Hollow Organ Overdistension

Drug	Target(s)	Target Relevant in Mechanotransduction?
Gleevec	c-kit/PDGFR	Yes [120]
Iressa	EGFR/ErbB1	Yes [58]
Herceptin	HER2/ErbB2	Yes [52]
Sutent	VEGFR2	ND
Rapamycin	mTOR	Yes [109]
Y27632	ROCK	Yes [74,123]

ND, not determined.

guinea-pig bladder muscle strips [120]. In that study, the authors concluded that the observed effects were mediated through inhibition of c-kit. However, Gleevec targets multiple kinases, including the PDGFR, recently implicated by us as activated following mechanical injury in the intact bladder. Thus, the ability of Gleevec to attenuate aberrant contractility may result from targeting multiple discrete mediators of bladder mechanotransduction.

Experimental data from our group implicated the EGFR and ErbB2 as mediators of distension-induced gene expression and growth in BSMC [51,57], and showed that intravesical instillation of Iressa could suppress distension-stimulated phosphorylation of the EGFR and BrdU uptake. To our knowledge, this was the first demonstration of intravesical administration of a small molecule kinase inhibitor and suggests a novel therapeutic strategy to mitigate the deleterious effects of pathologic fibroproliferative remodeling subsequent to chronic bladder wall overdistension.

Several non-receptor kinases have also proved central to the process of bladder muscle mechanotransduction, and are attractive candidates for therapeutic intervention since they often represent points of convergence for multiple upstream signals. As noted in Section 15.3, the Rho A-dependent kinase, ROCK has been implicated in regulation of smooth muscle tone within the lower urinary tract, and has been shown to increase within the bladder wall following outlet obstruction (reviewed in [121]). In studies on the vasculature in the mid-1990s, the pyridine derivative Y-27632 was identified as a potent relaxant for vascular smooth muscle, through its ability to inhibit ROCK-mediated Ca^{2+} sensitization [122]. More recently, Y-27632 was also found to inhibit phosphorylation of CPI-17, a protein kinase C target that regulates SM contraction, and that has been implicated in bladder contractility [74,123]. Y-27632 has been used successfully in rodent models of bladder dysfunction secondary to denervation or diabetic insult to attenuate carbachol-stimulated BSM contractions and promote smooth muscle relaxation. Although Y-27632 has not been used clinically, another ROCK inhibitor, HA-1077 (fasudil) has been shown to suppress arterial spasms in human patients with angina [124,125]. Thus, targeting of ROCK is feasible in patients and may have utility in the treatment of bladder dysfunction characterized by aberrant contractility.

Another non-receptor kinase that has emerged as a viable therapeutic target for the treatment of fibroproliferative bladder disease is the mammalian target of rapamycin mTOR. Using the Ingenuity Pathway Analysis (IPA) software suite, Aitken and colleagues mapped a series of stretch- or hypoxia-responsive genes identified from prior analyses of BSMC to gene identifiers within the IPA database [109]. This led to the identification of networks and signaling pathways that linked the observed gene expression changes, and also identified putative chemical inhibitors that could be tested experimentally. One such agent was rapamycin, the immunosuppressant and inhibitor of mTOR. The authors proceeded to validate the relevance of the mTOR pathway to BSMC proliferation in response to both mechanical strain and hypoxia *in vitro* and in the *ex vivo* bladder injury model described earlier. Because rapamycin and its analogs (temsirolimus, everolimus) are FDA-approved and in current clinical use, it is tempting to speculate that these agents could be moved quickly into clinical use for treatment of obstructive uropathies.

15.6.2 MECHANOTRANSDUCTION AND TISSUE REGENERATION

Medical therapy is a mainstay of treatment for lower urinary tract dysfunction. However, in certain patients, surgical intervention in the form of bladder augmentation is necessary to restore bladder function. The standard approach involves the use of intestinal segments to increase bladder capacity, but is often hampered by serious complications related to the absorptive nature of the donor tissue. As a result, much effort has been expended in the search for appropriate substitutes, ranging from synthetic scaffolds to autologous neobladders with many of these efforts leading to the creation of a new field of functional tissue engineering (reviewed in [126]). It has become apparent in recent years that conditioning of engineered tissues with appropriate mechanical stimulation prior to implantation is likely to be critical for successful regeneration of functional tissues *in vivo* [127–129]. Mechanical stimuli, in the form of cyclical filling and emptying, are an important component of normal bladder development [130]. Indeed, the absence of the physical forces conferred by bladder cycling in utero is believed to contribute to the devastating failure of bladder development, as observed in classic exstrophy [131]. Understanding how to recapitulate appropriate mechanical signals during the process of engineering replacement tissues and how to integrate them with innervation and vasculogenesis will represent a significant breakthrough in functional tissue regeneration. This topic has been summarized in two excellent recent reviews [128,129] and is not considered further here.

15.7 SUMMARY

In this chapter, we have summarized the literature on various aspects of bladder smooth muscle mechanotransduction, from techniques used to probe the mechanical activity of BSMC and tissues, to signaling pathways activated by mechanical stimulation and potential strategies for therapeutic intervention. At the present time, biomedical researchers have access to an unprecedented array of scientific resources, including (a) genomic, proteomic, and metabolomic profiles from a range of tissues exposed to physiologically relevant stimuli; (b) information on specificity, selectivity, pharmacokinetics, and bioavailability for a host of pharmaceutical agents; (c) clinical trial data on FDA-approved as well as investigational agents; and (d) software such as the IPA suite, enabling identification and mapping of higher-order signaling networks onto genomic and other datasets. The challenge moving forward is to leverage these unparalleled resources to design agents with the desired specificity and selectivity for key signaling nodes that mediate pathologic mechanotransduction within the bladder wall.

REFERENCES

1. Zderic, SA, Wein, A, Rohrman, D, Gong, C, Nigro, D, Haugaard, N et al. 1998. Mechanisms of bladder smooth-muscle hypertrophy and decompensation: Lessons from normal development and the response to outlet obstruction. *World J Urol* 16:350–358.
2. Holmdahl, G, Sillen, U. 2005. Boys with posterior urethral valves: Outcome concerning renal function, bladder function and paternity at ages 31 to 44 years. *J Urol* 174:1031–1034; discussion 1034.
3. United States Renal Data System, Atlas of End-Stage Renal Disease in the United States Annual Data Report. 2006, National Institute of Diabetes and Digestive and Kidney Diseases: Bethesda, MD.
4. Ji, G, Barsotti, RJ, Feldman, ME, Kotlikoff, MI. 2002. Stretch-induced calcium release in smooth muscle. *J Gen Physiol* 119:533–544.
5. Wei, B, Chen, Z, Zhang, X, Feldman, M, Dong, XZ, Doran, R et al. 2008. Nitric oxide mediates stretch-induced Ca2+ release via activation of phosphatidylinositol 3-kinase-Akt pathway in smooth muscle. *PLoS One* 3:e2526.
6. Del Alamo, JC, Meili, R, Alonso-Latorre, B, Rodriguez-Rodriguez, J, Aliseda, A, Firtel, RA et al. 2007. Spatio-temporal analysis of eukaryotic cell motility by improved force cytometry. *Proc Natl Acad Sci USA* 104:13343–13348.
7. Sabass, B, Gardel, ML, Waterman, CM, Schwarz, US. 2008. High resolution traction force microscopy based on experimental and computational advances. *Biophys J* 94:207–220.
8. Schwarz, US, Balaban, NQ, Riveline, D, Bershadsky, A, Geiger, B, Safran, SA. 2002. Calculation of forces at focal adhesions from elastic substrate data: The effect of localized force and the need for regularization. *Biophys J* 83:1380–1394.
9. Dembo, M, Wang, YL. 1999. Stresses at the cell-to-substrate interface during locomotion of fibroblasts. *Biophys J* 76:2307–2316.
10. Harris, AK, Wild, P, Stopak, D. 1980. Silicone rubber substrata: A new wrinkle in the study of cell locomotion. *Science* 208:177–179.
11. Butler, JP, Tolic-Norrelykke, IM, Fabry, B, Fredberg, JJ. 2002. Traction fields, moments, and strain energy that cells exert on their surroundings. *Am J Physiol Cell Physiol* 282:C595–C605.
12. Wang, N, Naruse, K, Stamenovic, D, Fredberg, JJ, Mijailovich, SM, Tolic-Norrelykke, IM et al. 2001. Mechanical behavior in living cells consistent with the tensegrity model. *Proc Natl Acad Sci USA* 98:7765–7770.
13. Wang, N, Tolic-Norrelykke, IM, Chen, J, Mijailovich, SM, Butler, JP, Fredberg, JJ et al. 2002. Cell prestress. I. Stiffness and prestress are closely associated in adherent contractile cells. *Am J Physiol Cell Physiol* 282:C606–C616.

14. Engler, AJ, Sen, S, Sweeney, HL, Discher, DE. 2006. Matrix elasticity directs stem cell lineage specification. *Cell* 126:677–689.

15. Krishnan, R, Park, CY, Lin, YC, Mead, J, Jaspers, RT, Trepat, X et al. 2009. Reinforcement versus fluidization in cytoskeletal mechanoresponsiveness. *PLoS One* 4:e5486.

16. Tan, JL, Tien, J, Pirone, DM, Gray, DS, Bhadriraju, K, Chen, CS. 2003. Cells lying on a bed of microneedles: An approach to isolate mechanical force. *Proc Natl Acad Sci USA* 100:1484–1489.

17. du Roure, O, Saez, A, Buguin, A, Austin, RH, Chavrier, P, Siberzan, P et al. 2005. Force mapping in epithelial cell migration. *Proc Natl Acad Sci USA* 102:2390–2395.

18. Pelham, RJ Jr. Wang, Y. 1997. Cell locomotion and focal adhesions are regulated by substrate flexibility. *Proc Natl Acad Sci USA* 94:13661–13665.

19. An, SS, Fabry, B, Mellema, M, Bursac, P, Gerthoffer, WT, Kayyali, US et al. 2004. Role of heat shock protein 27 in cytoskeletal remodeling of the airway smooth muscle cell. *J Appl Physiol* 96:1701–1713.

20. An, SS, Fabry, B, Trepat, X, Wang, N, Fredberg, JJ. 2006. Do biophysical properties of the airway smooth muscle in culture predict airway hyperresponsiveness? *Am J Respir Cell Mol Biol* 35:55–64.

21. Parker, KK, Brock, AL, Brangwynne, C, Mannix, RJ, Wang, N, Ostuni, E et al. 2002. Directional control of lamellipodia extension by constraining cell shape and orienting cell tractional forces. *FASEB J* 16:1195–1204.

22. Stamenovic, D, Coughlin, MF. 1999. The role of prestress and architecture of the cytoskeleton and deformability of cytoskeletal filaments in mechanics of adherent cells: A quantitative analysis. *J Theor Biol* 201:63–74.

23. Stamenovic, D, Liang, Z, Chen, J, Wang, N. 2002. Effect of the cytoskeletal prestress on the mechanical impedance of cultured airway smooth muscle cells. *J Appl Physiol* 92:1443–1450.

24. Stamenovic, D, Suki, B, Fabry, B, Wang, N, Fredberg, JJ. 2004. Rheology of airway smooth muscle cells is associated with cytoskeletal contractile stress. *J Appl Physiol* 96:1600–1605.

25. Choquet, D, Felsenfeld, DP, Sheetz, MP. 1997. Extracellular matrix rigidity causes strengthening of integrin-cytoskeleton linkages. *Cell* 88:39–48.

26. Munevar, S, Wang, YL, Dembo, M. 2004. Regulation of mechanical interactions between fibroblasts and the substratum by stretch-activated Ca2+ entry. *J Cell Sci* 117:85–92.

27. Matthews, BD, Overby, DR, Mannix, R, Ingber, DE. 2006. Cellular adaptation to mechanical stress: Role of integrins, Rho, cytoskeletal tension and mechanosensitive ion channels. *J Cell Sci* 119:508–518.

28. Fabry, B, Maksym, GN, Butler, JP, Glogauer, M, Navajas, D, Fredberg, JJ. 2001. Scaling the microrheology of living cells. *Phys Rev Lett* 87:148102.

29. Alcaraz, J, Buscemi, L, Grabulosa, M, Trepat, X, Fabry, B, Farre, R et al. 2003. Microrheology of human lung epithelial cells measured by atomic force microscopy. *Biophys J* 84:2071–2079.

30. Mijailovich, SM, Kojic, M, Zivkovic, M, Fabry, B, Fredberg, JJ. 2002. A finite element model of cell deformation during magnetic bead twisting. *J Appl Physiol* 93:1429–1436.

31. Trepat, X, Deng, L, An, SS, Navajas, D, Tschumperlin, DJ, Gerthoffer, WT et al. 2007. Universal physical responses to stretch in the living cell. *Nature* 447:592–595.

32. Derec, C, Ducouret, G, Ajdari, A, Lequeux, F. 2003. Aging and nonlinear rheology in suspensions of polyethylene oxide-protected silica particles. *Phys Rev E Stat Nonlin Soft Matter Phys* 67:061403.

33. Mason, TG, Weitz, DA. 1995. Optical measurements of frequency-dependent linear viscoelastic moduli of complex fluids. *Phys Rev Lett* 74:1250–1253.

34. Haberstroh, KM, Kaefer, M, DePaola, N, Frommer, SA, Bizios, R. 2002. A novel in-vitro system for the simultaneous exposure of bladder smooth muscle cells to mechanical strain and sustained hydrostatic pressure. *J Biomech Eng* 124:208–213.

35. Borer, JG, Park, JM, Atala, A, Nguyen, HT, Adam, RM, Retik, AB et al. 1999. Heparin-binding EGF-like growth factor expression increases selectively in bladder smooth muscle in response to lower urinary tract obstruction. *Lab Invest* 79:1335–1345.

36. Capolicchio, G, Aitken, KJ, Gu, JX, Reddy, P, Bagli, DJ. 2001. Extracellular matrix gene responses in a novel ex vivo model of bladder stretch injury. *J Urol* 165:2235–2240.

37. Fry, CH. 2004. Experimental models to study the physiology, pathophysiology, and pharmacology of the lower urinary tract. *J Pharmacol Toxicol Methods* 49:201–210.

38. Langendorff, O. 1895. Examination of a surviving mammalian heart. *Pflugers Arch* 61:291–332.

39. Hutcheson, JC, Stein, R, Chacko, S, Carr, M, Canning, DA, Zderic, SA. 2004. Murine in vitro whole bladder model: A method for assessing phenotypic responses to pharmacologic stimuli and hypoxia. *Neurourol Urodyn* 23:349–354.

40. Drake, MJ, Harvey, IJ, Gillespie, JI. 2003. Autonomous activity in the isolated guinea pig bladder. *Exp Physiol* 88:19–30.
41. Drake, MJ, Hedlund, P, Harvey, IJ, Pandita, RK, Andersson, KE, Gillespie, JI. 2003. Partial outlet obstruction enhances modular autonomous activity in the isolated rat bladder. *J Urol* 170:276–279.
42. Zhang, EY, Stein, R, Chang, S, Zheng, Y, Zderic, SA, Wein, AJ et al. 2004. Smooth muscle hypertrophy following partial bladder outlet obstruction is associated with overexpression of non-muscle caldesmon. *Am J Pathol* 164:601–612.
43. Austin, JC, Chacko, SK, DiSanto, M, Canning, DA, Zderic, SA. 2004. A male murine model of partial bladder outlet obstruction reveals changes in detrusor morphology, contractility and myosin isoform expression. *J Urol* 172:1524–1528.
44. Gloeckner, DC, Sacks, MS, Fraser, MO, Somogyi, GT, de Groat, WC, Chancellor, MB. 2002. Passive biaxial mechanical properties of the rat bladder wall after spinal cord injury. *J Urol* 167:2247–2252.
45. Nagatomi, J, Gloeckner, DC, Chancellor, MB, DeGroat, WC, Sacks, MS. 2004. Changes in the biaxial viscoelastic response of the urinary bladder following spinal cord injury. *Ann Biomed Eng* 32:1409–1419.
46. Danzer, E, Schwarz, U, Wehrli, S, Radu, A, Adzick, NS, Flake, AW. 2005. Retinoic acid induced myelomeningocele in fetal rats: Characterization by histopathological analysis and magnetic resonance imaging. *Exp Neurol* 194:467–475.
47. Danzer, E, Kiddoo, DA, Redden, RA, Robinson, L, Radu, A, Zderic, SA et al. 2007. Structural and functional characterization of bladder smooth muscle in fetal rats with retinoic acid-induced myelomeningocele. *Am J Physiol Renal Physiol* 292:F197–F206.
48. Tugay, M, Tugay, S, Etus, V, Yazir, Y, Utkan, T. 2008. Alterations in the mechanical properties of bladder smooth muscle in hydrocephalus rat model. *J Pediatr Surg* 43:713–717.
49. Chai, TC, Zhang, CO, Shoenfelt, JL, Johnson, HW, Jr., Warren, JW, Keay, S. 2000. Bladder stretch alters urinary heparin-binding epidermal growth factor and antiproliferative factor in patients with interstitial cystitis. *J Urol* 163:1440–1444.
50. Park, JM, Borer, JG, Freeman, MR, Peters, CA. 1998. Stretch activates heparin-binding EGF-like growth factor expression in bladder smooth muscle cells. *Am J Physiol* 275:C1247–C1254.
51. Nguyen, HT, Adam, RM, Bride, SH, Park, JM, Peters, CA, Freeman, MR. 2000. Cyclic stretch activates p38 SAPK2-, ErbB2-, and AT1-dependent signaling in bladder smooth muscle cells. *Am J Physiol Cell Physiol* 279:C1155–C1167.
52. Adam, RM, Eaton, SH, Estrada, C, Nimgaonkar, A, Shih, SC, Smith, LE et al. 2004. Mechanical stretch is a highly selective regulator of gene expression in human bladder smooth muscle cells. *Physiol Genomics* 20:36–44.
53. Yang, R, Amir, J, Liu, H, Chaqour, B. 2008. Mechanical strain activates a program of genes functionally involved in paracrine signaling of angiogenesis. *Physiol Genomics* 36:1–14.
54. Orsola, A, Adam, RM, Peters, CA, Freeman, MR. 2002. The decision to undergo DNA or protein synthesis is determined by the degree of mechanical deformation in human bladder muscle cells. *Urology* 59:779–783.
55. Adam, RM, Roth, JA, Cheng, HL, Rice, DC, Khoury, J, Bauer, SB et al. 2003. Signaling through PI3K/Akt mediates stretch and PDGF-BB-dependent DNA synthesis in bladder smooth muscle cells. *J Urol* 169:2388–2393.
56. Galvin, DJ, Watson, RW, Gillespie, JI, Brady, H, Fitzpatrick, JM. 2002. Mechanical stretch regulates cell survival in human bladder smooth muscle cells in vitro. *Am J Physiol Renal Physiol* 283:F1192–F1199.
57. Estrada, CR, Adam, RM, Eaton, SH, Bagli, DJ, Freeman, MR. 2006. Inhibition of EGFR signaling abrogates smooth muscle proliferation resulting from sustained distension of the urinary bladder. *Lab Invest* 86:1293–1302.
58. Halachmi, S, Aitken, KJ, Szybowska, M, Sabha, N, Dessouki, S, Lorenzo, A et al. 2006. Role of signal transducer and activator of transcription 3 (STAT3) in stretch injury to bladder smooth muscle cells. *Cell Tissue Res* 326:149–158.
59. Aitken, KJ, Block, G, Lorenzo, A, Herz, D, Sabha, N, Dessouki, O et al. 2006. Mechanotransduction of extracellular signal-regulated kinases 1 and 2 mitogen-activated protein kinase activity in smooth muscle is dependent on the extracellular matrix and regulated by matrix metalloproteinases. *Am J Pathol* 169:459–470.
60. Kushida, N, Kabuyama, Y, Yamaguchi, O, Homma, Y. 2001. Essential role for extracellular Ca(2+) in JNK activation by mechanical stretch in bladder smooth muscle cells. *Am J Physiol Cell Physiol* 281:C1165–C1172.

61. Higashiyama, S, Abraham, JA, Miller, J, Fiddes, JC, Klagsbrun, M. 1991. A heparin-binding growth factor secreted by macrophage-like cells that is related to EGF. *Science* 251:936–939.
62. Elenius, K, Paul, S, Allison, G, Sun, J, Klagsbrun, M. 1997. Activation of HER4 by heparin-binding EGF-like growth factor stimulates chemotaxis but not proliferation. *EMBO J* 16:1268–1278.
63. Fujita, O, Asanuma, M, Yokoyama, T, Miyazaki, I, Ogawa, N, Kumon, H. 2006. Involvement of STAT3 in bladder smooth muscle hypertrophy following bladder outlet obstruction. *Acta Med Okayama* 60:299–309.
64. Park, JM, Adam, RM, Peters, CA, Guthrie, PD, Sun, Z, Klagsbrun, M et al. 1999. AP-1 mediates stretch-induced expression of HB-EGF in bladder smooth muscle cells. *Am J Physiol* 277:C294–C301.
65. Persson, K, Dean-Mckinney, T, Steers, WD, Tuttle, JB. 2001. Activation of the transcription factors nuclear factor-kappaB and activator protein-1 in bladder smooth muscle exposed to outlet obstruction and mechanical stretching. *J Urol* 165:633–639.
66. Ramachandran, A, Ranpura, SA, Gong, EM, Mulone M., Cannon Jr., GM, Adam, RM. An Akt- and Fra-1-dependent pathway mediates PDGF-induced expression of thrombomodulin, a novel regulator of smooth muscle cell migration. *Am J Pathol* 177:119–131.
67. Damaser, MS, Kim, KB, Longhurst, PA, Wein, AJ, Levin, RM. 1997. Calcium regulation of urinary bladder function. *J Urol* 157:732–738.
68. Masters, JG, Neal, DE, Gillespie, JI. 1999. The contribution of intracellular Ca2+ release to contraction in human bladder smooth muscle. *Br J Pharmacol* 127:996–1002.
69. Visser, AJ, van Mastrigt, R. 2000. The role of intracellular and extracellular calcium in mechanical and intracellular electrical activity of human urinary bladder smooth muscle. *Urol Res* 28:260–268.
70. Rivera, L, Brading, AF. 2006. The role of Ca2+ influx and intracellular Ca2+ release in the muscarinic-mediated contraction of mammalian urinary bladder smooth muscle. *BJU Int* 98:868–875.
71. Somlyo, AP, Somlyo, AV. 2003. Ca2+ sensitivity of smooth muscle and nonmuscle myosin II: Modulated by G proteins, kinases, and myosin phosphatase. *Physiol Rev* 83:1325–1358.
72. Stanton, MC, Clement, M, Macarak, EJ, Zderic, SA, Moreland, RS. 2003. Partial bladder outlet obstruction alters Ca2+ sensitivity of force, but not of MLC phosphorylation, in bladder smooth muscle. *Am J Physiol Renal Physiol* 285:F703–F710.
73. Takahashi, R, Nishimura, J, Hirano, K, Seki, N, Naito, S, Kanaide, H. 2004. Ca2+ sensitization in contraction of human bladder smooth muscle. *J Urol* 172:748–752.
74. Petkov, GV, Bonev, AD, Heppner, TJ, Brenner, R, Aldrich, RW, Nelson, MT. 2001. Beta1-subunit of the Ca2+-activated K+ channel regulates contractile activity of mouse urinary bladder smooth muscle. *J Physiol* 537:443–452.
75. Tanaka, Y, Okamoto, T, Imai, T, Yamamoto, Y, Horinouchi, T, Tanaka, H et al. 2003. Bk(Ca) channel activity enhances with muscle stretch in guinea-pig urinary bladder smooth muscle. *Res Commun Mol Pathol Pharmacol* 113–114:247–252.
76. Thorneloe, KS, Knorn, AM, Doetsch, PE, Lashinger, ES, Liu, AX, Bond, CT et al. 2008. Small-conductance, Ca(2+)-activated K+ channel 2 is the key functional component of SK channels in mouse urinary bladder. *Am J Physiol Regul Integr Comp Physiol* 294:R1737–R1743.
77. Nozaki, K, Tomizawa, K, Yokoyama, T, Kumon, H, Matsui, H. 2003. Calcineurin mediates bladder smooth muscle hypertrophy after bladder outlet obstruction. *J Urol* 170:2077–2081.
78. Clement, MR, Delaney, DP, Austin, JC, Sliwoski, J, Hii, GC, Canning, DA et al. 2006. Activation of the calcineurin pathway is associated with detrusor decompensation: A potential therapeutic target. *J Urol* 176:1225–1229.
79. Layne, JJ, Werner, ME, Hill-Eubanks, DC, Nelson, MT. 2008. NFATc3 regulates BK channel function in murine urinary bladder smooth muscle. *Am J Physiol Cell Physiol* 295:C611–C623.
80. Rohrmann, D, Levin, RM, Duckett, JW, Zderic, SA. 1996. The decompensated detrusor I: The effects of bladder outlet obstruction on the use of intracellular calcium stores. *J Urol* 156:578–581.
81. Zderic, SA, Rohrmann, D, Gong, C, Snyder, HM, Duckett, JW, Wein, AJ et al. 1996. The decompensated detrusor II: Evidence for loss of sarcoplasmic reticulum function after bladder outlet obstruction in the rabbit. *J Urol* 156:587–592.
82. Stein, R, Hutcheson, JC, Gong, C, Canning, DA, Carr, MC, Zderic, SA. 2001. The decompensated detrusor IV: Experimental bladder outlet obstruction and its functional correlation to the expression of the ryanodine and voltage operated calcium channels. *J Urol* 165:2284–2288.
83. Wu, C, Thiruchelvam, N, Sui, G, Woolf, AS, Cuckow, P, Fry, CH. 2007. Ca2+ regulation in detrusor smooth muscle from ovine fetal bladder after in utero bladder outflow obstruction. *J Urol* 177:776–780.
84. Sui, G, Fry, CH, Malone-Lee, J, Wu, C. 2009. Aberrant Ca2+ oscillations in smooth muscle cells from overactive human bladders. *Cell Calcium* 45:456–464.

85. Lassmann, J, Sliwoski, J, Chang, A, Canning, DA, Zderic, SA. 2008. Deletion of one SERCA2 allele confers protection against bladder wall hypertrophy in a murine model of partial bladder outlet obstruction. *Am J Physiol Regul Integr Comp Physiol* 294:R58–R65.

86. Wibberley, A, Chen, Z, Hu, E, Hieble, JP, Westfall, TD. 2003. Expression and functional role of Rho-kinase in rat urinary bladder smooth muscle. *Br J Pharmacol* 138:757–766.

87. Bing, W, Chang, S, Hypolite, JA, DiSanto, ME, Zderic, SA, Rolf, L et al. 2003. Obstruction-induced changes in urinary bladder smooth muscle contractility: A role for Rho kinase. *Am J Physiol Renal Physiol* 285:F990–F997.

88. Poley, RN, Dosier, CR, Speich, JE, Miner, AS, Ratz, PH. 2008. Stimulated calcium entry and constitutive RhoA kinase activity cause stretch-induced detrusor contraction. *Eur J Pharmacol* 599:137–145.

89. Chiquet, M, Tunc-Civelek, V, Sarasa-Renedo, A. 2007. Gene regulation by mechanotransduction in fibroblasts. *Appl Physiol Nutr Metab* 32:967–973.

90. Guven, A, Onal, B, Kalorin, C, Whitbeck, C, Chichester, P, Kogan, B et al. 2007. Long term partial bladder outlet obstruction induced contractile dysfunction in male rabbits: A role for Rho-kinase. *Neurourol Urodyn* 26:1043–1049.

91. Takahashi, N, Shiomi, H, Kushida, N, Liu, F, Ishibashi, K, Yanagida, T et al. 2009. Obstruction alters muscarinic receptor-coupled RhoA/Rho-kinase pathway in the urinary bladder of the rat. *Neurourol Urodyn* 28:257–262.

92. Bagli, DJ, Joyner, BD, Mahoney, SR, McCulloch, L. 1999. The hyaluronic acid receptor RHAMM is induced by stretch injury of rat bladder in vivo and influences smooth muscle cell contraction in vitro [corrected]. *J Urol* 162:832–840.

93. Herz, DB, Aitken, K, Bagli, DJ. 2003. Collagen directly stimulates bladder smooth muscle cell growth in vitro: Regulation by extracellular regulated mitogen activated protein kinase. *J Urol* 170:2072–2076.

94. Upadhyay, J, Aitken, KJ, Damdar, C, Bolduc, S, Bagli, DJ. 2003. Integrins expressed with bladder extracellular matrix after stretch injury in vivo mediate bladder smooth muscle cell growth in vitro. *J Urol* 169:750–755.

95. Coplen, DE, Macarak, EJ, Howard, PS. 2003. Matrix synthesis by bladder smooth muscle cells is modulated by stretch frequency. *In Vitro Cell Dev Biol Anim* 39:157–162.

96. Hubschmid, U, Leong-Morgenthaler, PM, Basset-Dardare, A, Ruault, S, Frey, P. 2005. In vitro growth of human urinary tract smooth muscle cells on laminin and collagen type I-coated membranes under static and dynamic conditions. *Tissue Eng* 11:161–171.

97. Wognum, S, Schmidt, DE, Sacks, MS. 2009. On the mechanical role of de novo synthesized elastin in the urinary bladder wall. *J Biomech Eng* 131:101018.

98. Nagatomi, J, Toosi, KK, Chancellor, MB, Sacks, MS. 2008. Contribution of the extracellular matrix to the viscoelastic behavior of the urinary bladder wall. *Biomech Model Mechanobiol* 7:395–404.

99. Tamura, I, Rosenbloom, J, Macarak, E, Chaqour, B. 2001. Regulation of Cyr61 gene expression by mechanical stretch through multiple signaling pathways. *Am J Physiol Cell Physiol* 281:C1524–C1532.

100. Chaqour, B, Yang, R, Sha, Q. 2006. Mechanical stretch modulates the promoter activity of the profibrotic factor CCN2 through increased actin polymerization and NF-kappaB activation. *J Biol Chem* 281:20608–20622.

101. Chaqour, B, Whitbeck, C, Han, JS, Macarak, E, Horan, P, Chichester, P et al. 2002. Cyr61 and CTGF are molecular markers of bladder wall remodeling after outlet obstruction. *Am J Physiol Endocrinol Metab* 283:E765–E774.

102. Chen, CC, Chen, N, Lau, LF. 2001. The angiogenic factors Cyr61 and connective tissue growth factor induce adhesive signaling in primary human skin fibroblasts. *J Biol Chem* 276:10443–10452.

103. Chen, CC, Mo, FE, Lau, LF. 2001. The angiogenic factor Cyr61 activates a genetic program for wound healing in human skin fibroblasts. *J Biol Chem* 276:47329–47337.

104. Quan, T, He, T, Shao, Y, Lin, L, Kang, S, Voorhees, JJ et al. 2006. Elevated cysteine-rich 61 mediates aberrant collagen homeostasis in chronologically aged and photoaged human skin. *Am J Pathol* 169:482–490.

105. Brigstock, DR. 2002. Regulation of angiogenesis and endothelial cell function by connective tissue growth factor (CTGF) and cysteine-rich 61 (CYR61). *Angiogenesis* 5:153–165.

106. Ghafar, MA, Anastasiadis, AG, Olsson, LE, Chichester, P, Kaplan, SA, Buttyan, R et al. 2002. Hypoxia and an angiogenic response in the partially obstructed rat bladder. *Lab Invest* 82:903–909.

107. Sabha, N, Aitken, K, Lorenzo, AJ, Szybowska, M, Jairath, A, Bagli, DJ. 2006. Matrix metalloproteinase-7 and epidermal growth factor receptor mediate hypoxia-induced extracellular signal-regulated kinase 1/2 mitogen-activated protein kinase activation and subsequent proliferation in bladder smooth muscle cells. *In Vitro Cell Dev Biol Anim* 42:124–133.

108. Aitken, KJ, Tolg, C, Panchal, T, Leslie, B, Yu, J, Elkelini, M et al. 2010. Mammalian target of rapamy-cin (mTOR) Induces proliferation and de-differentiation responses to three coordinate pathophysiologic stimuli (mechanical strain, hypoxia, and extracellular matrix remodeling) in rat bladder smooth muscle. *Am J Pathol* 176:304–319.
109. Balestreire, EM, Apodaca, G. 2007. Apical epidermal growth factor receptor signaling: Regulation of stretch-dependent exocytosis in bladder umbrella cells. *Mol Biol Cell* 18:1312–1323.
110. Yu, W, Khandelwal, P, Apodaca, G. 2009. Distinct apical and basolateral membrane requirements for stretch-induced membrane traffic at the apical surface of bladder umbrella cells. *Mol Biol Cell* 20:282–295.
111. Araki, I, Du, S, Kobayashi, H, Sawada, N, Mochizuki, T, Zakoji, H et al. 2008. Roles of mechanosensi-tive ion channels in bladder sensory transduction and overactive bladder. *Int J Urol* 15:681–687.
112. Lazzeri, M. 2006. The physiological function of the urothelium–more than a simple barrier. *Urol Int* 76:289–295.
113. Ikeda, Y, Kanai, A. 2008. Urotheliogenic modulation of intrinsic activity in spinal cord-transected rat bladders: Role of mucosal muscarinic receptors. *Am J Physiol Renal Physiol* 295:F454–F461.
114. Araki, I, Du, S, Kamiyama, M, Mikami, Y, Matsushita, K, Komuro, M et al. 2004. Overexpression of epithelial sodium channels in epithelium of human urinary bladder with outlet obstruction. *Urology* 64:1255–1260.
115. Du, S, Araki, I, Mikami, Y, Zakoji, H, Beppu, M, Yoshiyama, M et al. 2007. Amiloride-sensitive ion channels in urinary bladder epithelium involved in mechanosensory transduction by modulating stretch-evoked adenosine triphosphate release. *Urology* 69:590–595.
116. Gevaert, T, Vriens, J, Segal, A, Everaerts, W, Roskams, T, Talavera, K et al. 2007. Deletion of the transient receptor potential cation channel TRPV4 impairs murine bladder voiding. *J Clin Invest* 117:3453–3462.
117. Sellers, DJ, McKay, N. 2007. Developments in the pharmacotherapy of the overactive bladder. *Curr Opin Urol* 17:223–230.
118. Hegde, SS. 2006. Muscarinic receptors in the bladder: From basic research to therapeutics. *Br J Pharmacol* 147 Suppl 2:S80–S87.
119. Biers, SM, Reynard, JM, Doore, T, Brading, AF. 2006. The functional effects of a c-kit tyrosine inhibitor on guinea-pig and human detrusor. *BJU Int* 97:612–616.
120. Christ, GJ, Andersson, KE. 2007. Rho-kinase and effects of Rho-kinase inhibition on the lower urinary tract. *Neurourol Urodyn* 26:948–954.
121. Uehata, M, Ishizaki, T, Satoh, H, Ono, T, Kawahara, T, Morishita, T et al. 1997. Calcium sensitization of smooth muscle mediated by a Rho-associated protein kinase in hypertension. *Nature* 389:990–994.
122. Chang, S, Hypolite, JA, DiSanto, ME, Changolkar, A, Wein, AJ, Chacko, S. 2006. Increased basal phos-phorylation of detrusor smooth muscle myosin in alloxan-induced diabetic rabbit is mediated by upregu-lation of Rho-kinase beta and CPI-17. *Am J Physiol Renal Physiol* 290:F650–F656.
123. Masumoto, A, Mohri, M, Shimokawa, H, Urakami, L, Usui, M, Takeshita, A. 2002. Suppression of coro-nary artery spasm by the Rho-kinase inhibitor fasudil in patients with vasospastic angina. *Circulation* 105:1545–1547.
124. Shimokawa, H, Hiramori, K, Iinuma, H, Hosoda, S, Kishida, H, Osada, H et al. 2002. Anti-anginal effect of fasudil, a Rho-kinase inhibitor, in patients with stable effort angina: A multicenter study. *J Cardiovasc Pharmacol* 40:751–761.
125. Wood, D, Southgate, J. 2008. Current status of tissue engineering in urology. *Curr Opin Urol* 18:564–569.
126. Kim, BS, Nikolovski, J, Bonadio, J, Mooney, DJ. 1999. Cyclic mechanical strain regulates the develop-ment of engineered smooth muscle tissue. *Nat Biotechnol* 17:979–983.
127. Korossis, S, Bolland, F, Ingham, E, Fisher, J, Kearney, J, Southgate, J. 2006. Review: Tissue engineering of the urinary bladder: Considering structure-function relationships and the role of mechanotransduction. *Tissue Eng* 12:635–644.
128. Farhat, WA, Yeger, H. 2008. Does mechanical stimulation have any role in urinary bladder tissue engineering? *World J Urol* 26:301–305.
129. Baskin, L, Meaney, D, Landsman, A, Zderic, SA, Macarak, E. 1994. Bovine bladder compliance increases with normal fetal development. *J Urol* 152:692–695.
130. Gearhart, JP, Mathews, R, 2007. Exstrophy-epispadias complex. In *Campbell-Walsh Urology*, eds. AJ Wein, LR Kavoussi, AC Novick, AW Partin, CA Peters, Chap. 123, p. 3497. Saunders-Elsevier, Philadelphia, PA.

16 Mechanobiology of Bladder Urothelial Cells

Shawn Olsen, Kevin Champaigne, and Jiro Nagatomi

CONTENTS

16.1 OVERVIEW

The lower urinary tract, which consists of the bladder, urethra, and urethral sphincters, is responsible for the storage and timely elimination of urine. According to the definitions of the International Continence Society, lower urinary tract symptoms (LUTS) can be categorized as storage, voiding, or post-micturition symptoms [1]. LUTS include frequency, urgency, incontinence, intermittency, and the feeling of incomplete emptying, among others. The EpiLUTS study [2] estimated the prevalence of LUTS, occurring at least "often" in individuals aged 40 and older, to be 47.9% for men and 52.5% for women in the United States, the United Kingdom, and Sweden. These symptoms are considered bothersome and affect the quality of life of the patient as well as the patient's family [3]. The proper functioning of the lower urinary tract is dependent upon complex, coordinated interactions between the structural, muscular, and nervous system components [4]. Unlike most other visceral organs, the bladder and urethra are under both involuntary and voluntary control, providing for both reflexive responses and the conscious control of micturition [5]. Mechanical parameters such as bladder wall tension or intravesical pressure due to bladder filling or shear stress due to urine flow through the urethra, are sensed at the cellular level and contribute to both conscious sensation and reflex responses. Malfunctions in these sensing responsibilities are therefore likely to cause lower urinary tract dysfunction [6,7]. However, the mechanisms by which sensory transduction is accomplished at the cellular level within the lower urinary tract are not well understood.

The urothelium is the epithelial lining of the urinary bladder, and recent discoveries have suggested that in addition to providing a barrier to urine, the urothelium actively participates in sensory functions related to thermal, chemical, and mechanical stimuli and releases chemical signals in response [8–11]. Ion channels have been proposed as part of a potential transduction mechanism for this function of the urothelium [12]. Thus far, despite the identification of numerous proteins as potentially mechanosensitive, the mechanism by which these proteins transduce mechanical stimuli into biochemical signals that produce a cellular response is still unknown [13]. In this chapter, we will first discuss the role that urothelial cells play in normal bladder physiology. We will then review the current theories of urothelial cell mechanotransduction involving mechanically triggered adenosine triphosphate (ATP) release and activation of membrane-bound ion channels, which may contribute to the bladder's ability to sense fullness and communicate with the nervous system. By determining the specific mechanisms by which urothelial cells participate in bladder physiology and pathophysiology, it is believed that further strides can be made toward the development of effective treatments for LUTS.

16.2   UROTHELIAL ANATOMY AND PHYSIOLOGY

16.2.1   STRUCTURE AND FUNCTION OF THE UROTHELIUM

The urothelium is comprised of at least three layers of cells (Figure 16.1): the basal cell layer attached to the basement membrane, an intermediate cell layer, and a superficial (apical) layer composed of large, hexagonally shaped cells, also called umbrella cells [14]. Basal cells are the smallest cells (~10 μm in diameter) that are in contact with the underlying connective tissue and capillaries. Intermediate cells (10–25 μm in diameter) are pyriform in shape and sit on the basal layer to form a cross section that can be from one to several cell layers thick [9]. The large umbrella cells compose the superficial layer and normally range from 25 to 250 μm in diameter [9]. The umbrella cells contain tight junctions that reduce the passage of ions and other solutes from urine to the bloodstream

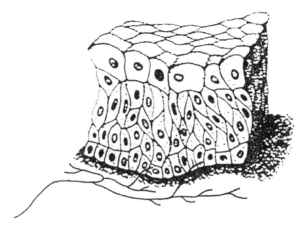

FIGURE 16.1   Illustrated cross section of the bladder urothelium. The basal cells are the small cells in contact with afferent and efferent nerves. As shown, the umbrella cells in contact with the urine are much larger than the rest of the urothelium in order to provide minimum points of entry for ions to pass through the junction. (From de Groat, W.C., *Urology*, 64(6A), 7, 2004. With permission.)

during the storage phase [14]. Upon bladder filling, urothelial cells also increase apical surface area to support an increased bladder volume [9].

16.2.2 NEUROANATOMY OF THE BLADDER

Unlike most other visceral organ systems, the lower urinary tract is under both voluntary and involuntary control (Figure 16.2). The normal adult can control the micturition reflex, while infants display reflexive control of bladder voiding [6]. As such, the integration of the sensory and motor control components of both the autonomic and somatic nervous systems yields additional complexity and opportunities for bladder nerves to interact with urothelial cells for added mechanosensing ability through multiple efferent and afferent pathways.

16.2.2.1 Efferent Pathways

The autonomic nervous system is divided into the parasympathetic and sympathetic nervous systems, which function in a complementary manner. Parasympathetic control of lower urinary tract function originates in the sacral spinal cord in the S2–S4 region [5,15]. Activation of the parasympathetic efferent nerves simultaneously causes bladder detrusor contraction (through the activation of M3 receptors by the neurotransmitter acetylcholine) and internal urethral smooth muscle relaxation (through the release of nitric oxide [NO]), thereby allowing voiding to occur. Sympathetic nerves related to the control of the lower urinary tract, which promote the storage of urine, emerge from the lower thoracic and upper lumbar spinal cord in the T10–L2 region. These postganglionic nerves release noradrenaline, which inhibits contraction of the detrusor through β3 receptors and induces contraction of the internal urethral sphincter through α1 receptors [5]. Somatic motor nerves also originate from the S2–S4 region and release acetylcholine, activating nicotinic receptors and inducing contraction of the striated muscle fibers of the external urethral sphincter.

16.2.2.2 Afferent Pathways

Sensations of bladder fullness or irritation are carried to the central nervous system by afferent nerve fibers [5]. The highest density of afferent axons appears in the mucosa, mostly within a few micrometers of the urothelium, with numerous afferent axons additionally penetrating some

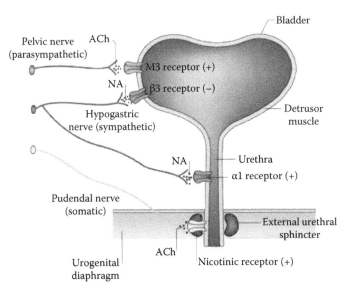

FIGURE 16.2 Efferent pathways of the lower urinary tract. (Adapted from Fowler, C.J. et al., *Nat. Rev. Neurosci.*, 9(6), 453, 2008. With permission.)

distance into the urothelium [16]. The distribution of afferent axons throughout the bladder is most dense in the trigone and neck regions, with progressively fewer afferent axons appearing toward the bladder dome [16].

Bladder afferent nerves have been shown to express multiple receptor types that are activated by chemical signaling mechanisms as well as various mechanical, chemical, or thermal stimuli that either enhance or inhibit afferent activity upon activation [17]. The expression of such a wide variety of receptors illustrates that bladder sensations and reflexes are likely modulated by a complex combination of simultaneous signals that are integrated to provide overall sensory functionality. Possibly due to the number of signal sources that are integrated to produce multiple modes of bladder sensation, the task of identifying and fully characterizing the molecular mechanisms responsible for sensing mechanical stimuli associated bladder filling has been elusive. This yet unidentified mechanotransduction mechanism may be inherent in the afferent axons themselves, possibly due to a mechanical coupling of the axon to the extracellular matrix or adjacent cells. Alternatively, the axons could be responsive to communication from nearby cells such as detrusor smooth muscle cells, myofibroblasts, or urothelial cells that actually possess the mechanosensitive mechanism.

16.2.3 PHYSIOLOGY OF BLADDER FILLING

The bladder has two distinct functions for normal urological activity: urine storage and voiding. During storage, urine flows from the kidney, passes through the ureters, and is collected into the bladder; the urethra is closed and the bladder smooth muscle is relaxed, which allows for the bladder pressure to remain low (<10 cm H_2O) over a large range of bladder volumes [18]. When the bladder fills, the urothelium becomes thinner as the basal membrane begins to stretch. The underlying cells that were once multilayered begin to be pushed laterally into a monolayer to accommodate the increased urine volume. As a result, the umbrella cells undergo a large change in shape from cuboidal to a flat, squamous morphology [9]. The conventional view is that the afferent nerves of the parasympathetic nervous system are activated via mechanosensitive Aδ-fibers when the bladder is stretched due to filling [19]. These signals are interpreted by the central nervous system as indicative of the fullness of the bladder and thus promote the desire to void.

Until recently, it was considered that the storage and voiding functions of the bladder were controlled strictly by the nervous system. Recent research, however, has implicated the urothelium as a possible component of bladder sensory function [20–23]. Multiple studies suggest that the urothelium may communicate bladder fullness by transmitting signals via a paracrine signaling pathway to the underlying nervous system by means of ATP release [9,11]. These findings that the urothelium releases signaling molecules in response to bladder filling and noxious stimuli such as mechanical damage or chemical irritants have replaced the previous perception of the urothelium as merely a passive barrier.

16.3 BLADDER MECHANOSENSING EVENTS MEDIATED BY UROTHELIAL CELLS

For the bladder to function normally, sensory mechanisms must exist within the lower urinary tract that are responsible for the transduction of mechanical stimuli (such as hydrostatic pressure, stretch, tension, and/or shear stress) into biochemical or bioelectrical signals. Two main theories exist regarding the generation of afferent signals via the mechanotransduction of stimuli associated with bladder filling: (1) the mechanosensory function is provided directly by the afferent nerves, or (2) other cell types transduce the mechanical stimuli into chemical signals that are received by afferent nerves, which then generate appropriate signals to the central nervous system [24]. Output signals due to bladder filling may be transmitted via neurotransmitter molecules and received as

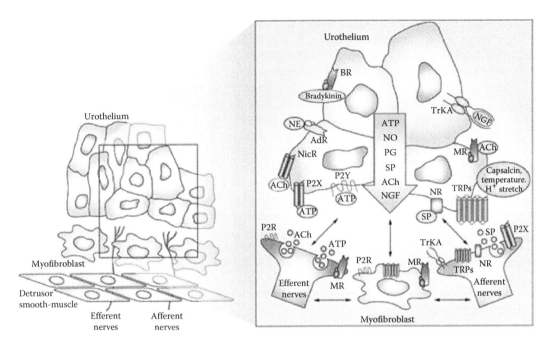

FIGURE 16.3 Possible signaling pathways between bladder afferent and efferent nerves, urothelial cells, and myofibroblasts. *Abbreviations:* ACh, acetylcholine; AdR, adrenergic receptor; BR, bradykinin receptor; H+, proton; MR, muscarinic receptor; NE, norepinephrine; NGF, nerve growth factor; NR, neurokinin receptor; NicR, nicotinic receptor; NO, nitric oxide; P2R, purinergic 2 receptor unidentified subtype; P2X and P2Y, purinergic receptors; PG, prostaglandin; SP, substance P; Trk-A, receptor tyrosine kinase A; TRPs, transient receptor potential channels. (From Birder, L.A. and de Groat, W.C., *Nat. Clin. Pract. Urol.*, 4(1), 46, January 2007. With permission.)

input signals by nearby components to cause downstream effects such as increased bladder surface area and afferent action potential generation. An increasingly accurate model of lower urinary tract function can be obtained through the understanding of such parameters as the sources of a particular biochemical transmitter, the factors or stimuli that cause its release, potential targets such as receptor molecules, modulators of the sensitivity of the receptors, and the downstream stimulatory or inhibitory effects on the targets.

Birder and de Groat [7,14] described a number of potential autocrine and paracrine signaling interactions between urothelial cells, myofibroblasts, smooth muscle cells, afferent nerves, and efferent nerves involved in bladder function, which were described as being part of a "uroepithelial-associated sensory web" (Figure 16.3). Each communication pathway is likely modulated by multiple factors, which may increase or decrease neurotransmitter release, increase or decrease receptor sensitivity, or modify neurotransmitter degradation rate. This chapter will focus on three widely observed responses by urothelial cells during bladder filling: (1) ATP is released by urothelial cells, which presumably activates purinergic receptors on nearby cells; (2) the internal concentration of calcium in urothelial cells increases; and (3) the surface area of the umbrella cells of the urothelium increases [25]. The following sections review the current literature on some of these bladder responses to mechanical stimuli.

16.3.1 ATP is Released by Urothelial Cells in Response to Mechanical Stimuli

In 1972, Burnstock first proposed the existence of nerves activated by ATP functioning as a neurotransmitter in various organs, including the urinary bladder [26]. As the main energy source for most cellular functions, ATP is ubiquitously present in cells and actively participates in such crucial intracellular processes as ion transport, muscle contraction, and cytoskeletal organization [27]. However, in the extracellular domain, ATP possesses characteristics typically associated with signaling molecules such as being released in response to particular stimuli, interacting with receptors, and being rapidly inactivated by enzymatic reactions [28].

Ferguson et al. first demonstrated using an in vitro model that ATP is released from the urothelial cell layer following the application of hydrostatic pressure (3.5 cm H_2O) to the mucosal (or apical) surface of the bladder wall in rabbits [22]. The results of this seminal work provided the first evidence that the release of ATP may be part of a mechanosensory pathway to detect bladder filling. The experiments were conducted using a modified Ussing chamber, where epithelial tissue was mounted between two fluid-filled compartments. Isolation of the apical and basolateral sides of the tissue enabled measurement of the ion transport across the epithelium, and in this test configuration, enabled the measurement of the concentration of released signaling molecules. Hydrostatic pressure was applied differentially to the two chambers, causing the mounted urothelium tissue to be stretched toward the chamber experiencing the lower pressure, which produced the subsequent release of ATP. Membrane stretch–induced ATP release has since been elicited using similar techniques from urothelial tissues in pig and human [29], guinea pig [30], mouse [31,32], and rat [33]. In addition to stretch, ATP has been shown to be released from rat urothelium upon application of acid, capsaicin, or electrical field stimulation, all of which generate a larger ATP response than stretch [33].

16.3.2 Mechanisms of ATP Release

Although the release of ATP by urothelial cells in response to mechanical stimuli has been widely observed, the mechanism by which it occurs has yet to be elucidated. Potential mechanisms that have been proposed include vesicle exocytosis, connexin or pannexin hemichannels, or transporters such as members of the ATP-binding cassette (ABC) transporter family. For example, an in vitro study provided evidence that ATP was released by both the mucosal and serosal surfaces of isolated rabbit bladder tissue when stretched by the application of hydrostatic pressure to the apical surface of the umbrella cells [23]. This release of ATP was selectively inhibited through the application of a wide variety of substances, with differing results depending upon whether the drug was applied to the mucosal or serosal side. Inhibitory effects were obtained using secretion inhibitors, gap junction antagonists, or ABC transporter protein blockers, which, together with the heterogeneous response of the two sides of the tissue, suggest a multifaceted and polar sensory cascade. Further, ATP signaling pathways were shown to regulate both exocytosis and endocytosis rates in umbrella cells, thereby controlling apical membrane dynamics. Specifically, exposure of the urothelium to increased hydrostatic pressure favored exocytosis, which was shown to produce a net increase in plasma membrane surface area through fusion of discoidal/fusiform-shaped vesicles from the cytoplasm with the plasma membrane. This observation is consistent with the theory that the urothelium surface area increases during bladder filling through urothelial cell remodeling.

16.3.3 ATP Release via Connexin/Pannexin Hemichannels

Another potential mechanism for mechanically induced ATP release involves the transport of the molecule through connexin or pannexin hemichannels [34,35]. Connexins are transmembrane proteins that form gap junctions with adjacent cells and allow the passage of ions and small molecules of less than approximately 1 kDa directly between the cytoplasmic volumes of connected cells [27]. In addition to the traditional paired channels of a gap junction, the existence of nonjunctional connexin hemichannels on the surface of cells has been observed, where no corresponding channel on a neighboring cell is present [36]. Although normally closed, under certain circumstances including a low extracellular calcium concentration, these channels can open to provide a route for solutes including ATP to pass between the cytoplasm and the extracellular space [37]. Connexin hemichannels may thereby participate in intercellular signaling processes by providing similar functionality to traditional ion channels, but for larger molecules such as ATP.

Although multiple isoforms of connexin have been shown to be mechanosensitive [35], and multiple isoforms are known to be expressed in the rat bladder [38], the potential role of connexins in mechanosensory events within the lower urinary tract remains to be verified. Multiple studies, related both specifically to the urothelium and to other epithelial tissues, illustrate a potential mechanosensitive role for connexin hemichannels or gap junctions.

Results of an in vitro study demonstrated a decrease in stretch-evoked ATP release from rat urothelium upon the application of the gap junction antagonists anandamide and flufenamic acid [23]. Additionally, the expression of connexin itself has been shown to be controlled by one or more unidentified mechanical stimuli sensing mechanisms. In a rat model of bladder outlet obstruction, which was designed to cause increased bladder intraluminal pressure and stretch, in situ hybridization and immunofluorescence techniques showed that connexin 43 (Cx43) was upregulated in smooth muscle cells and Cx26 (but not Cx43) was upregulated in the urothelium. This increase in Cx43 and Cx26 was postulated to be involved in bladder tissue responses to increased wall stretch. In a separate study, ATP release from renal tubular epithelial cells, which was triggered by tubular flow or hypotonicity in wild-type mice, was shown to be absent in Cx30-deficient mice. This suggests that Cx30 proteins may form mechanosensitive connexin hemichannels in epithelial cells that directly control the efflux of ATP in response to mechanical stimuli [39].

Closely related to the connexin proteins are the recently identified pannexins. Members of this second family of hemichannel-forming proteins in mammalian cells have been shown to be widely expressed in many cell types, with often partially overlapping expression and functionality with connexins [40]. Unlike connexin hemichannels, which generally require extremely low concentrations of extracellular calcium to enable activation, pannexin hemichannels can be activated in more physiologically relevant conditions in vitro [41]. Pannexin hemichannels have been shown to be both mechanosensitive to membrane stretch and permeable to ATP [34], but specific roles for pannexins within the lower urinary tract have yet to be identified.

16.3.4 ATP-SENSITIVE PURINERGIC RECEPTORS IN THE BLADDER

For ATP signaling to be involved in mechanotransduction within the bladder as currently assumed, not only must the release of ATP occur, but appropriate receptors must also be incorporated in nearby cells to receive those signals. Multiple subtypes of membrane-bound receptors that are sensitive to ATP, known as purinergic receptors, have been discovered in numerous organs and tissues [28], including the urothelial, myofibroblast, smooth muscle, and nerve cells of the urinary bladder [7,42]. Burnstock and colleagues first proposed that the purinergic P2 receptor family (consisting of P2X and P2Y subfamilies), which are sensitive to both ATP and ADP, played a significant role in transmitting painful sensations [26]. P2X receptors are typically located in sensory neurons and activated by ATP through intrinsic ion channel gating, whereas the P2Y family of receptors are G-protein-coupled receptors that perform various biological functions based on G-protein coupling [28].

All currently known subtypes of the P2X channels ($P2X_1$–$P2X_7$) and three subtypes of the P2Y receptors ($P2Y_1$, $P2Y_2$, and $P2Y_4$) have been identified in the urothelium of the cat [7,43]. It appears, however, that $P2X_1$ and $P2X_3$ play the most important role in normal bladder function. Specifically, $P2X_1$ has been found to be expressed in the efferent nerves leading to the bladder, whereas $P2X_3$ is expressed in the afferent nerves leading away from the bladder, especially in the pelvic nerve near the urothelium [23,44]. Patch-clamp studies on cell bodies of primary afferent nerves have demonstrated two kinds of responses to ATP [45]. The first response is a rapidly rising inward current, which desensitizes within 100 ms, and the second is a slower rising inward current, which decreases very little after 1–2 s [45]. These desensitizing and immediate nondesensitizing characteristics of the afferent pathways are believed to be responsible for the activation of neuronal pathways leading

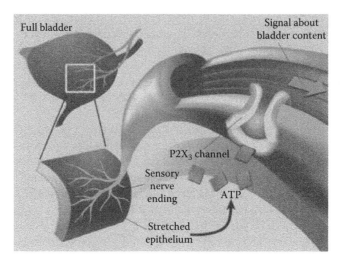

FIGURE 16.4 Proposed mechanism of P2X$_3$ receptor in bladder mechanosensation. ATP is released from the stretched urothelium and binds to the P2X$_3$ receptor. The bound ATP activates a neuronal pathway that signals bladder fullness. (From Rapp, D.E. et al., *Eur. Urol.*, 48(2), 303, 2005. With permission.)

from the bladder to the spinal cord. The involvement of P2X$_3$ receptors in bladder mechanosensation has been further confirmed in studies using P2X$_3$ knockout mice. Compared to wild-type mice, the knockout animals exhibited marked urinary bladder hyporeflexia with reduced voiding frequency and increased voiding volume [46], as well as an increased bladder capacity [47]. These findings suggest that P2X$_3$ receptors participate in the activation of afferent nerves due to bladder fullness in mice and that these receptors are most likely activated by ATP that is released from the bladder urothelium (Figure 16.4).

16.4 ROLE OF ION CHANNELS IN UROTHELIAL MECHANOTRANSDUCTION

16.4.1 MECHANOSENSITIVE ION CHANNELS

Mechanosensing by a cell can conceivably be achieved through a number of cellular proteins including the cytoskeleton [48], integrins [49], focal adhesion complexes [50,51], and a particular focus of this chapter, mechanosensitive ion channels (MSCs) [52]. Ion channels are generally considered to be either "open" or "closed," where the ionic conductance can switch rapidly between nearly zero and a consistent higher value [27] in a binary manner. MSCs are specifically defined as having the ability to modulate their gating (i.e., the probability of being in an open or closed state) in response to mechanical deformation of the cell membrane [53,54] and thereby provide mechanosensitivity. More generally, however, any change in the flow of ions such as Na$^+$, Ca^{2+}, and K$^+$ across the cell membrane (or across internal membrane structures within the cell) due to mechanical stimuli (e.g., fluid flow or membrane stretch) could cause an alteration of cell function, including the release of regulating molecules such as NO [55] and ATP [22].

16.4.2 ELECTROPHYSIOLOGY OF MECHANOSENSITIVE CHANNELS

The flow of calcium ions in particular is often studied due to its potential contribution to various signaling processes. In general, the ion current (in coulombs) carried out of the cell by a single calcium permeable channel when in the open state can be represented by the following equation:

$$i_{Ca} = \gamma_{Ca}(V_m - V_{C2}) \tag{16.1}$$

where

γ_{Ca} is the conductance for calcium (in siemens or Ω^{-1})

V_m is the membrane potential

V_{Ca} is the equilibrium potential for calcium (both in volts) [56]

Multiplying this current by the total number of the particular species of ion channel within the cell membrane (N) and the probability of each channel being open (p) yields the relationship for the overall cellular current given by

$$I_{Ca} = Np\gamma_{Ca}(V_m - V_{C2}) \tag{16.2}$$

The current (as well as the related calcium ion flux) due to a particular type of ion channel under particular circumstances is therefore proportional to the number of channels within the membrane, the open probability of those channels, the channel conductance, and the electrochemical potential difference [57]. Each of these factors is potentially modulated by mechanical stimuli, as discussed in the following text:

Open probability (p): The primary short-term modulating factor is open probability, which can be affected by local chemical, electrical, or mechanical conditions [27]. Chemical conditions include the presence of ligands (such as those associated with G-protein-coupled receptors), ions or small molecule concentrations, or conditions such as temperature or pH. Electrical factors are primarily membrane voltage related, which modulate voltage-sensitive ion channels, but externally applied electric fields can also have an effect. Mechanical factors may include strain in the nearby areas of the membrane or potentially hydrostatic pressure.

Number of channels (N): The number of channels on the cell membrane can be affected within short time periods by membrane dynamics or vesicle trafficking, where channels are added or removed from the membrane through endocytosis or exocytosis [58]. Over longer time frames, the number of channels may also be modulated by protein expression changes induced via modified gene transcription, translation, or posttranslational effects.

Conductance (γ_{Ca}): Although the conductance of a particular channel type is generally considered nearly constant, it can be affected by structural changes in the protein and electrostatic forces [56].

Equilibrium potential (V_{Ca}): The ($V_m - V_{Ca}$) factor in Equations 16.1 and 16.2 corresponds to the driving force provided by the electrochemical potential difference [59]. The equilibrium potential for calcium, V_{Ca}, can be calculated by the Nernst equation

$$V_{Ca} = \frac{RT}{Z_{Ca}F} \ln\left[\frac{[C_a]_0}{[C_a]_1}\right] \tag{16.3}$$

where

valence term Z_{Ca} is +2 for the divalent cation calcium

R is the universal gas constant

T is absolute temperature

F is Faraday's constant

Therefore, the calcium flux is dependent upon the relative concentrations of Ca^{2+} on the inside and outside of the cell membrane, in addition to temperature.

Membrane potential (V_m): The V_m term is related to the intracellular concentrations, extracellular concentrations, and permeabilities of K$^+$, Na$^+$, and Cl$^-$, as defined by the Goldman–Hodgkin–Katz (GHK) voltage equation

$$V_m = \frac{RT}{F} \ln\left[\frac{P_K\left[K\right]_0 + P_{Na}\left[Na\right]_0 + P_{Cl}\left[Cl\right]_i}{P_K\left[K\right]_i + P_{Na}\left[Na\right]_1 + P_{Cl}\left[Cl\right]_0}\right] \tag{16.4}$$

where
   R is the universal gas constant
   T is absolute temperature
   F is Faraday's constant [56]

Therefore, the ion flux for every ion including calcium can be modulated by changes in concentrations of each ion species present, their permeabilities, or temperature. Although every permeable ion species contributes to V_m, only K$^+$, Na$^+$, and Cl$^-$ are included in Equation 16.4, as other ions contribute a negligible amount to the determination of V_m due to low concentrations and permeabilities except in unique situations [56].

Cellular mechanosensitivity via ion channels may therefore be enabled through a wide array of physical mechanisms and is not necessarily dependent upon a membrane channel that is directly gated by a mechanical force. The combination of the diversity of mechanosensitive functions and the identification thus far of only a few such mechanically activated channels would appear to support this observation. Therefore, experiments to isolate and investigate the complex and interrelated factors associated with ion channel–related mechanosensitivity are needed to enable a fundamental understanding of these processes.

16.4.3 MECHANOSENSITIVE ION CHANNELS IN BLADDER UROTHELIAL CELLS

As stated earlier, urothelial cells have been found to express several types of ion channels that are considered mechanosensitive (Table 16.1). These channels include members of the epithelial sodium channel (ENaC) family [10] and members of the transient receptor potential (TRP) family of channels, including specifically TRPV1 [60] and TRPV4 [61,62]. TRP channels are a family of over 20 members prevalent in many mammalian tissues that have been demonstrated to respond to a variety of stimuli [60,63–65]. TRPV1 and TRPV4 channels, which are in the vanilloid subfamily, are typically activated by heat, mechanical stimuli, and naturally occurring vanilloids [66]. While several studies have successfully demonstrated that some or all of these ion channels appear to be activated during bladder filling, the precise role that these ion channels play in urothelial mechanotransduction has not yet been determined. The following sections review the current literature on TRPV1, TRPV4, and ENaC channels and their role in bladder mechanotransduction.

16.4.3.1 TRPV1

TRPV1 channels are nonselective cation channels that are highly permeable to Ca^{2+} and allow for the influx of Ca^{2+} in response to the depletion of intracellular Ca^{2+} stores [67]. Capsaicin, the active component in chili peppers, has been identified as a ligand that activates the TRPV1 channel [68]. In the bladder, TRPV1 is expressed in sensory nerve fibers, urothelial cells, myofibroblasts, and probably smooth muscle cells [12]. Previous studies have shown that capsaicin-invoked desensitization affects neurons important for bladder mechanosensation. These studies suggest that capsaicin-sensitive neurons participate in bladder physiology, specifically for bladder hyperreactivity [69].

The role of TRPV1 in the bladder mechanosensory function was first demonstrated using TRPV1 knockout mice [68]. When the mucosal surface of the bladder tissue was exposed to a fixed pressure (30 cm H$_2$O) using a custom-designed pressure chamber in vitro, the level of stretch-evoked

TABLE 16.1
Properties of Mechanosensitive Ion Channels Possibly Related to Bladder Sensory Transduction

Ion Channel	Activator	Blocker	Modulator	Location
ENaC	Constitutive, mechanical, cold	Amiloride	Aldosterone, serine protease, CAP-1, trypsin, Na^+, pH, Ca^{2+}, Nedd4	Urothelium Nerve endings DRG
TRPV1	Heat ($\geq 43°C$), low pH, voltage, anandamide vanilloids, OEA, eicosanoids, AA	Capsazepine, BCTC	PKC, Na^+, Ca^{2+}, Mg^{2+}, adenosine, $PI(4,5)P_2$	Urothelium Nerve endings Detrusor Myofibroblasts DRG
TRPV2	Noxious heat ($>53°C$), mechanical, growth factors			DRG Urothelium Nerve endings? Myofibroblasts?
TRPV4	Moderate heat ($>24°C$), cell swelling, mechanical (shear stress), anandamide, 4a-PDD, $5',6'EET$, AA			Urothelium DRG
TRPA1	Mechanical, noxious cold?, cinnamaldehyde, isothiocyanate, garlic, marijuana, bradykinin		$[Ca^{2+}]_i$, voltage	Urothelium Nerve endings DRG
TRPM8	Cold ($8°C–28°C$), menthol, icilin, voltage		pH, Ca^{2+}, $PI(4,5)P_2$	Urothelium Nerve endings DRG

Source: Araki, I. et al., *Int. J. Urol.*, 15(0), 681, 2008. With permission.
Notes: 4a-PDD, 4a-phorbol 12,13-didecanoate; $5',6'EET$, $5',6'$-epoxyeicosatrienoic acid; BCTC, N-(4-tertiarybutylphenyl)-4-(3-chloropyridin-2-yl) tetrahydropyrazine-1(2H)-2-carboxamide; CAP-1, channel activating protein-1; DRG, dorsal root ganglia; ENaC, epithelial Na+ channel; nerve endings, suburothelial nerve endings; OEA, oleoylethanol-amide; $PI(4,5)P_2$, phosphatidylinositol (4,5) bisphosphate; PKC, protein kinase C; TRPA1, transient receptor potential, subfamily A (Ankyrin), member 1; TRPM8, transient receptor potential, subfamily M (Melastatin), member 8; TRPV1, transient receptor potential, subfamily V (Vanilloid), member 1; TRPV2, transient receptor potential, subfamily V (Vanilloid), member 2; TRPV4, transient receptor potential, subfamily V (Vanilloid), member 4.

ATP release from urothelial cells in the knockout mice was significantly lower in comparison to the wild type. In addition, the TRPV1 knockout mice exhibited a higher frequency of low-amplitude, nonvoiding bladder contractions as compared to the TRPV1 wild-type littermates. These results suggest that TRPV1 plays a significant role in detection of mechanical stimuli as well as in maintenance of bladder stability. It should be noted, however, that the pressure that the bladders were exposed to in this in vitro experiment was much higher than the normal sensing range for sensory nerve Aδ-fibers in bladders (10–20 cm H_2O) [70], indicating that TRPV1-mediated ATP release may be more of a safeguard for abnormal bladder pressures.

16.4.3.2 TRPV4

TRPV4 is another nonselective cation channel that shares approximately 40% amino acid identity with TRPV1 [71]. TRPV4 has been found in epithelial cells in the renal tubule, trachea, and recently in the bladder [72]. A study that investigated the bladder function of TRPV4 knockout mice demonstrated an increase in nonvoiding contractions similar to TRPV1 knockout mice, but also found an increase in time between voids [73]. This suggests potentially overlapping but distinct roles for TRPV4 and TRPV1 in bladder sensation. TRPV4 was first discovered to be activated by hypotonicity-induced cell swelling and thus membrane stretch [74], but more recently has been associated with mechanical stimuli such as shear stress and pressure in mice [12]. Once activated in response to cell swelling and membrane stretch, TRPV4 mediates a chain of events that leads to a release of cellular signaling molecules (e.g., ATP, acetylcholine, NO).

The results of a recent study provide additional evidence that TRPV4 may be involved in mechanosensory function of the bladder. First, it was demonstrated that rat urothelial cells expressed TRPV4 channels and that in vitro exposure of these cells to synthetic phorbol ester, 4α-phorbol 12, 13-didecanoate (4α-PDD), led to a pronounced increase in Ca^{2+} influx and ATP release [63]. Second, cystometry performed on anesthetized rats revealed that continuous intravesical infusion of 4α-PDD (100 μM) resulted in increased bladder voiding pressure [63]. Together these results suggest that artificial activation of TRPV4 may generate a false sense of bladder fullness, which may be explained as follows. When urothelial cells are stimulated by the fullness of the bladder or by a chemical such as 4α-PDD, TRPV4 is activated and ATP is released from these cells, which in turn triggers the afferent input via the purinergic pathway. Constant activation of purinergic receptors under the stimulation of TRPV4 by 4α-PDD results in sustained firing of efferent nerve pathways leading to sustained bladder contraction even if the bladder is empty. The sensation felt is similar to the feeling of incomplete bladder voiding in humans with urinary tract disorders (e.g., urethral outlet obstruction or voiding dysfunction associated with multiple sclerosis) [63]. While the exact mechanism for mechanical activation of TRPV4 on the bladder urothelium is still unknown, these results provide a basis for a hypothesis that TRPV4 plays an active role in urothelial mechanotransduction.

16.4.3.3 Other TRP Channels

In addition to TRPV1 and TRPV4, several other TRP ion channels have been identified as mechanosensitive. To date, however, their role in urothelial mechanotransduction has not been sufficiently investigated. For example, TRPM8, a channel known for its sensitivity to cold temperature, is thought to play a role in the bladder reflex to cooling [75]. While TRPM8 is readily expressed in the bladder mucosa, it is not as prevalent as other MSCs [12]. TRPA1 is another TRP family channel that has recently been found in rodent and human bladder urothelium and its sensitivity to chemical and thermal stimuli has been demonstrated [12]. Moreover, TRPV2, TRPC1, and TRPM3 have all been proposed as mechanosensitive channels in mammals, but of these, only TRPV2 has been found in the urinary tract [21,76].

16.4.3.4 Epithelial Sodium Channels

ENaCs are highly selective cation channels that are considered mechanosensitive and known to be involved in the reabsorption of Na^+ ions into the epithelia of the colon, airways, distal nephron, and bladder [77]. Using patch-clamp techniques, ENaCs have been demonstrated to transport Na^+ ions using a two-step process involving both ENaCs and a basolateral Na^+–K^+ pump [77]. The ENaC has been studied extensively as a mechanosensory transducer in rats [78,79], especially as a baroreceptor and a cutaneous sensory structure [21].

Studies have shown that the ENaC is present in the human urothelium [80] and that it has the ability to change its sodium transport properties after exposure to changes in hydrostatic pressure [10]. Recently, the effects of amiloride (a diuretic compound and an ENaC blocker) on the micturition reflex of rat bladders in vivo as well as the stretch-evoked ATP release from bladder strips in vitro were examined [21]. By measuring the time interval between two consecutive bladder contractions, the authors found that urothelial cells exposed to amiloride (1 mM) increased the pressure threshold for inducing micturition (i.e., the pressure at the point preceding a bladder contraction) but did not affect the micturition pressure. From these results it was concluded that the voiding efficiency did not change, but the sensitivity to bladder fullness decreased due to ENaC inhibition. In other words, only the bladder afferent pathways, but not efferent pathways, were inhibited by the infusion of amiloride. It was also found that amiloride (1 mM) blocked ATP release from a full-thickness rat bladder in response to stretch (up to 50%) in vitro but did not affect basal ATP release. Since more than 90% of the ATP was released from the bladder mucosal layer, the authors concluded that stretch-evoked ENaC activation triggered the ATP release from the bladder mucosa. These results concur with an earlier study that pretreatment with amiloride (10 μM) suppressed ATP release by feline urothelial cells in response to a hypo-osmotic solution in vitro [81]. While several studies have successfully examined specific ion channels that appear to be activated during bladder filling, the precise role that these ion channels play in urothelial mechanotransduction has not been fully identified.

16.5 SUMMARY

Mechanical parameters such as bladder wall tension or intravesical pressure due to bladder filling are sensed at the cellular level and contribute to both conscious sensation and reflex responses. The urothelium has been shown to respond to mechanical stimuli through the release of ATP, which, in turn, can activate purinergic receptors on bladder afferent nerves. Urothelial cells have also been shown to actively respond to elevated pressure by increasing the surface area of the umbrella cells that line the bladder. The underlying mechanisms for the conversion of mechanical stimuli to these cellular responses and mechanotransduction by urothelial cells are being rigorously investigated. Current theories include activation of MSCs by mechanical deformation of the urothelial cell membrane and subsequent changes in intracellular ion concentration, tension-induced exocytosis/endocytosis, and direct mechanical activation of ATP release.

The results of research to date suggest that the mechanosensation process of the bladder is likely a multifaceted process that exploits an unconventional sensing mechanism. Interruption of particular stages of the signaling cascade has been achieved, for example, through ion channel blockers and cytoskeletal disruption. Indeed, a number of therapeutic drugs that modulate ion channel function are clinically administered for overactive bladder or similar conditions. However, identification of the inherent cause and effect relationships and the fundamental mechanical transduction processes involved in the bladder has thus far been elusive. The need to advance the understanding of previously discovered urothelial membrane channels as well as the possibility of discovering unknown mechanotransduction mechanisms for improved treatment of LUTS will continue to be the motivation for additional research in the mechanobiology of bladder urothelial cells.

REFERENCES

1. P. Abrams, L. Cardozo, M. Fall, D. Griffiths, P. Rosier, U. Ulmsten, P. van Kerrebroeck, A. Victor, and A. Wein, The standardisation of terminology of lower urinary tract function: Report from the of the International Standardisation Sub-Committee Continence Society, *Neurourol. Urodyn.*, 21(2), 167–178, 2002.
2. K.S. Coyne, C.C. Sexton, C.L. Thompson, I. Milsom, D. Irwin, Z.S. Kopp, C.R. Chapple, S. Kaplan, A. Tubaro, L.P. Aiyer, and A.J. Wein, The prevalence of lower urinary tract symptoms (LUTS) in the USA, the UK and Sweden: Results from the Epidemiology of LUTS (EpiLUTS) study, *BJU Int.*, 104(3), 352–360, August 2009.
3. K.S. Coyne, L.S. Matza, and J. Brewster-Jordan, 'we have to stop again?!': The impact of overactive bladder on family members, *Neurourol. Urodyn.*, 28(8), 969–975, 2009.
4. K.E. Andersson and A. Arner, Urinary bladder contraction and relaxation: Physiology and pathophysiology, *Physiol. Rev.*, 84(3), 935–986, July 2004.
5. C.J. Fowler, D. Griffiths, and W.C. de Groat, The neural control of micturition, *Nat. Rev. Neurosci.*, 9(6), 453–466, June 2008.

6. M.F. Campbell, A.J. Wein, and L.R. Kavoussi, *Campbell-Walsh Urology*. W.B. Saunders, Philadelphia, PA, 2007.

7. G. Apodaca, E. Balestreire, and L.A. Birder, The uroepithelial-associated sensory web, *Kidney Int.*, 72(9), 1057–1064, November 2007.

8. L.A. Birder, More than just a barrier: Urothelium as a drug target for urinary bladder pain, *Am. J. Physiol. Renal Physiol.*, 289(3), F489–F495, September 2005.

9. G. Apodaca, The uroepithelium: Not just a passive barrier, *Traffic*, 5(3), 117–128, 2004.

10. D.R. Ferguson, Urothelial function, *BJU Int.*, 84(3), 235–242, August 1999.

11. W.C. de Groat, The urothelium in overactive bladder: Passive bystander or active participant?, *Urology*, 64(6A), 7–11, December 2004.

12. I. Araki, S. Du, H. Kobayashi, N. Sawada, T. Mochizuki, H. Zakoji, and M. Takeda, Roles of mechanosensitive ion channels in bladder sensory transduction and overactive bladder, *Int. J. Urol.*, 15(0), 681–687, 2008.

13. M. Chalfie, Neurosensory mechanotransduction, *Nat. Rev. Mol. Cell Biol.*, 10(1), 44–52, January 2009.

14. L.A. Birder and W.C. de Groat, Mechanisms of disease: Involvement of the urothelium in bladder dysfunction, *Nat. Clin. Pract. Urol.*, 4(1), 46–54, January 2007.

15. D. Purves, G.J. Augustine, L.C. Katz, A.-S. LaMantia, and S.M. Williams, *Neuroscience*. Sinauer Associates, Inc., Sunderland, MA, 2001.

16. G. Gabella and C. Davis, Distribution of afferent axons in the bladder of rats, *J. Neurocytol.*, 27(3), 141–155, March 1998.

17. T. Gevaert, J. Vriens, A. Segal, W. Everaerts, T. Roskams, K. Talavera, G. Owsianik, W. Liedtke, D. Daelemans, I. Dewachter, F. Van Leuven, T. Voets, D. De Ridder, and B. Nilius, Deletion of the transient receptor potential cation channel TRPV4 impairs murine bladder voiding, *J. Clin. Invest.*, 117(11), 3453–3462, November 2007.

18. W.C. de Groat and N. Yoshimura, Pharmacology of the lower urinary tract, *Annu. Rev. Pharmacol. Toxicol.*, 41, 691–721, 2001.

19. W.C. de Groat, A neurologic basis for the overactive bladder, *Urology*, 50(6A Suppl), 36–52; discussion 53–6, December 1997.

20. S. Barrick, H. Lee, S. Meyers, M. Caterina, M. Zeidel, B. Chopra, W. De Groat, A. Kanai, and L. Birder, Receptors and channels: TRPV4 receptors in urinary bladder urothelium: Involvement in urinary bladder function, *J. Pain*, 5(3, Suppl 1), S10, 2004.

21. S. Du, I. Araki, Y. Mikami, H. Zakoji, M. Beppu, M. Yoshiyama, and M. Takeda, Amiloride-sensitive ion channels in urinary bladder epithelium involved in mechanosensory transduction by modulating stretch-evoked adenosine triphosphate release, *Urology*, 69(3), 590–595, 2007.

22. D. Ferguson et al., ATP is released from rabbit urinary bladder epithelial cells by hydrostatic pressure changes—A possible sensory mechanism?, *J. Physiol.*, 505(2), 503–511, 1997.

23. E.C. Wang, J.M. Lee, W.G. Ruiz, E.M. Balestreire, M. von Bodungen, S. Barrick, D.A. Cockayne, L.A. Birder, and G. Apodaca, ATP and purinergic receptor-dependent membrane traffic in bladder umbrella cells, *J. Clin. Invest.*, 115(9), 2412–2422, September 2005.

24. V.P. Zagorodnyuk, S.J.H. Brookes, N.J. Spencer, and S. Gregory, Mechanotransduction and chemosensitivity of two major classes of bladder afferents with endings in the vicinity to the urothelium, *J. Physiol. London*, 587(14), 3523–3538, July 2009.

25. W.Q. Yu, P. Khandelwal, and G. Apodaca, Distinct apical and basolateral membrane requirements for stretch-induced membrane traffic at the apical surface of bladder umbrella cells, *Mol. Biol. Cell*, 20(1), 282–295, January 2009.

26. G. Burnstock, Purinergic nerves, *Pharmacol. Rev.*, 24(3), 509–581, 1972.

27. B. Alberts, A. Johnson, J. Lewis, M. Raff, K. Roberts, and P. Walter, *Molecular Biology of the Cell*. Garland Science, London, U.K. 2001.

28. G. Burnstock, Physiology and pathophysiology of purinergic neurotransmission, *Physiol. Rev.*, 87(2), 659–797, April 2007.

29. V. Kumar, C.C. Chapple, and R. Chess-Williams, Characteristics of adenosine triphosphatase release from porcine and human normal bladder, *J. Urol.*, 172(2), 744–747, August 2004.

30. G.E. Knight, P. Bodin, W.C. De Groat, and G. Burnstock, ATP is released from guinea pig ureter epithelium on distension, *Am. J. Physiol. Renal Physiol.*, 282(2), F281–F288, February 2002.

31. K. Matsumoto-Miyai, A. Kagase, Y. Murakawa, Y. Momota, and M. Kawatani, Extracellular Ca^{2+} regulates the stimulus-elicited ATP release from urothelium, *Auton. Neurosci. Basic Clin.*, 150(1–2), 94–99, October 2009.

32. T. Mochizuki, T. Sokabe, I. Araki, K. Fujishita, K. Shibasaki, K. Uchida, K. Naruse, S. Koizumi, M. Takeda, and M. Tominaga, The TRPV4 cation channel mediates stretch-evoked Ca2+ influx and ATP release in primary urothelial cell cultures, *J. Biol. Chem.*, 284(32), 21257–21264, August 2009.

33. P. Sadananda, F. Shang, L. Liu, K.J. Mansfield, and E. Burcher, Release of ATP from rat urinary bladder mucosa: Role of acid, vanilloids and stretch, *Br. J. Pharmacol.*, 158(7), 1655–1662, December 2009.

34. L. Bao, S. Locovei, and G. Dahl, Pannexin membrane channels are mechanosensitive conduits for ATP, *FEBS Lett.*, 572(1–3), 65–68, August 2004.

35. L. Bao, F. Sachs, and G. Dahl, Connexins are mechanosensitive, *Am. J. Physiol.-Cell Physiol.*, 287(5), C1389–C1395, November 2004.

36. D.A. Goodenough and D.L. Paul, Beyond the gap: Functions of unpaired connexon channels, *Nat. Rev. Mol. Cell Biol.*, 4(4), 285–294, April 2003.

37. M.L. Cotrina, J.H.C. Lin, A. Alves-Rodrigues, S. Liu, J. Li, H. Azmi-Ghadimi, J. Kang, C.C.G. Naus, and M. Nedergaard, Connexins regulate calcium signaling by controlling ATP release, *Proc. Natl. Acad. Sci. USA*, 95(26), 15735–15740, December 1998.

38. J.A. Haefliger, P. Tissieres, T. Tawadros, A. Formenton, J.L. Beny, P. Nicod, P. Frey, and P. Meda, Connexins 43 and 26 are differentially increased after rat bladder outlet obstruction, *Exp. Cell Res.*, 274(2), 216–225, April 2002.

39. S. Arnold, S.L. Vargas, I. Toma, F. Hanner, K. Willecke, and P.P. Janos, Connexin 30 deficiency impairs renal tubular ATP release and pressure natriuresis, *J. Am. Soc. Nephrol.*, 20(8), 1724–1732, August 2009.

40. E. Scemes, D.C. Spray, and P. Meda, Connexins, pannexins, innexins: Novel roles of "hemi-channels" *Pflugers Arch.*, 457(6), 1207–1226, April 2009.

41. H.A. Praetorius and J. Leipziger, ATP release from non-excitable cells, *Purinergic Signal.*, 5(4), 433–446, December 2009.

42. G. Burnstock, Purinergic mechanosensory transduction and visceral pain, *Mol. Pain*, 5, 12, November 2009.

43. L.A. Birder, H.Z. Ruan, B. Chopra, Z. Xiang, S. Barrick, C.A. Buffington, J.R. Roppolo, A. Ford, W.C. de Groat, and G. Burnstock, Alterations in P2X and P2Y purinergic receptor expression in urinary bladder from normal cats and cats with interstitial cystitis, *Am. J. Physiol.-Renal Physiol.*, 287(5), F1084–F1091, November 2004.

44. B.A. O'Reilly, A.H. Kosaka, G.F. Knight, T.K. Chang, A.P.D.W. Ford, J.M. Rymer, R. Popert, G. Burnstock, and S.B. McMahon, P2X receptors and their role in female idiopathic detrusor instability, *J. Urol.*, 167(1), 157–164, 2002.

45. R.A. North, P2X3 receptors and peripheral pain mechanisms, *J. Physiol.*, 554(2), 301–308, 2004.

46. D.A. Cockayne, S.G. Hamilton, Q.M. Zhu, P.M. Dunn, Y. Zhong, S. Novakovic, A.B. Malmberg, G. Cain, A. Berson, L. Kassotakis, L. Hedley, W.G. Lachnit, G. Burnstock, S.B. McMahon, and A. Ford, Urinary bladder hyporeflexia and reduced pain-related behaviour in P2X(3)-deficient mice, *Nature*, 407(6807), 1011–1015, October 2000.

47. M. Vlaskovska, L. Kasakov, W. Rong, P. Bodin, M. Bardini, D.A. Cockayne, A.P.D.W. Ford, and G. Burnstock, P2X3 knock-out mice reveal a major sensory role for urothelially released ATP, *J. Neurosci.*, 21(15), 5670–5677, August 1, 2001.

48. F.J. Alenghat and D.E. Ingber, Mechanotransduction: All signals point to cytoskeleton, matrix, and integrins, *Sci. STKE*, 2002(119), pe6, February 12, 2002.

49. F.J. Alenghat, S.M. Nauli, R. Kolb, J. Zhou, and D.E. Ingber, Global cytoskeletal control of mechanotransduction in kidney epithelial cells, *Exp. Cell Res.*, 301(1), 23–30, 2004.

50. K. Burridge, K. Fath, T. Kelly, G. Nuckolls, and C. Turner, Focal adhesions: Transmembrane junctions between the extracellular matrix and the cytoskeleton, *Annu. Rev. Cell Biol.*, 4, 487–525, 1988.

51. Y. Sawada and M.P. Sheetz, Force transduction by triton cytoskeletons, *J. Cell Biol.*, 156(4), 609–615, February 18, 2002.

52. P.G. Gillespie and R.G. Walker, Molecular basis of mechanosensory transduction, *Nature*, 413(6852), 194–202, 2001.

53. C. Kung, A possible unifying principle for mechanosensation, *Nature*, 436(7051), 647–654, August 2005.

54. F. Sachs and C.E. Morris, Mechanosensitive ion channels in nonspecialized cells, *Rev. Physiol. Biochem. Pharmacol.*, 1321–77, 1998.

55. X. Yao and C.J. Garland, Recent developments in vascular endothelial cell transient receptor potential channels, *Circ. Res.*, 97(9), 853–863, October 28, 2005.

56. R. Plonsey and R.C. Barr, *Bioelectricity: A Quantitative Approach.* Springer, New York, 2007.

57. B. Nilius and G. Droogmans, Ion channels and their functional role in vascular endothelium, *Physiol. Rev.*, 81(4), 1415–1459, October 2001.

58. S.J. Royle and R.D. Murrell-Lagnado, Constitutive cycling: A general mechanism to regulate cell surface proteins, *Bioessays*, 25(1), 39–46, January 2003.

59. T.F. Weiss, *Cellular Biophysics, Volume 2: Electrical Properties*, The MIT Press, Cambridge, MA, 1996.

60. L.A. Birder, A.J. Kanai, W.C. de Groat, S. Kiss, M.L. Nealen, N.E. Burke, K.E. Dineley, S. Watkins, I.J. Reynolds, and M.J. Caterina, Vanilloid receptor expression suggests a sensory role for urinary bladder epithelial cells, *Proc. Nat. Acad. Sci.*, 98(23), 13396–13401, 2001.

61. W. Liedtke, D.M. Tobin, C.I. Bargmann, and J.M. Friedman, Mammalian TRPV4 (VR-OAC) directs behavioral responses to osmotic and mechanical stimuli in *Caenorhabditis elegans*, *Proc. Natl. Acad. Sci.*, 100(90002), 14531–14536, 2003.

62. M. Suzuki, A. Mizuno, K. Kodaira, and M. Imai, Impaired pressure sensation in mice lacking TRPV4, *J. Biol. Chem.*, 278(25), 22664–22668, June 20, 2003.

63. L. Birder, F.A. Kullmann, H. Lee, S. Barrick, W. de Groat, A. Kanai, and M. Caterina, Activation of urothelial transient receptor potential vanilloid 4 by 4{alpha}-phorbol 12,13-didecanoate contributes to altered bladder reflexes in the rat, *J. Pharmacol. Exp. Ther.*, 323(1), 227–235, 2007.

64. D. Daly, W. Rong, R. Chess-Williams, C. Chapple, and D. Grundy, Bladder afferent sensitivity in wild-type and TRPV1 knockout mice, *J. Physiol.*, 583(2), 663–674, 2007.

65. T. Rosenbaum, A. Gordon-Shaag, M. Munari, and S.E. Gordon, Ca2+/calmodulin modulates TRPV1 activation by capsaicin, *J. Gen. Physiol.*, 123(1), 53–62, 2003.

66. R.A. Ross, Anandamide and vanilloid TRPV1 receptors, *Br. J. Pharmacol.*, 140(5), 790–801, 2003.

67. A. Apostolidis, C.M. Brady, Y. Yiangou, J. Davis, C.J. Fowler, and P. Anand, Capsaicin receptor TRPV1 in urothelium of neurogenic human bladders and effect of intravesical resiniferatoxin, *Urology*, 65(2), 400–405, February 2005.

68. L.A. Birder et al., Altered urinary bladder function in mice lacking the vanilloid receptor, *Nat. Neurosci.*, 5, 856–860, 2002.

69. Y.C. Chuang, M.O. Fraser, Y. Yu, J.M. Beckel, S. Seki, Y. Nakanishi, H. Yokoyama, M.B. Chancellor, N. Yoshimura, and W.C. de Groat, Analysis of the afferent limb of the vesicovascular reflex using neurotoxins, resiniferatoxin and capsaicin, *Am. J. Physiol. Regul. Integr. Comp. Physiol.*, 281(4), R1302–R1310, October 2001.

70. K.-E. Andersson, Bladder activation: Afferent mechanisms, *Urology*, 59(5, Suppl 1), 43–50, 2002.

71. A.D. Guler, H. Lee, T. Iida, I. Shimizu, M. Tominaga, and M. Caterina, Heat-evoked activation of the ion channel, TRPV4, *J. Neurosci.*, 22(15), 6408–6414, 2002.

72. N.S. Delany, M. Hurle, P. Facer, T. Alnadaf, C. Plumpton, I. Kinghorn, C.G. See, M. Costigan, P. Anand, C.J. Woolf, D. Crowther, P. Sanseau, and S.N. Tate, Identification and characterization of a novel human vanilloid receptor-like protein, VRL-2, *Physiol. Genomics*, 4(3), 165–174, January 19, 2001.

73. M. Born, I. Pahner, G. Ahnert-Hilger, and T. Jons, The maintenance of the permeability barrier of bladder facet cells requires a continuous fusion of discoid vesicles with the apical plasma membrane, *Eur. J. Cell Biol.*, 82(7), 343–350, July 2003.

74. B. Nilius, J. Vriens, J. Prenen, G. Droogmans, and T. Voets, TRPV4 calcium entry channel: A paradigm for gating diversity, *Am. J. Physiol. Cell Physiol.*, 286(2), C195–C205, 2004.

75. R.J. Stein, S. Santos, J. Nagatomi, Y. Hayashi, B.S. Minnery, M. Xavier, A.S. Patel, J.B. Nelson, W.J. Futrell, N. Yoshimura, M.B. Chancellor, and F. De Miguel, Cool (TRPM8) and hot (TRPV1) receptors in the bladder and male genital tract, *J. Urol.*, 172(3), 1175–1178, September 2004.

76. W. Everaerts, J. Vriens, G. Owsianik, G. Appendino, T. Voets, D. De Ridder, and B. Nilius, Functional characterization of transient receptor potential channels in mouse urothelial cells, *AJP—Renal Physiol.*, 298(3), F692–F701, 2010.

77. L. Schild, E. Schneeberger, I. Gautschi, and D. Firsov, Identification of amino acid residues in the alpha, beta, and gamma subunits of the epithelial sodium channel (ENaC) involved in amiloride block and ion permeation, *J. Gen. Physiol.*, 109(1), 15–26, 1997.

78. H.A. Drummond, F.M. Abboud, and M.J. Welsh, Localization of beta and gamma subunits of ENaC in sensory nerve endings in the rat foot pad, *Brain Res.*, 884(1–2), 1–12, November 24, 2000.

79. B. Fricke, R. Lints, G. Stewart, H. Drummond, G. Dodt, M. Driscoll, and M. von During, Epithelial Na+ channels and stomatin are expressed in rat trigeminal mechanosensory neurons, *Cell Tissue Res.*, 299(3), 327–334, March 2000.

80. I. Araki, S. Du, M. Kamiyama, Y. Mikami, K. Matsushita, M. Komuro, Y. Furuya, and M. Takeda, Overexpression of epithelial sodium channels in epithelium of human urinary bladder with outlet obstruction, *Urology*, 64(6), 1255–1260, December 2004.

81. L.A. Birder, S.R. Barrick, J.R. Roppolo, A.J. Kanai, W.C. de Groat, S. Kiss, and C.A. Buffington, Feline interstitial cystitis results in mechanical hypersensitivity and altered ATP release from bladder urothelium, *Am. J. Physiol. Renal Physiol.*, 285(3), F423–F429, September 2003.

82. D.E. Rapp, M.B. Lyon, G.T. Bales, and S.P. Cook, A role for the P2X receptor in urinary tract physiology and in the pathophysiology of urinary dysfunction, *Eur. Urol.*, 48(2), 303–308, 2005.

17 The Mechanobiology of Aqueous Humor Transport across Schlemm's Canal Endothelium

Darryl R. Overby

CONTENTS

17.1 AQUEOUS HUMOR OUTFLOW AND INTRAOCULAR PRESSURE

Intraocular pressure (IOP) is controlled by the turnover of aqueous humor, a clear fluid that fills the anterior segment of the eye and serves to nourish the avascular tissues of the cornea, lens, and trabecular meshwork (Figure 17.1). Aqueous humor is secreted into the posterior chamber by the ciliary processes, enters the anterior chamber through the pupil, and drains through one of two outflow pathways. The conventional or trabecular outflow pathway carries the majority of outflow and includes the trabecular meshwork, Schlemm's canal, and downstream collecting vessels that lead to the episcleral veins. The secondary uveoscleral outflow pathway appears to account for only 3%–35% of total outflow in the human eye (as surveyed by Nilsson [1]) and likely contributes little toward IOP regulation, although uveoscleral outflow does become important for glaucoma therapies involving prostaglandins and its analogues.

As aqueous drains through the conventional outflow pathway, it experiences a pressure drop from IOP in the anterior chamber to venous pressure in the episcleral veins, which corresponds to

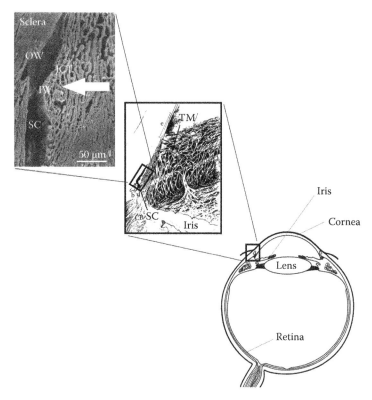

FIGURE 17.1 Overview of the aqueous humor outflow pathway, showing the position of Schlemm's canal near the root of the iris. Aqueous humor drains from the anterior chamber by passing through the porous trabecular meshwork (TM), which includes the juxtacanalicular connective tissue (JCT), before entering Schlemm's canal (SC). Aqueous enters SC by crossing its inner wall endothelium (IW) in the basal-to-apical direction. Not shown are collector channels, which open into the SC lumen and lead to veins within the sclera. Arrow indicates direction of flow. (Reproduced from Ethier, C.R. et al., *Biophys. J.*, 87(4), 2828, 2004. With permission; where the middle panel is adapted from Hogan, M.J. et al., *Histology of the Human Eye*, W.B. Saunders, Philadelphia, PA, 1971. With permission.)

a difference of nearly 5 mmHg in the normal human eye [2]. Tissues within the outflow pathway must mechanically support this pressure drop, which is maintained by the hydraulic resistance generated within the outflow pathway. In glaucoma, outflow resistance becomes elevated and leads to increased IOP [3].

Conventional wisdom attributes the bulk of outflow resistance generation to the inner wall endothelium of Schlemm's canal and its subjacent extracellular matrix, including the inner wall basement membrane and the juxtacanalicular connective tissue or JCT (Figure 17.1). Importantly, the source of elevated resistance in glaucoma is attributed to these same tissues [4]. However, it remains to be determined exactly how these tissues function to generate outflow resistance. This is an important issue because outflow resistance is the principal determinant of IOP and reducing IOP is the only successful strategy to treat glaucoma and curtail the progression of blindness [5]. Extensive reviews of these topics are included elsewhere [6–8].

Regardless of the precise mechanism of outflow resistance generation, practically all of the aqueous humor passing through the conventional outflow pathway must cross the continuous endothelial monolayer of Schlemm's canal before entering the canal lumen. The exact transport mechanism across this endothelium has been a matter of longstanding debate (see summary by Johnson and Erickson [8]); however it is now accepted that aqueous crosses the endothelium through transendothelial channels, and these channels appear to open and close in response to biomechanical stress.

Because these channels influence the rate and pattern of flow across the inner wall and through the JCT, they may serve as important modulators of outflow resistance generation. In the remainder of this chapter, we examine the mechanisms of aqueous humor transport across the endothelium of Schlemm's canal.

17.2   FUNCTIONAL ANATOMY OF SCHLEMM'S CANAL ENDOTHELIUM

Schlemm's canal is an endothelial-lined vessel that encircles the iris and, together with the trabecular meshwork, serves as the primary drainage route for aqueous humor from the anterior chamber of the eye. After passing through the trabecular meshwork, aqueous enters Schlemm's canal by crossing the inner wall endothelium in the "backward" basal-to-apical direction (Figure 17.1). Aqueous then flows circumferentially through the canal lumen to drain into one of approximately 30 collector channels or aqueous veins [9,10] that lead to the episcleral veins where aqueous mixes with the systemic circulation.

The ultrastructure of Schlemm's canal was first described using transmission electron microscopy by Garron et al. [11], and soon after by other investigators [12–18], who reported that Schlemm's canal is lined by a continuous endothelial monolayer.* Later studies using scanning electron microscopy to visualize the inner wall en face reported that approximately 23,000 endothelial cells make up the inner wall of Schlemm's canal [19]. The cells are long (~100μm) and slender (4–8μm) and tend to be oriented along the axis of the canal [19], with a projected area of 408μm^2 [20] to 480μm^2 [19].

A peculiar feature of the inner wall endothelium is the presence of cellular outpouchings known as *giant vacuoles* that bulge into the canal lumen, leaving a space between the cell and underlying matrix (Figure 17.2A). While giant vacuoles were originally suspected to be post-mortem artifacts [21–23], early concerns were resolved [24,25], and it is now fully accepted that giant vacuoles are true physiological structures that occur naturally in vivo [26]. A second peculiar feature of Schlemm's canal endothelium is the presence of micron-sized *transendothelial pores* that pass through the inner wall (Figure 17.2). These are not tears or ruptures in the cell, but are rather membrane-lined passageways through the endothelium that pass either between or through individual cells [27]. Current thinking attributes great importance to these pores, since these are the putative sites where aqueous humor crosses the inner wall endothelium, and recent data [28,29] show that pore numbers are reduced in glaucoma. Giant vacuoles and pores may be interrelated, because they often co-localize, and the formation of both structures appears to be coupled to biomechanical stress. Giant vacuoles and pores are discussed further in Sections 17.2.1 and 17.2.2.

The endothelium of Schlemm's canal exhibits similarities to both vascular and lymphatic endothelia (see review by Ramos et al. [30]). Like vascular endothelium, Schlemm's canal endothelium is a continuous monolayer, and although it possesses pores, it does not exhibit large gaps as often seen in lymphatic endothelium [31]. Schlemm's canal endothelial cells also express markers in common with vascular endothelial cells, such as von Willibrand factor [32], CD31/PECAM-1 and VE-cadherin [33], and neighboring cells are connected to one another by tight junctions [34–36], which are virtually absent from lymphatic endothelia [31]. In contrast, the basement membrane underlying the inner wall is discontinuous [13,18,37,38] as observed in lymphatic endothelia [31], which may be related to the basal-to-apical direction of transendothelial flow experienced by both endothelial types. Consistent with this view is the fact that the basement membrane underlying the outer wall, which is usually not heavily involved in filtration, is continuous and more clearly defined [13,34,37].

---

* In an early report, Holmberg [11] described the endothelium of Schlemm's canal as a double endothelial layer with fine fibrillar material lying between the two layers. However, this is not accurate, and Holmberg himself revised this conclusion after further study revealed that the underlying layer was discontinuous [15]. This second layer has been attributed to JCT cells aligned parallel to the inner wall [18].

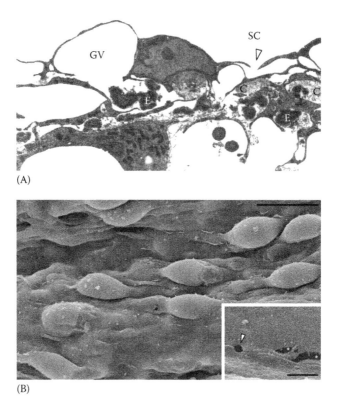

(A)

(B)

FIGURE 17.2 The inner wall endothelium of Schlemm's canal as seen by transmission and scanning electron microscopy. (A) shows a giant vacuole (GV) and a pore (arrowhead) along the inner wall, along with extracellular matrix (E, elastin; C, collagen) in the underlying juxtacanicular connective tissue. (Adapted from Gong, H., Tripathi, R.C., and Tripathi, B.J., *Microsc. Res. Tech.*, 33(4), 336, 1996; magnification in source article was ×89,100.) (B) shows the inner wall endothelium en face. Slender inner wall cells are generally aligned along the axis of the canal, and bulging structures may either be nuclei or giant vacuoles. A B-pore is shown at higher magnification (inset) along with artifactual "tears" (asterisks) in the cells that may be caused by fixative-induced stresses within the endothelium. Bars in (B) are 10 and 1 μm in the inset. (Images in (B) were kindly provided by Drs. A. Thomas Read and Darren W. Chan, University of Toronto, Toronto, Ontario, Canada.)

The basal-to-apical direction of aqueous humor flow across the inner wall strongly influences the biomechanical stress experienced by this endothelium. Flow across Schlemm's canal endothelium is driven by a pressure gradient that exerts forces on the endothelial cells. These forces act in the apical direction and tend to push the cells away from their basement membrane. The exact value of the pressure drop depends on the resistance generated by the inner wall, which is unknown, but estimates of ~1 mmHg are typical for the normal eye [39]. Taking this value and assuming an inner wall cell area of 450 μm^2 [19,20] yields a force estimate of 60 nN per cell, and larger forces would be expected with elevated IOP. These forces are significant on cellular terms, as can be appreciated by comparing against the forces necessary to initiate cell peeling from a substrate [40–43] or the forces exerted at focal adhesions during cell contraction [44,45] both of which are typically on the order of 10 nN. These pressure-derived forces likely deform and strain the endothelial cells, influencing the formation of giant vacuoles and pores that affect the fluid transport mechanism. Giant vacuoles and pores are described next.

17.2.1 GIANT VACUOLES

Giant vacuoles are outpouchings of Schlemm's canal endothelial cells that bulge into the lumen of the canal, leaving a fluid-filled cavity between the basal cell membrane and the underlying matrix

(Figure 17.2A). Garron [11] and Speakman [21,46] provided the first descriptions of these structures, although Speakman used light microscopy and misinterpreted giant vacuoles as large pores passing through the endothelium. Despite the name "vacuole" and their typical intracytoplasmic "signet ring" appearance on electron micrographs [11], the giant vacuole cavity is entirely extracellular, and serial sectioning has revealed that most, if not all, giant vacuoles have basal openings to the JCT [12,37,47–49].

Giant vacuoles are oriented along the long axis of the cell, stretching up to 10 μm long and 3–5 μm wide [19]; however these dimensions are sensitive to pressure, as described below. Giant vacuoles are generally located near the cell nucleus, and multiple giant vacuoles can occur within the same cell where the cavities often intercommunicate through openings in thin cellular partitions [37,50,51]. The internal cavity of giant vacuoles is often not a simple hemispherical dome shape, but is more convoluted with the cavity extending beyond the basal opening size like an invagination or tube through the cell. This often contributes to the false impression on single micrographs that the giant vacuole is entirely surrounded by cytoplasm.

Giant vacuole formation is a pressure-dependent process that occurs independently of cellular metabolism because they persist despite hypothermic conditions [52]. The size and density of giant vacuoles increase with IOP [53,54], and very few giant vacuoles are observed if the eye is fixed at 0 mmHg or by immersion [52] or if IOP becomes negative during retroperfusion [23]. This pressure sensitivity allows the inner wall to function as a one-way valve, allowing aqueous humor drainage when IOP is positive, but preventing the reflux of blood or plasma from the episcleral veins when IOP decreases [53].

Some, but not all, giant vacuoles possess transendothelial pores on their apical surface. These pores open into the lumen of Schlemm's canal and are thought to provide a route for aqueous humor flow across the endothelium. The inner wall pores are discussed next.

17.2.2 Transendothelial Pores

Transendothelial pores are elliptically shaped openings, on average about 1 μm in size, that pass through the inner wall endothelium and are entirely lined by cell membrane (Figure 17.2). There are two populations of pores: *intra*-cellular "I-pores" that pass transcellularly through individual cells and *inter*-cellular "B-pores" or border pores that pass between neighboring cells [27]. B-pores are likely the terminations of the paracellular aqueous flow pathway described by Epstein and Rohen [55]. I-pores are three to four times more plentiful and tend to be smaller than B-pores (major axis: 0.97 ± 0.23 μm vs. 1.64 ± 0.36 μm) [27]. Pores are generally associated with giant vacuoles, however a small population of pores, perhaps as large as 30% [56], is often found along flat, non-vacuolated regions of the inner wall [37].

The total pore density of the inner wall is generally reported to range from 1000–2000 pores/ mm^2 [19,27,57], corresponding to roughly one pore for each cell (assuming an area of 450 μm^2/cell, which was chosen as a mid-point between published values [19,20]). However, the exact in vivo pore density is unknown because the process of aldehyde fixation causes an artifactual increase in pore numbers [57]. Accounting for this artifact, Johnson et al. [29] demonstrated that glaucomatous eyes have a reduced inner wall pore density, confirming an earlier report [28]. This suggests that the inner wall endothelium of glaucomatous eyes may have an impaired ability to form pores, and this may contribute to elevated outflow resistance. The relationship between pores and outflow resistance is discussed in Section 17.3.2.

Before proceeding, it is worth pointing out one additional peculiarity of the inner wall. Inomata et al. [37] noted the presence of smaller pore-like structures that they referred to as "mini-pores," approximately 0.06 μm in diameter, that were typically bridged by a thin diaphragm. Mini-pores were often observed in the walls of giant vacuoles and in regions of the endothelium that appeared taut and thin [37,50]. Very few mini-pores were observed in eyes fixed at 0 or 8 mmHg, but they become increasingly prevalent as IOP was increased above 15 mmHg [49,50]. Mini-pores and their potential contribution to formation of "true" pores will be discussed further in Section 17.5.

17.2.3 Pores and Giant Vacuoles as Transendothelial Flow Channels

Holmberg [12] was the first to use electron microscopy to confirm the presence of open flow channels across the endothelium of Schlemm's canal. These channels were described to pass through the basal opening of giant vacuoles, through the vacuole cavity, and through the apical opening or "pore" to enter the lumen of Schlemm's canal (Figure 17.3). The existence of these channels was challenged in some early reports [11,13] that were largely based on individual sections. However, more detailed studies using serially sectioned transmission electron micrographs [12,15,17,18,50] and scanning electron microscopy [19,58,59] have firmly established that transendothelial channels indeed exist across the inner wall, although the exact number of these channels has been a matter of debate.

Using serial sectioning transmission electron microscopy, Kayes [17] and Tripathi [18] reported that only 1%–2% of giant vacuoles are true transendothelial channels containing both pores and basal openings. Using similar techniques, however, Grierson and Lee [49] reported that 8%–14% of giant vacuoles (between 15 and 30 mmHg) were true transendothelial channels. Higher estimates up to ~30% were reported using scanning electron microscopy (29% [58], ~30% [37], 15% [56]) that can survey a much larger area of the inner wall, suggesting that the lower estimates obtained using transmission electron microscopy may be attributed to sampling errors. The highest estimates are consistent with the expected prevalence of transendothelial channels, based on typical measures of

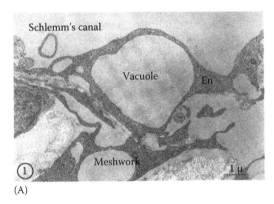

(A)

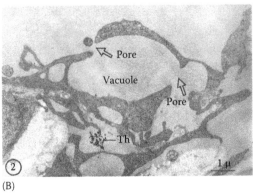

(B)

FIGURE 17.3 A giant vacuole in the endothelial lining of Schlemm's canal. In (A), the giant vacuole appears entirely intracytoplasmic. However, in a neighboring section (B), the giant vacuole exhibits both a basal opening and an apical pore, and therefore serves as a transendothelial flow channel (arrows). Bars are 1 μm. (Reproduced from Inomata, H., Bill, A., and Smelser, G.K., *Am, J. Ophthalmol.*, 73(5), 760, 1972. With permission.)

TABLE 17.1
Table Listing Giant Vacuole Dimensions and Estimates of Areal Strain Imposed on Inner Wall Cells

IOP (mmHg)	GV Width (μm)	GV Length (μm)	No. of GVs per Canal Length (mm⁻¹)	No. of GVs per Cell	Mean GV Surface Area (μm²)	Areal Strain per GV (%)	Areal Strain per Cell (%)
8	1.6±0.2	3.8±0.5	803±364	0.5	16	2–10	1–5
15	2.4±0.3	5.2±0.3	2284±705	1.5	33	4–20	6–30
22	3.3±0.3	6.4±0.5	3679±971	2.5	57	7–34	17–84
30	5.1±0.4	7.3±1.0	5384±1177	3.6	106	14–64	51–228

Notes: Areal strain is defined as the change in cell area divided by the original cell area. The left side of the table lists data reproduced from Grierson and Lee [51] in rhesus monkey, while the right side represents calculations based upon their data. We emphasize that these data should be used only for relative comparison, and not for absolute values. All values are mean ± SD where indicated. GV refers to giant vacuole. See Section 17.3.1 for further detail.

pore density (1500 pores/mm²), cell area (450 μm²), and number of giant vacuoles (1.5 per cell, see Table 17.1) that would predict a ratio of 0.45 pores per giant vacuole.* It therefore seems likely that a significant fraction, approximately one-third, of giant vacuoles has pores and may thereby serve as transendothelial channels for aqueous humor flow across the inner wall.

What evidence is there that aqueous humor crosses the inner wall by flowing through these transendothelial channels? To answer this question, we first consider two lines of evidence in support of aqueous humor flow through pores. First, 200–500 nm particles pass easily through the outflow pathway [61], where they would otherwise be blocked if flow passed largely through tight junctions that typically have a cut-off size of less than 10 nm [62]. To the contrary, however, it may be possible that particulates themselves induce pore formation, as has been observed for colloidal iron particles in the microvascular endothelium [63]. Secondly, as pointed out by Johnson [7,8], even if the entire bulk of outflow resistance were directly attributable to the inner wall (which almost certainly overestimates its resistance), the hydraulic conductivity of this tissue (≥3 μL/min/mmHg/cm²) is at least 10-fold larger than any other non-fenestrated endothelium possessing tight junctions, implying the existence of a highly conductive pathway for flow across the inner wall, namely through pores. Unfortunately, we still do not know whether flow predominately crosses the inner wall through transcellular routes involving I-pores or paracellular routes involving B-pores [55] or whether both routes are involved.

In regards to giant vacuoles, there is ample evidence demonstrating tracer accumulation in giant vacuole cavities, strongly suggestive of flow through these structures [37,47,64,65]. However, giant vacuoles do not appear to be an absolute requirement for transendothelial flow since some pores are found on non-vacuolated regions of the inner wall [37]. However, as we describe in Section 17.5, there are good reasons to suspect that giant vacuoles may promote the local formation of pores, while not being an absolute requirement for pore formation.

17.3 MECHANOSENSITIVITY OF SCHLEMM'S CANAL ENDOTHELIUM

In Section 17.2, we discussed giant vacuoles and pores and their contribution to transendothelial flow across the inner wall of Schlemm's canal. In this section, we discuss how these structures respond to biomechanical stress created by the pressure drop across the endothelium (Section 17.3.1) and how these structures relate to the generation of aqueous humor outflow resistance (Section 17.3.2).

---

* We acknowledge Prof. Mark Johnson for clarifying this argument [60].

17.3.1 Influence of Pressure Drop on Giant Vacuoles and Pores

Johnstone and Grant [53] were the first to describe how pressure-gradients affect the morphology of the inner wall endothelium of Schlemm's canal. When the pressure in the anterior chamber was larger than in the episcleral veins, the inner wall progressively became distended and distorted, with larger and more numerous giant vacuoles, expansion of the sub-endothelial space, and progressive collapse of the canal as the inner wall was displaced outward to approach the outer wall [53] (Figure 17.4). Studies by other groups soon confirmed these findings [25,49–51,54]. Distension of the endothelium was postulated to facilitate "opening" of the inner wall to allow drainage of aqueous humor and particulates, such as blood cells and pigment, from the anterior chamber [53]. No vacuoles were observed when anterior chamber pressure was lowered to episcleral venous pressure [53,54], and no "reverse" vacuoles were observed when anterior chamber pressure was dropped below venous pressure [25,53]. The inner wall therefore appears to function as a one-way valve that allows efflux of aqueous and particulates when anterior chamber pressure exceeds venous pressure but prevents the reflux of plasma and blood cells from the episcleral veins when anterior chamber pressure decreases* [53,54,67]. A similar unidirectional valve mechanism was described for cerebrospinal fluid (CSF) outflow [68] (see Section 17.4.1).

In a meticulous study using serial sectioning light microscopy, Grierson and Lee [51] reported changes in giant vacuole dimensions and numbers as a function of IOP in rhesus monkey eyes fixed in vivo. These data, reproduced on the left side of Table 17.1, reveal that between 15 and 30 mmHg, vacuole width (measured in the anterior-posterior direction) increased twofold, while length (measured along the axis of the canal) increased by 40%. Likewise, for the same range of IOP, there was nearly a 2.5-fold increase in vacuole numbers. In the original report, Grierson and Lee [51] noted a sharp decline in vacuole numbers at 50 mmHg; however this was an artifact due to collapse of Schlemm's canal that distorted and flattened vacuoles against the outer wall and to breaks that appeared in the inner wall at pressures above 30 mmHg [67,69]. In addition, Grierson and Lee were keen to point out that their data were originally intended to "indicate trends" and were "not intended

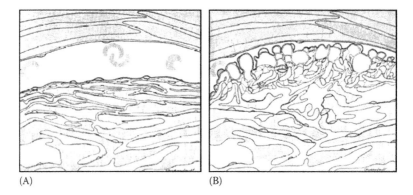

| (A) | (B) |

FIGURE 17.4 The inner wall of Schlemm's canal responds to changes in pressure drop. (A) For a pressure drop near 0 mmHg, the inner wall is flat with very few vacuoles and a compressed subendothelial tissue. Occasional red blood cells can be seen within the canal lumen. (B) For elevated pressure drops, the inner wall becomes highly vacuolated, with expansion of the subendothelial tissue and progressive collapse of the canal lumen. (Reproduced from Johnstone, M.A. and Grant, W.G., *Am. J. Ophthalmol.*, 75(3), 365, 1973. With permission.)

---

* As pointed out by Svedbergh on page 24 of his doctoral thesis [66], "closure" of the inner wall in response to low anterior chamber pressures may be facilitated by collapse of giant vacuoles where apical pores and basal openings are unlikely to overlap. This model explains how the valve function may be maintained without requiring rapid disappearance of pores from the inner wall.

to provide absolute values" [51]. For these reasons, we will restrict our analysis to relative changes and focus only on changes that occur with IOP less than 30 mmHg.

The strain experienced by the inner wall cells is an important factor controlling deformation during giant vacuole formation and may contribute to the formation of pores. To estimate the strain imposed on inner wall cells during giant vacuole formation, we follow Grierson and Lee [51] and assume that the giant vacuole may be represented as a prolate spheroid with equatorial and polar diameters equal to the vacuole width and length, respectively. Our approach is to calculate the relative change in cell surface area caused by the addition of the giant vacuole.

We consider two geometrical cases. For case A, the giant vacuole is considered to be a complete spheroid that rests atop the cell such that the surface area change is exactly equal to the spheroid's total surface area.* For case B, the giant vacuole is considered to be exactly half of a total spheroid, such that the plane of the cell perfectly bisects the spheroid. The surface area change for case B is then given by the difference between exactly half of the total spheroid surface area and its projected area.† Both cases neglect the small area contribution of the basal opening and apical pore, and neglect any strain experienced by the inner wall cell apart from giant vacuole formation. If the resting cell area is assumed to be flat with no invaginations and is taken to be $167 \mu m^2$, which is less than the typical cell area reported in monkeys ($253 \mu m^2$ [19,58]) but is chosen to be consistent with Grierson and Lee's [51] data,‡ then the areal strain can be calculated as the ratio of the change in cell area on account of the giant vacuole over the resting area. The strain estimated from cases A and B provide upper and lower bounds, shown in column 7 of Table 17.1. This analysis assigns a single giant vacuole to each cell and assumes that the strain is uniformly distributed throughout the entire cell, as if the non-vacuolated parts of the cell "donate" membrane and cytoplasm during the formation of the vacuole. These strain estimates would increase if the deformation were restricted to a smaller resting cell area in the immediate vicinity of the giant vacuole.

What if multiple giant vacuoles form within individual cells? Using data from Grierson and Lee [51], we can estimate the number of vacuoles per cell based upon the reported number of vacuoles per length of canal (column 4 of Table 17.1) and the number of reported nuclei per length (1500 per mm§; page 244 of Grierson and Lee [51]). The corresponding estimates, given in column 5 of Table 17.1, reveal that under normal pressure conditions, a single inner wall cell can contain multiple giant vacuoles. Multiplying the number of giant vacuoles per cell by the strain per vacuole yields estimates of strain experienced by inner wall cells in vivo (column 8 of Table 17.1), where again we have considered cases A and B to establish bounding estimates, and we have assumed a uniform strain distribution throughout the entire cell. Here, we point out that the lower estimates corresponding to case B in column 8 of Table 17.1 exactly match the calculations used by Ethier (Prof. C. R. Ethier, personal communication) to produce his reported 50% change in cell area between 8 and 30 mmHg [70]. This analysis predicts that inner wall cells experience extremely large changes in strain on account of giant vacuole formation, with areal strain increasing by threefold to fourfold between 15 and 30 mmHg for a single giant vacuole per cell. Much larger changes in strain, perhaps up to sevenfold or eightfold, are predicted when multiple giant vacuoles are allowed per cell as occurs in vivo. Again, we re-emphasize that Grierson and Lee [51] explicitly stated that their data

---

* The total surface area of a prolate spheroid is given by $2\pi R_e^2 + 2\pi R_e R_p \arcsin(\phi)/\phi$, where R_p and R_e are the polar and equatorial radii, respectively, and $\phi = \sqrt{R_p^2 - R_e^2}/R_p$.

† The projected surface area of a horizontally-oriented prolate spheroid is given by $\pi R_e R_p$.

‡ This cell area was estimated using the reported nuclear count (1500 per mm canal length) from Grierson and Lee [51] and assuming an anterior-posterior canal width of 0.25 mm. See following footnote for additional information.

§ In calculating the number of nuclei per unit length of inner wall, Grierson and Lee [51] divided the average number of nuclei per section (8.9; see page 238 of their report) by the average nuclear length (6 μm; page 244) measured by serial sectioning. This, however, neglects the contribution of the finite section thickness (1.5 μm; page 235), and accounting for this yields a slightly lower estimate of nuclear density (1200/mm) and a larger cell area ($208 \mu m^2$; see previous footnote). While this affects the absolute values of our estimates in Table 17.1, it has no effect upon relative differences, and so we continue to use the value of 1500 per mm for our analysis.

should be used only for relative comparisons and not for absolute values, and therefore our strain calculations (Table 17.1) based on their data should be used only for relative comparisons between different pressure levels.

How do inner wall cells accommodate such large changes in strain? Here it is important to recognize that few cells in the body are likely to experience changes in strain approaching the values listed in Table 17.1 as part of their normal function, with the possible exception of muscle cells, which are quite well adapted biomechanically, or cells undergoing migration or diapedesis (see Section 17.4.3). Cell death by loss of plasma membrane integrity typically occurs when cells are stretched by up to 50% [71], so it would appear that Schlemm's canal endothelial cells must possess some special mechanism that would allow them to withstand large mechanical strains, particularly at the level of the cell membrane.

Most cells posses a large reservoir of excess membrane stored in ruffles, folds or vesicles [72,73] that allow the cell to rapidly change shape without altering the total membrane surface area. However, once this excess membrane reservoir is depleted, the membrane becomes taut and stiff and typically ruptures with strains less than 5% [74]. Using electron microscopy, Grierson and Lee [50,75] twice reported a perceptible decline in micropinocytotic vesicles with increasing IOP between 8 and 30 mmHg, as well as a decrease in number of marginal folds. Unfortunately, quantitative values were not provided to compare against the changing giant vacuole surface area. Interestingly, the authors [50,75] originally attributed the change in vesicle numbers to an increase in receptor-mediated vesicle formation induced by the presence of blood plasma in the canal at lower pressures. However, it seems reasonable that these vesicles may also serve a mechanical role, and that the "additional" cell membrane area necessary for giant vacuole formation may be provided by evaginating vesicles and membrane folds, with additional contributions from simplifications occurring at the cell junctions [76]. If this is true, then this suggests that the cell membrane may contribute significantly to the mechanics of giant vacuole formation.

Evidence for the effects of IOP on transendothelial pores is less well established. While several authors have reported an increase in pore density with IOP [49,50,59,77], these effects may be attributed to an artifact whereby pore density increases in proportion to the volume of fixative passing through the outflow pathway [27,57,78]. Accounting for this effect eliminated the relationship between *total* pore density and fixation pressure [57]. However, one study still revealed a significant relationship between fixation pressure and density of B-pores, but not I-pores [27], but this was not observed in a second study by the same group [78]. In terms of pore size, Grierson and Lee [49] demonstrated an approximately linear relationship between pore size and giant vacuole size, both of which increased with IOP, but no such relationship was observed for pores that were not associated with giant vacuoles. In summary, while it may be possible that some portion of inner wall pores may respond to changes in pressure, data in this area are inconclusive and we currently do not fully understand how inner wall pores are influenced by pressure gradients across the inner wall.

17.3.2 Relationship of Pores and Giant Vacuoles to Outflow Resistance Generation

In a quantitative study of the inner wall of Schlemm's canal using scanning electron microscopy, Bill and Svedbergh [19] concluded that only a small portion of total outflow resistance (perhaps less than 10%) is directly attributable to transendothelial pores in human eyes. This finding was generally confirmed by later studies in man [56,79,80] and monkeys [77,79,80], although slightly higher estimates of inner wall pore resistance (perhaps up to 24% of total resistance) were reported by one group [79].

The outcome from these studies [19,56,77,79,80] has typically been used to argue against the functional significance of pores toward the generation of outflow resistance. However, there are at least three counter arguments against this view. First, pore densities reported in many previous studies are now known to have been affected by fixation artifact that increases the apparent pore density in proportion to the volume of fixative passing through the outflow pathway [27,57]. Accounting for this effect reveals that the true in vivo pore density may be two- to threefold less than reported by Bill and Svedbergh [19], suggesting that the inner wall may contribute more than 10% to total outflow resistance. Secondly, disrupting the continuity of the inner wall with pharmacologic agents (e.g., EDTA [81], α-chymotrypsin [82], cytochalasin-D [83], H-7 [84], but curiously not RGD [84]) causes a decrease in outflow resistance that is too large to be explained if only 10% of outflow resistance were attributed to the inner wall. Third, accounting for the fixative effect, a recent study [29] concluded that pore density is reduced in glaucomatous eyes, confirming an earlier report [28]. The decreased pore density in glaucoma may be as large as fivefold, based upon extrapolations to zero fixative volume as best estimates of true in vivo pore density [29].

We still lack a complete picture of how pores and transendothelial channels affect outflow resistance generation. Recent thinking, however, considers pores as modulators of resistance rather than direct generators of the bulk of resistance. As described by the funneling model [85], inner wall pores may act to confine aqueous drainage patterns in JCT, effectively increasing the resistance generated by extracellular matrix therein. This topic was extensively covered in a recent review article [6], and we will not repeat this discussion here except to point out that while progress is still being made, no existing model, including funneling, has yet explained how pores contribute to the generation of aqueous humor outflow resistance, either in normal or glaucomatous eyes.

We would be remiss here if we did not at least recognize the existence of an alternative hypothesis, namely that inner wall pores may be altogether separate from the mechanism of outflow resistance generation. In this scenario, the site of bulk resistance generation would lie elsewhere, and giant vacuoles and pores would be mere epiphenomena that form in response to hydrodynamic conditions imposed on them by the more dominant flow limiting structure. Possibilities include extracellular matrix within the JCT (e.g., glycosaminoglycans that are not well preserved in most ultrastructural studies [86]) and the basement membrane of the inner wall. A recent review [6] weighs the pros and cons of these alternatives.

17.4   GIANT VACUOLE AND PORE FORMATION IN OTHER ENDOTHELIA

The inner wall endothelium of Schlemm's canal is not unique in its ability to form giant vacuoles and pores. In this section, we explore giant vacuole and pore formation in three other endothelial systems, in the hopes that by illuminating these functional analogies, we may achieve a broader perspective and perhaps gain some specific insights into the mechanisms governing transport across Schlemm's canal endothelium.

17.4.1   CEREBROSPINAL FLUID TRANSPORT THROUGH ARACHNOID GRANULATIONS

As early as 1914, researchers were drawing analogies between the aqueous humor and CSF outflow pathways [87]. This latter pathway involves CSF drainage through *arachnoid granulations*, or their microscopic manifestations known as *arachnoid villi*, which are sites where the arachnoid mater herniates through the dura to bulge into the venus sinuses of the brain [88]. Like Schlemm's canal, arachnoid granulations are lined by a continuous endothelial monolayer containing tight junctions [89], where flow crosses the endothelium in the basal-to-apical direction in response to a pressure drop [89]. Also like Schlemm's canal, the arachnoid granulations function as a one-way valve, allowing efflux from the arachnoid mater but preventing reflux of venous contents into the meninges when the pressure gradient reverses [53,68].

Most strikingly, the endothelial lining of the arachnoid granulations/villi exhibits structures that appear in many ways identical to the giant vacuoles and pores of the inner wall of Schlemm's canal (Figure 17.5) [89,90]. This interesting functional analogy has been reviewed elsewhere [91], and we will not repeat the discussion here except to point out that the size and density of giant vacuoles and pores on the arachnoid granulations are, like the inner wall, highly sensitive to pressure [90]. However, unlike aqueous humor outflow, CSF outflow resistance *decreases* with increasing pressure once the pressure difference increases beyond a critical "opening" value, which in

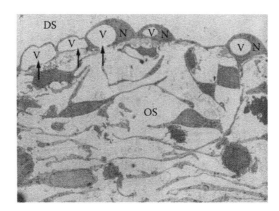

FIGURE 17.5 A transmission electron micrograph of an arachnoid villus showing the continuous endothelial lining of the dural sinus (DS) that is analogous to Schlemm's canal endothelium. Several giant vacuoles (V) can be seen, three of which exhibit basal openings to the subjacent tissue (arrows). (Reproduced from Tripathi [91] who reported an original magnification of ×2,880. Tripathi, R.C., *Exp. Eye Res.*, 25(Suppl), 65, 1977. With permission.)

man is approximately 3 mmHg above resting pressure [92], and this resistance decline has been described in terms of an increasing number of giant vacuoles and pores that form transendothelial channels in response to increasing pressure [90,92]. In contrast, aqueous humor outflow resistance *increases* with increasing IOP [93–96]. While this effect is attributed to pressure-induced collapse of Schlemm's canal [96,97], the fact that aqueous humor outflow resistance increases with increasing pressure is seemingly at odds with the fact that more giant vacuoles (and presumably more transendothelial channels) form along the inner wall at higher pressure, if indeed these transendothelial channels were important modulators of outflow resistance generation. Perhaps, transendothelial channels have a larger influence on resistance generation in the CSF compared to the aqueous humor outflow pathways. This would be supported by findings from Van Buskirk [95] who reported that aqueous outflow resistance in postmortem human eyes does not change with increasing IOP up to 20 mmHg if canal collapse is prevented by maximal lens depression.

At the very least, however, the similar morphologies between the aqueous humor and CSF outflow pathways clearly demonstrate that giant vacuoles and pores are not unique properties of Schlemm's canal endothelial cells, but rather may be a more general response of endothelial cells (or a broader subset of endothelial cells) to basal-to-apical directed pressure gradients. Recently developed ex vivo perfusion models of arachnoid granulations that allow microscopic visualization may provide new avenues to explore these exciting functional similarities [98].

17.4.2 Transcellular Gap Formation in the Microvasculature

Studies by Majno and Palade in 1961 [99,100] established that increased vascular permeability during acute inflammation coincides with the formation of gaps or openings in the endothelium of postcapillary venules. These gaps are lined by a smooth cell membrane border and range from 0.1 to 0.8 μm in width [99], and appear to function as transendothelial flow channels because intravascular tracer particles are found decorating gap borders, penetrating the vascular wall, and accumulating along the basement membrane in the vicinity of the gaps [63,99].

Since Majno's early work [99,100], there has been the general consensus that gaps through the microvascular wall pass entirely through the *paracellular* route after widening of intercellular clefts between cells. However, careful reconstruction of gap ultrastructure using serial section transmission electron microscopy has revealed that many gaps are actually *transcellular*, passing through rather than between individual cells, with the border of these gaps never contacting the intercellular cleft

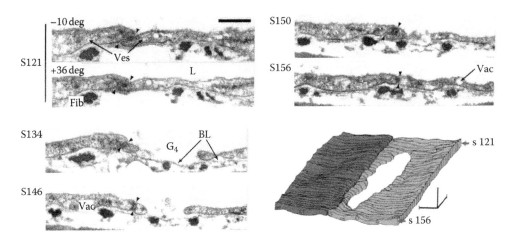

FIGURE 17.6 Serial transmission electron micrographs through a transcellular gap (G4) in a rat venule that had been perfused with ionophore A23187. The section numbers are given to the left of each panel. The intercellular junctions (arrowheads) appear intact and separate from the gap in each section. A three-dimensional reconstruction of the transcellular gap is shown at the lower right, where the two endothelial cells are shaded differently to identify the intercellular cleft, which appears intact and separate from the gap in all sections. Scale bars are 500 nm. BL, basal lamina; Ves, vesicle; Vac, intracytoplasmic vacuole. (Reproduced from Michel, C.C. and Neal, C.R., *Microcirculation*, 6(1), 45, 1999. With permission.)

(Figure 17.6) [63,101–103]. While some investigators maintain that gaps are almost exclusively intercellular in response to particular inflammatory stimuli or in particular vascular locations [104], other investigators have reported a significant proportion of transcellular gaps in the microvasculature of several species (frog [101–103,105], rat [63,105], mouse [63], guinea-pig [63], human [106]) that form in response to diverse stimuli (mild temperature elevation [101], calcium ionophore [103,105], histamine [63,106], serotonin [63], vascular endothelial growth factor [63], intravascular pressure elevation [102]), suggesting that transcellular gap formation is a common mechanism shared by many endothelial subtypes. Data from these studies reveal that transcellular gaps tend to be 0.3–0.8 μm in size, which is somewhat smaller than typical pore dimensions reported for the inner wall (~1 μm [27]). These gaps persist despite fixation with either glutaraldehyde or osmium tetroxide [103], suggesting that they are unlikely to be fixation artifacts, as discussed by Michel [62,107].

The formation of transcellular gaps tends to occur in the vicinity of cell margins, often within 1 μm of the margin, in a region where the cell is highly attenuated (see review by Michel and Neal [107]). In a study of transcellular gaps formed during elevated intravascular pressure in frog mesenteric capillaries, Neal and Michel [102] reported that the cell thickness is approximately 60 nm near the border of transcellular gaps, but may be somewhat thicker (~80 nm) near intercellular gaps. However, transcellular gaps do not always appear at the thinnest part of the cell, which may be as little as 15–30 nm [102]. Here, it is interesting to point out that pores in the inner wall of Schlemm's canal also tend to occur in thin regions of the cell, and I-pores that are not associated with giant vacuoles (or the *non-vacuolar transcellular channels*, as referred to by Grierson [49,50]) were "invariably ... found in the narrow marginal regions of the endothelial cells" (page 866–867 of Grierson and Lee [50]). This hints at a common mechanism of transcellular gap and I-pore formation that may occur in the peripheral cytoplasm in response to cell thinning, perhaps caused by biomechanical stimuli such as stretch or contraction. This will be discussed further in Section 17.5.

In contrast to intercellular gaps that are thought to form by unzipping or "stress failure" of cell–cell junctions, the formation of transcellular gaps appears to depend upon an unknown cellular mechanism that functions to relieve endothelial tension at sites where the peripheral cytoplasm is weaker than at the junction [101]. Michel and Neal [107] speculate that transcellular gaps could arise from rapidly enlarging vesicles, intracytoplasmic vacuoles or "vesiculo-vacuolar organelles"

Hypotheses for transcellular pore formation

Mediator induced   Pressure induced

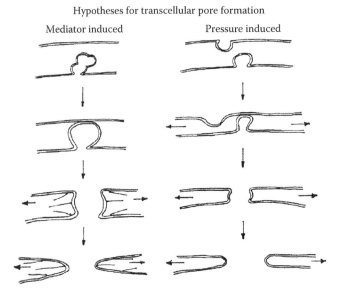

FIGURE 17.7 Possible mechanisms for the formation of transcellular gaps in the microvasculature as described by Michel and Neal [107]. The sequence on the left suggests that mediator-induced gaps may arise from intracytoplasmic vacuoles which first fuse with the basal and apical cell membrane to form vacuolar channels. Contraction of the cytoskeleton around the vacuoles then leads to the formation of gaps. The sequence on the right suggests that pressure-induced gaps may form from caveolae, which first form transendothelial channels and then gaps as the endothelium is thinned by stretch. (Reproduced from Michel, C.C. and Neal, C.R., *Microcirculation*, 6(1), 45, 1999. With permission.)

or VVOs (as described by the Dvorak laboratory [108,109]) in the cytoplasm that make contact between the apical and basal cell membranes, perhaps promoted by cytoplasmic thinning and cellular contractility, or from rupture of fenestrae or vacuoles that are bridged by a thin diaphragm (Figure 17.7). In studies of frog and rat mesentery, a much higher than normal incidence of fenestrae and vacuoles was reported in the attenuated regions of the cell surrounding transcellular gaps [101,102,105]. These structures may be analogous to the "mini-pores" observed in the inner wall of Schlemm's canal that occur in the thin regions of the cell and appear to increase with pressure [49,50] and may represent an early stage in the formation of "true" pores [37].

Unfortunately, there is little knowledge of the cellular and molecular mechanisms underlying transcellular gap formation. However, burgeoning interest in a parallel field is making headway uncovering the mechanisms of leukocyte diapedesis through transcellular routes. This is described next.

17.4.3 Transcellular Openings during Leukocyte Diapedesis

Leukocyte extravasation involves a highly coordinated process of leukocyte rolling, activation, and arrest on the vascular wall that is followed by *diapedesis* or migration of the leukocyte across the endothelial monolayer. Dating back to the early work of Julius Arnold [110], many investigators have maintained that leukocyte diapedesis occurs through paracellular openings between individual endothelial cells. However, this traditional view is currently being challenged by recent in vivo and in vitro studies that show leukocytes passing transcellularly through individual cells (see Carman [111] and colleagues [112] for an excellent background on this topic).

Using serial section transmission electron microscopy, Feng et al. [113] provided conclusive evidence for transcellular diapedesis in vivo, as was observed in skin venules of guinea pigs treated with the inflammatory peptide FMLP. Their [113] data show three-dimensional reconstructions of neutrophils passing through micron-sized transcellular openings (Figure 17.8). Like the transcellular

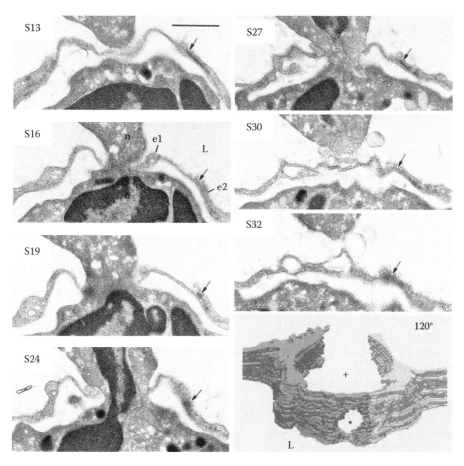

FIGURE 17.8   Transcellular diapedesis of a neutrophil (n) from the vascular lumen (L) across a thinned portion of venular endothelium in guinea pig skin after local intradermal injection of 10^{-5} M FMLP. Seven sections are shown (taken from 12 sections in the original publication [113]), where the section numbers are indicated on the left of each panel. The transcellular opening passes through a single endothelial cell (e1), but an intact junction (arrows) with a second endothelial cell (e2) is near to, but entirely separate from, the transcellular opening. A 3-dimensional reconstruction (lower right) reveals the location of the transcellular opening (*) with respect to the junction and shows a dome-shaped space (+) beneath the endothelial cells. Two cells are included in the 3-d reconstruction, and the cells are shaded differently to emphasize the location of the intercellular cleft. Note that the 3-d reconstruction is rotated to give a nearly en face perspective of the luminal cell surface (L). The dome-shaped space below the cells bears close resemblance to a giant vacuole on the inner wall of Schlemm's canal. See text for further discussion. The bar is 1 μm. (Reproduced from Feng, D. et al., *J. Exp. Med.*, 187(6), 903, 1998. With permission.)

gaps described in Section 17.4.2, these openings were typically within 2–3 μm of an intercellular cleft in regions where the endothelial cell was highly attenuated [113], but these openings tended to be somewhat larger (1.2 ± 0.3 μm; mean $\pm$ SEM) than the previously discussed gaps, possibly because the neutrophil contributes to the widening of the opening.

Before proceeding, it is worth noting that Figure 17.8 shows a striking similarity to giant vacuoles of the inner wall of Schlemm's canal, and similar dome structures are commonly observed when extravasating leukocytes become held up at the basal lamina, forcing the overlying endothelium into a thin dome-like structure that bulges into the vessel lumen, as described in early ultrastructural reports [114,115]. While common mechanisms may indeed exist between domes and openings and giant vacuole and pores, speculation at this point would be premature for the primary reason

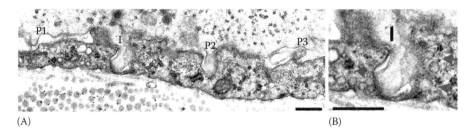

FIGURE 17.9 Multiple podosomes ("P1–P3" ~0.2–0.5 μm) and an invasive podosome ("I," ~0.8 μm) project from a lymphocyte into the endothelium of an inflamed venule in a guinea pig model of dermatitis (A). The invasive podosome has traversed the entire thickness of the endothelial cell, causing the apical and basal plasma membranes to be placed in close proximity, as can be seen by the higher magnification image in (B). The leukocyte is shown in the top part of the image. Arrowheads indicate endothelial vesicles and VVOs. Scale bars are 500 nm. (Reproduced from Carman, C.V. et al., *Immunity*, 26(6), 784, 2007. With permission.)

that leukocytes extravasate by moving from the luminal to abluminal compartments, and therefore formation of the transcellular opening likely precedes formation of the dome. This sequence is opposite that which we expect for the inner wall, where giant vacuole formation likely precedes pore formation (see Section 17.5).

As was reported in several studies [113–116], adherent leukocytes extend numerous cytoplasmic processes into the underlying endothelium, and these protrusive processes always precede, and are functionally required for, transcellular diapedesis. Carmen and colleagues [116] identified these processes as podosomes; cylindrical protrusive organelles ~0.5 μm in length and diameter that are enriched in F-actin and focal adhesion proteins. They observed that these structures were highly dynamic, with dozens forming and disappearing within minutes, as if the leukocyte was *palpating* the endothelium to identify a suitable location for diapedesis. This palpation occurred stochastically over the endothelium, but transcellular openings only formed in regions where podosomes could extend deeply, at which point they lengthened into *invasive podosomes* (>1 μm) that often spanned the full cell thickness, placing the apical and basal membranes in close proximity (Figure 17.9) [116]. At the site of invasive podosomes, the endothelial cell was enriched in vesicles and VVOs that were fused to the endothelial plasma membrane and there was a local increase in fusogenic proteins VAMP2 and 3 [116] involved in regulating membrane fusion events. Inhibition of fusogenic activity by NEM or by intracellular calcium chelation markedly reduced transcellular diapedesis, suggesting that vesicle fusion is necessary to form transcellular openings [116]. The process also required leukocyte protrusion because WASP-deficient leukocytes that are unable to form podosomes exhibited more than an 80% reduction in transcellular diapedesis [116].

In summary, the emerging model suggests that leukocyte protrusion, combined with a thinning endothelium, vesicles, and fusogenic protein activity promotes transcellular diapedesis in microvascular endothelia. It is tempting to speculate whether the leukocyte protrusion or "palpation" step could be functionally replaced by a biomechanical stimulus, such as the transendothelial pressure drop or endothelial stretch, that may initiate a similar mechanism for pore formation in Schlemm's canal endothelium.

17.5 SPECULATIONS ON THE MECHANISMS OF GIANT VACUOLE AND PORE FORMATION

The mechanisms underlying pore and giant vacuole formation in the inner wall of Schlemm's canal are unknown. In this section, we review two hypothetical models of giant vacuole and pore formation from the literature, and speculate on mechanistic details in light of our previous discussions of transcellular gap formation in microvascular endothelia in Section 17.4.

In 1971, Tripathi [47] proposed a cyclic, pressure-dependent model of giant vacuole and pore formation for the inner wall of Schlemm's canal. This model describes an "endothelial vacuolization cycle" that begins with a small indentation on the basal cell surface that enlarges on account of the transendothelial pressure drop and grows to form a giant vacuole that eventually develops a pore and ruptures into Schlemm's canal, depositing its contents and creating a temporary transendothelial channel [47,117]. The channel is then rapidly closed by the cell cytoplasm that moves to occlude the basal opening [47], thereby completing the cycle. A key idea in this model is that at any given instant, only 2% of giant vacuoles are in the transendothelial channel stage, which Tripathi based on his earlier work using serial section transmission electron microscopy [18]. As discussed in Section 17.2.3, this 2% estimate is at the low end of published values and is inconsistent with estimates of pore counts and with more extensive assessments of giant vacuole and pore morphology obtained using scanning electron microscopy [19,37,58,59].

Inomata, Bill and Smelser [37] proposed an alternative and more detailed model of giant vacuole and pore formation, which we repeat here in verbatim (see page 783 of their report).

> We start with an endothelial cell which develops an invagination near the nucleus at a place where the attachment to the subjacent tissue is poor. The invagination is partly occluded by the protoplasmic processes of the cells in the endothelial trabeculum, and fibrils and ground substance limit the rate of inflow into the vacuole which is forming. The vacuole enlarges and the wall toward the canal of Schlemm becomes thinner and thinner. Pinocytosis in this wall produces one or several membrane-covered minipores. These permit some flow of water, but retain high-molecular-weight substances. The diaphragm in the minipores breaks when the pressure difference between the vacuole and the canal of Schlemm is great enough, and aqueous humor then passes freely through the vacuole into Schlemm's canal and any flocculent material is washed away. At this stage several new vacuoles may have formed, and these vacuoles tend to open into each other and into the old vacuole.

There are at least two advantages of Inomata's model. First, it provides clear and testable descriptions of the mechanisms responsible for initiating giant vacuole and pore formation (i.e., rupture of relatively weak perinuclear cell-matrix attachments and mini-pores, respectively) and their relationships to the transendothelial pressure drop. Secondly, a higher percentage (~30% [37]) of giant vacuoles exist in the transendothelial channel state for the Inomata model, which is quantitatively consistent with predictions based upon extensive pore counting (see Section 17.2.3). This implies that transendothelial channels, once formed, likely remain patent and allow aqueous humor flow for a much longer duration in the Inomata compared to the Tripathi model. Inomata did not propose a mechanism for closure of transendothelial channels, but admitted to the possibility that "… pores into the canal of Schlemm possibly open and close quite frequently" (page 785 [37]). Finally, while the Inomata model as dictated above does not explicitly account for pores located in non-vacuolated regions of the inner wall, Inomata [37] suspected a "… similarity in origin and function" (page 781) based upon the similar size between vacuolar and non-vacuolar pores, and pointed out that precursor mini-pores were "… often found both in the walls of vacuoles and in the flat portion of the endothelium as well" (page 781).

Formation of I-pores in the inner wall may occur through mechanisms that are similar to those involved in transcellular gap formation in the microvasculature. Based upon the discussion in Section 17.4, consensus factors associated with transcellular gap formation include endothelial thinning, driven either by extracellular pressure-derived forces or intracellular contractile forces, combined with local accumulation of vesicles with fusogenic potential or diaphragms that rupture in response to stress (summarized in Figure 17.7). This is consistent with the Inomata model whereby cell tension, created either by the pressure drop across a giant vacuole wall or by tension within the endothelium as a whole, leads to cell thinning and a transition of local vesicles to a transcellular pore. In this model, giant vacuoles may promote pore formation on account of the biomechanical stress that they impose on the cell making up the giant vacuole wall, while not being an absolute requirement for pore formation since similar levels of stress may be imposed on non-vacuolated regions that are in tension as a result of stresses within the tissue as a whole.

The success of any one model of I-pore formation hinges on its ability to describe that critical instant when apical and basal plasma membranes somehow fuse together and rupture to create a membrane-lined transcellular pore. While these details remain largely unclear, it seems attractive to speculate on a role for fusogenic proteins such as VAMP2 and 3 that may be involved in bridging the apical and basal membranes through fusion of one or more vesicles. It is not difficult to imagine how such a process, when combined with a thinning endothelium under stress, would produce a structure that resembles the diaphragmatic "mini-pores" observed on the inner wall (Section 17.2.2), and rupture of mini-pores under prolonged or elevated loads may lead to the formation of a transcellular channel or I-pore.

Regardless of the precise mechanism by which apical and basal membranes fuse together to create the transcellular pore, the hypothetical mechanism described above suggests at least one important corollary. Namely, because giant vacuoles and pores result from biomechanical deformation within the endothelium, and because they in turn affect the transendothelial pressure drop through their influence on transport, the hypothesized transport mechanism implies the existence of a two-way coupling between flow crossing the endothelium and cellular deformation. Importantly, this predicts that the biomechanical properties of Schlemm's canal endothelial cells are important determinants of endothelial transport and may thereby act to modulate outflow resistance and IOP. This suggests the potential for pharmacological targeting of the inner wall cell biomechanical properties as new treatment for glaucoma.

Similar to the formation of I-pores, B-pores may form as a result of tensional forces acting on intercellular junctions that undergo "stress failure" or adaptive disassembly of their molecular adhesions. A diverse group of molecules regulate junctional assembly/disassembly, including PECAM-1/CD31 and VE-cadherin that are present on the inner wall [33] and are known to be mechanosensitive [118]. Reviews of these molecular mechanisms in the context of the microvasculature are given elsewhere [119–121], and for space limitations were not extensively covered in this chapter.

Before concluding, it is worth considering the relationship between pore density and fixation conditions on the inner wall of Schlemm's canal. As reported by Sit [57], and supported by others [27,29,78], inner wall pore density increases in proportion to the volume of fixative perfused through the outflow pathway. This has two important consequences. First, it seems almost certain that some fraction of pores is artifactual, but how many pores and which populations of pores (I-pores, B-pores, or both) are affected remains unclear. Second, we do not know the true in vivo pore density for the inner wall, although estimates have been generated by extrapolating to zero fixative volume from a regression of pore density versus fixative volume measured over several eyes [29,57]. Because earlier pore investigators were largely unaware of this artifact, readers must be somewhat cautious when interpreting data that precedes the study by Sit et al [57]. Furthermore, much of our understanding of inner wall transport may be changed if it turns out that a significant majority of pores are actually artifacts. For example, nearly the entire bulk of outflow resistance can be attributed to the inner wall if I-pores are found to be entirely artifactual and not present in vivo [27]. The reason why pores are sensitive to fixative volume is not clear, but this may be result of a "time-exposure" image of an inner wall that continues to form physiologic pores throughout the fixation process [57] or the result of artifactual pore formation driven by biomechanical stresses within the tissue during fixation (Figure 17.2B). Much of these issues would be resolved if we could find better ways to preserve inner wall ultrastructure or if selective markers could be identified for pores.

17.6   SUMMARY

In this chapter, we discussed the role of biomechanical forces in the transport of aqueous humor across Schlemm's canal endothelium. The primary points of this chapter may be summarized as follows:

- The inner wall endothelium of Schlemm's canal, together with its subjacent extracellular matrix, is the putative site of aqueous humor outflow resistance generation in the normal eye. The elevated outflow resistance characteristic of glaucoma is attributable to these same tissues.

- Aqueous humor likely crosses the inner wall through transendothelial pores that may or may not be associated with giant vacuoles. Approximately 30% of giant vacuoles appear to have pores, and this population of giant vacuoles combined with the non-vacuole-associated pores serves to provide channels for transendothelial flow.
- Significant biomechanical forces act on the inner wall on account of the transendothelial pressure drop. These forces influence the size and number of giant vacuoles, leading to significant mechanical strain on the inner wall cells. For a single giant vacuole, this strain may increase by as much as threefold to fourfold between 15 and 30 mmHg. Pores also appear to be sensitive to pressure drop, but studies in this area are largely inconclusive.
- Several other endothelial tissues form structures resembling giant vacuoles and pores. Most notably, micron-sized transcellular gaps are observed in microvascular endothelia during inflammation and leukocyte diapedesis. These gaps consistently form in attenuated regions of the endothelium at sites where vesicles and vesicle fusion may bridge apical and basal cell membranes. The mechanism of transcellular gap formation in the microvasculature is an area of intense research that may provide important clues to the mechanism of pore formation in Schlemm's canal endothelium.
- We speculate that the mechanism of intracellular "I-pore" formation on the inner wall of Schlemm's canal involves endothelial thinning, either caused by the pressure drop across a giant vacuole or tension within the endothelium as a whole, combined with local vesicle fusion or rupture of diaphragmatic vesicle borders that results in bridging of the apical and basal cell membranes. This mechanism is similar to the Inomata model [37] and may involve mini-pores and fusogenic proteins, such as VAMP2 and 3 that have been shown to be necessary for transcellular diapedesis in the microvasculature.
- The proposed mechanism of pore and giant vacuole formation describes a coupled interaction where biomechanical stresses acting on the endothelium drive the formation of transendothelial channels that feedback to influence the stresses by affecting the transendothelial pressure drop. This mechanism predicts that the biomechanical properties of Schlemm's canal endothelial cells are principal determinants of the transport mechanism and may thereby modulate outflow resistance generation.

ACKNOWLEDGMENTS

We thank Prof. C. Ross Ethier for valuable discussions and for reviewing a draft of this manuscript. We are grateful for support provided by donors of *National Glaucoma Research*, a program of the American Health Assistance Foundation (grants numbers G2006-057 and G2009-032), and the National Eye Institute (grant number EY018373).

REFERENCES

1. Nilsson, S.F., The uveoscleral outflow routes. *Eye*, 1997, **11**(Pt 2): 149–154.
2. McEwen, W., Application of Poiseuille's law to aqueous outflow. *AMA Arch Ophthalmol*, 1958, **60**(2): 290–294.
3. Grant, W.M., Clinical measurements of aqueous outflow. *Am J Ophthalmol*, 1951, **34**(11): 1603–1605.
4. Grant, W.M., Experimental aqueous perfusion in enucleated human eyes. *Arch Ophthalmol*, 1963, **69**: 783–801.
5. The Advanced Glaucoma Intervention Study (AGIS): 7. The relationship between control of intraocular pressure and visual field deterioration. The AGIS Investigators. *Am J Ophthalmol*, 2000, **130**(4): 429–440.
6. Overby, D.R., W.D. Stamer, and M. Johnson, The changing paradigm of outflow resistance generation: Towards synergistic models of the JCT and inner wall endothelium. *Exp Eye Res*, 2009, **88**(4): 656–670.

7. Johnson, M., What controls aqueous humour outflow resistance?, *Exp Eye Res*, 2006, **82**: 545–557.

8. Johnson, M. and K. Erickson, Mechanisms and routes of aqueous humor drainage, in *Principles and Practices of Ophthalmology*, D.M. Albert and F.A. Jakobiec, Eds. WB Saunders Co.: Philadelphia, PA, 2000, pp. 2577–2595.

9. Dvorak-Theobald, G., Schlemm's canal: Its anastomoses and anatomic relations. *Trans Am Ophthalmol Soc*, 1934, **32**: 574–595.

10. Dvorak-Theobald, G., Further studies of the canal of Schlemm: Its anastomoses and anatomic relations. *Am J Ophthalmol*, 1955, **39**: 65–89.

11. Garron, L.K. et al., Electron microscopic studies of the human eye. I. Preliminary investigations of the trabeculas. *Am J Ophthalmol*, 1958, **46**(1, Part 2): 27–35.

12. Holmberg, A., The fine structure of the inner wall of Schlemm's canal. *Arch Ophthalmol*, 1959, **62**: 956.

13. Garron, L.K. and M.L. Feeney, Electron microscopic studies of the human eye. II. Study of the trabeculae by light and electron microscopy. *Arch Ophthalmol*, 1959, **62**: 966–973.

14. Fine, B.S., Observations on the drainage angle in man and rhesus monkey: A concept of the pathogenesis of chronic simple glaucoma. A light and electron microscopic study. *Invest Ophthalmol*, 1964, **3**: 609–646.

15. Holmberg, Å., Schlemm's canal and the trabecular meshwork. An electron microscopic study of the normal structure in man and monkey (Cercopithecus aethiops). *Doc Ophthalmol*, 1965, **19**: 339.

16. Fine, B.S., Structure of the trabecular meshwork and the canal of Schlemm. *Trans Am Acad Ophthalmol Otolaryngol*, 1966, **70**(5): 777–790.

17. Kayes, J., Pore structure of the inner wall of Schlemm's canal. *Invest Ophthalmol Vis Sci*, 1967, **6**: 381.

18. Tripathi, R.C., Ultrastructure of Schlemm's canal in relation to aqueous outflow. *Exp Eye Res*, 1968, **7**(3): 335–341.

19. Bill, A. and B. Svedbergh, Scanning electron microscopic studies of the trabecular meshwork and the canal of Schlemm—An attempt to localize the main resistance to outflow of aqueous humor in man. *Acta Ophthalmol (Copenh)*, 1972, **50**(3): 295–320.

20. Lütjen-Drecoll, E. and J.W. Rohen, Endothelial studies of the Schlemm's canal using silver-impregnation technic. *Albrecht Von Graefes Arch Klin Exp Ophthalmol*, 1970, **180**(4): 249–266.

21. Speakman, J., Aqueous outflow channels in the trabecular meshwork in man. *Br J Ophthalmol*, 1959, **43**(3): 129–138.

22. Speakman, J., Endothelial cell vacuolation in the cornea. *Br J Ophthalmol*, 1959, **43**(3): 139–146.

23. Shabo, A.L., T.S. Reese, and D. Gaasterland, Postmortem formation of giant endothelial vacuoles in Schlemm's canal of the monkey. *Am J Ophthalmol*, 1973, **76**(6): 896–905.

24. Speakman, J.S., Drainage channels in the trabecular wall of Schlemm's canal. *Br J Ophthalmol*, 1960, **44**(9): 513–523.

25. Kayes, J., Pressure gradient changes on the trabecular meshwork of monkeys. *Am J Ophthalmol*, 1975, **79**(4): 549–556.

26. Grierson, I. and N.F. Johnson, The post-mortem vacuoles of Schlemm's canal. *Albrecht Von Graefes Arch Klin Exp Ophthalmol*, 1981, **215**(4): 249–264.

27. Ethier, C.R. et al., Two pore types in the inner-wall endothelium of Schlemm's canal. *Invest Ophthalmol Vis Sci*, 1998, **39**(11): 2041–2048.

28. Allingham, R.R. et al., The relationship between pore density and outflow facility in human eyes. *Invest Ophthalmol Vis Sci*, 1992, **33**(5): 1661–1669.

29. Johnson, M. et al., The pore density in the inner wall endothelium of Schlemm's canal of glaucomatous eyes. *Invest Ophthalmol Vis Sci*, 2002, **43**(9): 2950–2955.

30. Ramos, R.F. et al., Schlemm's canal endothelia, lymphatic or blood vasculature? *J Glaucoma*, 2007, **16**(4): 391–405.

31. Schmid-Schönbein, G.W., Microlymphatics and lymph flow. *Physiol Rev*, 1990, **70**(4): 987–1028.

32. Krohn, J., Expression of factor VIII-related antigen in human aqueous drainage channels. *Acta Ophthalmol Scand*, 1999, **77**(1): 9–12.

33. Heimark, R.L., S. Kaochar, and W.D. Stamer, Human Schlemm's canal cells express the endothelial adherens proteins, VE-cadherin and PECAM-1. *Curr Eye Res*, 2002, **25**(5): 299–308.

34. Grierson, I. et al., Associations between the cells of the walls of Schlemm's canal. *Albrecht Von Graefes Arch Klin Exp Ophthalmol*, 1978, **208**(1–3): 33–47.

35. Raviola, G. and E. Raviola, Paracellular route of aqueous outflow in the trabecular meshwork and canal of Schlemm. A freeze-fracture study of the endothelial junctions in the sclerocorneal angel of the macaque monkey eye. *Invest Ophthalmol Vis Sci*, 1981, **21**(1 Pt 1): 52–72.

36. Bhatt, K., H. Gong, and T.F. Freddo, Freeze-fracture studies of interendothelial junctions in the angle of the human eye. *Invest Ophthalmol Vis Sci*, 1995, **36**(7): 1379–1389.

37. Inomata, H., A. Bill, and G.K. Smelser, Aqueous humor pathways through the trabecular meshwork and into Schlemm's canal in the cynomolgus monkey (Macaca irus). An electron microscopic study. *Am J Ophthalmol*, 1972, **73**(5): 760–789.

38. Brilakis, H.S., C.R. Hann, and D.H. Johnson, A comparison of different embedding media on the ultrastructure of the trabecular meshwork. *Curr Eye Res*, 2001, **22**(3): 235–244.

39. Ethier, C.R., A.T. Read, and D. Chan, Biomechanics of Schlemm's canal endothelial cells: Influence on F-actin architecture. *Biophys J*, 2004, **87**(4): 2828–2837.

40. Lotz, M.M. et al., Cell adhesion to fibronectin and tenascin: Quantitative measurements of initial binding and subsequent strengthening response. *J Cell Biol*, 1989, **109**(4 Pt 1): 1795–1805.

41. Ward, M.D. and D.A. Hammer, A theoretical analysis for the effect of focal contact formation on cell-substrate attachment strength. *Biophys J*, 1993, **64**(3): 936–959.

42. Garcia, A.J., P. Ducheyne, and D. Boettiger, Quantification of cell adhesion using a spinning disc device and application to surface-reactive materials. *Biomaterials*, 1997, **18**(16): 1091–1098.

43. Ra, H.J. et al., Muscle cell peeling from micropatterned collagen: Direct probing of focal and molecular properties of matrix adhesion. *J Cell Sci*, 1999, **112** (Pt 10): 1425–1436.

44. Balaban, N.Q. et al., Force and focal adhesion assembly: A close relationship studied using elastic micropatterned substrates. *Nat Cell Biol*, 2001, **3**(5): 466–472.

45. Tan, J.L. et al., Cells lying on a bed of microneedles: An approach to isolate mechanical force. *Proc Natl Acad Sci U S A*, 2003, **100**(4): 1484–1489.

46. Speakman, J., The development and structure of the normal trabecular meshwork. *Proc R Soc Med*, 1959, **52**(1): 72–74.

47. Tripathi, R.C., Mechanism of the aqueous outflow across the trabecular wall of Schlemm's canal. *Exp Eye Res*, 1971, **11**(1): 116–121.

48. Fink, A.I., M.D. Felix, and R.C. Fletcher, Schlemm's canal and adjacent structures in glaucomatous patients. *Am J Ophthalmol*, 1972, **74**(5): 893–906.

49. Grierson, I. and W.R. Lee, Pressure effects on flow channels in the lining endothelium of Schlemm's canal. A quantitative study by transmission electron microscopy. *Acta Ophthalmol (Copenh)*, 1978, **56**(6): 935–952.

50. Grierson, I. and W.R. Lee, Pressure-induced changes in the ultrastructure of the endothelium lining Schlemm's canal. *Am J Ophthalmol*, 1975, **80**(5): 863–884.

51. Grierson, I. and W.R. Lee, Light microscopic quantitation of the endothelial vacuoles in Schlemm's canal. *Am J Ophthalmol*, 1977, **84**(2): 234–246.

52. Van Buskirk, E.M. and W.M. Grant, Influence of temperature and the question of involvement of cellular metabolism in aqueous outflow. *Am J Ophthalmol*, 1974, **77**(4): 565–572.

53. Johnstone, M.A. and W.G. Grant, Pressure-dependent changes in structures of the aqueous outflow system of human and monkey eyes. *Am J Ophthalmol*, 1973, **75**(3): 365–383.

54. Grierson, I. and W.R. Lee, Changes in the monkey outflow apparatus at graded levels of intraocular pressure: A qualitative analysis by light microscopy and scanning electron microscopy. *Exp Eye Res*, 1974, **19**(1): 21–33.

55. Epstein, D.L. and J.W. Rohen, Morphology of the trabecular meshwork and inner-wall endothelium after cationized ferritin perfusion in the monkey eye. *Invest Ophthalmol Vis Sci*, 1991, **32**(1): 160–171.

56. Grierson, I. et al., The trabecular wall of Schlemm's canal: A study of the effects of pilocarpine by scanning electron microscopy. *Br J Ophthalmol*, 1979, **63**(1): 9–16.

57. Sit, A.J. et al., Factors affecting the pores of the inner wall endothelium of Schlemm's canal. *Invest Ophthalmol Vis Sci*, 1997, **38**(8): 1517–1525.

58. Bill, A., Scanning electron microscopic studies of the canal of Schlemm. *Exp Eye Res*, 1970, **10**(2): 214–218.

59. Lee, W.R. and I. Grierson, Pressure effects on the endothelium of the trabecular wall of Schlemm's canal: A study by scanning electron microscopy. *Albrecht Von Graefes Arch Klin Exp Ophthalmol*, 1975, **196**(3): 255–265.

60. Johnson, M., Transport through the aqueous outflow system of the eye, PhD dissertation. Department of Mechanical Engineering, Massachusetts Institute of Technology: Cambridge, MA, 1987.

61. Johnson, M. et al., The filtration characteristics of the aqueous outflow system. *Exp Eye Res*, 1990, **50**(4): 407–418.

62. Michel, C.C. and F.E. Curry, Microvascular permeability. *Physiol Rev*, 1999, **79**(3): 703–761.

63. Feng, D. et al., Reinterpretation of endothelial cell gaps induced by vasoactive mediators in guinea-pig, mouse and rat: Many are transcellular pores. *J Physiol*, 1997, **504**(Pt 3): 747–761.

64. de Kater, A.W., S. Melamed, and D.L. Epstein, Patterns of aqueous humor outflow in glaucomatous and nonglaucomatous human eyes. A tracer study using cationized ferritin. *Arch Ophthalmol*, 1989, **107**(4): 572–576.

65. Ethier, C.R. and D.W. Chan, Cationic ferritin changes outflow facility in human eyes whereas anionic ferritin does not. *Invest Ophthalmol Vis Sci*, 2001, **42**(8): 1795–1802.

66. Svedbergh, B., Aspects of the aqueous humor drainage: Functional ultrastructure of Schlemm's canal, the trabecular meshwork and the corneal endothelium at different intraocular pressures. *Acta Univ Upsal*, 1976, **256**: 1–71.

67. Grierson, I. and W.R. Lee, The fine structure of the trabecular meshwork at graded levels of intraocular pressure. (2) Pressures outside the physiological range (0 and 50 mmHg). *Exp Eye Res*, 1975, **20**(6): 523–530.

68. Welch, K. and V. Friedman, The cerebrospinal fluid valves. *Brain*, 1960, **83**: 454–469.

69. Svedbergh, B., Effects of artificial intraocular pressure elevation on the outflow facility and the ultrastructure of the chamber angle in the vervet monkey (Cercopithecus ethiops). *Acta Ophthalmol (Copenh)*, 1974, **52**(6): 829–846.

70. Ethier, C.R., The inner wall of Schlemm's canal. *Exp Eye Res*, 2002, **74**(2): 161–172.

71. Tschumperlin, D.J. and S.S. Margulies, Equibiaxial deformation-induced injury of alveolar epithelial cells in vitro. *Am J Physiol*, 1998, **275**(6 Pt 1): L1173–L1183.

72. Lee, J. and G.W. Schmid-Schönbein, Biomechanics of skeletal muscle capillaries: Hemodynamic resistance, endothelial distensibility, and pseudopod formation. *Ann Biomed Eng*, 1995, **23**(3): 226–246.

73. Raucher, D. and M.P. Sheetz, Characteristics of a membrane reservoir buffering membrane tension. *Biophys J*, 1999, **77**(4): 1992–2002.

74. Needham, D. and R.S. Nunn, Elastic deformation and failure of lipid bilayer membranes containing cholesterol. *Biophys J*, 1990, **58**(4): 997–1009.

75. Grierson, I. and W.R. Lee, Pressure effects on the distribution of extracellular materials in the rhesus monkey outflow apparatus. *Albrecht Von Graefes Arch Klin Exp Ophthalmol*, 1977, **203**(3–4): 155–168.

76. Ye, W. et al., Interendothelial junctions in normal human Schlemm's canal respond to changes in pressure. *Invest Ophthalmol Vis Sci*, 1997, **38**(12): 2460–2468.

77. Svedbergh, B., Effects of intraocular pressure on the pores of the inner wall of Schlemm's canal: A scanning electron microscopic study. *Jpn J Ophthalmol*, 1976, **20**: 127–135.

78. Ethier, C.R. and F.M. Coloma, Effects of ethacrynic acid on Schlemm's canal inner wall and outflow facility in human eyes. *Invest Ophthalmol Vis Sci*, 1999, **40**(7): 1599–1607.

79. Moseley, H., I. Grierson, and W.R. Lee, Mathematical modelling of aqueous humour outflow from the eye through the pores in the lining endothelium of Schlemm's canal. *Clin Phys Physiol Meas*, 1983, **4**(1): 47–63.

80. Eriksson, A. and B. Svedbergh, Transcellular aqueous humor outflow: A theoretical and experimental study. *Albrecht Von Graefes Arch Klin Exp Ophthalmol*, 1980, **212**(3–4): 187–197.

81. Hamanaka, T. and A. Bill, Morphological and functional effects of Na2EDTA on the outflow routes for aqueous humor in monkeys. *Exp Eye Res*, 1987, **44**(2): 171–190.

82. Hamanaka, T. and A. Bill, Effects of alpha-chymotrypsin on the outflow routes for aqueous humor. *Exp Eye Res*, 1988, **46**(3): 323–341.

83. Johnson, D.H., The effect of cytochalasin D on outflow facility and the trabecular meshwork of the human eye in perfusion organ culture. *Invest Ophthalmol Vis Sci*, 1997, **38**(13): 2790–2799.

84. Bahler, C.K. et al., Pharmacologic disruption of Schlemm's canal cells and outflow facility in anterior segments of human eyes. *Invest Ophthalmol Vis Sci*, 2004, **45**(7): 2246–2254.

85. Johnson, M. et al., Modulation of outflow resistance by the pores of the inner wall endothelium. *Invest Ophthalmol Vis Sci*, 1992, **33**(5): 1670–1675.

86. Gong, H. et al., A new view of the human trabecular meshwork using quick-freeze, deep-etch electron microscopy. *Exp Eye Res*, 2002, **75**(3): 347–358.

87. Wegefarth, P. and L.H. Weed, Studies on cerebro-spinal fluid. No. VII: The analogous processes of the cerebral and ocular fluids. *J Med Res*, 1914, **31**(1): 167–176.

88. Upton, M.L. and R.O. Weller, The morphology of cerebrospinal fluid drainage pathways in human arachnoid granulations. *J Neurosurg*, 1985, **63**(6): 867–875.

89. Tripathi, B.J. and R.C. Tripathi, Vacuolar transcellular channels as a drainage pathway for cerebrospinal fluid. *J Physiol*, 1974, **239**(1): 195–206.

90. Levine, J.E., J.T. Povlishock, and D.P. Becker, The morphological correlates of primate cerebrospinal fluid absorption. *Brain Res*, 1982, **241**(1): 31–41.

91. Tripathi, R.C., The functional morphology of the outflow systems of ocular and cerebrospinal fluids. *Exp Eye Res*, 1977, **25**(Suppl): 65–116.

92. Mann, J.D. et al., Regulation of intracranial pressure in rat, dog, and man. *Ann Neurol*, 1978, **3**(2): 156–165.

93. Levene, R. and B. Hyman, The effect of intraocular pressure on the facility of outflow. *Exp Eye Res*, 1969, **8**(2): 116–121.

94. Brubaker, R.F., The effect of intraocular pressure on conventional outflow resistance in the enucleated human eye. *Invest Ophthalmol*, 1975, **14**(4): 286–292.

95. Van Buskirk, E.M., Changes in the facility of aqueous outflow induced by lens depression and intraocular pressure in excised human eyes. *Am J Ophthalmol*, 1976, **82**(5): 736–740.

96. Moses, R.A., The effect of intraocular pressure on resistance to outflow. *Surv Ophthalmol*, 1977, **22**(2): 88–100.

97. Van Buskirk, E.M., Anatomic correlates of changing aqueous outflow facility in excised human eyes. *Invest Ophthalmol Vis Sci*, 1982, **22**(5): 625–632.

98. Grzybowski, D.M. et al., In vitro model of cerebrospinal fluid outflow through human arachnoid granulations. *Invest Ophthalmol Vis Sci*, 2006, **47**(8): 3664–3672.

99. Majno, G. and G.E. Palade, Studies on inflammation. 1. The effect of histamine and serotonin on vascular permeability: An electron microscopic study. *J Biophys Biochem Cytol*, 1961, **11**: 571–605.

100. Majno, G., G.E. Palade, and G.I. Schoefl, Studies on inflammation. II. The site of action of histamine and serotonin along the vascular tree: A topographic study. *J Biophys Biochem Cytol*, 1961, **11**: 607–626.

101. Neal, C.R. and C.C. Michel, Transcellular openings through microvascular walls in acutely inflamed frog mesentery. *Exp Physiol*, 1992, **77**(6): 917–920.

102. Neal, C.R. and C.C. Michel, Openings in frog microvascular endothelium induced by high intravascular pressures. *J Physiol*, 1996, **492**(Pt 1): 39–52.

103. Neal, C.R. and C.C. Michel, Transcellular openings through frog microvascular endothelium. *Exp Physiol*, 1997, **82**(2): 419–422.

104. Baluk, P. et al., Endothelial gaps: Time course of formation and closure in inflamed venules of rats. *Am J Physiol*, 1997, **272**(1 Pt 1): L155–170.

105. Neal, C.R. and C.C. Michel, Transcellular gaps in microvascular walls of frog and rat when permeability is increased by perfusion with the ionophore A23187. *J Physiol*, 1995, **488**(Pt 2): 427–437.

106. Braverman, I.M. and A. Keh-Yen, Three-dimensional reconstruction of endothelial cell gaps in psoriatic vessels and their morphologic identity with gaps produced by the intradermal injection of histamine. *J Invest Dermatol*, 1986, **86**(5): 577–581.

107. Michel, C.C. and C.R. Neal, Openings through endothelial cells associated with increased microvascular permeability. *Microcirculation*, 1999, **6**(1): 45–54.

108. Dvorak, A.M. and D. Feng, The vesiculo-vacuolar organelle (VVO). A new endothelial cell permeability organelle. *J Histochem Cytochem*, 2001, **49**(4): 419–432.

109. Feng, D. et al., Ultrastructural studies define soluble macromolecular, particulate, and cellular transendothelial cell pathways in venules, lymphatic vessels, and tumor-associated microvessels in man and animals. *Microsc Res Tech*, 2002, **57**(5): 289–326.

110. Arnold, J., Ueber diapedesis. *Virch Arch*, 1873, **58**(2): 203–230.

111. Carman, C.V. and T.A. Springer, Trans-cellular migration: Cell-cell contacts get intimate. *Curr Opin Cell Biol*, 2008, **20**(5): 533–540.

112. Sage, P.T. and C.V. Carman, Settings and mechanisms for trans-cellular diapedesis. *Front Biosci*, 2009, **14**: 5066–5083.

113. Feng, D. et al., Neutrophils emigrate from venules by a transendothelial cell pathway in response to FMLP. *J Exp Med*, 1998, **187**(6): 903–915.

114. Marchesi, V.T. and H.W. Florey, Electron micrographic observations on the emigration of leucocytes. *Q J Exp Physiol Cogn Med Sci*, 1960, **45**: 343–348.

115. Williamson, J.R. and J.W. Grisham, Electron microscopy of leukocytic margination and emigration in acute inflammation in dog pancreas. *Am J Pathol*, 1961, **39**: 239–256.

116. Carman, C.V. et al., Transcellular diapedesis is initiated by invasive podosomes. *Immunity*, 2007, **26**(6): 784–797.

117. Cole, D.F. and R.C. Tripathi, Theoretical considerations on the mechanism of the aqueous outflow. *Exp Eye Res*, 1971, **12**(1): 25–32.

118. Tzima, E. et al., A mechanosensory complex that mediates the endothelial cell response to fluid shear stress. *Nature*, 2005, **437**(7057): 426–431.
119. Schnittler, H.J., Structural and functional aspects of intercellular junctions in vascular endothelium. *Basic Res Cardiol*, 1998, **93** (Suppl 3): 30–39.
120. Bazzoni, G. and E. Dejana, Endothelial cell-to-cell junctions: Molecular organization and role in vascular homeostasis. *Physiol Rev*, 2004, **84**(3): 869–901.
121. Vandenbroucke, E. et al., Regulation of endothelial junctional permeability. *Ann N Y Acad Sci*, 2008, **1123**: 134–145.
122. Hogan, M.J., J.A. Alvarado, and J.E. Weddel, *Histology of the Human Eye*. W. B. Saunders: Philadelphia, PA, 1971.
123. Gong, H., R.C. Tripathi, and B.J. Tripathi, Morphology of the aqueous outflow pathway. *Microsc Res Tech*, 1996, **33**(4): 336–367.

18 Mechanobiology in Health and Disease in the Central Nervous System

Theresa A. Ulrich and Sanjay Kumar

CONTENTS

18.1 INTRODUCTION

Over 2300 years have passed since Aristotle first identified "touch" as one of the five exteroceptive senses, marking one of the first discussions in the academic literature of mechanosensation, the ability to sense mechanical forces. In the intervening time, we have learned a great deal about how cells convert mechanical sensations into chemical signals, a process known as mechanotransduction. Importantly, we have learned that mechanotransduction serves not only as a sensory mechanism for cells and tissues, but as a regulatory mechanism as well. Numerous studies over the last several decades have revealed the robust sensitivity of mammalian cells to mechanical cues such as shear flow, cyclic strain, and microenvironmental stiffness, demonstrating that the mechanical microenvironment participates centrally in the homeostasis of cells and tissues, and that disruptions of these mechanical cues can contribute to the onset and progression of disease [1,2]. The field of mechanobiology, which examines how cells sense, process, and respond to mechanical stimuli, has emerged at the intersection of the physical and biological sciences to examine the often subtle yet complex relationship between mechanical force and cell and tissue behavior.

The central nervous system (CNS) has been the focus of extensive tissue biomechanics research for over three decades, spurred in part by widespread public interest in brain and spinal cord injury in automobile accidents, sports injuries, and other traumatic settings. Yet, despite this history of pioneering work in tissue-level biomechanics, investigation of cellular mechanobiology in the CNS has been comparatively limited. While the earliest work in this field stemmed directly from studies of neuronal trauma and regeneration (e.g., cell-level studies of axonal repair following injury), the last decade has seen rapid acceleration in the pace of research exploring how the properties of the normal (nontraumatized) mechanical microenvironment affect the health and disease of CNS cells and tissues.

Here, we provide an overview of the nascent field of cellular mechanobiology in the CNS, beginning with a description of the structure and mechanical microenvironment within the brain and spinal cord, continuing with an overview of the mechanobiological characteristics of cells in the CNS, and concluding with demonstrated and potential roles for how dynamic interactions between cells and their mechanical microenvironment can contribute to the onset or exacerbation of disease in the nervous system.

18.2   THE MECHANICAL MICROENVIRONMENT OF CNS TISSUES

No other organ system in the body is as carefully protected from external mechanical forces as the CNS by its encasement within the bony skull and vertebral column (and as a consequence, no other organ system is as vulnerable to the buildup of excess pressure during disease or injury, as we will discuss in the last section of this chapter). The soft tissues inside the CNS are further protected by three membranes known as meninges (the superficial dura mater, central arachnoid mater, and deep pia mater) and by the serum-like cerebrospinal fluid (CSF) that bathes both the brain and the spinal cord. A network of cavities deep within the brain, known as ventricles, is lined by a population of epithelial cells whose main function is to secrete CSF and to aid its circulation throughout the ventricles, the subarachnoid space (located between the arachnoid mater and pia mater), and the central canal of the spinal cord.

The brain parenchyma is organized into distinct anatomical structures (the cerebral hemispheres, diencephalon, brain stem, and cerebellum), which are crisscrossed by blood vessels and defined by a complex and highly organized neuroarchitecture of white matter (composed of neuronal axons) and gray matter (composed of nerve cell bodies). Far less is known about the structure and function of the extracellular matrix (ECM) in CNS tissues than in most other organ systems in the body. A well-defined ECM containing large amounts of collagen, fibronectin, and laminin exists in the basement membrane of the cerebral vasculature and in the meninges surrounding the brain cortex; however, these structures stand in stark contrast to the rest of the brain parenchyma, which consists of a relatively amorphous, anisotropic, and heterogeneous matrix containing mainly hyaluronic acid as well as various other glycosaminoglycans and proteoglycans [3,4].

The spinal cord is similarly heterogeneous, consisting of longitudinally oriented white matter tracts surrounding a central, butterfly shaped region of gray matter. Like the brain, the spinal cord is organized into distinct anatomical structures with specialized functions (e.g., the ascending and descending sensory and motor white matter tracts), and contains an anisotropic ECM that consists mostly of hyaluronic acid [5]. The structural inhomogeneities within both the brain and the spinal cord lead to extensive heterogeneity in the mechanical and biochemical microenvironment of resident cells, however, cyclic mechanical strain is generated throughout the CNS by pulsatile flow of CSF at a rate of approximately 1 Hz [6–8].

The mechanical stiffness of human tissues varies widely, ranging from very soft brain, fat, and mammary tissues with elastic moduli of less than 1 kPa to calcified bone tissues with elastic moduli near 10,000 kPa (10 MPa). Reported measurements of the mechanical properties of human and animal brain and spinal cord tissues have varied by more than an order of magnitude from several hundred Pa to several kPa, likely due to variations in species, age, sample size, and anatomic origin, sample preparation, testing conditions, and time postmortem [9–13].

The spectrum of available methods and published results for mechanical analysis of CNS tissues has recently been reviewed in considerable detail [14,15]. Here, we will focus our discussion on the two methods that are most commonly used today: rotational rheometry and magnetic resonance elastography (MRE). Rotational rheometry has been employed in materials characterization for decades and generally involves the small-amplitude oscillatory deformation of a sample of known geometry, measurement of the resulting stress within the sample, and calculation of material properties, such as the shear modulus and elastic modulus (see [16] for a straightforward and detailed explanation of rheometry theory and methods in the context of biological samples). MRE, on the other hand, is a noninvasive alternative that allows determination of tissue stiffness in living subjects, and therefore has become increasingly popular for measuring the mechanical properties of brain tissue [17–20]. This technique involves application of acoustic shear waves to the tissue of interest; a phase-sensitive magnetic resonance imaging (MRI) sequence is then used to visualize and quantitatively measure propagating strain waves throughout the tissue, and an algorithm is subsequently used to generate a map of the elastic modulus of the imaged region [21].

A recent report by Vappou et al. demonstrated a surprisingly strong agreement between the linear viscoelastic behavior of brain tissues measured by rotational rheometry and MRE [22], which may facilitate longitudinal or comparative studies of brain tissue mechanics in healthy and diseased individuals in the future. However, an important caveat of both rotational rheometry and MRE is that both measurements require a macroscopic sample of tissue that may contain multiple cell types and ECM with microscale heterogeneities, thereby complicating inference of the local mechanical microenvironment of single cells. This limitation was illustrated by a recent atomic force microscopy (AFM) indentation study by Elkin et al., which measured the apparent elastic modulus of the cell-scale microenvironment in anatomical subregions of the rat hippocampus, revealing spatial heterogeneity in local tissue mechanics (Figure 18.1) [11].

The brain is generally considered to be a nonlinear viscoelastic material with mechanical properties that exhibit relatively low but measurable interspecies variability. These mechanical properties appear to be independent of perfusion pressure and cranial confinement as long as strains are modest. Importantly, the mechanics of brain tissue change rapidly and dramatically within hours after death [23], which may account for some of the variability in reported measurements of brain tissue elasticity. Rheological studies of immature and adult porcine and rat brain tissue have revealed that an additional source of variability may be a decrease in the mechanical stiffness of the brain parenchyma with age, an observation that has been attributed in part to the increasing lipid content associated with myelination of rapidly branching axonal and dendritic arbors during development as well as reduction in water content with age [24,25]. Dynamic MRI studies appear to support the idea of age-dependent alterations in the mechanical properties of mature CNS tissues [10]. While dynamic-phase contrast MRI is not traditionally thought of as a measurement of tissue mechanics, it can be used to measure the coupling of brain, spinal cord, and CSF pulsations to the driving vascular pulsations; because the biomechanical properties of CNS tissues exert strong influence over this coupling, observed differences between adult and

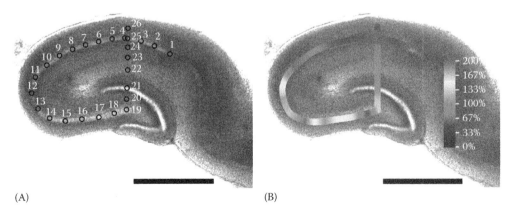

(A) (B)

FIGURE 18.1 (See color insert.) Biomechanical heterogeneity of the rat hippocampus. (A) Spatial distribution of indentation measurements. The elastic modulus of the rat hippocampus was measured via atomic force microscopy (AFM) indentation along a layer of pyramidal neuron cell bodies at the depicted locations. (B) Elasticity map of the hippocampus. Apparent elastic modulus normalized to the mean apparent elastic modulus of all indentations is depicted by the color bar. The material properties of the hippocampus were spatially heterogeneous. Scale bar is 1 mm. (Reprinted from Elkin, B.S. et al., *J. Neurotrauma.*, 24, 812, 2007. With permission.)

elderly individuals provide indirect evidence that human CNS tissue mechanics progressively change with age.

In one of the earliest studies of brain tissue biomechanics, Metz et al. compared the viscoelastic properties of living tissue in anesthetized animals with postmortem and postfixation tissue samples; although the authors reported significant increases in the elastic modulus of the fixed tissue, they surprisingly observed very little change in the nonlinearity of the stress–strain relationship [26]. Numerous studies have since reported nonlinear stress–strain responses in mammalian brain tissue, similar to the characteristic strain-stiffening observed in collagenous soft tissues [27–32]; these nonlinearities feature prominently in models of traumatic injury, but their significance to cellular mechanobiology remains incompletely understood because the range of frequencies at which cells probe their environment is not yet well determined.

18.3 MECHANOBIOLOGY IN CNS DEVELOPMENT

The mechanical properties of the ECM can direct a wide range of cellular properties, including cell shape and cytoarchitecture [33–35], motility [36,37], matrix remodeling [38], differentiation [39–42], and the extension of functional cellular projections [43–46]. As the human brain develops, billions of cells are generated in the proliferative tissues lining the lateral ventricles of the brain. These cells migrate throughout the developing CNS, differentiate into neurons or glial cells, and establish a diverse array of organized structures with distinctive shapes and intricate internal architecture [47]. Neurogenesis continues, albeit in a much more limited way, into adulthood through the self-renewal and differentiation of adult neural stem cells (aNSCs) found within the hippocampus and subventricular zone [48]. The mechanosensitivity of aNSCs was recently explored by Saha et al., who demonstrated that the differentiation and self-renewal of aNSCs can be modulated by controlling the mechanical stiffness of the surrounding microenvironment [42]. In particular, culturing aNSCs on the surface of soft polymeric substrates with stiffnesses close to living brain tissue (~100–500 Pa) favored differentiation into neurons, whereas culturing aNSCs on polymeric substrates with identical surface chemistry but much greater mechanical rigidity (~1–10 kPa) favored differentiation into glial cell types. The latter finding raises the intriguing possibility that tissue stiffening may play an instructive role in glial scar formation rather than merely serving as a passive consequence of the process. This mechanosensitivity, coupled with the mechanical heterogeneity present throughout the normal and diseased brain, lends support to the hypothesis that mechanical cues may be dynamically involved in CNS development, in the maintenance of homeostasis, and in the development of disease.

The mechanisms governing morphogenesis (the development of shape, organization, and structure) in CNS tissues have long been the subject of vigorous investigation and debate, but one of the most elegant and robust hypotheses that has emerged is the tension-based theory of morphogenesis proposed by David Van Essen in 1997 [47]. This theory holds that mechanical tension along axons, neurites, and glial cell extensions is sufficient to generate many of the structural features of the mammalian CNS, including the folding patterns of the cortex and the observed compactness of the neural circuitry. This hypothesis is supported by the unique mechanical properties of neurons (described below), which would be expected to facilitate regulation of steady tension in neuronal processes during development through both passive and active mechanisms.

18.4 MECHANOBIOLOGY OF NEURONS IN THE CNS

The notion that mechanosensation is a central function of many neurons, such as the somatosensory neurons that transduce tactile and sound cues, was clear long before the recent surge of interest in cellular mechanotransduction mechanisms. It was less obvious whether differentiated neurons in the CNS, which are largely shielded from external mechanical loading by the presence of the cranium and vertebral column, should be expected to retain similar sensitivity to the mechanical properties of their microenvironment. Yet, a growing body of work has indicated that the relationship between mature neurons and microenvironmental mechanics is as dynamic and delicate in the CNS as in tissues that routinely experience mechanical loading. For example, the growth and functionality of neurons appears to be tightly coupled to microenvironmental rigidity. Two recent studies by Janmey and coworkers revealed that substrate rigidity can modulate the outgrowth of cells from explants of the spinal cord [45] and cortical brain tissue [49] in a manner that mirrors the stiffness-dependence of aNSC differentiation discussed previously [42]. Specifically, explanting CNS tissues onto substrates that match the approximate stiffness of CNS tissues (several hundred Pa) was found to optimally support both neuronal and glial survival while suppressing overgrowth of the cultures by astrocytes, whereas explanting onto softer substrates selected for neuronal survival and growth and explanting onto stiffer substrates selected for glial survival and proliferation.

The extension and branching of neurites (thin projections from the cell body, including axons and dendrites) in culture can be similarly regulated by the rigidity of the cellular microenvironment (Figure 18.2) [43–46]; substrates that are either much softer or much stiffer than the normal brain

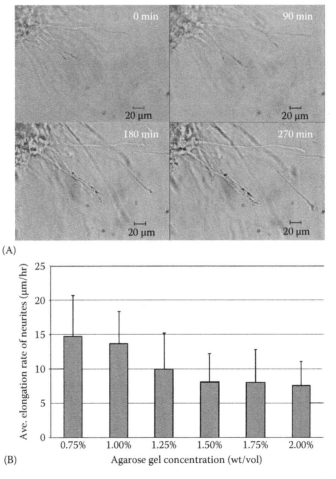

(A)

(B)

FIGURE 18.2   Neurite extension is regulated by microenvironmental mechanics. (A) Phase-contrast images of dorsal root ganglion neurons extending neurites into a 1% agarose hydrogel. Images were acquired at 90 min intervals. (B) Effect of agarose concentration on neurite extension. The rate of neurite extension is modulated by agarose gel stiffness, controlled by varying the concentration of agarose from 0.75% to 2.0% w/v. (Reprinted from Balgude, A.P. et al., *Biomaterials*, 22, 1077, 2001. With permission.)

microenvironment often do not support robust neurite extension in vitro, although the details of this relationship appear to depend upon the cell source and substrate geometry and composition. This correlation may have implications for neuroregeneration and tumorigenesis; for example, it was recently proposed that the softening of reactive astrocytes following mechanical injury may provide a compliant, brain-like mechanical substrate that promotes neurite extension [50]. We recently showed that the potency with which retinoic acid can induce neurite extension, reduce proliferation and suppress N-Myc expression in neuroblastoma tumor cells all depend strongly on ECM stiffness [51]. Given the substantial mechanical stress exerted by neurites on adhesive substrates [52–54], this phenomenon is likely related to the capacity of the underlying substrate to support generation of contractile forces within elongating projections. Interestingly, neurites have been elicited in vitro by direct application of tensile forces with glass microneedles, where active elongation is observed when tension is maintained above a threshold value, and active retraction is observed when tension is released [53–56]. Follow-up studies using magnetic beads to apply external loads to elongating neurites showed that forces on the order of 1.5 nN are required to elicit neurites, and that force-induced neurite initiation and elongation appears to be a highly conserved property that is largely independent of cell age and synaptic phenotype [57,58].

The initiation, extension, and maintenance of neurites is key to the functional integrity of the CNS, and importantly, these processes are intimately related to the mechanical properties of neurons and their subcellular components. AFM characterization of individual neurons showed that retinal neurons display the rheological characteristics of elastic solids, with cell processes that are often softer than the cell body [59]. Individual neurites display simple elastic behavior under transient stretching (neurite length increases in proportion to applied tension) [54] and viscoelastic properties under sustained stretching [60]; specifically, the initial tension relaxes passively to a lower level on the timescale of minutes, and when resting tension is released, the neurite shortens slightly prior to active retraction. Interestingly, cytoskeletal elements within the cell are known to behave in similar fashion; actin stress fibers in living cells also retract as viscoelastic cables following incision with a femtosecond laser nanoscissor [61], and curved microtubules have been observed to straighten briefly following nanoscissor incision due to release of elastic energy prior to rapid depolymerization [62]. This is also consistent with the viscoelastic behavior that has been measured in a wide variety of cultured non-neuronal cell types by magnetic twisting cytometry [35,63–65], AFM [66–69], micropipette aspiration [70–72], optical trapping [73,74], cellular de-adhesion measurements [75], and a variety of other methods [76].

Biochemical modifications to cytoskeletal elements may play a key role in neurite mechanics as well; in particular, it has long been known that cytoskeletal dynamics play a central role in driving the directionality of axonal growth [77–79]. More recently, however, contributions of cytoskeletal networks to the shape and mechanics of mature axons have begun to emerge. For example, repulsive forces between phosphate groups on neurofilament sidearm domains contribute to the organization of neurofilament networks within axons (Figure 18.3) [80–82]. Indeed, enzymatic dephosphorylation of neurofilaments significantly alters the mechanical properties of both purified neurofilament gels and single neurofilament proteins [83,84]. Conversely, traumatic nerve injury is sometimes accompanied by altered posttranslational modification and organization of neurofilaments and other cytoskeletal proteins [85,86]. Similar mechanisms have been proposed for the organization of microtubule bundles, which are commonly found in both axons and dendrites, by microtubule-associated proteins [87]. Modulators of the actin cytoskeleton also play a prominent role in regulating axonal growth and guidance; for example, activation of myosin motors and Rho-family GTPases, (e.g., Rho, Rac, Cdc42) play vital roles in regulating axon growth dynamics, and dysfunctional Rho family GTPase signaling has been tied to a surprisingly wide variety of adult and congenital neurological disorders [88–90]. Mechanosensitive ion channels have also been implicated in neurite growth kinetics [91–93] in addition to their well-documented role in fast sensation of stretch and other forces [94–96]. For example, inhibition of Ca^{2+} influx through stretch-activated ion channels in neurons from explanted *Xenopus laevis* spinal cord tissue was found to dramatically accelerate

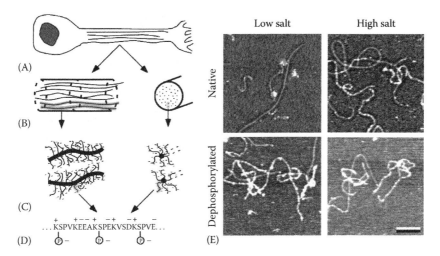

FIGURE 18.3   Axonal neurofilament (NF) networks. (A, B) Parallel orientation of NFs along the length of the axon. (C) Role of phosphorylated NF C-terminal sidearm domains in modulating interactions between adjacent NFs. (D) Representative amino acid sequence of sidearm domain of NF heavy chain showing cationic, anionic, and phosphorylated residues. (E) Effect of salt concentration and phosphorylation on structure of NF sidearms in vitro. When purified and imaged by AFM, native (i.e., phosphorylated) bovine NFs are surrounded by dark "exclusion zones" due to the thermal motion of the extended, unstructured sidearms. These zones persist whether the NFs are imaged in a low-salt buffer (top left) or a high-salt buffer (top right) in which long-range electrostatic interactions are screened. When the NFs are enzymatically dephosphorylated, the exclusion zones are no longer observed in either buffer, suggesting sidearm collapse (bottom left, right). (A–D: Reprinted from Kumar, S. et al., *Biophys. J.*, 82, 2360, 2002. With permission; E: Reprinted from Kumar, S. and Hoh, J.H., *Biochem. Biophys. Res. Commun.*, 324, 489, 2004. With permission.)

the rate of neurite extension, implying that calcium influx through mechanosensitive channels can inhibit outgrowth [92].

18.5   MECHANOBIOLOGY OF GLIAL CELLS IN THE CNS

Neurons have traditionally commanded the vast majority of attention in neurobiology research due to their propagation of action potentials, electrical impulses which travel along axons to facilitate quick and efficient transmission of signals over distances that can be greater than a meter. Non-neuronal cells within the CNS, known as glia (derived from the Greek word for "glue"), have historically been relegated to the accessory role of "support cells," with known functions ranging from the secretion and cilia-driven circulation of CSF by ependymal cells, to the myelination of axons by oligodendrocytes, to the structural support presumably provided by astrocytes, the most abundant cell type in the CNS. Over the last 20 years, new discoveries have challenged these old stereotypes and have stimulated renewed interest in the biology of glial cells, particularly astrocytes. Far from serving simply as the "glue of the nervous system," astrocytes are now thought to regulate adult neurogenesis [97–99], learning, and synaptic plasticity [100–105], and orchestration of the host response to injury in addition to protecting the CNS via maintenance of homeostasis and induction and maintenance of the blood–brain barrier [106].

The first direct viscoelastic characterization of individual glia and neurons, published in 2006, directly challenged the dogma of glia as structural support cells. These AFM indentation measurements revealed that glial cells are softer than neurons (and would therefore provide poor structural support) and that elastic forces dominate viscous forces in glial mechanics (therefore making them a poor glue) [59]. Yet, in 2008, direct probing of this hypothesis in situ via comparative tensile testing of spinal cord explants with an intact or disrupted glial matrix demonstrated that glia do provide significant mechanical support to spinal cord tissue under uniaxial tension [107]. While this is still an active area of debate, it is possible that the high compliance of individual glial cells allows them to protect neurons by cushioning them during trauma, whereas the architectural arrangement of star-shaped astrocytes into a cellular scaffold that physically couples blood vessels, neurons, and other glia may be critical in providing mechanical support to the tissue, especially given the absence of a robust ECM.

The idea that astrocytes may be ideally situated to sense and resist mechanical disruption in the brain through their unique scaffold architecture is not new; in fact, it has been over two decades since Alen Mathewson and Martin Berry first hypothesized that "architectural disruption" in the brain may be responsible for the phenomenon of astrocyte activation [108]. This activation, also referred to as reactive gliosis or astrogliosis, involves both astrocyte hypertrophy (abnormal enlargement of cell size) and hyperplasia (increase in cell number) in response to CNS pathologies ranging from neurodegenerative diseases to direct trauma, and often results in the formation of a glial scar [109]. Because astrocytes function as a syncytium of interconnected cells, mechanical deformation in one area of the brain due to primary stress (e.g., the mass effect of a tumor or direct stress due to trauma) or secondary stress (e.g., increasing pressure due to edema, the buildup of fluid following tissue insult) could quickly be biochemically and mechanically communicated to distant astrocytes, allowing rapid induction of reactive gliosis and other host response mechanisms.

Mathewson and Berry's architectural disruption hypothesis has been supported by in vitro observations of strain rate-dependent gliosis in three-dimensional (3D) cell culture models [110] and also by growing evidence that the mechanosensory machinery of astrocytes is particularly robust. For example, astrocytes directly convert mechanical stimuli into chemical signals using stretch-activated or stretch-inactivated ion channels. Ostrow et al. used micropipette indentation to directly demonstrate this phenomenon, showing that deformation of a single astrocyte induces transmembrane flux of Ca^{2+}, which propagates as a transcellular wave through gap junctions to neighboring astrocytes in confluent culture [111]. Chemical stimuli such as glutamate can similarly trigger the initiation and propagation of calcium waves in cultured astrocytes [112,113]; these calcium waves are believed to constitute a key signaling mechanism to orchestrate astrocytic functions ranging from guidance of CNS growth cones [114] to alterations in cell structure, gene expression, and proliferation [115,116]. Importantly, mechanical induction of intracellular second-messengers (e.g., inositol triphosphate and release of intracellular calcium) have been observed in cultured astrocytes as well [111].

Mounting evidence suggests that the components of the cellular contractile machinery regulate calcium signaling via gap junctions in addition to traditional mechanotransductive pathways. A relationship between a functional cytoskeleton and active calcium signaling in astrocytes was clearly established in 1998 by Cotrina et al., who demonstrated that neonatal astrocytes are unable to propagate calcium waves until the actin cytoskeleton is fully developed (a process that takes several hours), despite the existence of extensive gap junctional coupling shortly after subculture [117]. Furthermore, the radius of propagated calcium waves increases in direct proportion to the percentage of cells exhibiting a well-organized actin cytoskeleton (measured as the fraction of cells with visible actin stress fibers). Importantly, associated pharmacologic inhibition experiments revealed that calcium wave propagation is significantly attenuated by inhibition of myosin light chain kinase activity or actin polymerization but does not require microtubule organization.

In addition to these specialized mechanotransductive mechanisms, glia also exhibit many of the mechanobiological properties that are typical of cell types outside of the CNS. For example, mechanical stress induces rapid reorganization of both the intermediate filament network and the actin cytoskeleton [118,119]. The unique coupling of cytoskeletal elements across the astrocyte syncytium (e.g., organization of stress fibers into parallel bundles spanning multiple cells), however, suggests that transmission of mechanical signals in the brain may act across an unexpectedly long range [120]. This is potentially important as a mechanism for the dynamic production and regulation of coordinated responses to CNS injury and disease, as we will explore next.

18.6   CNS MECHANOBIOLOGY IN INJURY AND DISEASE

Many CNS diseases are intimately associated with structural changes that would be expected to alter the mechanical properties of the ECM and resident cells. For example, Alzheimer's Disease involves the gradual buildup of amyloid plaques and neurofibrillary tangles in the brain [121], a process which may affect both macroscopic and microenvironmental mechanics. Enhanced cell proliferation, de novo secretion of ECM proteins, and increased interstitial pressure within tumors increase the mechanical rigidity of tumor tissues relative to normal brain, as visualized by MRI and ultrasound elastography imaging of CNS malignancies [17,122–125]. It has even been hypothesized that some diseases may actually be *caused* by changes in the brain's mechanical properties; for example, it is thought that loss of tissue tensile strength following infarction can lead to physical obstruction of CSF flow, resulting in normal pressure hydrocephalus [126,127].

On the cellular level, the innate physiological response to CNS trauma or disease is related to mechanobiological phenomena in a number of interesting ways. For example, one of the hallmarks of astrocyte activation is increased expression of cytoskeletal intermediate filaments, including glial fibrillary associated protein, vimentin, and nestin, in addition to upregulation of focal adhesion proteins, such as vinculin, talin, and paxillin, and the actin-crosslinking protein alpha-actinin, implying that activated astrocytes should express a highly contractile phenotype [118,119,128]. The increase in tissue volume accompanying astrocyte hypertrophy and hyperplasia increases the stress on surrounding cells, as does secretion of additional ECM proteins, such as collagen IV and laminin [129], which subsequently form a scar of collagenous basement membrane that is thought to be one of the major impediments to axonal regeneration [130]. This local increase in stress can produce positive feedback to initiate further pathological changes, including enhanced expression of endothelin, a potent vasoconstrictor and astrocytic mitogen that is also associated with astrocyte activation in response to a variety of pathologies [116].

There are few other organ systems in which changes in local stress or tissue volume have the same implications as in the brain, where the maximum tissue volume is resolutely fixed by the encasing skull and local mechanical disturbances can be rapidly transmitted over long distances by the unique architecture of the astrocyte syncytium. As a result, secondary increases in mass or pressure that accompany many CNS diseases and injuries (e.g., from edema or the mass effects of tumor growth) can often contribute more to morbidity and mortality than the primary insult. The distinction between primary and secondary mechanical effects is especially significant in the case of traumatic brain injury, where direct trauma often imparts large, transient forces to CNS soft tissues, resulting in a complex cascade of secondary mechanical and biochemical responses. Investigations of tissue and cell-scale responses to traumatic CNS injury have broadly sought to understand not only how the transient forces present during the injury are directly transmitted to cells and tissues, but also what acute and long-term host response mechanisms are subsequently activated, and how these mechanisms can be exploited to enhance repair and the return of function [131].

Having reviewed general concepts in CNS mechanobiology, we now turn to a specific CNS pathology for which we and others have recently begun to elucidate the role of mechanobiology: the growth and invasion of the brain tumor glioblastoma multiforme (GBM).

CASE STUDY   Mechanobiology of Glioblastoma Multiforme

Primary brain tumors are abnormal masses of tissue that originate in the brain; they can be malignant (cancerous) or benign (noncancerous, i.e., not recurrent or progressive). Tumors arising from glia or their progenitors are called gliomas and are clinically divided into four grades according to the level of malignancy at diagnosis [132]. Grade IV gliomas, also known as GBM, represent the most common, aggressive, and neurologically destructive primary brain tumors. As the name would imply, GBM tumors are grossly heterogeneous (both inter- and intratumorally and at all levels, from tissue to cell to molecular and genetic), which may help explain why they are remarkably refractory to therapy. It has been over 80 years since the *Journal of the American Medical Association* (JAMA) published neurosurgeon Walter Dandy's report of what is perhaps the most radical GBM therapy to date: surgical hemispherectomy—literally, removal of an entire hemisphere of brain cortex [133]. Yet, Dandy concluded that even this radical procedure is ineffective at preventing rapid recurrence of the tumor, and sadly, an editorial published in JAMA almost 80 years later still cites a brain tumor diagnosis as one of the most feared by patients, physicians, and oncologists alike [134]. This fearsome reputation is well-deserved: despite extensive clinical and biological research efforts over the past several decades, there are still few proven risk factors for the development of GBM and little hope for long-term survival [134,135]. Even with the best available surgical care, chemotherapy, and radiation therapy, the average life expectancy at diagnosis is 12–15 months [136].

Clearly, the factors driving GBM progression are tightly woven into a complex network that has not yet been adequately dissected from either a basic science or therapeutic perspective. Tremendous effort has been devoted to elucidating the genetic and biochemical underpinnings of GBM over the last several decades; however, poor translation of candidate therapies from animal models to human patients has only increased the sense of urgency for the development of new approaches in both the laboratory and the clinic. Using words that could as easily have been written by Walter Dandy in 1928, the aforementioned 2005 JAMA editorial ended with a warning: "Advancements for patients with malignant glioma have been negligible, and there is a real risk of going nowhere by simply continuing to travel the same path" [134]. Indeed, the survival time for GBM has increased only incrementally over the past 25 years, with the most substantial recent advance (due to the alkylating agent temozolomide) improving survival by only an additional few months.

One novel path that has recently proved promising is investigation of GBM tumor cell mechanobiology. It has been known for decades that glioma cells retain many of the mechanosensory abilities of their nonmalignant counterparts, including stretch-activated ion channels and the ability to communicate transcellularly via gap junction-mediated calcium signaling [137–140]. Increased expression of connexins, the molecular building blocks of gap junctions, is correlated with enhanced calcium signaling between glioma cells and host astrocytes in rat xenograft models, and results in glioma cell invasion through a greater volume of brain parenchyma [141]. This implies that functional integration into the astrocytic syncytium, which itself is ideally posed to sense and transmit mechanical and biochemical signals over long distances in the brain, may constitute a significant support system for invasive GBM cells as they migrate away from the tumor mass into the surrounding tissue.

The remarkable capacity of single GBM tumor cells to diffusely infiltrate the surrounding brain parenchyma prior to diagnosis and following treatment is often cited as one of the key factors driving the uncommon aggressiveness of GBM. This infiltration ultimately renders surgical debulking and tumor bed irradiation ineffective, as was powerfully demonstrated by the failure of even Walter Dandy's hemispherectomies to prevent tumor recurrence. A central therapeutic goal in GBM has therefore been to develop new strategies to limit tumor cell invasion, thereby rendering the tumor more susceptible to anatomically directed therapies. While it is clear that biochemical signaling from the ECM is an important regulator of GBM tumor cell invasion, the biophysical components of this crosstalk have received comparatively little attention to date.

Hints that glioma invasion may be partly regulated by cell and tissue biomechanics have been inconspicuously scattered throughout the academic literature for over half a century. For example, neuropathologist Hans Scherer published extensive observations in 1940 describing the organization of invasive glioma cells into distinctive and predictable patterns that radiate away from the main tumor mass into the surrounding brain parenchyma [142]. Importantly, he noted that invasive cells spread preferentially along the surface of anatomical structures in the brain, including the basement membrane of blood vessels, white matter tracts, and the pia mater.

These infiltrative patterns are now commonly known as the "secondary structures of Scherer," and their association with biomechanically distinct components of the brain architecture may be especially informative from a mechanobiological perspective.

Several recent studies have highlighted the importance of microenvironmental mechanics to glioma cell physiology and invasion. For example, we recently cultured a panel of GBM cell lines on biochemically identical polymeric ECM substrates of defined mechanical rigidity (ranging from 0.08 kPa to 119.0 kPa). Our studies revealed stark rigidity-dependent differences in cell structure, motility, and proliferation (Figure 18.4) [143]. Specifically, tumor cells cultured on highly rigid ECMs spread extensively, formed prominent actomyosin stress fibers and mature focal adhesions, and migrated rapidly, whereas cells cultured on the most compliant ECMs (with rigidities comparable to normal brain tissue) appeared uniformly rounded and failed to productively migrate. We have subsequently explored the role of the focal adhesion protein α-actinin in glioma cell mechanobiology, motivated by the observations that α-actinin structurally couples the cellular adhesive and contractile machineries [144] and is significantly upregulated in high-grade astrocytomas [145]. We found that suppression of either α-actinin isoform (1 and 4) in human glioma reduces cell motility and traction forces and compromises the ability of cells to mechanically adapt to changes in ECM stiffness [146]. Importantly, glioma

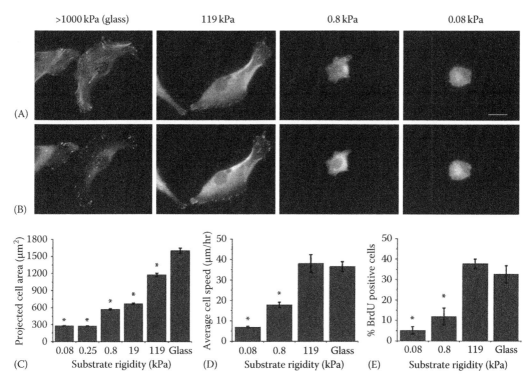

FIGURE 18.4 (See color insert.) The mechanical rigidity of the ECM regulates glioblastoma multiforme (GBM) tumor cell structure, motility, and proliferation. (A) Cell shape and cytoarchitecture. Human glioma cells cultured on fibronectin-conjugated glass and polyacrylamide gels of three different stiffnesses were fixed and stained for F-actin (green), nuclear DNA (blue), and the focal adhesion protein vinculin (red). Cells on glass and 119 kPa substrates exhibit robust focal adhesions and a well-defined cytoskeletal architecture, whereas cells on 0.80 and 0.08 kPa polyacrylamide gels are rounded with cortical rings of F-actin and small, punctate vinculin-positive focal complexes. Bar is 25 μm. (B) Isolated view of vinculin signal only, showing structure and distributions of cell-ECM adhesions. Effect of ECM mechanical rigidity on (C) Cell spreading area; (D) Migration rate; and (E) Cell proliferation, as measured by bromodeoxyuridine (BrdU) incorporation. (Adapted from Ulrich, T.A. et al., *Cancer Res.*, 69, 4167, 2009.)

cell rigidity-sensitivity can be blunted by direct or indirect pharmacologic inhibition of myosin-based contractility, providing support for a model in which ECM rigidity provides a transformative, microenvironmental cue that acts through actomyosin contractility to regulate the invasive properties of GBM tumor cells.

This model is consistent with the results of previous 3D in vitro studies of GBM invasion. While microenvironmental mechanics are more difficult to control in 3D cell culture models without altering integrin ligand density or microstructure, studies in which the stiffness of collagen I matrices was increased by increasing the concentration of collagen suggest that biophysical and biomolecular factors are both crucial regulators of glioma invasiveness [147,148]. We recently investigated 3D motility in collagen I matrices stiffened through the progressive addition of agarose, which we found restricted invasion by increasing steric barriers to motility and reducing the ability of tumor cells to bundle and remodel the collagen fibers [149]. Consistent with this finding, Kaufman et al. manipulated pore size in collagen I matrices by controlling the temperature of gelation and showed that pore sizes below 4–6 µm strongly limited glioma cell invasion speed [150]. To more carefully analyze microenvironmental mechanics during tumor invasion, Gordon et al. embedded 1 µm latex beads within Matrigel-based in vitro spheroid invasion assays, utilizing particle-tracking methods to analyze the spatial displacement of the tumor microenvironment at all stages of spheroid growth and invasion [151]. Their studies yielded a surprising juxtaposition of forces within the matrix surrounding the spheroid: volumetric expansion of the main tumor spheroid pushes the bulk of the gel outward, even as the matrix at the invasive front is pulled inward due to the localized generation of cell traction forces.

Importantly, there is evidence that the mechanobiological machinery of glioma cells differs from that of their nonmalignant counterparts, a crucial prerequisite for the development of mechanobiologically inspired therapeutics. For example, glioma cells exhibit reduced expression of cadherins (calcium-dependent transmembrane glycoproteins that facilitate cell–cell adhesion), enhanced expression of matrix metalloproteinases, increased expression of focal adhesion proteins, such as focal adhesion kinase (FAK), and altered expression of integrins compared to normal astrocytes [4,152–156]. These differences are potentially significant from the standpoints of both basic pathophysiology and therapeutics. For example, integrin-mediated adhesion of tumor cells to ECM proteins has been associated with greater resistance to ionizing radiation and chemotherapies, a phenomenon known as cell adhesion-mediated radioresistance/drug resistance (CAM-RR/CAM-DR) [157]. Recent studies have linked $\beta 1$ integrin signaling in particular with inhibition of drug-induced apoptosis [158] and promotion of radioresistance [159]. Enhanced expression of integrins $\alpha 2$, $\alpha 3$, $\alpha 5$, and $\beta 1$ in drug-resistant glioma cells has been correlated with enhanced adhesivity to ECM proteins such as fibronectin and collagen as well [160]; these proteins are more commonly found in tumor tissue and basement membrane than normal brain parenchyma, suggesting that CAM-DR may also promote tumor progression and invasiveness. Surprisingly, a recent study showed that pharmacologic inhibition of fibronectin assembly in the ECM can enhance sensitivity of GBM cells to nitrosourea chemotherapy in vitro and in vivo [161]; however, the dynamics of this relationship and the mechanisms driving ECM-derived chemosensitivity are not yet understood. Nevertheless, promising new chemotherapeutics are already beginning to target components of the contractility and adhesion machinery, including the potent integrin antagonists Cilengitide and SJ749 [162,163], radioiodinated antibodies directed against tumor-secreted ECM proteins [164,165], drugs that inhibit Rho GTPase-based signaling [166], and small-molecule inhibitors of FAK and other focal adhesion proteins [167].

The dynamic range of glioma cell mechanosensitivity may differ from that of their nonmalignant counterparts as well; for example, while highly compliant substrates (~100–500 Pa) have been found to select against the survival and proliferation of astrocytes in both explant studies and aNSC differentiation studies [42,49], cultured glioma cells survive and proliferate even on 80 Pa polyacrylamide gels [143]. Interestingly, however, the rate of incorporation of bromodeoxyuridine (BrdU), a labeled nucleotide taken up only by dividing cells, on these soft gels is reduced approximately fivefold over BrdU incorporation on a stiff (119 kPa) polyacryl-amide gel or glass. While the mechanism governing this effect is not yet known, ECM rigidity has been previously observed to modulate cell growth in other systems, including cultured fibroblasts [168], hepatocytes [169], and a variety of adult stem cells [40,42].

One potential explanation for the correlation between microenvironmental rigidity and cell proliferation is that changes in ECM rigidity might alter the speed of progression through the cell cycle by altering mechanochemical feedback during mitosis. Indeed, direct application of mechanical force has been observed to slow cytokinesis and induce shape asymmetries in *Dictyostelium discoideum* cells, which is actively corrected via mobilization of nonmuscle myosin II (NMMII) to produce a restoring force [170]. Second, ECM rigidity might regulate mitosis by synergistically triggering mechanotransductive and mitogenic signaling pathways, as has been suggested by recent studies from the breast cancer community, which reveal that modest increases in ECM rigidity can transform cultured breast epithelial cells from a benign, highly differentiated phenotype into a dysplastic and proliferative one [171,172]. Importantly, this matrix-driven transformation is accompanied by activation of extracellular signal-regulated kinase (ERK)-mediated proliferative signaling and activation of the contractile markers Rho GTPase and NMMII, which enables enhanced generation of contractile forces. Importantly for therapeutic applications, this rigidity-dependent phenotype can be reversed by pharmacological inhibition of Rho-associated kinase (ROCK) or ERK activity.

These contractility-mediating pathways are also intimately related to cell migration and inva-sion, which depend on actomyosin-generated contractile forces and involve a variety of dynamic and spatially regulated changes to both the cytoskeleton and the adhesion complexes that mediate interactions with the surrounding ECM. In an important study that has general implications for cell migration through 3D ECMs, Rosenfeld and coworkers recently demonstrated that NMMII is needed to deform the nucleus of glioma cells to enable amoeboid motion through ECM pores, and invading tumor cells in vivo significantly upregulate NMMII expression relative to endog-enous brain cells (Figure 18.5) [173]. Other recent work has shown that pharmacologic inhibition of myosin light chain kinase results in dramatic inhibition of glioma cell motility [174] and that ROCK-dependent mechanisms are important in GBM cell migration and therapeutic sensitivity [175–182]. Rho/ROCK signaling is thought to be especially important in regulating cell survival and tumorigenesis as well; for example, the ROCK inhibitor Y-27632 and transfection with dominant negative RhoA and ROCK were each found to induce apoptosis in vitro and resulted in significantly smaller tumor mass following tumor inoculation in vivo [183].

In summary, the emerging field of GBM mechanobiology has begun to infuse an apprecia-tion for mechanics into our overall understanding of GBM pathophysiology. These promising and surprising early results suggest that further exploration of the mechanobiological aspects of GBM tumor cells may constitute a new and valuable path toward the identification of novel therapeutic targets, and that these paradigms and approaches might be productively extended to other CNS pathologies.

18.7 CONCLUSION

The last several decades have seen an emerging appreciation for the complex and unexpected ways in which mechanobiology can regulate the CNS in health and disease. Here, we have reviewed this growing field, starting with the structure and mechanical microenvironment of the brain and the spinal cord, continuing with the mechanobiological characteristics of cells in the CNS, and concluding with the importance of mechanobiology to the progression of specific CNS disease states. Despite remarkable progress, many key challenges remain. In particular, it will be critical to determine how mechanobiological signaling in the CNS fits into the context of traditionally

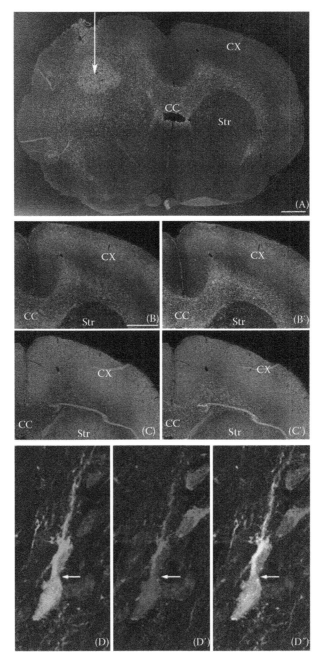

FIGURE 18.5 (See color insert.) Invasive glioma cells demonstrate enhanced myosin IIA expression. (A) Growth and spread of implanted tumor cells. Rat brain slice stained for human nuclear antigen (green) shows that implanted primary human glioma cells spread from the site of tumor inoculation (arrow) across the corpus collosum (CC) to the contralateral white matter, between the cortex (CX) and striatum (Str). Bar is 1 mm. (B) Enhanced immunofluorescence localization of myosin IIA in invasive cells (red). Bar is 1 mm. (B′) Colocalization of myosin IIA and human nuclear antigen. (C, C′) Corresponding immunofluorescence localization of myosin IIB (red), demonstrating equivalent or reduced expression in invasive glioma cells (green) relative to the surrounding normal brain tissue. (D) Nuclear deformation of invasive cells. A GFP-expressing human glioma cell (green) requires significant nuclear deformation (arrow; nucleus stained blue with DAPI) to infiltrate the surrounding normal brain tissue. (D′) Expression of myosin IIA (red) in infiltrative cells. (D″) Colocalization of myosin IIA and GFP. (Adapted from Beadle, C. et al., *Mol. Biol. Cell*, 19, 3357, 2008. With permission.)

understood genetic and biochemical control of neurobiology; cells in vivo are simultaneously subjected to space- and time-dependent mechanical and soluble/matrix-bound biochemical signals, and their response to this constellation of inputs undoubtedly depends on cell-intrinsic factors (e.g., regulation of gene expression) that may dramatically change from one cell to another in the same microenvironment. Understanding which signals dominate cell behavior in specific physiological settings is key to dissecting these cues' relative contributions to normal function and pathophysiology in the CNS. Related to this, it will be essential to extend the many elegant tools that have been developed for studying the mechanics and mechanobiology of single cells in culture to living tissues and organisms. As discussed earlier, noninvasive imaging methodologies that derive contrast from variations in tissue mechanics, such as ultrasound and MRE, hold great promise along these lines. One would also expect that an increased use of mechanosensitive optical probes and fluorescently labeled mechanosensory proteins in animal models may also enable in vivo visualization of CNS mechanotransduction. We anticipate that careful attention to these and other challenges will bring us closer to a day when both the CNS mechanical microenvironment and cellular mechanotransductory signaling systems can be exploited to regenerate nerve tissue, combat invasive brain tumors, and attack complex neuropathologies.

REFERENCES

1. Suresh S. Biomechanics and biophysics of cancer cells. *Acta Biomater* 2007; 3: 413–438.
2. Discher D, Dong C, Fredberg JJ et al. Biomechanics: Cell research and applications for the next decade. *Ann Biomed Eng* 2009; 37: 847–859.
3. Rao JS. Molecular mechanisms of glioma invasiveness: The role of proteases. *Nat Rev Cancer* 2003; 3: 489–501.
4. Bellail AC, Hunter SB, Brat DJ, Tan C, Van Meir EG. Microregional extracellular matrix heterogeneity in brain modulates glioma cell invasion. *Int J Biochem Cell Biol* 2004; 36: 1046–1069.
5. Jaime Struve PCM, Li Y-Q, Kinney S, Fehlings MG, Kuntz C, IV, Sherman LS. Disruption of the hyaluronan-based extracellular matrix in spinal cord promotes astrocyte proliferation. *Glia* 2005; 52: 16–24.
6. Greitz D, Franck A, Nordell B. On the pulsatile nature of intracranial and spinal CSF-circulation demonstrated by MR imaging. *Acta Radiol* 1993; 34: 321–328.
7. Soellinger M, Rutz AK, Kozerke S, Boesiger P. 3D cine displacement-encoded MRI of pulsatile brain motion. *Magn Reson Med* 2009; 61: 153–162.
8. Wagshul ME, Chen JJ, Egnor MR, McCormack EJ, Roche PE. Amplitude and phase of cerebrospinal fluid pulsations: Experimental studies and review of the literature. *J Neurosurg* 2006; 104: 810–819.
9. Fallenstein GT, Hulce VD, Melvin JW. Dynamic mechanical properties of human brain tissue. *J Biomech* 1969; 2: 217–226.
10. Uftring SJ, Chu D, Alperin N, Levin DN. The mechanical state of intracranial tissues in elderly subjects studied by imaging CSF and brain pulsations. *Magn Reson Imag* 2000; 18: 991–996.
11. Elkin BS, Azeloglu EU, Costa KD, Morrison B, III. Mechanical heterogeneity of the rat hippocampus measured by atomic force microscope indentation. *J Neurotrauma* 2007; 24: 812–822.
12. Prange MT, Margulies SS. Regional, directional, and age-dependent properties of the brain undergoing large deformation. *J Biomech Eng* 2002; 124: 244–252.
13. Prange MT, Meaney DF, Margulies SS. Defining brain mechanical properties: Effects of region, direction, and species. *Stapp Car Crash J* 2000; 44: 205–213.
14. Hrapko M, van Dommelen JA, Peters GW, Wismans JS. The influence of test conditions on characterization of the mechanical properties of brain tissue. *J Biomech Eng* 2008; 130: 031003.
15. Cheng S, Clarke EC, Bilston LE. Rheological properties of the tissues of the central nervous system: A review. *Med Eng Phys* 2008; 30: 1318–1337.
16. Janmey PA, Georges PC, Hvidt S. Basic rheology for biologists. *Methods Cell Biol* 2007; 83: 3–27.
17. Vappou J, Breton E, Choquet P, Willinger R, Constantinesco A. Assessment of in vivo and post-mortem mechanical behavior of brain tissue using magnetic resonance elastography. *J Biomech* 2008; 41: 2954–2959.
18. Kruse SA, Rose GH, Glaser KJ et al. Magnetic resonance elastography of the brain. *Neuroimage* 2008; 39: 231–237.

19. Green MA, Bilston LE, Sinkus R. In vivo brain viscoelastic properties measured by magnetic resonance elastography. *NMR Biomed* 2008; 21: 755–764.
20. Atay SM, Kroenke CD, Sabet A, Bayly PV. Measurement of the dynamic shear modulus of mouse brain tissue in vivo by magnetic resonance elastography. *J Biomech Eng* 2008; 130: 021013.
21. Manduca A, Oliphant TE, Dresner MA et al. Magnetic resonance elastography: Non-invasive mapping of tissue elasticity. *Med Image Anal* 2001; 5: 237–254.
22. Vappou J, Breton E, Choquet P, Goetz C, Willinger R, Constantinesco A. Magnetic resonance elastography compared with rotational rheometry for in vitro brain tissue viscoelasticity measurement. *MAGMA* 2007; 20: 273–278.
23. Gefen A, Margulies SS. Are in vivo and in situ brain tissues mechanically similar? *J Biomech* 2004; 37: 1339–1352.
24. Gefen A, Gefen N, Zhu Q, Raghupathi R, Margulies SS. Age-dependent changes in material properties of the brain and braincase of the rat. *J Neurotrauma* 2003; 20: 1163–1177.
25. Thibault KL, Margulies SS. Age-dependent material properties of the porcine cerebrum: Effect on pediatric inertial head injury criteria. *J Biomech* 1998; 31: 1119–1126.
26. Metz H, McElhaney J, Ommaya AK. A comparison of the elasticity of live, dead, and fixed brain tissue. *J Biomech* 1970; 3: 453–458.
27. Hrapko M, van Dommelen JA, Peters GW, Wismans JS. The mechanical behaviour of brain tissue: Large strain response and constitutive modelling. *Biorheology* 2006; 43: 623–636.
28. Maikos JT, Elias RA, Shreiber DI. Mechanical properties of dura mater from the rat brain and spinal cord. *J Neurotrauma* 2008; 25: 38–51.
29. Darvish KK, Crandall JR. Nonlinear viscoelastic effects in oscillatory shear deformation of brain tissue. *Med Eng Phys* 2001; 23: 633–645.
30. Takhounts EG, Crandall JR, Darvish K. On the importance of nonlinearity of brain tissue under large deformations. *Stapp Car Crash J* 2003; 47: 79–92.
31. Miller K, Chinzei K, Orssengo G, Bednarz P. Mechanical properties of brain tissue in-vivo: Experiment and computer simulation. *J Biomech* 2000; 33: 1369–1376.
32. Storm C, Pastore JJ, MacKintosh FC, Lubensky TC, Janmey PA. Nonlinear elasticity in biological gels. *Nature* 2005; 435: 191–194.
33. Yeung T, Georges PC, Flanagan LA et al. Effects of substrate stiffness on cell morphology, cytoskeletal structure, and adhesion. *Cell Motil Cytoskeleton* 2005; 60: 24–34.
34. Pelham RJ, Jr., Wang Y. Cell locomotion and focal adhesions are regulated by substrate flexibility. *Proc Natl Acad Sci USA* 1997; 94: 13661–13665.
35. Wang N, Ingber DE. Control of cytoskeletal mechanics by extracellular matrix, cell shape, and mechanical tension. *Biophys J* 1994; 66: 2181–2189.
36. Lo CM, Wang HB, Dembo M, Wang YL. Cell movement is guided by the rigidity of the substrate. *Biophys J* 2000; 79: 144–152.
37. Peyton SR, Putnam AJ. Extracellular matrix rigidity governs smooth muscle cell motility in a biphasic fashion. *J Cell Physiol* 2005; 204: 198–209.
38. Halliday NL, Tomasek JJ. Mechanical properties of the extracellular matrix influence fibronectin fibril assembly in vitro. *Exp Cell Res* 1995; 217: 109–117.
39. Engler AJ, Sen S, Sweeney HL, Discher DE. Matrix elasticity directs stem cell lineage specification. *Cell* 2006; 126: 677–689.
40. Rowlands AS, George PA, Cooper-White JJ. Directing osteogenic and myogenic differentiation of MSCs: Interplay of stiffness and adhesive ligand presentation. *Am J Physiol Cell Physiol* 2008; 295: C1037–C1044.
41. Winer JP, Janmey PA, McCormick ME, Funaki M. Bone marrow-derived human mesenchymal stem cells become quiescent on soft substrates but remain responsive to chemical or mechanical stimuli. *Tissue Eng Part A* 2008; 15: 147–154.
42. Saha K, Keung AJ, Irwin EF et al. Substrate modulus directs neural stem cell behavior. *Biophys J* 2008; 95: 4426–4438.
43. Balgude AP, Yu X, Szymanski A, Bellamkonda RV. Agarose gel stiffness determines rate of DRG neurite extension in 3D cultures. *Biomaterials* 2001; 22: 1077–1084.
44. Leach JB, Brown XQ, Jacot JG, Dimilla PA, Wong JY. Neurite outgrowth and branching of PC12 cells on very soft substrates sharply decreases below a threshold of substrate rigidity. *J Neural Eng* 2007; 4: 26–34.
45. Flanagan LA, Ju YE, Marg B, Osterfield M, Janmey PA. Neurite branching on deformable substrates. *Neuroreport* 2002; 13: 2411–2415.

46. Willits RK, Skornia SL. Effect of collagen gel stiffness on neurite extension. *J Biomater Sci Polym Ed* 2004; 15: 1521–1531.
47. Van Essen DC. A tension-based theory of morphogenesis and compact wiring in the central nervous system. *Nature* 1997; 385: 313–318.
48. Zhao C, Deng W, Gage FH. Mechanisms and functional implications of adult neurogenesis. *Cell* 2008; 132: 645–660.
49. Georges PC, Miller WJ, Meaney DF, Sawyer ES, Janmey PA. Matrices with compliance comparable to that of brain tissue select neuronal over glial growth in mixed cortical cultures. *Biophys J* 2006; 90: 3012–3018.
50. Miller WJ, Levental I, Scarsella D, Haydon PG, Janmey PA, Meaney D. Mechanically induced reactive gliosis causes atp-mediated alterations in astrocyte stiffness. *J Neurotrauma* 2009; 26: 789–797.
51. Lam WA, Cao L, Umesh V, Keung AJ, Sen S, Kumar S. Extracellular matrix rigidity modulates neuroblastoma cell differentiation and N-myc expression. *Mol Cancer* 2010; 9: 35.
52. Lamoureux P, Buxbaum RE, Heidemann SR. Direct evidence that growth cones pull. *Nature* 1989; 340: 159–162.
53. Zheng J, Lamoureux P, Santiago V, Dennerll T, Buxbaum RE, Heidemann SR. Tensile regulation of axonal elongation and initiation. *J Neurosci* 1991; 11: 1117–1125.
54. Dennerll TJ, Joshi HC, Steel VL, Buxbaum RE, Heidemann SR. Tension and compression in the cytoskeleton of PC-12 neurites. II: Quantitative measurements. *J Cell Biol* 1988; 107: 665–674.
55. Bray D. Axonal growth in response to experimentally applied mechanical tension. *Dev Biol* 1984; 102: 379–389.
56. Chada S, Lamoureux P, Buxbaum RE, Heidemann SR. Cytomechanics of neurite outgrowth from chick brain neurons. *J Cell Sci* 1997; 110(Pt 5): 1179–1186.
57. Fischer TM, Steinmetz PN, Odde DJ. Robust micromechanical neurite elicitation in synapse-competent neurons via magnetic bead force application. *Ann Biomed Eng* 2005; 33: 1229–1237.
58. Fass JN, Odde DJ. Tensile force-dependent neurite elicitation via anti-beta1 integrin antibody-coated magnetic beads. *Biophys J* 2003; 85: 623–636.
59. Lu YB, Franze K, Seifert G et al. Viscoelastic properties of individual glial cells and neurons in the CNS. *Proc Natl Acad Sci USA* 2006; 103: 17759–17764.
60. Dennerll TJ, Lamoureux P, Buxbaum RE, Heidemann SR. The cytomechanics of axonal elongation and retraction. *J Cell Biol* 1989; 109: 3073–3083.
61. Kumar S, Maxwell IZ, Heisterkamp A et al. Viscoelastic retraction of single living stress fibers and its impact on cell shape, cytoskeletal organization, and extracellular matrix mechanics. *Biophys J* 2006; 90: 3762–3773.
62. Heisterkamp A, Maxwell IZ, Mazur E et al. Pulse energy dependence of subcellular dissection by femtosecond laser pulses. *Opt Express* 2005; 13: 3690–3696.
63. Puig-De-Morales M, Grabulosa M, Alcaraz J et al. Measurement of cell microrheology by magnetic twisting cytometry with frequency domain demodulation. *J Appl Physiol* 2001; 91: 1152–1159.
64. Maksym GN, Fabry B, Butler JP et al. Mechanical properties of cultured human airway smooth muscle cells from 0.05 to 0.4 Hz. *J Appl Physiol* 2000; 89: 1619–1632.
65. Wang N, Ingber DE. Probing transmembrane mechanical coupling and cytomechanics using magnetic twisting cytometry. *Biochem Cell Biol* 1995; 73: 327–335.
66. Alcaraz J, Buscemi L, Grabulosa M et al. Microrheology of human lung epithelial cells measured by atomic force microscopy. *Biophys J* 2003; 84: 2071–2079.
67. Park S, Koch D, Cardenas R, Käs J, Shih CK. Cell motility and local viscoelasticity of fibroblasts. *Biophys J* 2005; 89: 4330–4342.
68. Mahaffy RE, Shih CK, MacKintosh FC, Kas J. Scanning probe-based frequency-dependent microrheology of polymer gels and biological cells. *Phys Rev Lett* 2000; 85: 880–883.
69. Mahaffy RE, Park S, Gerde E, Kas J, Shih CK. Quantitative analysis of the viscoelastic properties of thin regions of fibroblasts using atomic force microscopy. *Biophys J* 2004; 86: 1777–1793.
70. Evans E, Yeung A. Apparent viscosity and cortical tension of blood granulocytes determined by micropipet aspiration. *Biophys J* 1989; 56: 151–160.
71. Discher DE, Mohandas N, Evans EA. Molecular maps of red cell deformation: Hidden elasticity and in situ connectivity. *Science* 1994; 266: 1032–1035.
72. Chien S, Schmid-Schonbein GW, Sung KL, Schmalzer EA, Skalak R. Viscoelastic properties of leukocytes. *Kroc Found Ser* 1984; 16: 19–51.
73. Mills JP, Qie L, Dao M, Lim CT, Suresh S. Nonlinear elastic and viscoelastic deformation of the human red cell with optical tweezers. *Mech Chem Biosyst* 2004; 1: 169–180.

74. Laurent VM, Henon S, Planus E et al. Assessment of mechanical properties of adherent living cells by bead micromanipulation: Comparison of magnetic twisting cytometry vs optical tweezers. *J Biomech Eng* 2002; 124: 408–421.

75. Sen S, Kumar S. Cell-matrix de-adhesion dynamics reflect contractile mechanics. *Cell Mol Bioeng* 2009; 2: 218–230.

76. Janmey P, Schmidt C. Experimental measurements of intracellular mechanics. In: Mofrad MR, Kamm RD, eds. *Cytoskeletal Mechanics: Models and Measurements*. New York: Cambridge University Press; 2006, pp. 18–49.

77. Bouquet C, Nothias F. Molecular mechanisms of axonal growth. *Adv Exp Med Biol* 2007; 621: 1–16.

78. Wen Z, Zheng JQ. Directional guidance of nerve growth cones. *Curr Opin Neurobiol* 2006; 16: 52–58.

79. Pak CW, Flynn KC, Bamburg JR. Actin-binding proteins take the reins in growth cones. *Nat Rev Neurosci* 2008; 9: 136–147.

80. Kumar S, Hoh JH. Modulation of repulsive forces between neurofilaments by sidearm phosphorylation. *Biochem Biophys Res Commun* 2004; 324: 489–496.

81. Kumar S, Yin X, Trapp BD, Hoh JH, Paulaitis ME. Relating interactions between neurofilaments to the structure of axonal neurofilament distributions through polymer brush models. *Biophys J* 2002; 82: 2360–2372.

82. Kumar S, Yin X, Trapp BD, Paulaitis ME, Hoh JH. Role of long-range repulsive forces in organizing axonal neurofilament distributions: Evidence from mice deficient in myelin-associated glycoprotein. *J Neurosci Res* 2002; 68: 681–690.

83. Guo X, Rueger D, Higgins D. Osteogenic protein-1 and related bone morphogenetic proteins regulate dendritic growth and the expression of microtubule-associated protein-2 in rat sympathetic neurons. *Neurosci Lett* 1998; 245: 131–134.

84. Aranda-Espinoza H, Carl P, Leterrier JF, Janmey P, Discher DE. Domain unfolding in neurofilament sidearms: Effects of phosphorylation and ATP. *FEBS Lett* 2002; 531: 397–401.

85. Okonkwo DO, Pettus EH, Moroi J, Povlishock JT. Alteration of the neurofilament sidearm and its relation to neurofilament compaction occurring with traumatic axonal injury. *Brain Res* 1998; 784: 1–6.

86. Smith DH, Meaney DF. Axonal damage in traumatic brain injury. *Neuroscientist* 2000; 6: 483–495.

87. Mukhopadhyay R, Hoh JH. AFM force measurements on microtubule-associated proteins: The projection domain exerts a long-range repulsive force. *FEBS Lett* 2001; 505: 374–378.

88. Linseman DA, Loucks FA. Diverse roles of Rho family GTPases in neuronal development, survival, and death. *Front Biosci* 2008; 13: 657–676.

89. Giniger E. How do Rho family GTPases direct axon growth and guidance? A proposal relating signaling pathways to growth cone mechanics. *Differentiation* 2002; 70: 385–396.

90. Brown ME, Bridgman PC. Myosin function in nervous and sensory systems. *J Neurobiol* 2004; 58: 118–130.

91. Imai K, Tatsumi H, Katayama Y. Mechanosensitive chloride channels on the growth cones of cultured rat dorsal root ganglion neurons. *Neuroscience* 2000; 97: 347–355.

92. Jacques-Fricke BT, Seow Y, Gottlieb PA, Sachs F, Gomez TM. Ca^{2+} influx through mechanosensitive channels inhibits neurite outgrowth in opposition to other influx pathways and release from intracellular stores. *J Neurosci* 2006; 26: 5656–5664.

93. Sigurdson WJ, Morris CE. Stretch-activated ion channels in growth cones of snail neurons. *J Neurosci* 1989; 9: 2801–2808.

94. Kamkin A, Kiseleva I, eds. *Mechanosensitivity of the Nervous System*. Springer: the Netherlands; 2009.

95. Bianchi L. Mechanotransduction: Touch and feel at the molecular level as modeled in Caenorhabditis elegans. *Mol Neurobiol* 2007; 36: 254–271.

96. Tang Y, Yoo J, Yethiraj A, Cui Q, Chen X. Mechanosensitive channels: Insights from continuum-based simulations. *Cell Biochem Biophys* 2008; 52: 1–18.

97. Barkho BZ, Song H, Aimone JB et al. Identification of astrocyte-expressed factors that modulate neural stem/progenitor cell differentiation. *Stem Cells Dev* 2006; 15: 407–421.

98. Ma DK, Ming GL, Song H. Glial influences on neural stem cell development: Cellular niches for adult neurogenesis. *Curr Opin Neurobiol* 2005; 15: 514–520.

99. Song H, Stevens CF, Gage FH. Astroglia induce neurogenesis from adult neural stem cells. *Nature* 2002; 417: 39–44.

100. Araque A, Sanzgiri RP, Parpura V, Haydon PG. Astrocyte-induced modulation of synaptic transmission. *Can J Physiol Pharmacol* 1999; 77: 699–706.

101. Bacci A, Verderio C, Pravettoni E, Matteoli M. The role of glial cells in synaptic function. *Philos Trans R Soc Lond B Biol Sci* 1999; 354: 403–409.

102. Wenzel J, Lammert G, Meyer U, Krug M. The influence of long-term potentiation on the spatial relationship between astrocyte processes and potentiated synapses in the dentate gyrus neuropil of rat brain. *Brain Res* 1991; 560: 122–131.
103. Araque A, Carmignoto G, Haydon PG. Dynamic signaling between astrocytes and neurons. *Annu Rev Physiol* 2001; 63: 795–813.
104. Anderson BJ, Li X, Alcantara AA, Isaacs KR, Black JE, Greenough WT. Glial hypertrophy is associated with synaptogenesis following motor-skill learning, but not with angiogenesis following exercise. *Glia* 1994; 11: 73–80.
105. Ullian EM, Sapperstein SK, Christopherson KS, Barres BA. Control of synapse number by glia. *Science* 2001; 291: 657–661.
106. Volterra A, Meldolesi J. Astrocytes, from brain glue to communication elements: The revolution continues. *Nat Rev Neurosci* 2005; 6: 626–640.
107. Shreiber DI, Hao H, Elias RA. Probing the influence of myelin and glia on the tensile properties of the spinal cord. *Biomech Model Mechanobiol* 2008; 8: 311–321.
108. Mathewson AJ, Berry M. Observations on the astrocyte response to a cerebral stab wound in adult rats. *Brain Res* 1985; 327: 61–69.
109. Milos Pekny MN. Astrocyte activation and reactive gliosis. *Glia* 2005; 50: 427–434.
110. Cullen DK, Simon CM, LaPlaca MC. Strain rate-dependent induction of reactive astrogliosis and cell death in three-dimensional neuronal-astrocytic co-cultures. *Brain Res* 2007; 1158: 103–115.
111. Ostrow LW, Langan TJ, Sachs F. Stretch-induced endothelin-1 production by astrocytes. *J Cardiovasc Pharmacol* 2000; 36: S274–S277.
112. Cornell-Bell AH, Finkbeiner SM, Cooper MS, Smith SJ. Glutamate induces calcium waves in cultured astrocytes: Long-range glial signaling. *Science* 1990; 247: 470–473.
113. Charles AC, Merrill JE, Dirksen ER, Sanderson MJ. Intercellular signaling in glial cells: Calcium waves and oscillations in response to mechanical stimulation and glutamate. *Neuron* 1991; 6: 983–992.
114. Hung J, Colicos MA. Astrocytic Ca(2+) waves guide CNS growth cones to remote regions of neuronal activity. *PLoS ONE* 2008; 3: e3692.
115. Agulhon C, Petravicz J, McMullen AB et al. What is the role of astrocyte calcium in neurophysiology? *Neuron* 2008; 59: 932–946.
116. Ostrow LW, Sachs F. Mechanosensation and endothelin in astrocytes–hypothetical roles in CNS pathophysiology. *Brain Res Rev* 2005; 48: 488–508.
117. Cotrina ML, Lin JH, Nedergaard M. Cytoskeletal assembly and ATP release regulate astrocytic calcium signaling. *J Neurosci* 1998; 18: 8794–8804.
118. Abd-El-Basset EM, Fedoroff S. Upregulation of F-actin and alpha-actinin in reactive astrocytes. *J Neurosci Res* 1997; 49: 608–616.
119. Takamiya Y, Kohsaka S, Toya S, Otani M, Tsukada Y. Immunohistochemical studies on the proliferation of reactive astrocytes and the expression of cytoskeletal proteins following brain injury in rats. *Brain Res* 1988; 466: 201–210.
120. Abd-el-Basset EM, Fedoroff S. Contractile units in stress fibers of fetal human astroglia in tissue culture. *J Chem Neuroanat* 1994; 7: 113–122.
121. Parihar MS, Hemnani T. Alzheimer's disease pathogenesis and therapeutic interventions. *J Clin Neurosci* 2004; 11: 456–467.
122. Scholz M, Noack V, Pechlivanis I et al. Vibrography during tumor neurosurgery. *J Ultrasound Med* 2005; 24: 985–992.
123. Xu L, Lin Y, Han JC, Xi ZN, Shen H, Gao PY. Magnetic resonance elastography of brain tumors: Preliminary results. *Acta Radiol* 2007; 48: 327–330.
124. Unsgaard G, Rygh OM, Selbekk T et al. Intra-operative 3D ultrasound in neurosurgery. *Acta Neurochir (Wien)* 2006; 148: 235–253; discussion 53.
125. Selbekk T, Bang J, Unsgaard G. Strain processing of intraoperative ultrasound images of brain tumours: Initial results. *Ultrasound Med Biol* 2005; 31: 45–51.
126. Bradley WG Jr, Whittemore AR, Watanabe AS, Davis SJ, Teresi LM, Homyak M. Association of deep white matter infarction with chronic communicating hydrocephalus: Implications regarding the possible origin of normal-pressure hydrocephalus. *AJNR Am J Neuroradiol* 1991; 12: 31–39.
127. Del Bigio MR. Neuropathological changes caused by hydrocephalus. *Acta Neuropathol* 1993; 85: 573–585.
128. Kalman M, Szabo A. Immunohistochemical investigation of actin-anchoring proteins vinculin, talin and paxillin in rat brain following lesion: A moderate reaction, confined to the astroglia of brain tracts. *Exp Brain Res* 2001; 139: 426–434.

129. Liesi P, Kauppila T. Induction of type IV collagen and other basement-membrane-associated proteins after spinal cord injury of the adult rat may participate in formation of the glial scar. *Exp Neurol* 2002; 173: 31–45.

130. Hermanns S, Reiprich P, Müller HW. A reliable method to reduce collagen scar formation in the lesioned rat spinal cord. *J Neurosci Meth* 2001; 110: 141–146.

131. Spaethling JM, Geddes-Klein DM, Miller WJ et al. Linking impact to cellular and molecular sequelae of CNS injury: Modeling in vivo complexity with in vitro simplicity. *Progr Brain Res* 2007; 161: 27–39.

132. Furnari FB, Fenton T, Bachoo RM et al. Malignant astrocytic glioma: Genetics, biology, and paths to treatment. *Genes Dev* 2007; 21: 2683–2710.

133. Dandy WE. Removal of right cerebral hemisphere for certain tumors with hemiplegia: Preliminary report. *JAMA* 1928; 90: 823–825.

134. Fisher PG, Buffler PA. Malignant gliomas in 2005: Where to GO from here? *JAMA* 2005; 293: 615–617.

135. Holland EC. Glioblastoma multiforme: The terminator. *Proc Natl Acad Sci USA* 2000; 97: 6242–6244.

136. Stupp R, Hegi ME, van den Bent MJ et al. Changing paradigms—An update on the multidisciplinary management of malignant glioma. *Oncologist* 2006; 11: 165–180.

137. Charles AC, Naus CC, Zhu D, Kidder GM, Dirksen ER, Sanderson MJ. Intercellular calcium signaling via gap junctions in glioma cells. *J Cell Biol* 1992; 118: 195–201.

138. Zhang W, Couldwell WT, Simard MF, Song H, Lin JH, Nedergaard M. Direct gap junction communication between malignant glioma cells and astrocytes. *Cancer Res* 1999; 59: 1994–2003.

139. Yamasaki T, Enomoto K, Moritake K, Maeno T. Analysis of intra- and intercellular calcium signaling in a mouse malignant glioma cell line. *J Neurosurg* 1994; 81: 420–426.

140. Fry T, Evans JH, Sanderson MJ. Propagation of intercellular calcium waves in C6 glioma cells transfected with connexins 43 or 32. *Microsc Res Tech* 2001; 52: 289–300.

141. Lin JH, Takano T, Cotrina ML et al. Connexin 43 enhances the adhesivity and mediates the invasion of malignant glioma cells. *J Neurosci* 2002; 22: 4302–4311.

142. Scherer HJ. The forms of growth in gliomas and their practical significance. *Brain* 1940; 63: 1–35.

143. Ulrich TA, Juan Pardo EM, Kumar S. The mechanical rigidity of the extracellular matrix regulates the structure, motility, and proliferation of glioma cells. *Cancer Res* 2009; 69: 4167–4174.

144. Otey CA, Carpen O. Alpha-actinin revisited: A fresh look at an old player. *Cell Motil Cytoskeleton* 2004; 58: 104–111.

145. Quick Q, Skalli O. Alpha-actinin 1 and alpha-actinin 4: Contrasting roles in the survival, motility, and RhoA signaling of astrocytoma cells. *Exp Cell Res* 2010; 316: 1137–1147.

146. Sen S, Dong M, Kumar S. Isoform-specific contributions of alpha-actinin to glioma cell mechanobiology. *PLoS ONE* 2009; 4: e8427.

147. Kaufman LJ, Brangwynne CP, Kasza KE et al. Glioma expansion in collagen I matrices: Analyzing collagen concentration-dependent growth and motility patterns. *Biophys J* 2005; 89: 635–650.

148. Hegedus B, Marga F, Jakab K, Sharpe-Timms KL, Forgacs G. The interplay of cell-cell and cell-matrix interactions in the invasive properties of brain tumors. *Biophys J* 2006; 91: 2708–2716.

149. Ulrich TA, Jain A, Tanner K, Mackay JL, Kumar S. Probing cellular mechanobiology in three-dimensional culture with collagen-agarose matrices. *Biomaterials* 2010; 31: 1875–1884.

150. Yang YL, Motte S, Kaufman LJ. Pore size variable type I collagen gels and their interaction with glioma cells. *Biomaterials* 2010; 31: 5678–5688.

151. Gordon VD, Valentine MT, Gardel ML et al. Measuring the mechanical stress induced by an expanding multicellular tumor system: A case study. *Exp Cell Res* 2003; 289: 58–66.

152. Rutka JT, Muller M, Hubbard SL et al. Astrocytoma adhesion to extracellular matrix: Functional significance of integrin and focal adhesion kinase expression. *J Neuropathol Exp Neurol* 1999; 58: 198–209.

153. Friedlander DR, Zagzag D, Shiff B et al. Migration of brain tumor cells on extracellular matrix proteins in vitro correlates with tumor type and grade and involves alphaV and beta1 integrins. *Cancer Res* 1996; 56: 1939–1947.

154. Paulus W, Baur I, Schuppan D, Roggendorf W. Characterization of integrin receptors in normal and neoplastic human brain. *Am J Pathol* 1993; 143: 154–163.

155. Lefranc F, Brotchi J, Kiss R. Possible future issues in the treatment of glioblastomas: Special emphasis on cell migration and the resistance of migrating glioblastoma cells to apoptosis. *J Clin Oncol* 2005; 23: 2411–2422.

156. Belot N, Rorive S, Doyen I et al. Molecular characterization of cell substratum attachments in human glial tumors relates to prognostic features. *Glia* 2001; 36: 375–390.

157. Hehlgans S, Haase M, Cordes N. Signalling via integrins: Implications for cell survival and anticancer strategies. *Biochim Biophys Acta* 2007; 1775: 163–180.

158. Aoudjit F, Vuori K. Integrin signaling inhibits paclitaxel-induced apoptosis in breast cancer cells. *Oncogene* 2001; 20: 4995–5004.
159. Cordes N, Seidler J, Durzok R, Geinitz H, Brakebusch C. Beta1-integrin-mediated signaling essentially contributes to cell survival after radiation-induced genotoxic injury. *Oncogene* 2006; 25: 1378–1390.
160. Hikawa T, Mori T, Abe T, Hori S. The ability in adhesion and invasion of drug-resistant human glioma cells. *J Exp Clin Cancer Res* 2000; 19: 357–362.
161. Yuan L, Siegel M, Choi K et al. Transglutaminase 2 inhibitor, KCC009, disrupts fibronectin assembly in the extracellular matrix and sensitizes orthotopic glioblastomas to chemotherapy. *Oncogene* 2007; 26: 2563–2573.
162. Reardon DA, Nabors LB, Stupp R, Mikkelsen T. Cilengitide: An integrin-targeting arginine-glycine-aspartic acid peptide with promising activity for glioblastoma multiforme. *Expert Opin Investig Drugs* 2008; 17: 1225–1235.
163. Maglott A, Bartik P, Cosgun S et al. The small alpha5beta1 integrin antagonist, SJ749, reduces proliferation and clonogenicity of human astrocytoma cells. *Cancer Res* 2006; 66: 6002–6007.
164. Akabani G, Reardon DA, Coleman RE et al. Dosimetry and radiographic analysis of 131I-labeled anti-tenascin 81C6 murine monoclonal antibody in newly diagnosed patients with malignant gliomas: A phase II study. *J Nucl Med* 2005; 46: 1042–1051.
165. Leins A, Riva P, Lindstedt R, Davidoff MS, Mehraein P, Weis S. Expression of tenascin-C in various human brain tumors and its relevance for survival in patients with astrocytoma. *Cancer* 2003; 98: 2430–2439.
166. Ying H, Biroc SL, Li WW et al. The Rho kinase inhibitor fasudil inhibits tumor progression in human and rat tumor models. *Mol Cancer Ther* 2006; 5: 2158–2164.
167. Shi Q, Hjelmeland AB, Keir ST et al. A novel low-molecular weight inhibitor of focal adhesion kinase, TAE226, inhibits glioma growth. *Mol Carcinog* 2007; 46: 488–496.
168. Wang HB, Dembo M, Wang YL. Substrate flexibility regulates growth and apoptosis of normal but not transformed cells. *Am J Physiol Cell Physiol* 2000; 279: C1345–C1350.
169. Semler EJ, Ranucci CS, Moghe PV. Mechanochemical manipulation of hepatocyte aggregation can selectively induce or repress liver-specific function. *Biotechnol Bioeng* 2000; 69: 359–369.
170. Effler JC, Kee YS, Berk JM, Tran MN, Iglesias PA, Robinson DN. Mitosis-specific mechanosensing and contractile-protein redistribution control cell shape. *Curr Biol* 2006; 16: 1962–1967.
171. Paszek MJ, Zahir N, Johnson KR et al. Tensional homeostasis and the malignant phenotype. *Cancer Cell* 2005; 8: 241–254.
172. Huang S, Ingber DE. Cell tension, matrix mechanics, and cancer development. *Cancer Cell* 2005; 8: 175–176.
173. Beadle C, Assanah MC, Monzo P, Vallee R, Rosenfeld SS, Canoll P. The role of myosin II in glioma invasion of the brain. *Mol Biol Cell* 2008; 19: 3357–3368.
174. Gillespie GY, Soroceanu L, Manning TJ, Jr, Gladson CL, Rosenfeld SS. Glioma migration can be blocked by nontoxic inhibitors of myosin II. *Cancer Res* 1999; 59: 2076–2082.
175. Lepley D, Paik JH, Hla T, Ferrer F. The G protein-coupled receptor S1P2 regulates Rho/Rho kinase pathway to inhibit tumor cell migration. *Cancer Res* 2005; 65: 3788–3795.
176. Manning TJ, Jr, Parker JC, Sontheimer H. Role of lysophosphatidic acid and rho in glioma cell motility. *Cell Motil Cytoskeleton* 2000; 45: 185–199.
177. Goldberg L, Kloog Y. A Ras inhibitor tilts the balance between Rac and Rho and blocks phosphatidylinositol 3-kinase-dependent glioblastoma cell migration. *Cancer Res* 2006; 66: 11709–11717.
178. Zhai GG, Malhotra R, Delaney M et al. Radiation enhances the invasive potential of primary glioblastoma cells via activation of the Rho signaling pathway. *J Neurooncol* 2006; 76: 227–237.
179. Chan AY, Coniglio SJ, Chuang YY et al. Roles of the Rac1 and Rac3 GTPases in human tumor cell invasion. *Oncogene* 2005; 24: 7821–7829.
180. Ader I, Delmas C, Bonnet J et al. Inhibition of Rho pathways induces radiosensitization and oxygenation in human glioblastoma xenografts. *Oncogene* 2003; 22: 8861–8869.
181. Salhia B, Rutten F, Nakada M et al. Inhibition of Rho-kinase affects astrocytoma morphology, motility, and invasion through activation of Rac1. *Cancer Res* 2005; 65: 8792–8800.
182. Senger DL, Tudan C, Guiot MC et al. Suppression of Rac activity induces apoptosis of human glioma cells but not normal human astrocytes. *Cancer Res* 2002; 62: 2131–2140.
183. Rattan R, Giri S, Singh AK, Singh I. Rho/ROCK pathway as a target of tumor therapy. *J Neurosci Res* 2006; 83: 243–255.

Section II (Part 4)

Literature Review of Mechanobiology Research Findings and Theories

Frontiers of Mechanobiology

19 Mechanostimulation in Bone and Tendon Tissue Engineering

Samuel B. VanGordon, Warren Yates, and Vassilios I. Sikavitsas

CONTENTS

19.1 INTRODUCTION

Defects of bone and tendon are widespread, debilitating ailments. In mild cases, the symptoms of such musculoskeletal disorders may be limited to sporadic aches and pains. When more severe, these conditions could manifest themselves as limited mobility or brittle bones, which may lead to chronic breakages or other potentially life-threatening consequences. Though treating musculoskeletal injuries is of paramount importance, currently available remedies are still flawed. Disease and infection are associated with autografts, allografts, and xenografts. Autografts also result in donor site morbidity. The use of artificial materials to replace lost bone or ligament results in pain for the patient. These shortcomings have led researchers to pursue other methods to fix damaged or failing tissues. Interest in tissue engineering and the use of bioreactors has grown to address the deficits of current clinical practice.

19.2 TISSUE ENGINEERING

The major goal of tissue engineering is to produce tissue grafts for the replacement or regeneration of damaged or failing tissue. This goal can be accomplished by combining cells with a carrier on which cell and tissue growth can occur and by applying mechanical and chemical stimuli to promote functional tissue growth [1]. The newly formed tissue construct could then be implanted and integrated into the site of injury. This makes scaffolds, cells, and stimuli from mechanical and chemical means major components in tissue engineering strategies. Scaffolds provide a surface for the attachment and growth of cells and tissue. Ideal scaffolds provide proper mechanical support while allowing for cellular attachment, proliferation, and extracellular matrix (ECM) deposition [2]. The material and structure that a scaffold is composed of depends on the tissue application. Harder materials such as titanium are used in bone replacement while elastic materials are used for tendon grafts. The elicitation of an immune response from a scaffold material should also be minimal. The cells used for tissue regeneration should be able to proliferate and deposit tissue-specific ECM. Stem cells (SCs) are usually favored due to their ability to proliferate at a high rate and differentiate into different cell types. Stimuli used in tissue engineering try to mimic the natural environment that the designated engineered tissue would experience *in vivo*. Chemical stimuli can be provided by the use of growth and differentiation factors specific to particular cellular responses. Growth factors can influence ECM synthesis, cell proliferation, and cell differentiation. Mechanical forces can be applied to tissue constructs by compression, stretching, and fluid shear.

19.2.1 BONE TISSUE ENGINEERING

Bone is a hard mineralized tissue that supports the structure and motility of the body. While minor fractures of bone can be healed naturally by the body, larger nonunion fractures require supplemental reinforcement from implantable hardware or a material that can be placed into the fracture to fuse the disjointed bone. Bone tissue engineering pursues new alternatives to fill nonunion fractures and allow for natural healing and remodeling of bone tissue.

19.2.1.1 Scaffolds

Scaffolds for bone tissue engineering must provide ample mechanical support to allow for cellular attachment, cellular proliferation, ECM production, ECM attachment, and when in conjunction with preosteoblastic cells, osteogenic differentiation. Common materials for scaffolds for bone

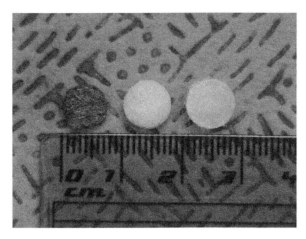

FIGURE 19.1 Three popular types of scaffolds with different pore architecture: titanium mesh, porous poly(L-lactic acid), and nonwoven fibrous poly(L-lactic acid) mesh (from left to right).

tissue engineering are ceramics [3], natural polymers [4–6], synthetic polymers [6], and metals like titanium [7]. Natural polymers used for scaffolding are commonly composed of fibrin and collagen while some synthetic polymers that have been used are polyanhydrides, polycarbonates, polyphosphazene, polyfumarates, and poly(butylenes terephthalate) poly(ethylene oxide) [6]. Synthetic polymers that are popular for their biodegradability and FDA approval are poly(lactic acid) (PLA), poly(glycolic acid) (PGA), and their copolymers, poly(lactic-co-glycolic acid) (PLGA) [6]. Scaffolds for bone tissue engineering commonly have high porosity (~70%–90%) to allow for proper oxygen and nutrient transport. Woo et al. found that the incorporation of a scaffold nanostructure promotes osteogenesis [8]. Porous foams, fiber meshes (Figure 19.1), and structures from rapid prototyping are commonly used scaffold structures. Common fabrication techniques include solvent casting, particulate leaching, melt blowing, electrospinning, fiber bonding, melt molding, and membrane lamination [9].

19.2.1.2 Cells

Cells commonly considered for bone tissue engineering are preosteoblastic SCs, primary osteoblasts, and osteocytes. This has to do, in part, with their roles in bone formation and remodeling [10]. While osteoclasts play an integral role in bone remodeling by breaking down bone tissue [10], the goal of bone tissue engineering is the formation of bone tissue. Preosteoblastic SCs are favored for bone tissue generation because of their higher proliferation rate over osteoblasts and osteocytes. Mesenchymal stem cells (MSCs) which have the capacity to differentiate into osteoblasts have been found to be osteoinductive when transplanted into a bone injury site [11] and have demonstrated accelerated osteoblastic differentiation when exposed to mechanical stimulation, especially fluid shear forces [12,13]. Stiehler et al. [12] found that when MSCs were cultured on 75:25 porous PLGA scaffolds in spinner flasks, shear increased DNA content, alkaline phosphatase (ALP) activity, and transcription levels of osteogenic markers like collagen I, bone morphogenic protein 2 (BMP-2) and runt-related transcription factor 2 (Runx2) were observed for up to 3 week cultures. Holtorf et al. cultured bone marrow stromal cells (BMSCs) on titanium fiber mesh scaffolds under flow perfusion without specialized media containing dexamethasone in order to investigate the effectiveness of flow perfusion as an osteogenic stimulant [13]. It was found that BMSCs cultured under continuous flow perfusion had greater scaffold cellularity, ALP activity, osteopontin secretion, and mineral deposition when compared to BMSCs cultured statically in the presence or absence of dexamethasone [13]. These findings exhibited the ability of perfusion culturing alone to enhance osteogenic differentiation in adult SCs from the bone marrow.

19.2.1.3   Chemical Stimuli

To assist the osteoblastic differentiation of osteoprogenitors, cell culture medias are often supplemented with chemicals and growth factors [14]. The most commonly used agents that promote osteogenic differentiation are dexamethasone and BMP-2. Dexamethasone is a corticosteroid that promotes osteoblast differentiation in several cell culture systems. Dexamethasone is also implicated in chondrocytic differentiation via the enhancement of SOX9 expression [15] while inducing fully differentiated chondrocytes to secrete ALP, suggesting an important role in cartilage calcification [16]. BMP-2 is a part of the transforming growth factor β (TGFβ) superfamily that stimulates proliferation and differentiation of both chondrocytes and osteoblasts and causes increased matrix production in each cell type. BMP-2 has been found to and is commonly used to induce MSC differentiation into osteoblasts [16,17]. Other members of the TGFβ superfamily like BMP-4, BMP-6, and BMP-7 and different TGFβ isoforms have also been implicated in osteoblastic differentiation [18]. The incorporation of differentiating growth factors or peptides into scaffolds by physical or chemical means is a common tissue engineering strategy that results in enhanced osteoinductivity [7,19].

19.2.1.4   Mechanical Stimuli

Bone formation and remodeling are considered sensitive to mechanical forces exerted on bone tissue [10,16,20,21]. Compression and tension is created from the loading and unloading of bone. Loading results in bone deformation that leads to pressure differences in the interstitial fluid leading to flow. Coupled with vascular transport there is movement of fluid through the canaliculi and lacunae of bone [22–27]. These compressions, tensions, and fluid movements translate forces onto the major bone cells (i.e., osteoblasts and osteocytes). The stresses and strains cause the deformation of the cell bodies [28,29]. How mechanical stimuli are translated into biochemical pathways is not fully understood. Some hypothesized mechanisms of mechanical signal conversion involve membrane ion channels [30–37], focal adhesions [36,38–44], intracellular junctions [45–52], and cilia [53–57]. Mechanical stimulation seems to regulate osteoblast function, proliferation and differentiation [58]. Interstitial flow plays a major role in bone homeostasis [25] and has been reported to enhance *in vitro* osteogenesis [25,59,60]. It is proposed that engineered bone grafts, if adequately perfused, can lead to enhanced cell stimulation, nutrient transportation, and bone regeneration [25].

19.2.1.5   Osteogenic Differentiation Markers

The progression of preosteoblastic cells toward their mature phenotype can be monitored by an array of common markers. Some of these markers are ALP production, prostaglandins production (predominantly prostaglandin E_2, PGE_2), and nitric oxide production. Proteins and biomolecules essential for the creation of ECM are also applicable for neotissue evaluation. Popular ECM markers include collagen I, osteopontin, osteocalcin, osteonectin, bone sialoproteins (bone sialoprotein 1 is osteocalcin), and ECM mineral deposition.

As knowledge of osteogenic signaling pathways grow, common pathway markers are often evaluated relative to "housekeeping" genes (e.g., glyceraldehyde 3-phosphate dehydrogenase), which exhibit constant mRNA expression. Examples of some common pathway markers include cyclooxygenases (Cox-1, Cox-2), SMAD-1, SMAD-5, Runx2, Sp7 transcription factor (Osx), and extracellular-signal regulated protein kinases (ERK-1, ERK-2). These insights have spurred a shift in tissue engineering research practice, causing researchers to value mRNA expression data in addition to mere osteogenic differentiation biochemical measurements.

19.2.2   Tendon Tissue Engineering

Tendons serve to transmit forces between skeletal muscle tissue and bone. Tendons stabilize our joints and permit locomotion. Tendon injuries affect many people each year and can be debilitating to the extent that the sufferer cannot work or even walk until recovered. The purpose of tendon tissue engineering is to remedy tendon ailments for which physical therapies are insufficient. In most cases, this means replacing lengths of tendon or bridging the gap incurred by a laceration.

19.2.2.1 Scaffolds

By far the most popular scaffold chosen for tendon tissue engineering is some form of collagen matrix, mimicking tenocytes' natural environment. Collagen hydrogels and sponges are common options. Biodegradable polymers such as PGA, PLA, polyurethanes, and polyanhydrides, favored in many tissue applications, are relegated to static tendon tissue engineering studies due to their mechanical properties [61]. In some cases, natural materials such as chitosan, silk, small intestine submucosa (SIS), or decellularized tissue (e.g., the human umbilical vein [HUV], rabbit patellar tendon, rat tail tendon) have been investigated and shown great promise [62–66]. Finally, some research has been performed on animal tendons or tendon fascicles directly, though these studies are more exploratory in nature and not intended for translation to clinical practice [67–69].

19.2.2.2 Tenocytes as Mechanosensory Cells

Cells are known to respond to physical stimuli. One suggested mechanism for transducing forces to cells is tensegrity in which the nucleus of the cell is connected to the ECM through the cytoskeleton [70–74]. Gap junctions play an important role in propagating mechanostimulation signals. Specifically, connexin 32 stimulates load response and connexin 43 is inhibitory [75]. Tensile forces applied to fibroblasts via β_1 integrins and the actin microtubule network modulates mechanotranscriptional coupling of filamin A [73]. A strong, significant inverse correlation exists between the load applied to tenocytes and the observed expression of metalloproteinases (MMP). This correlation is not evident when the tenocyte's cytoskeleton is disrupted with cytochalasin D, which corroborates the tensegrity model of mechanotransduction [72,74]. It is thought that $\alpha_1\beta_1$ and $\alpha_2\beta_1$ integrins are the major receptors responsible for regulating ECM remodeling: $\alpha_1\beta_1$ mediates the signals inducing downregulation of collagen gene expression, whereas the $\alpha_2\beta_1$ integrin mediates induction of MMP-1 [76]. Mechanical stimulation, whether acting via integrins, stretch-activated ion channels, or another method (G-protein receptors, stress- or mitogen-activated protein kinases), may be initiated by different means but trigger the activation of many of the same intracellular responses within minutes (release of chemical second messengers, protein phosphorylation, secretion of growth factors, cytoskeleton restructuring, attachment or release of integrins-ECM adhesions, changes in gene expression) [70,72,77–79].

Cyclic strain (specifically cyclic stretching), when applied to tendon cells, has been associated with several effects. There are instances in which the observed effect is dependent on frequency, magnitude, duration of stretch or rest, cell concentration, or some combination thereof. Cyclic strain for 10 min inhibits collagen synthesis, whereas strain for 24 h increases it [68]. Both strain rate and strain amplitude may independently alter collagenase gene expression in rat tenocytes through increases in shear stress and cell deformation, respectively [70]. Cyclic stretching has been shown to activate stress-activated protein kinase (SAPK) c-Jun N-terminal kinase (JNK) in a magnitude- but not frequency-dependent manner [77]. In this particular study, activation of JNK peaked at 30 min and returned to baseline levels within 2 h [77]. Although transient activation is normal, the persistent activation of JNK has been implicated in the initiation of the apoptotic cascade, which could be similar to the mechanism by which tendon overuse injuries (which include local apoptosis of tenocytes) occur [77]. Cyclic mechanical loading combined with PDGF-BB has a synergistic effect on DNA synthesis [71]. IGF-1 combined with cyclic mechanical loading produces a modestly synergistic effect on DNA synthesis [71]. Cyclic stretching in human patellar tendon fibroblasts (HPTFs) increases gene expression and protein production of collagen I and TGFβ-1 in a magnitude-dependent manner [80]. Cyclic stretching has been shown to upregulate Cox-1 and Cox-2 expressions in a magnitude- and frequency-dependent manner as well as PGE_2 and leukotriene B_4, which have an inverse relationship [80–83]. Cyclic strain has been shown to significantly increase mRNA expression of procollagen I, TGFβ-1, and connective tissue growth factor when compared to static controls [84].

Tensile stimulation over extended periods of time promotes dramatic upregulation of collagen I gene expression [85]. The shape factor of tenocytes influences collagen I expression, with more elongation corresponding to greater expression (even when the total area of cellular spreading is constant) [86]. Tenocyte production of IGF-I is significantly increased during tendinosis and may be accompanied by phosphorylation of insulin receptor substrate 1 and ERK-1/2 [87]. Fibroblasts in stressed collagen gels increased production of tenascin-C and collagen XII compared to cells on nonstressed gels, which have been shown to increase MMP-1 synthesis and decrease collagen I mRNA expression and protein synthesis [88–93]. Under conditions of mechanical unloading, tenocytes in rat tail tendon fascicle explants have been shown to rapidly and dramatically increase gene expression for MMP-3 and MMP-13 for sustained periods (at least 48 h), whereas the same conditions produced a short-term upregulation of collagen I, decorin, cathepsin K, and MMP-2 gene expression that dwindled to baseline levels or even decreased within a few hours [67]. It has been suggested that the impetus for tendinopathy and the pathological symptoms associated (collagen disruption, higher MMP expression, apoptosis) is mechanobiological understimulation of tenocytes [94].

19.3 BIOREACTORS

Bioreactors are devices designed to culture cells seeded on three-dimensional (3D) constructs in a controlled and sterile environment [95]. Bioreactors provide cellular constructs or tissues with mechanical and chemical stimuli to achieve a desired tissue phenotype. Mechanical stimuli used in bioreactors try to mimic those found in physiological systems. The types of forces generated vary with the specific bioreactor design. Popular bioreactor designs for engineering bone and tendon tissues make use of dynamic flow, cyclic compression, and cyclic stretching (Figure 19.2). The following are descriptions of various bioreactors that have been used for the engineering of bone and tendon tissues.

19.3.1 BONE TISSUE BIOREACTORS

19.3.1.1 Static Culture

The simplest and most widely used tissue engineering cell culturing technique is the static culture. In static cultures, cells are seeded on scaffolds, often dropwise, supplied with sufficient cell culture media and placed in standard incubators. Static cell culturing is often used as a control

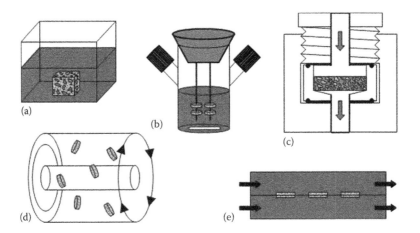

FIGURE 19.2 Common bioreactors used in bone tissue engineering. (a) static culture, (b) spinner flask, (c) scaffold perfusion, (d) rotation wall; (e) perfused column.

culture [96,97] and in scaffold biocompatibility testing [98].Initial efforts to produce *in vitro* generated bone grafts involved seeding of BMSCs on 75:25 PLGA foams with varying pore sizes and seeding densities cultured with media supplemented with dexamethasone [99]. The cells seeded on the scaffolds proliferated, had high ALP activity, and calcified matrix was observed. The penetration of the tissue into the porous constructs was limited to 240 μm, producing a non-uniform cell and ECM distribution. A similar study was performed using rat calvarial osteoblasts with comparable results [100] demonstrating the need for a more complex culture method for bone tissue engineering. Culturing cell/scaffold constructs statically has often resulted in low cell seeding efficiencies, uneven cell distributions [59,101], poor nutrient transport, moderate cell proliferation, slow mineralized ECM growth [99], and an obvious inability to take advantage of fluid shear-induced mechanical stimulation [102,103].

19.3.1.2   Spinner Flask

The spinner flask is one of the simplest bioreactor designs. Scaffolds threaded onto needles are suspended from the cover of a cylindrical flask. The scaffolds are submerged in culture media that is agitated to produce convective flow by a magnetic stir bar at the base of the vessel. The convection is induced in order to promote a homogeneous distribution of oxygen and nutrients over scaffolds [102,104]. The fluid environment at the external surface of scaffolds in the system is turbulent and characterized by the existence of eddies. The presence of convective flow is considered to enhance nutrient transport into pores and expose external cells to relatively high shear forces. The spinner flask has shown improved cell seeding [105], proliferation, and osteoblastic differentiation over static cultures [102]. A drawback of the design is that the shear stress magnitudes are not homogeneous over the scaffolds and cells at different locations are exposed to a wide range of stresses. Also, in larger 3D constructs, while external convective forces create improved nutrient and oxygen transport on the outside of scaffolds, internal porous networks still have to rely on diffusion along concentration gradients [59,106]. Shea et al. demonstrated this effect when porous PLA constructs statically seeded with MC3T3-E1 preosteoblastic-like cells cultured in spinner flasks could not penetrate the constructs beyond the first 200 μm from the external surface even after 12 weeks of culture [107]. While the spinner flask is simple in design, the more complex flow perfusion systems have shown greater enhancement of *in vitro* cell proliferation and osteogenesis [25,59,60,103].

19.3.1.3   Rotating-Wall Vessels

The rotating-wall vessel was designed by NASA to study cell cultures in a microgravity environment [102,108]. The reactor design consists of two concentric cylinders whose annular space acts as a culture chamber. The inner cylinder is stationary and gas permeable while the outer cylinder rotates to create dynamic laminar flow. Adjustment of the rotation speed of the outer cylinder can create a pseudo microgravity environment that can suspend scaffolds in the annular culture space [109]. In contrast to the spinner flask, rotating-wall vessel fluid flows are moderate and shear forces exerted on cells residing on the exterior of scaffolds are of a relatively narrow range. With a stagnation zone at the upstream edge, shear stresses decrease in the direction of flow with no variations from scaffold to scaffold [110]. Spinner flasks have shown to support increased osteogenesis over rotating-wall vessels with increased ALP activity, osteocalcin secretion, and calcium deposition from rat MSCs seeded on 3D porous 75:25 PLGA biodegradable scaffolds [102]. Rotating-wall vessels have been widely utilized in cartilage [110–113] and osteochondral tissue growth [114] contrasting its modest use in bone tissue engineering.

19.3.1.4   Mechanical Loading

Mechanical loading of cellular constructs have been shown to modulate osteogenic gene expression [115,116], cell proliferation [116,117], ECM production [116,118], and ECM calcification [116,118]. Designs associated with mechanical loading are commonly based on the compression of the cell cultures or the deformation of cell-seeded membranes or scaffolds. A major drawback in many mechanical loading systems for bone tissue engineering is the restrictive use of monolayer cultures rather than in 3D cell/scaffold constructs. More experiments utilizing mechanical loading applied on 3D constructs are needed to allow for a detailed investigation of the effects of mechanical loading on osteoblastic differentiation and bone tissue engineering.

19.3.1.4.1 Compression

Compression systems operate by the loading of cultured cells by either a physical application of force on the culture [115,119] or by the use of hydrostatic pressure [120]. Compression loading is sometimes performed in cycles to mimic physiological conditions.

Cyclic compression promotes increased ECM production, osteogenic differentiation, and remodeling. Rath et al. [115] seeded electrospun poly(ε-caprolactone) scaffolds statically with calvarial osteoblasts and initially cultured the constructs statically for 4 weeks. The cellular scaffolds were then exposed to 10% cyclic compressive strain that resulted in increased expression of BMP-2, Runx2, and MAD homolog 5 (SMAD5) in comparison to static cultures. Increased expression of ALP, collagen I, osteocalcin, osteonectin, and osteopontin were also observed. When the compressive strain was increased to 20%, the observed stimulation in comparison to 10% strained samples was diminished, implying that 20% compressive strain is not beneficial if osteogenic differentiation is desired. This demonstrates the existence of a threshold for compressive forces beneficial for engineered bone tissue engineering constructs.

Nagatomi et al. exposed calvarial osteoblasts to cyclic hydrostatic pressure for 1 h per day for up to 19 days [118]. They found that compressive loading resulted in increased collagen production and ECM mineral deposition. Mitsui et al. found that human osteoblastic cells that experienced direct physical compression of 1 g/cm^2 caused increased expression of Runx2 and ostenix along with mineralization of ECM [121].

Mitsui et al. found that human osteoblastic cells that experienced direct physical compression of 1 g/cm^2 caused increased expression of Cox-2, osteocalcin, osteopontin, and PGE$_2$. The same compressive stimulation administered in conjunction with the inhibition of PGE$_2$ resulted in diminished expression of osteocalcin [119]. Mitsui et al. found that human osteoblastic cells that experienced direct physical compression of 1 and 3 g/cm^2 resulted in the expression of different matrix metalloproteinases implying that the amount of force loaded on the cells could control the remodeling process [122]. Liu et al. also exposed MSCs to dynamic and continuous hydrostatic compressive loading for up to 7 days. It was found that MSCs promoted osteoclastogenesis by upregulation of the ratio of receptor activator of nuclear factor-jB ligand (RANKL) to osteoprotegerin genes indicating signaling for osteoclast recruitment and bone remodeling [120].

19.3.1.4.2 Substrate Deformation

Substrate deformation systems come in an array of designs. Common substrate deformation systems use axial stretching [117,123] and membrane or scaffold bending [124]. Basic axial stretching systems use membrane dishes clamped at opposite ends. While one clamped end is stationary, the other clamped end is attached to a motor system that elongates the membrane dish. Bending systems use opposing pressure points on the top and bottom of the construct to deflect the construct. Similar to compression systems, substrate deformation is sometimes performed cyclically to mimic the physiological loading of bone.

Using a uniaxial membrane stretching bioreactor, proliferation rate of human osteoblast-like cells cultured on silicone membranes increased significantly after being subjected to 1% cyclic strain for 15 min/day for 3 days [117]. When the strain was increased above 1%, cell proliferation was reduced. It was also observed that ALP activity was not affected by the applied cyclic strain. Using a similar system, Winter et al. found that osteoblasts seeded on silicone membranes strained with 1000 microstrains at 1 Hz cyclically for 15-min intervals for 1 h, experienced a decrease in ALP activity and increased ECM mineral deposition and cellular DNA in comparison to cells strained continuously for 1 h [123].

Kaspar et al. [116] found that applying 1% cyclic strain to osteoblast-seeded silicone membranes for 30 min/day for 2 days resulted in an increase in osteoblast proliferation and collagen I expression. It was also observed that ALP activity and osteocalcin secretion were reduced. From these findings it can be concluded that cyclic strain at physiological levels leads to an increase in osteoblast activities related to matrix production but a decrease in activities related to matrix mineralization [116].

Jagodzinski et al. cultured human bone MSCs on a silicone dish and subjected it to 8% cyclic strain at 1 Hz for a total of 6 h per day for up to 7 days [125]. Increased ALP activity and osteocalcin secretion along with increased expression of collagen I and II were observed.

Rat bone MSCs were cultured on porous PLA scaffolds and subjected to four-point bending deformation with 2500 microstrains at 0.004 Hz for 16 days [124]. The mechanical loading resulted in a decrease in ECM deposition and no noticeable increase in ALP activity. This deviation of the results from previous studies' results could be attributed to the use of frequency differences and scaffolding material and architecture.

19.3.1.5 Perfusion Fluid Flow

Perfusion systems drive fluid flow through porous scaffolds. Flow through the scaffold porosity exposes cells seeded at the interior of scaffolds to fluid shear which is stimulatory to bone cells [97,126,127]. Fluid shear also improves cell seeding, increases nutrient transfer, promotes uniform cell distribution, and improves ECM growth throughout porous constructs [97,101,103,127–130]. Although flow perfusion bioreactors are based on the same set of fundamentals, their specific designs vary. The following is a description of two types of popular perfusion bioreactors that have been shown to drive fluid flow through constructs.

19.3.1.5.1 Perfusion Column

Perfusion column bioreactors use continuous fluid flow though sealed culture chambers that contain scaffolds in the flow path. This causes the recirculated culture media flowing in the chamber to flow through the interconnected pores of the scaffolds [103,131–133]. While this increases nutrient and oxygen transport for the scaffolds, the flow is divided between the porous network of the scaffolds and the surrounding chamber. This division of the media flow reduces the convective flow through constructs [96].

19.3.1.5.2 Scaffold Perfusion

Scaffold perfusion systems are similar to perfusion columns but instead of having flow of media through and around the scaffolds, the flow is restricted so it only passes through the interconnected porous network of a scaffold (Figure 19.3) [96,134,135]. By perfusing the media through the scaffolds, the flow can be better controlled and nutrient and oxygen transport limitations can be mitigated or eliminated [130].

19.3.1.5.3 Effects of Flow Perfusion

MSCs cultured in a scaffold perfusion bioreactor exhibited a more uniform distribution of cells and ECM growth throughout the porous construct when compared to cells grown on a construct statically [60]. Although the cell proliferation rate and ALP activity were comparable in the static and perfused cultures, the scaffolds cultured under perfusion demonstrated increased mineral deposition.

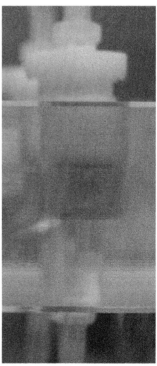

FIGURE 19.3 Scaffold perfusion chamber used for both seeding and culturing on poly(L-lactic acid) porous scaffolds for bone tissue engineering in Alvarez-Baretto et al. [128].

Osteoblastic cells statically seeded on PLGA foam scaffolds were used to compare the use of spinner flask, rotating wall, and flow perfusion bioreactors [59]. The scaffolds cultured in the flow perfusion system resulted in a more uniform distribution of cells over the scaffolds' porous networks and higher ALP activity [59].

Meinel et al. cultured human MSCs on collagen scaffolds in a flow perfusion chamber and a spinner flask. The scaffolds cultured with flow perfusion showed a more uniform distribution of the mineralized ECM over the entire construct in comparison to scaffolds grown in the spinner flask. The scaffolds cultured in the spinner flask produced a greater amount of deposited mineral than the perfusion cultured scaffolds [134]. The collagen scaffolds used in this experiment degraded with time causing loss in cell numbers during the culturing process. This may have attributed to the increased amount of mineral observed in the spinner flask cultures [134].

Bancroft et al. [126] studied the effects of shear rate on the mineralized matrix growth by varying the flow rate when culturing rat BMSCs on titanium fiber meshes in a flow perfusion system. It was found that as the flow rate increased, distribution of ECM and deposition of mineral too increased. The increasing mineral deposition was linked to increasing shear forces or increasing chemotransport.

To determine if shear forces or mass transport effects control the growth and mineralization of MSCs, rat BMSCs seeded on titanium fiber mesh scaffolds were cultured under flow perfusion with a constant flow rate and variable cell culture media viscosity [97]. As viscosity of media increased, increased amounts of mineralized matrix and increased distribution of ECM were observed. The increase in viscosity correlates to increased shear forces under comparable chemotransport conditions denoting that fluid shear is an important stimulatory factor that can be utilized in the creation of engineered bone grafts.

Holtorf et al. cultured BMSCs on titanium fiber mesh scaffolds under flow perfusion without osteogenic media (no dexamethasone) in order to investigate the effectiveness of flow perfusion as an osteogenic stimulant [13]. It was found that the cells cultured under flow perfusion conditions had greater scaffold cellularity, ALP activity, osteopontin secretion, and mineral deposition when compared to static controls cultured with and without dexamethasone [13]. These findings demonstrated the ability of perfusion culture alone to promote MSC osteogenic differentiation.

Although fluid shear forces increase cell proliferation in general [130], this increase is not dose dependent [127]. Human trabecular bone seeded with MC3T3-E1osteoblastic-like cells showed higher proliferation under low flow rates of about 0.01 mL/min when compared to static cultures and higher flow rates of 0.20–1.0 mL/min. Trabecular bone seeded with MC3T3-E1 cells cultured under higher flow rates showed upregulation of Runx2, osteocalcin, and ALP in comparison to low flow rate and static cultures [127].

19.3.2 TENDON TISSUE BIOREACTORS

Because tendons have no significant vasculature, they are easily classified as structural tissues, whose functional efficacy is predicated on their structure [95]. Consequently, efforts to fabricate a tissue-engineered tendon model *in vitro* aim to recreate the natural, hierarchical mostly collagenous ECM structure found *in vivo* [136–146]. Influential *in vivo* work by Ohno et al., Yamamato et al., and Yamato et al. considered stress shielding of the rabbit patellar tendon and revealed subsequent dramatic reduction of mechanical properties, which has guided much of today's bioreactor designs [79,144,147,148]. As mentioned earlier, cyclic stretching has been shown to have multiple effects on tendon cells and is thus the preferred method of stimulation employed by bioreactors for tendon tissue engineering. Though bioreactors designed for tendon tissue engineering that do not implement cyclic stretching are not as common, there are designs that apply static tensile stress, as well as some that incorporate compression using pneumatic elements [62,64,71,84,85,139,145,149–158].

Altman et al. developed a novel bioreactor that simultaneously applied translational strain (10%, 2 mm) and rotational strain (25%, 90°) simultaneously at a frequency of 0.0167 Hz (one full cycle per min) to account for the helical orientation of natural collagen fibers [159,160]. They found that unlike tissue engineering for other cell types, mechanical stimulation alone could direct the differentiation of mesenchymal progenitor cells from bovine bone marrow seeded on collagen matrices into a ligament cell lineage (which has similarities to tendon) [159].

Androjna studied the effects of cyclic loading on canine tenocytes seeded on a SIS substrate and found that stretching resulted in significantly improved construct stiffness (129.1% ± 10.2%), specifically cyclic stretching at 8% strain and stiffness [64].

Joshi and Webb found that human dermal fibroblasts seeded on porous polyurethane scaffolds cyclically strained at low or subphysiological ranges (2.5% strain, 0.1–0.5 Hz, 7,200–28,000 cycles/day) resulted in the highest values of elastic modulus [152].

Riboh et al. employed a Flexcell Strain Unit (Flexcell Int., Hillsborough, NC) and found that cyclic uniaxial strain provided by the bioreactor promoted collagen I production, cell proliferation, and tenocyte-like morphology in rabbit epitenon tenocytes and sheath fibroblasts, as well as BMSCs and adipose tissue derived SCs [154].

Arnoczky et al., using biaxial cyclic strain on canine patellar tenocytes, found that cyclic strain produced an immediate, significant activation of JNK1 and JNK2 that lessened with time. This activation was highly dose-dependent with respect to strain, but dose-independent with respect to frequency [77].

Maeda et al. used a custom-made cyclic strain bioreactor in conjunction with rat fascicles and found that after 6 h, strained fascicles exhibited significantly upregulated collagen III mRNA expression and 24 h of cyclic strain resulted in a small but significant downregulation of decorin expression compared with unstrained controls [69]. The same group found previously that 24 h of cyclic strain upregulated collagen synthesis, whereas only 10 min of strain inhibited it [68]. Cyclic strain enhanced the retention of newly synthesized collagen [68].

Lavagnino et al. also used a custom-made cyclic strain bioreactor with rat tail tendons, finding that tendon cells exposed to higher strain rates and strain amplitudes expressed significantly less MMP-13 mRNA. However, increased strain amplitude in conjunction with reduced strain rate had little effect on MMP-13 mRNA expression. Thus, they concluded that increased shear stress, rather than cell membrane strain, was the primary factor in the inhibition of MMP-13 mRNA expression [70].

Using collagen-coated magnetite beads to apply cyclic tensile forces to human gingival fibroblasts, D'Addario et al. found that the microtubule network modulates mechanotranscriptional coupling of filamin A, an actin-crosslinking protein [73].

Bagnaninchi et al. found that the use of 250 μm chitosan channels seeded with porcine tenocytes could be used to allow the primary tenocytes to proliferate in a bundle-like structure when using a perfusion bioreactor. Also, the use of flow perfusion increased ECM production compared to static culture conditions. Optical coherence tomography was also shown to be an effective, nondestructive, on-line method for semiquantitatively observing the culture's progress [62].

Butler et al. designed a bioreactor to apply tensile stimulation to rabbit MSCs on collagen I scaffolds and found that stimulated constructs exhibited 3–4 times more collagen I and III gene expression as well as 2.5 times the linear stiffness and 4 times the linear modulus. Decorin and fibronectin gene expression showed no significant change versus nonstimulated samples [150].

Although the overwhelming majority of tendon tissue engineering research has been performed using dynamic loading techniques, it has been suggested that statically loaded fibroblast-collagen constructs display superior mechanical properties when comparedwith dynamically loaded constructs [139]. In fact, static loading has received ample attention from several research groups [74,145,155–158]. Arnoczky et al. found that static tensile loading inhibits MMP-1 mRNA expression in rat tail tenocytes in a nonlinear dose-dependent manner [74].

19.4 BONE ENGINEERING STUDY

19.4.1 Materials and Methods

19.4.1.1 Cell Culture

Adult MSCs were extracted from 175–200 g Wistar rats (Harlan Laboratories) using well-established methods [161,162]. Briefly, rats were euthanized and their tibias and femurs were extracted. The epiphyses of the bones were cut off, and the bone marrow was flushed out and suspended in α-minimum essential media (α-MEM, Invitrogen) supplemented with 10% fetal bovine serum (FBS) (Atlanta Biologicals) and 1% antibiotic-antimycotic (Invitrogen). Suspended cells were plated on polystyrene 75 cm² culture flasks (BD). Cells were incubated at 37°C and 5% CO_2. Nonadherent cells were discarded after 3 days of culture. When cell cultures reached ~80% confluency, cells were washed with phosphate-buffered saline (PBS, Atlanta Biologicals), lifted with trypsin (Invitrogen), centrifuged at 400g for 5 min, re-suspended in α-MEM, and re-plated. Once these flasks became ~80% confluent, the cells were lifted using fresh α-MEM and used for seeding. This method was also used for the tendon engineering study, presented later.

19.4.1.2 Scaffold Fabrication

Scaffolds were prepared using solvent casting particulate leaching method [6,163]. Briefly, poly(L-lactic acid) (114,500 MW, 1.87 PDI, Birmingham Polymers) was dissolved into chloroform 5% w/v. The solution was then poured over a bed of sodium chloride crystals (NaCl, 250–350 nm). Solvent was allowed to evaporate away for 24 h. The remaining salt-polymer was placed into an 8 mm diameter cylindrical mold and compressed at 500 psi. While compressed, the composite was heated to 130°C and held for a duration of 30 min. The resulting composite rod was cut into discs using a diamond wheel saw (Model 650, South Bay Technology, Inc.) Resulting discs were placed into deionized water under agitation for 2 days to leach out NaCl. Water was replaced twice per day. Discs were then removed and placed into vacuum to dry water from the scaffolds. The resulting product was 8 mm diameter, 2.3 mm thick discs that were ~85% porous. Porosity was determined from polymer density compared to weight and dimensions of discs.

19.4.1.3 Scaffold Pre-Wetting

Before seeding, scaffolds were placed into a beaker containing ethanol and sealed with a septum stopper. Vacuum was applied to the vessel until air was no longer visibly escaping the scaffolds. This was performed in order to pre-wet the scaffolds to allow for their ubiquitous penetration by cells and media during seeding and culturing [163]. The scaffolds were immediately removed from the beaker and press-fitted into cassettes.

19.4.1.4 Static Seeding and Culture

Pre-wet scaffolds were placed in a low-attachment 6-well culture plate (Corning) and washed twice with 10 mL of PBS. MSCs suspended at 500,000 cells in 0.25 mL α-MEM were pipetted over each scaffold and allowed to penetrate the scaffolds. Over a period of 2 h, cell suspensions were re-passed through scaffolds every 15 min. After seeding, scaffolds were submerged in 10 mL of osteogenic media (α-MEM supplemented with 10% FBS, 1% antibiotic-antimycotic, 50 g/mL ascorbic acid (Sigma), 10 mM β-glycerophosphate (Sigma), and 10^{-8} M dexamethasone (Sigma)). Scaffolds were cultured statically under incubation for 2, 5, and 9 days. Media was replaced every 2 days. After the culture period, scaffolds were removed from their culture cassettes and placed in 600 mL of RNAlater® (Ambion) and stored at −20°C for later RNA analysis.

19.4.1.5 Dynamic Seeding and Culture

Pre-wet scaffolds in cassettes were placed into a flow perfusion bioreactor [96]. The perfusion bioreactor has been previously used for culture of MSCs in 3D scaffolds. 200 mL of α-MEM was perfused through the scaffolds and bioreactor for 1 h to rinse and cure the system. 500,000 MSCs per scaffold were then seeded using oscillatory flow perfusion [128]. Briefly, MSCs were suspended at 500,000 cells per 0.25 mL of α-MEM. Cell suspensions of 0.25 mL were then pipetted over each scaffold. Cells were then perfused though the scaffolds at a flow rate of 0.15 mL/min for 5 min until the entire volume was passed through the scaffold. The flow direction was then reversed and the cells were allowed to pass back through the scaffold. After 2 h of this procedure being repeated, flow was stopped and the cells were allowed to rest and adhere for 2 h. After the resting period, a flow rate of 0.15 mL/min was applied to the seeded scaffolds for 1 day to condition the cells. After conditioning, α-MEM media was exchanged with osteogenic media and a flow rate of 1 mL/min was applied. After the culture period, scaffolds were removed and placed in 600 mL of RNAlater® and stored at −20°C for later RNA analysis.

19.4.1.6 RNA Extraction and Analysis

PLLA scaffolds are removed from RNAlater®. Scaffolds were then broken down and RNA isolated using RNAqueous®-4PCR Kit (Ambion) in accordance with the manufacturer's directions. cDNA was then synthesized from RNA using TaqMan® Reverse Transcription Reagents (Applied Biosystems, Austin, TX) in accordance to manufacturer's directions using random hexamers as the primers for reverse transcription. Gene expression levels of collagen I (Coll-a1), collagen III (Coll-a3), and Cox-2) were relatively quantified using a Power SYBR® Green PCR Master Mix kit (Applied Biosystems) using glyceraldehyde 3-phosphate dehydrogenase (GAPDH) as reference. Primer sequences were as follows: Coll-a1 forward 5′-GGA GAG TAC TGG ATC GAC CCT AAC-3′ and reverse 5′-CTG ACC TGT CTC CAT GTT GCA-3′; Coll-a3 forward 5′-CAG CTG GCC TTC CTC AGA CTT-3′ and reverse 5′-GCT GTT TTT GCA GTG GTA TGT AAT GT-3′; Cox-2 forward 5′-TAC TGT GTA GCT CCC CTT CG-3′ and reverse 5′-TGC CCA GAA CTA CCC ACTA A-3′; GAPDH forward 5′-AAC TCC CTC AAG ATT GTC AGC AA-3′ and reverse 5′-GTG GTC ATG AGC CCT TCC A-3′. Quantitative real-time PCR was conducted as follows: 10 min at 25°C, followed by 45 cycles of amplification where each cycle consisted of 15 s at 95°C followed by 1 min at 60°C. Relative gene expression (compared to GAPDH) was analyzed using the $\Delta\Delta C_T$ method [164].

19.4.2 RESULTS

Scaffolds grown under flow perfusion showed increased expression of collagen I and III in comparison to scaffolds cultured under static conditions for the same time period (Figure 19.4).

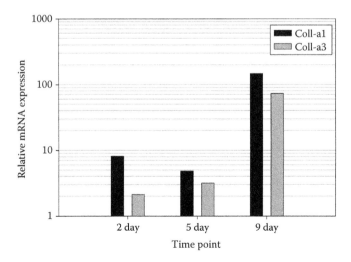

FIGURE 19.4 Relative collagen I (Coll-a1) and collagen III (Coll-a3) mRNA expression of MSCs seeded on PLLA foams and cultured in osteogenic media in a flow perfusion bioreactor compared to statically cultured MSCs for the same length of time.

Perfusion cultures also exhibited a minimum of 100-fold increase in the expression of Cox-2 (data not shown).

19.4.3 DISCUSSION

Cox-2 is responsible for the production of PGE_2. An increase in the production of PGE_2 is associated with osteogenic differentiation of MSCs. Collagen I and III are two major components of bone ECM. Increases in the expression of Coll-a1 and Coll-a3 show increases in the production of bone ECM from shear flow stimulation. Increases in the expression of bone ECM proteins and PGE_2 demonstrate fluid shear stress as a stimulatory mechanism for preosteoblastic differentiation and osteoblastic cell signaling.

19.5 TENDON ENGINEERING STUDY

19.5.1 MATERIALS AND METHODS

19.5.1.1 Scaffold Preparation

HUVs were obtained and prepared as described elsewhere with minor modifications as detailed below [65,66]. Cords were separated from full-term placentas obtained from Norman Regional Hospital (Norman, OK) within 5 days of delivery. Cords were trimmed to 8.5 cm in length, rinsed with distilled water, and mounted through the HUV on stainless steel mandrels (McMaster-Carr, Atlanta, GA) prior to freezing. The mounted cords were frozen to −80°C at a rate of −2.5°C/min. An automated lathe was used to dissect the HUV from the rest of the cord (while frozen). The resulting HUV scaffold had a wall thickness of 0.75 mm. The scaffolds were allowed to thaw, inverted (so that lumenal and ablumenal sides were reversed), and then trimmed to 6.5 cm in length prior to decellularization. For decellularization, scaffolds were washed for 24 h in 1% w/v sodium dodecyl sulfate solution, followed by a thorough (multistage) wash using distilled water. Scaffolds were then washed for 24 h in 95% ethanol and again thoroughly washed using distilled water. HUV scaffolds were then sterilized in 0.2% v/v peracetic acid solution. The scaffolds were then washed extensively in sterile water and pH-balanced (7.2–7.4) in sterile PBS prior to storage at −4°C in sterile PBS. Scaffolds were used within 1 week of storage.

19.5.1.2 Scaffold Seeding

Scaffolds were fitted with steel adapters at each end and immersed in α-MEM 1 h prior to seeding. In addition to steel adapters attached at each end of the HUV construct (Figure 19.5), a steel screw was inserted into the adapter on one end. The other end's screw was not inserted until after the constructs were seeded.

   Cells were obtained and cultured as described in the bone engineering study, presented earlier. Cells were rinsed with PBS, trypsinized (0.25% trypsin-EDTA, Invitrogen), and subsequently removed from the culture flask using α-MEM. Then the cell solution was centrifuged at 2200 rpm (IEC Model HNS2). The supernatant was discarded and the cell pellet was resuspended in a collagen I (Gibco, Cat. No. A10644) solution such that the resultant cell solution contained 1 million cells/mL. Using 1 mL pipettes, 0.6 mL of the collagen/cell suspension was injected inside each HUV. This volume of collagen solution completely fills the void space inside the HUV. The adapter through which the scaffold was seeded then had its respective screw inserted, effectively sealing the collagen/cell solution inside the scaffold. HUVs were incubated at 37°C for at least 40 min while a firm collagen gel formed.

19.5.1.3 Bioreactor Design and Mechanical Stimulation

As described previously, the bioreactor employed for this study is a custom design, specifically created to culture multiple tissue constructs simultaneously in a sterile environment while applying

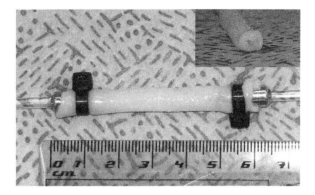

FIGURE 19.5 A decellularized HUV with steel adapters at each end attached using cable ties. A glass rod assists in the attachment of the adapters. Inset: Decellularized HUV, end view.

controlled mechanical stimuli at various frequencies, strain levels, and waveforms to cell-seeded scaffolds [66]. The bioreactor is capable of mechanically stimulating three constructs simultaneously, each cultured in a separate chamber. All experimental groups were therefore tested in triplicate. The constructs are stimulated using a bidirectional linear actuator (capable of applying controlled loads in both positive and negative directions from the baseline position) controlled by a signal converter (Wavetec model 185, San Diego, CA). The output signal from the signal converter is amplified before reaching the actuator. The specifications of the actuator are as follows: linear travel zone = 5 in. (12.7 cm), peak force sustained for 10 s = 49 lb (22.22 kg), and maximum force sustained continuously = 14 lb (6.35 kg) (Figure 19.6).

19.5.1.4 RNA Extraction and Analysis

RNA extraction and RT-PCR procedures were executed as described previously [66], though they will be reproduced here for clarity with changes described below. Tissue samples (≤25 mg) dedicated for gene expression analysis were stored at −20°C in RNAlater® (Ambion, Austin, TX) to preserve RNA until analysis could be performed. RNA was isolated from the tissue using an RNAqueous®-4PCR Kit (Ambion) in accordance with the manufacturer's directions. cDNA was synthesized from RNA using TaqMan® Reverse Transcription Reagents (Applied Biosystems, Austin, TX) in accordance with manufacturer's directions. Gene expression levels of collagen I (Coll-a1), collagen III (Coll-a3), tenascin-C, and GAPDH were quantified using a Power SYBR® Green PCR Master Mix kit (Applied Biosystems). Primer sequences for Coll-a1, Coll-a3, and GAPDH were the same as those used in bone tissue engineering study (4.1.6). Tenascin-C primers were forward 5′-CAG AAG CCT TGG CCA TGT G-3′ and reverse 5′-GTG GTC ATG AGC CCT TCC A-3′. Quantitative real-time RT-PCR was conducted as follows: 10 min at 25°C, followed by 45 cycles of amplification where each cycle consists of 15 s at 95°C followed by 1 min at 60°C. Relative gene expression (compared to GAPDH) was analyzed using the $\Delta\Delta C_T$ method [164].

19.5.2 Results

After 3 days of culture, qRT-PCR indicated that compared to the static control group, Group 2 (stretched twice per day) exhibited a downregulation of collagen I (>100-fold), collagen III (>30-fold), and tenascin-C (>50%) (Figure 19.7). Fully contrasting with Group 2 was Group 3, which displayed gene upregulation across the board: over 10-fold for collagen I, almost 4-fold for collagen III, and nearly 40-fold for tenascin-C (Figure 19.7). Group 4 showed the mildest changes on average, including merely doubled collagen I gene expression, collagen III expression diminished by less than half, and tenascin-C expression just over three times that of the control group.

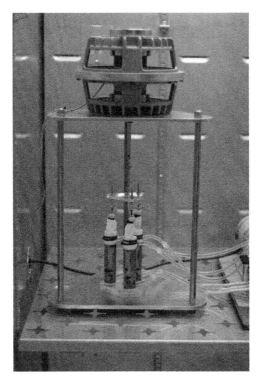

FIGURE 19.6 (See color insert.) In this study, mechanical forces were delivered to the constructs using a bidirectional motor (upper portion of image) that was connected to three culture chambers via a distributor plate. The constructs may be seen inside the glass culture chambers. Also visible are the send and return lines for media circulation.

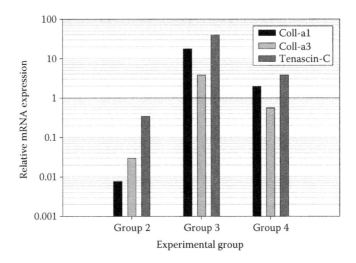

FIGURE 19.7 Relative mRNA expression of MSCs seeded on decellularized HUV and cultured in a mechanical stimulator compared to static controls (Group 1). Group 1 was cultured inside the bioreactor but subjected to no mechanical stimulation (static control). Group 2 was subjected to stimulation of 1 h stretching cycles, 2 cycles per day (morning and evening), 0.01667 Hz. Group 3 was subjected to stimulation of 1 h stretching cycles, 1 cycle per day, 0.01667 Hz. Group 4 was subjected to stimulation of 1 h stretching cycles, 1 cycle per day, 0.03333 Hz.

19.5.3 Discussion

The three genes considered here are all positively associated with tenocyte lineage. The conditions of experimental Group 2, which entail a significantly reduced resting period between applications of tensile stimulation, appear to have a profoundly adverse effect on MSCs' tendency toward tenocyte differentiation. The conditions of experimental Group 4, which entail stretching at twice the frequency of the other groups, appear to have a limited positive effect on tenocyte differentiation. Comparing Groups 2 and 4, which received approximately the same number of "total stretches," suggests the primary detrimental condition appears to be the rest period granted to the constructs. The secondary detrimental condition, as observed by comparing Groups 3 and 4, appears to be the rate of stretching (where the lower frequency is preferable). These results are corroborated by the previous research of Abousleiman et al. [65,66].

Clearly, more thorough research is required to fully understand the nature of tenocyte differentiation. Here, through a rudimentary analysis, two characteristics of mechanical stimulation were considered. Both of them appear to have substantial effects on tenocyte-associated gene expression of MSCs.

19.6 FUTURE

Mechanical stimulation has been shown to be critical in the differentiation of bone and tendon cells making mechanotransduction a useful tool in the development of novel tissue engineering strategies. Different types of mechanical stimuli appear to influence osteoblastic and tenocytic cells. The proper manipulation of mechanical stimuli in conjunction with their respective biochemical signals create new possibilities in the development of functioning tissue engineered constructs. Better understanding of the cellular mechanism with which cells transduce mechanical signals into intracellular processes will further enrich the tools tissue engineering utilizes in the regeneration of mechanosensitive tissues like bone and tendons. It is expected that in the near future, specific mechanical signals will be utilized on a regular basis in the differentiation of SCs toward a specific mechanosensitive phenotype complementing or even replacing biochemical agents. The creation of each specific mechanical stimulus will require the use of bioreactors making them a requirement toward the development of successful bone and tendon grafts.

REFERENCES

1. Langer, R. and J.P. Vacanti, Tissue engineering. *Science*, 1993, **260**(5110): 920–926.
2. Bonassar, L.J. and C.A. Vacanti, Tissue engineering: The first decade and beyond. *Journal of Cellular Biochemistry Supplement*, 1998, **30–31**: 297–303.
3. Hollinger, J.O. and G.C. Battistone, Biodegradable bone repair materials. Synthetic polymers and ceramics. *Clinical Orthopaedics and Related Research*, 1986, **207**: 290–305.
4. Hadjipanayi, E., R.A. Brown, and V. Mudera, Interface integration of layered collagen scaffolds with defined matrix stiffness: Implications for sheet-based tissue engineering. *Journal of Tissue Engineering Regenerative Medicine*, 2009, **3**(3): 230–241.
5. Glowacki, J. and S. Mizuno, Collagen scaffolds for tissue engineering. *Biopolymers*, 2008, **89**(5): 338–344.
6. Liu, X. and P.X. Ma, Polymeric scaffolds for bone tissue engineering. *Annals of Biomedical Engineering*, 2004, **32**(3): 477–486.
7. Nienhuijs, M.E. et al., Bone-like tissue formation using an equine COLLOSS E-filled titanium scaffolding material. *Biomaterials*, 2006, **27**(16): 3109–3114.
8. Woo, K.M. et al., Nano-fibrous scaffolding promotes osteoblast differentiation and biomineralization. *Biomaterials*, 2007, **28**(2): 335–343.
9. Hutmacher, D.W., Scaffolds in tissue engineering bone and cartilage. *Biomaterials*, 2000, **21**(24): 2529–2543.
10. Parfitt, A.M., The cellular basis of bone remodeling: The quantum concept reexamined in light of recent advances in the cell biology of bone. *Calcified Tissue International*, 1984, **36** (Suppl 1): S37–S45.

11. Rosenbaum, A.J., D.A. Grande, and J.S. Dines, The use of mesenchymal stem cells in tissue engineering: A global assessment. *Organogenesis*, 2008, **4**(1): 23–27.
12. Stiehler, M. et al., Effect of dynamic 3-D culture on proliferation, distribution, and osteogenic differentiation of human mesenchymal stem cells. *Journal of Biomedical Materials Research part A*, 2009, **89**(1): 96–107.
13. Holtorf, H.L., J.A. Jansen, and A.G. Mikos, Flow perfusion culture induces the osteoblastic differentiation of marrow stroma cell-scaffold constructs in the absence of dexamethasone. *Journal of Biomedical Materials Research. Part A*, 2005, **72**(3): 326–334.
14. Kern, S. et al., Comparative analysis of mesenchymal stem cells from bone marrow, umbilical cord blood, or adipose tissue. *Stem Cells*, 2006, **24**(5): 1294–1301.
15. Sekiya, I. et al., Dexamethasone enhances SOX9 expression in chondrocytes. *Journal of Endocrinology*, 2001, **169**(3): 573–579.
16. Sikavitsas, V.I., J.S. Temenoff, and A.G. Mikos, Biomaterials and bone mechanotransduction. *Biomaterials*, 2001, **22**(19): 2581–2593.
17. Marie, P.J., F. Debiais, and E. Hay, Regulation of human cranial osteoblast phenotype by FGF-2, FGFR-2 and BMP-2 signaling. *Histology and Histopathology*, 2002, **17**(3): 877–885.
18. Yamaguchi, A., T. Komori, and T. Suda, Regulation of osteoblast differentiation mediated by bone morphogenetic proteins, hedgehogs, and Cbfa1. *Endocrine Review*, 2000, **21**(4): 393–411.
19. Li, C. et al., Electrospun silk-BMP-2 scaffolds for bone tissue engineering. *Biomaterials*, 2006, **27**(16): 3115–3124.
20. Wolff, J., *The Law of Bone Remodeling*. 1986, New York, Berlin, Heidelberg: Springer.
21. Buckwalter, J.A. et al., Bone biology. II: Formation, form, modeling, remodeling, and regulation of cell function. *Instructional Course Lectures*, 1996, **45**: 387–399.
22. Cowin, S.C., Bone poroelasticity. *Journal of Biomechanics*, 1999, **32**(3): 217–238.
23. Bonewald, L.F. and M.L. Johnson, Osteocytes, mechanosensing and Wnt signaling. *Bone*, 2008, **42**(4): 606–615.
24. Riddle, R.C. and H.J. Donahue, From streaming-potentials to shear stress: 25 years of bone cell mechanotransduction. *Journal of Orthopaedic Research*, 2009, **27**(2): 143–149.
25. Hillsley, M.V. and J.A. Frangos, Bone tissue engineering: The role of interstitial fluid flow. *Biotechnology and Bioengineering*, 1994, **43**(7): 573–581.
26. Knothe Tate, M.L., U. Knothe, and P. Niederer, Experimental elucidation of mechanical load-induced fluid flow and its potential role in bone metabolism and functional adaptation. *The American Journal of the Medical Sciences*, 1998, **316**(3): 189–195.
27. Burger, E.H. and J. Klein-Nulend, Mechanotransduction in bone–role of the lacuno-canalicular network. *FASEB Journal*, 1999, **13** (Suppl): S101–S112.
28. Basso, N. and J.N. Heersche, Characteristics of in vitro osteoblastic cell loading models. *Bone*, 2002, **30**(2): 347–351.
29. Mullender, M. et al., Mechanotransduction of bone cells in vitro: Mechanobiology of bone tissue. *Medical and Biological Engineering and Computing*, 2004, **42**(1): 14–21.
30. Yamaguchi, D.T. et al., Parathyroid hormone-activated calcium channels in an osteoblast-like clonal osteosarcoma cell line. cAMP-dependent and cAMP-independent calcium channels. *Journal of Biological Chemistry*, 1987, **262**(16): 7711–7718.
31. Davidson, R.M., D.W. Tatakis, and A.L. Auerbach, Multiple forms of mechanosensitive ion channels in osteoblast-like cells. *Pflugers Archive*, 1990, **416**(6): 646–651.
32. Ypey, D.L. et al., Voltage, calcium, and stretch activated ionic channels and intracellular calcium in bone cells. *Journal of Bone and Mineral Research*, 1992, **7** (Suppl 2): S377–S387.
33. Ryder, K.D. and R.L. Duncan, Parathyroid hormone enhances fluid shear-induced [Ca2+]i signaling in osteoblastic cells through activation of mechanosensitive and voltage-sensitive Ca2+ channels. *Journal of Bone and Mineral Research*, 2001, **16**(2): 240–248.
34. Rawlinson, S.C., A.A. Pitsillides, and L.E. Lanyon, Involvement of different ion channels in osteoblasts' and osteocytes' early responses to mechanical strain. *Bone*, 1996, **19**(6): 609–614.
35. Li, J. et al., L-type calcium channels mediate mechanically induced bone formation in vivo. *Journal of Bone and Mineral Research*, 2002, **17**(10): 1795–1800.
36. Pavalko, F.M. et al., A model for mechanotransduction in bone cells: The load-bearing mechanosomes. *Journal of Cellular Biochemistry*, 2003, **88**(1): 104–112.
37. Liu, D. et al., Activation of extracellular-signal regulated kinase (ERK1/2) by fluid shear is Ca(2+)- and ATP-dependent in MC3T3-E1 osteoblasts. *Bone*, 2008, **42**(4): 644–652.

38. Ingber, D.E., Tensegrity: The architectural basis of cellular mechanotransduction. *Annual Review of Physiology*, 1997, **59**: 575–599.

39. Pavalko, F.M. et al., Fluid shear-induced mechanical signaling in MC3T3-E1 osteoblasts requires cytoskeleton-integrin interactions. *American Journal of Physiology*, 1998, **275**(6 Pt 1): C1591–C1601.

40. Bidwell, J.P. et al., Nuclear matrix proteins and osteoblast gene expression. *Journal of Bone and Mineral Research*, 1998, **13**(2): 155–167.

41. Maniotis, A.J., C.S. Chen, and D.E. Ingber, Demonstration of mechanical connections between integrins, cytoskeletal filaments, and nucleoplasm that stabilize nuclear structure. *Proceedings of the National Academy of Sciences of the United States of America*, 1997, **94**(3): 849–854.

42. Weyts, F.A. et al., ERK activation and alpha v beta 3 integrin signaling through Shc recruitment in response to mechanical stimulation in human osteoblasts. *Journal of Cellular Biochemistry*, 2002, **87**(1): 85–92.

43. Di Palma, F. et al., Modulation of the responses of human osteoblast-like cells to physiologic mechanical strains by biomaterial surfaces. *Biomaterials*, 2005, **26**(20): 4249–4257.

44. Jaasma, M.J. et al., Adaptation of cellular mechanical behavior to mechanical loading for osteoblastic cells. *Journal of Biomechanics*, 2007, **40**(9): 1938–1945.

45. Cheng, B. et al., PGE(2) is essential for gap junction-mediated intercellular communication between osteocyte-like MLO-Y4 cells in response to mechanical strain. *Endocrinology*, 2001, **142**(8): 3464–3473.

46. Genetos, D.C. et al., Oscillating fluid flow activation of gap junction hemichannels induces ATP release from MLO-Y4 osteocytes. *Journal of Cellular Physiology*, 2007, **212**(1): 207–214.

47. Cherian, P.P. et al., Mechanical strain opens connexin 43 hemichannels in osteocytes: A novel mechanism for the release of prostaglandin. *Molecular Biology of the Cell*, 2005, **16**(7): 3100–3106.

48. Cherian, P.P. et al., Effects of mechanical strain on the function of Gap junctions in osteocytes are mediated through the prostaglandin EP2 receptor. *Journal of Biological Chemistry*, 2003, **278**(44): 43146–43156.

49. Saunders, M.M. et al., Gap junctions and fluid flow response in MC3T3-E1 cells. *American Journal of Physiology Cell Physiology*, 2001, **281**(6): C1917–C1925.

50. Saunders, M.M. et al., Fluid flow-induced prostaglandin E2 response of osteoblastic ROS 17/2.8 cells is gap junction-mediated and independent of cytosolic calcium. *Bone*, 2003, **32**(4): 350–356.

51. Ziambaras, K. et al., Cyclic stretch enhances gap junctional communication between osteoblastic cells. *Journal of Bone and Mineral Research*, 1998, **13**(2): 218–228.

52. Lozupone, E. et al., Intermittent compressive load stimulates osteogenesis and improves osteocyte viability in bones cultured "in vitro". *Clinical Rheumatology*, 1996, **15**(6): 563–572.

53. Jacobs, C.R., Primary cilia. *Journal of Musculoskeletal and Neuronal Interactions*, 2007, **7**(4): 297–298.

54. Whitfield, J.F., The solitary (primary) cilium—A mechanosensory toggle switch in bone and cartilage cells. *Cellular Signalling*, 2008, **20**(6): 1019–1024.

55. Whitfield, J.F., Primary cilium—Is it an osteocyte's strain-sensing flowmeter? *Journal of Cellular Biochemistry*, 2003, **89**(2): 233–237.

56. Malone, A.M. et al., Primary cilia mediate mechanosensing in bone cells by a calcium-independent mechanism. *Proceedings of the National Academy of Sciences of the United States of America*, 2007, **104**(33): 13325–13330.

57. Malone, A.M. et al., Primary cilia in bone. *of Musculoskeletal and Neuronal Interactions*, 2007, **7**(4): 301.

58. Pavlin, D. et al., Mechanical loading stimulates differentiation of periodontal osteoblasts in a mouse osteoinduction model: Effect on type I collagen and alkaline phosphatase genes. *Calcified Tissue International*, 2000, **67**(2): 163–172.

59. Goldstein, A.S. et al., Effect of convection on osteoblastic cell growth and function in biodegradable polymer foam scaffolds. *Biomaterials*, 2001, **22**(11): 1279–1288.

60. Gomes, M.E. et al., Effect of flow perfusion on the osteogenic differentiation of bone marrow stromal cells cultured on starch-based three-dimensional scaffolds. *Journal of Biomedical Materials Research part A*, 2003, **67**(1): 87–95.

61. Ahmed, T., E. Dare, and M. Hincke, Fibrin: A versatile scaffold for tissue engineering applications. *Tissue Engineering: Part B*, 2008, **14**(2): 199–215.

62. Bagnaninchi, P.O. et al., Chitosan microchannel scaffolds for tendon tissue engineering characterized using optical coherence tomography. *Tissue Engineering*, 2007, **13**(2): 323–331.

63. Kardestuncer, T. et al., RGD-tethered silk substrate stimulates the differentiation of human tendon cells. *Clinical Orthopaedics and Related Research*, 2006, **448**: 234–239.

64. Androjna, C., R. Spragg, and K. Derwin, Mechanical conditioning of cell-seeded small intestine submucosa: A potential tissue-engineering strategy for tendon repair. *Tissue Engineering*, 2007, **13**(2): 233–243.

65. Abousleiman, R.I. et al., The human umbilical vein: A novel scaffold for musculoskeletal soft tissue regeneration. *Artificial Organs*, 2008, **32**(9): 735–742.

66. Abousleiman, R.I. et al., Tendon tissue engineering using cell-seeded umbilical veins cultured in a mechanical stimulator. *Tissue Engineering: Part A*, 2009, **15**(4): 787–795.

67. Leigh, D.R., E.L. Abreu, and K. Derwin, Changes in gene expression of individual matrix metalloproteinases differ in response to mechanical unloading of tendon fascicles in explant culture. *Journal of Orthopaedic Research*, 2008, **26**(10): 1306–1312.

68. Maeda, E. et al., Time dependence of cyclic tensile strain on collagen production in tendon fascicles. *Biochemical and Biophysical Research Communications*, 2007, **362**: 399–404.

69. Maeda, E. et al., Differential regulation of gene expression in isolated tendon fascicles exposed to cyclic tensile strain in vitro. *Journal of Applied Physiology*, 2009, **106**: 506–512.

70. Lavagnino, M. et al., A finite element model predicts the mechanotransduction response of tendon cells to cyclic tensile loading. *Biomechanics and Modeling in Mechanobiology*, 2008, **7**: 405–416.

71. Banes, A.J. et al., PDGF-BB, IGF-I and mechanical load stimulate DNA synthesis in avian tendon fibroblasts in vitro. *Journal of Biomechanics*, 1995, **28**(12): 1505–1513.

72. Ingber, D.E., Tensegrity: The architectural basis of cellular mechanotransduction. *Annual Review of Physiology*, 1997, **59**: 575–599.

73. D'Addario, M. et al., Regulation of tension-induced mechanotranscriptional signals by the microtubule network in fibroblasts. *The Journal of Biological Chemistry*, 2003, **278**(52): 53097.

74. Arnoczky, S.P. et al., Ex vivo static tensile loading inhibits MMP-1 expression in rat tail tendon cells through a cytoskeletally based mechanotransduction mechanism. *Journal of Orthopaedic Research*, 2004, **22**: 328–333.

75. Waggett, A.D., M. Benjamin, and J.R. Ralphs, Connexin 32 and 43 gap junctions differentially modulate tenocyte response to cyclic mechanical load. *European Journal of Cell Biology*, 2006, **85**: 1145–1154.

76. Langholz, O. et al., Collagen and collagenase gene expression in three-dimensional collagen lattices are differentially regulated by $\alpha 1 \beta 1$ and $\alpha 2 \beta 1$ integrins. *Journal of Cell Biology*, 1995, **131**(6 pt 2): 1903–1915.

77. Arnoczky, S.P. et al., Activation of stress-activated protein kinases (SAPK) in tendon cells following cyclic strain: The effects of strain frequency, strain magnitude, and cytosolic calcium. *Journal of Orthopaedic Research*, 2002, **20**: 947–952.

78. Martineau, L.C. and P.F. Gardiner, Insight into skeletal muscle mechanotransduction: MAPK activation is quantitatively related to tension. *Journal of Applied Physiology*, 2001, **91**: 693–702.

79. Butler, D.L., N. Juncosa-Melvin, and M.R. Dressler, Functional efficacy of tendon repair processes. *Annual Review of Biomedical Engineering*, 2004, **6**: 303–329.

80. Wang, J.H.-C., Mechanobiology of tendon. *Journal of Biomechanics*, 2006, **39**: 1563–1582.

81. Wang, J.H. et al., Cyclic mechanical stretching of human tendon fibroblasts increases the production of prostaglandin E2 and levels of cyclooxygenase expression: A novel in vitro model study. *Connective Tissue Research*, 2003, **44**: 128–133.

82. Almekinders, L.C., A.J. Banes, and C.A. Ballenger, Effects of repetitive motion on human fibroblasts. *Medicine and Science in Sports and Exercise*, 1993, **25**: 603–607.

83. Almekinders, L.C., A.J. Baynes, and L.W. Bracey, An in vitro investigation into the effects of repetitive motion and nonsteroidal antiinflammatory medication on human tendon fibroblasts. *American Journal of Sports Medicine*, 1995, **23**: 119–123.

84. Webb, K. et al., Cyclic strain increases fibroblast proliferation, matrix accumulation, and elastic modulus of fibroblast-seeded polyurethane constructs. *Journal of Biomechanics*, 2006, **39**: 1136–1144.

85. Chokalingam, K. et al., Tensile stimulation of murine stem cell-collagen sponge constructs increases collagen type i gene expression and linear stiffness. *Tissue Engineering: Part A*, 2009, **15**(9): 2561–2570.

86. Li, F. et al., Cell shape regulates collagen type i expression in human tendon fibroblasts. *Cell Motility and the Cytoskeleton*, 2008, **65**: 332–341.

87. Scott, A. et al., Tenocyte responses to mechanical loading in vivo: A role for local insulin-like growth factor 1 signaling in early tendinosis in rats. *Arthritis and Rheumatism*, 2007, **56**(3): 871–881.

88. Chiquet-Ehrismann, R. et al., Tenascin-C expression by fibroblasts is elevated in stressed collagen gels. *Journal of Cell Biology*, 1994, **127**: 2093–2101.

89. Chiquet, M. et al., Regulation of extracellular matrix synthesis by mechanical stress. *Biochemistry and Cell Biology*, 1996, **74**: 737–744.

90. Eckes, B. et al., Downregulation of collagen synthesis in fibroblasts within three-dimensional collagen lattices involves transcriptional and posttranscriptional mechanisms. *FEBS Letters*, 1993, **318**: 129–133.

91. Lambert, C.A. et al., Pretranslational regulation of extracellular matrix macromolecules and collagenase expression in fibroblasts by mechanical forces. *Laboratory Investigation*, 1992, **66**(4): 444–451.

92. Hatamochi, A. et al., Regulation of collagen VI expression in fibroblasts, effects of cell density, cell-matrix interactions, and chemical transformation. *Journal of Biological Chemistry*, 1989, **264**: 3494–3499.

93. Mauch, C. et al., Collagenase gene expression in fibroblasts is regulated by a three-dimensional contact with collagen. *FEBS Letters*, 1989, **250**: 301–305.

94. Arnoczky, S.P., M. Lavagnino, and M. Egerbacher, The mechanobiological aetiopathogenesis of tendinopathy: Is it the over-stimulation or the under-stimulation of tendon cells? *International Journal of Experimental Pathology*, 2007, **88**(4): 217–226.

95. Freed, L.E. et al., Advanced tools for tissue engineering: Scaffolds, bioreactors, and signaling. *Tissue Engineering*, 2006, **12**(12): 3285–3305.

96. Bancroft, G.N., V.I. Sikavitsas, and A.G. Mikos, Design of a flow perfusion bioreactor system for bone tissue-engineering applications. *Tissue Engineering*, 2003, **9**(3): 549–554.

97. Sikavitsas, V.I. et al., Mineralized matrix deposition by marrow stromal osteoblasts in 3D perfusion culture increases with increasing fluid shear forces. *Proceedings of the National Academy of Sciences of the United States of America*, 2003, **100**(25): 14683–14688.

98. Meinel, L. et al., Engineering bone-like tissue in vitro using human bone marrow stem cells and silk scaffolds. *Journal of Biomedical Materials Research Part A*, 2004, **71**(1): 25–34.

99. Ishaug, S.L. et al., Bone formation by three-dimensional stromal osteoblast culture in biodegradable polymer scaffolds. *Journal of Biomedical Materials Research*, 1997, **36**(1): 17–28.

100. Ishaug-Riley, S.L. et al., Three-dimensional culture of rat calvarial osteoblasts in porous biodegradable polymers. *Biomaterials*, 1998, **19**(15): 1405–1412.

101. Wendt, D. et al., Oscillating perfusion of cell suspensions through three-dimensional scaffolds enhances cell seeding efficiency and uniformity. *Biotechnology and Bioengineering*, 2003, **84**(2): 205–214.

102. Sikavitsas, V.I., G.N. Bancroft, and A.G. Mikos, Formation of three-dimensional cell/polymer constructs for bone tissue engineering in a spinner flask and a rotating wall vessel bioreactor. *Journal of Biomedical Materials Research*, 2002, **62**(1): 136–148.

103. Zhao, F. and T. Ma, Perfusion bioreactor system for human mesenchymal stem cell tissue engineering: Dynamic cell seeding and construct development. *Biotechnology and Bioengineering*, 2005, **91**(4): 482–493.

104. Darling, E.M. and K.A. Athanasiou, Articular cartilage bioreactors and bioprocesses. *Tissue Engineering*, 2003, **9**(1): 9–26.

105. Vunjak-Novakovic, G. et al., Dynamic cell seeding of polymer scaffolds for cartilage tissue engineering. *Biotechnology Progress*, 1998, **14**(2): 193–202.

106. Sucosky, P. et al., Fluid mechanics of a spinner-flask bioreactor. *Biotechnology and Bioengineering*, 2004, **85**(1): 34–46.

107. Shea, L.D. et al., Engineered bone development from a pre-osteoblast cell line on three-dimensional scaffolds. *Tissue Engineering*, 2000, **6**(6): 605–617.

108. Vunjak-Novakovic, G. et al., Microgravity studies of cells and tissues. *Annals of the New York Academy of Sciences*, 2002, **974**: 504–517.

109. Begley, C.M. and S.J. Kleis, The fluid dynamic and shear environment in the NASA/JSC rotating-wall perfused-vessel bioreactor. *Biotechnology and Bioengineering*, 2000, **70**(1): 32–40.

110. Williams, K.A., S. Saini, and T.M. Wick, Computational fluid dynamics modeling of steady-state momentum and mass transport in a bioreactor for cartilage tissue engineering. *Biotechnology Progress*, 2002, **18**(5): 951–963.

111. Duke, P.J., E.L. Daane, and D. Montufar-Solis, Studies of chondrogenesis in rotating systems. *Journal of Cellular Biochemistry*, 1993, **51**(3): 274–282.

112. Freed, L.E. et al., Chondrogenesis in a cell-polymer-bioreactor system. *Experimental Cell Research*, 1998, **240**(1): 58–65.

113. Vunjak-Novakovic, G. et al., Bioreactor studies of native and tissue engineered cartilage. *Biorheology*, 2002, **39**(1–2): 259–268.

114. Yoshioka, T. et al., Repair of large osteochondral defects with allogeneic cartilaginous aggregates formed from bone marrow-derived cells using RWV bioreactor. *Journal of Orthopaedic Research*, 2007, **25**(10): 1291–1298.

115. Rath, B. et al., Compressive forces induce osteogenic gene expression in calvarial osteoblasts. *Journal of Biomechanics*, 2008, **41**(5): 1095–1103.

116. Kaspar, D. et al., Dynamic cell stretching increases human osteoblast proliferation and CICP synthesis but decreases osteocalcin synthesis and alkaline phosphatase activity. *Journal of Biomechanics*, 2000, **33**(1): 45–51.

117. Neidlinger-Wilke, C., H.J. Wilke, and L. Claes, Cyclic stretching of human osteoblasts affects proliferation and metabolism: A new experimental method and its application. *Journal of Orthopaedic Research*, 1994, **12**(1): 70–78.

118. Nagatomi, J. et al., Cyclic pressure affects osteoblast functions pertinent to osteogenesis. *Annals of Biomedical Engineering*, 2003, **31**(8): 917–923.

119. Mitsui, N. et al., Optimal compressive force induces bone formation via increasing bone sialoprotein and prostaglandin E(2) production appropriately. *Life Sciences*, 2005, **77**(25): 3168–3182.

120. Liu, J. et al., Pressure-loaded MSCs during early osteodifferentiation promote osteoclastogenesis by increase of RANKL/OPG ratio. *Annals of Biomedical Engineering*, 2009, **37**(4): 794–802.

121. Mitsui, N. et al., Optimal compressive force induces bone formation via increasing bone morphogenetic proteins production and decreasing their antagonists production by Saos-2 cells. *Life Sciences*, 2006, **78**(23): 2697–2706.

122. Mitsui, N. et al., Effect of compressive force on the expression of MMPs, PAs, and their inhibitors in osteoblastic Saos-2 cells. *Life Sciences*, 2006, **79**(6): 575–583.

123. Winter, L.C. et al., Intermittent versus continuous stretching effects on osteoblast-like cells in vitro. *Journal of Biomedical Materials Research Part A*, 2003, **67**(4): 1269–1275.

124. Shimko, D.A. et al., A device for long term, in vitro loading of three-dimensional natural and engineered tissues. *Annals of Biomedical Engineering*, 2003, **31**(11): 1347–1356.

125. Jagodzinski, M. et al., Effects of cyclic longitudinal mechanical strain and dexamethasone on osteogenic differentiation of human bone marrow stromal cells. *European Cells and Materials*, 2004, **7**: 35–41; discussion 41.

126. Bancroft, G.N. et al., Fluid flow increases mineralized matrix deposition in 3D perfusion culture of marrow stromal osteoblasts in a dose-dependent manner. *Proceedings of the National Academy of Sciences of the United States of America*, 2002, **99**(20): 12600–12605.

127. Cartmell, S.H. et al., Effects of medium perfusion rate on cell-seeded three-dimensional bone constructs in vitro. *Tissue Engineering*, 2003, **9**(6): 1197–1203.

128. Alvarez-Barreto, J.F. et al., Flow perfusion improves seeding of tissue engineering scaffolds with different architectures. *Annals of Biomedical Engineering*, 2007, **35**(3): 429–442.

129. Wang, Y. et al., Application of perfusion culture system improves in vitro and in vivo osteogenesis of bone marrow-derived osteoblastic cells in porous ceramic materials. *Tissue Engineering*, 2003, **9**(6): 1205–1214.

130. Chung, C.A. et al., Enhancement of cell growth in tissue-engineering constructs under direct perfusion: Modeling and simulation. *Biotechnology and Bioengineering*, 2007, **97**(6): 1603–1616.

131. Mizuno, S., F. Allemann, and J. Glowacki, Effects of medium perfusion on matrix production by bovine chondrocytes in three-dimensional collagen sponges. *Journal of Biomedical Materials and Research*, 2001, **56**(3): 368–375.

132. Zhao, F. et al., Effects of oxygen transport on 3-d human mesenchymal stem cell metabolic activity in perfusion and static cultures: Experiments and mathematical model. *Biotechnology Program*, 2005, **21**(4): 1269–1280.

133. Zhao, F., R. Chella, and T. Ma, Effects of shear stress on 3-D human mesenchymal stem cell construct development in a perfusion bioreactor system: Experiments and hydrodynamic modeling. *Biotechnology and Bioengineering*, 2007, **96**(3): 584–595.

134. Meinel, L. et al., Bone tissue engineering using human mesenchymal stem cells: Effects of scaffold material and medium flow. *Annals of Biomedical Engineering*, 2004, **32**(1): 112–122.

135. Xie, Y. et al., Three-dimensional flow perfusion culture system for stem cell proliferation inside the critical-size beta-tricalcium phosphate scaffold. *Tissue Engineering*, 2006, **12**(12): 3535–3543.

136. Annovazzi, L. and F. Genna, An engineering, multiscale constitutive model for fiber-forming collagen in tension. *Journal of Biomedical Materials Research Part A*, 2010, **92**(1): 254–266.

137. Bailey, A.J. and N.D. Light, Intermolecular cross-linking in fibrotic collagen. *Ciba Foundation Symposium*, 1985, **114**: 80–96.

138. Eyre, D.R., M.A. Paz, and P.M. Gallop, Cross-linking in collagen and elastin. *Annual Review of Biochemistry*, 1984, **53**: 717–748.

139. Feng, Z. et al., Construction of fibroblast-collagen gels with orientated fibrils induced by static or dynamic stress: Toward the fabrication of small tendon grafts. *Journal of Artificial Organs*, 2006, **9**: 220–225.
140. Fratzl, P. et al., Fibrillar structure and mechanical properties of collagen. *Journal of Structural Biology*, 1997, **112**: 119–122.
141. Harada, I. et al., A simple combined floating and anchored collagen gel for enhancing mechanical strength of culture system. *Journal of Biomedical Materials Research Part A*, 2007, **80**(1): 123–130.
142. Józsa, L. et al., Alterations in dry mass content of collagen fiber in degenerative tendinopathy and tendon-rupture. *Matrix*, 1989, **9**: 140–146.
143. Puxkandl, R. et al., Viscoelastic properties of collagen: Synchrotron radiation investigations and structural model. *Philosophical Transactions of the Royal Society B Biological Sciences*, 2002, **357**(1418): 191–197.
144. Yamato, M. et al., Condensation of collagen fibrils to the direct vicinity of fibroblasts as a cause of gel contraction. *Journal of Biochemistry*, 1995, **117**(5): 940–946.
145. Young, R.G. et al., Use of mesenchymal stem cells in a collagen matrix for achilles tendon repair. *The Journal of Bone and Joint Surgery*, 1998, **16**: 406–413.
146. Screen, H.R.C. et al., An investigation into the effects of the hierarchical structure of tendon fascicles on micromechanical properties. *Proceedings of IMechE Part H: Journal of Engineering in Medicine*, 2004, **218**: 109–119.
147. Ohno, K. et al., Effects of complete stress-shielding on the mechanical properties and histology of in situ frozen patellar tendon. *Journal of Orthopaedic Research*, 1993, **11**(4): 592–602.
148. Yamamoto, N. et al., Effects of stress shielding on the mechanical properties of rabbit patellar tendon. *Journal of Biomechanical Engineering*, 1993, **115**(1): 23–28.
149. Butler, D.L. et al., Functional tissue engineering for tendon repair: A multidisciplinary strategy using mesenchymal stem cells, bioscaffolds, and mechanical stimulation. *Journal of Orthopaedic Research*, 2008, **26**(1): 1–9.
150. Butler, D.L. et al., Using functional tissue engineering and bioreactors to mechanically stimulate tissue-engineered constructs. *Tissue Engineering: Part A*, 2009, **15**.
151. Garvin, J. et al., Novel system for engineering bioartificial tendons and application of mechanical load. *Tissue Engineering*, 2003, **9**(5): 967–979.
152. Joshi, S.D. and K. Webb, Variation of cyclic strain parameters regulates development of elastic modulus in fibroblast/substrate constructs. *Journal of Orthopaedic Research*, 2008, **26**(8): 1105–1113.
153. Nirmalanandhan, V.S. et al., Combined effects of scaffold stiffening and mechanical preconditioning cycles on construct biomechanics, gene expression, and tendon repair biomechanics. *Tissue Engineering: Part A*, 2009, **15**(8): 2103–2111.
154. Riboh, J. et al., Optimization of flexor tendon tissue engineering with a cyclic strain bioreactor. *Journal of Hand Surgery*, 2008, **33**(8): 1388–1396.
155. Awad, H. et al., Repair of patellar tendon injuries using a cell-collagen composite. *Journal of Orthopaedic Research*, 2003, **21**: 420–431.
156. Awad, H. et al., In vitro characterization of mesenchymal stem cell-seeded collagen scaffolds for tendon repair: Effects of initial seeding density on contraction kinetics. *Journal of Biomedical Materials Research*, 2000, **51**: 233–240.
157. Huang, D. et al., Mechanisms and dynamics of mechanical strengthening in ligament-equivalent fibroblast-populated collagen matrices. *Annals of Biomedical Engineering*, 1993, **21**: 289–305.
158. Bell, E., B. Ivarsson, and C. Merrill, Production of a tissue-like structure by contraction of collagen lattices by human fibroblasts of different proliferative potential in vitro. *Proceedings of the National Academy of Sciences*, 1979, **76**: 1274–1278.
159. Altman, G.H. et al., Cell differentiation by mechanical stress. *FASEB Journal*, 2002, **16**: 270–272.
160. Altman, G.H. et al., Advanced bioreactor with controlled application of multidimensional strain for tissue engineering. *Journal of Biomechanical Engineering*, 2002, **124**: 742–749.
161. Datta, N. et al., Effect of bone extracellular matrix synthesized in vitro on the osteoblastic differentiation of marrow stromal cells. *Biomaterials*, 2005, **26**(9): 971–977.
162. Holtorf, H.L., J.A. Jansen, and A.G. Mikos, Ectopic bone formation in rat marrow stromal cell/titanium fiber mesh scaffold constructs: Effect of initial cell phenotype. *Biomaterials*, 2005, **26**(31): 6208–6216.
163. Mikos, A.G. et al., Wetting of poly(L-lactic acid) and poly(DL-lactic-co-glycolic acid) foams for tissue culture. *Biomaterials*, 1994, **15**(1): 55–58.
164. Livak, K.J. and T.D. Schmittgen, Analysis of relative gene expression data using real-time quantitative PCR and the $2^{-\Delta\Delta CT}$ method. *Methods*, 2001, **25**: 402–408.

20 Mechanobiology in the Stem Cell Niche: Integrating Physical and Chemical Regulation of Differentiation

Wilda Helen and Adam J. Engler

CONTENTS

20.1 INTRODUCTION

Since the mid-2000s, there has been increasing interest in understanding stem cell biology, particularly as it relates to the control of stem cell fate, so that cells can be engineered for therapeutic applications of tissue and organ regeneration as well as the treatment of human diseases such as heart disease, diabetes, Parkinson's disease, and pathologies related to trauma [1–5]. However, successful cell-based treatment regimes require a more thorough understanding of what environmental signals stem cells recognize that influence their differentiation, how this recognition occurs, and which pathways become active as a result, in order to direct uncommitted stem cells toward a specific phenotype. These key environmental signals include both soluble cues as well as biochemical and biophysical stimuli presented by the extracellular matrix (ECM), a three-dimensional (3D) fibrillar protein assembly ubiquitously expressed throughout the body to which cells adhere. In this chapter, we will focus on (1) the interplay between matrix mechanics, forces, and chemistry, (2) their role in differentiating stem cells into various cell types, and (3) their application in regenerative medicine. We will also identify a few of the many open issues remaining in the field. To begin such a discussion however, a brief review for the non-stem cell or matrix biologist will put the classification of stem cells and ECM properties into their proper biological context.

20.1.1 THE TYPES OF STEM CELLS

Stem cells are unique cell populations, defined by their ability to undergo both self-renewal and differentiation into a variety of mature cell lineages. Stem cells can be obtained from two major sources: (1) embryonic stem cells (ESCs), derived from the inner cell mass of the blastocyst [6,7], and (2) adult stem cells, derived from many sources, such as bone marrow, peripheral blood, adipose tissue and umbilical cord [8]. Though their origins may be dramatically different, the primary means of classifying stem cells has been to use their "potency," i.e., the range of cell types into which the cell can be differentiated. Immediately upon fertilization, the zygote is presumed to be "totipotent," meaning that it can give rise to all tissue types, both those that comprise the organism as well as those that go to support the organism, e.g., placenta. For the sake of this discussion however, we will limit ourselves to ESCs and mesenchymal stem cells (MSCs), a specific type of adult bone marrow–derived stem cell.

ESCs are now attracting interest as a promising cell source for regenerative medicine, since they are "pluripotent," meaning that they are capable of being induced to develop into all cell types of the body, including neural cells, heart muscle cells, blood cells, endothelial cells and chondrocytes [6]. Undifferentiated ESCs are typically recognized by their ability to express self-renewal markers such as STAT-3, Nanog, and Oct3/4 among others [9], and, to maintain these markers,. murine ESCs are commonly cultured on top of a feeder layer of mouse embryonic fibroblasts, which are thought to secrete factors that inhibit differentiation and maintain self-renewal [10]. One essential fibroblast-provided factor that has been identified is the cytokine leukemia inhibitory factor (LIF), which activates the transcription factor STAT-3 to maintain self-renewal [11] and has enabled researchers to culture murine ESCs on gelatin-coated dishes in medium supplemented with LIF rather than growing them on a feeder cell layer. On the other hand, human ESCs were initially cultured in the presence of LIF [7], but alternative feeder cell sources, including human embryo and foreskin fibroblast, are now conventionally used to support human ESC culture based on the observation that self-renewal is more easily maintained in human ESCs grown on feeder layers [12,13]. Use of human feeder cells could limit cross-species contamination for cell and tissue therapy, but such applications require large quantities of undifferentiated ESCs, that are not practical for feeder cells and may produce variable results. Thus, ideal culture conditions for human ESCs' use as a tissue therapy cell type would be defined as culture conditions free of animal components and feeder layers [14]. Withdrawal of these self-renewing conditions, along with the addition of stimuli specific to the desired differentiation pathway, e.g., activin A to promote endoderm expression [15], will begin the process of inducing lineage expression in a stem cell. Differentiation stimuli, however, come in a variety of forms and will be the focus of our subsequent discussion.

In addition to ESCs, adult stem cells are another class of stem cells that have shown therapeutic potential, despite their more limited differentiation capacity compared to ESCs. Various types of adult stem cells exist, including MSCs, hematopoietic stem cells, and more tissue-specific cells, such as endothelial progenitor cells and muscle satellite cells [16–18], though their differentiation potential varies widely. MSCs are bone marrow–derived "multipotent" stem cells, retaining the ability to self renew and to differentiate into diverse lineages, including osteocytes, chondrocytes, adipocytes, and myocytes [8,19]. Prior to their differentiation into these cell types, MSCs are identified by their expression of CD71, CD105, CD166, Thy-1, STRO-1, β1 integrin, and hyaluronic acid receptor as well as their lack of expression of CD14, CD34, and CD45 [8,20,21]. Although MSCs represent only 0.01% of the total nucleated cell population in the marrow, MSCs can be expanded in vitro nearly 1 million times while maintaining their multilineage potential [8,22], making them a suitable regenerative candidate. Unlike other cell types, MSCs also do not express major histocompatibility complex II (MHC II) antigens, which are responsible for immune rejection, making MSCs an ideal candidate cell source for allogeneic cell transplantation [23]. Additionally, the ease of isolation from the iliac crest, ability to form a genetically stabile and reproducible cell population, slower proliferation compared to ESCs, and differentiation potential into a variety of tissues that often require repair, e.g., muscle and bone, also make MSCs a suitable regenerative cell type [17,24]. From these reasons, bone marrow–derived MSCs gained significant interest beginning over a decade ago, especially as concerted effort began on better characterizing the differentiation pathways and the marrow and host tissue microenvironments [8,25]. That said, many of these initial studies involved chemically induced MSC into cells resembling muscle, bone, and cartilage and did not utilize ECM-based methods as a primary means of differentiating cells. As a result, some improvement has been observed in a subset of disease models [26–29], but it is uncertain whether such improvements come from MSC growth factor and cytokine secretion, MSC differentiation, or MSC interaction with other cell types at the site of injury. Moreover, a near equal number of studies in disease models in mechanically active tissues, e.g., myocardial infarcts, muscular dystrophy, etc., have not shown significantly improved outcomes [4,28,30,31], suggesting that further understanding of the microenvironment, especially as it varies with disease, is critical for successful outcomes.

20.1.2 THE "BIG PICTURE": THE CLINICAL ROLE MECHANOBIOLOGY SHOULD PLAY IN STEM CELL-BASED REGENERATIVE MEDICINE

The potential for stem cells to differentiate into almost all cell types, in addition to providing unlimited number of cells, has stirred interest in their use as integral part of modern clinical treatment, but, to realize their therapeutic potential, understanding how microenvironmental cues modulate and guide stem cell differentiation is critical. Moreover when stem and progenitor cells are transplanted, they are placed into a microenvironment where cells are subjected not only to chemical signals from growth factors and ECM proteins, which can modulate cell survival, organization, migration, proliferation and differentiation [32–34], but also to the physical microenvironment that ECM provides, which are discussed below and include matrix structure [35–37], surface properties [38–40], matrix elasticity and cell shape [41,42], and mechanical strain [43–45]. Knowing these stimuli, two important considerations should be addressed: (1) these signals are presented to ESCs and MSCs during development and in the mature organism, respectively, in a dynamic environment where the mechano-chemical environment is constantly changing, e.g., gastrulation, and (2) the influence and interactions of these signals with one another may be additive or synergetic making the total effect on differentiation greater than [42] or different from [40] the individual cues. Moreover, cells direct and remodel their environment to a degree by the application of mechanical forces generated by actin and myosin within the cell, eliciting a molecular response, both in themselves as well as adjacent stem cells. Overall these considerations lead to changes in form and/or function of these cells, ultimately affecting their ability to differentiate and assemble into larger structures, e.g., multi-nucleated muscle fibers. This realization, coupled with the complexity of tissues, emphasizes the importance of creating a set of design principles that incorporate as many biomimetic properties mentioned above as possible; to date however, such principles, from the standpoint of mechanobiology, have been underappreciated. Through the examples provided in subsequent sections, we propose that the field of tissue engineering must first develop a better understanding of relevant mechanical and chemical properties of native and diseased tissues, such as altered matrix composition and elasticity in fibrotic muscle [4], as well as how they change with time and location in the organism, in order to then develop design criteria to that will improve biomimicry in materials. Yet it follows from these tenets that in using stem cells or uncommitted cells for tissue engineering, knowledge of mechanotransduction pathways (biochemical signals that arise from mechanical forces) may be a crucial step in promoting appropriate differentiation pathway(s) for stem cells and can feed back in material design. In light of this approach, this chapter summarizes where the field has come in terms of understanding the effect that the biochemical and physical microenvironment created by ECM as well as the effect that surface properties of matrix proteins has on the differentiation of stem and progenitor cells.

20.2 THE INFLUENCE OF PHYSICAL MICROENVIRONMENT ON STEM CELL DIFFERENTIATION

One of cellular physiology's most important lessons from in the past decade has been the realization that the behavior of cells can be modulated by the physical properties of the adjacent environment or the culture surface to which they are attached. One of the notable discoveries has been the influence that these properties, i.e., (1) passive mechanical elasticity or "stiffness," (2) matrix topography, (3) active mechanical forces from adjacent cells and/or matrix, and (4) matrix structure or organization, have on the differentiation of stem cells. While this section provides a partial list of all of key ECM physical properties, it is by no means exhaustive, and there are likely to be many unknown parameters that also induce such effects but remain to be discovered.

20.2.1 PASSIVE MATRIX ELASTICITY OR "STIFFNESS"

The elasticity, or Young's modulus, is a material parameter described as the relationship between the force applied to a material's face and the elongation of a material in response to this force, i.e., $E = \sigma/\varepsilon$, and is measured in Pascal. Tissues, which are full of matrix, have long had an underappreciated variation in their stiffness, though their differences are clear from manual palpation: fat is soft and bones are rigid (Figure 20.1A). However, many tissues in the body have subtle mechanical differences that give rise to a variety of changes in structure and function, i.e., muscle is somewhat compliant, $E \sim 5$–$20\,kPa$ depending on muscle type and species [46,47], but not compliant to the point where cells cannot "feel" their matrix when pulling on it nor stiff to the point where cells cannot deform the substrate [48] (Figure 20.1B and C). Initial studies using acrylamide gels among other materials helped determine the influence of matrix elasticity on a variety of cell behaviors including cell proliferation, contractility, motility, and spreading. For example, spread cell area increased with substrate elasticity [49–52] and their spreading [51] and contractility [52] are highly sensitive to ECM ligand density but not necessarily to specific cytoskeletal elements. In parallel to these observations, matrix stiffness has been reported to influence focal adhesion structure [53,54], cytoskeletal organization [49,55,56], and the strength of the integrin cytoskeleton linkages [57] as well as feed back on matrix assembly [58]. Moreover, not only do cells respond to local stiffness, but they can also detect the strain fields of adjacent cells and align themselves with them [47,59] as well as detect spatial variations in elasticity and migrate in the direction of an increasing elasticity gradient [50] with gradients as small as $12\,Pa/\mu m$ [60] in a behavior termed "durotaxis." From these observations, one may predict that cells are capable of responding to substrate elasticity through a truly active tactile exploration process, by exerting contractile forces and then interpreting substrate deformation to formulate a response, rather than responding directly to the force [61].

Though many cell behaviors have been demonstrated to be elasticity dependent, the relationship between matrix stiffness and the differentiation of stem cells was unknown until recent studies began relating tissue properties, e.g., soft brain, $E_{brain} < 1\,kPa$, and decalcified but stiff bone, $E_{bone} \sim 25$–$40\,kPa$, to cell fate. Differentiated cells such as fibroblasts, muscle cells, neurons and epithelial cells all exert different levels of traction forces in response to the stiffness of different substrates, resulting in various cellular morphologies and force balance between cells and the substrates [56]. It was subsequently shown that human MSC differentiation can be modulated by varying mechanical properties of the substrates [42]. When seeded on soft collagen-coated gel ($E_{brain} \sim 0.1$–$1\,kPa$), the majority of human MSCs tend to resemble neural-like cells, whereas most MSCs express early muscle proteins and adopt a spindle-shaped morphology on intermediate matrices ($E_{muscle} \sim 8$–$17\,kPa$) (Figure 20.2). On stiffer substrates ($E_{bone} \sim 25$–$40\,kPa$), most of the MSCs

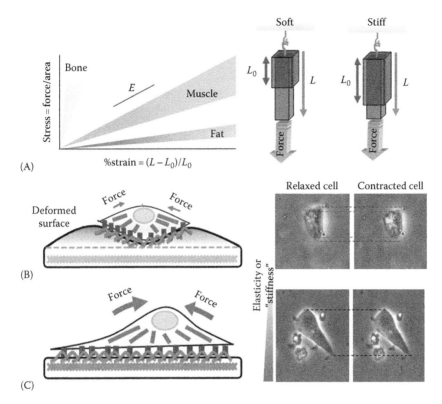

FIGURE 20.1 Tissue stiffness, forces, and cells. (A) Schematic of a stress versus strain plot for several types of tissues extended by forces (per cross-sectional area at right). The slope of the stress–strain plot for these tissues, when subjected to a small strain, gives the Young's elastic modulus, E, for each tissue. (B) The cell's interaction with soft matrix; cells generate some force and undergo large deformation where as (C) on stiff matrix, cell-generated forces cannot deform the material as readily. Lines in the images at right indicate the amount of cell shortening during a cardiomyocyte contraction. Note that there is noticeable movement of the cardiomyocyte on soft matrix (dark gray line: relaxed edge of the cell; light gray line: contracted edge of the cell) versus the lack of deformation on stiff matrix (black line).

tend to make osteoblast-like cells and express initial bone proteins. Furthermore, phenotype commitment was "plastic": Initial lineage commitment to one cell type could be modified with soluble factors of another cell type whereas cells plated on the matrix for longer time were essentially defined only by matrix elasticity. Transduction of these physical cues from matrices is thought to occur by the cells generating traction forces on the matrix [41], as inhibition of nonmuscle myosin II can block the ability of cells to differentiate based on matrix elasticity [42]. The ultimate transduction of this signal into a biochemical read out for stem cell differentiation, however, remains undetermined but could rely on one or more of the proposed mechanotransduction mechanisms discussed later. It should be noted that these observations were made on flat substrates, which may limit its applicability to in vivo situations where fibrillar, 3D tissue structures are present.

In addition to the matrix elasticity effects on MSCs, it's effects on ESC behavior, including differentiation, have only recently been explored. Human ESCs cultured on a synthetic polymer hydrogel composed of a semi-interpenetrating polymer network, maintained their expression of the self-renewal markers Oct-4, SSEA-3, and SSEA-4 only on the softest materials, $E < 0.5\,\mathrm{kPa}$ [62], which mimic native ESC mechanical properties [63]. However, initial evidence indicates that crosslinking matrix to make it 20-fold less compliant than the matrix from Li et al. [63], i.e., close to E_{muscle}, may in fact hinder murine ESC differentiation towards any one specific lineage [64]. It

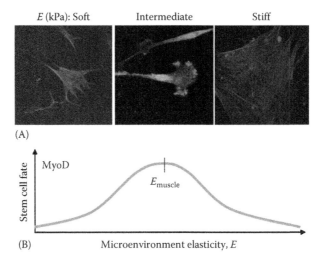

FIGURE 20.2 (See color insert.)   Stiffness-dependent differentiation of stem cells. (A) Immunofluorescence images of the actin cytoskeletal (red) and transcription factor MyoD1 (green) in MSCs plated across a range of matrix stiffness. On the soft matrices, MSCs have long extensions (red). On intermediate matrices, MSCs upregulate the transcription factor MyoD1 (green). On the stiff matrices, MSCs do not express significant MyoD1 marker but are more spread. (B) Schematic of the fluorescent intensity of MSCs versus substrate elasticity indicates that maximal myogenic expression occurs at E typical of muscle type.

would seem then that perhaps a combination of cues in addition to stiffness may be important for ESCs, which may be plausible as ESCs do not encounter a significant amount of assembled matrix until the time leading up to gastrulation.

20.2.2 MATRIX TOPOGRAPHY

As another important matrix cue, topography (e.g., aligned fibrillar matrices, shape, texture, etc.) has been reported to play an important role in cell organization, alignment, and differentiation. Cells can sense topographical changes in the substrate over several length scales, ranging from features of a few nanometers to those that are tens of micrometers, depending on the material and cell type. More surprising though is the ability of these features to promote topographic guidance and direct cell fate as well as the fact that cells simply do not migrate away from such patterns if possible [14,23,65]. Several studies revealed that not only are feature dimensions important but also their conformation—whether they are ridges, groove, whorls, pits, pores or steps—and, more intriguingly, even their symmetry [66–68]; all of which likely dictate how cells generate forces in these geometries.

Using technologies borrowed from the semiconductor and microelectronic industries, a plethora of techniques has been developed for creating patterned surfaces to investigate cellular behavior as diverse as cell–matrix and cell–cell interactions, cell migration, cell differentiation in response to matrix topography, mechanotransduction, and cell response to gradient effect of surface-bound ligands. However, such techniques generally create materials that rely on either (1) interwoven or spun fibers (Figure 20.3A) or (2) micropatterned surfaces of a solid material (Figure 20.3B). With polyamide-based electrospun nanofibers, significant enhancement of murine ESC expression of the self-renewal marker Nanog as well as increased proliferation was observed versus flat substrates [37]. Poly(ε-caprolactone)-based scaffolds have also been used to differentiate human MSCs, specifically to fat, cartilage, and bone. This material, however, is very tough and nylon-like, and thus only when provided with the appropriate inductive agents did differentiation occur [69]. As with the previous section, this perhaps suggests that a more compliant material be used, such as spider silk [70] in which stem cells have been coaxed into skin, bone, cartilage. Moreover,

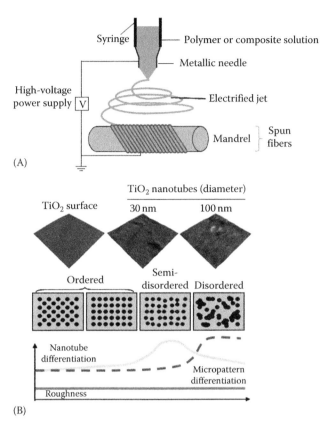

FIGURE 20.3 (See color insert.) Structural and topographical cues that regulate cell-generated forces and differentiate stem cells. (A) Schematic of the fabrication of interwoven or spun fibers. (B) The top row shows AFM topography images of a titanium substrate and TiO_2 nanotubes of different diameter. Scans are $5 \times 5\,\mu m$. The second row shows of nanofeatures having ordered, semi-disordered, and disordered symmetries. Materials typical of this fabrication scheme involve polymer matrices such as PMMA. Surface roughness (red) does not change significantly as shown in the bottom schematic, though both nanotube (blue) and polymer surfaces (green) induce differentiation when specific topography is recognized by MSCs. (From Dalby, M.J. et al., *Nat. Mater.*, 6, 997, 2007.)

new synthetic silks made from diblock copolymers [71] suggest that this material may be made as a "smart" engineered product that could encompass several important design parameters discussed here.

Micropatterned substrates and nanostructures have also been observed to stimulate behavioral change in cells and tissues [35,36,72] (Figure 20.3B). In particular, nanoscale structures have played a critical role in accelerating the rate of proliferation and enhancing tissue acceptance with a reduced immune response [73]. It should be noted that nanotopography effects need not increase substrate roughness to have an impact: osteogenic differentiation of human MSCs increased 50% when cultured on 100 nm diameter TiO_2 nanotubes compared with those cultured on 30 nm TiO_2 nanotubes or a flat TiO_2 substrate without osteogenic inducing media [36]. Though patterning might be independent of roughness changes, the order or orientation of the pattern appears to be critical for polymer [35] and alumina materials [74]: nanopatterns showed negligible osteogenic differentiation unless the topographical pattern is semi-disordered [35]. Though synthetic materials provide an appropriate testing ground for these ideas, disordered nanotopographies are likely present in a variety of natural matrices [75] which appear to support not only the generation of sufficient forces for differentiation but also the ability to unfold and remodel themselves to potentially initiate other extracellular signals [76]. In the appropriate context, changes in topography, i.e., surface coating versus fibrillar structure, could enhance a differentiation signal to a stem cell, and initial evidence of this appears to support such a hypothesis [64].

20.2.3 ACTIVE MECHANICAL FACTORS INFLUENCE STEM CELLS

Unlike the passive elasticity of tissues or the topography of the matrix, two components that do not require applications of force directly onto the cells, there is an important subset of mechanically-based environmental cues that originate from active processes. During tissue development and remodeling especially, cells are subjected to various forms of mechanical stimulation, which in turn shapes and regulates a large number of physiological processes and is one component of the "tensional homeostatis" maintained by multicellular structures [77]. Upon mechanical stimulation, cells convert these mechanical signals into biochemical factors through the process known as mechanotransduction [78] that involve mechanisms that are still being debated as illustrated in Section 20.4. Though the effect of mechanical factors on cell function has been well documented [79–82], their ability to stimulate stem cell differentiation has only recently been investigated, as in the stimulation of cardiomyogenesis in murine ESCs via a matrix stretch-induced increase of reactive oxygen species generation and expression of nicotinamide adenine dinucleotide phosphate oxidase [43]. Murine ESC differentiation into smooth muscle, as assessed by increased expression of smooth muscle α-actin (α-SMA) and smooth muscle myosin heavy chain, has also shown to be modulated by cyclic strain [83]. Unlike passive elasticity though, how the strain is actively applied to the stem cells appears to be critical: human ESC differentiation was inhibited by biaxial cyclic strain over a range of frequencies in conditioned but not unconditioned media [45]. Again this suggests that mechanical stress plays a role in regulating the self-renewal and differentiation via a yet to be determined mechanism, and that role acts synergistically with chemical signals to result in a more fully committed cell.

Active mechanical forces can also be transduced from forces applied to the apical surface of the cell. Though cells cultured on surfaces do not engage matrix there, fluid shear stress is often used as this scenario mimics the environment that endothelial cells would encounter as they line blood vessels. Huang et al. seeded murine ESC-derived Flk1-positive mesoderm cells into a microporous tube that matched the compliance of human arteries and found that after 2 days under a 0.22 Pa pulsatile flow, ES cells covering the superficial layer of the microporous tube appeared endothelial-like, expressed platelet endothelial cell adhesion molecule 1, and were regularly oriented in the direction of the pulsatile flow [44]. Expression of mature endothelial cell-specific markers in MSCs occurred only after the application of higher shear, 1.5 Pa, for 12 days [84]. In addition, Wang et al. found that shear stress induces uptake of acetylated low-density lipoprotein and formation of tube like structure on Matrigel. When cultured in a 3D porous scaffold, MSCs experience flow rates within the material that likely drop below a cell response threshold, e.g., 10 μPa, and the flow becomes a purely convective component.

Mechanotransduction of actively applied forces on stem cells is not limited to matrix and fluid forces and has been extended to hydrostatic pressure as well. MSC differentiation into the chondrogenic pathway was correlated to compressive hydrostatic pressure whereas osteogenic pathway was correlated to low hydrostatic stress [85], an observation that correlates nicely to the hydrostatic pressures found in a variety of tissues. Such pressure appears to tightly regulate the formation of intramembranous ossification if local strains are lower (approximately ±5%) and hydrostatic pressure smaller than ±0.15 MPa, endochondral ossification for compressive hydrostatic pressures greater than −0.15 MPa and local strains smaller than ±15%, and all other mechanical environments favored differentiation of fibrous connective tissue [86]. Taken together, these results indicate that actively applied physical factors can regulate cells and induce biophysical responses that result in biochemical changes. Again, how such cues are sensed by the cells to commit to a lineage remains to be determined.

20.2.4 MATRIX STRUCTURES AND ORGANIZATION

Various matrix structures have been shown to support stem cell organization and differentiation, and recent studies demonstrated that the selective advantage of 3D porous structures over 2D substrates, especially as it relates to the biodegradability and growth factor release of polymer scaffolds. What is lacking, however, is a better characterization of force production and mechanics for stem cells. This is critically important as cell-generated forces, cytoskeletal organization, and elasticity are known to be dramatically different in scaffolds than on surfaces [87] and this can give rise to complex 3D structures [88]. What is known, however, is that the formation of these complex structures is highly coupled to ligand presentation, geometry, and growth factors: binding site orientation directly influences cell adhesion strength [89] whereas the structure of multicellular networks within the material influences stem cell response to growth factors and tension development [90]. That said, the majority of observations in this field are mechanically driven, even if not attributed to that: human ESCs seeded into 3D structures differentiate and form a vessel-like network that appear to anastomose, yet, on surfaces, the ESCs failed to organize and could only support moderate differentiation [32]. This suggests that matrices provided a more favorable physical structure for differentiation than flat substrates, but again quantitative models of this are necessary.

One such reason for this may revolve around the cell-material interaction and the resulting cell shape. For instance, 3D matrices have greater surface area, so contact inhibition and the cellular processes modulated by contact inhibition may be minimized [91]. Such a point may be the reason for differences in ECM secretion by human MSCs on 3D and 2D poly(ethylene terephthalate) substrates [92] where minimal ECM deposition occurred on 2D substrates but an organized fibrillar network formed and substantial amounts of collagen and laminin were secreted by MSCs on the 3D scaffold. Most relevant, however, is the implication that the amount of contact between the cell and its matrix, coupled with growth factors, can differentially modulate differentiation toward fat or bone [41]. Underlying this process is a tension-sensitive switch that likely is dependent on higher expression of integrins [92] to facilitate the transmission of Rho-mediated forces.

The organization of 3D matrices, which are inherently susceptible to cell-generated forces [76], can also act as sinks of growth factors that can be liberated and signal to stem cells only under the appropriate force conditions. The transition of fibroblasts to contractile myofibroblasts has been shown to be dependent on force-induced release of the growth factor TGF-β from stores of the protein within the ECM [93]. As for stem cells specifically, Mizuno and Kuboki cultured bone marrow cells on type I collagen matrix including TGF-β and found that after 10 days, bone marrow cells on the TGF-β containing matrix formed mineralized nodules, whereas cells on the collagen matrix alone did not form them [94]. Since mineralized bone is much more rigid than decalcified, collagenous bone, this could form a potential feedback loop that could induce further calcification due to the physical change in the organization of the environment along with many other biochemical changes. 3D structures also provide a porous structure for growth factors to diffuse, which can induce differentiation into contractile phenotypes as well [95,96]. Though these large, interconnected pores promote cell colonization [97] and structure formation [95], further exploration of the inducing or resulting forces would greatly advance our understanding of the physical role that ECM structure plays in stem cell differentiation.

20.3 SURFACE CHEMISTRY AND ITS AFFECT ON STEM CELL DIFFERENTIATION

Unlike the physical properties of the ECM, the influence of surface ECM properties on cellular growth, movement, and orientation has been previously recognized [67,98]. However, their ability to stimulate stem cell differentiation has only recently been investigated, and some controversy exists in the field as not all chemical stimuli appear to be effective differentiation stimuli. For this reason, it is important to discuss current efforts to understand how specific surface properties influence stem cells as well as how those properties combine with physical ones and cell-generated forces to affect mechanobiology.

20.3.1 THE CHEMISTRY OF CELL-MATERIAL INTERACTIONS: SURFACE STRUCTURE AND CHARGE

Recent studies on surface structures and charge have revealed the importance of surface properties in affecting stem cell differentiation. Various chemical patterning techniques including chemical vapor deposition, electrochemical deposition, electron-beam lithography, and layer by layer micro-fluidic patterning have been developed for creating patterned surfaces, and due to their cell adhesive properties, such substrates have been used to investigate cellular behavior [99]. While these techniques can create a range of surface topographies reviewed above, questions of varying (i.e., between different fabrication techniques) or unsuitable chemistry for cells (i.e., non-biocompatible materials) have arisen. For example, microstructured poly(methyl methacrylate) (PMMA) substrates had lower osteogenic expression in the absence of osteogenic supplement compared to flat PMMA, which was reversed in the presence of osteogenic supplement [39]. However, that chemistry is also used in the semi-disordered topography of Dalby et al. who observed differentiation in the absence of inducing factors [35] (Figure 20.3B), suggesting that the specific interaction between the cells and the material may be governed by the exact presentation of the structure: round patterns versus shallow pits. Such varied observations are not limited to the presentation of synthetics: smooth muscle cells in micropatterned collagen grooves had enhanced proliferation and altered shape [100] whereas collagen channels kept cells in place and allowed them to simply survive [101]. Subtle differences in material preparation, e.g., thick collagen-coated grooves versus channels cast in a collagen gel, may change fiber alignment and thus mechanical properties, but it will also affect the distribution of the charges on the material's surface. This could lead to changes in shear forces that cells can withstand and is likely due to both, changes in structure, e.g., fibronectin coated surfaces enable cells to withstand eightfold higher shear than fibrillar fibronectin matrices [102], as well as changes in charge [89]. There is an equally important set of non-fouling materials which do not support proteins [103], though discussion here is limited to surfaces which support protein adsorption, cell attachment, and ultimately cell-generated forces.

The concern about material interactions and charge arises from the fact that matrix itself remodels substantially during development [75], and with this comes a likely change in matrix composition and density as well as charge, all of which could dramatically dictate stem cell differentiation. Since these parameters are all interdependent properties in naturally-occurring ECM, deciphering a specific role for the environment's charge, especially as it relates to the cell's adhesion, is difficult. That said, when protein adsorption is minimized, stromal cells attach more readily to positively charged substrates than to neutral or negative charged surfaces [38]. Though cell spreading and differentiation remains minimal, such surfaces did not specifically inhibit differentiation when stimulated with the broad-spectrum glucocorticoid dexamethasone [104]. Though these results hint at a potential role for modulating stem cells, the exact regulation and interplay with biophysical stimuli that basic surface structure and charge has remains an unresolved issue.

20.3.2 THE CHEMISTRY OF CELL-MATERIAL INTERACTIONS: SURFACE CHEMISTRY

Surface chemistry and composition has also been shown to direct stem cell differentiation into specific lineages. Natural variations in composition exists throughout the body [75], and changes in ECM ligand and density are known to influence the murine ESC self-renewal versus differentiation switch [40]. Synthetic materials, however, offer the ability to directly mimic specific adult environments to understand stem cells' regenerative properties, such as MSCs in non-mineralized bone. Murphy et al. examined MSC differentiation on mineralized and non-mineralized poly(lactide-co-glycolide) (PLG) and found that only on non-mineralized PLG substrates and in the presence of osteogenic supplements did human MSCs exhibit enhanced expression of osteogenic markers [105].

Comparisons of alumina, titania, and hydroxyapatite substrates indicate that the titanium surface was most adhesive, especially those with grain size below 50 nm, suggesting that surface chemistry and nanophase topography could direct human MSC adhesion [106].

Hydrophilicity is also another important surface chemistry concern as osteoblast differentiation and mineralization is known to be directly dependent on hydrophilicity and charge via adsorption of fibronectin and β_1 integrins [107]. Osteoblast-specific gene expression and mineralization is dependent on $\alpha_5\beta_1$ integrin binding to FN [108], and $\alpha_V\beta_3$ integrin overexpression in osteoblast-like cells down-regulates differentiation and mineralization [109]. Together this suggests a role for surface chemistry regulating cell-matrix contact and thus cell function. On the other hand, actual contact with a specific substrate is not required to be influential: dissolved calcium carbonate from gold-coated coral substrates promoted osteogenesis while the gold coating prevented cell-material contact which instead also induced expression of chondrogenic genes. Whether in direct contact or not, it appears that there is cause for biomaterial chemistry regulating differentiation, but while physical cues likely couple to soluble growth factors [41,42], such a coupling for the surface chemistry of matrix remains an open question.

20.4 MECHANISMS FOR STEM CELL MECHANOBIOLOGY: SEEING THE "FORCE" IN STEM CELLS

After a cell adheres to and spreads on a material, regardless of whether it inhabits a flat or rough ligand-coated surface, compliant hydrogel, or a 3D fibrillar matrix, it contracts against the material, exerting a force brought on by myosin sliding against filamentous actin. In the case of stem cells, the cell-substrate interaction can impart vital information to the cell as described above, but most importantly, it provides resistance against which the cell pulls. This resistance is based on the mechanical properties of the material. A muscle analogy is perhaps insightful: when lifting a weight that is too light or too heavy, little muscle mass is built as no work can be done. When lifting a weight that is neither light nor heavy, the muscle can move the weight and feel sufficient resistance to develop additional muscle. So for cells, matrix elasticity becomes the "weight" that the cell must be able to contract against. If the material allows for sufficient deformation at a given contractile force, then the cell can "feel" the material [61], transmit this signal to the nucleus, and respond via differentiation toward a particular lineage. Though the exact mechanism behind this process is still hotly debated, we present here three mechanisms well-supported by recent findings (Figure 20.4), each of which can be modulated by physical and chemical properties of the matrix.

In a first mechanism ([50]; Figure 20.4A), soft microenvironments induce very little positive mechanical feedback as the focal contacts-ECM complex cannot generate sufficient tension as the matrix deforms too much (Figure 20.1B). In the presence of a stiff substrate, smaller displacement is realized, though stronger mechanical feedback may then activate stress-sensitive ion channels [110] and influx of calcium among other ions to promote a change in cell behavior. Since cells directly sense the distance of receptor movement as a result of exerted probing forces [61], such a mechanism could be plausible, but while this mechanism can begin to explain some changes in myocytes behavior [111], it may not fit well for stem cells, where multiple lineages emerge from a common cell type. Another proposed mechanism relies on the influence of the Rho GTPase family and their effectors (Figure 20.4B), such as the Rho/ROCK signaling pathway [112], to regulate cell contractility and balance the adhesive strength of mature focal contacts. On a soft matrix [42] or when not sufficiently spread [41], cells cannot generate a suitable amount of force, though overexpression of the actin cytoskeleton but not adhesions [51] can rescue this phenotype. A rigid substrate, or perhaps a material with suitable charge density to promote cell adhesion, can provide sufficient resistance to balance large traction forces, and as a result, adhesions become large and robust [49]. However, stem cell differentiation likely requires many intermediate forms contributed by both biophysical and biochemical signals, and, with such a model, intermediate levels of Rho/ROCK signaling simulated

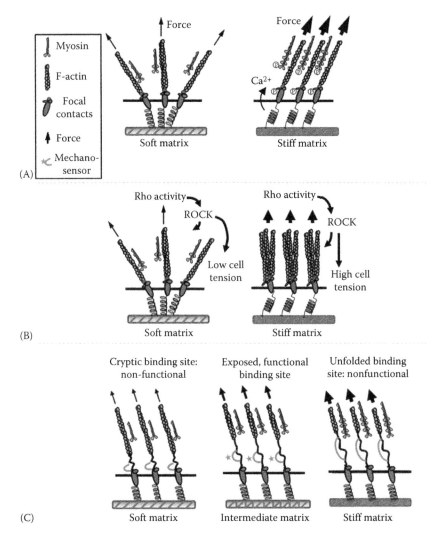

FIGURE 20.4 Mechanotransduction mechanisms for matrix elasticity. (A) In the first proposed mechanism, cell on soft matrix have tension at the anchorage site that is weak while the receptor ligand complex is somewhat labile. Cells on stiff matrix exhibit the increase of calcium to cause myosin phosphorylation, which leads to an increased energy consumption and further increase in tension. (B) In the second proposed mechanism, cells on soft matrix possess low RhoA activity, which prevents assembly of the actin cytoskeleton and focal adhesion maturation; both are required to propel the cell body forward and maintain adhesion to the substrate. Cells on stiff matrix possess high RhoA activity, which promotes the formation of robust actin stress fibers and focal adhesion via ROCK, resulting in more stationary, contractile cells. (C) In a final mechanism presented here, changes in protein folding change as forces are exerted to expose binding sites. Cells on soft matrix with weak intracellular forces cannot sufficiently change the conformation of a mechanically sensitive protein of interest to expose a cryptic binding site, making it nonfunctional. By comparison, cells on stiff matrix generate high tension, which causes the protein to unfold to such a degree that the binding site is rendered non functional. However, cells on matrix with optimal elastic properties may apply the appropriate amount of forces such that the cell may change the conformation of the protein, making the cryptic binding site accessible.

via soluble growth factors combined with spreading-induced tension may promote labile adhesions and optimal force generation for differentiation [41,90].

Another recently outlined mechanism involves force-induced conformational changes. Though in vitro forced unfolding of proteins has been shown for the past decade or more, recent evidence supports receptor-like [113] or adhesion molecules [114] acting as molecular "strain gauges," where the appropriate application of traction forces changes the conformation of proteins to turn a non-functional binding site into a functional one. After the ends of the protein bind to adjacent proteins, the protein of interest is subjected to traction forces. Too little or too much intracellular force will cause the protein either to remain folded and inaccessible or to unfold completely rendering the binding site nonfunctional, respectively. However, application of the appropriate level of force may change the conformation of the protein of interest, making a cryptic binding site accessible (Figure 20.4C, middle). In situ changes in force have recently been measured in MSCs using a cysteine shotgun method [115], indicating changes in cytoskeletal proteins in response to force though such a model is predicated on mechano-sensitive proteins residing between the force-generating actomyosin and the resisting matrix. Though each model presents a plausible mechanism to convert a biophysical signal into a useable biochemical stimulus for stem cell differentiation, a combination of all or some of these along with other mechanisms yet to be elucidated is likely to be the case.

20.5 CONCLUSIONS AND PERSPECTIVES ON STEM CELL MECHANOBIOLOGY

Although the pace of identifying the mechanical signals that regulate stem cell differentiation is accelerating, our knowledge about the microenvironment and the molecular mechanisms and signaling pathways that lead to efficient stem cell differentiation and tissue formation is still limited to those presented here. Though biochemical assays have been used to localize the activity of signaling molecules in living cells, molecular and imaging techniques need to be developed to elucidate the increasing number of signaling molecules involved in stem cell survival, differentiation, organization, and morphogenesis into tissue like structure as well as determine their mechanical role, if any, in determining cell fate. Much promise exists for elucidating the biophysical and biochemical pathways involved in mechanical signals transduction by stem cells in situ, should the necessary, high-throughput methods be developed and widely used. Such new approaches will hopefully be developed in concert with the development of new materials that push the envelop from widely-studied 2D stem cell differentiation into the realm of differentiation in more biomimetic 3D matrices that have the appropriate biophysical and biochemical signals as outlined here. All of these and many more considerations will help push the field of tissue engineering and regenerative medicine forward as the community develops new therapies based on our improved understanding of the biophysical and biochemical pathways involved in mechanical signal transduction.

REFERENCES

1. Roy, N.S. et al., Functional engraftment of human ES cell-derived dopaminergic neurons enriched by coculture with telomerase-immortalized midbrain astrocytes. *Nature Medicine*, 2006, **12**(11), 1259–1268.
2. Strauer, B.E., M. Brehm, and C.M. Schannwell, The therapeutic potential of stem cells in heart disease. *Cell Proliferation*, 2008, **41**(Suppl 1), 126–145.
3. Ringe, J. and M. Sittinger, Tissue engineering in the rheumatic disease. *Arthritis Research and Therapy*, 2009, **11**(1), 211.
4. Berry, M.F. et al., Mesenchymal stem cell injection after myocardial infarction improves myocardial compliance. *American Journal of Physiology: Heart and Circulatory Physiology*, 2006, **290**(6), H2196–H2203.
5. Porada, C.D., E.D. Zanjani, and G. Almeida-Porad, Adult mesenchymal stem cells: A pluripotent population with multiple applications. *Current Stem Cell Research and Therapy*, 2006, **1**(3), 365–369.

6. Keller, G., Embryonic stem cell differentiation: Emergence of a new era in biology and medicine. *Genes and Development*, 2005, **19**(10), 1129–1155.
7. Thomson, J.A. et al., Embryonic stem cell lines derived from human blastocysts. *Science*, 1998, **282**, 1145–1147.
8. Pittenger, M.F. et al., Multilineage potential of adult human mesenchymal stem cells. *Science*, 1999, **284**(5411), 143–147.
9. Ivanova, N. et al., Dissecting self-renewal in stem cells with RNA interference. *Nature*, 2006, **442**(7102), 533–538.
10. Evans, M.J. and M.H. Kaufmann, Establishment in culture of pluripotential cells from mouse embryos. *Nature*, 1981, **292**(5819), 154–156.
11. Niwa, H. et al., Self-renewal of pluripotent embryonic stem cells is mediated via activation of STAT3. *Genes and Development*, 1998, **12**, 2048–2060.
12. Hovatta, O. et al., A culture system using human foreskin fibroblasts as feeder cells allows production of human embryonic stem cells. *Human Reproduction*, 2003, **18**(7), 1404–1409.
13. Richards, M. et al., Human feeders support prolonged undifferentiated growth of human. *Nature Biotechnology*, 2002, **20**, 933–936.
14. Hwang, N.S., S. Varghese, and J. Elisseeff, Controlled differentiation of stem cells. *Advanced Drug Delivery Reviews*, 2008, **60**, 199–214.
15. Yasunaga, M. et al., Induction and monitoring of definitive and visceral endoderm differentiation of mouse ES cells. *Nature Biotechnology*, 2005, **23**(12), 1542–1550.
16. Berardi, A.C. et al., Functional isolation and characterization of human hematopoietic stem cells. *Science*, 1995, **267**(5194), 104–108.
17. Pittenger, M.F. and B.J. Martin, Mesenchymal stem cells and their potential as cardiac therapeutics. *Circulation Research*, 2004, **95**(1), 9–20.
18. Fuch, S. et al., Transendocardial delivery of autologous bone marrow enhances collateral perfusion and regional function in pigs with chronic experimental myocardial ischemia. *Journal of the American College of Cardiology*, 2001, **37**(6), 1726–1732.
19. Xu, W. et al., Mesenchymal stem cells from adult human bone marrow differentiate into a cardiomyocyte phenotype in vitro. *Experimental Biology and Medicine*, 2004, **229**(7), 623–631.
20. Simmons, P.J. and B. Torok-Storb, Identification of stromal cell precursors in human bone marrow by a novel monoclonal antibody, STRO-1. *Blood*, 1991, **78**, 55–62.
21. Deans, R.J. and A.B. Moseley, Mesenchymal stem cells: Biology and potential clinical uses. *Experimental Hematology*, 2000, **28**(8), 875–884.
22. Wagner, W. et al., Comparative characteristics of mesenchymal stem cells from human bone marrow, adipose tissue, and umbilical cord blood. *Experimental Hematology*, 2005, **33**(11), 1402–1416.
23. Huang, N.F., R.J. Lee, and S. Li, Chemical and physical regulation of stem cells and progenitor cells: Potential for cardiovascular tissue engineering. *Tissue Engineering*, 2007, **13**, 1809–1823.
24. Stolzing, A. et al., Age-related changes in human bone marrow-derived mesenchymal stem cells: Consequences for cell therapies. *Mechanisms of Ageing and Development*, 2008, **129**(3), 163–173.
25. Pereira, R.F. et al., Cultured adherent cells from marrow can serve as long-lasting precursor cells for bone, cartilage, and lung in irradiated mice. *Proceedings of the National Academy of Sciences of the United States of America*, 1995, **92**(11), 4857–4861.
26. Chopp, M. and Y. Li, Treatment of neural injury with marrow stromal cells. *Lancet Neurology*, 2002, **1**(2), 92–100.
27. Hofstetter, C.P. et al., Marrow stromal cells form guiding strands in the injured spinal cord and promote recovery. *Proceedings of the National Academy of Sciences of the United States of America*, 2002, **99**, 2199–2204.
28. Toma, C. et al., Human mesenchymal stem cells differentiate to a cardiomyocyte phenotype in the adult murine heart. *Circulation*, 2002, **105**(1), 93–98.
29. Shake, J.G. et al., Mesenchymal stem cell implantation in a swine myocardial infarct model: Engraftment and functional effects. *The Annals of Thoracic Surgery*, 2002, **73**(6), 1919–1925.
30. Murry, C.E. et al., Haematopoietic stem cells do not transdifferentiate into cardiac myocytes in myocardial infarcts. *Nature*, 2004, **428**(6983), 664–668.
31. Breitbach, M. et al., Potential risks of bone marrow cell transplantation into infarcted hearts. *Blood*, 2007, **110**(4), 1362–1369.
32. Levenberg, S. et al., Differentiation of human embryonic stem cells on three-dimensional polymer scaffolds. *Proceedings of the National Academy of Sciences of the United States of America*, 2003, **100**(22), 12741–12746.

33. Sauer, H. et al., Redox control of angiogenic factors and CD31-positive vessel-like structures in mouse embryonic stem cells after direct current electrical field stimulation. *Experimental Cell Research*, 2005, **304**(2), 380–390.

34. Park, J.S. et al., Differential effects of equiaxial and uniaxial strain on mesenchymal stem cells. *Biotechnology and Bioengineering*, 2004, **88**(3), 359–368.

35. Dalby, M.J. et al., The control of human mesenchymal cell differentiation using nanoscale symmetry and disorder. *Nature Materials*, 2007, **6**, 997–1003.

36. Oh, S. et al., Stem cell fate dictated solely by altered nanotube dimension. *Proceedings of the National Academy of Sciences of the United States of America*, 2009, **106**(7), 2130–2135.

37. Nur, E.K.A. et al., Three-dimensional nanofibrillar surfaces promote self-renewal in mouse embryonic stem cells. *Stem Cells*, 2006, **24**(2), 426–433.

38. Qiu, Q. et al., Attachment, morphology, and protein expression of rat marrow stromal cells cultured on charged substrate surfaces. *Journal of Biomedical Materials Research*, 1998, **42**, 117–127.

39. Engel, E. et al., Mesenchymal stem cell differentiation on microstructured poly (methyl methacrylate) substrates. *Annals of Anatomy*, 2009, **191**(1), 136–144.

40. Flaim, C.J., S. Chien, and S.N. Bhatia, An extracellular matrix microarray for probing cellular differentiation. *Nature Methods*, 2005, **2**(2), 119–125.

41. McBeath, R. et al., Cell shape, cytoskeletal tension, and RhoA regulate stem cell lineage commitment. *Developmental Cell*, 2004, **6**, 483–495.

42. Engler, A.J. et al., Matrix elasticity directs stem cell lineage specification. *Cell*, 2006, **126**, 677–689.

43. Schmelter, M. et al., Embryonic stem cells utilize reactive oxygen species as transducers of mechanical strain-induced cardiovascular differentiation. *The FASEB Journal*, 2006, **20**, 1182–1184.

44. Huang, H. et al., Differentiation from embryonic stem cells to vascular wall cells under the in vitro pulsatile flow loading. *The International Journal of Artificial Organs*, 2005, **8**, 110–118.

45. Saha, S. et al., Inhibition of human embryonic stem cell differentiation by mechanical strain. *Journal of Cellular Physiology*, 2006, **206**, 126–137.

46. Collinsworth, A.M. et al., Apparent elastic modulus and hysteresis of skeletal muscle cells throughout differentiation. *American Journal of Cell Physiology*, 2002, **283**(4), C1219–C1227.

47. Engler, A.J. et al., Myotubes differentiate optimally on substrates with tissue-like stiffness: Pathological implications for soft or stiff microenvironments. *The Journal of Cell Biology*, 2004, **166**, 877–887.

48. Engler, A.J. et al., Embryonic cardiomyocytes beat best on a matrix with heart-like elasticity: Scar-like rigidity inhibits beating. *Journal of Cell Science*, 2008, **121**(22), 3794–3802.

49. Pelham, R.J. Jr. and Y. Wang, Cell locomotion and focal adhesions are regulated by substrate flexibility. *Proceedings of the National Academy of Sciences of the United States of America*, 1997, **94**(25), 13661–13665.

50. Lo, C.M. et al., Cell movement is guided by the rigidity of the substrate. *Biophysical Journal*, 2000, **79**(1), 144–152.

51. Engler, A.J. et al., Substrate compliance versus ligand density in cell on gel responses. *Biophysical Journal*, 2004, **86**, 617–628.

52. Reinhart-King, C.A., M. Dembo, and D.A. Hammer, The dynamics and mechanics of endothelial cell spreading. *Biophysical Journal*, 2005, **89**(1), 676–689.

53. Bershadsky, A.D., N.Q. Balaban, and B. Geiger, Adhesion-dependent cell mechanosensitivity. *Annual Review of Cell and Developmental Biology*, 2003, **19**, 677–695.

54. Cukierman, E. et al., Taking cell-matrix adhesions to the third dimension. *Science* 2001, **294**(5547), 1708–1712.

55. Yeung, T. et al., Effects of substrate stiffness on cell morphology, cytoskeletal structure, and adhesion. *Cell Motility and the Cytoskeleton*, 2005, **60**(1), 24–34.

56. Discher, D.E., P. Janmey, and Y.L. Wang, Tissue cells feel and respond to the stiffness of their substrate. *Science*, 2005, **310**(5751), 1139–1143.

57. Choquet, D., D.P. Felsenfeld, and M.P. Sheetz, Extracellular matrix rigidity causes strengthening of integrin-cytoskeleton linkages. *Cell*, 1997, **88**(1), 39–48.

58. Halliday, N.L. and J.J. Tomasek, Mechanical properties of the extracellular matrix influence fibronectin fibril assembly in vitro. *Experimental Cell Research*, 1995, **217**(1), 109–117.

59. Reinhart-King, C.A., M. Dembo, and D.A. Hammer, Cell-cell mechanical communication through compliant substrates. *Biophysical Journal*, 2008, **95**(12), 6044–6051.

60. Zaari, N. et al., Photopolymerization in microfluidic gradient generators: Microscale control of substrate compliance to manipulate cell response. *Advanced Materials*, 2004, **16**(23–24), 2133–2137.

61. Saez, A. et al., Is the mechanical activity of epithelial cells controlled by deformations or forces? *Biophysical Journal*, 2005, **89**(6), L52–L54.
62. Li, Y.J. et al., Hydrogels as artificial matrices for human embryonic stem cell self-renewal. *Journal of Biomedical Materials Research. Part A*, 2006, **79**(1), 1–5.
63. Davidson, L.A. et al., Measurements of mechanical properties of the blastula wall reveal which hypothesized mechanisms of primary invagination are physically plausible in the sea urchin Strongylocentrotus purpuratus. *Developmental Biology*, 1999, **209**(2), 221–238.
64. Helen, W., J.E. Schwarzbauer, and A.J. Engler, unpublished.
65. Li, S., J.L. Guan, and S. Chien, Biochemistry and biomechanics of cell motility. *Annual Review of Biomedical Engineering*, 2005, **7**, 105–150.
66. Teixeira, A.I. et al., Epithelial contact guidance on well-defined micro- and nanostructured substrates. *Journal of Cell Science*, 2003, **116**(10), 1881–1892.
67. Curtis, A. and C. Wilkinson, Topographical control of cells. *Biomaterials*, 1997, **18**(24), 1573–1583.
68. Flemming, R.G. et al., Effects of synthetic micro- and nano-structured surfaces on cell behavior. *Biomaterials*, 1999, **20**(6), 573–588.
69. Li, W.J. et al., Multilineage differentiation of human mesenchymal stem cells in a three-dimensional nanofibrous scaffold. *Biomaterials*, 2005, **25**(25), 5158–5166.
70. Wang, Y. et al., Stem cell-based tissue engineering with silk biomaterials. *Biomaterials*, 2006, **27**(36), 6064–6082.
71. Rabotyagova, O.S., P. Cebe, and D.L. Kaplan, Self-assembly of genetically engineered spider silk block copolymers. *Biomacromolecules*, 2009, **10**(2), 229–236.
72. Jung, D.R. et al., Topographical and physicochemical modification of material surface to enable patterning of living cells. *Critical Reviews in Biotechnology*, 2001, **21**(2), 111–154.
73. Mahdavi, A. et al., A biodegradable and biocompatible gecko-inspired tissue adhesive. *Proceedings of the National Academy of Sciences of the United States of America*, 2008, **105**(7), 2307–2312.
74. La Flamme, K.E. et al., Biocompatibility of nanoporous alumina membranes for immunoisolation. *Biomaterials*, 2007, **28**(16), 2638–2645.
75. Hay, E.D., *Cell Biology of Extracellular Matrix*. 2nd edn. New York: Plenum Press. 1991: 468.
76. Smith, M.L. et al., Force-induced unfolding of fibronectin in the extracellular matrix of living cells. *PLoS Biology*, 2007, **5**(10), e268.
77. Paszek, M.J. et al., Tensional homeostasis and the malignant phenotype. *Cancer Cell*, 2005, **8**(3), 241–254.
78. Orr, A.W. et al., Mechanisms of mechanotransduction. *Developmental Cell*, 2006, **10**(3), 11–20.
79. Kurpinski, K. et al., Regulation of vascular smooth muscle cells and mesenchymal stem cells by mechanical strain. *Molecular and Cellular Biomechanics*, 2006, **3**(1), 21–34.
80. Li, S., N.F. Huang, and S. Hsu, Mechanotransduction in endothelial cell migration. *Journal of Cellular Biochemistry*, 2005, **96**(6), 1110–1126.
81. Lammerding, J., R.D. Kamm, and R.T. Lee, Mechanotransduction in cardiac myocytes. *Annals of the New York Academy of Sciences*, 2004, **1015**, 53–70.
82. Stegemann, J.P., H. Hong, and R.M. Nerem, Mechanical, biochemical, and extracellular matrix effects on vascular smooth muscle cell phenotype. *Journal of Applied Physiology*, 2005, **98**(6), 2321–2327.
83. Riha, G.M. et al., Cyclic strain induces vascular smooth muscle cell differentiation from murine embryonic mesenchymal progenitor cells. *Surgery*, 2007, **141**, 394–402.
84. Wang, H. et al., Shear stress induces endothelial differentiation from a murine embryonic mesenchymal progenitor cell line. *Arteriosclerosis, Thrombosis, and Vascular Biology*, 2005, **25**(9), 1761–1762.
85. Carter, D.R., P.R. Blenman, and G.S. Beaupré, Correlations between mechanical stress history and tissue differentiation in initial fracture healing. *Journal of Orthopaedic Research*, 1988, **6**(5), 736–748.
86. Claes, L.E. and C.A. Heigele, Magnitudes of local stress and strain along bony surfaces predict the course and type of fracture healing. *Journal of Biomechanics*, 1999, **32**(3), 255–266.
87. Panorchan, P. et al., Microrheology and ROCK signaling of human endothelial cells embedded in a 3D matrix. *Biophysical J*, 2006, **91**(9), 3499–3507.
88. Nelson, C.M. et al., Tissue geometry determines sites of mammary branching morphogenesis in organotypic cultures. *Science*, 2006, **314**(5797), 298–300.
89. Friedland, J.C., M.H. Lee, and D. Boettiger, Mechanically activated integrin switch controls alpha5beta1 function. *Science*, 2009, **323**(5914), 642–644.
90. Ruiz, S.A. and C.S. Chen, Emergence of patterned stem cell differentiation within multicellular structures. *Stem Cells*, 2008, **26**(11), 2921–2927.
91. Fagotto, F. and B.M. Gumbiner, Cell contact-dependent signaling. *Developmental Biology*, 1996, **180**(2), 445–454.

92. Grayson, W.L. and T. Ma, Human mesenchymal stem cells tissue development in 3D PET matrices. *Biotechnology Progress*, 2004, **20**(3), 905–912.

93. Goffin, J.M. et al., Focal adhesion size controls tension-dependent recruitment of alpha-smooth muscle actin to stress fibers. *Journal of Cell Biology*, 2006, **172**(2), 259–268.

94. Mizuno, M. and Y. Kuboki, TGF-beta accelerated the osteogenic differentiation of bone marrow cells induced by collagen matrix. *Biochemical and Biophysical Research Communications*, 1995, **211**(3), 1091–1098.

95. Gerecht-Nir, S. et al., Three-dimensional porous alginate scaffolds provide a conducive environment for generation of well-vascularized embryoid bodies from human embryonic stem cells. *Biotechnology and Bioengineering*, 2004, **88**(3), 313–320.

96. Kasten, P. et al., Porosity and pore size of beta-tricalcium phosphate scaffold can influence protein production and osteogenic differentiation of human mesenchymal stem cells: an in vitro and in vivo study. *Acta Biomaterialia*, 2008, **4**(6), 1904–1915.

97. Karande, T.S., J.L. Ong, and C.M. Agrawal, Diffusion in musculoskeletal tissue engineering scaffolds: Design issues related to porosity, permeability, architecture, and nutrient mixing. *Annals of Biomedical Engineering*, 2004, **32**(12), 1728–1743.

98. Rosenberg, M.D., Cell guidance by alterations in monomolecular films. *Science*, 1963, **139**, 411–412.

99. Whitesides, G.M. et al., Soft lithography in biology and biochemistry. *Annual Review of Biomedical Engineering*, 2001, **3**, 335–373.

100. Thakar, R.G. et al., Regulation of vascular smooth muscle cells by micropatterning. *Biochemical and Biophysical Research Communications*, 2003, **307**, 883–890.

101. Liu, Y. et al., 3D femtosecond laser patterning of collagen for directed cell attachment. *Biomaterials*, 2005, **26**(22), 4597–605.

102. Engler, A.J. et al., A Novel Mode of Cell Detachment from Fibrillar Fibronectin Matrix under Shear. *Journal of Cell Science*, 2009, **122**, 1647–53.

103. Falconnet, D. et al., Surface engineering approaches to micropattern surfaces for cell-based assays. *Biomaterials*, 2006, **27**(16), 3044–3063.

104. Guo, L. et al., Osteogenic differentiation of human mesenchymal stem cells on chargeable polymer-modified surfaces. *Journal of Biomedical Materials Research A*, 2008, **87**(4), 903–912.

105. Murphy, W.L. et al., Effects of a bone-like mineral film on phenotype of adult human mesenchymal stem cells in vitro. *Biomaterials*, 2005, **26**, 303–310.

106. Dulgar-Tulloch, A.J., R. Bizios, and R.W. Siegel, Human mesenchymal stem cell adhesion and proliferation in response to ceramic chemistry and nanoscale topography. *Journal of Biomedical Materials Research. Part A*, 2009, **90**(2), 586–594.

107. Keselowsky, B.G., D.M. Collard, and A.J. Garcia, Integrin binding specificity regulates biomaterial surface chemistry effects on cell differentiation. *Proceedings of the National Academy of Sciences of the United States of America*, 2005, **102**(17), 5953–5957.

108. Moursi, A.M., R.K. Globus, and C.H. Damsky, Interactions between integrin receptors and fibronectin are required for calvarial osteoblast differentiation in vitro. *Journal of Cell Science*, 1997, **110**(18), 2187–2196.

109. Cheng, S.L. et al., Bone mineralization and osteoblast differentiation are negatively modulated by integrin alpha(v)beta3. *Journal of Bone and Mineral Research*, 2001, **16**(2), 277–288.

110. Lee, J. et al., Regulation of cell movement is mediated by stretch-activated calcium channels. *Nature*, 1999, **400**(6742), 382–386.

111. Jacot, J.G., A.D. McCulloch, and J.H. Omens, Substrate stiffness affects the functional maturation of neonatal rat ventricular myocytes. *Biophysics Journal*, 2008, **95**(7), 3479–3487.

112. Peyton, S.R. and A.J. Putnam, Extracellular matrix rigidity governs smooth muscle cell motility in a biphasic fashion. *Journal of Cellular Physiology*, 2005, **204**(1), 198–209.

113. von Wichert, G. et al., RPTP-alpha acts as a transducer of mechanical force on alphav/beta3-integrin-cytoskeleton linkages. *Journal of Cell Biology*, 2003, **161**(1), 143–153.

114. del Rio, A. et al., Stretching single talin rod molecules activates vinculin binding. *Science*, 2009, **323**(5914), 638–641.

115. Johnson, C.P. et al., Forced unfolding of proteins within cells. *Science*, 2007, **317**, 663–666.

116. Diridollous, S. et al., In vivo model of the mechanical properties of the human skin under suction. *Skin Research and Technology*. 2000, **6**(4), 214–221.

21 Mesenchymal Stem Cell Mechanobiology

Wen Li Kelly Chen and Craig A. Simmons

CONTENTS

21.1 INTRODUCTION

Mesenchymal stem cells (MSCs) are a population of plastic-adherent cells that has clonogenic capacity and multilineage differentiation potential in vitro. They can be isolated from bone marrow, adipose tissue, umbilical cord, and several other postnatal connective tissues [1]. Unlike the hematopoietic system, where the positive and negative selection of specific surface markers facilitates the isolation of hematopoietic stem cells (HSCs) from which a *single* HSC is capable of reconstituting the multilineage blood system in irradiated mice, the ability of a *single* MSC to reconstitute different mesenchymal compartments in vivo has not been demonstrated. Thus, the usage of "mesenchymal stem cell" to describe these cells is somewhat controversial since "stemness" implies self-renewal and multipotent capabilities, neither of which has been rigorously substantiated at the clonal level in vivo [1]. Nevertheless, MSCs are a promising cell source for regenerative medicine owing to their high ex vivo expansion rate, low immunogenicity, and their ability to give rise to committed progenies relevant for clinical applications, such as those in the mesenchymal (adipogenic, chondrogenic, osteogenic, tenogenic, myogenic, or endothelial) or even non-mesenchymal (neurogenic) lineages.

In the body, MSCs exist in specialized niches where the integration of biochemical and mechanical cues is essential for proper cell function. MSCs in vivo are subjected both to forces applied externally due to fluid flow or tissue deformation and to forces generated endogenously by the actinomyosin cytoskeleton as the cells interact physically with other cells and their adhesion substrate. Mechanical forces are increasingly recognized as a potent modulator of MSC commitment, particularly to cell lineages that are normally exposed to mechanical forces in vivo, such as those of the cardiovascular and musculoskeletal systems. In fact, alterations in the mechanical environment have been linked to aberrant stem cell differentiation in the development of several diseases. For example, in the absence of significant gravitational forces during space flight, bone loss is accelerated and changes in bone mineral density are similar to those in osteoporosis [2,3]. In fact, 1 month spent in a microgravity environment is equivalent to 1 year worth of bone loss in postmenopausal women [2], suggesting that mechanics, independent of age-related hormonal effects, is a major regulator in bone homeostasis. Pathological development under microgravity conditions may be the result of suppression of human MSC (hMSC) osteogenic differentiation and the simultaneous promotion of adipogenesis [2]. As another example, it has been suggested that calcification and fibrosis of cardiac valves may be a result of pathological differentiation of a population of tissue-resident MSCs with rich osteogenic and myofibrogenic differentiation potential [4]. Lineage specification of these cells is modulated by the stiffness of the extracellular matrix (ECM) [5], leading to the suggestion that changes in ECM stiffness, which are a hallmark of valve disease, may influence MSC differentiation [5]. These studies point to the importance of mechanical cues (both externally applied and intracellularly generated) in regulating MSC function and tissue homeostasis versus disease.

In light of the importance of mechanical signals in maintaining proper stem cell function, the study of mechanical regulation of MSC specification may broadly advance understanding of stem cell (patho)physiology and lead to new developments in tissue engineering and regenerative medicine strategies. In this review, we highlight the importance of various mechanical/physical cues in regulating MSC function and discuss the potential mechanisms involved.

21.2 REGULATION OF MSC FATE BY EXTERNALLY APPLIED FORCES

Dynamic, externally applied forces are important regulators of MSC differentiation towards cell types of the musculoskeletal and cardiovascular systems. MSCs sense and respond to mechanical stimuli of different types, magnitudes, directionalities, frequencies, and durations. In order to study the effects of external loading in MSC lineage commitment, a number of in vitro systems have been used to apply a variety of mechanical stimuli, in attempts to recapitulate the dominant mechanical forces in vivo, including fluid flow-induced shear stress, and compressive and tensile loading.

21.2.1 Cell Culture Systems

21.2.1.1 Fluid Flow-Induced Shear Stress

The effect of fluid flow on MSC differentiation is investigated most frequently in two-dimensional (2D) parallel plate flow chambers and in three-dimensional (3D) spinner-flask and perfusion bioreactors. Parallel plate flow chambers can introduce defined levels of fluid shear stress across the apical surface of a cell monolayer via a pressure gradient generated from active pumping or hydrostatic head [6]. The ability to generate precise and uniform flow patterns in parallel plate flow chambers facilitates the mechanistic study of flow-activated signaling pathways. In contrast, perfusion bioreactors are mostly used for mechanical conditioning and improving mass-transport of engineered tissue constructs, but less commonly used for mechanistic investigations. Three-dimensional fluid flow can cause highly complex spatial shear stress distribution as a result of the heterogeneity of scaffold properties (e.g., pore size, structure, and distribution). In attempts to address these uncertainties, computational models have been developed to predict shear stress profiles in perfusion bioreactors, and they may be used to estimate the level of shear stress required to elicit the desired cell responses in different scaffold and perfusion environments [7].

21.2.1.2 Substrate Deformation

Stretching of a monolayer of cells on a deformable substrate is a common way to study cellular response to mechanical distension. Uniaxial stretch is usually achieved by gripping and elongating two ends of a flexible membrane or by four-point bending in which the deformable substrate is flexed upwards to create tensile strain on a convex surface on which cells are seeded [6]. The commercially available Flexercell system is a commonly used apparatus for studying equibiaxial deformation because the newer versions are able to deliver uniform equibiaxial strain by the displacement of a cylindrical post against the flexible substrate. This setup can be adapted to impart uniaxial or biaxial strain with different strain levels by simply changing the geometry of the loading post.

21.2.1.3 Compressive Loading

Compressive loading of cells or tissues can be achieved by hydrostatic pressurization or platen displacement [8]. The caveat of hydrostatic pressure is twofold. First, the pressurization of the gas phase increases solubility of O_2 and CO_2 in the media, thereby confounding experimental interpretation of mechanics-induced and metabolite-induced cell response [6]. Second, hydrostatic pressurization does not cause net cellular deformation; therefore, the means by which this signal is conveyed remains unclear. Compression via platen displacement can be subdivided into confined or unconfined compression [8]. Confined compression, in which the cell/biomaterial composite is compacted uniaxially while constraining lateral expansion, is similar to the type of physiological loading in cartilage. In practice, however, the spatial constraints in confined compression may obstruct nutrient transport [8], limiting its use in cell studies. Unconfined compression is simpler to implement, and therefore, despite its complex strain field, is widely used for mechanical loading of cell-laden scaffolds, mostly in cartilage tissue engineering.

21.2.2 Osteogenic Differentiation

MSC osteogenic differentiation can be modulated by different types of dynamic forces. Tensile strain is a potent inducer of MSC osteogenesis. Low levels of cyclic, equibiaxial or biaxial strain, albeit at different magnitudes (0.8%–5%), frequencies (0.25–1 Hz), and durations, stimulated MSC osteogenic differentiation [9–14], whereas high-level cyclic biaxial strain (10%–15%) showed an inhibitory effect [12]. Substrate deformation alone, in the absence of soluble supplements, was sufficient to upregulate osteogenic lineage markers with concomitant reduction of chondrogenic, adipogenic, and neurogenic gene expression (this differential genetic regulation was termed "gene focusing") [9]. The functional competence of the resulting hMSC-derived osteoblasts was confirmed by the presence of a mineralized matrix [9]. Additionally, stretching suppressed lipid droplet formation and preserved the osteogenic capacity of C3H10T1/2 cells (a mouse MSC line) in adipogenic supplements [10,11], suggestive of the dominant influence of physical cues over chemical signaling in osteogenic commitment. In fact, mechanical stimulation can modulate biochemical signaling.

For example, cyclic uniaxial tensile loading of hMSC-laden collagen gels induced upregulation of gene expression of BMP-2 (a potent osteogenic stimulus) in serum-free conditions, suggesting a role for tensile loading-induced BMP-2 autocrine signaling in hMSC osteogenic differentiation [15].

Fluid flow-induced shear stress also influences MSC osteogenic gene expression, but the role of flow in osteogenesis requires further confirmation by functional assessment (i.e., mineralization). Oscillatory [16–18], pulsatile [19,20], steady [20,21], and intermittent flow [21] of different magnitudes and frequencies have been shown to enhance expression of certain bone-related genes, but their relative effectiveness is unknown. Oscillatory flow patterns [16] and shear stress in the range of 8–30 dyn/cm² [22] were thought to be most reflective of the load-induced interstitial flow in the lacuna-canalicular system. However, unlike low-magnitude substrate deformation, which restricts MSC lineage selection by gene focusing [9], oscillatory flow induced nonspecific differentiation by simultaneously upregulating Runx2 (bone), PPARγ (fat), and Sox9 (cartilage) mRNA expression [18]. Therefore, oscillatory flow alone cannot direct MSCs osteogenic specification. Moreover, the majority of planar flow studies focused on the study of early signaling pathways (e.g., Ca$^+$ wave [16], Wnt [17], mitogen activated protein kinases (MAPKs) [23], nitric oxide ([19], and cyclooxygenase-2/prostaglandin E$_2$ signaling [19,21]) and their correlations with osteogenic gene transcription, which may not necessarily reflect functional competence at the protein level. For instance, though both continuous and intermittent flow induced similar enhancement of osteopontin mRNA expression [21], only the former induced osteopontin protein secretion [21]. Moreover, continuous flow activated both the extracellular signal regulated protein kinase (ERK1/2) and p38 signaling pathways [21], which can have opposite influences on osteogenic differentiation [13]. Thus the influence of flow stimulation on MSC differentiation is ambiguous without further validation with functional bone markers. In fact, the only flow study on MSC-mediated mineralization revealed that although osteocalcin protein expression was increased in the sheared samples, bone nodule formation was suppressed [24]. Thus, further long-term investigations of flow induced MSC osteogenesis are required to establish links between early molecular events and their functional outcomes.

Bioreactors are not only important for developing tissue-engineered bone grafts, but are also useful for the study of fluid stress in a more physiological 3D environment. Flow-conditioning of encapsulated MSCs enhanced tissue mineralization in perfusion chambers and spinner flask systems [25–29]. The stimulatory effect of fluid shear, however, is confounded with improved mass transport in dynamically loaded systems. To resolve this ambiguity, it was demonstrated that by changing the viscosity of culture media, one can manipulate the level of fluid shear stress while maintaining a constant perfusion rate, thereby decoupling mechanical effects from chemotransport influence [30]. Elevated shear stress, as a result of higher media viscosity, enhanced calcium deposition [30]. Mechanical stress can advance osteogenic differentiation independent of molecular transport. In fact, it has been speculated that elevated fat content, and by association reduced marrow viscosity, may diminish shear stress in osteoporotic bones, and consequently exacerbate osteoporosis [31].

21.2.3 CHONDROGENIC DIFFERENTIATION

Compressive loading regulates in vivo cartilage homeostasis and in vitro chondrogenic differentiation. Cells in matrices subjected to unconfined compressive loading experience strain, fluid shear stress, and hydrostatic pressure [32,33]. Finite element modeling revealed that these forces are heterogeneously distributed within the loaded constructs [32], and therefore the encapsulated cells are exposed to different mechanical stimuli depending on their spatial location within the scaffolds. Generally under dynamic compression, the loaded construct is subjected to predominantly hydrostatic pressure and strain with increased fluid shear at the construct periphery [32]. When subjected to cyclic, unconfined compression (10% strain, 1 Hz), bovine MSCs increased aggrecan gene expression in both the central and the annular regions of the hydrogel, which suggests that strain and hydrostatic pressure, but not fluid flow, are the dominant mechanical inducers of MSC chondrogenic commitment [32]. In a separate study, when agarose-encapsulated equine MSCs were dynamically compressed between a porous platen and an impermeable base, enhanced proteoglycan deposition was found near the base where there was greater hydrostatic pressure and low fluid flow [34]. The spatial profile of mechanical signals is sensitive to changes in loading parameters. For instance, the region of high hydrostatic pressure and strain can be enlarged by increasing loading frequency [32]. Therefore, there is the opportunity to optimize MSC responses by fine-tuning the loading conditions.

On the whole, aggrecan and collagen type II gene expression and/or proteoglycan deposition are responsive to hydrostatic pressure [35–40] and cyclic platen compression [41–47], but their effectiveness can be modulated by biochemical cues. The degree of MSC chondrogenic response under intermittent hydrostatic pressure is dose dependent [39]. Increasing loading pressure from 0.1 to 10 MPa enhanced cartilaginous protein deposition [39]. Interestingly, whereas TGF-β and *hydrostatic pressure* have additive effects on chondrogenic differentiation [37,39], high dose of *exogenous* TGF-β (~10 ng/mL) can mask or even inhibit *compression-mediated* chondrogenic differentiation [34,45–48]. On the other hand, *endogenous* TGF-β secretion was thought to contribute to compression-induced chondrogenic commitment, suggestive of a critical level of TGF-β signaling required for mechanoregulation of MSC chondrogenesis. Dynamic unconfined compression (10% strain, 1 Hz) applied to rabbit MSCs-seeded agarose gel in *serum-free media* resulted in upregulation of cartilage-specific markers and TGF-β gene expression [47], proving that mechanics alone is capable of initiating chondrogenic specification. The mechanism involved the upregulation of endogenous TGF-β expression and heightened TGF-β sensitivity, as indicated by elevated TGF-β types I and II receptor expressions [49]. Inhibition of TGF-β1 receptor with LY364947 significantly attenuated chondrogenic gene expression [45]. Compression-mediated TGF-β autocrine signaling [49] is an example of how mechanoregulation of biochemical signaling can influence MSC differentiation.

21.2.4 Smooth Muscle Differentiation

Mechanical stimulation can induce smooth muscle cell-related marker expression in MSCs; however, the mode of mechanical stimulus conducive to robust smooth muscle cell differentiation is unclear and ECM protein type appears to be the determining factor in stretch-induced myogenic differentiation. Uniaxial but not equibiaxial stretching (10% strain, 1 Hz), simulating the anisotropic strain experienced by smooth muscle cells in vivo, promoted brief upregulation of contractile markers (smooth muscle α-actin [α-SMA] and smooth muscle-22α mRNA) in hMSCs on collagen I-coated substrates [50]. However, the observed myogenic gene expression was transient, possibly as a result of the non-physiological cellular alignment typical of many in vitro stretching systems [50]. High-magnitude 2D substrate deformation (both uniaxial and biaxial) can cause cells to re-orient perpendicular to the direction of strain to minimize cytoskeleton stress buildup [50–57]. In the native vasculature, however, smooth muscle cells align parallel to cyclic strain in the circumferential direction [50,58]. Micro-grooved elastomeric substrates were used to restrain cell elongation parallel to unidirectional strain, better mimicking the physiological condition [52]. However, despite upregulation of an intermediate smooth muscle marker (calponin mRNA), no terminal differentiation was observed using the combination of contact guidance and uniaxial strain, suggesting that additional cues, such as different ECM proteins and/or soluble factors, may be required to further MSC myogenic differentiation [52]. Indeed, when hMSCs were subjected to 8%–12% equibiaxial strain, α-SMA and calponin proteins were upregulated on fibronectin but downregulated on collagen I or non-coated substrates, demonstrating the indispensible contribution of ECM cues in the correct mechanoregulation of myogenic differentiation. In addition, other types of mechanical forces, such as the combination of pulsatile shear (14 dyn/cm²) and compression (pressure at 120/60 mmHg) have been shown to further myogenic specification, evidenced by the expression of a late myogenic marker, smooth muscle myosin heavy chain [59]. Shear/compression conditioning on mouse MSCs demonstrated superior effectiveness over substrate stretching in directing MSC myogenic maturation. However, the observed improvement could be the result of differences in cell type- and/or species-dependent myogenic potential and mechanosensitivity.

Based on the accumulated knowledge from 2D myogenic differentiation studies, vascular engineering using hMSCs-laden biodegradable scaffolds requires the optimal combination of biochemical and mechanical conditioning in a carefully orchestrated temporal sequence. MSCs seeded in fibronectin-coated polyglycolic acid (PGA) polymers were first allowed to proliferate in the presence of platelet-derived growth factor-BB to obtain the appropriate cellularity in static culture, and were subsequently subjected to pulsatile perfusion (5% cyclic strain, shear) in the presence of TGF-β1 to promote myogenic maturation, resulting in vessel grafts with improved tissue organization and burst pressure [60].

21.2.5 TENOGENIC DIFFERENTIATION

Tendons and ligaments are dense and fibrous connective tissues that regularly undergo cyclic tensile loading in vivo. Dynamic stretching in both 2D and 3D systems have been shown to upregulate tendon/ligament-related markers in MSCs. While low-magnitude (3% strain) cyclic biaxial substrate deformation favored osteogenic differentiation, high-magnitude (10% strain) biaxial or uniaxial planar stretching upregulated ligament and tendon fibroblast marker expression, including collagen types I and III, and tenascin-C [55–57]. Strategies to tissue engineer tendons/ligament replacements involved uniaxial tensile of MSC-seeded scaffolds (collagen gel or sponge), some of which have resulted in cellular and ECM protein alignment along the loading axis [61,62] and improved structural and material properties of the loaded constructs [63,64].

21.2.6 ENDOTHELIAL CELL DIFFERENTIATION

The effect of mechanical regulation of MSC endothelial differentiation is controversial. Application of laminar flow at 15 dyn/cm^2 induced gene and protein expression of endothelial cell-related markers (CD31 and von Willebrand factor) and tubular formation of C3H10T1/2 cells without exogeneous factors [65]. The mechanism was associated with shear-induced upregulation of angiogenic factor gene expression, VEGF (and its receptor), and the concurrent reduction of smooth muscle cell stimulator, TGFβ1 (and its receptor) [66]. However, when human placenta-derived MSCs were subjected to 12 dyn/cm^2 of shear, pre-differentiation in endothelial supplements was required to prevent cell detachment [67]. The combination of chemical treatment and shear stress substantially enhanced MSC differentiation into endothelial cells as indicated by increased PECAM-1 and vWF protein expression, and tubule formation on Matrigel [67]. The indispensable role of soluble factor induction in the latter study suggests that shear stress may only enhance terminal differentiation of precommitted progenitors but cannot initiate early lineage commitment [67]. The discrepancies between the efficiency of shear-induced MSC endothelial differentiation can be attributed to the use of different ECM protein coating (collagen I [65] vs. fibronectin [67]) and cell source (mouse MSC cell line [65] vs. human placenta-derived MSC [67]) in different studies.

21.3 REGULATION OF MSC FATE BY INTRACELLULARLY GENERATED FORCE

In biological systems, form often correlates with function. Cell shape modulation has functional consequences in many differentiated cell types. For instance, hepatocytic albumin secretion correlates inversely with cell spreading [68], adipocytes adopt a spherical morphology to maximize fat storage [69], and osteoblast flattening is conducive to bony matrix deposition [69]. Similarly, as reviewed below, cell shape modulation of MSCs influences lineage commitment. Cell shape is a function of cell contractility, generated when adherent cells interact with their immediate environment (e.g., neighboring cells, and/or substrates with different geometry, topography, and/or material stiffness). Thus, by altering cell–cell interactions or the characteristics of the substrate, one can investigate cellular response to alterations in intracellularly generated forces. Usually this represents a more accessible model for studying mechanoregulation of MSCs, and can allow specific definition of single cell mechanical environments. For example, precise manipulation of substrate adhesivity via micropatterning techniques allows clonal investigation of MSC mechanobiology, thereby reducing the effects of heterogeneity and contamination from differentiated cells that are common in population-based dynamic loading studies.

21.3.1 Cell–Cell Interactions and Geometry Sensing

MSC lineage commitment towards the adipogenic and osteogenic fate can be controlled by the degree of cell spreading. At high cell density, densely packed, rounded hMSCs became adipocytes, but sparsely plated hMSCs spread and acquired an osteogenic fate [69]. Cell shape-mediated adipo/osteogenic lineage specification was validated on singly seeded hMSCs confined to micropatterned fibronectin islands of increasing sizes, which eliminated the potential confounding contributions from cell–cell interaction and paracrine signaling at high cell density [69].

Cell contractility (cytoskeleton tension) is regulated by RhoA GTPase/RhoA-associated kinase (ROCK) signaling. Constitutively active RhoA overcame soluble factor stimulation and promoted ALP expression in adipogenic media, substantiating the importance of cytoskeletal tension in lineage commitment [69]. Indeed, when grown on 2D patterned substrates or encapsulated in 3D matrices, hMSCs selectively underwent osteogenic differentiation in regions of high tension near the edge of the geometry and adipogenesis at regions of lower tension close to the center [70]. Interestingly, this tension-induced spatial segregation of MSC commitment was most evident in mixed osteogenic/adipogenic media, suggesting that the different tensional environment modulates the specificity of cellular response towards different chemical factors. This phenomenon shed light on how different tissues evolve in a localized area with similar bulk morphogen composition but different tensional state. The osteo/adipogenic lineage segregation observed in the 3D matrices has similar anatomical form and tension distribution of that in a long bone, where rigid bone encloses a cavity of soft fat tissues [70]. It seems that tension, generated through either dynamic stretching or passive guidance (via constrained 2D adhesive islands or 3D matrices), has a selective role towards MSC osteogenic specification.

21.3.2 Nanotopography

Cells respond to nanoscale features of the substrates to which they are attached. In physiological conditions, cells can sense and respond to subtle changes in protein conformation, which occurs at the length scale of nanometers [71]. Advances in nanofabrication have facilitated the study of cellular function on surfaces with precise and reproducible nano-topographical features [72,73].

Both the arrangement and dimensions of nanotopographical features are important design parameters to achieve the desired biological responses. It has been reported that semi-ordered arrays of 120 nm nanopits induced more hMSC osteogenic differentiation than highly ordered or randomly indented nano-surfaces [72]. Semi-ordered nanoporous (79 nm pore size) alumina substrates enhanced cell adhesion, proliferation, and osteogenic differentiation of mouse MSCs over amorphous substrates [74]. A 100 nm titanium (Ti) nanotube-assembled surface showed superior osteogenic ability than nanotubes with smaller diameters (30–60 nm) [73]. hMSC osteogenesis can be achieved in the absence of chemical stimuli on nano-featured surfaces that promoted fibrillar extension [72,73], suggesting that nano-topographical cues are conveyed through adhesion- and cytoskeletal tension-mediated signaling pathways. Interestingly, the nano-topographical effect on MSC differentiation has been attributed to the extent of serum protein retention on nanotube surfaces [73]. Scanning electron microscopy revealed that agglomerated serum proteins, comparable in size to 30–60 nm nanotubes, were more easily trapped at the surface features of similar dimensions, resulting in a higher density of adhesive ligands for cell attachment [73]. The increased proximity of ligand presentation (nanometers apart) produced a highly adhesive surface that promoted greater initial cell adhesion and restricted cell spreading, which together contributed to the rounding of hMSC that had analogous effect to cell shape-mediated suppression of osteogenesis on small micropatterned surfaces [69,73]. Nanotopography has also been shown to induce MSC commitment to non-mesenchymal lineage with higher efficiency than chemically induced differentiation. MSCs plated on nanogratings expressed higher levels of the mature neuronal markers MAP2 and β-tubullin III in growth media than MSCs grown on flat surfaces with neurogenic supplements [75]. This response was modulated by feature dimensions: increasing grating dimensions (300 nm, 1 μm, 10 μm) positively regulated MSC proliferation, but adversely affected MAP2 expression [75]. Disruption of the actin with cytochalasin-D reduced neurogenic tyrosine hydroxylase mRNA expression, alluding to the importance of an intact cytoskeleton in mediating physical cues such as nanotopography.

21.3.3 EXTRACELLULAR MATRIX MECHANICS

All adherent cells in the body interact with their microenvironment via surface receptor binding to various extracellular components. The ECM is not only a reservoir of biochemical factors but it is also responsible for the structural integrity of tissues. This integration of biochemical and physical cues manifests in specialized ECM niches that drive diverse tissue-specific functions.

Substrate (ECM) stiffness is an important physical regulator of a range of cell functions (adhesion, migration, proliferation, differentiation, and apoptosis) in many *differentiated* cell types including neurons, astrocytes, cardiomyocytes, smooth muscle cells, hepatocytes, chondrocytes, skeletal myoblasts, fibroblasts, and pre-osteoblasts [76]. It has been shown that matrix mechanics alone is capable of guiding early hMSC lineage specification [77]. On 2D collagen I-coated substrates, hMSCs differentiated towards the neurogenic, myogenic, and osteogenic fate on substrates with elasticities corresponding to those of brain, muscle and pre-calcified bone, respectively [77] (Figure 21.1). In combination with chemical induction, differentiation was further enhanced in an additive manner with lineage-specific marker expression levels reaching those of murine myoblasts and human osteoblasts (hFOBs).

Although there is strong evidence supporting the effect of substrate stiffness in regulating MSC commitment, mechanics alone is insufficient to achieve terminal differentiation. Further, many tissues in the body have comparable elasticities [78]; for example, the tissue modulus of brain, fat and the bone marrow all fall around 0.2 kPa [78], but these cells/tissues have vastly different phenotypic and functional characteristics. Clearly, additional micro-environmental cues are needed to definitively drive MSC commitment. All anchorage-dependent cells sense their microenvironment via integrin–ECM interactions, and since ECM composition varies greatly in different tissues, ECM proteins likely have modulating effects on how MSCs respond to the matrix stiffness. For example, on substrates of comparable rigidities (~0.5 kPa), hMSCs showed characteristic neurite branching and upregulation of early neurogenic marker on collagen I [77], but remained rounded, quiescent, and retained their multipotency on 90% collagen I/10% fibronectin-coated surfaces [78]. Reportedly, fibronectin has an inhibitory effect on neural stem cell differentiation and neurite extension, and hence could be accountable for the differences observed in cell response [78]. Furthermore, on the substrates with stiffness comparable to that of trabecular bone (~80 kPa), hMSCs cultured on collagen I-coated surface expressed high level of Runx2, which was 2-, 7-, and 14-fold greater than those on fibronectin-, laminin-, and collagen IV-coated substrates respectively [79]. However, the same stiffness/ECM (80 kPa/collagen I) condition also favored MyoD1 expression which contradicted a previous study [77] where myogenic commitment on collagen I-coated substrates was initiated at a lower elasticity, distinct from the optimal range for osteogenic differentiation. It is possible that the difference in type I collagen coating density in the two independent studies contributed to the

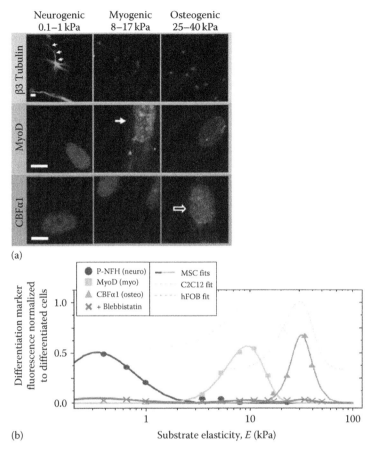

FIGURE 21.1 (See color insert.) Matrix stiffness modulates human mesenchymal stem cell lineage specification. (a) hMSCs expressed early neurogenic, myogenic, and osteogenic markers on substrates with stiffness value resembling the tissue elasticity of brain, muscle, and pre-calcified bone, respectively. (b) Quantification of the lineage marker intensities normalized to the respective positive control cells (C2C12 myoblasts and hFOB osteoblasts) as a function of substrate stiffness. Specific lineage commitment was optimized at distinct range of substrate stiffness and inhibition of non-muscle myosin II activity with blebbistatin abolished the stiffness-mediated hMSC specification. (From Engler, A.J. et al., *Cell*, 126(4), 677, 2006.)

discrepancy. Subtle tuning of ECM composition, in addition to ECM type and mechanics, may have modulating effects on MSC lineage segregation.

21.4 MSC MECHANOTRANSDUCTION

Mechanical/physical regulation of cell and tissue function is ubiquitous; however, little is known about the mechanisms by which cells sense and convert physical signals into biochemical processes that can impact changes in gene and protein expression, and how these molecular events reciprocally alter the local mechanical environment resulting in global changes in tissue mechanics and function. Mechanistic studies on MSCs are limited mainly due to the lack of definitive MSC markers for purification. Thus the inherent heterogeneity in a progenitor population undermines the accurate delineation of signaling pathways. Nonetheless, the study of mechanotransduction in cell lines and other homogenous cell populations may shed light on how similar mechanical signals are transduced in progenitor cells given that they all possess similar mechanosensitive machinery.

21.4.1 MECHANORECEPTORS

Transmembrane proteins such as integrins, ion channels, and G-protein-coupled receptors (GPCRs), have been proposed as mechanoreceptors, and they can be activated by force-induced changes in protein conformation. However, evidence suggests that ion channels [80,81] and GPCRs do not respond to mechanical stimuli directly, but instead react to intracellular mechanical stress relayed through their association with the cytoskeleton. Since all adherent cells require substrate anchorage to remain viable, and cell adhesion involves ECM–integrin–cytoskeleton coupling, integrins are likely the front line mechanoreceptors, upstream of both ion channels and GPCRs. Indeed, increased tension at integrin–ECM adhesions induced extracellular calcium influx [82,83], and disruption of actin cytoskeleton could influence ion channel activities [84]. GPCR subunit localization within focal adhesions alludes to their potential interaction with integrin and cytoskeleton components. Cells respond to mechanical stimulation by manipulating their integrin receptor profile as well as ECM protein secretion [85]. The following section focuses on integrins due to their role as the front line mechanosensors and the fact that they are 10–100 times more abundant than other surface receptors [71].

21.4.1.1 Integrin–ECM Interaction

Cells interact with their environment via specific integrin–ECM binding. Integrins link different ECM components to the actin cytoskeleton at sites called focal adhesions. The cytoplasmic clustering of focal adhesion, proteins, and their mechanical coupling to the cytoskeleton network facilitates force propagation across the cell.

Activation of $\alpha_5\beta_1$ integrin interaction entails the adaptation from a relaxed state to a tensioned state via myosin II force generation [86]. Disruption of myosin II activity post-seeding did not affect the number of total fibronectin-$\alpha_5\beta_1$ bonds but significantly reduced the frequency of tensioned bond (cross-linked), which were essential for focal adhesion kinase (FAK) Y397 autophosphorylation [86] and thus MAPK signaling [87]. Given that substrate rigidity modulates intracellular tension generation, the number of cross-linked β_1 integrins and level of FAK pY397 activation increased linearly with increasing substrate stiffness (0.2, 20, 100 kPa) [86]. The link between tensioned bond formation and FAK activation indicates the importance of mechanical stress in FAK activation and provides a mechanism through which physical stimuli can be converted to chemical signals, thereby influencing downstream signal transduction.

Although the RGD binding motif was responsible for ECM–integrin binding (relaxed bond), synergy sites in the fibronectin molecules are crucial for bond stabilization and therefore tensioned bond formation [86]. Selective mutation at the two synergy sites of fibronectin led to 90% reduction in adhesion strength and FAKpY397 phosphorylation [86]. While RGD peptides are frequently used to improve bioadhesivity of artificial scaffolds, it is unclear whether they undermine the mechanosensivities of implanted cells due to their oversimplified molecular structure. It has been shown that RGD-coated surfaces required 10-fold more ligand density than fibronectin-coated surfaces to achieve similar cell spreading area, and even then the traction force generated was still fivefold lower on the RGD-surfaces, resulting in diffuse actin fiber and focal adhesion organization [88]. These results suggest that RGD motifs alone are not conducive to cytoskeletal coupling. It is unclear if other ECM protein–integrin combinations require similar synergy binding sites to reinforce mechanical coupling to the cytoskeleton.

21.4.1.2 MSC Mechanics and Mechanosensitivity

The differentiation stage of a cell affects its sensitivity towards mechanical (e.g., cyclic strain [89] and pulsatile fluid flow [19]) and physical cues (substrate stiffness and nano spacing of adhesive ligands) [89,90]. In general, primitive cells appeared to be less mechano-responsive than their mature counterparts, but can acquire increased mechanosensitivity if they were pre-differentiated with chemical supplements (based on observations from osteogenic cells) [90].

MSC lineage commitment is accompanied by major changes in membrane mechanics [91,92], cytoskeleton organization [91,93], and integrin mobility in the plasma membrane [92]. Membrane tether length measurement showed that undifferentiated MSCs have a higher membrane storage and weak membrane-cytoskeleton interaction [91]. The plasma membrane of MSCs is anchored to its cytoskeleton through only focal complexes in contrast to mature cell types, in which the membrane cytoskeleton interaction is stabilized by focal complexes as well as numerous linker proteins such as ezrin, radixin, moesin, and myosin-I, contributing to a significantly shorter membrane tether and stronger membrane–cytoskeleton association [91]. Given that the plasma membrane contains putative mechanosensors such as ion channels, G-protein, and integrins, the weaker membrane–cytoskeleton interaction may affect the efficiency of signal transduction.

Quantum dot tracking of integrin diffusion across the plasma membrane showed that integrin movement is heavily constrained in hMSCs in comparison to differentiated cells [92]. Cytoskeleton tension may be a regulator of integrin diffusion, owing to the fact that hMSCs are twice as stiff as differentiated osteoblasts [91] and that actin disruption increased integrin diffusion rate by almost 10-fold [92]. Knowing the importance of integrins as mechanosensors, the immobility and therefore inaccessibility of integrins to engage in dynamic integrin–ECM binding may adversely impact mechanosensitivity.

There is no direct evidence that weak membrane–cytoskeleton interaction and the reduced integrin diffusion rate are responsible for the low MSC mechanosensitivity. However, as MSCs underwent chemically induced osteogenic differentiation, integrin mobility increased significantly [92] and the actin cytoskeleton remodeled extensively to become a thinner meshwork with higher density of membrane–cytoskeleton interactions. Consequently, cell elasticity converged to that of differentiated osteoblast [90,93]. Future studies can look into establishing a causal relationship between these observations and MSC differentiation.

21.4.2 Intracellular Signal Transduction

Following activation of mechanosensors, the incoming signals must be relayed to the appropriate intracellular targets in order to elicit a change in cell function. Intracellular signal propagation occurs via two plausible mechanisms: *cytoskeleton-driven signal transduction* and *biochemically mediated signal transduction*. These two mechanisms are not mutually exclusive, but synergize to coordinate global cell function.

21.4.2.1 Cytoskeleton-Driven Signal Transduction

In the tensegrity model, a cell is thought to maintain its shape by establishing a force equilibrium where continuous tension in the actin cytoskeleton and to a lesser extent the intermediate filaments is counterbalanced by compression of microtubules and of matrix adhesion sites [94]. Mechanical coupling of surface mechanoreceptors to the internal cytoskeleton network is required for long-range stress propagation [94]. For instance, when stress was applied locally to β_1 integrin receptors, focal adhesion assembly was induced and cytoskeleton stiffness was increased to resist deformation, demonstrating that force transmitted from a discrete location can cause a global change in cell stiffness due to the structural continuity of the cytoskeleton [95]. On the other hand, when force was applied to a membrane metabolic receptor that is not stably linked to the internal cytoskeleton (e.g., acetylated-low density lipoprotein receptor), resistance to deformation was not observed [95]. It was deduced that the separation of the metabolic receptor from the cytoskeleton prevented force propagation and therefore failed to cause strain-stiffening of the cytoskeleton. Various transmembrane proteins have differential cytoskeleton coupling capacities in the following order: integrin β_1 > PECAM > integrin $\alpha_v\beta1$ > E-selectin > integrin α_5, α_2, α_v >>> ALDL receptor > HLA antigen (from [96]). Therefore, the diversity in ECM–integrin mediated mechano-regulation of cell functions is partly attributable to differences in integrin mechanical coupling strengths.

The cytoskeleton network envelops and anchors intracellular organelles; force transmitted from integrins via the cytoskeleton deforms distant cellular structures (i.e., mitochondria and the nucleus), thereby impacting cell function [97,98]. Stress applied to a single RGD-coated bead induced rapid (0.3 s) long-range (60 μm) Src activation, which correlated with deformations on microtubules and endosomes [99]. Force application to integrins has been shown to cause mitochondria displacement and alignment [98], as well as nucleus deformation [100]. Stretch-induced opening of nuclear pores may enhance transport of mRNA and/or other regulatory proteins [97]. Nuclear deformation may cause unwinding of DNA and exposure of promoter binding site to facilitate its interactions with transcriptional factors in the nucleus matrix [97,101]. Some have argued that nuclear shape rather than cell shape is a modulator of stem cell fate, as differential nuclear shape index (nuclear area to height ratio) has been shown to correlate closely with osteogenic-specific gene and protein expression changes in MSCs with *similar* spreading area [101]. For further information on nuclear mechanotransduction, readers can refer to excellent reviews by Dahl et al. [102] and Wang et al. [97].

It was postulated that stress propagation via the cytoskeleton is a more rapid transduction pathway than biochemically mediated signal transduction, via the second messenger system, for example [97]. It was estimated that stress waves move through tensed cytoskeletal filaments at 30 m/s; motor-assisted molecular translocation occurs at 1 μm/s; and chemicals (e.g., calcium) diffuse at 2 μm/s [97].

In addition, force preferentially propagates through stiff tensile structures to reduce stress dissipation [97]. Similarly, given the high nucleus/cytoplasm stiffness ratio (ninefold) [100], stress propagation may preferentially target the nucleus [97]. The lack of stress fiber formation in cells plated on poly-L-lysine-coated (as opposed to collagen I-coated) dishes resulted in a 10-fold reduction in stress transmission efficiency to the nucleolus [103]. Long range Src activation was attenuated on soft gels (0.7 kPa) [99]. Collectively, these data reiterated the importance of integrin–ECM interaction and intracellular tension in mechanotransduction. Therefore, mechanoregulation of cell behavior can be realized by fine-tuning substrate biochemistry and mechanics.

In summary, while there is a disconnect in the current literature between early stress-induced signal propagation (milliseconds to minutes) and downstream cellular responses (days to weeks), there is one common theme: the reliance on cytoskeletal tension in early signaling and long-term cell function.

21.4.2.2 RhoA/ROCK Mediated Signaling

Cytoskeletal tension-mediated cell regulation is mediated by coupling of RhoA and its downstream effector, Rho-associated kinase (ROCK) [104]. ROCK activation increases myosin light chain (MLC) phosphorylation and in turn myosin II activity, thereby generating contractility [105]. Though cytoskeletal tension has been shown to impact cell proliferation, differentiation and apoptosis, there is no direct evidence that stress propagation alone, in the absence of biochemical cascades, independently elicits these responses. Contractility-induced MSC differentiation is usually coupled with downstream biochemical signaling pathways.

Different types of stress, both externally applied (stretch, shear, and compression) or intracellularly generated (geometric guidance and substrate stiffness) differentially modulate Rho activity in many cell types [18,69,105–107]. MSC lineage specification has been associated with Rho activity [69,108]. Disruption of actin [69], ROCK, MLC phosphorylation, and myosin II activity [70] independently eradicated MSC osteogenic differentiation. Constitutively active RhoA activity partially rescued stress fiber disruption in microgravity conditions and recovered hMSC osteogenic capacity [108]. Interestingly, though constitutively active RhoA was able to reverse a chemical-induction effect by encouraging osteogenic differentiation in adipogenic media, it failed to rescue osteogenesis in rounded cells constrained to small adhesive patterns [69]. Cell adhesion, shape and cytoskeletal tension were indispensible for RhoA regulation of ROCK [104]. In contrast, overexpression of ROCK was able to regulate myosin II activity and MSC commitment independent of cell shape [69].

Substrate stiffness modulation of pre-osteoblast differentiation required FAK and RhoA/ROCK activities, which could subsequently influence downstream MAPK signaling [106,109,110]. Inhibition of non-muscle myosin II activity, downstream of ROCK, abolished stiffness-mediated hMSC specification [77]. Inhibition of RhoA or ROCK attenuated ERK phosphorylation, resulting in diminished osteogenic differentiation [106]. RhoA overexpression marginally increased ERK activation and Runx2 expression on the compliant matrices by day 14 [106], confirming that elevated RhoA activity, in the absence of cytoskeletal tension, could not revert to the osteogenic fate, similar to that observed in rounded cells [69]. These studies reaffirmed the importance of cytoskeleton tension in RhoA-regulation of ROCK activity, MAPK signaling, and ultimately, forced-mediated differentiation. The connection between tensegrity-based tension generation and biochemical-based MAPK cascades helps to bridge the gap between early cell signaling events and long-term cell behavior by linking surface receptors (integrins) to cytoskeletal components, intracellular signaling proteins (MAPKs), and DNA binding molecules (transcription factors, e.g., Runx2). In general, many signaling molecules (glycolytic enzymes, protein kinases, lipid kinases, and phospholipases GTPases) [84] interact with the cytoskeletal network; therefore, stress propagation through the cytoskeleton may induce activation of various signaling molecules by changing their conformation and/or location relative to their binding targets [111].

21.4.2.3 Calcium-Mediated Signaling

Calcium ions are common second messengers implicated in a range of signaling events, and differential Ca^+ oscillation patterns have been detected during MSC commitment. Second messengers are small, abundant signaling molecules generated in response to receptor binding. They can diffuse efficiently across the cytoplasm to activate downstream signaling by targeting key signaling proteins. Oscillatory fluid flow-induced cytosolic Ca^{2+} mobilization has been associated with increased proliferation and upregulation of osteogenic mRNA expression in hMSC [16]. hMSCs have substantially different Ca^{2+} oscillation profiles from their differentiated progenies, and imposed calcium oscillation using electrical stimulation could accelerate osteogenic differentiation, establishing a link between short-term Ca^{2+} activity with long-term MSC function (10 days) [112]. Mechanics-induced Ca^{2+} oscillation is likely to have similar impact on MSC terminal differentiation.

Spontaneous Ca^{2+} oscillation in hMSCs involved extracellular Ca^{2+} influx and intracellular Ca^{2+} recruitment from the endoplasmic reticulum, and these events were *independent* of actin and microtubule integrity and myosin contractility [113], suggesting that second messenger-based Ca^{2+} propagation may be separate from cytoskeleton-mediated signal transduction. However, the Ca^{2+} oscillation pattern was sensitive to substrate stiffness, and reduced Ca^{2+} activity correlated with reduced RhoA on compliant matrices [113]. This is counterintuitive since rigidity sensing normally requires an intact cytoskeleton and myosin activity. Nevertheless, it is possible that other pathways that have not been previously implicated may be responsible for stiffness sensing. In addition, though a compromised cytoskeleton did not seem to affect second messenger (calcium signaling) activation, it might affect the targets of Ca^{2+}, thereby adversely impacting downstream cell responses.

21.4.2.4 Mitogen-Activated Protein Kinase (MAPK) Pathways

Many types of mechanical stimuli can modulate MSC function by converging to the MAPK pathways, specifically ERK1/2 and p38 signaling. The loss of ERK activity under microgravity condition suppressed hMSCs osteogenic differentiation and resulted in marked reduction in bone-related gene expressions [2]. Similarly, inhibition of ERK1/2 activity with U0126 blocked oscillatory fluid flow-induced MSC proliferation [114], and abolished 3D fluid shear- and substrate strain-induced matrix mineralization [13,25]. Interestingly, suppression of ERK1/2 with PD98059 (another ERK1/2 inhibitor) in a separate study only marginally reduced matrix mineralization [9], possibly due to differential specificities of chemical inhibitors. Future mechanistic studies may benefit from the use of more rigorous blocking methods, such as small interfering RNAs (siRNAs) or genetic knock-downs. p38 MAPK signaling negatively regulates strain-induced MSC osteogenic differentiation, as evidenced by the accelerated calcium deposition in p38-inhibited cultures [13]. Moreover, despite the reciprocal relationship between adipogenesis and osteogenesis, stretch-induced inhibition of hMSC adipogenesis is independent of ERK1/2 signaling [115]. Collectively, these studies suggest that mechanical stimulation can simultaneously activate multiple signaling pathways and the resultant MSC response depends on the net outcome of these signaling events.

21.4.2.5 Wnt Signaling

Both canonical and non-canonical Wnt pathways play an important role in developmental pathways in vivo and have been implicated in MSC lineage specification in vitro [116]. However, the mechanoregulation of Wnt signaling in MSC commitment has not been studied extensively. Mechanically induced activation of the canonical pathway has been reported in strain-induced MSC osteogenic differentiation. Mechanical strain induced β-catenin nuclear translocation and suppressed β-catenin degradation in adipogenic media by deactivating glycogen synthase kinase-3β [1]. Oscillatory fluid flow, on the other hand, stimulated Runx2 gene transcription by weakening the β-catenin/N-cadherin association at adherens junctions, thereby facilitating β-catenin nuclear translocation independent of canonical signaling [17]. Moreover, oscillatory fluid flow activated noncanonical signaling by inducing Wnt protein (Wnt5a) and receptor kinase (Ror2) expression, which in turn may stimulate downstream RhoA activation [17]. Given the importance of RhoA in mechanotransduction discussed previously (Figure 21.2), the noncanonical Wnt signaling must play an important role in cytoskeleton-mediated MSC commitment. Also,

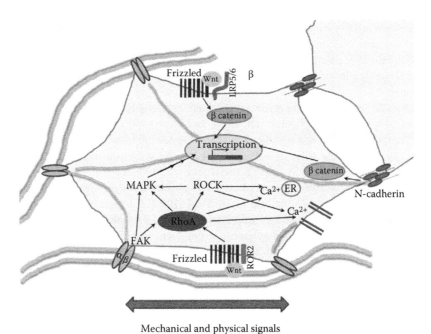

Mechanical and physical signals
(Compression, tension, shear, substrate stiffness, topography,...)

FIGURE 21.2 A diagram illustrating the sensory and signaling pathways implicated in mechanics-induced mesenchymal stem cell (MSC) differentiation (mainly osteogenic). Many signaling pathways (Wnt, MAPKs, Ca^{2+}) have been either directly or indirectly associated with RhoA/ROCK activities, suggesting that cytoskeletal contractility plays an important role in MSC mechanotransduction. ER, endoplasmic reticulum; MAPK, mitogen-activated protein kinase, FAK, focal adhesion kinase; ROCK, RhoA-associated kinase.

since cell–cell contact (density) modulates cell shape and actin filaments interact with adherens junctions, stress propagation to the adherens junctions may trigger β-catenin translocation [17], thereby transducing cell shape signals to MSC commitment.

21.5 SUMMARY AND FUTURE DIRECTIONS

This review highlights recent progress in the study of mechanoregulation of mesenchymal stem cell differentiation and the potential underlying mechanisms involved. Though understanding of mechanotransduction is still limited, it appears that most physical signals are sensed and transduced via similar molecular machinery, and that mechanical and biochemical signaling pathways collaboratively contribute to the overall signaling network.

Mechanical forces are important regulators of MSC-mediated tissue homeostasis. The study of MSC mechanobiology furthers our understanding of how mechanical stress dysregulation impacts MSC-mediated pathological development and facilitates the engineering of the appropriate microenvironmental cues to direct MSC specification for tissue engineering applications. However, the molecular details of how physical forces are sensed and subsequently transduced into biochemical cascades that impact MSC fate decision is ill defined. The delineation of complex cell behavior arisen from the interactions of multiple signals (i.e., biochemical and physical) necessitates a paradigm shift from reductionism to a "systems biology" approach. In order to realize the regenerative potential of MSCs, many challenges need to be overcome and the important few are outlined below.

21.5.1 INTEGRATION OF MULTIPLE SIGNALS

A deeper understanding of how mechanical loading, material properties and biochemical factors collectively regulate MSC differentiation necessitates a systematic investigation of large permutations of experimental conditions. High-throughput array platforms with the capacity to manipulate various cell-biomaterial [117–120] and cell-ECM protein [121] interactions have been accomplished. The development of high-density miniaturized devices capable of delivering defined dynamic mechanical stimulation (2D and 3D) in combination with biochemical factors could elucidate the interplay between multiple stimuli. These screening studies may help to identify the optimal combination of microenvironmental cues that can predictably guide stem cell differentiation, and the knowledge obtained will be applicable for the macroscale generation of functional tissue constructs.

21.5.2 MECHANOSENSING AND MECHANOTRANSDUCTION

Molecular activation or inhibition of signaling pathways is often conveyed through protein conformational changes; therefore, protein mechanics—the spatial and temporal regulation of force-induced structural changes—is crucial to the mechanistic understanding of mechanotransduction. The extent of protein deformation depends on protein structure. Most of the currently known force sensitive proteins have β-sheets for secondary structures. For example, both tensile and contractile forces have been implicated in the unfolding of fibronectin, thereby exposing cryptic binding sites for ECM–integrin interactions [122]. The understanding of protein mechanics is important for the design of ECM-mimetic motifs that are permissive to MSC mechanosensing.

Nano-topographical cues, which are comparable in size to proteins, may also regulate MSC responses by modulating protein conformation. The study of protein mechanics at this physiologically relevant length scale may provide added insights into how mechanics-mediated changes in protein conformation can activate downstream signaling pathways. However, the probing of subtle molecular deformation (0.1–1 nm) in a short reaction time (nanoseconds to seconds) without compromising the native protein conformation is extremely challenging [122] and requires the development of sophisticated molecular probes and imaging techniques.

REFERENCES

1. Bianco, P., P.G. Robey, and P.J. Simmons, Mesenchymal stem cells: Revisiting history, concepts, and assays. *Cell Stem Cell*, 2008, **2**(4): 313–319.
2. Zayzafoon, M., W.E. Gathings, and J.M. McDonald, Modeled microgravity inhibits osteogenic differentiation of human mesenchymal stem cells and increases adipogenesis. *Endocrinology*, 2004, **145**(5): 2421–2432.
3. Meyers, V.E. et al., Modeled microgravity disrupts collagen I/integrin signaling during osteoblastic differentiation of human mesenchymal stem cells. *J Cell Biochem*, 2004, **93**(4): 697–707.
4. Chen, J.H. et al., Identification and characterization of aortic valve mesenchymal progenitor cells with robust osteogenic calcification potential. *Am J Pathol*, 2009, **174**(3): 1109–1119.
5. Yip, C.Y. et al., Calcification by valve interstitial cells is regulated by the stiffness of the extracellular matrix. *Arterioscler Thromb Vasc Biol*, 2009, **29**(6): 936–942.
6. Ethier, C.R. and C.A. Simmons, *Introductory Biomechanics: From Cells to Organisms. Cambridge Texts in Biomedical Engineering*, 2007, Cambridge: Cambridge University Press.
7. Porter, B. et al., 3-D computational modeling of media flow through scaffolds in a perfusion bioreactor. *J Biomech*, 2005, **38**(3): 543–549.
8. Brown, T.D., Techniques for mechanical stimulation of cells in vitro: A review. *J Biomech*, 2000, **33**(1): 3–14.
9. Ward, D.F. Jr. et al., Mechanical strain enhances extracellular matrix-induced gene focusing and promotes osteogenic differentiation of human mesenchymal stem cells through an extracellular-related kinase-dependent pathway. *Stem Cells Dev*, 2007, **16**(3): 467–480.
10. David, V. et al., Mechanical loading down-regulates peroxisome proliferator-activated receptor gamma in bone marrow stromal cells and favors osteoblastogenesis at the expense of adipogenesis. *Endocrinology*, 2007, **148**(5): 2553–2562.
11. Sen, B. et al., Mechanical strain inhibits adipogenesis in mesenchymal stem cells by stimulating a durable beta-catenin signal. *Endocrinology*, 2008, **149**(12): 6065–6075.
12. Koike, M. et al., Effects of mechanical strain on proliferation and differentiation of bone marrow stromal cell line ST2. *J Bone Miner Metab*, 2005, **23**(3): 219–225.
13. Simmons, C.A. et al., Cyclic strain enhances matrix mineralization by adult human mesenchymal stem cells via the extracellular signal-regulated kinase (ERK1/2) signaling pathway. *J Biomech*, 2003, **36**(8): 1087–1096.
14. Qi, M.C. et al., Mechanical strain induces osteogenic differentiation: Cbfa1 and Ets-1 expression in stretched rat mesenchymal stem cells. *Int J Oral Maxillofac Surg*, 2008, **37**(5): 453–458.
15. Sumanasinghe, R.D., S.H. Bernacki, and E.G. Loboa, Osteogenic differentiation of human mesenchymal stem cells in collagen matrices: Effect of uniaxial cyclic tensile strain on bone morphogenetic protein (BMP-2) mRNA expression. *Tissue Eng*, 2006, **12**(12): 3459–3465.
16. Li, Y.J. et al., Oscillatory fluid flow affects human marrow stromal cell proliferation and differentiation. *J Orthop Res*, 2004, **22**(6): 1283–1289.
17. Arnsdorf, E.J., P. Tummala, and C.R. Jacobs, Non-canonical Wnt signaling and N-cadherin related beta-catenin signaling play a role in mechanically induced osteogenic cell fate. *PLoS ONE*, 2009. **4**(4): e5388.
18. Arnsdorf, E.J. et al., Mechanically induced osteogenic differentiation—The role of RhoA, ROCKII and cytoskeletal dynamics. *J Cell Sci*, 2009, **122**(Pt 4): 546–553.
19. Knippenberg, M. et al., Adipose tissue-derived mesenchymal stem cells acquire bone cell-like responsiveness to fluid shear stress on osteogenic stimulation. *Tissue Eng*, 2005, **11**(11–12): 1780–1788.
20. Sharp, L.A., Y.W. Lee, and A.S. Goldstein, Effect of low-frequency pulsatile flow on expression of osteoblastic genes by bone marrow stromal cells. *Ann Biomed Eng*, 2009, **37**(3): 445–453.
21. Kreke, M.R. et al., Effect of intermittent shear stress on mechanotransductive signaling and osteoblastic differentiation of bone marrow stromal cells. *Tissue Eng Part A*, 2008, **14**(4): 529–537.
22. Weinbaum, S., S.C. Cowin, and Y. Zeng, A model for the excitation of osteocytes by mechanical loading-induced bone fluid shear stresses. *J Biomech*, 1994, **27**(3): 339–360.
23. Grellier, M. et al., Responsiveness of human bone marrow stromal cells to shear stress. *J Tissue Eng Regen Med*, 2009, **3**(4): 302–309.
24. Kreke, M.R. and A.S. Goldstein, Hydrodynamic shear stimulates osteocalcin expression but not proliferation of bone marrow stromal cells. *Tissue Eng*, 2004, **10**(5–6): 780–788.
25. Kim, S.H. et al., ERK 1/2 activation in enhanced osteogenesis of human mesenchymal stem cells in poly(lactic-glycolic acid) by cyclic hydrostatic pressure. *J Biomed Mater Res A*, 2007, **80**(4): 826–836.

26. Meinel, L. et al., Bone tissue engineering using human mesenchymal stem cells: Effects of scaffold material and medium flow. *Ann Biomed Eng*, 2004, **32**(1): 112–122.
27. Datta, N. et al., In vitro generated extracellular matrix and fluid shear stress synergistically enhance 3D osteoblastic differentiation. *Proc Natl Acad Sci USA*, 2006, **103**(8): 2488–2493.
28. Stiehler, M. et al., Effect of dynamic 3-D culture on proliferation, distribution, and osteogenic differentiation of human mesenchymal stem cells. *J Biomed Mater Res A*, 2009, **89**(1): 96–107.
29. Zhao, F., R. Chella, and T. Ma, Effects of shear stress on 3-D human mesenchymal stem cell construct development in a perfusion bioreactor system: Experiments and hydrodynamic modeling. *Biotechnol Bioeng*, 2007, **96**(3): 584–595.
30. Sikavitsas, V.I. et al., Mineralized matrix deposition by marrow stromal osteoblasts in 3D perfusion culture increases with increasing fluid shear forces. *Proc Natl Acad Sci USA*, 2003, **100**(25): 14683–14688.
31. Gurkan, U.A. and O. Akkus, The mechanical environment of bone marrow: A review. *Ann Biomed Eng*, 2008, **36**(12): 1978–1991.
32. Mauck, R.L. et al., Regulation of cartilaginous ECM gene transcription by chondrocytes and MSCs in 3D culture in response to dynamic loading. *Biomech Model Mechanobiol*, 2007, **6**(1–2): 113–125.
33. Urban, J.P., The chondrocyte: A cell under pressure. *Br J Rheumatol*, 1994, **33**(10): 901–908.
34. Kisiday, J.D. et al., Dynamic compression stimulates proteoglycan synthesis by mesenchymal stem cells in the absence of chondrogenic cytokines. *Tissue Eng Part A*, 2009, **15**(10): 2817–2824.
35. Wagner, D.R. et al., Hydrostatic pressure enhances chondrogenic differentiation of human bone marrow stromal cells in osteochondrogenic medium. *Ann Biomed Eng*, 2008, **36**(5): 813–820.
36. Angele, P. et al., Cyclic hydrostatic pressure enhances the chondrogenic phenotype of human mesenchymal progenitor cells differentiated in vitro. *J Orthop Res*, 2003, **21**(3): 451–457.
37. Miyanishi, K. et al., Effects of hydrostatic pressure and transforming growth factor-beta 3 on adult human mesenchymal stem cell chondrogenesis in vitro. *Tissue Eng*, 2006, **12**(6): 1419–1428.
38. Finger, A.R. et al., Differential effects on messenger ribonucleic acid expression by bone marrow-derived human mesenchymal stem cells seeded in agarose constructs due to ramped and steady applications of cyclic hydrostatic pressure. *Tissue Eng*, 2007, **13**(6): 1151–1158.
39. Miyanishi, K. et al., Dose- and time-dependent effects of cyclic hydrostatic pressure on transforming growth factor-beta3-induced chondrogenesis by adult human mesenchymal stem cells in vitro. *Tissue Eng*, 2006, **12**(8): 2253–2262.
40. Elder, S.H. et al., Influence of hydrostatic and distortional stress on chondroinduction. *Biorheology*, 2008, **45**(3–4): 479–486.
41. Angele, P. et al., Cyclic, mechanical compression enhances chondrogenesis of mesenchymal progenitor cells in tissue engineering scaffolds. *Biorheology*, 2004, **41**(3–4): 335–346.
42. Terraciano, V. et al., Differential response of adult and embryonic mesenchymal progenitor cells to mechanical compression in hydrogels. *Stem Cells*, 2007, **25**(11): 2730–2738.
43. Pelaez, D., C.Y. Huang, and H.S. Cheung, Cyclic compression maintains viability and induces chondrogenesis of human mesenchymal stem cells in fibrin gel scaffolds. *Stem Cells Dev*, 2008, **18**(1): 93–102.
44. Haudenschild, A.K. et al., Pressure and distortion regulate human mesenchymal stem cell gene expression. *Ann Biomed Eng*, 2009, **37**(3): 492–502.
45. Li, Z. et al., Mechanical load modulates chondrogenesis of human mesenchymal stem cells through the TGF-beta pathway. *J Cell Mol Med*, 2009, **14**(6A): 1338–1346.
46. Campbell, J.J., D.A. Lee, and D.L. Bader, Dynamic compressive strain influences chondrogenic gene expression in human mesenchymal stem cells. *Biorheology*, 2006, **43**(3–4): 455–470.
47. Huang, C.Y. et al., Effects of cyclic compressive loading on chondrogenesis of rabbit bone-marrow derived mesenchymal stem cells. *Stem Cells*, 2004, **22**(3): 313–323.
48. Thorpe, S.D. et al., Dynamic compression can inhibit chondrogenesis of mesenchymal stem cells. *Biochem Biophys Res Commun*, 2008, **377**(2): 458–462.
49. Huang, C.Y., P.M. Reuben, and H.S. Cheung, Temporal expression patterns and corresponding protein inductions of early responsive genes in rabbit bone marrow-derived mesenchymal stem cells under cyclic compressive loading. *Stem Cells*, 2005, **23**(8): 1113–1121.
50. Park, J.S. et al., Differential effects of equiaxial and uniaxial strain on mesenchymal stem cells. *Biotechnol Bioeng*, 2004, **88**(3): 359–368.
51. Wang, J.H. and B.P. Thampatty, Mechanobiology of adult and stem cells. *Int Rev Cell Mol Biol*, 2008, **271**: 301–346.
52. Kurpinski, K. et al., Anisotropic mechanosensing by mesenchymal stem cells. *Proc Natl Acad Sci USA*, 2006, **103**(44): 16095–16100.

53. Hamilton, D.W., T.M. Maul, and D.A. Vorp, Characterization of the response of bone marrow-derived progenitor cells to cyclic strain: Implications for vascular tissue-engineering applications. *Tissue Eng*, 2004, **10**(3–4): 361–369.
54. Neidlinger-Wilke, C., E.S. Grood, J. H.-C. Wang, R.A. Brand, and L. Claes, Cell alignment is induced by cyclic changes in cell length: Studies of cells grown in cyclically stretched substrates. *J Orthop Res*, 2001, **19**(2): 286–293.
55. Zhang, L. et al., Time-related changes in expression of collagen types I and III and of tenascin-C in rat bone mesenchymal stem cells under co-culture with ligament fibroblasts or uniaxial stretching. *Cell Tissue Res*, 2008, **332**(1): 101–109.
56. Zhang, L. et al., Effect of uniaxial stretching on rat bone mesenchymal stem cell: Orientation and expressions of collagen types I and III and tenascin-C. *Cell Biol Int*, 2008, **32**(3): 344–352.
57. Chen, Y.J. et al., Effects of cyclic mechanical stretching on the mRNA expression of tendon/ligament-related and osteoblast-specific genes in human mesenchymal stem cells. *Connect Tissue Res*, 2008, **49**(1): 7–14.
58. Kurpinski, K. et al., Regulation of vascular smooth muscle cells and mesenchymal stem cells by mechanical strain. *Mol Cell Biomech*, 2006, **3**(1): 21–34.
59. Kobayashi, N. et al., Mechanical stress promotes the expression of smooth muscle-like properties in marrow stromal cells. *Exp Hematol*, 2004, **32**(12): 1238–1245.
60. Gong, Z. and L.E. Niklason, Small-diameter human vessel wall engineered from bone marrow-derived mesenchymal stem cells (hMSCs). *FASEB J*, 2008, **22**(6): 1635–1648.
61. Kuo, C.K. and R.S. Tuan, Mechanoactive tenogenic differentiation of human mesenchymal stem cells. *Tissue Eng Part A*, 2008, **14**(10): 1615–1627.
62. Altman, G.H. et al., Cell differentiation by mechanical stress. *FASEB J*, 2002, **16**(2): 270–272.
63. Butler, D.L. et al., Functional tissue engineering for tendon repair: A multidisciplinary strategy using mesenchymal stem cells, bioscaffolds, and mechanical stimulation. *J Orthop Res*, 2008, **26**(1): 1–9.
64. Juncosa-Melvin, N. et al., Mechanical stimulation increases collagen type I and collagen type III gene expression of stem cell-collagen sponge constructs for patellar tendon repair. *Tissue Eng*, 2007, **13**(6): 1219–1226.
65. Wang, H. et al., Shear stress induces endothelial differentiation from a murine embryonic mesenchymal progenitor cell line. *Arterioscler Thromb Vasc Biol*, 2005, **25**(9): 1817–1823.
66. Wang, H. et al., Fluid shear stress regulates the expression of TGF-beta1 and its signaling molecules in mouse embryo mesenchymal progenitor cells. *J Surg Res*, 2008, **150**(2): 266–270.
67. Wu, C.C. et al., Synergism of biochemical and mechanical stimuli in the differentiation of human placenta-derived multipotent cells into endothelial cells. *J Biomech*, 2008, **41**(4): 813–821.
68. Bhadriraju, K. and L.K. Hansen, Hepatocyte adhesion, growth and differentiated function on RGD-containing proteins. *Biomaterials*, 2000, **21**(3): 267–272.
69. McBeath, R. et al., Cell shape, cytoskeletal tension, and RhoA regulate stem cell lineage commitment. *Dev Cell*, 2004, **6**(4): 483–495.
70. Ruiz, S.A. and C.S. Chen, Emergence of patterned stem cell differentiation within multicellular structures. *Stem Cells*, 2008, **26**(11): 2921–2927.
71. Sniadecki, N.J. et al., Nanotechnology for cell-substrate interactions. *Ann Biomed Eng*, 2006, **34**(1): 59–74.
72. Dalby, M.J. et al., The control of human mesenchymal cell differentiation using nanoscale symmetry and disorder. *Nat Mater*, 2007, **6**(12): 997–1003.
73. Oh, S. et al., Stem cell fate dictated solely by altered nanotube dimension. *Proc Natl Acad Sci USA*, 2009, **106**(7): 2130–5.
74. Popat, K.C. et al., Osteogenic differentiation of marrow stromal cells cultured on nanoporous alumina surfaces. *J Biomed Mater Res A*, 2007, **80**(4): 955–964.
75. Yim, E.K., S.W. Pang, and K.W. Leong, Synthetic nanostructures inducing differentiation of human mesenchymal stem cells into neuronal lineage. *Exp Cell Res*, 2007, **313**(9): 1820–1829.
76. Yip, C.Y., J.H. Chen, and C.A. Simmons, Engineering substrate mechanics to regulate cell response, in *Micro- and Nanoengineering of the Cell Microenvironment: Technologies and Applications*, A. Khademhosseini, Editor, 2008, London: Artech House, pp. 161–178.
77. Engler, A.J. et al., Matrix elasticity directs stem cell lineage specification. *Cell*, 2006. **126**(4): 677–689.
78. Winer, J.P. et al., Bone marrow-derived human mesenchymal stem cells become quiescent on soft substrates but remain responsive to chemical or mechanical stimuli. *Tissue Eng Part A*, 2009, **15**(1): 147–154.

79. Rowlands, A.S., P.A. George, and J.J. Cooper-White, Directing osteogenic and myogenic differentiation of MSCs: Interplay of stiffness and adhesive ligand presentation. *Am J Physiol Cell Physiol*, 2008, **295**(4): C1037–C1044.

80. Ingber, D.E., Tensegrity II. How structural networks influence cellular information processing networks. *J Cell Sci*, 2003, **116**(Pt 8): 1397–1408.

81. Vogel, V. and M. Sheetz, Local force and geometry sensing regulate cell functions. *Nat Rev Mol Cell Biol*, 2006, **7**(4): 265–275.

82. Lo, C.M. et al., Cell movement is guided by the rigidity of the substrate. *Biophys J*, 2000, **79**(1): 144–152.

83. Lee, J. et al., Regulation of cell movement is mediated by stretch-activated calcium channels. *Nature*, 1999, **400**(6742): 382–386.

84. Janmey, P.A., The cytoskeleton and cell signaling: Component localization and mechanical coupling. *Physiol Rev*, 1998, **78**(3): 763–781.

85. Wozniak, M. et al., Mechanically strained cells of the osteoblast lineage organize their extracellular matrix through unique sites of alphavbeta3-integrin expression. *J Bone Miner Res*, 2000, **15**(9): 1731–1745.

86. Friedland, J.C., M.H. Lee, and D. Boettiger, Mechanically activated integrin switch controls alpha5beta1 function. *Science*, 2009, **323**(5914): 642–644.

87. Plotkin, L.I. et al., Mechanical stimulation prevents osteocyte apoptosis: Requirement of integrins, Src kinases, and ERKs. *Am J Physiol Cell Physiol*, 2005, **289**(3): C633–C643.

88. Rajagopalan, P. et al., Direct comparison of the spread area, contractility, and migration of balb/c 3T3 fibroblasts adhered to fibronectin- and RGD-modified substrata. *Biophys J*, 2004, **87**(4): 2818–2827.

89. Jansen, J.H. et al., Stretch-induced phosphorylation of ERK1/2 depends on differentiation stage of osteoblasts. *J Cell Biochem*, 2004, **93**(3): 542–551.

90. Hsiong, S.X. et al., Differentiation stage alters matrix control of stem cells. *J Biomed Mater Res A*, 2008, **85**(1): 145–156.

91. Titushkin, I. and M. Cho, Modulation of cellular mechanics during osteogenic differentiation of human mesenchymal stem cells. *Biophys J*, 2007, **93**(10): 3693–3702.

92. Chen, H. et al., Altered membrane dynamics of quantum dot-conjugated integrins during osteogenic differentiation of human bone marrow derived progenitor cells. *Biophys J*, 2007, **92**(4): 1399–1408.

93. Rodriguez, J.P. et al., Cytoskeletal organization of human mesenchymal stem cells (MSC) changes during their osteogenic differentiation. *J Cell Biochem*, 2004, **93**(4): 721–731.

94. Ingber, D.E., Tensegrity I. Cell structure and hierarchical systems biology. *J Cell Sci*, 2003, **116**(Pt 7): 1157–1173.

95. Wang, N., J.P. Butler, and D.E. Ingber, Mechanotransduction across the cell surface and through the cytoskeleton. *Science*, 1993, **260**(5111): 1124–1127.

96. Ingber, D.E., Mechanical signaling and the cellular response to extracellular matrix in angiogenesis and cardiovascular physiology. *Circ Res*, 2002, **91**(10): 877–887.

97. Wang, N., J.D. Tytell, and D.E. Ingber, Mechanotransduction at a distance: Mechanically coupling the extracellular matrix with the nucleus. *Nat Rev Mol Cell Biol*, 2009, **10**(1): 75–82.

98. Wang, N. et al., Mechanical behavior in living cells consistent with the tensegrity model. *Proc Natl Acad Sci USA*, 2001, **98**(14): 7765–7770.

99. Na, S. et al., Rapid signal transduction in living cells is a unique feature of mechanotransduction. *Proc Natl Acad Sci USA*, 2008, **105**(18): 6626–6631.

100. Maniotis, A.J., C.S. Chen, and D.E. Ingber, Demonstration of mechanical connections between integrins, cytoskeletal filaments, and nucleoplasm that stabilize nuclear structure. *Proc Natl Acad Sci USA*, 1997, **94**(3): 849–854.

101. Thomas, C.H. et al., Engineering gene expression and protein synthesis by modulation of nuclear shape. *Proc Natl Acad Sci USA*, 2002, **99**(4): 1972–1977.

102. Dahl, K.N., A.J. Ribeiro, and J. Lammerding, Nuclear shape, mechanics, and mechanotransduction. *Circ Res*, 2008, **102**(11): 1307–1318.

103. Hu, S. et al., Prestress mediates force propagation into the nucleus. *Biochem Biophys Res Commun*, 2005, **329**(2): 423–428.

104. Bhadriraju, K. et al., Activation of ROCK by RhoA is regulated by cell adhesion, shape, and cytoskeletal tension. *Exp Cell Res*, 2007, **313**(16): 3616–3623.

105. Olson, M.F., Contraction reaction: Mechanical regulation of Rho GTPase. *Trends Cell Biol*, 2004, **14**(3): 111–114.

106. Khatiwala, C.B. et al., ECM compliance regulates osteogenesis by influencing MAPK signaling downstream of RhoA and ROCK. *J Bone Miner Res*, 2009, **24**(5): 886–898.

107. Haudenschild, D.R., D.D. D'Lima, and M.K. Lotz, Dynamic compression of chondrocytes induces a Rho kinase-dependent reorganization of the actin cytoskeleton. *Biorheology*, 2008, **45**(3–4): 219–228.
108. Meyers, V.E. et al., RhoA and cytoskeletal disruption mediate reduced osteoblastogenesis and enhanced adipogenesis of human mesenchymal stem cells in modeled microgravity. *J Bone Miner Res*, 2005, **20**(10): 1858–1866.
109. Khatiwala, C.B. et al., The regulation of osteogenesis by ECM rigidity in MC3T3-E1 cells requires MAPK activation. *J Cell Physiol*, 2007, **211**(3): 661–672.
110. Khatiwala, C.B., S.R. Peyton, and A.J. Putnam, Intrinsic mechanical properties of the extracellular matrix affect the behavior of pre-osteoblastic MC3T3-E1 cells. *Am J Physiol Cell Physiol*, 2006, **290**(6): C1640–C1650.
111. Chen, C.S. and D.E. Ingber, Tensegrity and mechanoregulation: From skeleton to cytoskeleton. *Osteoarthr Cartilage*, 1999, **7**(1): 81–94.
112. Sun, S. et al., Physical manipulation of calcium oscillations facilitates osteodifferentiation of human mesenchymal stem cells. *FASEB J*, 2007, **21**(7): 1472–1480.
113. Kim, T.J. et al., Substrate rigidity regulates Ca2+ oscillation via RhoA pathway in stem cells. *J Cell Physiol*, 2009, **218**(2): 285–293.
114. Riddle, R.C. et al., MAP kinase and calcium signaling mediate fluid flow-induced human mesenchymal stem cell proliferation. *Am J Physiol Cell Physiol*, 2006, **290**(3): C776–C784.
115. Turner, N.J. et al., Cyclic stretch-induced TGFbeta1/Smad signaling inhibits adipogenesis in umbilical cord progenitor cells. *Biochem Biophys Res Commun*, 2008, **377**(4): 1147–1151.
116. Davis, L.A. and N.I. Zur Nieden, Mesodermal fate decisions of a stem cell: The Wnt switch. *Cell Mol Life Sci*, 2008, **65**(17): 2658–2674.
117. Neuss, S. et al., Assessment of stem cell/biomaterial combinations for stem cell-based tissue engineering. *Biomaterials*, 2008, **29**(3): 302–313.
118. Anderson, D.G., S. Levenberg, and R. Langer, Nanoliter-scale synthesis of arrayed biomaterials and application to human embryonic stem cells. *Nat Biotechnol*, 2004, **22**(7): 863–866.
119. Anderson, D.G. et al., Biomaterial microarrays: Rapid, microscale screening of polymer-cell interaction. *Biomaterials*, 2005, **26**(23): 4892–4897.
120. Albrecht, D.R. et al., Photo- and electropatterning of hydrogel-encapsulated living cell arrays. *Lab Chip*, 2005, **5**(1): 111–118.
121. Flaim, C.J., S. Chien, and S.N. Bhatia, An extracellular matrix microarray for probing cellular differentiation. *Nat Methods*, 2005, **2**(2): 119–125.
122. Bao, G., Protein mechanics: A new frontier in biomechanics. *Exp Mech*, 2009, **49**(1): 153–164.

22 The Use of Microfluidic Technology in Mechanobiology Research

Brittany Ho McGowan, Sachin Jambovane,
Jong Wook Hong, and Jiro Nagatomi

CONTENTS

22.1 INTRODUCTION

While the amount of research examining the effects of mechanical force stimuli on biological responses of various cells is growing, the understanding is incomplete due to limited approaches. To perform rigorous investigations of complex biological systems, it is desirable to examine the cells in question under a wide range of experimental conditions. Due to the limited capabilities of commercially available systems, investigators often design and fabricate custom setups, which can be costly and difficult to replicate. Another issue associated with the investigation of mechanobiology is that the assessment of molecular-level responses using techniques, such as immunoblotting, requires a large quantity of protein samples from each experiment and thus requires a large number of cells. This too can be costly with large quantities of cells and culture reagents needed. Because of the parallel processing ability and high-throughput nature at relatively lower costs of microfluidics, several attempts have been made in applying microscale fluid formats toward mechanobiology research. While it is still relatively new technology, researchers are beginning to take advantage of the numerous appealing features of microfluidics in the culturing of hepatocytes, epithelial cells, and bone cells, as well as in the investigation of the effects of fluid flow on vascular, renal, skeletal, and stem cells. This chapter reviews the basics of microfluidics, design criteria for devices in biological applications, and the current literature in mechanobiology research using microfluidic devices.

22.2 MICROFLUIDIC DEVICES

22.2.1 Definition and History

Microfluidics is a multidisciplinary area, encompassing physics, chemistry, engineering, and biotechnology that studies the behavior of fluids at volumes thousands of times smaller than a common droplet. More precisely, microfluidics is the science of designing, manufacturing, and formulating devices and processes that deal with volumes of the order of nano-(10^{-9}) to femto-(10^{-18}) L. Typical advantages of microfluidics include lower consumption of sample volumes, reduction in the cost of reagents, faster reaction times, higher process control, and the ability to combine multiple steps on a single device [1,2]. Integration is a key part of microfluidic devices, allowing the automation and incorporation of many biological or chemical processes using only a small amount of material [3]. However, the field of microfluidics is still relatively young and exploiting the advantages above will require further technology growth [3].

The development of microfluidics started three decades ago with the development of gas chromatography and the inkjet printer [1]. Since then, it has grown rapidly to provide applications in chromatography, electrophoretic separation systems, electroosmotic pumping systems, diffusive separation systems, micromixers, DNA amplification, cytometers, and chemical microreactors [2]. In recent time, microfluidic systems have also attracted diverse biological applications including flow cytometry [4], protein crystallization [5], DNA extraction [6], digital PCR amplification [7], stem cell culture [8], synthetic ecosystems consisting of *Escherichia coli* populations [9], and the determination of enzyme kinetic parameters [10].

22.2.2 Physics and Chemistry of Microfluidics

Because of the small dimensions of the devices, physical and chemical processes that take place on microfluidic circuits/chips exhibit unique characteristics in terms of fluid and mass transport, as well as fluid interactions with material surfaces. This section will review the basic features of physics and chemistry in microfluidics. It is the combination of these traits that provide the many benefits, discussed previously, of this technology in simplifying procedures and reducing overall experimental costs.

22.2.2.1 Laminar Flow

One of the main characteristics of microfluidic devices is the presence of laminar flow. Laminar flow is a well-streamlined flow pattern in which particles move in parallel to the direction of flow. It is distinguished by a parabolic velocity profile (Figure 22.1a) and a Reynolds number less than 2100, which is calculated from a ratio of the inertial force to viscous force. The small dimensions of microfluidic channels decrease the inertial forces and allow viscous forces to dominate resulting in laminar flow. One advantage of laminar flow is that it allows two different fluids to move through the same channel with negligible integration. However, this trait can make mixing liquids a challenge [11].

22.2.2.2 Surface-Area-to-Volume Ratio

The surface-area-to-volume ratio (SAV) is another important parameter in microfluidic devices. In microfluidics, the SAV increases by orders of magnitude with small decreases in channel dimensions. The high SAV helps speed up reaction rates and heat transfer. A high SAV also makes microfluidic devices advantageous for cell culture because the in vivo SAV is normally much higher than that in conventional culture flasks and dishes. However, the efficiency of fluid movements can be reduced at this small scale due to the dominance of surface effects on fluid behavior. For example, a high SAV allows molecules to easily adsorb to the surface of channels.

22.2.2.3 Diffusion

Diffusion is one of the most fundamental molecular processes observed in mass transport. Diffusion describes the dispersion of a group of molecules along a concentration gradient to, over time, become uniformly distributed. In microfluidic devices, mixing of two or more fluids occurs only by the process of diffusion across the boundary of two streams (Figure 22.1b) due to the absence of turbulence in laminar flow conditions. For applications such as analytical chemistry, the ability to make diffusion coefficient measurements for different solutes, by taking advantage of this physical phenomenon is an important technical advantage of microfluidic devices.

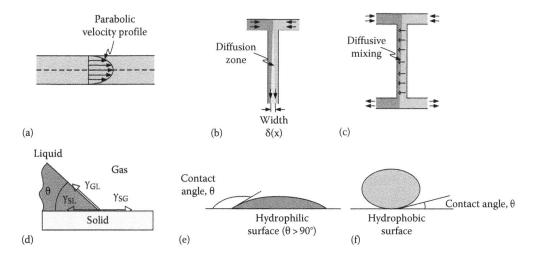

FIGURE 22.1 (See color insert.) Physical basis of microfluidic devices: (a) laminar flow in a microchannel (b) diffusion between two coflowing fluids, (c) diffusive mixing between two coflowing laminar flows, (d) solid–liquid–gas interface, (e) droplet on a hydrophilic surface, and (f) droplet on a hydrophobic surface.

22.2.2.4 Mixing

For the successful development of a fully integrated microfluidic device, often referred to as a "lab-on-a-chip," it is necessary to allow rapid mixing of fluids that are analyzed. Due to the characteristic laminar flow profile, the mixing of fluids is dominated by diffusion, occurring at a slow rate (Figure 22.1c). In the last decade, various mixing approaches, such as chaotic flow micromixers, electrokinetic micromixers, peristaltic micromixers, etc. [11–15] have been designed and implemented to exploit the potential of the microfluidic systems [16].

22.2.2.5 Wetting

Wetting describes the interfacial behavior of a liquid in contact with a solid caused by a balance between the surface energies at the gas liquid and solid interfaces. At a microscale, the high SAV makes the wettability of the microfluidic material very important in the fluid behavior and must be characterized carefully. Fundamentally, wetting occurs at three interfaces: the solid–liquid, gas–liquid, and solid–gas interfaces (Figure 22.1d) with the surface energy corresponding to each interface represented as γ_{SL}, γ_{SG}, and γ_{GL}, respectively. The contact angle, θ, determines the hydrophilicity/hydrophobicity of the material surface (Figure 22.1e and f). The principle of wetting is a critical concept used in digital (droplet) microfluidic devices.

22.2.3 Flow Control in Microfluidic Devices

Fluid flow control is one of the key issues for the successful integration of multiple functions onto a single microfluidic chip. While a number of new approaches are constantly developed, electrokinetic, droplet-based, and pneumatic valve-based strategies are some of the most widely used strategies in highly integrated microfluidic devices.

Electrokinetic flow control consists of two main strategies: electrophoresis and electroosmosis. Electrophoresis describes the migration of charged particles in a fluid under the influence of a uniformly applied electrical field. Electroosmosis, on the other hand, describes the motion of polar liquid due to the combined effect of charged walls and an applied electrical field. In addition, other forms of electrokinetic phenomena such as sedimentation and streaming potentials have been utilized to move particles and fluids in microfluidics [17,18].

On digital microfluidic devices, liquid droplets are manipulated on dielectrically coated surface electrodes by applying an electric field. The electrode surface is usually coated with a hydrophobic material, such as Teflon, to ensure partial wetting with a very small contact angle. When an electric field is applied between two electrodes, charges are created and the electric force moves the droplets from one side to the other. In principle, digital microfluidics involves movement of droplets by the combined effect of electrostatic and dielectrophoresis forces [19,20].

Pneumatic based flow control relies on sequential opening and closing of very small valves ($100\,\mu m \times 100\,\mu m$ area) to move fluids through microfluidic devices. Rigorous characterization and performance improvement of elastomeric microvalves over the past few years [21–23] have demonstrated that the shape of the microchannels plays a critical role in the success of pneumatic valve-based flow control. For example, it has been observed that rectangular and trapezoidal-shaped channels would not permit complete closure of elastomeric microvalves, but a round-shaped cross section has been proven more effective.

22.2.4 Fabrication of Microfluidic Devices

The emergence of microfabrication began with the revolution in the fabrication of integrated circuits [24]. Techniques such as micromachining, soft lithography, negative mold fabrication, and multilayer soft lithography have since then been heavily applied for the fabrication of microfluidic devices.

22.2.4.1 Bulk Micromachining versus Surface Micromachining

The two most widely used methods for the fabrication of microdevices are bulk micromachining and surface micromachining. Bulk micromachining is a subtractive fabrication method whereby silicon is lithographically patterned and material is then etched away to form three-dimensional (3D) structures. Surface micromachining, in contrast, is an additive method where layers of semiconductor-type materials (e.g., polysilicon, metals, silicon oxide, silicon nitride) are sequentially added and patterned to make 3D structures. However, these methods are limited in their capabilities, as it becomes very difficult to accurately machine small-sized channels in rigid materials, like silicon oxide and metals. In addition, the use of multiple layers commonly seen in microfluidic devices requires strong adhesion between layers to prevent separation during machining [23]. For bulk micromachining a wafer-bonding technique is used to hold the layers together. In surface micromachining, the thermal stress created between layers restricts the total device thickness to about 20 μm.

22.2.4.2 Soft Lithography

An alternative microfabrication technique based on replication molding is widespread among many areas of micro- and nanotechnology for microfluidic device development. This technique is commonly known as soft lithography, due to the use of a soft polymer base material [25]. The advantages of lithography include its application in rapid prototyping, ease of fabrication without expensive equipment, and flexible process parameters. The process can be divided into two steps: negative mold fabrication where the structure pattern is established, and replica casting where the polymer is cured over the pattern and later removed to create the final device.

22.2.4.3 Negative Mold Fabrication and Replica Casting

Similar to how a plastic is made in standard manufacturing, a mold is created that is a negative of the final channel design, to allow the polymer to cure as the (positive) structure. The mold can be fabricated by micromachining of silicon [26], LIGA [27], or photolithographic patterning of a photoresist [28]. The creation of inexpensive molds in combination with short polymer curing times allow for rapid fabrication of a single microdevice using soft lithography.

22.2.4.4 Multilayer Soft Lithography

An advanced form of soft lithography known as multilayer soft lithography is a technique to combine multiple layers of patterned structures together by varying the relative composition of a two-component silicone rubber from one layer to another [23]. This process allows multilayered microfluidic devices (up to seven patterned layers, each approximately 40 μm in thickness in the most current study) to be fabricated more easily than conventional silicon-based microfabrication. Another advantage of this technique is that because the layered device is monolithic (same material for all of the layers) interlayer adhesion failures and thermal stress problems are avoided [23].

22.3 DESIGN CRITERIA FOR MICROFLUIDIC DEVICES IN BIOLOGICAL APPLICATIONS

The advantages of microfluidic technology described previously can simplify laboratory techniques and assays and offer similar benefits in the studies of biological systems. Due to the flexibility in soft lithography and other microfluidic fabrication techniques, a large amount of unique designs have been developed for the culturing and manipulation of various cell types (Table 22.1). Although each of these designs is different, they must all consider a few common design criteria that are critical for the successful maintenance of cells and exploration of cell behavior.

TABLE 22.1
Summary of Microfluidic Cell Culture Designs

Cell Type	General Purpose	Controlled Parameters (Ranges)	Advantages (Positives)	Disadvantages (Negatives)	Reference
Hep G2 (human hepatoblastoma), C2C12 (murine myoblasts), endothelial cells (human), MC3T3-E1 (murine preosteoblasts)	Explore hypoxic culture conditions	Channel size: 30–200 μm height Cell density: unknown	Combines rigid and elastic materials, lower media evaporation	Requires precise layer alignment	[34]
Hep G2 (human hepatoblastoma), hepatocytes (human)	Growth/expansion of cells	Channel size: 400×150×400 μm Cell density: 40,000 cells/cm²	Promotes 3D cell arrangement	Overgrowth of cells, occlusion of channels	[36]
Hepatocytes (rat)	Enhance cell functions	Chamber size: 10 mm×10 mm×100 μm Cell density: 500,000 cells/cm²	Scaffold provides attachment site for cell aggregates, media perfusion on both sides of the cells	Tedious fabrication process	[37]
A549 (human epithelial cells), C3A (human liver cells), bone marrow stem cells (rat)	Promote native ECM production	Channel size: 1 cm×600 μm×100 μm Cell density: 5,000,000 cells/cm²	Encourages ECM formation, no need for hydrogel scaffold	Requires tight control of cell seeding density	[38]
MC3T3-E1 (murine osteoblasts)	Enhance cell functions	Channel size: 400×150×100 μm Cell density: 80,000 cells/device	Controllable fluid flow rate and shear stress, higher ALP expression compared with Petri dishes	Initial nonuniform cell growth, cell clusters can block fluid flow	[39]

Cell type	Purpose	Specifications	Advantages	Disadvantages	Ref.
Endothelial cells (human)	Mimic vasculature geometry and produce uniform seeding	Channel size: 175–200 μm diameter Cell density: unknown	Smooth bifurcations, rounded channels, evenly spread cell monolayer	Hard to align channels, more complicated fabrication	[42]
Neural stem cells (human)	Growth and differentiation of cells	Chamber size: 2.4 mm × 100 μm Cell number: 50,000 cells	Applies multiple gradients, cell surface pattern capability	Does not prevent cell crosstalk, limited control over cell migration	[48]
NMuMG (murine epithelial cells)	Growth/expansion of cells	Channel size: 20 mm × 1,000 μm × 250 μm Cell density: 40,000 cells/cm²	Cells demonstrated more 3D-like structures, better representation of in vivo ECM conditions	Structural instability for long-term culture	[57]
Chondrocytes (bovine)	Growth/expansion of cells	Chamber size: 2 mm dia, 200 μm depth Cell density: 8,000 cells/chamber	3D agarose cell scaffold	Response not significantly different from macroculture	[58]
L2 (rat epithelial cells), H4IIE (rat hepatocytes), HepG2 (human hepatoblastoma)	Investigate chemical toxicity	Chamber size: 0.4–3.5 mm × 2–109 mm × 20–100 μm Cell number: 8,000–1,000,000 cells	Mimics different organ systems and circulatory design, simple coculture system	Cell lines do not provide complete organ response	[65]
Hep G2 (human hepatoblastoma)	Large-scale expansion of cells	Channel size: 400 × 400 × 100 μm Cell density: unknown	Incorporation of an oxygen chamber, low media flow rate	More cells and reagents needed, larger sized device	[80]
MLO-Y4 (murine osteocytes)	Mimic microarchitecture of bone	Chamber size: 20 μm dia, 10 μm depth Cell density: 1 cell/chamber	Allows individual cell compartmentalization, mimics in vivo conditions	Time consuming, tedious process	[81]

22.3.1 Selection of Material

The choice of materials for the fabrication of microfluidic devices is very wide. The most common materials used in biological applications include polydimethylsiloxane (PDMS), polystyrene, cyclic olefin copolymers, and borosilicate glass.

22.3.1.1 Polydimethylsiloxane

The most commonly used material in the fabrication of a microfluidic device is PDMS [29]. As one of the simplest and most common silicones, PDMS is comprised of a silicon–oxygen backbone, with side methyl groups (Figure 22.2) [30]. PDMS has been shown to exhibit good insulating properties, is chemically inert, and is nonbiodegradable. Due to these features and its versatility, PDMS is utilized in a wide variety of applications. Low molecular weight PDMS can be used as an antifoamer, surfactant, or lubricant in industrial processes, while high-molecular-weight PDMS is cured to form a rubber, used as an electrical cable insulator, medical tubing, and in cosmetic surgery [31]. Changing the number of repeating units and the degree of cross-linking allows the properties of PDMS to be closely controlled [28,32]. The polymer can be modified to tailor its normally low strength, to be used in a number of biomedical applications including finger joint replacements, prosthetic heart valves, and nose reconstruction [33]. Today, it is the standard material for microfluidics due to its low cost, easy material modification, and ability to cure to other materials. The material transparency of PDMS and its high permeability to gas (including oxygen and carbon dioxide) make it advantageous for cell culture in particular because it simplifies image capturing and cell environment maintenance [34,35]. Due to the presence of the CH_3-group, however, the PDMS surface is inherently hydrophobic and has poor wettability for aqueous solutions. To overcome this issue, the PDMS surface may be treated with oxygen plasma or NaOH solution. Once the PDMS surface is made hydrophilic by these treatments, it can be further treated with adhesive proteins to support culturing of adherent cells.

The use of PDMS microfluidic culture systems has been shown to enhance the cellular activities of hepatocytes [36,37], epithelial cells [38], and osteoblasts [39] when compared with standard well plates. Recently however, the use of this material in long-term cell culture has been challenged. It was found that fibroblasts cultured in PDMS microdevices had significantly lower cell proliferation numbers when compared with cultures on well plates [40]. Observing other activities, such as cell division and extracellular signal-regulated kinase phosphorylation, it was suggested that the use of PDMS microfluidic culture devices can result in cell behavior that is significantly different from that observed in traditional cultures using polystyrene [40]. This change in behavior is possibly due to the absorption of small hydrophobic molecules into the bulk of the polymer. This was evidenced when PDMS channels were flushed with Nile Red (a hydrophobic fluorophore) and washed multiple times with Alconox® and deionized water, and the channels still demonstrated fluorescence, indicating possible absorption/adsorption of the molecules [41]. It can be speculated that the removal of certain critical proteins from culture media could affect the overall cell behavior.

22.3.1.2 Polystyrene

Polystyrene has been investigated as a material for the fabrication of microdevices because it is a common material for conventional cell culture flasks and well plates. Polystyrene (Figure 22.3)

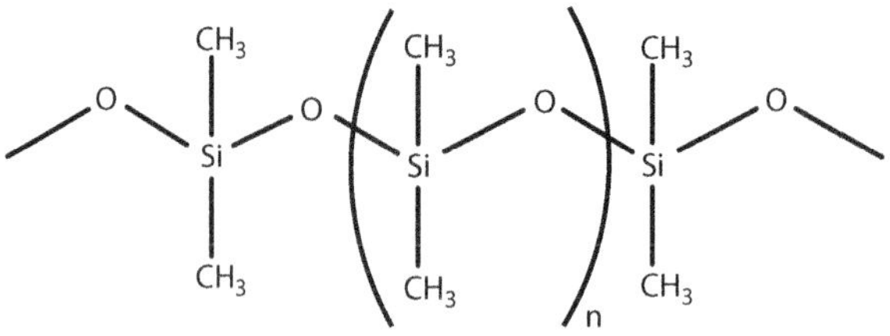

FIGURE 22.2 Chemical structure of PDMS.

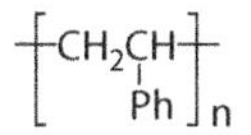

$$\left[\begin{array}{c} CH_2CH \\ | \\ Ph \end{array} \right]_n$$

FIGURE 22.3 Chemical structure of polystyrene. (From *PolymersnetBASE*. Taylor & Francis, Boca Raton, FL, 2010. With permission.)

is a thermoplastic that can be manufactured to be clear or opaque and is typically hard and brittle. In addition to use in research, this polymer is used in everyday products such as electronics, appliances, toys, and kitchenware. It is also frequently used to aid in packaging when converted to a foam-like structure [31]. Within scientific research products, most cell culture and assay substrates are made using polystyrene. Manufacturers of cell culture products (e.g., Integrated BioDiagnostics) have already begun prototyping and developing microfluidic devices using polystyrene to meet the needs of researchers who are increasingly finding the benefits of microculture technology [29]. While not as commonly used as PDMS, polystyrene has been used successfully in microfluidic cultures today. For example, a device to mimic the bifurcated network of vasculature has been fabricated using polystyrene [42]. Channels (~200 μm) with eight "generations" of bifurcations, were successfully incorporated into the polystyrene material by embossing, and primary human endothelial cells formed a viable monolayer within the channels, demonstrating that this material can be used to culture cells on a microlevel [34,42]. Unlike PDMS, interactions of cells with polystyrene have already been well characterized due to its frequent use. There are, however, a few drawbacks to utilizing this material for culture devices as well. The fabrication process such as hot embossing or injection molding is more expensive and complicated than the standard lithography techniques used for PDMS devices [34,41,42]. Polystyrene is also more rigid than the extracellular matrix that many cells normally attach to in the body, limiting the cell response implications for in vivo behaviors [42].

22.3.1.3 Cyclic Olefin Copolymers

Another material used in the fabrication of microfluidic cell culture devices is cyclic olefin copolymer. It is formed from the polymerization of ethylene and a cyclic olefin monomer (e.g., cyclopentene, cyclobutene, norbornene; Figure 22.4) through one of two different processes (Figure 22.5) [43,44]. Some of its properties include a high glass-transition temperature, low density, and high chemical resistance [44,45]. This polymer is routinely used in plastic lenses, medical lab equipment, and electronic data storage media [44,45], but can be fabricated into a microfluidic chip through injection molding or photolithography. Like PDMS, a cyclic olefin copolymer offers low water absorption and good optical properties making it an ideal material for biological applications [46]. Although cyclic olefin copolymers have been utilized more frequently as a material to fabricate biosensors (e.g., for monitoring oxygen levels in blood), they have also been successfully molded into fluidic channels to maintain endothelial cell cultures for up to 24 h [34,46].

22.3.1.4 Borosilicate Glass

Like polystyrene, borosilicate glass is a material frequently utilized in conventional cell culture studies. It can be used in combination with PDMS microfluidic channels, providing the bottom surface to which cells can attach [47,48] or it can be fabricated into microfluidic devices to culture cells [35,46,49]. For example, the use of a glass microchannel has been reported for the culturing of

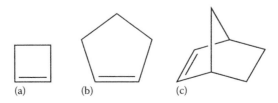

FIGURE 22.4 Examples of cyclic olefin monomers: (a) cyclobutene, (b) cyclopentene, and (c) norbornene. (From Tritto, I. et al., *Coordin. Chem. Rev.*, 250, 212, 2006. With permission.)

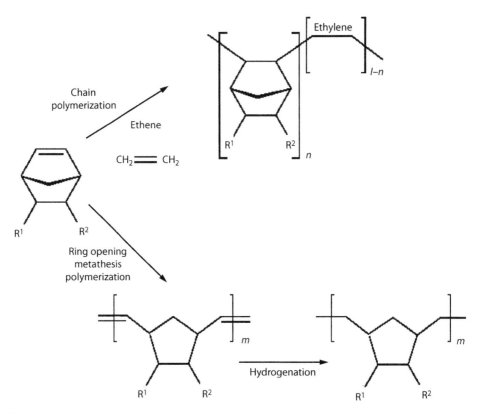

FIGURE 22.5 Polymerization processes for cyclic olefin copolymer. (From Shin, J.Y. et al., *Pure Appl. Chem.*, 77, 801, 2005. With permission.)

osteoblasts and hepatocytes [50,51]. While borosilicate glass provides a biocompatible, rigid, and optically transparent material, its disadvantages include a high cost and need for advanced clean room facilities in device fabrication [29,52,53]. Channels are created by deep etching, or other complex micromachining techniques and the device is sealed with high-temperature treatment [49,54]. The rigidity of borosilicate glass makes maintaining the device shape easier than with other softer materials, such as PDMS. However, like polystyrene, it does not mimic the mechanical properties of the in vivo extracellular matrix well [55]. Another issue associated with the use of borosilicate glass is that, unlike PDMS, it is not gas permeable. For this reason, oxygen must be provided through the culture media to promote cell survival [35].

22.3.1.5 Biologically Based Polymers

In addition to these synthetic materials, biological polymers have been used in microfluidic devices as these materials provide a culture environment similar to the extracellular matrix. For example, alginate, a polysaccharide extracted from brown algae, has been successfully fabricated into microfluidic channels [56]. This polymer is advantageous because it is biocompatible, biodegradable, and has high diffusive permeability [56]. However, the use of alginate requires more complex fabrication steps to bond and seal multiple layers of the polymer structures [56]. Additionally, the degradation of alginate over time may prove to be problematic for long-term maintenance of cells.

Gelatin, a hydrolyzed form of collagen typically extracted from the skin of cows and pigs, has also been used as a material for microfluidic devices. By cross-linking with microbial transglutaminase, gelatin has been shaped into a device to culture murine epithelial cells [57]. The cells cultured in the gelatin microchannels were shown to migrate into the floor of the material and form

3D-like cell extensions after 24 h of culture [57]. Although the use of biologically based polymers in microfluidic devices allows cells to be cultured in a more in vivo-like environment, the requirement for additional fabrication steps to maintain the long-term structural integrity of devices continues to be a challenge.

22.3.1.6 Summary

A number of different materials are currently being investigated for the fabrication of microfluidic devices in biological applications, from hard materials like borosilicate glass and polystyrene, to those that better mimic in vivo conditions, like alginate and cross-linked gelatin. While both borosilicate glass and various polymers have shown success in the culture of different cell types in microfluidic devices, the use of soft polymers, especially PDMS, is preferred due to cost and manufacturability considerations [49]. Despite the controversies that PDMS may influence the experimental outcomes with cells, this polymer continues to be the material of choice for most microfluidic cell culture devices. Some of the recently attempted approaches include, but are not limited to, pretreatment of the cell culture surface with a surfactant solution or extracellular matrix protein in order to control the adsorption of unwanted hydrophobic molecules (e.g., hormones, lipids) that may influence the cell fate [58]. The most common surface modification method consists of an oxygen plasma treatment followed by a coating of fibronectin [49,59–61]. It is important to keep in mind that although culturing of cells on PDMS is just as artificial as that in tissue culture plastic, because of the unique physical features of microfluidic devices it may provide better milieu to elicit the cell responses similar to those under in vivo conditions.

22.3.2 Geometry

Another critical aspect that must be considered when designing a cell culture area is the geometry of the microfluidic device as channel dimensions can have a significant effect on the success of the device. Size of the cell types being investigated, necessary culture media volume, and the fabrication process all must be considered when designing the size and shape of the microfluidic device. Due to the ease of soft lithography techniques, a large number of design options can be implemented into the geometry of a microfluidic device. Most designs fall under two generic geometry categories: channels and chambers.

22.3.2.1 Microfluidic Channels

Microfluidic channel designs can be as simple as one single channel in which cells are seeded, to a whole system of channels with varying diameters and connected through bifurcations. Channel dimensions typically range from 1–6 cm in length by 200–1000 μm in width and 25–250 μm in height [50,57,62]. These channels are frequently rectangular in shape, but some work has introduced round channels to encourage a more uniform cell distribution and to better mimic vasculature geometry [42]. However, fabricating round channels is difficult to achieve using soft lithography techniques, which typically create rectangular cross sections [42]. Instead, an electroplating and embossing method has been used to form half-cylindrical channels with polystyrene, which were then aligned to form the fully cylindrical channels [42]. Although this method allows the formation of round channels with smooth bifurcations, aligning the two layers of half-cylindrical channels with one another presents technical challenges in microfabrication [42].

In addition to varying size and channel shape, channels can incorporate other structures, such as micropillars to aid in entrapment of 3D cell aggregates (Figure 22.6) [38,55]. These were only select examples of experiments that were performed to encourage in vivo-like cell responses through the culture of cells in all three dimensions. The use of microfluidic channels allows researchers to develop a number of other geometrical changes that can enhance cellular functions and the expansion of cells.

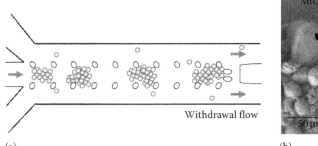
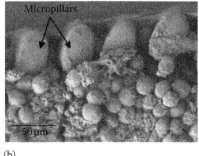

(a)   (b)

FIGURE 22.6 Microfluidic channel incorporating micropillars: (a) schematic of the channel design and (b) SEM of cells cultured in the microfluidic device. (From Ong, S.M. et al., *Biomaterials*, 29(22), 3237, 2008. With permission; Toh, Y.C. et al., *Lab. Chip.*, 7(3), 302, 2007. With permission.)

22.3.2.2 Microfluidic Chambers

In addition to fluidic channels, microfabrication techniques allow the formation of small wells/ chambers in which cells of various organs can be cultured under a range of controlled conditions. These areas typically range in size from 5–10 mm × 5–10 mm × 50–100 μm [37,63,64]. By connecting multiple chambers that carry cells harvested from different tissues/organs, representing different functions, these chambers can be used to model an in vivo system for numerous applications. For example, the effects of a toxin on lung and liver functions can be studied by creating multiple chambers containing a combination of lung and liver cells [65]. Moreover, by mimicking the compartmentalization of different cells, interactions between these cells through endocrine or other signaling molecules could be explored. Subsections can also be formed within a single chamber to encourage cell attachment in certain locations. Microholes and microchambers may be incorporated within a larger microfluidic chamber, allowing cells to grow and attach over one another rather than in a single monolayer. Studies have demonstrated that human and rat hepatocytes cultured in chambers containing microholes, formed into aggregates and a more organized network when compared with hepatocytes in a conventional well plate within 10 days [36,37]. As with the use of microfluidic channels, a number of different designs have been tested utilizing microfluidic chambers. Compared with traditional well plates, the microfluidic chambers allow the experimenters to better tailor the cell culture environment to the specific cells being tested and therefore enhance their cellular function.

22.3.3 NUTRIENT SUPPLY

Cells require a supply of nutrients and removal of metabolic waste for survival. For this reason, culture media and any secreted waste surrounding the cells must be moved regularly through the channels of the microfluidic device. Typically small pumps, including rotary and ultrasonic pumps ranging in rates between 10 μL/min to several mL/min, are used to drive fluid transfer [54]. Other approaches to help move fluid through the microfluidic device include the use of a system of Braille pins [29,34]. This method uses a number of small pins, acting as pneumatic valves, to deform the fluidic membrane and push fluids through the channels (Figure 22.7). A Braille pin array offers more flexibility in controlling fluid movements as it is not limited to flow in one direction [61]. The design of the channel geometries is more restricted though, because the pins are fixed in specific locations in the Braille display and cannot be spaced closely together [61]. Passive methods of fluid movement can also be used, relying on the surface tension of the liquids to move fluids between the ends of the channels. By creating droplets of different sizes, a pressure gradient is created that moves the fluids through the channel. This offers the benefit of little equipment and requires no

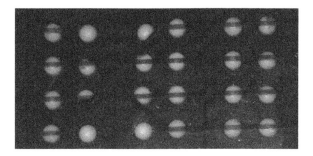

FIGURE 22.7 (See color insert.) A Braille pin system with microfluidic channels colored by green dye. (From Mehta, G. et al., *Anal. Chem.*, 81(10), 3714, 2009. With permission.)

tubing to perform fluidic experiments. However, this technique only works effectively for low flow rates and low pressures [29].

Constant perfusion of culture media is necessary not only to supply nutrients to cells but also to ensure fluid is not lost through evaporation within the microfluidic device over time. Because microfluidic devices utilize such a small amount of fluid, if the culture environment is not properly humidified, a significant amount of moisture can be lost [29,40]. One way to minimize fluid loss through evaporation and to ensure successful cell culture is to create sacrificial water pools as a constant source of humidity within the device [40].

As much as the humidity, the constant supply of oxygen is also crucial for the survival of cells in cultures. When the device is made of PDMS, this factor is not often an area of concern due to its high gas permeability. However, the construction of microfluidic devices with multiple layers of PDMS can limit the diffusion of oxygen and other gases. Therefore, the incorporation of a specific oxygen chamber can increase the survival and activity of cells [36]. This was evidenced by a study that cultured hepatocytes in microfluidic devices consisting of eight layers of PDMS and increased the cell numbers and albumin production when a specific chamber to allow air exchange was incorporated into the system [36].

22.3.4 APPLICATION OF MECHANICAL FORCE STIMULI TO CELLS

Mechanobiology research requires additional design criteria for applying controlled mechanical force stimuli to the cell culture. The simplest and most popular stimulus applied to cells cultured in microfluidic devices is shear stress. The media perfusion to supply nutrients and remove wastes also induces shear stress on the cells, which can be controlled through a change in perfusion rate, channel geometry, or viscosity of the fluids.

For example, by using a peristaltic pump, medium reservoir, and bubble trap in series with a PDMS microfluidic chip, osteoblastic cell cultures were exposed to two different flow rates and multiple shear stress profiles [39]. Another approach utilized a grid of Braille pins to generate different shear stresses by changing the frequency at which the pins would move and pump fluids through the device [60]. The applied shear stress can also be varied by the geometry of the channels in which the cells are cultured. For example, a constant flow rate of 1.25 mL/min applied to channel widths of 500, 750, and 1000 μm resulted in a shear stress of 4000, 2700, and 2000 dyn/cm^2, using Equation 22.1 where τ is the shear stress, Q is the volumetric flow rate, μ is the fluid viscosity, and h and w are the channel height and width, respectively [62]. Another study demonstrated that a gradient of shear stress can be created even within a single channel, by varying the width of the channel across its length (Figure 22.8) [66]. Varying the overall microfluidic chamber and channel geometry, allows a wide range of shear magnitudes to be applied to the cultured cells. By seeding cells well below the flow of media for instance, extremely low shear stresses were reached

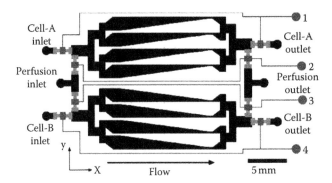

FIGURE 22.8 Geometry design that varies channel width across the device length. (From Gutierrez, E. and Groisman, A., *Anal. Chem.*, 79(6), 2249, 2007. With permission.)

(0.0001 dyn/cm²) compared with the shear stress on cells attached to substrates directly in line with media flow (0.01 dyn/cm²) [47].

$$\tau = \frac{6Q\mu}{h^2 w} \tag{22.1}$$

When the cellular response to fluid flow is studied, it is often difficult to separate the effects of shear stress from the effects of a change in chemotransport. This problem may be partially solved by keeping the media flow rate constant, but varying the viscosity of the fluid by adding dextran in different amounts to create different shear stresses on the cells under investigation [51]. This polysaccharide, naturally produced by bacteria, has shown good biocompatibility and could be utilized in experiments to explore the effects of shear stress without damaging the cells [51,67]. However, there are still limitations to this method because dextran is slowly metabolized by the cells, eventually causing an increase in sugar concentration in the media [67].

In addition to flow-induced shear stress, a mechanical force stimuli commonly investigated in studies of mechanobiology is plane strain. One example is a microdevice that utilized moving pins to deform the substrate on which porcine aortic valvular interstitial cells were seeded (Figure 22.9). Strain was created by inducing culture substrate deformation through the application of pressure or a vacuum [68]. By seeding cells on a thin PDMS membrane, a change in surrounding pressure resulted in flexing of the membrane, exposing the cells to controlled stretch [64].

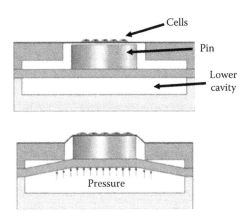

FIGURE 22.9 Microfluidic pin design to create substrate strain. (From Moraes, C. et al., *Lab Chip.*, 10(2), 227. With permission.)

22.4 MECHANOBIOLOGY RESEARCH USING MICRODEVICES

Although microfluidic cell culture technology is relatively new, researchers have already begun taking advantage of its numerous appealing features in order to improve understanding of the mechanobiology of various cells. This technology offers the advantages of requiring fewer cells and lower amounts of expensive reagents than conventional cell cultures, in addition to greater control of mechanical stimulus application. Most studies thus far have explored the effects of fluid flow and the resulting shear stress on the cells, although a few are beginning to investigate the role of strain and pressure in mechanotransduction. Example mechanobiology studies and their advantages over conventional techniques are summarized in Table 22.2.

22.4.1 VASCULAR CELLS

Because of its importance in the understanding of cardiovascular diseases, the effect of fluid shear stress on endothelial cells is one of the most studied areas of research in mechanobiology [53,60]. The conventional approaches to expose cultured endothelial cells to shear stress are parallel plate flow chambers and cone-and-plate viscometers [69]. As described earlier, it is relatively easy to replicate this mechanical environment at a microscale, by creating a PDMS microfluidic device that allows exposure of cultured endothelial cells to different levels of shear stress [60]. One of the first studies in endothelial mechanobiology using microdevices, demonstrated that the application of a shear stress of $9\,dyn/cm^2$ at 1 Hz resulted in cell elongation and alignment in the direction of flow, but this was not seen at a stress of approximately $1\,dyn/cm^2$ [60]. Additionally, stresses between 2.5 and $12\,dyn/cm^2$ at frequencies from 0.25 to 2 Hz also resulted in a change in cell morphology. These results, demonstrating that pulsatile flow may be a key factor in vascular endothelial cell alignment, are in agreement with earlier studies at a macroscale indicating that microsystems provide a powerful and more economic new tool for mechanobiology research [70].

Another mechanosensitive cell in the cardiovascular system is the cardiomyocyte. Like the vascular endothelial cells, it is suggested that fluid flow and shear stress are critical for the proper development of these cells [64]. Moreover, since the heart muscle constantly experiences large deformation, it is also important to examine the effect of stretch. A microfluidic device allows application of both strain and shear stress to determine how the combination of these two stimuli affects cardiomyocyte structure and protein expression [64]. Exposure of cardiomyocytes to the combination of these stimuli was found to increase the number of contractions up to 80 beats per minute (bpm) from the static conditions (30 bpm), but no difference in hypertrophy protein markers was observed between these conditions. Conventional in vitro studies have demonstrated an increase in hypertrophic genetic markers (e.g., atrial natriuretic factor, brain natriuretic peptide) when cardiomyocytes were exposed to 5%–20% stretch [71]. Although the conventional and microfluidic studies indicate contrasting results, the latter examined a different set of markers (e.g., α-smooth muscle actin and phosphorylated phospholamban) and, thus, it is possible that more physiologically relevant signaling pathways are activated when the cells are stimulated at the microscale. It has been argued that the cardiovascular system in vivo is much more complex than can be modeled in vitro [71]. However, microfluidic systems may provide a novel way to model more physiological systems with finer control over the cell culture mechanical environments.

22.4.2 RENAL CELLS

Because of the organ's main function to continuously filter blood, epithelial cells in the kidney are also subjected to constant fluid flow in vivo. Thus investigation of renal cell mechanobiology has been an important topic of research. The application of shear stress to these renal epithelial cells in vitro can lead to the activation of multiple signaling pathways through an increase in intracellular calcium levels. It is hypothesized that the uptake of calcium is activated by the opening of

TABLE 22.2
Summary of Mechanobiology Research Using Microfluidics

Cell Type	Substrate (2D, 3D, etc.)	Stimulus (Ranges)	Output (Cellular/ Molecular Responses)	Advantage over Conventional Setups	Limitations	Reference
MC3T3-E1 (murine osteoblasts)	Fibronectin coated PDMS (3D-like)	Shear stress (5–70 mPa)	ALP expression	Low cell number required	Nonuniform cell spreading, cell overgrowth	[39]
C2C12 (murine myoblasts), myocytes (rat), H09, and H13 (human embryonic stem cells)	Glass (2D), collagen coated glass (2D), hyaluronic acid (3D)	Shear stress (<0.0001–0.1 dyn/cm²)	α-Smooth muscle actin expression	Close control over shear, no interaction between wells	More risk of contamination	[47]
Colla1GFP MC3T3-E1 (murine osteoblasts)	Matrigel modified Tempex borosilicate glass (3D-like)	Shear stress (0.07 dyn/cm²)	ALP expression	3D-like environment, easy application of dynamic culture conditions	Cells form monolayer (still 2D)	[50]
Hep G2 (human hepatoblastoma)	Type 1 collagen-coated quartz glass (2D)	Shear stress (0.14–6 Pa)	Albumin production, cell morphology	Manipulation of shear stress without change in nutrient flow	Use of dextran may lead to change in glucose level	[51]
HDMEC (human endothelial cells)	Fibronectin coated PDMS (2D)	Shear stress (1–12 dyn/ cm² at 0.25–2 Hz)	Cell alignment and shape	Easy seeding of cells into multiple chambers, multiple independent flow loops	Restricted shear level (up to 12 dyn/cm²)	[60]
C2C12 (murine myoblasts)	Fibronectin coated PDMS (2D)	Media flow rate (3–370 nL/min)	Myosin heavy chain response	Flexible fluid movement control, compact equipment	Pins cannot be tightly spaced	[61]
WT NR6 (murine fibroblasts)	Fibronectin coated PDMS (2D)	Shear stress (2000, 2700, and 4000 dyn/ cm²)	Cell detachment	Video capture of cell behavior	High shear values	[62]
Aortic valvular interstitial cells (porcine)	Type I collagen coated PDMS (2D)	Strain (2%–15% at 1 Hz)	β-catenin expression	Uniform strain field, wide range of strains	Potential crosstalk between cells, not suited for real-time imaging	[68]
MDCK (canine epithelial cells)	Glass (2D)	Shear stress (0.54–6 dyn/cm²)	Actin filament assembly, intracellular calcium levels	Shear stress gradient	Potential crosstalk between cells, potential data artifacts	[72]
L2 (rat alveolar), H4IIE (rat hepatoma)	Fibronectin and type I collagen-coated silicon (2D)	Shear stress (0.8– 6.7 dyn/cm²)	Oxygen concentration	Integrated real-time sensor, mimics different organ systems	Cell lines do not provide complete organ response, increased cell detachment	[82]

mechanosensitive ion channels through cytoskeleton filament reorganization. Renal epithelial cells exposed to a gradient of shear stresses ($0.54–6 \, dyn/cm^2$) through a tapered microfluidic channel geometry, demonstrated reorganization of fluorescently stained actin filaments over the course of 18 h [72]. Fluorescent imaging of calcium sensitive dye indicated that a 3 h exposure of kidney epithelial cells to shear stress led to a decrease in calcium sensitivity of these cells to acute flow changes [72]. The results of the microfluidic study were in agreement with previous findings of macroscale cultures that the response of renal epithelial cells to a mechanical signal, including the influx of calcium, is influenced by the prestressed state of the cytoskeleton [73]. More importantly, microfluidic cultures provided the advantage of cost effectiveness and better control of the mechanical parameters for further research on the mechanotransduction pathways of kidney cells.

22.4.3 SKELETAL CELLS

Another area of intense research in mechanobiology has focused on skeletal cells due to the importance of mechanical loading on bone structural integrity. Bone cells have been shown to demonstrate enhanced bone-forming activities in response to a variety of mechanical stimuli (e.g., strain [74], pressure [75]) in conventional in vitro studies. Recently, the research focus has shifted to investigation of the effects of fluid flow on critical cells like osteoblasts and osteocytes [76]. Since the variation of shear stress with high precision is a relatively simple task to implement, microfluidic devices have been used in the more recent exploration of bone cell mechanotransduction mechanisms. When compared between murine osteoblasts cultured in static well plates, versus cells in perfusion microfluidic cultures exposed to shear stress ($0.07 \, dyn/cm^2$), alkaline phosphatase (ALP) expression was ten times higher in the perfusion group [50]. These results were in agreement with another study in which ALP expression was 2.5 times higher for a microfluidic perfusion culture than for a microfluidic static culture [39]. However, it is likely that there is an optimum range of applied shear stress, as a higher perfusion rate ($35 \, \mu L/min$) resulted in cell detachment after 2 days of culture [39]. Studies performed on various skeletal cells in conventional cultures have also shown enhanced expression of osteoblast markers when exposed to fluid flow [76]. The greater control over the fluidic environment in microfluidic cultures provides an advantage over conventional mechanobiology studies, as skeletal cells are also shown to be sensitive to the specific fluid flow profile applied [76,77].

22.4.4 STEM CELLS

The role of mechanical stimuli in stem cell function is frequently investigated as it has been hypothesized that differentiation into specific cell types is regulated by mechanical signals. One study found an increase in vascular differentiation markers in human embryonic stem cells after 4 days of culture in a perfusion micro-bioreactor with a maximum applied shear of $0.1 \, dyn/cm^2$ [47]. These findings are similar to the current understanding in vascular mechanics that suggest applied shear is important for the proper function and differentiation of vascular cells from stem cells [47]. Another study explored the effects of cyclic strain on the expression of β-catenin in aortic valvular interstitial cells using a microfabricated array of pins [68]. These cells contain mesenchymal stem cells in their population and were found to increase their nuclear β-catenin expression when they were subjected to 15% strain for 3 h, compared with static controls [68]. Additionally, when these valvular interstitial cells were exposed to lower strain magnitudes for longer exposure time a significant increase in the nuclear β-catenin expression was observed while the response to 15% strain diminished at this time point. Since this protein has been found to be important in the differentiation of these progenitor cells into osteoblasts, the authors concluded that stem cell differentiation may be enhanced by applied strain in a time- and magnitude-dependent fashion [68]. The use of microdevices to culture stem cells will provide researchers with more control over the environment of these sensitive cells, which will prove to be critical in the progress of the field of tissue engineering.

22.4.5 HEPATOCYTES

While it is relatively easy to design microfluidic devices to expose cultured cells to perfusion, initial work with hepatocytes suggests that shear stress is not always beneficial for cellular functions. An increase in flow-induced shear stress from 0.14 to 6 Pa in a microfluidic liver cell culture led to a decrease in albumin production over a culture of 3 days [51], indicating that beyond a certain level, shear stress can exhibit adverse effects on hepatocyte function. Additionally, the same study demonstrated that exposure of human hepatocytes to shear stresses above 2 Pa led to significant cell loss, suggesting that once the shear stress reaches a critical level the cells are no longer able to remain attached to the substrate [51]. Another study using conventional techniques to expose rat liver tissue to shear stress between 0.5 and 20 dyn/cm^2 (0.05–2 Pa) have demonstrated similar results resulting in higher albumin production under this low range of shear compared with no shear and shear stress higher than 20 dyn/cm^2 [78]. Together, these studies indicate that the use of microfluidic devices is highly effective in exploring liver cell functions under mechanical signals, but may be more powerful than the use of traditional culture setups considering all the positive features of the microscale format.

22.4.6 FIBROBLASTS

The application of shear stress to cultured cells can be important not only for the investigation of its effects on cell function, but can be useful for studying the strength of cell attachment to substrates. By combining shear stress with the use of video microscopy to capture fibroblast motions in real time, cell detachment under different shear stress levels has been investigated [62]. The highest stress of 4000 dyn/cm^2 resulted in a loss of approximately 90% of the cells within 12 min, while only 10% were lost with a shear stress of 2000 dyn/cm^2 [62]. The stress magnitudes recorded to cause cell detachment in microfluidic channels confirm findings using traditional setups [62]. Compared with conventional techniques, which can have a limited range of stress or be labor-intensive, these microdevices allow a simpler and faster way of performing analysis on cell cultures in response to a wide range of shear stress magnitudes [62].

22.5 FUTURE CONSIDERATIONS

In the past decade, a number of technological advancements in the field of microfluidics have led to an increase in molecular analysis and cell culture techniques at the microscale. The current trend with increasing demand for total analysis systems and assays using microfluidic devices, specifically within the medical world, is expected to continue [79]. Although significant progress has been made for microfluidic cell cultures and devices for studies of mechanobiology in the recent years, there is still a large potential for growth. For example, the supplementary equipment (e.g., bubble trap, pumps, valves) for experiments with microfluidic devices is often very bulky and, thus, simplification or miniaturization of the microfluidic system would be highly beneficial [69]. Moreover, very few techniques for quantitative analysis of cell behavior within microfluidic channels currently exist [40]. The most widely used techniques in conventional cell behavior analysis often require a large number of cells and are less sensitive to small changes in expression. Therefore, the small sample size becomes a limiting factor in the analysis of microfluidic cultures. Improvements in analysis techniques for cell cultures of this size will be essential for the progress of this technology. For example, an In Cell Western technique is being developed to quantify proteins in situ using a small number of cells [40]. Further development and standardization of these techniques and general fabrication techniques will help progress this technology into more frequent use in the future.

  Microfluidic technology is still relatively new and further studies to validate the results of experiments using cells cultured in microfluidic devices will be needed [69]. As improvements are made in the design of devices, this technology will increase their utility in the exploration of mechanotransduction pathways for a variety of cell types. While the studies to date have indeed demonstrated that experiments using conventional and microfluidic techniques produce similar cellular responses, the unique features of microfluidics will allow more sophisticated experiments to be performed in a high-throughput and cost-effective manner. Large advances in microfluidic technology have been demonstrated in the recent years and the near future holds exciting promise in its capabilities for studies in mechanobiology.

REFERENCES

1. Gravesen, P., J. Branebjerg, and O. Jensen, Microfluidics—A review. *Journal of Micromechanics and Microengineering*, 1993, **3**: 168.
2. Tabeling, P., *Introduction to Microfluidics*. Oxford University Press, Oxford, U.K., 2005.
3. Whitesides, G., The origins and the future of microfluidics. *Nature*, 2006, **442**(7101): 368–373.
4. Fu, A.Y. et al., An integrated microfabricated cell sorter. *Analytical Chemistry*, 2002, **74**(11): 2451–2457.
5. Hansen, C.L. et al., A robust and scalable microfluidic metering method that allows protein crystal growth by free interface diffusion. *Proceedings of the National Academy of Sciences*, 2002, **99**(26): 16531–16536.
6. Hong, J.W. et al., A nanoliter-scale nucleic acid processor with parallel architecture. *Nature Biotechnology*, 2004, **22**(4): 435–439.
7. Warren, L. et al., Transcription factor profiling in individual hematopoietic progenitors by digital RT-PCR. *Proceedings of the National Academy of Sciences*, 2006, **103**(47): 17807–17812.
8. Gomez-Sjoberg, R. et al., Versatile, Fully automated, microfluidic cell culture system. *Analytical Chemistry*, 2007, **79**(22): 8557–8563.
9. Balagadde, F.K. et al., A synthetic *Escherichia coli* predator-prey ecosystem. *Molecular System Biology*, 2008, **4**: 1–8.
10. Jambovane, S. et al., Determination of kinetic parameters, Km and kcat, with a single experiment on a chip. *Analytical Chemistry*, 2009, **81**(9): 3239–3245.
11. Stroock, A. et al., Chaotic mixer for microchannels. *Science*, 2002, **295**(5555): 647–651.
12. Nguyen, N. and Z. Wu, Micromixers—A review. *Journal of Micromechanics and Microengineering*, 2005, **15**(2): R1–R16.
13. Oddy, M., J. Santiago, and J. Mikkelsen, Electrokinetic instability micromixing. *Analytical Chemistry*, 2001, **73**(24): 5822–5832.
14. Chou, H.P., M.A. Unger, and S.R. Quake, A Microfabricated rotary pump. *Biomedical Microdevices*, 2001, **3**(4): 323–330.
15. Tai, C.H. et al., Micromixer utilizing electrokinetic instability-induced shedding effect. *Electrophoresis*, 2006, **27**(24): 4982–4990.
16. Haeberle, S. and R. Zengerle, Microfluidic platforms for lab-on-a-chip applications. *Lab on a Chip*, 2007, **7**(9): 1094–1110.
17. Li, D., *Electrokinetics in Microfluidics*. Academic Press, Burlington, Germany, 2004.
18. Li, D., *Encyclopedia of Microfluidics and Nanofluidics*. Springer Verlag, Berlin, 2008.
19. Abdelgawad, M. and A. Wheeler, The digital revolution: A new paradigm for microfluidics. *Advanced Materials*, 2009, **21**(8): 920–925.
20. Pollack, M., R. Fair, and A. Shenderov, Electrowetting-based actuation of liquid droplets for microfluidic applications. *Applied Physics Letters*, 2000, **77**: 1725.
21. Kartalov, E.P., W.F. Anderson, and A. Scherer, The analytical approach to polydimethylsiloxane microfluidic technology and its biological applications. *Journal of Nanoscience and Nanotechnology*, 2006, **6**: 2265–2277.
22. Studer, V. et al., Scaling properties of a low-actuation pressure microfluidic valve. *Journal of Applied Physics*, 2004, **95**(1): 393–398.
23. Unger, M.A. et al., Monolithic microfabricated valves and pumps by multilayer soft lithography. *Science*, 2000, **288**(5463): 113.
24. Madou, M.J., *Fundamentals of Microfabrication: The Science of Miniaturization*. CRC Press, Boca Raton, FL, 2002.
25. Xia, Y. and G.M. Whitesides, Soft lithography. *Annual Review of Materials Science*, 1998, **28**(1): 153–184.
26. Madou, M.J., *Fundamentals of Microfabrication*, 1st edn. CRC Press, Boca Raton, 1998, p. 589.
27. Hruby, J., LIGA technologies and applications. *MRS Bulletin*, 2001, **26**(4): 337–340.

28. McDonald, J.C. et al., Fabrication of microfluidic systems in poly(dimethylsiloxane). *Electrophoresis*, 2000, **21**(1): 27–40.
29. Young, E.W. and D.J. Beebe, Fundamentals of microfluidic cell culture in controlled microenvironments. *Chemical Society Reviews*, 2010, **39**(3): 1036–1048.
30. Colas, A. and J. Curtis, Silicone biomaterials: History and chemistry. *Biomaterials Science—An Introduction to Materials in Medicine*, 2nd edn., eds. B.D. Ratner et al.: Elsevier Academic Press, San Diego, CA, 2004.
31. *PolymersnetBASE*. 2010, Taylor & Francis, Boca Raton, FL.
32. Sia, S.K. and G.M. Whitesides, Microfluidic devices fabricated in poly(dimethylsiloxane) for biological studies. *Electrophoresis*, 2003, **24**(21): 3563–3576.
33. Cooper, S.L. et al., Polymers. *Biomaterials Science—An Introduction to Materials in Medicine*, 2nd edn., eds. B.D. Ratner et al.: Elsevier Academic Press, San Diego, CH, 2004.
34. Mehta, G. et al., Hard top soft bottom microfluidic devices for cell culture and chemical analysis. *Analytical Chemistry*, 2009, **81**(10): 3714–3722.
35. Vanapalli, S.A., M.H. Duits, and F. Mugele, Microfluidics as a functional tool for cell mechanics. *Biomicrofluidics*, 2009, **3**(1): 12006.
36. Leclerc, E., Y. Sakai, and T. Fujii, Perfusion culture of fetal human hepatocytes in microfluidic environments. *Biochemical Engineering Journal*, 2004, **20**: 143–148.
37. Ostrovidov, S. et al., Membrane-based PDMS microbioreactor for perfused 3D primary rat hepatocyte cultures. *Biomedical Microdevices*, 2004, **6**(4): 279–287.
38. Ong, S.M. et al., A gel-free 3D microfluidic cell culture system. *Biomaterials*, 2008, **29**(22): 3237–3244.
39. Leclerc, E. et al., Study of osteoblastic cells in a microfluidic environment. *Biomaterials*, 2006, **27**(4): 586–595.
40. Paguirigan, A.L. and D.J. Beebe, From the cellular perspective: Exploring differences in the cellular baseline in macroscale and microfluidic cultures. *Integr Biol (Camb)*, 2009, **1**(2): 182–195.
41. Toepke, M.W. and D.J. Beebe, PDMS absorption of small molecules and consequences in microfluidic applications. *Lab on a Chip*, 2006, **6**(12): 1484–1486.
42. Borenstein, J.T. et al., Functional endothelialized microvascular networks with circular cross-sections in a tissue culture substrate. *Biomedical Microdevices*, 2010, **12**(1): 71–79.
43. Tritto, I., L. Boggioni, and D.R. Ferro, Metallocene catalyzed ethene- and propene co-norbornene polymerization: Mechanisms from a detailed microstructural analysis. *Coordination Chemistry Reviews*, 2006, **250**: 212–241.
44. Shin, J.Y. et al., Chemical structure and physical properties of cyclic olefin copolymers. *Pure and Applied Chemistry*, 2005, **77**: 801–814.
45. *Cyclic olefin co-polymers*. Dow Aromatics Co-Products 2010 [cited April 5, 2010]; Available from: http://www.dow.com/aromatics/prod/app/cyclic.htm
46. Ahn, C.H. et al., Disposable smart lab on a chip for point-of-care clinical diagnostics. *Proceedings of IEEE*, 2004, **92**(1): 154–173.
47. Figallo, E. et al., Micro-bioreactor array for controlling cellular microenvironments. *Lab on a Chip*, 2007, **7**(6): 710–719.
48. Chung, B.G. et al., Human neural stem cell growth and differentiation in a gradient-generating microfluidic device. *Lab on a Chip*, 2005, **5**(4): 401–406.
49. Matsubara, Y. et al., Application of on-chip cell cultures for the detection of allergic response. *Biosensors and Bioelectronics*, 2004, **19**(7): 741–747.
50. Jang, K. et al., Development of an osteoblast-based 3D continuous-perfusion microfluidic system for drug screening. *Analytical and Bioanalytical Chemistry*, 2008, **390**(3): 825–832.
51. Tanaka, Y. et al., Evaluation of effects of shear stress on hepatocytes by a microchip-based system. *Measurement Science and Technology*, 2006, **17**: 3167–3170.
52. El-Ali, J., P.K. Sorger, and K.F. Jensen, Cells on chips. *Nature*, 2006, **442**(7101): 403–411.
53. van der Meer, A.D. et al., Microfluidic technology in vascular research. *Journal of Biomedicine and Biotechnology*, 2009, **2009**: 823148.
54. Franke, T.A. and A. Wixforth, Microfluidics for miniaturized laboratories on a chip. *Chemphyschem*, 2008, **9**(15): 2140–2156.
55. Toh, Y.C. et al., A novel 3D mammalian cell perfusion-culture system in microfluidic channels. *Lab on a Chip*, 2007, **7**(3): 302–309.
56. Cabodi, M. et al., A microfluidic biomaterial. *Journal of the American Chemical Society*, 2005, **127**(40): 13788–13789.

57. Paguirigan, A. and D.J. Beebe, Gelatin based microfluidic devices for cell culture. *Lab on a Chip*, 2006, **6**(3): 407–413.
58. Wu, M.H. et al., Development of PDMS microbioreactor with well-defined and homogenous culture environment for chondrocyte 3-D culture. *Biomedical Microdevices*, 2006, **8**(4): 331–340.
59. Leclerc, E., Y. Sakai, and T. Fujii, Cell culture in 3-dimensional microfluidic structure of PDMS (polydimethylsiloxane). *Biomedical Microdevices*, 2003, **5**(2): 109–114.
60. Song, J.W. et al., Computer-controlled microcirculatory support system for endothelial cell culture and shearing. *Analytical Chemistry*, 2005, **77**(13): 3993–3999.
61. Gu, W. et al., Computerized microfluidic cell culture using elastomeric channels and Braille displays. *Proceedings of the National Academy of Sciences of the United States of America*, 2004, **101**(45): 15861–15866.
62. Lu, H. et al., Microfluidic shear devices for quantitative analysis of cell adhesion. *Analytical Chemistry*, 2004, **76**(18): 5257–5264.
63. Ni, X.F. et al., On-chip differentiation of human mesenchymal stem cells into adipocytes. *Microelectronic Engineering*, 2008, **85**: 1330–1333.
64. Nguyen, M.D. et al., Microfluidic cardiac circulation model (microCCM) for functional cardiomyocyte studies. *Conference Proceedings of the IEEE Engineering in Medicine and Biology Society*, 2009, **2009**: 1060–1063.
65. Viravaidya, K., A. Sin, and M.L. Shuler, Development of a microscale cell culture analog to probe naphthalene toxicity. *Biotechnology Progress*, 2004, **20**(1): 316–323.
66. Gutierrez, E. and A. Groisman, Quantitative measurements of the strength of adhesion of human neutrophils to a substratum in a microfluidic device. *Analytical Chemistry*, 2007, **79**(6): 2249–2258.
67. Baldwin, A.D. and K.L. Kiick, Polysaccharide-modified synthetic polymeric biomaterials. *Biopolymers*, 2010, **94**(1): 128–140.
68. Moraes, C. et al., Microfabricated arrays for high-throughput screening of cellular response to cyclic substrate deformation. *Lab on a Chip*, 2010, **10**(2): 227–234.
69. Young, E.W. and C.A. Simmons, Macro- and microscale fluid flow systems for endothelial cell biology. *Lab on a Chip*. 2010, **10**(2): 143–160.
70. Hsiai, T.K. et al., Endothelial cell dynamics under pulsating flows: Significance of high versus low shear stress slew rates ($d(tau)/dt$). *Annals of Biomedical Engineering*, 2002, **30**(5): 646–656.
71. Shyu, K.G., Cellular and molecular effects of mechanical stretch on vascular cells and cardiac myocytes. *Clinical Science (London)*, 2009, **116**(5): 377–389.
72. Wang, J., J. Heo, and S.Z. Hua, Spatially resolved shear distribution in microfluidic chip for studying force transduction mechanisms in cells. *Lab on a Chip*, 2010, **10**(2): 235–239.
73. Alenghat, F.J. et al., Global cytoskeletal control of mechanotransduction in kidney epithelial cells. *Experimental Cell Research*, 2004, **301**(1): 23–30.
74. Kusumi, A. et al., Passage-affected competitive regulation of osteoprotegerin synthesis and the receptor activator of nuclear factor-kappaB ligand mRNA expression in normal human osteoblasts stimulated by the application of cyclic tensile strain. *Journal of Bone and Mineral Metabolism*, 2009, **27**(6): 653–662.
75. Nagatomi, J. et al., Cyclic pressure affects osteoblast functions pertinent to osteogenesis. *Annals of Biomedical Engineering*, 2003, **31**(8): 917–923.
76. Riddle, R.C. and H.J. Donahue, From streaming-potentials to shear stress: 25 years of bone cell mechanotransduction. *Journal of Orthopaedic Research*, 2009, **27**(2): 143–149.
77. Jacobs, C.R. et al., Differential effect of steady versus oscillating flow on bone cells. *Journal of Biomechanics*, 1998, **31**(11): 969–976.
78. Torii, T., M. Miyazawa, and I. Koyama, Effect of continuous application of shear stress on liver tissue: continuous application of appropriate shear stress has advantage in protection of liver tissue. *Transplantation Proceedings*, 2005, **37**(10): 4575–4578.
79. Ohno, K., K. Tachikawa, and A. Manz, Microfluidics: Applications for analytical purposes in chemistry and biochemistry. *Electrophoresis*, 2008, **29**(22): 4443–4453.
80. Leclerc, E., Y. Sakai, and T. Fujii, Microfluidic PDMS (polydimethylsiloxane) bioreactor for large-scale culture of hepatocytes. *Biotechnology Progress*, 2004, **20**(3): 750–755.
81. You, L. et al., 3D Microfluidic approach to mechanical stimulation of osteocyte processes. *Cellular and Molecular Bioengineering*, 2008, **1**(1): 103–107.
82. Sin, A. et al., The design and fabrication of three-chamber microscale cell culture analog devices with integrated dissolved oxygen sensors. *Biotechnology Progress*, 2004, **20**(1): 338–345.

23 Design of Abdominal Wall Hernioplasty Meshes Guided by Mechanobiology and the Wound Healing Response

Shawn J. Peniston, Karen J.L. Burg, and Shalaby W. Shalaby

CONTENTS

23.1 INTRODUCTION

The use of meshes to repair abdominal wall defects is believed to reduce the overall recurrence rate; however, its presence in the abdominal wall has resulted in short- and long-term complications. The etiology of these complications is, in part, the result of a mismatch in biomechanics that leads to less than optimal biocompatibility. The word biocompatibility is used extensively, yet a great deal of uncertainty exists about what it actually means and about the mechanisms that collectively control the concept. The study of mechanobiology holds promise to improve the knowledge base from the molecular to the organ level, whereby mechanical stimulation influences the nature of continuous cellular remodeling to maintain tissues and during the accelerated remodeling process characteristic of the wound healing process. As such, careful examination of the biomechanical environment and the temporal needs of the wound healing process are necessary to illicit a desirable response from a surgically implanted device. Specifically, this investigation reflects on the current state of mesh hernioplasty from a mechanobiology perspective with emphasis on the wound healing process, abdominal wall anatomy and biomechanics, etiology of abdominal wall hernias, biomechanics in the treatment of hernias, and the biomechanical environment of healing wounds. Additionally, a future mesh design concept is presented that takes into consideration the mechanobiology of healing wounds within the abdominal wall.

23.1.1 Wound Healing Process

23.1.1.1 Stages of Wound Healing

Wound healing is characterized by several overlapping, predictable stages that are strongly interlinked and contribute to the common goals of the elimination of foreign microorganisms, wound debridement, cell proliferation, matrix deposition, contraction, and maturation resulting in mechanical integrity. The major stages of wound healing include inflammation (generally 2–5 days post insult), cue proliferation (generally 2 days to 3 weeks post insult), and remodeling/maturation (generally 3 weeks to 2 years post insult). The time frames vary depending on the complex and often case-specific nature of wound healing. Many factors affect the timing of the wound healing process, including the location and extent of injury, level of wound contamination and infection, rate of fluid perfusion, level of local pH, presence of foreign bodies, host comorbidities, and the regenerative capacity of proximal cells [1,2]. The wound healing process is sequentially similar throughout the body with differences specific to the length of each stage and cell autocrine and paracrine signaling often resulting in anatomical site–specific cell responses such as atrophy, hypertrophy, hyperplasia, metaplasia, and/or phenotype changes. The end result of orderly and timely wound healing is ideally tissue with minimal fibrosis yet structural integrity, minimal to no wound contraction, and preinjury function. Often with the implantation of medical devices, the measure of their biocompatibility is graded on the local resolution of the wound healing process in a timely manner. To better illustrate this process a brief review of the typical, wound healing event is presented herein.

Injury to vascularized tissue triggers coagulation and the initiation of the inflammation process of wound healing. The inflammation process rapidly increases to a maximum at 2–3 days and then gradually resolves over the next couple of weeks. Overall, the inflammatory process can be divided into two stages of (a) vasomotor-vasopermeability and (b) leukocyte signaling and infiltration [3]. The initial vascular response to injury is a short-lived vasoconstriction intended to minimize blood loss followed by a period of local vasodilation. Vasodilation increases the pressure and flow of blood to the area and causes the release of serum fluid through permeable vascular walls resulting in tissue edema. Within the newly formed interstitial space created by the exudate, proteins such as fibronectin are deposited, which will create the initial scaffolding for subsequent cellular migration and locomotion into and within the wound [3]. Within the wound site, the formation of a blood clot via the coagulation cascade blocks the continued loss of blood and provides additional scaffolding for cellular activity during the remodeling phase. The blood clot is constructed from fibrin, an insoluble, cross-linked product of fibrinogen, and activated platelets. During the formation of the blood clot, activated platelets release a myriad of cytokines including platelet derived growth factors (PDGF) [4] and transforming growth factor-beta (TGF-β) [5], which result in the chemotaxis of leukocytes and fibroblasts. Figure 23.1 depicts the subsequent temporal cellular and extracellular matrix (ECM) formation of the wound healing response.

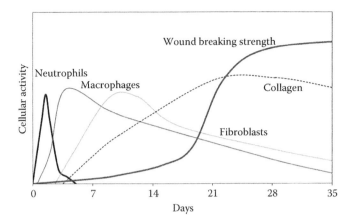

FIGURE 23.1 Temporal cellular activity and ECM formation of the wound healing response. (Adapted from Robson, M.C. et al., *Curr. Prob. Surg.*, 38(2), 72, 2001. With permission from Elsevier.)

The first leukocyte to aggregate at the wound site is the neutrophil. The primary role of the neutrophil is to clean the wound by removing bacteria and initiating debridement [3,7]. Neutrophils accomplish this goal through the release of free oxygen radicals and lysosomal enzymes from within their many intracellular granules. Neutrophils are short lived within wounds, especially when bacterial infiltration is minimal, and quickly (generally by day 3) become secondary in number to monocytes differentiating into macrophages. Macrophages continue to phagocytize tissue and bacterial debris taking over where the neutrophils left off; however, macrophages have the added responsibility of being crucial to cellular proliferation and recruitment. Once the wound is clear of bacteria and debris the rebuilding process can begin with the repopulation of cells.

The proliferation and remodeling phase is characterized by an increase in the number of fibroblasts followed by the deposition of ECM in the wound. Proliferation is driven by macrophages that secrete growth factors such as PDGF, TGF-β, interleukins, and tumor necrosis factor, which play key roles in the migration and activation of local fibroblasts [3]. Fibroblasts, which originate from mesenchymal cells located in loose tissue around blood vessels and fat, differentiate and migrate in response to these cytokines [8]. These newly formed fibroblasts use the fibrin/fibronectin network previously established for locomotion within the wound. In addition to collagen, fibroblasts produce glycoaminoglycans and proteoglycans, which form the tissue "ground substance." Initially, fibroblasts are primarily focused on replication and recruitment with minimal collagen synthesis. Within the first 4–6 days following insult, fibroblasts become the primary cell type, and the deposition of collagen increases rapidly for the next 3 weeks [7]. Concurrently, during the 3 week time period of collagen deposition, angiogenesis progresses to provide the wound with nourishment and oxygen while removing metabolic and waste products. Macrophages and fibroblasts provide the stimulus for the progression of capillaries, arterioles, and venules toward the wound space. Fibroblast growth factor and vascular endothelial growth factor released by macrophages are the primary cytokines responsible for angiogenesis [9]. Over time, capillaries slowly regress as collagen fills the wound space and a mature, avascular scar tissue is formed.

Wound remodeling and maturation are characterized by a gradual strengthening and reorganization of the collagen matrix that follow the initial wound healing events. During this time, Type III collagen, so-called immature collagen because of its smaller fiber diameter, greater elasticity, and lower strength, is replaced by Type I collagen. Prior to the deposition of Type I collagen the wound has essentially no mechanical integrity. During the wound maturation/remodeling process no net gain in collagen content is achieved; instead, the production of collagen is matched by the degradation of collagen by matrix metalloproteinases (MMPs). With the turnover of Type III collagen to Type I collagen the wound site develops mechanical strength and the cells attempt to replicate the preinjury tissue characteristics by contributing to the developing structure and orientation along the lines of tension. However, typically the wound site never fully obtains the original tissue structure, while values of approximately 80% of preinjury mechanical strength are reported [5].

23.1.1.2 Role of Collagen Microstructure in Soft Tissue Mechanical Behavior

Collagen is the most abundant protein present in the human body and consists of fibrils embedded in an amorphous gel-like matrix composed of proteoglycans and water. Fiber forming collagen, such as Types I and III, function to transmit and dissipate loads and store elastic strain energy applied to the joints of the body [10]. The mechanical characteristics of collagen lie in its unique and somewhat complex structure.

The development of collagen, the structural component of soft tissue, is paramount to effective and efficient resolution in wound healing. Collagen is synthesized by fibroblasts in a multistep process that starts with intracellular assimilation of peptide chains and is completed with extracellular collagen fiber and fiber bundle formation into macroscopic structures such as tendons or fascia.

Within wounds, collagenous tissue function is highly dependent on its structure. Collagen fibrils are highly oriented and thus stronger in the direction parallel to the fiber orientation, but they can be considered essentially inextensible [11]. As such, the conformation of collagen fibrils is essential to determining tissue function. For example, highly oriented structures such as tendons transmit force quickly and efficiently while randomly oriented fibers in the dermis allow considerable extension before resistance is achieved. Therefore, the aggregation of fibrils into fibers and their resultant diameter as well as the construction of the aggregate fibers gives soft tissue its mechanical properties and anatomically specific characteristics. The stress–strain curve for collagenous tissue under tensile loading is nonlinear and can be divided into three distinct regions, each of which can be attributed to a different structural element (Figure 23.2).

In tendon and ligament the curve has been characterized by a low strain nonlinear toe region, a curved mid-region, and then a linear yield and failure region [12]. In the toe region, low stress is required to remove macroscopic crimp in the form of gradual straightening of collagen fibers that have varying degrees of undulations [13]. The second region is characterized by collagen fibers that begin to line up in the direction of the load and provide increasing resistance. In addition, disordered molecules in the lateral gap region between fibrils reorganize [14]. In the third region, stretching of the triple helices and cross-links between helices produce side-by-side gliding of neighboring collagen molecules leading to structural changes at the level of the collagen fibrils [15]. Under typical conditions, physiologic levels are within the toe region of the stress–strain curve, resulting in a shock absorbing system, where low levels of stress are required to achieve deformations in the absence of significant molecular stretching of the fibers [11].

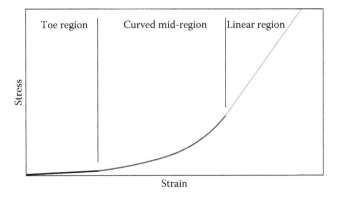

FIGURE 23.2 Stress–strain relationship for highly oriented collagenous tissue (rat tail tendon). (Adapted from Fratzl, P. et al., *J. Struct. Biol.*, 122(1–2), 119, 1998. With permission from Elsevier.)

Like synthetic polymers, biopolymers such as collagen exhibit viscoelastic mechanical behavior. Collagen exhibits the viscoelastic phenomenon of creep and stress relaxation. In other words, the mechanical properties of collagen have a time-dependent nature under continuous stress or strain conditions. The hierarchical structure of collagen, from triple helix to tissue, provides structural components that have elastic and viscous characteristics. For example, ligament viscoelasticity under uniaxial tensile loading is attributed to the inherent viscoelasticity of the collagen fibrils (bond rotation and stretching), the local ECM, interfibrillar cross-linking, and the movement of fluid within and in/out of the tissue [16]. The degree of cross-linking between fibrils has been linked to viscoelastic properties, with low levels dominating viscous behavior and high levels dominating elastic behavior through the stretching of nonhelical ends, cross-links, and the triple helix [10]. Furthermore, it has been determined that the overall strain in a tendon is always larger than the strain of individual fibrils, indicating that some of the viscoelastic deformation takes place in the proteoglycan-rich interfibrillar matrix [15].

23.1.1.3 Biomechanical Influences on Tissue Remodeling and the Wound Healing Process

Mechanical forces are fundamental in maintaining and regulating the structure and function of tissue. The importance of the mechanical loading of bone has been realized for many years; more recently, the importance of mechanical stimulation on other tissues such as ligament, tendon, skeletal muscle, intervertebral disc, and meniscus is being realized [17]. The influence of mechanotransduction in cellular signaling is still not fully understood. However, the influence of mechanical stimulation on fibroblasts has been investigated and determined to affect cell proliferation [18,19], collagen deposition [20,21], phenotype [22], apoptosis [23], cell spreading [24], orientation [25], and the release of MMPs [26]. From a biochemical aspect, observed changes in animal tendons and ligaments from joint immobilization studies include increases in the rate of collagen turnover, reduced levels of cross-linking, slight mass loss, a reduced amount of proteoglycans and hyaluronic acid, and decreased water content [27]. Biomechanically, the result of these changes is a reduction in tangent modulus, cross-sectional area, and ultimate strength [27]. Hannafin and coworkers compared static and mechanically cycled canine flexor digitorum profundus tendons in vitro and demonstrated that cells in static samples had altered morphology and decreased number; furthermore, cell and collagen alignment was modified, resulting in decreases in tensile modulus over an 8-week period [28]. Fibroblast–matrix interactions control cell shape and orientation and also directly regulate cellular functions, primarily through integrin receptors that cells use to adhere to and receive mechanical energy from the ECM [29]. In addition, it has been shown that elongated tendon fibroblasts, as they appear in a homeostatic, mechanically stressed ECM, produce greater amounts of collagen Type I as compared to less elongated cells [30].

Mechanical stimulation also has a role in wound contraction. A cell subpopulation will differentiate into myofibroblasts as fibroblasts increase in number within the wound site. Myofibroblasts express different sets of cytoskeletal proteins, such as α-smooth muscle actin (α-SMA), that play an important role in contraction. Myofibroblasts are stimulated to differentiate by a combination of TGF-β1 and ED-A fibronectin (ED-A FN), both of which are critical to the induction of α-SMA expression but are not sufficient to maintain myofibroblast differentiation in the absence of mechanical stimulation [31]. It has been suggested that, for normally strained tissues such as tendons, wound contraction is an attempt to restore the physiologic condition of tension [32]. However, excessive contraction can distort and disrupt tissue structure, resulting in undesirable consequences. The tissue contraction of a rat wound healing model occurred in a three-stage process [31]. Slow contraction occurs from 1 to 6 days post-wounding, which is characterized by an increase in the expression of ED-A FN and is said to be independent of myofibroblast influence [33]. Next, a steep increase in contraction is realized, with the increased expression of α-SMA. This phase lasts during a period from 3 to 10 days, initiating at earlier timepoints and persisting longer when the wound site is mechanically stressed. After approximately 10 days, a reduction in the α-SMA expression occurs, followed by contraction. Tissue contraction correlates with the level of α-SMA expression being higher when granulation tissue is subjected to greater levels of tension [31]. Tension within the wound site is said to prevent apoptosis of myofibroblasts [34], but once forces are relieved, apoptosis of myofibroblasts will occur even if growth factors are added to the wound [35]. In summary, mechanical stimulation is crucial to the biomechanical quality of collagen and myofibroblast modulation of the wound contraction process.

23.1.2 COLLAGENOUS TISSUE WOUND HEALING IN THE ABDOMINAL WALL

23.1.2.1 Anatomical Features of the Abdominal Wall

The abdominal cavity is approximated by the spine and back muscles posteriorly, the pelvic cavity inferiorly, and the thoracic cavity superiorly. Theoretically, the abdominal cavity possesses the necessary support to resist herniation. However, herniation in the inguinal region occurs through the myopectineal orifice, as described by Fagan and Awad [36]. The myopectineal orifice is quadrangular in shape and is divided superiorly and inferiorly by the inguinal ligament, which runs from the anterior-superior iliac spine to the pubic tubercle (Figure 23.3). The myopectineal orifice is perforated in the medial-lateral triangle by the spermatic cord and in the femoral triangle by the femoral artery and vein.

The inguinal canal is created from the passage of the spermatic cord, including the vas deferens exiting the abdominal cavity and transcending to the testes in the scrotum. The location where the spermatic cord initially enters the abdominal wall is called the deep inguinal ring (Figure 23.4).

The rectus abdominis is comprised of two ventrally located vertical pillars segmented on the midline by the linea alba. Attached to the rectus abdominis is a triple layer of flat muscles extending laterally and creating a cylindrical abdominal cavity that withstands internal pressure as well as external insults [38]. From the superficial to deep layer, these muscles include the external oblique, internal oblique, and transversus abdominus. The transversus abdominus is the main muscle used to retain the abdominal contents [38].

The rectus sheath is divided by the posterior and anterior layer relative to the rectus abdominis muscle and is comprised of the aponeurosis from each layer of the triple flat abdominal muscles.

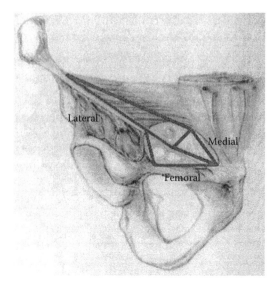

FIGURE 23.3 Myopectineal orifice depicting the medial, lateral, and femoral triangles. (Reprinted from Fagan, S.P. and Awad, S.S., *Am. J. Surg.*, 188(6A Suppl), 3S, 2004. With permission from Elsevier.)

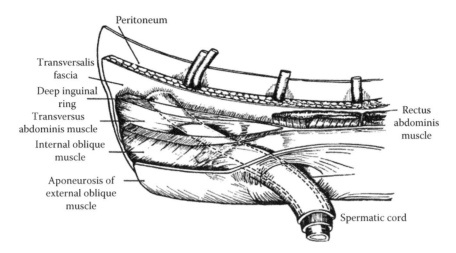

FIGURE 23.4 The anterior view of the inguinal region of the abdominal wall. (Adapted from Yeager, V., *Clin. Anat.*, 5(4), 289, 1992. With permission from John Wiley and Sons.)

The anterior layer of the rectus sheath is made up of primarily aponeurosis fiber from the external and internal oblique muscles and the posterior layer is comprised of aponeurosis fibers from the internal oblique and transversus muscles above the level of arcuate line. The arcuate line is generally located midway between the umbilicus and pubis and represents the transition zone in which the aponeurosis of the external oblique, the internal oblique, and the transversus abdominis muscles all pass anterior to the rectus muscle [39]. Below the arcuate line the posterior sheath of the rectus abdominis lacks strength as it is comprised of only transversalis fascia, areolar tissue, and peritoneum [39]. It should be noted that aponeurosis are like tendons or ligaments, with the major difference being that they originate from large flat muscles and thus take on the form of large, flat, thin sheets. Fascial layers on the other hand are considerably more extensible and primarily function to separate layers of tissue rather than provide load bearing structural support. For this reason the myopectineal orifice is susceptible to herniation.

23.1.2.2 Anatomical Classification of Hernias

The most common wound to the abdominal wall is inguinal hernia, which is surgically repaired at the annual rate of 800,000 in the United States, 200,000 in Germany, 100,000 in France, 80,000 in the United Kingdom, and 12,000 in Finland [40–42]. Hernias of the abdominal wall form at areas that are susceptible to loss of mechanical integrity through acquired or congenital pathologies. The weak points of the abdominal wall are the inguinal, umbilical, and femoral canal regions. From epidemiology data, it is known that the prevalence rates for abdominal wall hernias are approximately 73% inguinal, 9.5% umbilical, 6.2% incisional, 2.7% femoral, and 8.6% other types such as spigelian, hiatal, or epigastric [43]. Inguinal hernias are classified as direct or indirect.

Indirect inguinal hernias occur when a visceral sac leaves the abdominal cavity, enters the deep inguinal ring, and transcends the spermatic cord. The hernial sac contains peritoneum and viscera such as adipose tissue, intestinal loops, or omentum and is surrounded by all three fascial coverings of the spermatic cord. The hernia can traverse the entire inguinal canal and exit through the superficial inguinal ring. In severe cases the hernial sac enters the scrotum. Indirect inguinal hernias can occur in women, but they are twenty times more likely in males.

Direct inguinal hernias occur when the peritoneum with subperitoneal tissues and/or abdominal viscera herniate through a weak point in the abdominal wall. The typical location of direct inguinal hernias is within the confines of the medial triangle of the myopectineal orifice [36]. The hernial sac is formed by distention of the transversalis fascia lateral to the rectus abdominis muscle and it emerges to reach the superficial inguinal ring, gaining an outer covering of external spermatic fascia inside or parallel to that on the cord. It rarely enters into the scrotum. Direct hernias are most common in elderly men.

Femoral hernias occur within the femoral triangle of the myopectineal orifice and result from the distension or the rupture of the transversalis fascia [44]. It is generally accepted that they are the result of elevated intra-abdominal pressure and/or an enlarged femoral ring, which facilitates the peritoneum and preperitoneal adipose tissue to protrude through the femoral ring [45]. Subsequently, the hernial sac may travel along the femoral vessels and settle in the anterior thigh [46]. Femoral hernias are more frequent in women than in men (4:1) [47].

Umbilical hernia prevalence in adults has a female to male ratio of 3:1 with particular frequency in obese, multiparous women [48]. The etiology of umbilical hernias in adults is believed to initiate during embryonic development through defects in the closure of the embryo's abdominal orifice from which the umbilical cord emerges after the obliteration of the coelomic sac. Over time, this weakness manifests itself at the superior aspect of the umbilicus, becoming susceptible to increased intra-abdominal pressure, which may drive forward gobbets of preperitoneal fat or an incipient sac. This condition stretches the fascia before it into a funnel and eventually progresses into a hernia [49].

An incisional hernia is one that appears at the site of an incision from a previous abdominal operation. An incisional hernia can appear within months or take many years to become evident to the patient. Incisional hernias are sometimes referred to as ventral hernias due to their typical position between the rectus abdominal muscles and through the linea alba, the preferred midline incision used by surgeons for visceral access during laparotomy.

23.1.2.3 Etiology and Pathology of Hernia Development

Is the development of abdominal wall hernias the result of an anatomical defect or collagen disease? Historically, hernia genesis was attributed to a mechanical disparity between visceral pressure and resistance of the structures within the myopectineal orifice. Which of these factors are significant contributors? Increasingly, hernia etiopathology is described as a multifactorial process linking an evolutionary anatomical weakness, predisposed defects, and dynamic factors such as increased abdominal pressure. The influence of each of these factors in the primary formation and recurrence of hernias is an area of significant debate.

23.1.2.3.1 Mechanisms in the Development of Primary Hernias

Evolution has clearly left human beings with a section of the abdominal wall that is weaker in comparison to the rest of the abdominal wall: the majority of hernias occur in the myopectineal orifice of the inguinal region. The thin and weak transversalis fascia of the groin coupled with the lack of fascial sheath below the arcuate line together form the argument for an intrinsic defect in the human abdominal wall [50]. Many surgeons believe the transversalis fascia does not even resemble fascia or any tendinous-like structure; the transversalis fascia is a thin, fibro-membranous peritoneum with markedly reduced strength as compared with typical fascia [50]. The myopectineal orifice is sealed by the transversalis fascia; thus, all groin hernias are the result of the displacement of this fascia by a peritoneal sac.

Hernia development can be congenital in the case of indirect hernias. In this case a visceral sac leaves the abdominal cavity and transcends the spermatic cord. Congenital predisposition in males originates during the descent of the fetal testes into the scrotum. Interruptions in the closure of the deep inguinal ring can develop into a potential defect later in life [46]. Indirect hernias are more common on the right side than left. The right testicle descends from its position near the kidney into the scrotum after the left testicle has already completed its descent. The delay in the closure of the deep inguinal ring on the right side is believed to be responsible for its side-specific hernia predominance [50].

Increased intra-abdominal pressure is believed to be a contributing factor in the pathogenesis of herniation [51]. Risk factors include obesity and chronic constipation. Often hernias are thought to be the result of a single event (e.g., lifting a heavy object) but in fact repetitive mechanical strain is likely the damaging factor [52]. It is possible that chronic mechanical strain, not prior biologic defects, may induce secondary changes in structural tissue cellular and molecular function [53]. However, increased intra-abdominal pressure is speculative in nature with no clinical study to confirm its contribution to hernia formation [51]. Furthermore, no adequate animal model exists that can simulate hernia formation or replicate the increased intra-abdominal pressure from erect posture gravitational forces on the floor of the abdominal wall [50].

23.1.2.3.2 Causes of Recurrent Hernias

In approximately 60% of all excised meshes, recurrence is the reason for extraction [54]. The recurrence of hernia repairs has frustrated surgeons for many years, especially incisional hernia recurrence with rates of 11%–15% within the first year and a doubling of the high initial rates within the first 9 years [55]. The stated causes of recurrence include stress applied to the wound prior to the development of mechanical integrity, shear stresses at the margins of the mesh during collagen maturation, collagen metabolism disorders, infection, and surgical technical errors. Immediate gross failure in most cases is attributed to surgeon technical failure or infection, while longer term (>3 months) failure stems primarily from abnormal wound healing such as varied collagen metabolism and the progress of acute wound disruptions into symptomatic failures.

There is evidence that hernia recurrence of mesh-repaired laparotomies is the result of external stresses applied to the wound site prior to the development of tissue integration and wound strength (first 2–3 weeks post insult). As reviewed by Franz, one prospective study of primary repaired incisional hernias found that the total rate of acute wound disruption was about 11% at postoperative day 30 with the majority (94%) later developing into incisional hernias [53]. Primary repaired incisional hernias fail by wound dehiscence from sutures pulling through the wound edges. Similarly, mesh acute wound failures also occur from stressed suture lines at the margins of the mesh creating mesh-fascial dehiscence. The result of the failure is a loss in tension applied across the wound. The loss of mechanical load signaling may impair fibroblast biology, which promotes subsequent collagen abnormalities leading to the high rate of recurrent incisional hernia formation [53].

Recurrent hernias develop 99% of the time at the margins of the implanted mesh [54,56–59]. Owing to the significant strength of most meshes, central mesh ruptures are a documented but extremely rare occurrence [60,61]. The nonphysiologically low stretching capability of the mesh/tissue complex contrasts with the highly elastic abdominal wall, resulting in shear forces at the margins of the mesh. These forces overstress developing and maturing collagen, resulting in recurrence at the margins of the mesh [62].

Incisional hernias are also being investigated as an abnormal wound healing response with an inability to produce abundant, quality, strong collagen. As with primary hernias, metabolic factors are being investigated; however, more evidence supports collagenase as a central agent involved with the development of incisional hernias when compared to primary hernia formation [63]. Though this may be the case, it is hard to accept metabolic factors as the primary pathway to recurrent hernia when the majority of patients do not have a history of wound healing defects and do not express any defects in organs local to the surgical site or the vascular system.

23.2 BIOMECHANICAL CONSIDERATIONS FOR HERNIA TREATMENT

23.2.1 BIOMECHANICS OF THE ABDOMINAL WALL

The abdominal wall mechanics traditionally have been characterized by (1) the physiologic maximum force generated within the wall, and (2) the extension or strain associated with that maximum physiologic force. Peiper et al. [64] determined that loading of the inguinal region of the abdominal wall is predominately related to increases in intra-abdominal pressure and not muscular contraction. If one assumes that intra-abdominal pressure only governs the resistive strength required in the abdominal wall, then the required strength can be derived by Laplace's law as suggested by Klinge et al. [65]. Human abdominal pressures range from 0.2 kPa (resting) to 20 kPa maximum. According to *Laplace's* law, a thin-walled sphere where the total vessel wall tension [(pressure × vessel radius)/2] is independent of the layer thickness (wall thickness/vessel radius ≪ 1) can be described by, $F = p \times d/4$ (N/cm) where d is the diameter, p is the pressure, and F is the wall tension/cm of circumference. If the longitudinal diameter of the human abdominal wall is 32 cm, a tensile force of 16 N/cm is produced at the maximum pressure. To define the physiologic strain associated with a 16 N/cm load, Junge et al. [66] analyzed the abdominal wall of 14 fresh corpses and determined that longitudinally the average extension was 25% ± 7%. However, Cobb et al. [67] directly measured the intra-abdominal pressure of 10 healthy male and 10 healthy female subjects performing various activities including coughing and jumping, the two known activities that produce maximum intra-abdominal pressures. These measurements indicated that the maximum tensile force ranged from 11 to 27 N/cm. Wolloscheck et al. [68] investigated the tissue burst force of individual layers of the lower abdominal wall. Their findings include burst force values for the transversalis fascia, peritoneum including the pre and subperitoneal tissue, the aponeurosis of the internal oblique, and the aponeurosis of the external oblique measured as 10.5, 46.6, 51.7, and 92.6 N, respectively. This suggests that the transversalis fascia is the weakest of the load bearing tissues, further evidence of the minimal support provided by the transversalis fascia covering the myopectineal orifice.

23.2.2 MESH CHARACTERISTICS PERTINENT TO BIOCOMPATIBILITY

The most important modern advancement in hernia surgery has been the development of so-called tension-free repairs using meshes [69]. The design characteristics of meshes include construction and yarn type. Collectively, these characteristics significantly influence the biocompatibility of a mesh and its mechanical suitability for a particular application. However, although the design characteristics of meshes have produced significant debate with regard to their respective influences, no study has been completed that isolates each characteristic. For example, an in vivo study that compares the same construction with a different yarn type does not exist. To date, studies have been conducted using commercially available meshes, and each has a different construction, fiber form, and fiber chemistry making it difficult to isolate the influence of individual mesh design variables. However, observations about mesh design variables have provided insight into their general influence on biocompatibility.

23.2.2.1 Mesh Construction

Mesh construction can be extensively varied and has implications on pore size, area weight, drape, extensibility, and strength, all of which clinically translate into surgical handling characteristics, anatomical conformability, foreign body reaction, and the mechanical, cellular, and ECM characteristics of the mesh/tissue complex.

The porosity of a mesh is best described as the amount of open space in a unit area of mesh. However, this description does not provide a complete picture as the dimensions/area of individual pores, the distance between pores, and the size and quantity of interstitial pores are also contributing factors to in vivo performance. Porosity is a rarely measured characteristic. Recently, a study using digital imaging was conducted in order to better characterize the overall porosity and the distribution of different pore sizes for clinically relevant meshes [70]. The porosity of meshes is a primary determinant for tissue response, with pore size implicated in affecting long-term abdominal wall mobility. It has been suggested that pores smaller than 1 mm lead to extensive inflammation and fibrosis, resulting in a bridging of collagen between adjacent pores [54,62,71]. Ultimately, this occurrence produces a dense, continuous outer fibrotic capsule covering the whole mesh. Several investigations using a rat [72–74] and a porcine [75] animal model have concluded that larger pores result in increased abdominal wall mobility due to reduced fibrotic bridging. In addition, the absence of a complete capsule covering the mesh facilitates improved fluid transport through the mesh, as well as improved vascularization and organization of connective tissue [76]. Based on increasing evidence, porosity is a key factor in the incorporation of the mesh into the surrounding tissue and thus an important prerequisite to its biocompatibility.

Area weight, measured as the mass per area (g/m^2), is a determination of the total amount of biomaterial implanted for a given area. Theoretically, lower area weights induce a milder foreign body reaction, improved abdominal wall compliance, less contraction or shrinkage, and allow better tissue incorporation; as a result, the use of lighter weight or partially absorbable meshes has been investigated to improve hernia repair outcomes. However, the available data is contradictory and controversial. The outcomes from partially absorbable meshes are mixed. Early investigations in a rat animal model indicated long-term decreased inflammation as a result of the long-term decreased polypropylene content [74,77]. However, Weyhe et al. [78] have investigated the changes in foreign body reaction in both rat animal model and in vitro [79,80] and determined that simple reduction of the mesh mass was not the main determinant of biocompatibility; rather, pore size and material composition were better indicators. Clinical trials have produced conflicting outcomes as well. Several studies have found no significant differences between typical polypropylene meshes and partially absorbable meshes in multicenter studies using Lichtenstein hernioplasty [81,82]. Conze et al. [83] found similar outcomes when comparing partially absorbable and standard polypropylene meshes; however, partially absorbable mesh had a trend toward increased hernia recurrence. Other authors have reported improvements in some aspects of pain and discomfort using partially absorbable meshes [84], but in addition to less chronic pain, an increase in recurrence has also been reported [85]. In a recent review of current randomized controlled trials and retrospective studies, partially absorbable meshes seem to have some advantage with respect to postoperative pain and foreign body sensations, but their use is associated with increased recurrence rates [86]. The movement toward the use of lower weight meshes has shown researchers and clinicians that long-term, lower area weight meshes appear to reduce some complications but at the cost of an apparent increase in recurrence.

The mesh construction in vivo can influence the mechanical properties of the mesh/tissue complex. When a mesh is stressed, the mesh filaments show minimal if any elasticity, while the mesh itself produces extensive elongation from geometric deformation within the pores perpendicular to the applied stress. Thus, the elastic characteristics of a mesh and the ability to match those of the natural abdominal wall are predominately influenced by construction. In a study conducted by Greca et al., a polypropylene light weight, macroporous mesh ($19\,g/m^2$) was compared to a typical so-called "heavy weight" mesh ($85\,g/m^2$) in the repair of a full-thickness abdominal wall defect in a dog animal model. Although the heavy-weight mesh had 3.6 times the initial burst strength of the light-weight mesh, after 90 days in vivo the mesh/tissue complex for the light-weight mesh produced marginally greater average burst strengths than the heavy-weight mesh [87]. The authors attributed the increased strength to a higher concentration of mature Type I collagen developed around the light-weight mesh. The mechanical characteristics of the mesh construction are important factors as the maturation of the mesh/tissue complex is a critical indicator of biocompatibility.

The strength of a mesh is not only characterized by yarn diameter or denier, number of yarns, and tenacity of the yarn, but construction also affects the burst strength. The nature and extent of yarn looping within the mesh construction affects the burst strength due to stress concentrations. For a given mesh construction, stronger mesh yarn does not always translate linearly to increased mesh burst strength. Construction can play a limiting role in the strength of the construct.

23.2.2.2 Type of Yarn

Monofilament-based meshes have marked stiffness, whereas multifilament meshes have improved softness, less surface texture, and better drape characteristics for adaptation to anatomical curvatures. However, multifilaments physically have a pronounced increase in surface area, which influences their biocompatibility. In vivo testing in a rat animal model has suggested a more intense acute inflammatory response and increased fibrosis to multifilament yarn than monofilament yarn [88–90]. The authors attributed the upregulated response to the increased tissue/biomaterial surface contact area. Patients implanted with either multifilament or monofilament polypropylene meshes for inguinal hernioplasty were investigated for inflammatory response differences. Blood samples were collected before surgery and up to 168 h post-surgery and were analyzed for pro-inflammatory mediators. Results indicated a more intense acute inflammatory response to the multifilament mesh [91]. The increased response may be, in part, the result of the increased adsorption of host proteins to the implant, which in turn triggers the increased activation of inflammatory cells. Long-term multifilaments have not been identified as triggering an increased level of chronic inflammation as compared with monofilaments, in fact, more leukocytes and larger multinucleated giant cells have been observed to be associated with monofilament meshes [92]. This continuously stimulated long-term chronic inflammation may be the result of continuous mechanical irritation of the local tissue around the mesh. Currently, no studies investigate this phenomenon for meshes, although it is becoming increasingly understood that tissue response is, in part, driven by differences in mechanical stiffness between implant and surrounding tissue.

23.3 BIOMECHANICAL ENVIRONMENTS OF HEALING WOUNDS

23.3.1 EARLY PHASES OF WOUND HEALING AND MESH INTEGRATION

The early phases of wound healing and mesh integration are critical to the outcome of hernioplasty. The acute wound healing phase is when inflammatory and repair cells are recruited but during which no meaningful tissue integrity is yet established [93]. The early phases of normal wound healing include hemostasis and acute inflammation characterized by fibrin clot formation, bacterial clearing, and wound debridement followed by proliferation, which includes the formation of granulation tissue and wound contraction. Granulation tissue is made up of a cellular, randomly ordered ECM containing new blood vessels from angiogenesis. When the acute inflammation phase does not resolve in a timely manner it is said to enter a state of chronic inflammation.

23.3.1.1 Acute and Chronic Inflammation—Responses to Meshes

Tissue disruption and bleeding during mesh implantation results in fibrin clot formation and a burst of inflammatory cytokines that attract monocytes/macrophages, polymorphonuclear cells, and lymphocytes. Resident tissue cells such as mast cells and macrophages become activated and respond to the biomaterial. It is believed that these phagocytes attach and interact with the adsorbed proteins rather than with the material itself [54]. Phagocytes attempt, but are unable to remove the mesh due to its size and relatively inert nature. As a result, the wound healing response moves into a chronic inflammation phase due to the continued presence of the mesh. Consequently, the proliferation phase initiates, but with the continued presence of inflammatory cells as the host tissue attempts to isolate the mesh through encapsulation [94].

23.3.1.2 Proliferation—Tissue Integration and Collagen Development

The proliferation phase is primarily characterized by the increased presence of fibroblasts that produce significant amounts of ECM with maximum deposition at 2–3 weeks. Consequently, the bulk of the collagen is deposited in the proliferation stage; however, even under normal wound healing conditions, high-quality collagen is not formed until later in the maturation phase. Wound mechanical integrity begins to develop as the level of cross-linking increases. Fascial tissues within the abdominal wall have been shown to develop strength faster than dermal wounds. Fibroblasts extracted from rat fascial wounds show enhanced cell proliferation and increased wound collagen deposition compared to dermal fibroblasts [95]. Mechanisms for accelerated fascial healing include earlier activation of fascial fibroblasts and earlier induction of collagen synthesis compared to dermal wounds [96]. Unfortunately, the presence of the biomaterial means the acute wound healing trajectory may further delay the recovery of wound tensile strength for hernioplasty [97]. Wound inflammation is primarily a protective response; studies suggest that the mechanisms used to destroy bacteria and remove debris delay tissue repair [98]. However, in the case of prolonged inflammation excessive collagen production may prevail.

The inflammatory response to typical mesh biomaterials during the proliferation stage results in significant fibrosis that increases the rigidity and stiffness of the abdominal wall [99]. As reviewed by Welty et al. [100], 50% of patients report some form of physical restriction of the abdominal wall. Chronic inflammation is also responsible for slow and gradual secondary migration of mesh through trans-anatomical planes due to foreign-body reaction induced erosion of local tissue [101,102]. In addition, from a mechanical point of view, irritation and inflammation from a hard material in contact with soft tissue can induce erosion in the latter [103]. This is especially true in the case of stiff and rigid monofilament polypropylene mesh. Consequently, mesh can enter the abdominal cavity causing visceral adhesions and/or fistula formation [104].

Macrophages play a fundamental role in the process of ECM development through direct or indirect fibroblast signaling. Results from animal studies have suggested that the absence of macrophages results in defective scar formation [105]. On the other hand, increases in the time and intensity of inflammation are associated with increased levels of scarring [93]; however, the amount of collagen present at the wound site is not a good indicator of mechanical integrity. The bulk of ECM production occurs during the proliferation stage, but this does not represent the time at which the ECM is complete. Significant changes in orientation and conformation of fiber bundles have yet to occur during the remodeling and maturation stage in order to develop quality collagen.

23.3.1.3 Contraction of the Mesh/Tissue Complex

Collagen deposition is accompanied by wound contraction as a natural phenomenon that decreases the area of the wound defect. Mesh shrinkage has been explored extensively following observations during revision surgeries that significant folding and shrinking of the mesh was apparent. Mesh size reduction resulting from myofibroblast activity reduces the pore size and causes buckling and folding of the prosthesis. Pore size and overall mesh length has been observed to reduce by 20% in meshes explanted from patients [102]. Furthermore, a study conducted by Klinge et al. [106] using polypropylene mesh on the posterior sheath of the rectus fascia within the preperitoneal space of dogs, produced an overall reduction in area of 46% within 4 weeks of implantation. The extent of mesh shrinkage has been shown to be directly proportional to the degree of inflammatory response [59].

For mesh hernia repair, contraction is an undesired consequence that initiates during the second week and is associated with recurrence, migration, and pain [107,108]. The etiology is largely unknown but it has been suggested that during proliferative scars the impaired activity of myofibroblasts renders them unable to control normal fibrillar arrangement; in addition, excessive production of TGF-β during chronic inflammation increases contraction [6]. Early and complete integration of the mesh into the surrounding tissue may decrease mesh shrinkage. In a swine study to investigate the association between tissue ingrowth and mesh contraction, suture detachment from fixation points was observed, indicating that shrinkage occurred before the mesh had time to integrate into the tissue [107]. On the other hand, meshes that exhibited strong integration into the surrounding tissue exhibited the least amount of contraction. As reviewed by Garcia-Urena, in a study comparing mesh shrinkage with or without fixation after 90 days, the meshes in the fixation group shrank less and retained their original shape indicating the importance of mesh stability to shrinkage [108]. It is also known that extensive inflammation delays the proliferative phase of tissue deposition, which may reduce the integration of the mesh into the surrounding tissue. Together, the physical consequences of delayed tissue ingrowth due to inflammation and inadequate mesh stability have potential contributing roles in mesh shrinkage.

23.3.1.4 Mesh/Tissue Complex Stability and Applied Tension

The surgical repair of hernias with mesh results in a lower level of tension applied to the wound. Unlike primary repair, the edges of the hernia are not completely approximated; instead, the mesh is sutured in place with the defect forming what is known as a hernia ring that is spanned on all sides by the mesh. The mesh then becomes the scaffold, allowing ECM to develop within and around the hernia ring. The reduced wound tension of mesh repairs compared to primary repair has increased the clinical use of meshes for incisional hernia repair from 35% in 1987 to 66% in 1999 [109]. This shift is due to an observed reduction in recurrence, which has been attributed to lower levels of wound tension. The most common cause of wound failure in primary repair is dehiscence from the tearing of suture through the fascia. Although less common, the mechanism exists for meshes because of the endogenous tension of the tissue being repaired. However, early biomechanical wound disruption for meshes occurs almost exclusively at the suture line located at the margins of the mesh. For example, during the first few weeks after hernia repair, minimal ECM integrity has developed, and applied loads are concentrated at the points of suture attachment [110]. Concurrently, recovering surgical patients return to physical activity, placing increased loads across the acute wound during its weakest phase [64,111]. Therefore, suture integrity at the margins of the mesh must exclusively resist the applied stresses. Failure to resist the applied stress results in mesh-fascial dehiscence.

The general clinical consensus is that wound tension should be minimized during early postoperative periods [112]. Burger and coworkers, using computed tomography scans in a retrospective study, determined that, although incisional hernias in some cases are not realized until several months or years after surgery, the hernia process starts during the first postoperative month with observable separation of the rectus abdominal muscles [113]. These defects may initiate small and asymptomatic, but can steadily increase in size, and eventually result in the protrusion of abdominal contents and visible bulging. DuBay et al. [114] designed a rat surgical model that produces acute fascial separation following the rapid dissolution of cat gut suture. The ensuing incisional hernias have a well-defined hernia ring, protruding hernia sac, and visceral adhesions, providing all of the characteristics that clinically develop in humans. Following mesh repair of the induced hernia, animals were evaluated on postoperative days 7, 14, 28, and 60 for the development of recurrent incisional hernias. It was determined that 16% of the total 21% recurrences occurred by postoperative day 7 [114]. The authors concluded that recurrent incisional hernia formation is an early postoperative occurrence.

In addition to recurrence, wound biomechanical disruptions can affect angiogenesis. During the early stages of angiogenesis, capillary sprouts lack full thickness, which renders them delicate and easily disrupted. Budding vessels are so fragile that collagen fibers must surround them to prevent rupturing when blood pressure is imposed by arterial inflow [115]. Immobilization of the granulation tissue is essential through the proliferation phase to permit vascular regrowth and prevent microhemorrhages [8]. The prevention of these microhemorrhages facilitates the progression of the wound healing process without the release of additional pro-inflammatory cytokines. Wound stability during the early postoperative period is critical to preventing wound disruptions at the suture line that can evolve into recurrences and is critical for optimal wound site vascularization.

23.3.2 ADVANCED PHASES OF WOUND HEALING AND MATURATION

The so-called maturation phase of wound healing is characterized by a decrease in cell density and metabolic activity followed by balanced collagen degradation and deposition process whereby the wound site is remodeled through collagen fiber bundle organization based on chemical and mechanical stimulation.

The goal of hernioplasty should be to restore the morphology and functions of the abdominal wall such that the prosthesis provides a substitute with characteristics closest to the tissue's normal function, with a low elastic modulus more relevant to performance than the relative strength of the mesh [116]. Increasingly, it is being recognized that the growth and remodeling of fibrotic tissue that supports mechanical loads is governed by the same principles as Wolff's Law for bone [94]. A lack of stress produces atrophy while excessive stress results in necrosis. Ideally, low forces produce ECM distention, providing endogenous stress that fibroblasts maintain under tensional homeostasis [117]. Meshes begin to significantly decrease the mobility of the abdominal wall 2–3 weeks after implantation from the induction of the scar plate [116]. In the uniaxial direction, typical meshes have low initial elastic modulus due to deformations within the pores. However, biaxially, with fixation on all sides, current meshes have a high elastic modulus with very low elongation even at maximum physiologic conditions. Consequently, the mesh absorbs most of the mechanical load and the stress-deprived ECM will not efficiently remodel or mature. Matching of the implant load/extensional characteristics to the tissue should create a fibrotic capsule that is substantial and strong, with elastic properties that more closely match those of the surrounding tissue. Furthermore, when two contacting materials have marked differences in their elastic moduli, the result is high shear stress concentrations at the interface [94]. This difference is evidenced in reports of collagen fibers, which orientate along stress lines, aligning parallel to individual filaments of yarn within the mesh [118,119] with little or no orientation to the outer fibrotic capsule. Relative motion at the mechanical interface may also be responsible, in part, for the foreign body response to meshes. Increased shear stresses and a lack of mechanical stability at the interface may cause load-mediated cell necrosis and general irritation to the tissue stimulating inflammation [94].

A mesh/tissue complex that has firm attachment and close matching of the mesh and tissue elastic moduli is critical to the long-term mesh/tissue complex mechanical properties and the level of foreign body response to the mesh. Mechanical stimulation significantly influences the remodeling of the mesh/tissue complex and the development of proper physiology.

23.4 FUTURE HERNIOPLASTY MESH DESIGNS GUIDED BY MECHANOBIOLOGY

Since the development of absorbable sutures over three decades ago, there has been interest in using absorbable multifilament yarn for constructing absorbable meshes. However, this interest has been limited by unsuccessful attempts to repair load bearing soft tissues such as abdominal wall hernia repairs. Meshes constructed of only fast absorbing polyglycolide or 90/10 poly(glycolide-co-lactide) provide inadequate strength beyond 3–4 weeks of breaking strength retention. In addition, development of meshes constructed from relatively slow degrading high-lactide yarn has generated little to no interest. This situation has left the majority of soft tissue repair load bearing applications to be filled by non-absorbable materials, which suffer distinctly from undesirable features associated, in part, with their inability to (1) possess short-term stiffness to facilitate tissue stability during the development of wound strength; (2) gradually transfer the perceived mechanical loads as the wound is building mechanical integrity; (3) provide compliance with load transfer to the remodeling and maturing mesh/tissue complex; and (4) minimize the likelihood of long-term complications with their degradation and absorption at the conclusion of their intended functional performance. Wound healing is a dynamic process that results in different criteria at different phases for the optimal development of the mesh/tissue complex. Current nonabsorbable or partially absorbable meshes may not provide adequate early stability and then a sufficient level of long-term mechanical stimulation to the remodeling and maturing collagen.

Absorbable medical meshes in the form of multicomponent systems possessing different degradation rates can produce modulated mechanical, chemical, and physical properties, which individually or collectively may improve the wound healing process. Figure 23.5 illustrates the characteristics of a multicomponent mesh that has two different strength profiles compared to the wound strength profile and typical phases of wound healing. Each characteristic or phase is indicated with anticipated time intervals.

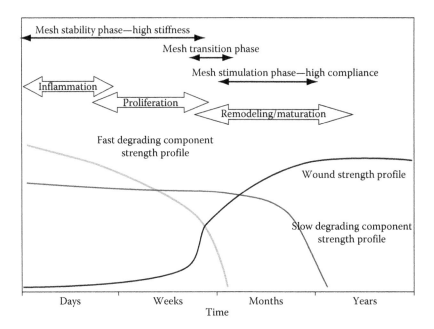

FIGURE 23.5 The modulated mechanical characteristics of a bicomponent mesh superimposed with the temporal wound healing response.

The mesh wound stability phase continues through the inflammation and proliferation phases. Early stiffness of the mesh may facilitate uninterrupted tissue integration and angiogenesis, while reducing the risk of recurrence from applied wound stresses prior to the development of wound strength. In addition, the added stiffness and stability may resist and/or minimize the wound contraction process. As the wound develops load bearing capability, stress is slowly transferred as the fast degrading component loses strength. Once the fast degrading component of the mesh is removed, the long lasting component is well encapsulated in the ECM. The mesh is positioned in a relaxed configuration such that the newly deposited collagen becomes load bearing and tensional homeostasis is returned to the abdominal wall. Over the ensuing months, the remodeling/maturing process of collagen degradation and synthesis adapts the tissue to the loading conditions. Finally, the slow degrading component loses its mechanical integrity by hydrolysis and is metabolized, leaving the wound site without the continuous foreign body response. Ultimately, the abdominal wall is left with a functional, sustainable layer of tissue that seals the defect area preventing future hernia formation.

REFERENCES

1. Dee, K.C. and D.A. Puleo, *An Introduction to Tissue–Biomaterials Interactions*. Wiley-Liss: Hoboken, NJ, 2002.
2. Mulder, G.D. et al., Factors influencing wound healing, in *Wound Biology and Management*, D.J. Leaper and K.G. Harding, Editors. Oxford University Press: Oxford, U.K., 1998, pp. 52–69.
3. Stadelmann, W.K., A.G. Digenis, and G.R. Tobin, Physiology and healing dynamics of chronic cutaneous wounds. *Am J Surg*, 1998, **176**(2A Suppl): 26S–38S.
4. Paul, M., Wound healing—Aiming for perfect skin regeneration. *Science*, 1997, **276**: 75–81.
5. Witte, M.B. and A. Barbul, General principles of wound healing. *Surg Clin North Am*, 1997, **77**(3): 509–528.
6. Robson, M.C., D.L. Steed, and M.G. Franz, Wound healing: Biologic features and approaches to maximize healing trajectories. *Curr Probl Surg*, 2001, **38**(2): 72–140.
7. Lorenz, H.P. and M.T. Longaker, Wounds: Biology, pathology, and management, in *Surgery: Basic Science and Clinical Evidence*, J.A. Norton et al., Editors. Springer-Verlag: New York, 2003, pp. 77–88.

8. Hardy, M.A., The biology of scar formation. *Phys Ther*, 1989, **69**(12): 1014–1024.
9. Yang, G.P. et al., From scarless fetal wounds to keloids: Molecular studies in wound healing. *Wound Repair Regen*, 2003, **11**(6): 411–418.
10. Silver, F.H. et al., Transition from viscous to elastic-based dependency of mechanical properties of self-assembled type I collagen fibers. *J Appl Polym Sci*, 2000, **79**: 134–142.
11. Bailey, A.J., R.G. Paul, and L. Knott, Mechanisms of maturation and ageing of collagen. *Mech Ageing Dev*, 1998, **106**(1–2): 1–56.
12. Fratzl, P. et al., Fibrillar structure and mechanical properties of collagen. *J Struct Biol*, 1998, **122**(1–2): 119–122.
13. Silver, F.H., A. Ebrahimi, and P.B. Snowhill, Viscoelastic properties of self-assembled type I collagen fibers: Molecular basis of elastic and viscous behaviors. *Connect Tissue Res*, 2002, **43**(4): 569–580.
14. Misof, K., G. Rapp, and P. Fratzl, A new molecular model for collagen elasticity based on synchrotron X-ray scattering evidence. *Biophys J*, 1997, **72**(3): 1376–1381.
15. Puxkandl, R. et al., Viscoelastic properties of collagen: Synchrotron radiation investigations and structural model. *Philos Trans R Soc Lond B Biol Sci*, 2002, **357**(1418): 191–197.
16. Bonifasi-Lista, C. et al., Viscoelastic properties of the human medial collateral ligament under longitudinal, transverse and shear loading. *J Orthop Res*, 2005, **23**(1): 67–76.
17. Benjamin, M. and B. Hillen, Mechanical influences on cells, tissues and organs—'Mechanical Morphogenesis'. *Eur J Morphol*, 2003, **41**(1): 3–7.
18. Screen, H.R. et al., Cyclic tensile strain upregulates collagen synthesis in isolated tendon fascicles. *Biochem Biophys Res Commun*, 2005, **336**(2): 424–429.
19. Webb, K. et al., Cyclic strain increases fibroblast proliferation, matrix accumulation, and elastic modulus of fibroblast-seeded polyurethane constructs. *J Biomech*, 2006, **39**(6): 1136–1144.
20. Yang, G., R.C. Crawford, and J.H. Wang, Proliferation and collagen production of human patellar tendon fibroblasts in response to cyclic uniaxial stretching in serum-free conditions. *J Biomech*, 2004, **37**(10): 1543–1550.
21. Park, S.A. et al., Biological responses of ligament fibroblasts and gene expression profiling on micropatterned silicone substrates subjected to mechanical stimuli. *J Biosci Bioeng*, 2006, **102**(5): 402–412.
22. Eckes, B. et al., Mechanical tension and integrin alpha 2 beta 1 regulate fibroblast functions. *J Investig Dermatol Symp Proc*, 2006, **11**(1): 66–72.
23. Bride, J. et al., Indication of fibroblast apoptosis during the maturation of disc-shaped mechanically stressed collagen lattices. *Arch Dermatol Res*, 2004, **295**(8–9): 312–317.
24. Puk, C.K. et al., The effects of short-term stimulation on fibroblast spreading in an in vitro 3D system. *J Biomed Mater Res A*, 2006, **76**(4): 665–673.
25. Eastwood, M. et al., Effect of precise mechanical loading on fibroblast populated collagen lattices: Morphological changes. *Cell Motil Cytoskeleton*, 1998, **40**(1): 13–21.
26. Prajapati, R.T. et al., Mechanical loading regulates protease production by fibroblasts in three-dimensional collagen substrates. *Wound Repair Regen*, 2000, **8**(3): 226–237.
27. Yasuda, K. and K. Hayashi, Changes in biomechanical properties of tendons and ligaments from joint disuse. *Osteoarthr Cartilage*, 1999, **7**(1): 122–129.
28. Hannafin, J.A. et al., Effect of stress deprivation and cyclic tensile loading on the material and morphologic properties of canine flexor digitorum profundus tendon: An in vitro study. *J Orthop Res*, 1995, **13**(6): 907–914.
29. Eckes, B. et al., Interactions of fibroblasts with the extracellular matrix: Implications for the understanding of fibrosis. *Springer Semin Immunopathol*, 1999, **21**(4): 415–429.
30. Li, F. et al., Cell shape regulates collagen type I expression in human tendon fibroblasts. *Cell Motil Cytoskeleton*, 2008, **65**(4): 332–341.
31. Hinz, B. et al., Mechanical tension controls granulation tissue contractile activity and myofibroblast differentiation. *Am J Pathol*, 2001, **159**(3): 1009–1020.
32. Faryniarz, D.A. et al., Myofibroblasts in the healing lapine medial collateral ligament: Possible mechanisms of contraction. *J Orthop Res*, 1996, **14**(2): 228–237.
33. Mirastschijski, U. et al., Matrix metalloproteinase inhibitor GM 6001 attenuates keratinocyte migration, contraction and myofibroblast formation in skin wounds. *Exp Cell Res*, 2004, **299**(2): 465–475.
34. Grinnell, F. et al., Release of mechanical tension triggers apoptosis of human fibroblasts in a model of regressing granulation tissue. *Exp Cell Res*, 1999, **248**(2): 608–619.
35. Burgess, L.P. et al., Wound healing. Relationship of wound closing tension to scar width in rats. *Arch Otolaryngol Head Neck Surg*, 1990, **116**(7): 798–802.

36. Fagan, S.P. and S.S. Awad, Abdominal wall anatomy: The key to a successful inguinal hernia repair. *Am J Surg*, 2004, **188**(6A Suppl): 3S–8S.
37. Yeager, V., Intermediate inguinal ring. *Clin Anat*, 1992, **5**(4): 289–295.
38. Park, A.E., J.S. Roth, and S.M. Kavic, Abdominal wall hernia. *Curr Probl Surg*, 2006, **43**(5): 326–375.
39. Flament, J., C. Avisse, and J. Delattre, Anatomy of the abdominal wall, in *Abdominal Wall Hernias: Principles and Management*, R. Bendavid et al., Editors. Springer-Verlag: New York, 2001, pp. 39–63.
40. Bhattacharjee, P., Surgical options in inguinal hernia: Which is the best. *Ind J Surg*, 2006, **68**(4): 191–200.
41. Paajanen, H., A single-surgeon randomized trial comparing three composite meshes on chronic pain after Lichtenstein hernia repair in local anesthesia. *Hernia*, 2007, **11**(4): 335–339.
42. Schwab, R. and U. Klinge, Principle actions for re-currences, in *Recurrent Hernia*, V. Schumpelick and R.J. Fitzgibbons, Editors. Springer-Verlag: Berlin, Germany, 2007.
43. Weber, A., D. Garteiz, and S. Valencia, Epidemiology of inguinal hernia: A useful aid for adequate surgical decisions, in *Abdominal Wall Hernias: Principles and Management*, R. Bendavid et al., Editors. Springer-Verlag: New York, 2001.
44. Palot, J. and C. Avisse, Open techniques of femoral hernia repair, in *Abdominal Wall Hernias: Principles and Management*, R. Bendavid et al., Editors. Springer-Verlag: New York, 2001.
45. Temiz, A. et al., A rare and frequently unrecognised pathology in children: Femoral hernia. *Hernia*, 2008, **12**(5): 553–556.
46. Bax, T., B.C. Sheppard, and R.A. Crass, Surgical options in the management of groin hernias. *Am Fam Physician*, 1999, **59**(1): 143–156.
47. Rodrigues, A.J. et al., Do the dimensions of the femoral canal play a role in the genesis of femoral hernia? *Hernia*, 2000, **4**: 45–51.
48. Velasco, M. et al., Current concepts on adult umbilical hernia. *Hernia*, 1999, **4**: 233–239.
49. Deysine, M., Umbilical hernias, in *Abdominal Wall Hernias: Principles and Management*, R. Bendavid et al., Editors. Springer-Verlag: New York, 2001.
50. McArdle, G., Is inguinal hernia a defect in human evolution and would this insight improve concepts for methods of surgical repair? *Clin Anat*, 1997, **10**(1): 47–55.
51. Rosch, R. et al., Hernia—A collagen disease? *Eur Surg*, 2003, **35**: 11–15.
52. Kang, S.K. et al., Hernia: Is it a work-related condition? *Am J Ind Med*, 1999, **36**(6): 638–644.
53. Franz, M.G., The biology of hernias and the abdominal wall. *Hernia*, 2006, **10**(6): 462–471.
54. Klosterhalfen, B., K. Junge, and U. Klinge, The lightweight and large porous mesh concept for hernia repair. *Expert Rev Med Devices*, 2005, **2**(1): 103–117.
55. Hollinsky, C. and S. Sandberg, Measurement of the tensile strength of the ventral abdominal wall in comparison with scar tissue. *Clin Biomech (Bristol, Avon)*, 2007, **22**(1): 88–92.
56. Junge, K. et al., Review of wound healing with reference to an unrepairable abdominal hernia. *Eur J Surg*, 2002, **168**(2): 67–73.
57. Schumpelick, V. et al., Light weight meshes in incisional hernia repair. *J Min Access Surg*, 2006, **2**(3): pNA.
58. Klinge, U. et al., Hernia recurrence as a problem of biology and collagen. *J Min Access Surg*, 2006, **2**(3): 151–154.
59. Kockerling, C. and C. Schug-Pass, Recurrence and mesh material, in *Recurrent Hernia*, V. Schumpelick and R.J. Fitzgibbons, Editors. Springer-Verlag: Berlin, Germany, 2007.
60. Schippers, E., Central mesh rupture—Myth or real concern? in *Recurrent Hernia*, V. Schumpelick and R.J. Fitzgibbons, Editors. Springer-Verlag: Berlin, Germany, 2007.
61. Langer, C. et al., Central mesh recurrence after incisional hernia repair with Marlex—Are the meshes strong enough? *Hernia*, 2001, **5**(3): 164–167.
62. Doctor, H., Evaluation of various prosthetic materials and newer meshes for hernia repairs. *J Min Access Surg*, 2006, **2**(3): 110–116.
63. Donahue, T.R., J.R. Hiatt, and R.W. Busuttil, Collagenase and surgical disease. *Hernia*, 2006, **10**(6): 478–485.
64. Peiper, C. et al., Abdominal musculature and the transversalis fascia: An anatomical viewpoint. *Hernia*, 2004. **8**(4): 376–380.
65. Klinge, U. et al., Modified mesh for hernia repair that is adapted to the physiology of the abdominal wall. *Eur J Surg*, 1998, **164**(12): 951–960.
66. Junge, K. et al., Elasticity of the anterior abdominal wall and impact for reparation of incisional hernias using mesh implants. *Hernia*, 2001, **5**(3): 113–118.
67. Cobb, W.S. et al., Normal intraabdominal pressure in healthy adults. *J Surg Res*, 2005, **129**(2): 231–235.
68. Wolloscheck, T. et al., Inguinal hernia: Measurement of the biomechanics of the lower abdominal wall and the inguinal canal. *Hernia*, 2004, **8**(3): 233–241.

69. Awad, S.S. and S.P. Fagan, Current approaches to inguinal hernia repair. *Am J Surg*, 2004, **188**(6A Suppl): 9S–16S.
70. Muhl, T. et al., New objective measurement to characterize the porosity of textile implants. *J Biomed Mater Res B Appl Biomater*, 2008, **84**(1): 176–183.
71. Schumpelick, V. and U. Klinge, Prosthetic implants for hernia repair. *Br J Surg*, 2003, **90**(12): 1457–1458.
72. Klinge, U. et al., Functional and morphological evaluation of a low-weight, monofilament polypropylene mesh for hernia repair. *J Biomed Mater Res*, 2002, **63**(2): 129–136.
73. Junge, K. et al., Influence of mesh materials on collagen deposition in a rat model. *J Invest Surg*, 2002, **15**(6): 319–328.
74. Klinge, U. et al., Influence of polyglactin-coating on functional and morphological parameters of polypropylene-mesh modifications for abdominal wall repair. *Biomaterials*, 1999, **20**(7): 613–623.
75. Cobb, W.S. et al., Textile analysis of heavy weight, mid-weight, and light weight polypropylene mesh in a porcine ventral hernia model. *J Surg Res*, 2006, **136**(1): 1–7.
76. Klinge, U. et al., Impact of polymer pore size on the interface scar formation in a rat model. *J Surg Res*, 2002, **103**(2): 208–214.
77. Rosch, R. et al., Vypro II mesh in hernia repair: Impact of polyglactin on long-term incorporation in rats. *Eur Surg Res*, 2003, **35**(5): 445–450.
78. Weyhe, D. et al., Experimental comparison of monofile light and heavy polypropylene meshes: Less weight does not mean less biological response. *World J Surg*, 2006, **30**(8): 1586–1591.
79. Weyhe, D. et al., The role of TGF-beta1 as a determinant of foreign body reaction to alloplastic materials in rat fibroblast cultures: Comparison of different commercially available polypropylene meshes for hernia repair. *Regul Pept*, 2007, **138**(1): 10–14.
80. Weyhe, D. et al., In vitro comparison of three different mesh constructions. *ANZ J Surg*, 2008, **78**(1–2): 55–60.
81. Bringman, S. et al., One year results of a randomised controlled multi-centre study comparing Prolene and Vypro II-mesh in Lichtenstein hernioplasty. *Hernia*, 2005, **9**(3): 223–227.
82. Smietanski, M., Randomized clinical trial comparing a polypropylene with a poliglecaprone and polypropylene composite mesh for inguinal hernioplasty. *Br J Surg*, 2008, **95**(12): 1462–1468.
83. Conze, J. et al., Randomized clinical trial comparing lightweight composite mesh with polyester or polypropylene mesh for incisional hernia repair. *Br J Surg*, 2005, **92**(12): 1488–1493.
84. Bringman, S. et al., Three-year results of a randomized clinical trial of lightweight or standard polypropylene mesh in Lichtenstein repair of primary inguinal hernia. *Br J Surg*, 2006, **93**(9): 1056–1059.
85. O'Dwyer, P.J. et al., Randomized clinical trial assessing impact of a lightweight or heavyweight mesh on chronic pain after inguinal hernia repair. *Br J Surg*, 2005, **92**(2): 166–170.
86. Weyhe, D. et al., Improving outcomes in hernia repair by the use of light meshes—A comparison of different implant constructions based on a critical appraisal of the literature. *World J Surg*, 2007, **31**(1): 234–244.
87. Greca, F.H. et al., The influence of porosity on the integration histology of two polypropylene meshes for the treatment of abdominal wall defects in dogs. *Hernia*, 2008, **12**(1): 45–49.
88. Krause, H.G. et al., Biocompatible properties of surgical mesh using an animal model. *Aust N Z J Obstet Gynaecol*, 2006, **46**(1): 42–45.
89. Klosterhalfen, B., U. Klinge, and V. Schumpelick, Functional and morphological evaluation of different polypropylene-mesh modifications for abdominal wall repair. *Biomaterials*, 1998, **19**(24): 2235–2246.
90. Conze, J. et al., Polypropylene in the intra-abdominal position: Influence of pore size and surface area. *Hernia*, 2004, **8**(4): 365–372.
91. Di Vita, G. et al., Impact of different texture of polypropylene mesh on the inflammatory response. *Int J Immunopathol Pharmacol*, 2008, **21**(1): 207–214.
92. Papadimitriou, J. and P. Petros, Histological studies of monofilament and multifilament polypropylene mesh implants demonstrate equivalent penetration of macrophages between fibrils. *Hernia*, 2005, **9**(1): 75–78.
93. Franz, M.G., D.L. Steed, and M.C. Robson, Optimizing healing of the acute wound by minimizing complications. *Curr Probl Surg*, 2007, **44**(11): 691–763.
94. Hilborn, J. and L.M. Bjursten, A new and evolving paradigm for biocompatibility. *J Tissue Eng Regen Med*, 2007, **1**(2): 110–119.
95. Dubay, D.A. et al., Fascial fibroblast kinetic activity is increased during abdominal wall repair compared to dermal fibroblasts. *Wound Repair Regen*, 2004, **12**(5): 539–545.
96. Franz, M.G. et al., Fascial incisions heal faster than skin: A new model of abdominal wall repair. *Surgery*, 2001, **129**(2): 203–208.

97. Franz, M.G. et al., Use of the wound healing trajectory as an outcome determinant for acute wound healing. *Wound Repair Regen*, 2000, **8**(6): 511–516.

98. Szpaderska, A.M. and L.A. DiPietro, Inflammation in surgical wound healing: Friend or foe? *Surgery*, 2005, **137**(5): 571–573.

99. Champault, G. et al., Inguinal hernia repair: The choice of prosthesis outweighs that of technique. *Hernia*, 2007, **11**(2): 125–128.

100. Welty, G. et al., Functional impairment and complaints following incisional hernia repair with different polypropylene meshes. *Hernia*, 2001, **5**(3): 142–147.

101. Agrawal, A. and R. Avill, Mesh migration following repair of inguinal hernia: A case report and review of literature. *Hernia*, 2006, **10**(1): 79–82.

102. Amid, P.K., Classification of biomaterials and their related complications in abdominal wall hernia surgery. *Hernia*, 1997, **1**: 15–21.

103. Klinge, U. and C.J. Krones, Can we be sure that the meshes do improve the recurrence rates? *Hernia*, 2005, **9**: 1–2.

104. Felemovicius, I. et al., Prevention of adhesions to polypropylene mesh. *J Am Coll Surg*, 2004, **198**(4): 543–548.

105. Bellon, J.M. et al., Integration of biomaterials implanted into abdominal wall: Process of scar formation and macrophage response. *Biomaterials*, 1995, **16**(5): 381–387.

106. Klinge, U. et al., Shrinking of polypropylene mesh in vivo: An experimental study in dogs. *Eur J Surg*, 1998, **164**(12): 965–969.

107. Gonzalez, R. et al., Relationship between tissue ingrowth and mesh contraction. *World J Surg*, 2005, **29**(8): 1038–1043.

108. Garcia-Urena, M.A. et al., Differences in polypropylene shrinkage depending on mesh position in an experimental study. *Am J Surg*, 2007, **193**(4): 538–542.

109. Flum, D.R., K. Horvath, and T. Koepsell, Have outcomes of incisional hernia repair improved with time? A population-based analysis. *Ann Surg*, 2003, **237**(1): 129–135.

110. Franz, M.G., The biological treatment of the hernia disease, in *Recurrent Hernia*, V. Schumpelick and R.J. Fitzgibbons, Editors. Springer-Verlag: Berlin, Germany, 2007.

111. Robson, M.C. et al., Effect of cytokine growth factors on the prevention of acute wound failure. *Wound Repair Regen*, 2004, **12**(1): 38–43.

112. Carlson, M.A., Acute wound failure. *Surg Clin North Am*, 1997, **77**(3): 607–636.

113. Burger, J.W. et al., Incisional hernia: Early complication of abdominal surgery. *World J Surg*, 2005, **29**(12): 1608–1613.

114. DuBay, D.A. et al., Mesh incisional herniorrhaphy increases abdominal wall elastic properties: A mechanism for decreased hernia recurrences in comparison with suture repair. *Surgery*, 2006, **140**(1): 14–24.

115. Hopf, H.W. et al., Hyperoxia and angiogenesis. *Wound Repair Regen*, 2005, **13**(6): 558–564.

116. Ferrando, J.M. et al., Experimental evaluation of a new layered prosthesis exhibiting a low tensile modulus of elasticity: Long-term integration response within the rat abdominal wall. *World J Surg*, 2002, **26**(4): 409–415.

117. Karamuk, E., J. Mayer, and G. Raeber, Tissue engineered composite of a woven fabric scaffold with tendon cells, response on mechanical simulation in vitro. *Comp Sci Tech*, 2004, **64**: 885–891.

118. Klinge, U. et al., Foreign body reaction to meshes used for the repair of abdominal wall hernias. *Eur J Surg*, 1999, **165**(7): 665–673.

119. Klosterhalfen, B. et al., Influence of implantation interval on the long-term biocompatibility of surgical mesh. *Br J Surg*, 2002, **89**(8): 1043–1048.

Index

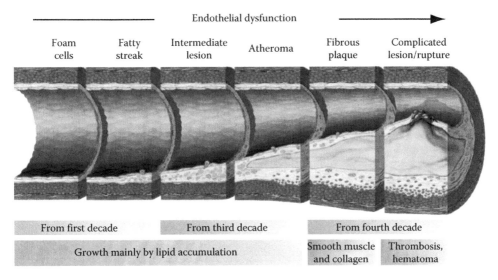

FIGURE 4.1 Progression of an atherosclerotic lesion. (From Pepine, C.J., *The American Journal of Cardiology*, 82(10), S24, 1998.)

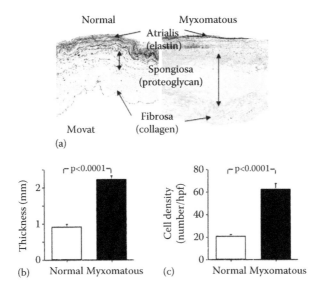

FIGURE 6.7 Comparison of normal and myxomatous mitral valves. (a) Myxomatous valves show a decrease in elastin, an expanded spongiosa, and a fibrosa with more loosely connected collaged fibers. (b) Myxomatous valves are significantly thicker and (c) contain a higher cell density than normal valves. (Adapted from Rabkin, E. et al., *Circulation*, 104, 2525, 2001.)

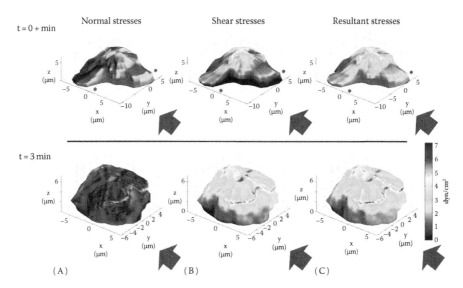

FIGURE 8.9   Shear-induced retraction of neutrophil pseudopods is independent of their orientation in the flow field and the stress distribution on the cell surface. Three-dimensional confocal microscopic images of HL-60-derived neutrophils exposed to 2.2 dyn/cm² wall shear stress (direction indicated by block arrows) was acquired at t=0+ and at t=3 min after the onset of flow of a parallel plate geometry over the cell. Variations in the cell surface geometry leads to a heterogeneous stress distribution over the cell. Within the time frame of shear experiments, the cell retracted existing pseudopods (*). Three-dimensional reconstructions of the cell at the two time points were used for computational fluid mechanics estimations of the distributions of the normal (A) and shear stress components (B) as well as the resultant (the vector combination of the normal and shear) stress (C) on the cell surface. Note that the shear stress component imposed by fluid flow is the dominant stress experienced by the cell compared to the normal component. (Image courtesy of Dr. Susan Su.)

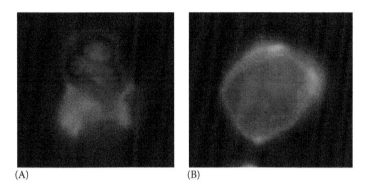

FIGURE 8.10 F-actin distribution in human neutrophils. Activated (A) and shear-inactivated (B) neutrophils were fixed, labeled with phalloidin (F-actin staining at the cell periphery) and propidium iodide (nuclear staining). *Note*: neutrophils are 10–15 μm in diameter.

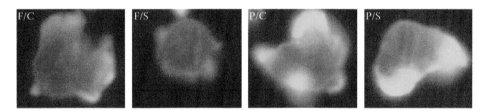

FIGURE 8.11 Deactivation of neutrophils by fluid shear stress involves redistribution of cytoskeletal F-actin. Neutrophils that had been prestimulated with either 10 nM fMLP (F) or 10 nM PMA (P) were exposed to cone-plate flow ($\tau \approx 5$ dyn/cm^2) for 5 min (S). Controls (C) were cells under no-flow, but otherwise, similar experimental conditions prevailed. After experiments, cells were fixed, labeled with phalloidin (F-actin staining in pseudopods and cell periphery) and propidium iodide (nuclear staining). *Note*: neutrophils are 10–15 μm in diameter. F/C = FMLP under control; F/S = FMLP + Shear; P/C = PMA under control; P/S = PMA + Shear.

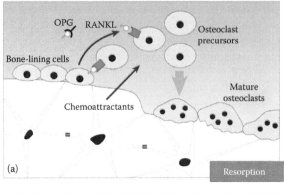

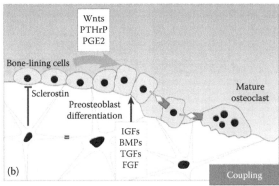

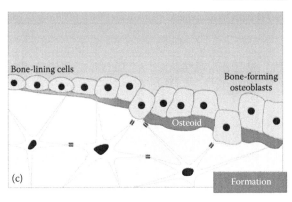

FIGURE 9.1 Bone remodeling is carried out by teams of osteoblasts and osteoclasts. (a) Osteoclast precursors (green) are recruited to bone surfaces via chemoattractants released (straight red arrow) from bone tissue (gray) into the marrow space (blue). RANKL (yellow) is expressed on the surface of osteoblasts and osteoblast progenitors, shown here as bone lining cells (pink). RANKL is also expressed in a soluble form. RANKL binds RANK on osteoclast precursors and initiates monocyte fusion and differentiation (blue arrow) into multinucleated mature osteoclasts. Osteoblasts and osteoblast progenitors also express osteoprotegerin (OPG), which acts as a decoy receptor for RANKL and blocks osteoclast differentiation. Mature osteoclasts adhere to bone surfaces and resorb matrix. (b) Coupling action of osteoclast and osteoblast activity occurs at bone surfaces where resorption of matrix by osteoclasts releases osteogenic growth factors (e.g., IGFs, BMPs, TGFs, and FGFs) thereby enhancing osteoblast differentiation. Wnts, PTHrP, and PGE2 released from neighboring osteoblastic cells (blue arrow) also enhance osteogenic activity. Cell–cell signaling (blue receptor/yellow ligand) can also play a role in osteoblast differentiation. Sclerostin, which is produced by osteocytes (white) embedded within the matrix, can negatively control this process by inhibiting osteoblast differentiation. (c) Mature cuboidal osteoblasts form new matrix or osteoid on previously resorbed bone surfaces. A fraction of the osteoblasts become embedded within the matrix, a process that is partially controlled by direct communication with osteocytes via gap junctions (red double dash).

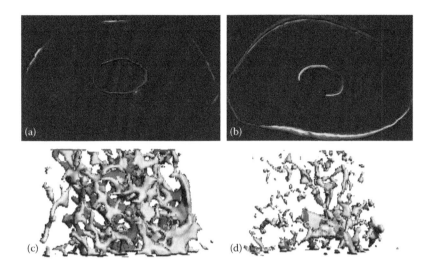

FIGURE 9.2 Mechanical adaptation of bone. Mechanical loading and disuse result in new bone formation and bone loss, respectively. Cyclic compressive loading of the mouse ulna results in significant bone formation on the periosteal and endosteal surfaces at midshaft (b) compared to the nonloaded contralateral limb (a). Fluorescent bone labels (green calcein, red alizarin) administered 7 days apart are incorporated into the newly forming matrix and allow the visualization of newly formed bone. Hind limb suspension in mice results in significant trabecular bone loss in the distal femur (d) as compared to normally ambulating controls (c) depicted by a reduction in trabecular number and thickness and an increase in spacing between individual trabeculae.

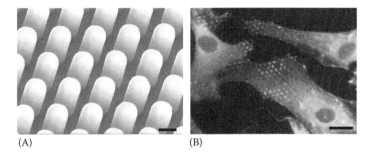

FIGURE 12.5 Micropost force sensor array. (A) SEM image of MFSA. Scale bar, 2 μm. (B) Fluorescence microscopy image of human skin fibroblasts cultured on a MFSA. TRITC-conjugated phalloidin (red), anti-vinculin monoclonal antibody and FITC-conjugated secondary antibody (green), and BisBenzimide H33258 (blue) were used to reveal actin filaments, focal contacts, and nuclei, respectively. Scale bar, 20 μm. (Adapted from Figs. 5 and 6 in Li, B., et al., *Cell Motil. Cytoskeleton*, 64, 509, 2007. With permission.)

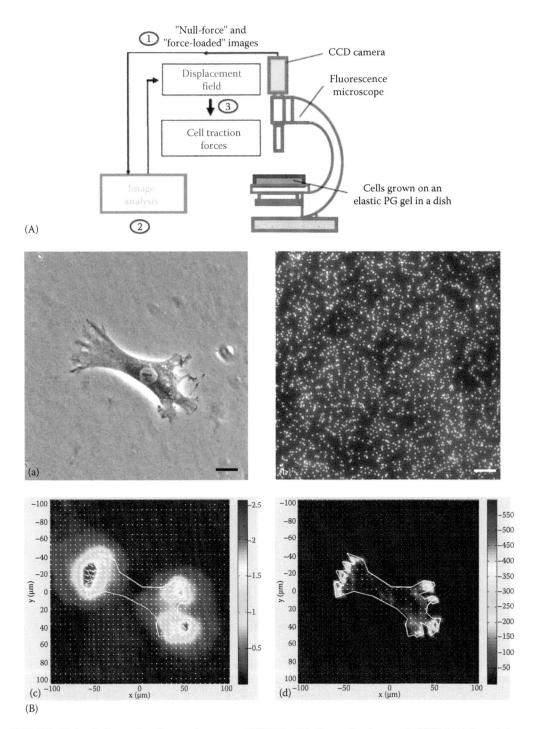

FIGURE 12.6 Cell traction force microscopy (CTFM). (A) General scheme of CTFM. (Adapted from Fig. 2 in Wang and Thampatty, *Int. Rev Cell Mol. Biol.*, 271, 301, 2008. With permission.) (B) Application of CTFM to determine CTFs of a human patellar tendon fibroblast (HPTF): (a) HPTF on a PAG, (b) fluorescence image of fluorescent beads embedded in gel, (c) substrate displacement field, and (d) cell traction force field. Scale bars, 20 μm.

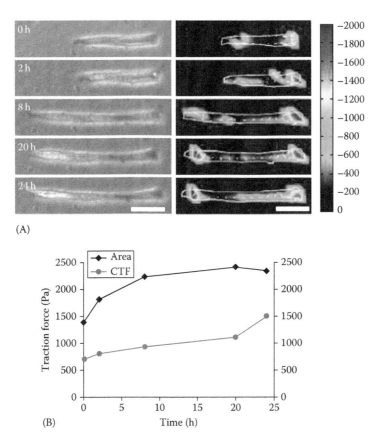

FIGURE 12.7 Traction force development of micropatterned C2C12 cells during differentiation. (A) Phase contrast images (left panel) and corresponding traction force maps (right panel), respectively. The color bar represents traction force level. Note that prior to plating on PAG, cells were pre-differentiated for 2 days in differentiation medium. Note the phase contrast images were slightly out of focus due to the requirement of in situ fluorescence image acquisition. Scale bars, 50 μm. (B) Changes in cell spreading area and cell traction force during C2C12 differentiation between first and second days of culture. (Reprinted from Fig. 5 in Li, B. et al., *J. Biomech.*, 41, 3349, 2008. With permission.)

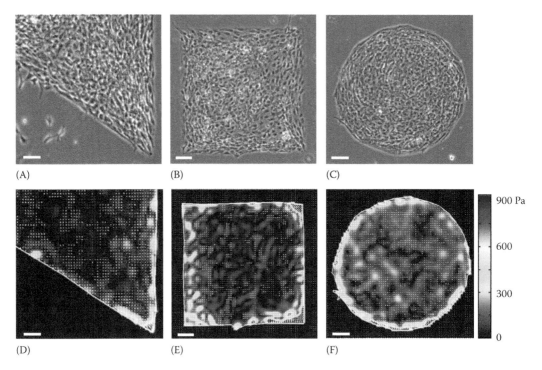

(A) (B) (C)

(D) (E) (F)

900 Pa

600

300

0

FIGURE 12.8 Mechanical stress distribution of patterned cell aggregates. (A–C) Micropatterned cell islands of distinctive shapes. (D–F) CTF distribution of cell islands determined by CTFM. Scale bars, 100 µm. (Adapted from Fig. 2 in Li, B. et al., *J. Biomech.*, 42, 1622, 2009. With permission.)

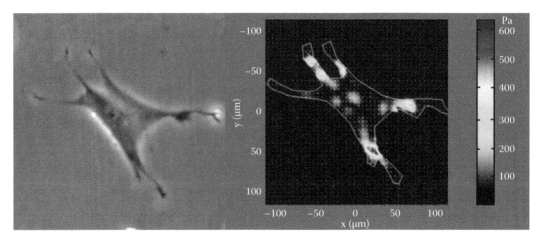

FIGURE 15.1 Traction force microscopy. Traction forces exerted by a BSMC, analyzed using traction force microscopy. *Left*, Phase contrast image. *Right*, Corresponding traction force map.

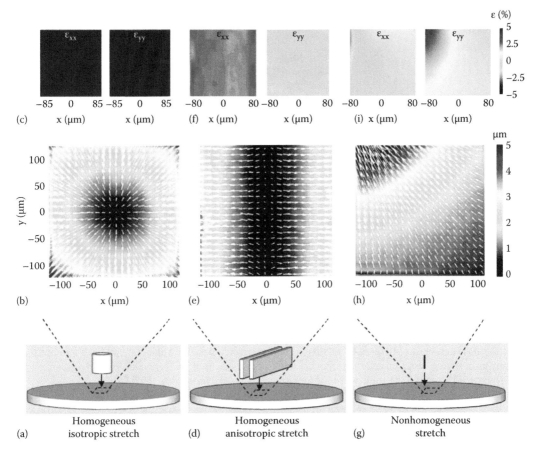

FIGURE 15.2 Cell mapping rheometry. (a–c) An annular punch indenter imposes a homogeneous isotropic biaxial stretch within the central region of the indenter. (d–f) Two parallel plates impose a homogeneous, anisotropic, and uniaxial stretch. (g–i) A microneedle imposes a nonhomogeneous stretch. Localized gel displacements are indicated by arrows, and their magnitude by color. (From Krishnan, R. et al., *PLoS One*, 4, e5486, 2009.)

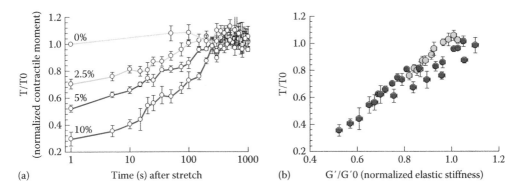

FIGURE 15.3 In response to an acute homogeneous stretch, traction forces evolve in concert with cell stiffness. (a) Traction forces, represented by the ratio of the contractile moment T relative to the pre-stretch baseline value, T0, plotted as a function of time after stretch cessation. The greater was the applied stretch, the greater were the reductions in cell traction forces. Peak strains of: 0% (gray; $n = 9$), 2.5% (green; $n = 12$), 5.0% (blue; $n = 11$) and 10.0% (red; $n = 14$). (b) Normalized traction force data correlate linearly with normalized cell elastic stiffness (G′/G′0). (From Krishnan, R. et al., *PLoS One*, 4, e5486, 2009.)

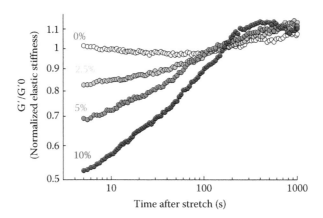

FIGURE 15.5 An acute homogeneous stretch fluidizes the airway SMC. Cell stiffness, represented by the ratio of the cell elastic stiffness G′ relative to the pre-stretch baseline value, G′0, reduces promptly and abruptly after stretch cessation and then recovers slowly. Peak strains of: 0% (gray; $n = 9$), 2.5% (green; $n = 12$), 5.0% (blue; $n = 11$), and 10.0% (red; $n = 14$). (From Trepat, X. et al. *Nature*, 447, 592, 2007.)

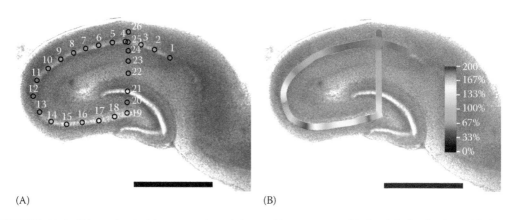

(A)   (B)

FIGURE 18.1 Biomechanical heterogeneity of the rat hippocampus. (A) Spatial distribution of indentation measurements. The elastic modulus of the rat hippocampus was measured via atomic force microscopy (AFM) indentation along a layer of pyramidal neuron cell bodies at the depicted locations. (B) Elasticity map of the hippocampus. Apparent elastic modulus normalized to the mean apparent elastic modulus of all indentations is depicted by the color bar. The material properties of the hippocampus were spatially heterogeneous. Scale bar is 1 mm. (Reprinted from Elkin, B.S. et al., *J. Neurotrauma.*, 24, 812, 2007. With permission.)

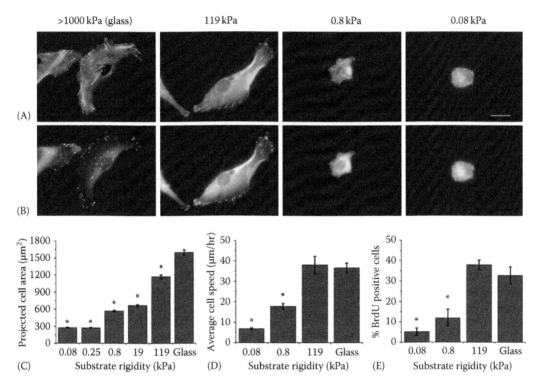

FIGURE 18.4 The mechanical rigidity of the ECM regulates glioblastoma multiforme (GBM) tumor cell structure, motility, and proliferation. (A) Cell shape and cytoarchitecture. Human glioma cells cultured on fibronectin-conjugated glass and polyacrylamide gels of three different stiffnesses were fixed and stained for F-actin (green), nuclear DNA (blue), and the focal adhesion protein vinculin (red). Cells on glass and 119 kPa substrates exhibit robust focal adhesions and a well-defined cytoskeletal architecture, whereas cells on 0.80 and 0.08 kPa polyacrylamide gels are rounded with cortical rings of F-actin and small, punctate vinculin-positive focal complexes. Bar is 25 μm. (B) Isolated view of vinculin signal only, showing structure and distributions of cell-ECM adhesions. Effect of ECM mechanical rigidity on (C) Cell spreading area; (D) Migration rate; and (E) Cell proliferation, as measured by bromodeoxyuridine (BrdU) incorporation. (Adapted from Ulrich, T.A. et al., *Cancer Res.*, 69, 4167, 2009.)

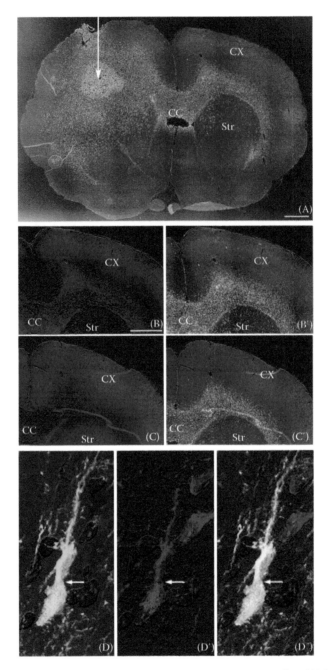

FIGURE 18.5 Invasive glioma cells demonstrate enhanced myosin IIA expression. (A) Growth and spread of implanted tumor cells. Rat brain slice stained for human nuclear antigen (green) shows that implanted primary human glioma cells spread from the site of tumor inoculation (arrow) across the corpus collosum (CC) to the contralateral white matter, between the cortex (CX) and striatum (Str). Bar is 1 mm. (B) Enhanced immuno-fluorescence localization of myosin IIA in invasive cells (red). Bar is 1 mm. (B′) Colocalization of myosin IIA and human nuclear antigen. (C, C′) Corresponding immunofluorescence localization of myosin IIB (red), dem-onstrating equivalent or reduced expression in invasive glioma cells (green) relative to the surrounding normal brain tissue. (D) Nuclear deformation of invasive cells. A GFP-expressing human glioma cell (green) requires significant nuclear deformation (arrow; nucleus stained blue with DAPI) to infiltrate the surrounding normal brain tissue. (D′) Expression of myosin IIA (red) in infiltrative cells. (D″) Colocalization of myosin IIA and GFP. (Adapted from Beadle, C. et al., *Mol. Biol. Cell*, 19, 3357, 2008. With permission.)

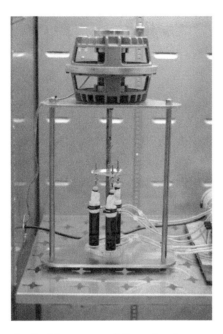

FIGURE 19.6 In this study, mechanical forces were delivered to the constructs using a bidirectional motor (upper portion of image) that was connected to three culture chambers via a distributor plate. The constructs may be seen inside the glass culture chambers. Also visible are the send and return lines for media circulation.

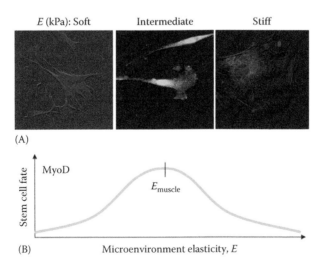

FIGURE 20.2 Stiffness-dependent differentiation of stem cells. (A) Immunofluorescence images of the actin cytoskeletal (red) and transcription factor MyoD1 (green) in MSCs plated across a range of matrix stiffness. On the soft matrices, MSCs have long extensions (red). On intermediate matrices, MSCs upregulate the transcription factor MyoD1 (green). On the stiff matrices, MSCs do not express significant MyoD1 marker but are more spread. (B) Schematic of the fluorescent intensity of MSCs versus substrate elasticity indicates that maximal myogenic expression occurs at E typical of muscle type.

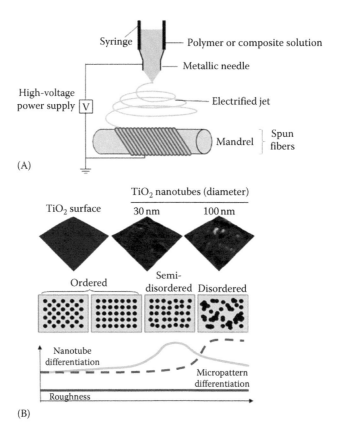

FIGURE 20.3 Structural and topographical cues that regulate cell-generated forces and differentiate stem cells. (A) Schematic of the fabrication of interwoven or spun fibers. (B) The top row shows AFM topography images of a titanium substrate and TiO_2 nanotubes of different diameter. Scans are $5 \times 5 \mu m$. The second row shows of nanofeatures having ordered, semi-disordered, and disordered symmetries. Materials typical of this fabrication scheme involve polymer matrices such as PMMA. Surface roughness (red) does not change significantly as shown in the bottom schematic, though both nanotube (blue) and polymer surfaces (green) induce differentiation when specific topography is recognized by MSCs. (From Dalby, M.J. et al., *Nat. Mater.*, 6, 997, 2007.)

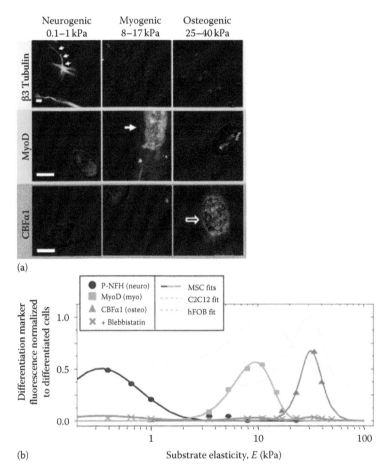

(a)

(b)

FIGURE 21.1 Matrix stiffness modulates human mesenchymal stem cell lineage specification. (a) hMSCs expressed early neurogenic, myogenic, and osteogenic markers on substrates with stiffness value resembling the tissue elasticity of brain, muscle, and pre-calcified bone, respectively. (b) Quantification of the lineage marker intensities normalized to the respective positive control cells (C2C12 myoblasts and hFOB osteoblasts) as a function of substrate stiffness. Specific lineage commitment was optimized at distinct range of substrate stiffness and inhibition of non-muscle myosin II activity with blebbistatin abolished the stiffness-mediated hMSC specification. (From Engler, A.J. et al., *Cell*, 126(4), 677, 2006.)

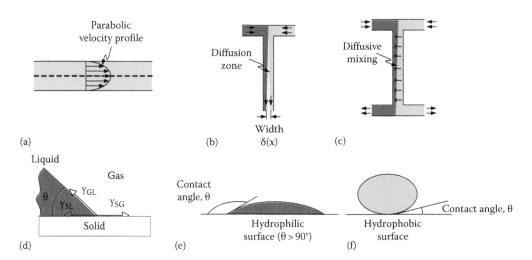

FIGURE 22.1 Physical basis of microfluidic devices: (a) laminar flow in a microchannel (b) diffusion between two coflowing fluids, (c) diffusive mixing between two coflowing laminar flows, (d) solid–liquid–gas interface, (e) droplet on a hydrophilic surface, and (f) droplet on a hydrophobic surface.

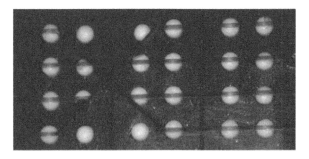

FIGURE 22.7 A Braille pin system with microfluidic channels colored by green dye. (From Mehta, G. et al., *Anal. Chem.*, 81(10), 3714, 2009. With permission.)

Printed and bound by CPI Group (UK) Ltd, Croydon, CR0 4YY

18/10/2024

01776210-0010